HANDBOOK OF EXPERIMENTAL POLLINATION BIOLOGY

HANDBOOK OF EXPERIMENTAL POLLINATION BIOLOGY

Edited By

C. Eugene Jones

Department of Biological Science
California State University, Fullerton
Fullerton, California

R. John Little

Rancho Santa Ana Botanic Garden
Claremont, California

 **SCIENTIFIC AND ACADEMIC EDITIONS**
DIVISION OF VAN NOSTRAND REINHOLD COMPANY INC
NEW YORK   CINCINNATI   TORONTO   LONDON   MELBOURNE

Manufactured in the United States of America

Published by Van Nostrand Reinhold Company Inc.
135 West 50th Street, New York, N.Y. 10020

Van Nostrand Reinhold Publishing
1410 Birchmount Road
Scarborough, Ontario MIP 2E7, Canada

Van Nostrand Reinhold
480 Latrobe Street
Melbourne, Victoria 3000, Australia

Van Nostrand Reinhold Company Limited
Molly Millars Lane
Wokingham, Berkshire, England

15 14 13 12 11 10 9 8 7 6 5 4 3 2 1

Library of Congress Cataloging in Publication Data

Main entry under title:
Handbook of experimental pollination biology.

   Includes index.
   1. Pollination. 2. Pollination—Research.
I. Jones, C. Eugene. II. Little, R. John.
QK926.H27 1983   582′.016   82-15998
ISBN 0-442-24676-5

199849

To Our Future,
Douglas, Philip, and Elizabeth Jones
and
Jeffery and Branden Little,
and to
Our Parents

PREFACE

The idea for this book grew out of a much simpler plan to print the results of a symposium on "Experimental Pollination Biology," held on October 25, 1980, at California State University, Fullerton, California (CSUF). The first coeditor (CEJ) was responsible for organizing this program and inviting the speakers: Herbert Baker, Lynn Carpenter, Gordon Frankie, Gene Jones, Peter Kevan, and Chris Plowright. The symposium was cosponsored by by Southern California Botanists and the Biological Sciences Department at CSUF.

After the symposium, we felt, as did Van Nostrand Reinhold editor Susan Munger, that the subject of experimental pollination biology required a more comprehensive treatment. Using the symposium speakers as a nucleus, we expanded the number of chapters and added several new topics (sections). Beginning in early 1981, contact was made with the appropriate authors, so nearly all of the chapters were written especially for this book.

The purpose of the *Handbook of Experimental Pollination Biology* is to critically review, examine, and discuss issues that are important to modern pollination research. One of the focal points of this volume is the presentation of experimental approaches and methodologies that can be useful or have potential for advancing the scope of pollination research. Basic "syndromes" of pollination (e.g., bird, insect, bat) have been well studied and described. However, recent pollination research has turned to investigating various components of plant–pollinator interactions.

The section topics reflect significant areas of current research. The chapters within a section vary in approach and emphasis: some are primarily comprehensive, whereas others interweave review materials with combinations of experimental and observational data, discussions of methodologies, model building, and tests of hypotheses. We wish to emphasize that there is considerable difference of opinion among pollination biologists on many issues. The discerning reader will note a fair number of differences between the authors in this book. We feel this is healthy. One of the authors even told us that he *hoped* his comments would stir up controversy. We agree wholeheartedly!

We especially acknowledge the Board of Directors of Southern California Botanists (of which we are Directors) for the opportunity to produce this volume in conjunction with their annual symposium series. We also appreciate the cooperation of the Department of Biological Science of California State University, Fullerton in cosponsoring the Pollination Biology Symposium. Special thanks to Cynthia Little, Cynthia Troyer, Coleen Cory, Mitch Cruzan, and Robert Fulton for helping to proofread various portions of the manuscript. The mention of a proprietary product in this book does not constitute a recommendation of this product by the USDA or any of the authors.

C. Eugene Jones
California State University, Fullerton
Fullerton, California 92634

R. John Little
Rancho Santa Ana Botanic Garden
Claremont, California

INTRODUCTION

In order to understand the pollination ecology of a species it is necessary to live with its populations and observe the plants at different times of day and night, under different weather conditions, and at different stages of the flowering season.... We have managed to be in the right place at the right time to see the course of pollination in full swing....

Verne and Karen Grant. 1965 *Pollination in the Phlox Family*. Columbia Univ. Press, New York.

This handbook is divided into eight sections, each consisting of two to four chapters. The sections are organized to emphasize particular topics in experimental pollination research. However, the information in a chapter is not necessarily limited in scope to the section in which it is placed. Indeed, most chapters encompass material from more than one section. This is a reflection of the interdisciplinary nature of many of these studies. In the following pages we give a brief overview of the material presented in the eight sections of this book, and a synopsis of each chapter.

SECTION I: FLORAL CUES AND ANIMAL BEHAVIOR

In the four chapters of this section many different aspects of floral cues and their relationship to pollinator behavior are discussed. These chapters, in dealing with the topics of color cues, post-pollination floral changes, floral fragrances, and buzz pollination, all address contemporary subjects of active research. Although the emphasis here is on bee behavior, the reader should be aware that floral cues are of equal importance with other vectors as well.

The adaptive significance of floral color as it relates to biotic pollinators and angiosperm diversity is discussed by Kevan. He first analyzes daylight in relation to color vision in insects, and then reviews the trichromaticity method he developed for determining, comparing, and graphically presenting the color of objects as seen through an insect's eye. A thorough review of the various techniques available for measuring color is presented, which will be valuable to those interested in methods used to analyze color. He emphasizes that when investigating the spectral reflectance of flowers, the investigator should not limit analysis to a single color (e.g., ultraviolet), but rather should consider the entire spectrum. Kevan completes his review with a discussion of the importance of floral color in relation to pollination communities, insect associations and behavior, biogeography, taxonomy, and evolution.

Gori begins with a review of the literature on the very interesting phenomenon of floral changes that occur in response to either pollination or aging. Such changes result in the flowers becoming less conspicuous, unattractive, or less accessible to the pollination vectors. Gori raises the question of the adaptive significance of such floral color modifications. Three alternative hypotheses are presented, and the critical predictions of each are outlined. Data gathered for a lupine, *Lupinus argenteus,* are used to test these hypotheses. Gori emphasizes that more empirical studies are needed to evaluate his three hypotheses relative to the importance of the male vs. female components of plant fitness in the evolution of post-pollination floral changes. He also suggests that comparative studies of congeneric species that do and do not undergo post-pollination color changes will be invaluable to our understanding of the relationship between flower retention and pollination behavior that results in floral changes in some species but not in others.

Floral odors are often extremely important as long-distance attraction and/or short-distance orientation cues for pollinators. Williams begins Chapter 3 with a review of the literature dealing with the chemical identification of floral fragrance components. A great many of these studies have been conducted with orchids, although other families have also been investigated. Even though the basic components of floral odors have been identified, relatively little work has been done regarding how pollinators actually respond to floral fragrances. Of interest to the pollination biologist, is an in-depth consideration of the methods used for collecting and identifying floral fragrances. Williams concludes with several excellent suggestions for future investigations that pertain to floral fragrances and their role in animal behavior and pollination biology.

The characteristics, distribution, and history of vibratile, (or buzz), pollination in flowering plants are reviewed by Buchmann. Vibratile pollination occurs in many different taxonomic groups. After discussing the various adaptations and contrivances of both "buzz pollinated" flowers and their pollinators, primarily bees, he presents two biophysical models to explain how pollen is ejected from poricidal anthers by buzzing. The topics of electrostatics and energetics in relation to buzz pollination are also presented. (An elaboration of the electrostatic aspects of pollination is presented in Chapter 8, Section II.) Buchmann concludes with some interesting suggestions for future research on buzz pollination.

SECTION II: CHEMICAL AND PHYSICAL ASPECTS OF POLLINATION

In this section, the chemical constituents, diversity, energetic potential and evolution of the various floral rewards, particularly nectar and pollen, are discussed. Physical aspects relating to pollination are reviewed in terms of the chemical basis of floral coloration. A consideration of the phenomena of electrostatics is presented. This new and potentially exciting approach may be of great value in understanding how pollen is transferred from plant to insect and vice versa, and how pollen is "attracted" to female flowers, cones, etc., in wind-pollinated plants.

In a thorough review of the literature dealing with sugars in floral nectar, Herbert and Irene Baker present an account of their own extensive work with some 765 species from around the world. Stressing that comparisons of results by different workers are very difficult, they present a concise explanation of their own methods for determining the composition of floral nectars. A discussion of the types of sugars found in nectar follows, and they report a general trend showing that species in the same family, and plants pollinated by the same class of pollination vectors, have similar sugar ratios, even, in the latter group, when the plants are taxonomically related. To illustrate the probable relationship between nectar sugar ratios and pollinator class, they review the experimental results of studies with honeybees and hummingbirds, and also report on observations involving bat flowers, and moth, butterfly, bee, and other insect flowers. They conclude with a consideration of some thoughts for future studies which will be of interest to workers in many fields.

Simpson and Neff have contributed an interesting chapter dealing with the diversity of floral rewards and the types of evolutionary constraints that appear important in their selection. Stating that these constraints have received comparatively little discussion, they contend that it is important to understand why some floral rewards are so successful and others so limited in distribution. They note that rewards that are easily handled or consumed, or that contain sufficient calories, appear to have been selectively favored over rewards that necessitate special adaptations. Ways by which reward substances might initially have been collected are examined in relation to physiological, structural, and behavioral adaptations of modern pollinators. In addition to pollen and nectar, which are the most common modern rewards, other substances such as oils, sexual attractants, resins and gums, food tissues, and brood places are also discussed. Some excellent suggestions for future studies are given in the concluding remarks.

Among the factors important for attracting pollinators are the visual signals presented by flowers and perceived as colors. Scogin reviews the relationships between floral colors and pollinator class and emphasizes that pollinators are not enticed to visit a flower by any single attractant, including color. The question of whether pollinator color preferences are genetically determined or are learned is raised. The three major chemical classes of floral pigments (flavonoids, carotinoids, and betalains) are reviewed, and the chemical and physical factors affecting the expression of flower color are discussed. A section on methods outlines basic procedures for extraction, purification, and identification of the three major classes of pigments. Scogin comments on the nonuniformity of floral colors in relation to geographical regions and phytosociological communities and concludes with some directions for future research.

In what may be considered a pioneering paper, Erickson and Buchmann review the evidence suggesting that the electrostatic forces surrounding plants and insect pollinators may be much more important in various pollination processes than most biologists realize. After introducing the subject of electrostatic potentials in plants and insects, they discuss the applicability of these properties in relation to pollinators. Because insects acquire significant positive electrical potentials during flight, and since plants bear a negative charge, the differences in electrical potential are known, or are believed to be implicated in a number of important biological phenomena. Pollen discharge and attraction in both wind- and insect-pollinated plants, spore discharge in fungi and nonvascular plants, and host-seeking behavior in certain insect species are just some of the areas that appear to be significantly influenced by electrostatic fields. This chapter should engender both controversy and numerous ideas for future research.

SECTION III: POLLINATOR ENERGETICS

In essence, pollinator energetics is a study in economics. Some have likened the "currency" in this system to plant products such as nectar and pollen. The plants invest in their production, which costs energy, while the pollinators that forage for them do so, also at an energetic cost. The plants and pollinators may be thought of as trying to maximize their investment by getting as much "return" as possible for the energy expended. However, as Hein-. rich points out, energy investment for pollinator and plant payoffs is only one of several possible ways to view these interactions, since neither the foragers nor the plants are governed solely by energetics. The approaches that an investigator can take to analyze and quantify these strategies are numerous.

Well-known for his work with pollinator energetics, Heinrich presents an overview of the subject and a synthesis of recent work on insect foraging energetics and models of optimal foraging. A useful discussion on methodologies pertinent to energetics studies is presented for such topics as measuring nectar volume, determining sugar concentrations, estimating energy budgets, and marking one or many bees. After reviewing numerous studies that involve optimal foraging, he points out that optimal criteria are established by the investigators themselves. He stresses the need to concentrate on learning more about pollinator behavior rather than on trying to make prior assumptions about what should be optimal. He cites several examples illustrating that some animals are better able to forage "optimally" than others because they have fewer conflicting constraints. An alternative "strategy" may be that of "inoptimal" foraging; and although few data are available on this subject, Heinrich suggests that floral mimicry may be one of the best examples.

Concentrating on the energetics of pollination in avian communities, Carpenter presents the results of a series of five energetics hypotheses (attractant, cornucopia, nectar timing, flower timing, and elevation) that she tested in the Sierra Nevada of California. Her hypotheses were based on the assumption that visitation rates limit reproductive output in hummingbird-pollinated plants. However, since the criterion that visitation rates may be limiting was not met, the energetics hypotheses were rejected. This study provides further insight into the possibility that "dependable and thorough pollination" may be a major advantage of bird-pollination. This contrasts with several previous studies demonstrating that insect pollination does limit reproductive output in plants. Carpenter calls for intercommunity studies that would compare systems in which pollinators were abundant to systems in which pollinators were limited.

In a Chapter presenting both new data and review material, Paton and Ford address energetics questions through a study of Australian honeyeater behavioral patterns in relation to their forage plants. Emphasizing that few detailed observations have been made on the foraging behavior of these birds, they test the changes in frequency and duration of foraging bouts to assess the influence of plant traits and honeyeater size on levels of pollination. The authors find that smaller species of honeyeaters effect higher amounts of cross-pollination than larger species, that cross-pollination is highest in plants with greater floral displays, and that cross-pollination may be lowest during peak flowering.

The final chapter in this section, by Evenson, discusses energetics from the viewpoint of energy allocation in plants, a subject applicable to many areas of botany, including studies dealing with pollinator energetics. Evenson notes that resource allocation strategies are acted upon by natural selection and are expressed in terms of not only the total energy allocated, but also in terms of other limiting resources such as time and material. Furthermore, pollinators may be a factor limiting reproductive effort rather than resources. Such possibilities, he emphasizes, must be carefully considered in the interpretation of resource allocation data. He reviews in considerable detail four major areas that relate to reproductive energy allocation, and concludes with suggestions for future work in this interesting field.

SECTION IV: COMPETITION AND POLLINATION

Many kinds of competition can occur as part of the pollination and fertilization process. For example, competition may exist between pollinators to obtain the available rewards of the flowers in a population; it may also exist between simultaneously blooming plants for the "services" of the available pollinators. Competition may or may not occur if plants share the same pollinators. Although a considerable body of literature has accumulated on the subject of competition and pollination, controversy remains over the causes and results of many plant-pollinator interactions. Studies in this field can be interesting, yet highly involved, and much remains to be learned. Competition between interspecific pollen tubes in terms of differential growth rates has recently been documented in a natural population (Chapter 16). The full implication of this remains to be worked out, but it gives an idea of the possibilities for future research. For example, competition *between* ovules for attraction of pollen tubes is a subject that seems not to have been pursued but should be considered in the future.

Waser's central thesis in Chapter 13 is that competition for pollination is probably common in nature, although there is little supportive evidence that it is important in producing or maintaining differences in floral characters among sympatric plant species. After defining competition for pollination as "any interaction in which co-occurring plant species (or phenotypes) suffer reduced reproductive success because they share pollinators," he introduces two different types of interaction: competition through pollinator preference and competition through interspecific pollen transfer. Acknowledging that the latter is the more common of the two, and "implied in producing or maintaining floral character differences among sympatric plant species," he reviews supporting evidence in terms of three different approaches that one can take in studying competition. To obtain better information on the importance of competition in maintaining floral diversity, he advocates that more experimental studies be made.

One of the potential outcomes of competition for pollinators is the development of situations of floral mimicry. Many of these examples involve complex pollinator behavior patterns and intriguing floral adaptations. In

Chapter 14, Little reviews the subject of floral mimicry concentrating on the topic of food deception mimicry. Beginning with a discussion of the definitions of mimicry, he suggests several reasons why the standard "Batesian" and "Mullerian" terminology is not appropriate for describing plant and floral mimicries. Food deception mimicries are considered in terms of three categories: two (or more) taxa mimicries, automimicry (conspecific), and mimicry based on naivete. After considering examples of these, he reviews the topic of "floral mutualisms" and discusses the relationship of these with deceptive mimicries, concluding that mutualisms are fundamentally different from deceptions although both are part of a continuum.

Powell and Jones report on a recently completed study dealing with floral mutualism between two herbaceous species, one an annual and the other perennial, pollinated by bumblebees and solitary bees. In this system the model and mimic roles are not clearly defined. They emphasize that this is similar to what are often called "Mullerian" mimicries, although they reject this terminology in favor of "floral mutualism" or "mutualistic mimicry." Finding evidence of convergence in floral and vegetative characters in sympatric populations of their study species, they suggest that this results in mutual pollinator attraction. An important result of their study was to demonstrate a close correlation between visitation ratios and ambient air temperatures. This finding was significant in view of the different types of rewards available from each plant species relative to the foraging strategies of the pollinators.

Mulcahy, Curtis, and Snow discuss the relatively unexplored subject of pollen tube competition. Based on data from earlier experiments showing that rapidly growing pollen tubes give rise to plants that themselves grow rapidly, the authors suggest that differential pollen tube growth rates have important implications in population biology. After demonstrating pollen tube competition in a natural population of *Geranium maculatum,* the authors conclude that gametophytic selection may indeed be an adaptively significant phenomenon in some populations.

SECTION V: GENE FLOW AND POLLINATION

Gene flow in plants is accomplished by movement of pollen and seeds. In this section, we are primarily concerned with gene flow of pollen in entomophilous plants, and some factors that influence the behavior of pollinators and, hence, the movement of pollen.

In a chapter dedicated to Fritz Müller, Wasner and Price critically examine the concept of genetic similarity between mates and its relationship to "optimal" and "actual" outcrossing distances. They show that genetic similarity between mates is the most important factor in defining optimal mating for plants. The authors predict that outbreeding depression can be detected between members of the same population for species that have localized gene flow. They suggest that a small physical distance between mates, known as an "optimal outcrossing distance" can be detected, and then demonstrate that this indeed occurs with their study plants. A thought-provoking discussion is provided in answer to the posed question: "Do plants achieve optimal outcrossing?"

Frankie and Haber examine the question of why bee pollinators move between mass-flowering trees in a tropical forest. After demonstrating that the nectar resource base is constantly changing in mass-flowering trees, the authors suggest that the bee pollinators, in the course of continuously sampling flowers for changes in flow rate (and probably concentration) of nectar, effect outcrossing.

SECTION VI: TEMPERATE AND TROPICAL PHENOLOGICAL STUDIES

Timing, duration, and frequency of flowering are important factors related to the behavior of pollinators. These factors are the essence of phenological studies. Phenological observations can be made at the community, population, or individual level. Plants exhibit a myriad of flowering patterns that, coupled with tremendous variation in sexuality and reproductive strategies, offer a complex array from which the pollination biologist tries to make sense. There is considerable challenge in trying to piece together such complex relationships. Phenological studies can also involve a consideration of competitive interactions between plants and pollinators, and analyses of gene flow. Presented in this section are three chapters that review, from different perspectives, temperate (Chapter 19) and tropical (Chapters 20 and 21) phenological studies. We learn that whereas many temperate studies have progressed into a hypothesis-testing stage, the majority of tropical studies are descriptive.

Pleasants has contributed an excellent review chapter in which he addresses the question, "To what extent can the structure of communities be explained by competition for pollinators?" Although this chapter could have been included in Section IV, it is placed here because it also considers phenological studies. Arguing that many papers dealing with floral phenology patterns lack statistical verification, Pleasants discusses some of the approaches that have been taken to provide statistical tests for regularity of flowering phenologies. He makes an important point in stating that sequentially flowering plants may be involved in mutualistic as well as competitive interactions, and he discusses the possible effects on floral phenology of two types of mutualism. After answering the questions " . . . do we really expect competition to produce a regular dispersion of flowering times?" and "Are rare species more attractive than common species?", he summarizes his interpretation of the competition hypothesis. Commenting on four factors that tend (or at least are believed) to reduce the interference component of pollinator competition, he states pragmatically that it is difficult to prove that interference competition was responsible for the evolution of these factors. The chapter concludes with a discussion of community structure, directed primarily toward bumblebee communities.

Bawa presents a well-organized and rather complete review of the literature that deals directly or peripherally with aspects of phenology in tropical flowering plants, particularly at the species and infraspecific level. After discussing the implications that flowering patterns have for many fields of botanical research, he emphasizes that the ecological basis of such variation has not received much attention. He reviews the topics of timing, duration, and frequency of flowering with intersexual variation in these factors and variation among sexual systems. He concludes with a succinct discussion of five factors that limit our current ability to understand different selective pressures.

Frankie, Haber, Opler, and Bawa have produced a "compendium" of information relating to the large bee pollination system in the Costa Rican dry forest. Characteristics of large bees and large bee flower (LBF) plants are discussed in relation to nine categories of phenological, reproductive, and pollination factors. Much useful natural history is revealed in the observations recorded here. Studies of this sort provide a model that others can use to make initial descriptive studies and surveys of plant communities. Such studies can be used as a springboard for more quantitative and analytical studies on phenological patterns in tropical dry forests. In their concluding remarks, the authors note the extreme difficulty of demonstrating experimentally the constraints to community organization that plant–pollinator interactions must pose in tropical forest ecosystems. We hope this will serve as a challenge to future workers.

SECTION VII: SYSTEMS ECOLOGY IN POLLINATION BIOLOGY

Systems ecology may be a new concept to some, but the idea behind a systems approach to research is certainly not new. The premise is that by subdividing a complex activity into small units, one can more easily study the separate parts of the whole process. Such approaches, for example, are common in business and engineering disciplines. The ultimate goal of the systems approach in ecology is to describe relationships mathematically. An intermediate step in this process involves model building. This is beginning to be done for certain aspects of pollination biology, but as the contributors to this section point out, the task is not easy.

Thompson has contributed a chapter that

serves as an "annotated checklist of complications to watch for" in relation to studies aimed at deciphering the question, "What is the nature of the interaction between two plant species populations that overlap substantially in time of bloom and in pollinator use?" Recognizing that pollination systems are often complex and cross-connected, Thompson identifies two extreme strategies for approaching community studies: "top-down" and "bottom-up." He argues for a bottom-up approach that identifies the components of the interaction. By specifying the behavior of the small system subunits, one can model system behavior either verbally or mathematically. Finally, in pointing out potential flaws of experimental design of past studies, he maintains that to find the truth about pollination systems one will usually have to dissect them and spend some time and effort on model building.

Waddington, after reviewing the literature on "flower-constancy" and mentioning several weaknesses in current techniques used for its analysis, suggests a more appropriate terminology and strategy for defining and describing a bee flower-visiting behavior: floral-visitation-sequence (F-V-S). This method, when combined with statistical tests and a probability matrix, can provide both qualitative and quantitative data. By understanding the relationship between foraging efficiency and all possible foraging behaviors, Waddington predicts that it will be possible to determine the "best" behavioral pattern. He discusses two examples of research currently in progress that attempts to test various components of the F-V-S system. For behavioral studies he prefers programs that combine laboratory experiments with field investigations. This is certainly a highly desirable approach. In closing, Waddington argues that "By formulating hypotheses that can be tested, the set can be constructed rule by rule." These rules in turn form the building blocks of models that can be used for predicting the F-V-S. This method of dissecting the hypothesis into smaller parts is also that advocated by Thompson.

Lertzman and Gass lucidly discuss the details of how one aspect of pollination, pollen transfer, can be studied through a systems approach. Beginning with a consideration of why models should be built for pollination studies, the authors mention several key points about the advantages of systems approaches in pollination ecology and the role of models. Two types of models are recognized, "pattern-oriented" and "process-oriented." Four alternative models of pollen transfer processes are discussed, which encompass both pattern and process models. Each more complex model adds new components to the determination of pollen transfer: reproductive state, finite stigmal surface area, variability in floral morphology, and depth of the pollen pool. The "most important and general conclusion" that they obtained is that "simple assumptions about pollen pool structure and the rules for pickup and dropoff" can greatly affect the computer-generated results for the amount of pollen actually carried over. Finding that all of their models predicted the majority of pollen will be deposited a short distance from the source plant, they empasize that significant amounts of pollen may, nevertheless, be carried quite far from the source, and may have important biological implications.

SECTION VIII: APPLIED POLLINATION ECOLOGY

Applied pollination ecology, when referring to entomophilous plants, refers to the study and use of animals, particularly bees, to effect cross-pollination for the purposes of breeding plants for crop improvement and hybrid seed, production of fruit crops, etc. Beekeeping and bee hybridization can also be subjects of great importance to applied pollination biologists. One reward to be realized from studies of pollination ecology is the opportunity to apply the study results to benefit agriculture. Contributions are made in a direct manner by pollination biologists working with agricultural crops, and indirectly by those who concentrate on various aspects of nonapplied pollination biology (e.g., experimentalists, theorists, and natural historians). Communication between applied and nonapplied pollination biologists, as pointed out in Chapter 28, is not as good as it could and should be. The purpose of this section, therefore, is to emphasize the advances and contributions that "applied" scientists have made to pollination biology, and to encourage an improved awareness of these benefits by scientists specializing in other fields.

Erickson's chapter consists of two parts: first a review, and then a collection of reports by other authors. He first presents an overview of the subject of pollination and hybridization in insect-pollinated crops, and makes an optimistic statement about the vast potential that exists for improving crop quantity and yield through modern methods of plant breeding. He also enumerates the many types of problems that plant breeders face when attempting to produce hybrid seed from hybrid seed parents. The underlying thesis of this chapter is that insect pollination of some hybrid crops may be exceedingly difficult to achieve. Such problems can have serious financial effects for those who produce the hybrids and for those who grow them for profit. In the second part, Erickson has presented a collection of short papers dealing with different hybrid crop plants (e.g., alfalfa, carrots, cotton, crucifers, and sunflowers). These reports illustrate the problems that can be encountered in attempting to cross-pollinate hybrid seed parents; they also are informative in terms of providing examples of experimental design, and they show the extent to which honeybees can be manipulated.

Estes, Amos, and Sullivan present a review that attempts to bring together some important contributions of both agricultural and experimental pollination biologists. Noting that these groups are concerned with precisely the same issues, they point out that unfortunately communication among pollination biologists in the agricultural and biological sciences is not as good as it ought to be. The same can be said for pollination biologists and plant systematists. Suggestions are made to remedy this situation. The review concentrates on the topics of foraging behavior and breeding systems because of their importance in agricultural systems. First, the potential for manipulation of honeybees is discussed in relation to agricultural crops. Second, various aspects of foraging for pollen by solitary bees and a consideration of oligolectic behavior are discussed in relation to pollinator behavior. The authors emphasize that a "cross-pollination" of ideas from agricultural and biological disciplines would be very productive. They call for a general model of optimal foraging strategy that would permit agricultural scientists "to test the effects of perturbations, singly and in tandem."

CONTENTS

Section I
FLORAL CUES AND ANIMAL BEHAVIOR

1
FLORAL COLORS THROUGH THE INSECT EYE: WHAT THEY ARE AND WHAT THEY MEAN

Peter G. Kevan

Department of Environmental Biology
University of Guelph
Guelph, Ontario, Canada

ABSTRACT

To understand the full meaning of floral colorations one must accept that they have an adaptive value, or had one in the past. This latter thought concerns the evolution of autogamy and agamospermy in plants that have retained floral characteristics of their entomophilous ancestry. The adaptive significance is in pollinator attraction, wherein evolutionary radiation of floral coloration through time would be expected as floras became increasingly rich and the need increased for pollinators to distinguish their resource flowers and for flowers to have accurate intraspecific pollen flow.

An analysis of floral coloration must take into account the spectral sensitivity of pollinators and their abilities to distinguish colors. At the same time, an understanding of the spectral sensitivites of insects must accommodate the photic environment in which they live. That is, there is relatively little ultraviolet as compared to the other insect primary colors (blue and yellow), which most insects studied use in trichromatic color vision. This explains why insects are so sensitive to ultraviolet. In an ecophysiological context, they boost the importance of the depauperate primary color. For this reason an equal-energy white is not a natural, or appropriate, starting point for colorimetry with respect to insect color vision. Equiproportionate reflectance in the insect visible spectrum is the meaningful datum.

Using that concept, color measurements with respect to insect color vision can be made. I have presented a simple trichromaticity method to this end (improvements are needed). Floral colors, and colors of other objects of interest to insects (e.g., prospective mates, predators, prey, oviposition sites, etc.) can be measured and plotted on spectral reflectance graphs, and more comparatively on trichromaticity triangles. These enable a direct comparison of the colors of objects and graphically present the differences in colors. The floral colors of entire floras can be presented and show much more spread and diversity of color than is evident to humans.

Although ultraviolet is invisible to humans and, in reflection and absorption patterns on flowers, provides much of the color diversity for insects, it is no more important than any other primary color in terms of floral attractiveness or plant reproductive strategy.

Floral colors must be taken in their full analysis to be understood in terms of function. Color preferences by different insects, color changes in flowers, and phenological floral colors in plant communities are all important aspects of pollination ecology. It is difficult to generalize on the importance of floral color polymorphisms. They may be related to pollination, or other factors, or both. Each case needs independent examination. Nevertheless, floral color polymorphisms may act as incipient isolating mechanisms, some of which have been suggested to encourage sympatric speciation. Floral color differences at the level of the pollination community must be important as a clue for pollinators to accurately forage and pollinate intraspecifically. As such the colors act as an isolating mechanism for the plants. Given that, it is understandable that floral pigments and coloration should be used as taxonomic characters and as clues to phylogenetic affinities of plants.

KEY WORDS: Floral colors, ultraviolet, insect color vision, white defined, insect colors, colorimetry, trichromatic color vision, pollination, plant communities, phenology, polymorphisms, isolating mechanisms, crypsis, mimicry, floral color changes, taxonomy, evolution.

CONTENTS

INTRODUCTION

The phenomenon of color has long captured people's imaginations. At first this interest was in arts and crafts, in painting and coloring. As Western civilization progressed, more analytical examinations of color came about (see Küppers, 1973; Gerritsen, 1975; Verity, 1980). Color classifications have been actively pursued, and with them attempts to measure color have been made. One of the most widely accepted schemes for color measurement has been developed through the trichromatic (i.e., three primary colors) theory of human color perception, as based on the works of Newton (1704), Young (1807), and Helmholtz (1852a,b,c, 1924–25). Representation of colors accordingly is on a trichromaticity chart that started as a color triangle (Mayer, 1745,

in Verity, 1980; Maxwell, 1855). The chart progressed as more understanding of the psycho-physiological aspects of human color perception was gained, so that now the chromaticity chart of the CIE (Commission International de l'Eclairage), adopted in 1931, is the standard. These matters are well discussed in Judd and Wyszecki (1963), Wyszecki and Stiles (1967), and Wright (1969).

This chapter is about colors and their significance to insects and pollination. It is not possible, at the moment, to produce a chart that reflects the psycho-physiological appreciations insects may have, although attempts have been made by Daumer (1956), Mazokhin-Porshnyakov (1969), and, with recent success from physiological studies, by Schlecht (1979). Furthermore, as will be shown, insect color vision is not the same in all Insecta, even where it exists. For the purposes of this chapter, I will restrict considerations to a simple triangle for representing color in the insect visual spectrum (IVS).

How colors may appear to insects is especially important in pollination biology. Insects and angiosperms are the major elements of the biota. They have coevolved since the Mesozoic, and burgeoned into prominence in Cretaceous time. Flowers attract insects by providing visual and olfactory stimuli to insects seeking food rewards in flowers. In return, the insect visitors transport pollen to subsequently visited flowers and so may bring about pollination. There is a wide variety of flowers and insects, and flower–pollinator combinations are often very precise and often quite general. Thus, insects should recognize the flowers that provide for their needs and the flowers, in turn, advertise to the insects that fit their needs. The ad-

vertisement to be discussed in this chapter is color. If one is to understand the role of color in pollination, one must try to observe like insects and understand what insects see and how they see. Our ability to do this is limited, but I hope this chapter will provide some insights, based on firm principles and approximate data, into measurement and meaning of floral colors with respect to insects.

To start, one must establish some ground rules. First the light received in the insect eye needs examining. This requires a knowledge of the photic environment of ambient light, and an understanding of the light reflected from objects of interest to insects. The visual appreciation that insects possess must be taken into account with respect to both ambient light and reflected rays to understand color vision in insects. All the above must be placed in ecological, evolutionary, and coevolutionary contexts to understand the workings and consequences of color to insects in the terrestrial biosphere.

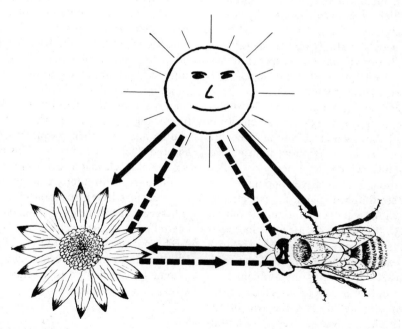

Figure 1-1. Pollination ecology and daylight. Dashed lines show the paths of light, emitted and reflected. From the sun and sky light makes up the photic environment for both animals and plants. Light is reflected from blossoms to the eyes of anthophiles, through which colored images are perceived. Solid lines represent paths of adaptation in an evolutionary sense. Both blossoms and anthophiles are adapted to the photic environment in terms of their reflectance and spectral sensitivity, respectively. The double-ended arrow indicates that co-adaptive relationships come about in pollination. The solid line to the insect can also represent physiological adaptation to diel changes in the quality of light.

Fig. 1-1 presents the relationships of daylight, insect color vision, and the colors of objects of interest to insects.

THE PHOTIC ENVIRONMENT: DAYLIGHT

Daylight is difficult to define precisely. Henderson (1970) has reviewed research into daylight and writes that determinations of daylight "have scarcely progressed beyond the collection of many spectral curves on different days and at different times, usually with normalization of the curves at a selected wavelength . . ." (pp. 187–188). Nevertheless, generalizations can be made. For color measurements, it suffices, within limits, to use the relative amounts of energy in daylight (i.e., normalized curves). Most (65%) of the solar energy reaching the surface of the earth is in the animal visual spectrum (i.e., about 300–780 nm). Also, the relative amounts of energy in that spectrum are more constant than in the infrared, which is greatly affected by atmospheric water vapor, CO_2, aerosols, dust, etc., which vary widely spatially and temporally.

Even so, in the animal visual spectrum the distribution of energy in daylight depends on various atmospheric factors. The air mass through which sunlight must travel is important and includes the angle of solar elevation and elevation above sea level. Sunlight reaches the surface of the earth as direct light and as scattered light, which together are called global radiation. The relative amount of energy in scattered light depends on aspect; e.g., southern skylight is paler blue than the darker northern skylight. When the air mass increases greatly, as at dawn and dusk, the relative amount of energy in the red part of the spectrum increases. This affects the ambient light and our perception of colors. In shade, ultraviolet is less attenuated than the rest of the spectrum. In cloud cover shorter wavelengths seem more attenuated (Möller, 1957).

A useful way of measuring the color of daylight (and other sources of radiation) is through correlated color temperature (CCT). The Planckian relationship provides a formal baseline for calculating the amount of energy at any wavelength being emitted from a fully radiating body at any given temperature. Thus, we can relate the temperature of a heated object as it goes from red-hot through to white-hot with its color. The surface temperatures of stars are measured in this way from the intensely hot blue-dwarfs to the relatively cool red-giants.

Generally, the CCT of daylight lies in the white-hot between 5500°K and 6200°K. A useful standard is 5800°K. Figure 1-2 shows the relative amounts of power in daylight for CCT of 4800°K to 10,000°K (a dark blue north skylight) compared with Planckian radiators from 4000°K to 10,000°K. The departure of the daylights from the Planckian curves at the shorter wavelengths (especially ultraviolet) should be kept in mind.

It is convenient to plot the chromaticity, or color of the radiation, on a chart in which the three primary colors are represented. This can be done on a trichromaticity chart or triangle, the apices of which represent the three primary colors. For humans, blue (436 nm), green (546 nm), and red (700 nm) are used. MacNichol (1964) and Rushton (1975) discuss three-pigment color vision and color-blindness in humans. Land (1977) has offered an alternate, retinex, theory to explain human color vision. Light, composed of an even mixture of all three, appears white. If the mixture is blue and red, then purple results; if green and red, then yellow results. This is the way in which color television sets operate, through illumination of differently colored phosphors on the screen. Close inspection of color printing will reveal the same principle, except that reflected light is perceived.

The mixtures of emitted or reflected wavelengths can be plotted according to the relative amounts in each of the primary colors. The proportions then are:

$$b = \frac{R_b}{R_b + R_g + R_r},$$

$$g = \frac{R_g}{R_b + R_g + R_r},$$

$$r = \frac{R_r}{R_b + R_g + R_r}$$

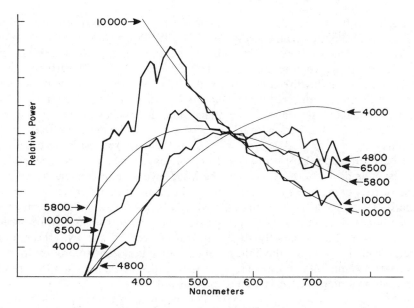

Figure 1-2. The relative power in daylights of different correlated color temperatures compared with Planckian radiators at 4000°K, 5800°K, and 10,000°K (smooth lines). Daylight curves for 4800°K and 10,000°K from Judd et al. (1964) and for 6500°K from Henderson and Hodgkiss (1963).

where R is the reflectance (or emittance) in *b*lue, *g*reen, and *r*ed. A color may be plotted by its coordinates (r, g) where r is the abscissa and g is the ordinate. The value of $b = 1 - (r + g)$. The coordinates can then be plotted on a triangular chromaticity diagram in which b has the coordinates $(0, 0)$; r is located at $(1, 0)$; and g is at $(0, 1)$. White will be located at $(0.333, 0.333)$ and yellow at $(0.5, 0.5)$.

Thus, daylight spectral curves can be used to plot the chromaticity of the daylight on a chromaticity diagram. For the immediate purpose (but see below) it is useful to define the white-point in terms of equal energy in the three primary wavelengths. That is, the white-point on the chromaticity triangle will fall at $(0.333, 0.333)$. As will be demonstrated, this is not a natural light, especially as far as insect vision is concerned. Figure 1-3 shows daylight chromaticities from a variety of sources ranging from dark blue at a CCT of 34,000°K to near white at 4800°K, together with the Planckian locus of CCT from below 2000°K to over 40,000°K. The equal-energy white-point E is also shown. From this figure one can see the blueness of daylight as the solid curves toward the blue corner of the chromaticity tri-

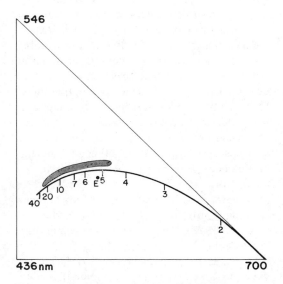

Figure 1-3. Area of chromaticities of daylights with correlated color temperatures from 4800°K to 34,000°K from various sources (see Kevan, 1978a) with the chromaticity locus for Planckian radiation from 2000°K to 40,000°K (large numbers along curve). E is the equal-energy white-point. These are presented on a color triangle for the human visual spectrum.

angle, arcing slightly above the Planckian locus which itself arcs above the equal-energy white-point E. An important point here is that in this chromaticity diagram, appropriate for human color vision (i.e., using the three human primary colors—blue, green, and red), the daylight loci and the equal-energy white-point are not far removed from each other, or from the Planckian locus.

Insects do not have the same three primary colors for color vision as do humans. Reviews on the spectral sensitivity of insects (e.g., Goldsmith and Bernard, 1974) point out the similarities among many and diverse insects. The insect visual spectrum is shifted entirely toward the shorter wavelengths of the daylight spectrum, from about 300 nm in the ultraviolet to 700 nm in the yellow orange. Research on insect color vision has been active for over 50 years. Honeybees [*Apis mellifera* (L.)] have been the most used experimental insect. Especially important to this chapter is the work of Karl Daumer (1956, 1958), who developed a technique whereby he could exclude different wavebands of light, or alter the brightness (or amount of energy) of those wavebands. He used the resulting colors of light to train honeybees, through carefully planned experiments using artificial feeding stations, differing (or not) only in color-discriminating abilities of bees. From these experiments he devised a trichromatic color perception model for honeybees. His results dovetail remarkably with electrophysiological studies (see Goldsmith and Bernard, 1974). Since then, color vision has been proposed for other insects, e.g., bumblebees (*Bombus* spp., Hymenoptera: Apidae) (Mazokhin-Porshnyakov, 1962), wasps (*Paravespula germanica* F., Hymenoptera: Vespidae) (Menzel, 1971), flies (Snyder and Miller, 1972), and various Lepidoptera (Strüwe 1972a,b, 1973; Swihart, 1972; Högland et al., 1973a,b; Schlecht, 1979). Not all have the same system as bees: some flies (Diptera) have only two-color vision (Burkhardt, 1964; Snyder and Miller, 1972); some butterflies may be red-sensitive (Strüwe, 1972 a,b; Swihart, 1972), and nocturnal insects may be color-blind (Moller-Räcke, 1953; Strüwe, 1973). Furthermore, there may be differences

depending on the structure of the compound eye, i.e., the apposition eye, such as that of bees, vs. the superposition eye (Kirschfeld, 1974; Schlecht, 1979). Nevertheless, most insects studied so far show three peaks of optical spectral sensitivity, one in each of the ultraviolet, blue, and yellow parts of the daylight spectrum (see Goldsmith and Bernard, 1974). Most of this research has been based on electrophysiology, especially from single cells of the insect eye. For understanding insect color vision in terms of its meaning to insects in nature, the behavioral approach is more appropriate. Therein all responses from the individual transducers of the eye through to the central nervous system and to resultant behavior are composed. Daumer's (1956) research is most useful.

For argument's sake Fig. 1-4 is a trichromaticity diagram with the three primary colors as used by Daumer (1956). Plotted on that figure are the same daylight chromaticities as used in Fig. 1-3, the same Planckian locus, and the equal-energy white-point. Bee white, as defined by Daumer's experiments, is also shown. It is clear that daylight is very different from equal-energy white, because of the attenuation of ultraviolet light by the oxygen

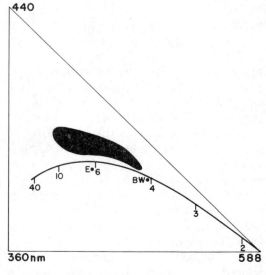

Figure 1-4. As Fig. 1-3 but plotted on a color triangle for the insect visual spectrum. BW is bee white, after Daumer (1956).

and ozone of the atmosphere. This raises the question of what color is natural light to insects, and what is white to insects.

THE PHOTIC ENVIRONMENT AND COLOR VISION: WHAT IS WHITE?

So far the discussion has been of daylight, with some consideration of spectral sensitivities of humans and insects. Humans are not equally sensitive to the three primary wavebands for their color vision, nor are insects. One might conjecture that the psycho-physiological phenomena in color vision would boost the organism's sensitivity to the wavelength(s) least represented. This would balance the three primary colors for color vision. Humans do this through adaptation as when viewing colored objects in rooms illuminated by relatively yellow incandescent lights, or by relatively blue fluorescent lights (Crawford, 1959; Ripps and Weale, 1969). Insects can also adapt to different photic environments (Thomas and Autrum, 1965; Högland et al., 1973a; Strüwe, 1973; Labhart, 1974; Menzel and Erber, 1978; Schlecht, 1979). More important is the general relative spectral sensitivity of insects. Insects are most sensitive to ultraviolet light, the light least abundant in their visual spectrum. That is, insects boost the importance of ultraviolet relative to the other primary wavebands for their color vision. I have pointed out (Kevan, 1978a) that a good relationship exists between insects' sensitivity, as represented by honeybees from Daumer's (1956) data and the amount of energy available in daylight in each of the primary wavebands. For daylight at $5500°K$, the bee's sensitivity multiplied by the percent of energy in daylight ($S_\lambda \cdot M_\lambda$) is close to one; for ultraviolet $S_\lambda \cdot M_\lambda = 1.08$, for blue $S_\lambda \cdot M_\lambda = 0.93$, and for yellow $S_\lambda \cdot M_\lambda = 1.05$. It is interesting that $S_\lambda \cdot M_\lambda$ for blue is the most constant for daylight CCT from $4800°K$ to $7500°K$ and ranges from 0.91 to 1.06. Blue is pivotal to adaptation (see above) (Menzel, 1971; Högland et al, 1973a; Menzel and Erber, 1978). One may conjecture that Daumer's experiments were performed with bees adapted to a daylight close to $5500°K$.

From the foregoing, it should be clear that equal-energy "white" is entirely inappropriate to insect color vision. It is far from a natural light and not part of the psycho-physiological processes of insects' seeing white. White must be redefined to incorporate those concepts and to make it a useful term ecologically. I have redefined white as follows: *"An object, the colour of which falls on the white locus of a chromaticity diagram constructed for any given colour sensor, is a diffuser which reflects in equal proportions all wavelengths in (i) a spectrum of natural light within the range of sensitivity of that given colour sensor or (ii) the primary colours of that colour sensor"* (Kevan, 1978a, p. 58). This definition is based on a white locus which is generated by equiproportionate reflectance across the visual spectrum. The color sensor and visual spectrum of concern here are of insects, although human or other organisms or machines are not excluded from this definition.

The concept of an equiproportionately reflecting white-point (which also grades through greys and includes black, as the latter is equiproportionately reflecting nothing) simplifies color measurement in the insect visual system immensely. The formulae given above are appropriate as they are. Further, color naming in the insect scheme becomes a matter of analogy.

MEASURING COLOR FOR INSECTS

It should be kept in mind that the chromaticity diagrams presented and used in this paper are simplistic. The CIE chart is curved at the apex, and those proposed by Daumer (1956, 1958) for bees and by Schlecht (1979) for *Deilephila elpenor* (Sphingidae) are also curved. The shapes of the CIE chromaticity chart for human colorimetry and the chromaticity diagram for color perception in *D. elpenor* (Schlecht, 1979) are more realistic as they include the effects of the overlapping curves of sensitivity for the three primary colors in humans and *D. elpenor* respectively. Research is under way to try to produce a generalized chromaticity chart for insects in which the three overlapping sensitivity curves calculated from nomograms (curves of receptivity with wavelength for visual pigments or cells) (Dartnall, 1953; Ebry and Honig, 1977), the relative sensitivity in each waveband, and the amount of light in each waveband in the photic environment, are all considered.

There are a variety of ways of measuring spectral reflectance. The most precise is by spectrophotometer with an integrating sphere (see Kevan et al., 1973; Kay, 1978). The most useful way to analyze floral colors and color patterns is by photography (see Daumer, 1958; Kevan, 1972, 1979; Mulligan and Kevan, 1973; Guldberg and Atsatt, 1975), or through video viewing (Eisner et al., 1969). Silbergleid (1976) discusses some techniques he has found useful. When one is using photographic or video images, there are important ideas to keep in mind. First, *ultraviolet reflectance alone does not give a full picture of the color of an object as it may appear in the insect visual spectrum.* I have pointed out (Kevan, 1979) that if one detects a pattern in ultraviolet, then a pattern may exist. Frohlich (1976) discusses artifacts in reflectance caused by flash photography. However, if one fails to detect a pattern in ultraviolet, then one cannot conclude that no pattern of interest to insects is present on the object (flower, insect, etc.). The other two primary colors are just as important as ultraviolet, as will be demonstrated below. Hence, measuring colors of objects as they may appear to insects must be made across the full gamut of wavebands representative of the insect visual spectrum. Daumer (1956, 1958) clearly appreciated this fact, yet it has been ignored by many contemporary researchers. Second, *one must know how much reflectance there is in each waveband.* To do this a grey-scale is needed. Kevan et al. (1973) described the construction of a grey-scale that is highly reflective in ultraviolet; commercially available grey-scales are not. Daumer (1958) used a scale built into his camera. Silberglied (1976, 1977) uses this technique and explains how to make a built-in grey-scale. This sort of grey-scale is not reflective, but is a transmitting scale made of increasingly darkening steps of neutral density filters. The importance of measuring reflectance is brought out in the work of Daumer (1958), Mazokhin-Porshnyakov

(1959), Kugler (1963), Kevan (1972, 1978a, 1979), Mulligan and Kevan (1973), and Guldberg and Atsatt (1975). I have shown (Kevan, 1979) that misinterpretation of photographic or video images can be made through overexposure. In that (1979) paper a properly exposed picture of the capitulum of a yellow composite shows hardly any ultraviolet reflectance, whereas overexposure (the control grey-scale in the comparison photograph is washed out) could lead to the conclusion that ultraviolet reflectance was strong. Third, *one must consider the background color against which the object of interest to insects should contrast if it is to be highly visible.* Frohlich (1976) and Kevan (1979) discuss this, and caution that color measurements be made of vegetation, soils, and so on. The light source is also important (see Frohlich, 1976), and I prefer daylight in shade (diffused light) to avoid problems caused by shadows in interpretation of reflectance measurements. Video cameras are sensitive to infrared, and ultraviolet-passing filters (and others) also pass infrared. Hence, video recording must be controlled to assure that one is not interpreting infrared images as those from the other part of the spectrum passed through the filters.

The aim of the technique I have outlined is to produce a spectral reflectance curve in the insect visual spectrum of the object of interest. This may be done as follows:

1. The object is set up for photography or video viewing with an ultraviolet and visible reflecting calibrated grey-scale alongside. Calibration of the chips or steps of the grey-scale requires the use of a reflectance spectrophotometer (see above).

2. The camera (photographic or video), preferably equipped with a quartz lens (glass lenses attenuate ultraviolet, especially so if coated for that purpose), is aimed at the object and grey-scale. Some quartz lenses are chromatically adjusted from ultraviolet through the human visual spectrum (Zeiss UV Sonnar); others are not (Pentax Quartz Takumar), and focus adjustments must be made for

short wavelengths (see Kevan, 1972). Ordinary glass lenses have been used successfully by various workers (Cruden, 1972; Horovitz and Cohen, 1972; Abrahamson and McCrea, 1977; Hill, 1977). Quartz lenses are expensive. Lutz (1924, 1933) avoided the problems by using a pin-hole camera. A bellows extension is often needed. A filter holder is attached to the front of the lens (these holders may not be available commercially and must be pieced together or built; see also Hill, 1977).

3. A series of broad-band monochromatic filters is used to capture the reflectance in each waveband. I have used square filters from Kodak and Ilford and recommend the following:

Kodak 18A	ultraviolet	(glass)	300–400 nm
Kodak 35	violet	(gelatin)	320–470 nm
Kodak 98	blue	(gelatin)	390–500 nm
Kodak 65	blue-green	(gelatin)	440–570 nm
Kodak 61	deep green	(gelatin)	480–610 nm
Kodak 90	dark greenish amber	(gelatin	540–650 nm
Kodak 25	red	(gelatin)	580 nm on
Kodak 70	dark red	(gelatin)	650 nm on

(See Kodak, 1970, for more information).

4. For each filter, a filter factor must be worked out. A through-the-lens light meter is helpful. Exposure bracketing from two times to eight times exposure values without a filter will provide, on one or two trial films, the best exposure of the grey-scale for each filter (see also Hill, 1977). Filter factors will differ with location too; at high elevations ultraviolet and blue exposures are not as prolonged as at sea level. Video cameras have built-in exposure systems so that the filter factor is less of a problem, but the machine will tend to give overexposures of dark images (see above).

5. Black and white films are sensitive in ultraviolet, and are not in infrared. I use Kodak Tri-X, which I force to ASA 1600 (rather than the ASA 400 designated) because I work outside in windy places and need fast shutter speeds to stop movement of the object in the wind. Also a small aperture is desirable to increase depth of field. Silberglied (1976, 1977) provides details on developing techniques.

6. Photographs, or video images, may now be made using no filter, and then progressing through the filter series adjusting the exposures as need be. A label for each filter should be included with the object as a record.

7. Once the image is recorded on film or video tape, one matches the reflectance of the object (or part of the object) with the reflectance of a chip (step) of the grey-scale in the same image. The matching can be done quite accurately by eye, although some sort of densitometry is more accurate. Also, an ordinary photographer's grey-scale can be used for comparisons on photographic prints. Thus, for each waveband one has the reflectance of the object of interest.

8. One then makes a table of wavelength and reflectance and plots a spectral reflectance curve (see below).

This process may seem rather involved, but is really no more tedious than many other scientific techniques. I am working toward improvements in the grey-scale construction and in reflectance measurements in step 6. These show promise and should hasten the process considerably.

Silbergleid's methods (1976, 1977, 1979) are designed especially for ultraviolet photography, but are readily applicable to spectral reflectance analysis with the series of filters I use. His techniques are also designed for use in the laboratory, as are those of Daumer (1958). The transmitting grey-scale dictates that a special background of white be used for at least part of the photographs. Eastman White Reflectance paint on metal is used, as it reflects close to 100% of the light from ultraviolet through red. This is necessary because it is the light reflected from the background that is transmitted through the grey-scale to produce the grey-scale on the photograph. My scheme imposes no limits on where a photo-graph can be taken, or on the background re-quirements: the reflectance grey-scale is merely placed with the object to be photo-graphed. My methods are designed for field work and at this time they are probably less accurate than those of Silberglied. However, improvements in the reflectance grey-scale, to-gether with video-enchanced and assisted den-sitometry of the photographic negatives, look promising as ways to increase the accuracy and speed of generating spectral reflectance curves.

NAMING COLORS FOR INSECTS

Figure 1-5 presents some spectral reflectance curves for a variety of colored surfaces, and Kevan (1972, 1978a) shows some curves taken directly from flowers. In Fig. 1-5 curve 5 shows equal reflectance across both the insect and human visual spectra, and is white (equipro-portionate white) to both. Curve 4 is white to humans, but is lacking ultraviolet. This is anal-ogous to curves 1, 2, and 3 to humans in that it lacks the primary color with the shortest wavelength (i.e., blue to humans; the ultravi-olet part of curves 1 and 2 is immaterial to

human color vision). Curves 1, 2, and 3 rep-resent yellow to humans. Analogously curve 4 represents insect-yellow. Curve 1 (and to a lesser extend curve 2) shows reflectance in ul-traviolet, not in blue, but again in green through red. The red is immaterial to insect color vision, so for insects these curves repre-sent reflectance in their primary colors with the shortest and longest wavelengths, ultravi-olet and yellow. This is analogous to curve 6 for humans, which shows reflectance in the blue and red primaries but not in green. This is purple. Thus, by analogy, ultraviolet (or in-sect-blue) and yellow (or insect-red) give in-sect-purple, a concept developed by Daumer (1958). Curve 6 to insects is lacking the pri-maries with the shortest and longest wave-lengths, having only reflectance as far as in-sects are concerned in blue (or insect-green by analogy).

Although we cannot tell how insects may see color, we know that they can distinguish colors. Naming the colors gives us an intuitive appreciation of how they may appear to insects.

So far I have discussed color names in a de-scriptive manner. By using the formulae (see above), one can take spectral reflectance data

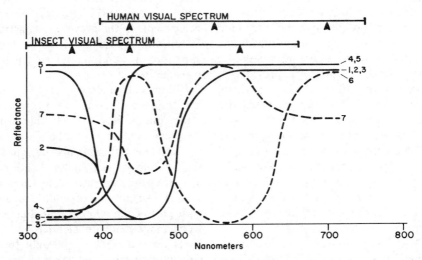

Figure 1-5. Spectral reflectance curves for some hypothetical colors. The human and insect visual spectra are given and the primary colors pointed out for each (arrowheads). Curve 1 is yellow, or insect-purple; curve 2 is yellow, or insect-reddish-purple; curve 3 is yellow, or insect-red; curve 4 is white, or insect-yellow; curve 5 is white, or insect-white; curve 6 is purple, or insect-green; curve 7 is greenish-yellow, or insect-mauve. (See also Figs. 1-6 and 1-7.)

and use them to plot colors on a chromaticity diagram as described above. The white-point is the equiproportionate reflectance white-point. This allows percent reflectance in the three primary wavebands to be used directly in the formulae. Figures 1-6 and 1-7 show the loci of the colors described by the curves in Fig. 1-5 in the human and insect chromaticity diagrams respectively. Many color loci can be plotted on the same diagram. This is useful for comparative study (see below).

A part of color measurement not considered in the chromaticity triangles is the combined effects of luminance factor and saturation, that is, the overall reflectance level (i.e., general reflectance may be low and one wavelength may show only a slightly higher level of reflectance; the color would be dull) and the amount to which the stronger wavelength(s) predominate(s) (i.e., if the stronger wavelength is not much stronger than the others, the color will be pale). The combined effects give rise to the paleness, dullness, brightness, and darkness of colors. To represent these factors in coordination with color, a color solid must be used (see Wyszecki and Stiles, 1967). Ideas of the importance of these factors in comparing colors can be gained from spectral reflectance curves.

Through plotting floral colors on a trichromaticity diagram for insects, and using analogous colors from the loci transferred to a diagram for humans, it is possible to make artistic

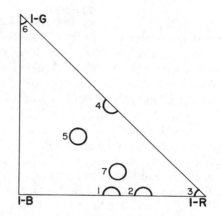

Figure 1-7. Trichromatic plots of the colors described in Fig. 1-5 on a color triangle for the insect visual spectrum.

approximations. G. Bennett and I have done this and published colored pictures of flowers as they appear to humans and may appear to insects (see Kevan, 1978a). The results are striking. Color contrasts within flowers and their color contrast against their background are heightened. Vegetation, which appears green to humans because of high absorption in red and reflectance in green, is fairly uniformly reflective across the insect visual spectrum, with a modest peak in insect-yellow (green). Thus, to insects vegetation is generally dull-yellow-grey (Kevan, 1972, 1978a), which we would consider a neutral background against which bright colors would contrast vividly.

ULTRAVIOLET: PATTERNS ON FLOWERS

That ultraviolet reflections exist on flowers has been known for nearly a century. Knuth (1891a,b) first photographed them. Lutz (1924, 1933) used a pin-hole camera for the same purpose. Richtmyer (1923), Lothmar (1933), and Seybold and Weissweiler (1944) used quartz spectrographs to detect ultraviolet and other reflectance, but could not discern patterns. For much of this early period, it must be remembered that the color vision of insects (especially of honeybees) was only beginning to be understood. Even after the classic studies of von Frisch (1913, 1914–15) on the color sensitivity of honeybees, and the work of Kühn

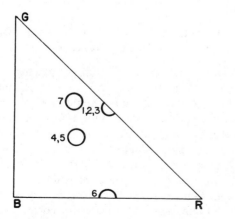

Figure 1-6. Trichromatic plots for the colors described in Fig. 1-5 on a color triangle for the human visual spectrum.

(1924, 1927) demonstrating that the honeybees' spectral sensitivity was entirely shifted to the shorter wavelengths (to include ultraviolet) compared with that of humans, color vision was not proven. The stumbling block was ruling out the importance of the levels of reflectance of differently colored objects being detected through tonal (i.e., black and white) vision. The sequence of investigations from von Frisch and Kühn to Lothmar (1933), Hertz (1937, 1939), and Daumer (1956) and Mazokhin-Porshnyakov (1962, 1969) has demonstrated trichromatic color vision in bees, for which ultraviolet is a primary waveband. Subsequently, color vision in other insects has been demonstrated. All, whether deuteranopic or including sensitivity in the red, show keen ultraviolet sensitivity (see Mazokhin-Porshnyakov, 1969; Goldsmith and Bernard, 1974).

Perhaps because ultraviolet is invisible to humans, it has been treated as something special and frequently investigated *in vacuo*. Under some circumstances this approach is valid; ultraviolet can be treated as a floral character. However, from an "ecobiophysical" viewpoint, that is severely limiting and prone to interpretational difficulties (Kevan, 1979). After all, examining the world through an ultraviolet passing filter to try to understand the insect's view, is like examining and interpreting the human world through blue spectacles. Silberglied (1979) provides an excellent review of "communication in ultraviolet" and gives a tabular overview of the works that consider ultraviolet reflections or patterns from flowers.

That ultraviolet is important cannot be denied, whether it is reflected, is absorbed, forms a pattern, or does not. Most examinations have been photographic (Kugler, 1947, 1963, 1971; Zeigenspeck, 1955; Daumer, 1958*; Mazokhin-Porshnyakov, 1959; Kullenberg, 1961; Mosquin, 1969; Kevan, 1970*, 1972*, 1978a*, 1979*; Ornduff and Mosquin, 1970; Thien, 1971; Thien and Marcks, 1972; Cruden, 1972; Horovitz and Cohen, 1972; Eisner et al., 1973a,b; Mulligan and Kevan, 1973*; Brehm and Krell, 1975; Guldberg and Atsatt, 1975;

*Indicates spectral analysis across the insect visual spectrum.

King and Krantz, 1975; Utech and Kawano, 1975; Frohlich, 1976; Kay, 1976, 1978; Abrahamson and McCrea, 1977; Hill, 1977; Eisikowitch, 1978; Silberglied, 1979; Parker, 1980; Schaal and Leverich, 1980). Recently video viewing has also been used (Eisner et al., 1969; Jones and Buchmann, 1974; Kevan and Laverty, unpub.*).

At present it is difficult to generalize about ultraviolet reflectance patterns. Guldberg and Atsatt (1975) sampled 300 plant species and indicated that size and ultraviolet patterning are positively correlated. They say that yellow and violet flowers are more likely to have ultraviolet patterning, whereas wind- and bird-pollinated flowers are less likely to be so adorned. (Interestingly, Goldsmith, 1980, has reported ultraviolet sensitivity in hummingbirds. Silberglied, 1979 reports on other vertebrates.) Guldberg and Atsatt show that some plant families—Amaryllidaceae, Geraniaceae, Fabaceae, Onagraceae, and Ranunculaceae—have a propensity for ultraviolet patterning, whereas others—Ericaceae and Polemoniaceae—show the opposite. Floral shape does not appear to be correlated with frequency of ultraviolet patterning (Kugler, 1963).

Very few flowers fall into the insect-blue (i.e., ultraviolet only) color category. Examples are *Papaver rhoeas* (Papaveraceae) (Lothmar, 1933; Kugler, 1947; Daumer, 1958) and a red horticultural variety of *Gaillardia aristata* (Asteraceae) (Daumer, 1958).

FLOWER COLORS IN POLLINATION COMMUNITIES

There have been few attempts at placing floral colors in a synecological context. Daumer (1958) examined a large number of species in Germany. I have studied most of the flora of a Canadian high arctic site (Kevan, 1972) and worked on Canadian weeds around Ottawa (Mulligan and Kevan, 1973). Now, I am completing research on the alpine flora of Colorado. Figures 1-8 to 1-11 respectively present the chromaticity coordinates for colors of arctic flowers in the human visual spectrum and the insect visual spectrum, and for Canadian weeds similarly. Comparison, for both habitats, of the loci of floral colors for humans and insects shows clearly that the diveristy, spread, and discreteness is much greater for insects than for humans.

Those features are important to anthophilous insects. The visual acuity of insects is much less than of humans (Wigglesworth, 1972; Goldsmith and Bernard, 1974) so that their ability to distinguish shape and form at a distance is relatively poor. The blotches of distinctive color that an insect would see at a distance would be an accurate clue for their recognition of the flower. This is important to the insects in foraging efficiency, in discrimination of flowers that they may easily manipulate or have learned to manipulate to extract rewards (see Laverty, 1980, for *Bombus*), and in flower constancy or fidelity through which the plant benefits by intraspecific pollen flow and greater efficiency in pollination.

FLORAL COLORATION AND INSECT ASSOCIATIONS

The color preferences of insects have been debated for over a century, and some generalizations can be made. Yellow flowers are often highly reflective and are visited by a large variety of insects. Unspecialized anthophiles, such as flies (Diptera) (Ilse, 1949; Kugler, 1950, 1951, 1956) and less advanced butterflies, such as Nymphalidae (Ilse, 1928; Tinbergen et al., 1942; Ilse and Vaidya, 1956), appear to prefer yellow flowers. Valentine (1975) concluded that yellow floral color is adaptively neutral. White flowers are also widely visited. Those with easily accessible nectar are visited by parasitic Hymenoptera (Kevan, 1973), although these insects also visit flowers of other colors. Nocturnally pollinated flowers are frequently pale, thereby being more visible to their moth pollinators than if colored otherwise (Baker, 1961). Ultraviolet reflection is often found in yellow flowers, making for insect-purple and insect-red patterns. Ultraviolet seems to impart no special attractiveness to flowers. From data in Kevan (1972) and Mulligan and Kevan (1973) there are highly visited flowers and rarely visited flowers with ultraviolet reflecting patterns, and there are flowers with

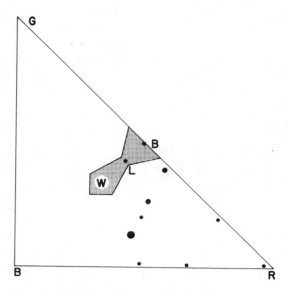

Figure 1-8. Trichromatic plots of the colors of flowers in the Canadian High Arctic on a color triangle for the human visual spectrum. *B* is Blue; *G* is Green; *R* is red; *W* is equiproportionate reflectance white. Diameters of spots outside shading are proportionate to number of species represented (1, 2, or 3). Width of the shaded area is proportionate to the number of species; 22 yellow, 6 pale yellow, and 20 white flowers or flower parts. (Data from Kevan, 1970, 1972.)

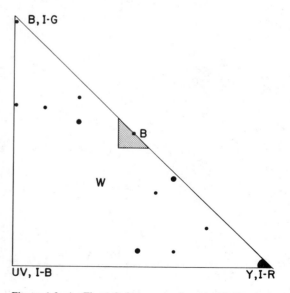

Figure 1-9. As Fig. 1-8, but on a color triangle for the insect visual spectrum. Insect-yellow (shaded area) shows 13 observations, and insect-red shows 14. *UV, I-B* is ultraviolet or insect-blue; *B, I-G* is blue or insect-green; and *Y, I-R* is yellow or insect-red; *W* is equiproportionate reflectance white.

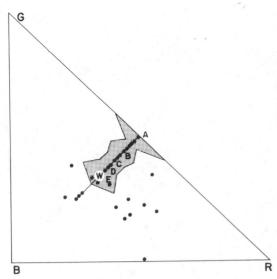

Figure 1-10. As Fig. 1-8, but for the weeds around Ottawa, Canada. The width of the shaded area at each letter represents for *A*, 37 observations; *B*, 21; *C*, 5; *D*, 6; *E*, about 22 (= *W*). Other spots represent single observations of flowers or floral parts. (Data from Mulligan and Kevan, 1973.)

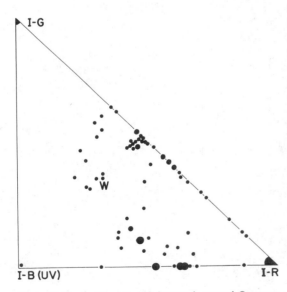

Figure 1-11. As Fig. 1-9, but for weeds around Ottawa, Canada. Points represent only 1, 2, or 3 observations of flowers or floral parts. At *I-R*, insect-red, there are 37 observations, many of floral parts in color combination with other loci.

weak or no ultraviolet patterns that are highly visited and others rarely visited (see Kevan, 1978a). There is no relationship between the reproductive strategies (i.e., outcrossers, selfers, or apomictics) of plants with white, yellow, or ultraviolet reflecting properties in the Canadian Arctic (Kevan, 1972), Canadian weeds (Mulligan and Kevan, 1973), and the Colorado alpine (Kevan, 1978a, 1980). Blue-reflecting flowers (i.e., those that are blue or purple, but not white) are frequented more by bees than any other insect group. Many of these flowers are zygomorphic or stereomorphic and adapted to pollination by bees. Lubbock (1881) concluded that blue was bees' favorite color. Knoll (1921) demonstrated the relationship of Bombyliidae (bee flies: Diptera) and blue flowers, especially with *Muscari* (Liliaceae). From studies on the reproductive strategies of blue-reflecting flowers, it appears that they tend to be obligate entomophiles, at least in the Colorado alpine (Kevan, 1980).

The association of butterflies with pinkish flowers was noted by Müller (1883a), and with red-flowered tropical and subtropical plants mentioned by Proctor and Yeo (1973). Some advanced Lepidoptera have been found to have red-sensitive vision (Goldsmith and Bernard, 1974). Red flowers are most often associated with hummingbird pollination (Grant, 1966; Grant and Grant, 1968; Raven, 1972). These flowers, not reflecting ultraviolet, would be very dull to insects with the type of color vision upon which this paper is based. That does not mean they would not be visited. Many insects (not just ants) forage from inconspicuous extrafloral nectaries, and other plant and plant pathogen exudates. Moreover, it is likely that the combination of dull color, often very deeply hidden nectar, and the nectar's being dilute (Baker, 1975) makes many red flowers undesirable to insects. Some red flowers which are visited by bees are reflective in ultraviolet, e.g., *Papaver rhoeas* (Papaveraceae) (Daumer, 1958). An interesting tropical forest tree endemic to Sulawesi, Indonesia, *Sarcotheca celebica* (Oxalidaceae), has flowers that are nonreflective across the insects' visual spectrum as so far defined. The flowers are small, red, and pollinated by a variety of bees, especially *Xylocopa* spp. (Hymenoptera: Xylocopidae) (Kevan and Lack, unpub.), which lack the red screening pigment of the eyes of familiar temperate bees and flies. It seems likely that these bees are also sensitive to red.

Mimetic coloration of dung, carrion, and other mimic-scented blossoms is well known (Proctor and Yeo, 1973; Faegri and van der Pijl, 1978), especially in *Stapelia* spp. (Asclepiadaceae), *Rafflesia* (Rafflesiaceae), and the Araceae (Knoll, 1926; Meeuse, 1966; Vogel, 1978). These have not been colorimetrically examined. The remarkable orchids *Ophrys* spp. are mimics of a number of bees and wasps. They mimic not only pheromones of the female insect to attract the males, but also mimic her shape and color. Kullenberg (1956, 1961) has studied *Gorytes* spp. (Sphecidae) and *Ophrys*. The color is mimicked to include ultraviolet, which is more reflective from the orchid flowers than from the female insect's wings. The flower represents a "super-normal visual stimulation" (Kullenberg, 1961, p. 294) coupled with olfactory enticement for the male wasps to attempt copulation with the flowers. Pollination takes place through this process of pseudocopulation. Recently Dafni et al. (1981) describe pollination in a Mediterranean orchid, *Serapias vomeracea*, which mimics sleeping holes for male solitary bees.

FLORAL, NECTAR, OR POLLINATION GUIDES

Floral guides are specialized patterns in blossoms that act as near-goal orientation cues to flower visitors to assist them in obtaining rewards while pollinating. Sprengel (1793) first described their function. Since then the functional significance of guides has been demonstrated (Kugler, 1943; Daumer, 1956, 1958; Manning, 1956a,b; Free, 1970; Jones and Buchmann, 1974; and others). Invisible guides, in ultraviolet, also have been studied. Müller (1876) noted that guides occur most frequently on flowers with hidden rewards. This relationship has been confirmed in more detailed studies by Kugler (1930–32, 1963),

who examined the occurrence of color pattern guides in some of the European flora; and similar data can be obtained from Kevan (1972) and Mulligan and Kevan (1973). Generally, butterfly flowers have the greatest frequency of guides (83% of the flowers), which are in both the visible and ultraviolet parts of the spectrum in about equal abundance (66% and 63%, respectively). The zygomorphic, or lipped, flowers show guides in 76%, but they predominate in the visible part of the spectrum (68% to 42% in ultraviolet). Capitulate blooms show fewer guides (67%), with ultraviolet guides rather than visible (56% to 30%) predominant. The more generalist flowers such as discs, bowls, bells, and funnels have the least frequency of guides (about 50%) with a slight tendency to ultraviolet in funnels and discs. From these data (from Kugler, 1963) it is clear that ultraviolet does not assume any special importance; it functions as a guide just like any other insect color.

The guide patterns themselves take on a variety of forms. In violets (*Viola* spp., Violaceae), many Fabaceae and Orchidaceae, the guides are lines that converge toward the entrance to the flower. Patterns of spots in many Fabaceae, Orchidaceae, Saxifragaceae, and Scrophulariaceae make a similar converging array toward the entrance of the flower. In some Scrophulariaceae (e.g., *Linaria vulgaris*, *Pedicularis groenlandica*, and others), some Fabaceae (e.g., *Caesalpinia eriostachys* and *Parkinsonia aculeata*), and other plants, the guide is a blotch or patch of a different color. Bull's-eye patterns are especially large and well represented in the Asteraceae and a wide variety of Rosaceae and less specialized Ranunculaceae. The bull's-eye patterns may also be small, as in the yellow centers of forget-me-nots (*Myosotis, Eritrichium*: Boraginaceae), *Primula* spp. (Primulaceae), some Brassicaceae (Horovitz and Cohen, 1972), and *Nymphoides indica* (Menyanthaceae) (Ornduff and Mosquin, 1970).

It is interesting to note that most guide patterns have colors reflecting shorter wavebands peripherally or as the background color. In target patterns, for example, the peripheral colors include ultraviolet or blue, while yellow (insect-red) makes up the center. The central color is also that for which the insects have the least ability to discriminate hue.

The presence of guides in flowers cannot be used to infer greater attractiveness to insects (Kevan, 1972; 1978a; Mulligan and Kevan, 1973) or to infer the reproductive strategy of the plant. *Taraxacum* spp. (Asteraceae), *Ranunculus* spp. (Ranunculaceae), and *Potentilla* spp. (Rosaceae) have well-developed bull's-eye (and other) patterns of insect-purple and insect-red, yet many are agamosperms. *Verbascum thapsus* (Scrophulariaceae) and *Eritrichium aretioides* are autogamous (Petersen, 1968; Mulligan, 1972). Various reproductive strategies are also found in *Viola,* and the *Lupinus nanus* (Fabaceae) complexes. (Horovitz and Harding, 1972) and *Nymphoides indica* (Ornduff and Mosquin, 1970). All the above have well-developed guides.

An intriguing suggestion by Thorp et al. (1975) is that nectar itself may absorb, reflect, or fluoresce daylight and act as a visual attractant to insects. Specular, or mirrorlike, reflection from glistening nectar in open flowers would be highly visible. However, I have suggested that fluorescence from nectar would be too weak for its additive or subtractive effects on color within the flower to be perceived by insects (Kevan, 1976). Nevertheless, the correlation of fluorescent nectar being more common in open flowers than in those with hidden nectar is interesting. Thorp et al. (1975) are continuing their research and may prove my criticism false (Thorp, 1979).

CHANGES IN FLORAL COLORS AND THEIR MEANINGS

During the lives of some flowers, color and other changes take place, which may be induced by pollination or just senescence (Gori, this book). Although this has been a well-known phenomenon (see Kerner and Oliver, 1904, Vol. II, pp. 191–192), its significance has not been well tested. Müller (1883b) documents that females of *Anthophora pilipes* (Hymenoptera: Anthophoridae) almost exclusively visit the pink flowers of *Pulmonaria of-*

ficinalis (Boraginaceae) while ignoring the blue, older flowers of the same plants. The pink to blue color change (or the reverse) in flowers as they age is well known in Boraginaceae, Polemoniaceae, Convolvulaceae, Onagraceae, Scrophulariaceae, Fabaceae, and others. From experience in Colorado, the older blue flowers are visited more often in *Mertensia* (Boraginaceae) and *Oxytropis* (Fabaceae) (see Kevan, 1978a). Clements and Long (1923) note the same effects. Müller (1883b) and Ludwig (1885, 1887) concluded that these color changes were used by insects foraging from the flowers as cues to the availability of rewards. In *Helleborus niger* (Ranunculaceae) the sepals are white when the nectaries (the reduced petals) are active. At this time the flowers are visited by bees. Later, the petals fall, and the sepals turn green. The flower is then not visited, and presumably the sepals contribute to the photosynthetic effort of the plant (see Kerner and Oliver, 1904, Vol. I, p. 376). Kugler (1936) showed that bumblebees *(Bombus)* ignore the older flowers of *Aesculus hippocastanum* (Hippocastanaceae), which are spotted red, while they visit the younger ones with yellow markings. Schaal and Leverich (1980) have shown the same effects of the change in the banner marking of *Lupinus texensis* (Fabaceae), in which over 90% of bumblebee and honeybee visits were to flowers with the conspicuous white to yellow banner spot. The spot changes to red-purple after five days; it then contrasts much less with the rest of the petals, and visitations all but cease. Pollen is produced most copiously, and is at its most viable, on the first days of opening, and fertilization is most successful on the later days (to day 5). Jones (unpub. ms.) has described the color change from yellow to burnt-orange in *Lotus scoparius* (Fabaceae), and its role in pollination.

In the Asteraceae, the older discs of the capitula frequently turn brown. Kugler (1950) has shown that *Eristalis tenax* (Diptera: Syrphidae) ignore the capitula of *Senecio jacobaea* in which that has happened, but visit those with yellow discs. The effect is important in the insects' recognition of prime flowers. Kikuchi (1962a,b, 1963a,b,c, 1964) has shown that more aggressive or dominant foragers on *Chrysanthemum leucanthemum* displace the less aggressive ones to older flowers with less reward. Kevan (1978a) and Kevan and Swift (unpub.) have shown that there are changes with age in the ultraviolet reflectance of the ray florets of *Solidago spathulata;* reflectance is greatest in prime flowers. Insect visitors interact, and the most dominant or aggressive are found on those capitula that offer the greatest amount of reward (nectar and pollen). Those insects are bumblebees, bristle flies (Tachinidae), and hover flies (Syrphidae) (Diptera); they are also the most active visitors and best pollen vectors. Other changes in ultraviolet patterns reported to influence pollinator visitation were found in comparing buds and flowers of *Chelidonium majus* (Papaveraceae) (Daumer, 1958) and in post-pollinated flowers of *Caesalpinea eriostachys* (Fabaceae) (Jones and Buchmann, 1974).

All these sorts of changes, including the change in nocturnal flowers from pale to darker, appear to function to increase the efficiency of pollination (see Gori, this book) by decreasing the conspicuousness of older flowers, or providing visual cues that visitors can learn to associate with depleted rewards, or both. The plant also benefits. However, in *Lantana* spp. (Verbenaceae) the story is less clear. F. Müller (1877) noted that in a Brazilian species, the flowers were yellow on the first day, orange on the second, and purple on the third. Some butterflies visited only the yellow flowers, some the yellow and orange, and none the purple. Dronamraju (1960) studied *L. camara* in Calcutta and noted the same phenomenon. There, *Precis almana* (Nymphalidae) preferred orange flowers; *Papilio demoleus* (Papilionidae), *Catopsila pyranthe* (Pieridae), and *Baoris mathias* (Hesperiidae) preferred pink flowers; and *Danaus chrysippus* (Danaidae) did not discriminate. The effects of such strong discrimination by at least some of the butterflies would favor disruptive selection and possibly sympatric speciation (Dronamraju, 1960). More research on pollinator specificity between species of *Lantana* (Schemske, 1976) and on nectar theft (Barrows, 1976) has not resolved the apparent paradox.

SEASONAL PROGRESSION OF COLOR IN PLANT COMMUNITIES

There is an old idea of floral color seasons (see Kerner and Oliver, 1904, Vol. II, pp. 197–198) that should be examined in an ecological context. It is known that some insects change their color preferences throughout the year [e.g., butterflies (Ilse, 1928; Ilse and Vaidya, 1956) and wasps (*Vespula squamosa,* Hymenoptera: Vespidae) (Sharp and James, 1979)] and that there are seasonal abundances of insects, some of which show color preferences (see above). Horovitz and Harding (1972) suggest a seasonally changing role for floral guides and coloration in *Lupinus nanus.* In comparing genetic stocks, they suggest that the relative lack of blue reflectance and of guide patterns on one accounts for the lack of outcrossing early in the season when attractants are critical. Later, when pollinators were more abundant, the other stock showed more outcrossing. Waser and Real (1979) discuss the cooperative activities of two hummingbird-pollinated flowers, *Ipomopsis aggregata* (Polemoniaceae) and *Delphinium nelsonii* (Ranunculaceae), which bloom at different times of the year in the montane regions of Colorado. In doing so, they provide a continuing resource for their long-lived pollinators. Stace and Fripp (1977) suggest that the phenology of the different color morphs of *Epacris impressa* (Epacridaceae) are correlated with pollinator abundances throughout the blooming season. They note that red-flowered plants bloom in winter when insect pollinators are not common but birds forage, and that white-flowered plants, which are spring bloomers, coincide with greater insect abundance. I believe that more research on the community dynamics of pollination throughout blooming seasons would divulge other interesting patterns in which coloration is probably important.

FLORAL COLOR POLYMORPHISM

That floral coloration is perceived by pollinators cannot be denied. Moreover, there has been a long debate as to their interest in responding differentially. Many of the early writers were also concerned with discrimination of colors by insects, before color vision had been proved (see Clements and Long, 1923). Darwin (1876, p. 416) wrote that bumblebees were good botanists, able to discriminate species with a variety of color morphs. Nevertheless, pollinators do discriminate between morphs of some flowers. Much of the research in this area has been in agriculture such as in cabbages (*Brassica oleraceae*) and swedes (*B. napus*) (Bateman, 1951); brussels sprouts *(B. oleraceae)* (Free and Williams, 1973; Faulkner, 1976) (Brassicaceae); alfalfa *(Medicago sativa)* (Pedersen and Todd, 1949; Clement, 1965; Pedersen, 1967; Kauffeld and Sorensen, 1971; Goplen and Brandt, 1975); cowpea *(Vigna sinensis)* (Leleji, 1973); and soybean *(Glycine max)* (Erickson, 1975 a,b) (Fabaceae).

Research on natural populations of plants has shown similar tendencies of pollinators to discriminate. Lloyd (1969) found that honeybees foraging on *Leavenworthia crassa* (Brassicaceae) did not discriminate between the two common forms (yellow-centered and yellow) but did show a preference for the rarest (intermediate) morph. By contrast, *Bombylius major* (Diptera: Bombyliidae) showed a strong preference for the most common, yellow-centered form. *Anthocharis genutiae* (Lepidoptera: Pieridae) did not discriminate. The plant is self-compatible. Levin (1969, 1972a,b), Levin and Kerster (1970), and Levin and Schaal (1970) have found clear differences in the visitation rates, as measured by pollen deposition, seed set, and observations, in color and shape morphs in *Phlox* spp. (Polemoniaceae). The pollinators are mostly Lepidoptera, which Levin and his co-workers conclude may act as strong selective agents for color. Floral colors act as isolating mechanisms reducing interspecific and intervarietal pollen flow. Cruden (1972) studied *Nemophilia menziesii* (Hydrophyllaceae) in California, where there are different geographic races pollinated by bees (mostly *Andrena* spp., Andrenidae). The bees discriminate among the flowers which differ in ultraviolet pattern. Cruden suggested that isolation of *N. menziesii* and pollinators in Miocene time broke down following

the retreat of the seas in northern California. There, hybridization takes place because flower constancy by the bees (which are different from more southerly places) is not pronounced. This series of events would not have taken place in the southern California ranges. Mogford (1974, 1978) studied two color morphs of *Cirsium palustre* (Asteraceae) in Wales. This thistle has white to purple capitula. He found that *Bombus lapponicus* (Apidae) would discriminate in some places, but not in others. He has noted that white morphs are more common at higher elevations, and that they are preferentially pollinated there. Kay (1976, 1978) has investigated color dimorphism in *Raphanus raphanistrum* (Brassicaceae) in relation to preferential and assortative pollination. The white morph is insect-white and the yellow morph insect-purple (i.e., both have ultraviolet reflectance). He showed conclusively that *Pieris* spp. (Lepidoptera: Pieridae) and *Eristalis* spp. (Diptera: Syrphidae) preferentially visit insect-purple flowers, whereas *Bombus* spp. preferentially visit insect-white ones. Rare morphs of some flowers seem to be at a disadvantage in pollination (Free, 1966). Scora (1964) noted that spotless mutants of *Monarda punctata* (Lamiaceae) are ignored by foraging bees. White variants in *Clarkia cylindrica* (Onagraceae) (Lewis, 1953) and *Pseudomuscari azureum* (Liliaceae) (Garbari, 1972) failed to set seed. Studies on Colorado *Delphinium nelsonii* (Ranunculaceae) suggest that the rare white morphs are less fecund owing to pollinator discrimination (Waser and Price, 1981).

Not all color polymorphisms in flowers seem to affect pollinator activity. Carter (1974) suggests a breakdown in isolating mechanisms in *Cercidium* spp. (Fabaceae) despite clear ultraviolet differences among the flowers. Galen and Kevan (1980) have shown that color polymorphism, from white (insect-yellow) to blue and purple (insect-green and insect-blue-green) in *Polemonium viscosum* (Polemoniaceae) in the Colorado Rockies does not influence the activity of pollinators [*Bombus* spp. and *Hyles lineata* (Lepidoptera: Sphingidae)]. The scent dimorphism (skunky and sweet) is the discriminator clue for the bumblebees.

Hannan (1981) concluded that color polymorphisms in *Platystemon californicus* (Papaveraceae) must be linked genetically to other adaptive characters as yet undetermined. Solitary bees (Andrenidae and Halictidae) are the most important pollen vectors, with wind ranking somewhat lower. Both pollination systems are more or less indiscriminate; so polymorphic populations show no intramorph differences in seedset. Woodson (1964) was unable to suggest any selective pressure to account for the floral color variation in *Asclepias tuberosus* (Asclepiadaceae). Similarly, Rafínski (1979) could not demonstrate any influence of selection in maintaining flower color variability in *Crocus scepusiensis* (Iridaceae) in the western Carpathian Mountains of Poland. In some cases, edaphic factors (minerals, moisture, etc.) may be important, as in *Anemone coronaria* (Horovitz, 1976), *Eschscholzia californica* in California and Chile (Frías et al., 1975) (Papaveraceae), and *Viola calminaria* (Kakes and Everards, 1976). In other cases physical factors may be involved, e.g., in *Epacris impressa* (Epacridaceae) in Australia, with white-flowered forms increasing with elevation and possibly cold (Stace and Fripp, 1977). *Abies concolor* (Pinaceae) has strobili that are darker at higher elevations and become preferentially warmed by comparison with lighter strobili under the same experimental conditions (Sturgeon and Mitton, 1980). Light intensity has also been suggested as important, forms of species with paler flowers being more prevalent in dimmer environments (such as deep forests) (see below, and Anderson, 1936). There are numerous examples of polymorphisms, but few have been explained in terms of their function (see below).

BIOGEOGRAPHY OF FLORAL COLORS

Biogeographical distributions of colors and color morphs of flowers of various species have been documented. A number of examples are given above. Weevers (1952) generally discredits any theory of biogeography for floral colors, but did note a greater proportion of blue flowers in high mountains as diverse as in Java and Switzerland as compared to those countries as a whole. The Arctic has a relative abundance of white and yellow flowers (Tikhomirov, 1959; Kevan, 1972). Weindorfer (1903) writes that the alpine flowers of Mt. Kosciusko in Australia are less colorful than those of the Alps. Mogford (1978) notes that *Cirsium palustre* is white at higher elevations. There are other examples of this sort of trend, e.g., *Trifolium pratense* (Fabaceae) (Müller, 1881), *Brimeura fastigiata* (Liliaceae) (Garbari, 1970), *Cypripedium acaule* (Orchidaceae) (Baldwin, 1970), *Campanula trachelium* (Campanulaceae) (Kerner and Oliver, 1904, Vol. II, p. 193) and *Linaria reflexa* (Scrophulariaceae) (Engler and Prantl, 1887). The reverse trend, to more intense coloration, is also noted for increasing elevation, e.g., *Primula* spp. (Primulaceae), *Daphne straita* (Thymelaeaceae), *Anacamptis pyramidalis* (Orchidaceae), *Crocus vernus* (Iridaceae), *Gymnadenia odoratissima* (Orchidaceae), *Euphrasia salisburgenis* (Scrophulariaceae), *Ageratum* spp. (Asteraceae), *Myosotis* spp. (Boraginaceae), and *Linaria aeruginea* (Scrophulariaceae) (Müller, 1883a; Knuth, 1906–9; Weevers, 1952; Tutin et al., 1964). The adaptive significance of this trend has been suggested to be for encouraging diurnal, visually oriented pollinators at high elevations, perhaps changing the pollinators from nocturnal moths to butterflies or diurnal moths, or to bees (Müller, 1881; Knuth, 1906–9). However, the idea of the lack of effectiveness of nocturnal pollination in the chill night alpine air may be untrue, as moths are very active at high elevations at night, at least in Colorado (Kendall et al., 1982). *Melampyrum pratense* (Scrophulariaceae) in mountains occurs as a form with crimson corolla tips as well as the usual yellow type (Smith, 1963). Kyhos (1971) notes that the typically yellow-flowered *Encelia farinosa farinosa* (Asteraceae) has proportionately more individuals with brownish-purple disc florets *(E. f. phenicodonta)* at higher elevations and to the north in the Sonoran desert. More important is the correlation between *E. f. phenicodonta* prevalence and proximity to water (coastal, river, rainfall). As yet, apart from Mogford's studies (1974, 1978), the involvement of pollinators has not been tested.

Certainly some of the coloration effects are temperature-induced (see Klebs, 1906; Baldwin, 1970; Garbari, 1970).

The habitat in which plants bloom is also important. Anderson (1936) noted that plants of *Dicentra cucullaria* (Fumariaceae) and *Hepatica acutiloba* (Ranunculaceae) of New England forests were more frequently white than in the less dense Ozark woodlands. He surmised that the paler flowers would be more visible in the dim light. More recently, Ostler and Harper (1978) in the Wasatch Mountains of Idaho and Utah showed the frequency of yellow-colored flowers in well-lighted communities decreased with increasing species richness, whereas blue flowers showed the opposite trend. They equated this with the greater importance of specialized (bee) pollinators in species-rich habitats. Del Moral and Standley (1979) and Baker (in Kevan and Baker, 1983) have shown that the flora of coniferous forests have pale flowers, whereas in the same region in open areas the flowers are yellow and on the forest margin, purple. In the more well-lighted forests of Colorado, *Ipomopsis aggregata* is red-flowered, whereas in more open habitats it is white or pink. This may correspond with hummingbird habitat in some way, but this idea has not been tested.

Utech and Kawano (1975) have shown some biogeographical trends in the Japanese flora, noting that white and yellow flowers are more common in more northerly places. They also discuss the reduction in the flora of plants with ultraviolet reflecting flowers. They suggest that the reduced ultraviolet intensity at high latitudes may have placed selective pressure for ultraviolet absorption. The selective pressures for floral coloration presumably result from pollination and the distinguishability of flowers of various species (see above). Furthermore, the amount of ultraviolet in arctic daylight does not seem much reduced as compared with elsewhere (Hisdal, 1967; Henderson, 1970; Caldwell, 1972). At high elevations ultraviolet may be biologically quite harmful; so its reflection or absorption by pigments could be protective (Caldwell, 1971). More data are needed from entire regional floras to further explore these ideas. Scogin (this book) discusses the biogeography of pigments in flowers and also concludes that there is a need for further work. It would seem appropriate that coloration and pigmentation studies be conducted together.

PROTECTIVE COLORATION AND CRYPSIS IN FLOWERS

Hinton (1973a,b) suggested that floral colorations may have had an aposematic (warning) role to discourage iguanodon and ceratopsian reptiles, which presumably had good color vision, from eating the plants. The flowers would also attract pollinators. He produced some interesting correlations between the incidence of distasteful or toxic secondary compounds in plants and the colors of the flowers from various parts of the world in climes where reptiles would have thrived. Many flowering plants are tender and edible as seedlings, but accumulate tannins, alkaloids, and other compounds as they age, flower, and fruit. These secondary compounds are interpreted as being protective of the reproductive investment made by the plant. The presence of flowers could be a clue to herbivores, large or small, to avoid that plant.

Crypsis by flower-visiting arthropods is not well studied, but has roles for both predators and potential prey in flowers. In temperate regions, crab spiders (Thomisidae) are frequently found in flowers, and they are cryptically colored (Gertsch, 1949). I found that when the arctic crab spiders, *Xysticus deichmannii*, were in the ultraviolet reflecting blossoms of *Arnica alpina* (Asteraceae), they were not on peripheral insect-purple parts of the ray florets, but on the central parts or on the disk florets, which are insect-red, the same color as the spiders themselves. On purple flowers the spiders assume a purple tinge (Kevan, 1972). Wickler (1968) describes the crab spider, *Misumena vatia*, as able to assume the color of its background in flowers. *Phymata pennsylvanica americana* (Hemiptera: Phymatidae), an ambush bug, has been studied as a predator using flowers (Balduf, 1941). These bugs lie in wait for their potential prey, most frequently in flowers with the greatest rewards (nectar and pollen). In tropical regions of Asia and Africa a number of cryptically colored mantids (Orthoptera: Mantidae) are known that mimic floral colors and ambush prey that visit the flowers they inhabit (Shelford, 1903; Williams, 1904; Wickler, 1968). The related orthopteroids, Phasmidae or stick insects, may also be flower mimics, but presumably gain protection (Annandale, 1902).

Other nonpredatory anthophiles seem to show protective coloration in flowers. Anthophilous thrips are frequently pale colored, as are many anthophilous bugs (Hemiptera). Kevan and Kevan (1970) and Kevan (1978b) suggest that anthophilous springtails (Collembola) are mostly pale-colored and visit flowers in which they are least conspicuous. Recently we (Kevan and co-workers) have recorded early stages of some Colorado grasshoppers (Orthoptera) as cryptic in the flowers of prickly-pear cacti (*Opuntia* spp.) (Cactaceae), where they eat anthers. However, as the grasshoppers grow and become darker, they are more conspicuous and are readily scared from the flowers. One of the most conspicuous cryptic colorations is of the moth *Schinia masoni* (Nocturnidae: Lepidoptera) in *Gaillardia aristata,* in which the yellow head and thorax of the moth match the disk florets of the blossom, and the red-brown wings match the ray florets. As the moth feeds around the disk florets, it remains inconspicuous (Ferner, 1980; Ferner and Rosenthal, 1981).

FLORAL COLORATION IN TAXONOMY AND EVOLUTION

The study of floral colors has a long history, not only with regard to the appearance of floras (see Allen, 1882), pollinator preference, and pollination (Plateau, 1899, 1900a,b; von Frisch, 1913; Knoll, 1921, 1922, 1926; Clements and Long, 1923; and see above), but also from the physical and chemical standpoint. Kraus (1872) had discovered ultraviolet-reflecting pigments by alcoholic extraction. Knuth (1891a,b) was the first to photographically demonstrate ultraviolet patterns in flowers. Exner and Exner (1910) made one of the first detailed explorations into the physical nature of floral colors. Weevers (1952) investigated various aspects of floral pigments, including their biochemistry.

Some generalizations can be made. White coloration may be due primarily to the physical effects of refraction at cell surfaces on the petals, and, within the tissue, refraction due to air spaces between the cells. (This is the same sort of physical effect as seen in white hair and snow.)

Floral pigments may be found in cell sap or in plastids. Scogin (this book) discusses the chemistry and effects of these pigments in flowers in the human visual spectrum. From the viewpoint of the insect visual spectrum, the effects of pigments in the ultraviolet must be considered. Generally, the anthoxanthins absorb ultraviolet (Weevers, 1952; Scogin, this book), as do the anthocyanins. Thus, the flavonoids, which are sometimes restricted in their location in flowers, may produce patterns. Thompson et al. (1972) showed that carotenoids throughout the rays of *Rudbeckia hirta* (Asteraceae) in concert with the flavonoids, depleted in the peripheral part of the rays, gave rise to the bull's-eye effect of insect-purple outer color and insect-red center rich in flavonoids. Since then, similar patterns have been reported in other *Rudbeckia* spp. (Abramhamson and McCrea, 1977), *Bidens* (Scogin and Zakar, 1976), *Coreopsis* (Scogin, 1976a; Scogin et al., 1977) (Asteraceae), *Oenothera* (Onagraceae) (Dement and Raven, 1974), *Antirrhinum* (Scrophulariaceae) (Scogin, 1976b), and some Apiaceae (Brehm and Krell, 1975), and with anthocyanins and flavonoids as in *Phlox* (Levy and Levin, 1974, 1975) and other flowers (Levy, 1978). These sorts of patterns have been studied most extensively in the Asteraceae (Harborne and Smith, 1978). In these (Baagoe, 1977a,b, 1978; Lane, 1979, 1980), as in Apiaceae (Brehm and Krell, 1975), epidermal papillae on the petals may or may not be involved (see Levy, 1978). Ultraviolet-reflecting trichomes have been demonstrated in *Limnanthes* flowers (Parker, 1980) (Limnanthaceae). The occurrence of yellow flavonoids in flowers may not correlate directly with ultraviolet patterns (Harborne and Smith, 1978).

Nevertheless, ultraviolet patterns, and floral coloration and pigmentation generally, can be used in taxonomic studies. They may be used as characters, plain and simple (e.g., Horovitz and Harding, 1972; Eisner et al., 1973b), or in attempts to understand phylogenetic relationships (e.g., Harborne, 1963, 1967; Arditti, 1969; Levy and Levin, 1974, 1975; Baagoe, 1977a,b, 1978; Harborne and Smith, 1978; Scogin, this book). As has been shown above, floral colorations can act as powerful isolating mechanisms (see also Grant, 1949, 1971; Heslop-Harrison, 1958; Free, 1966; Stebbins, 1974).

ACKNOWLEDGMENTS

I am grateful to a number of scientists with whom I have corresponded regarding the measurement of color: W. Budde, B. H. Crawford, and A. R. Robertson were especially helpful. I thank G. A. Mulligan for the use of the data on Canadian weeds. I have enjoyed fruitful discussions with a large number of colleagues from time to time, and would like to thank them all as a group.

Funding for the work described in this chapter has come from the National Research Council of Canada (NRC A-2560 to the late Professor B. Hocking), the Defense Research Board of Canada, The University of Alberta, the Plant Research Institute of Canada Agriculture, the University of Colorado at Colorado Springs and at Boulder (CRCW 1644266), and the National Science Foundation (NSF DEB 76-20125 and 79-10786). I am grateful to C. E. Jones and California State University, Fullerton for having given me the opportunity to present this material orally and in publication.

R. D. Leggett prepared the line illustrations; I should like to thank him along with the staff of the Biology Department at Fullerton for typing the manuscript.

LITERATURE CITED

Abrahamson, W. and K. D. McCrea. 1977. Ultraviolet light reflection and absorption patterns in populations of *Rudbeckia* (Compositae). *Rhodora* 79:269–277.

Allen, G. 1882. *The Colours of Flowers as Illustrated in the British Flora*. Macmillan, London.

Anderson, A. 1936. Color variation in eastern North American flowers as exemplified by *Hepatica acutiloba*. *Rhodora* 38:301–304.

Annandale, N. A. 1902. Notes on the habits of Malayan Phasmidae and of a flower like beetle larvae. *Proc. Phys. Soc. Edinburgh* 14:439–440.

Arditti, J. 1969. Floral anthocyanins in species and hybrids of *Broughtonia, Brassavola* and *Cattleyopsis* (Orchidaceae). *Am. J. Bot* 56:59–68.

Baagoe, J. 1977a. Taxonomical application of ligule microcharacters in Compositae. I. Anthemideae, Heliantheae, Tageteae. *Bot. Tidssk.* 71:193–224.

———. 1977b. Microcharacters in ligules of Compositae. *In* V. Heywood, J. Harborne, and B. L. Truner (eds.), *Biology and Chemistry of Compositae*. Academic Press, New York, Ch. 7.

———. 1978. Taxonomical application of ligule microcharacters in Compositae II. Arctotideae, Astereae, Calenduleae, Eremothamneae, Inuleae, Liabeae, Mutiseae and Senecioneae. *Bot. Tidssk.* 72:125–148.

Baker, H. G. 1961. The adaptation of flowering plants to nocturnal and crepuscular pollinators. *Q. Rev. Biol.* 36:64–73.

———. 1975. Sugar concentrations in nectars from hummingbird flowers. *Biotropica* 7:37–41.

Balduf, W. V. 1941. Life history of *Phymata pennsylvanica americana* Melin. *Ann. Ent. Soc. Am.* 34:204–214.

Baldwin, J. T. 1970. White phase in flower development in *Cypripedium acaule*. *Rhodora* 72:142–143.

Barrows, E. M. 1976. Nectar robbing and pollination in *Lantana camara* (Verbenaceae). *Biotropica* 8:132–135.

Bateman, J. 1951. The taxonomic discrimination of bees. *Heredity* 5:271–278.

Brehm, B. G. and D. Krell. 1975. Flavonoid localization in epidermal papillae of flower petals: a specialized adaptation for ultraviolet absorption. *Science (Washington)* 190:1221–1223.

Burkhardt, D. 1964. Colour discrimination in insects. *Adv. Insect Physiol.* 2:131–173.

Caldwell, M. 1971. Solar UV radiation and the growth and development of higher plants. *Photophysiology* 6:131–177.

———. 1972. Biologically effective solar ultraviolet irradiation in the Arctic. *Arct. Alp. Res.* 4:39–43.

Carter, A. M. 1974. Evidence for the hybrid origin of *Cercidium sonorae* (Leguminosae: Caesalpinoideae) of northwestern Mexico. *Madroño* 22:266–272.

Clement, M. W. 1965. Flower color, a factor in attractiveness on alfalfa clones for honeybees. *Crop. Sci.* 5:267–268.

Clements, F. C. and F. L. Long. 1923. *Experimental Pollination. An Outline of the Ecology of Flowers and Insects*. Carnegie Inst. (Washington) Publ. No. 336. vii + 274 pp.

Crawford, B. H. 1959. Measurement of color rendering tolerances. *J. Opt. Soc. Am.* 49:1147–1156.

Cruden, R. W. 1972. Pollination biology of *Nemophila menziesii* (Hydrophyllaceae) with comments on the evolution of oligolectic bees. *Evolution* 26:373–389.

Dafni, A., Y. Ivri, and N. B. M. Brantjes. 1981. Pollination of *Serapias vomeraceae* Brig. (Orchidaceae) by imitation of holes for sleeping solitary male bees (Hymenoptera). *Acta Bot. Neerl.* 30:69–73.

Dartnall, H. H. A. 1953. The interpretation of spectral sensitivity curves. *Brit. Med. Bull.* 9:24–30.

Darwin, C. R. 1876. *Cross and Self Fertilisation in the Vegetable Kingdom*. J. Murray, London.

Daumer, K. 1956. Reizmetrische Untersuchung des Farbensehens der Bienen. *Zeit. Vergl. Physiol.* 38:413–478.

———. 1958. Blumenfarben: wie sie die Bienen sehen. *Zeit. Vergl. Physiol.* 41:49–110.

Del Moral, R. and L. A. Standley. 1979. Pollination of angiosperms in contrasting coniferous forests. *Am. J. Bot.* 66:26–35.

Dement, W. A. and P. H. Raven. 1974. Pigments responsible for ultraviolet patterns in flowers of *Oenothera* (Onagraceae). *Nature (London)* 252:705–706.

Dronamraju, K. R. 1960. Selective visits of butterflies to flowers; a possible factor in sympatric speciation. *Nature (London) 186:178.*

Ebry, G. T. and B. Honig. 1977. New wavelength dependent visual pigment nomograms. *Vision Res.* 17:147–151.

Eisikowitch, D. 1978. Insect visiting of two subspecies of *Nigella arvensis* under adverse seaside conditions. *In* A. J. Richards (ed.), *The Pollination of Flowers by Insects* (Linnean Soc. Symp. No. 6). Academic Press, London, pp. 125–132.

Eisner, T., R. E. Silberglied, D. Aneshansley, J. E. Carrel, and H. C. Howland. 1969. Ultraviolet video-viewing: the television camera as an insect eye. *Science (Washington)* 166:1172–1174.

Eisner, T., M. Eisner, and D. Aneshansley. 1973a. Ultraviolet patterns on rear of flowers: basis of disparity of buds and blossoms. *Proc. Natl. Acad. Sci. U.S.A.* 70:1002–1004.

Eisner, T., M. Eisner, P. Hyypio, D. Aneshansley, and R. E. Silberglied. 1973b. Plant taxonomy: ultraviolet patterns of flowers visible as fluorescent patterns in pressed herbarium specimens. *Science (Washington)* 179:486–487.

Engler, A. and K. Prantl. 1887. *Die naturlichen Pflanzenfamilien.* Englemann, Liepzig.

Erickson, H. E. 1975a. Honeybees and soybeans. *Am. Bee J.* 115:351–353.

———. 1975b. Variability of floral characteristics influences honeybee visitation to soybean blossoms. *Crop Sci.* 15:767–771.

Exner, F. and S. Exner. 1910. Die physikalischen Grundlagen der Blütenfarbungen. *Sitzungber. Kais. Akad. Wiss., Math.-Naturwiss. Cl.* 119(1):191–245.

Faegri, K. and L. van der Pijl. 1978. *The Principles of Pollination Ecology,* 3rd ed. rev. Pergamon Press, Elmsford, N.Y.

Faulkner, G. J. 1976. Honeybee behaviour as affected by plant height and flower colour in brussels sprouts. *J. Apic. Res.* 15:15–18.

Ferner, J. W. 1980. "Cover photograph." *Bioscience* 30:75 + cover.

Ferner, J. W. and M. Rosenthal. 1981. A cryptic moth, *Schinia masoni* (Noctuidae) on *Gaillardia aristata* (Compositae) in Colorado. *Southwest. Natur.* 26:88–89.

Free, J. B. 1966. The foraging behaviour of bees and its effect on the isolation and speciation of plants. *In* J. G. Hawkes (ed.), *Reproductive Biology and Taxonomy of Vascular Plants.* Bot. Soc. Brit. Isles Conf. Rept. No. 9. Pergamon Press, Oxford.

———. 1970. The effect of flower shapes and nectar guides on the foraging behaviour of honeybees. *Behaviour* 23:269–286.

Free, J. B. and I. H. Williams. 1973. The foraging behaviour of honeybees *(Apis mellifera)* on brussels sprouts *(Brassica oleraceae* L.). *J. Appl. Ecol.* 10:489–499.

Frías, L. D., R. Godoy, P. Iturra, S. Koref-Santibanez, J. Navarro, N. Pacheco, and G. L. Stebbins. 1975. Polymorphism and geographic variation in flower color in Chilean populations of *Eschscholzia californica. Pl. Syst. Evol.* 123:185–198.

Frisch, K. von. 1913. Ueber den Farbesinn der Bienen und die Blumen farbe. *Munch. Med. Woch.* 60(1):15–18.

———. 1914–15. Der Farbensinn und Formensinn der Bienen. *Zool. Jahrb. Abt. Allg. Zool. Physiol. Tiere* 35:1–188.

Frohlich, M. W. 1976. Appearance of vegetation in ultraviolet light: absorbing flowers, reflecting backgrounds. *Science (Washington)* 194:839–841.

Galen, C. and P. G. Kevan. 1980. Scent and color, floral polymorphisms and pollination biology in *Polemonium viscosum* Nutt. *Am. Midl. Natur.* 104:281–289.

Garbari, F. 1970. Il genere Brimerua Salish. (Liliaceae). *Mem. Soc. Tosc. Sci. Nat. (Ser. B)* 77:12–36.

———. 1972. Note sul genere *"Pseudomuscari"* (Liliaceae). *Webbia* 27:369–381.

Gerritsen, F. 1975. *Theory and Practice of Color. A Color Theory Based on Laws of Perception.* Van Nostrand Reinhold Co., New York.

Gertsch, W. J. 1949. *American Spiders.* Van Nostrand Reinhold Co., New York.

Goldsmith, T. H. 1980. Hummingbirds see near ultraviolet light. *Science* 207:786–788.

Goldsmith, T. H. and G. D. Bernard. 1974. The visual system of insects. *In* M. Rockstein (ed.), *The Physiology of Insecta,* 2nd ed. Academic Press, New York and London, Ch. 5, Vol. II, pp. 165–272.

Goplen, B. P. and S. A. Brandt. 1975. Alfalfa flower color associated with differential seed set by leaf-cutter bees. *Agron. J.* 67:804–807.

Grant, K.A. 1966. A hypothesis concerning the prevalence of red coloration in California hummingbird flowers. *Am. Natur.* 100:85-98.

Grant, V. A. 1949. Pollination systems as isolating mechanisms in Angiosperms. *Evolution,* 3:82–97.

———. 1971. *Plant Speciation.* Columbia University Press, New York.

Grant, V. A. and V. Grant. 1968. *Hummingbirds and Their Flowers.* Columbia University Press, New York. vii + 115 pp.

Guldberg, L. D. and P. R. Atsatt. 1975. Frequency of reflection and absorption of ultraviolet light in flowering plants. *Am. Midl. Natur.* 93:35–43.

Hannan, G. L. 1981. Flower color polymorphism and pollination biology of *Platystemon californicus* Benth. (Papveraceae). *Am. J. Bot.* 68:233–243.

Harborne, J. B. 1963, Distribution of anthocyanins in higher plants. *In* T. Swain (ed.), *Chemical Plant Taxonomy.* Academic Press, New York, pp. 359–388.

———. 1967. *Comparative Biochemistry of the-Flavonoids.* Academic Press, New York.

Harborne, J. B. and D. H. Smith. 1978. Anthochlors and other flavonoids as honey guides in the Compositae. *Biochem. Syst. Ecol.* 6:287–291.

Helmholtz, H. von. 1852a. Ueber die Theorie der Zusammengeretzten Farben. *Poggendorf's Ann.* 87:45–66.

———. 1852b. On the theory of compound colours. *Phil. Mag.* 4:461–482.

———. 1852c. Ueber die Zusammensetzung der Spektralfarben. Poggendorff's Ann. 94: 1–28.

———. 1924–25. *Helmoltz's Treatise on Physiological Optics.* Trans. from 3rd German ed. (J. P. C. Southall, ed.), 3 vols. Opt. Soc. Am. (publisher). Reprinted 1962 by Dover Publ. Inc., New York.

Henderson, S. T. 1970. *Daylight and its Spectrum.* Adam Hilger, Ltd., London.

Henderson, S. T. and D. Hodgkiss. 1963. The spectral energy distribution of daylight. *Brit. J. Appl. Phys.* 14:125–131.

Hertz, M. 1937. Versuche über das Farbensystem der Bienen. *Naturwissenschaften* 25:492–493.

———. 1939. New experiments on color vision in bees. *J. Exp. Biol.* 16:1–8.

Heslop-Harrison, J. 1958. Ecological variation and ethological isolation. *In* O. Hedberg (ed.), *Systematics of Today. Uppsala Univ. Arsskr.* 1958(6):150–158.

Hill, R. J. 1977. Technical note: ultraviolet reflectance-absorbance photography; an easy, inexpensive research tool. *Brittonia* 29:382–390.

Hinton, H. E. 1973a. Natural deception. *In* R. Gregory (ed.), *Illusion in Nature and the Arts.* Duckworth Publ., London.

———. 1973b. Some recent work on the colors of insects and their likely significance. *Proc. Brit. Ent. Nat. His. Soc.* 6(2):43–54.

Hisdal, V. 1967. A comparative study of the spectral composition of the zenith sky radiation. *Årbok Norsk Polarinst. Oslo Ser.* 1967:7–27.

Högland, G., K. Hamdorf, and G. Rosner. 1973a. Trichromatic visual system in an insect and its sensitivity control by blue light. *J. Comp. Physiol.* 86:265–279.

Högland, G., K. Hamdorf, H. Langer, R. Paulsen, and J. Schwemer, 1973b. The photopigments in an insect retina. *In* H. Langer (ed.), *Photochemistry and Physiology of Visual Pigments.* Springer Verlag, Berlin, pp. 167–174.

Horovitz, A. 1976. Edaphic factors and flower colour distribution in *Anemone* (Ranunculaceae). *Pl. Syst. Evol.* 126:239–242.

Horovitz, A. and Y. Cohen. 1972. Ultraviolet reflectance characteristics in flowers of crucifers. *Am. J. Bot.* 59:706–713.

Horovitz, A. and J. Harding. 1972. Genetics of *Lupinus.* V. Intraspecific variability for reproductive traits in *Lupinus nanus. Bot. Gaz.* 133:155–165.

Ilse, D. 1928. Uber den Farbensinn der Tagfalter. *Zeit. Vergl. Physiol.* 8:658–691.

———. 1949. Colour discrimination in the dronefly, *Eristalis tenax. Nature (London)* 163:255.

Ilse, D. and V. G. Vaidya. 1956. Spontaneous feeding response to colors in *Papilio demoleus* L. *Proc. Indian Acad. Sci.* 43:23–31.

Jones, C. E. and S. L. Buchmann. 1974. Ultraviolet floral patterns as functional orientation cues in hymenopterous pollination systems. *Anim. Behav.* 22:481–485.

Judd, D. B., D. C. MacAdam, and G. Wyszecki. 1964. Spectral distribution of typical daylight as a function of correlated color temperature. *J. Opt. Soc. Am.* 54:1031–1040.

Judd, D. B. and G. W. Wyszecki. 1963. *Color in Business, Science and Industry.* Wiley, New York.

Kakes, P. and K. Everards. 1976. Genecological investigations in zinc plants. I. Genetics of flower color in crosses between *Viola calaminaria* Lej. and its subspecies *Westfalica* (Leg.) Ernst. *Acta Bot. Neerl.* 25:31–40.

Kauffeld, N. M. and E. L. Sorensen. 1971. Interrelations of honeybee preference of alfalfa clones and flower color, aroma, nectar volume and sugar concentrations. *Kansas Agric. Exp. Stn. Res. Publ.* No. 163, Manhattan, Kans. 14 pp.

Kay, Q. O. N. 1976. Preferential pollination of yellow-flowered morphs of *Raphanus raphinistrum* by *Pieris* and *Eristalis* spp. *Nature (London)* 261:230–232.

———. 1978. The role of preferential and assortative pollination in the maintenance of flower colour polymorphisms. *In* A. J. Richards (ed.), *The Pollination of Flowers by Insects* (Linnean Soc. Symp. No. 6). Academic Press, London, pp. 175–190.

Kendall, D. M. , P. G. Kevan, and J. D. Lafontaine. 1982. Nocturnal flight activity of moths in alpine tundras. *Can. Entomol.* 113:607–614.

Kerner von Marilaun, A. and F. W. Oliver, 1904. *The Natural History of Plants.* Gresham, London.

Kevan, P. G. 1970. High Arctic insect–flower relations: the inter-relationships of arthropods and flowers at Lake Hazen, Ellesmere Island, N.W.T., Canada. Unpub. Ph.D. thesis, University of Alberta, Edmonton, Canada.

———. 1972. Floral colors in the High Arctic with reference to insect flower relations and pollination. *Can. J. Bot.* 50:2289–2316.

———. 1973. Parasitoid wasps as flower visitors in the Canadian High Arctic. *Anz. Schadlingskunde, Pflanzen und Umweltschutz* 46:3–7.

———. 1976. Fluorescent nectar. (Technical comment.) *Science (Washington)* 194:341–342.

———. 1978a. Floral coloration, its colormetric analysis and significance in anthecology. *In* A. J. Richards (ed.), *The Pollination of Flowers by Insects* (Linnean Soc. Symp. No. 6). Academic Press, London, pp. 51–78.

———. 1978b. Anthophilous springtails (Collembola) from the Alaskan north slope, from Signy Island Antarctica, and from near Ottawa, Ontario. *Rev. Ecol. Biol. Sol.* 15:373–378.

———. 1979. Vegetation and floral colors using ultraviolet light: interpretational difficulties for functional significance. *Am. J. Bot.* 66:749–751.

———. 1980. Pollination in the alpine. *Abs. II Int. Cong. Syst. Evol.,* p. 58.

Kevan, P. G. and H. G. Baker. 1983. Insects as flower vistors and pollinators *Ann. Rev. Entomol.* 28:407–453.

Kevan, P. G. and D. K. Kevan. 1970. Collembola as pollen feeders and flower visitors with observations from the High Arctic. *Quaest. Ent.* 6:311–326.

Kevan, P. G., N. D. Grainger, G. A. Mulligan, and A. R. Robertson. 1973. A gray-scale for measuring reflectance and color in the insect and human visual spectra. *Ecology* 54:924–926.

Kikuchi, T. 1962a. Studies on the coaction among insects visiting flowers. I. Ecological groups in insects visiting the Chrysanthemum flower, *Chrysanthemum leucanthemum. Sci. Rep. Tohoku Univ. Ser. IV (Biol.)* 28:17–22.

———. 1962b. II. Dominance relationship in the so-called dronefly group. *Sci. Rep. Tohoku Univ. Ser. IV (Biol.)* 28:47–51.

———. 1963a. III. Dominance relationship among flower-visiting flies, bees and butterflies. *Sci. Rep. Tohoku Univ. Ser. IV (Biol.)* 29:1–8.

———. 1963b. IV. Preferring behaviour of some syrphid flies, *Eristalomyia tenax, Eristalis cerealis* and *Sphaeropharis cylindrica* in relation to the age of the chrysanthemum flower, *Chrysanthemum leucanthemum. Sci. Rep. Tohoku Univ. Ser. IV (Biol.)* 29:9–14.

———. 1963c. V. Effects of the dominant insects on the density of the subordinate ones. *Sci. Rep. Tohoku Univ. Ser. IV (Biol.)* 29:107–115.

———. 1964. VI. The distribution of the subordinate insects in one flower garden. *Sci. Rep. Tohoku Univ. Ser. IV (Biol.)* 30:143–149.

King, R. M. and V. E. Krantz. 1975. Ultraviolet reflectance patterns in the Asteraceae: I. Local and cultivated species. *Phytologia* 31:66–114.

Kirschfeld, K. 1974. The absolute sensitivity of lens and compound eyes. *Z. Naturforsch.* 29(c):592–596.

Klebs, G. 1906. Uber variationen der Blüten. *Fahrb. Wissensch. Bot.* 42:155–320.

Knoll, F. 1921. Insekten und Blumen. Experimentelle Arbeiten zur vertiefung unserer Kenntnisse uber die Wechselbeziehungen zwischen Pflanzen und Tieren. II. *Bombylius fuliginosus* und die Farbe der Blumen. *Abh. Zool.-Bot. Ges. Wien* 12(1):17–119.

———. 1922. III. Lichsinn und Blumenbesuch der Falters von *Macroglossa stellatarum. Abh. Zool.-Bot. Ges. Wien* 12(2):120–377.

———. 1926. IV. Die Arum-Blütenstände und ihre Besucher. *Abh. Zool.-Bot. Ges. Wien* 12:(3)378–481.

Knuth, P. 1891a. Die Einwirkung der Blütenfarben auf die photographische Platte. *Bot. Centralbl.* 48:160–165.

———. 1891b. Weitere Beobachtungen uber die Anlockungsmittel der Blüten von *Sycos angulata* L. und *Bryonia dioica* L. *Bot. Centralbl.* 48:314–318.

———. 1906–9. *Handbook of Flower Pollination.* Trans. J. R. Ainsworth Davis (3 vols.: I, 1906; II, 1908; III, 1909). Oxford.

Kodak. 1970. *Kodak Filters for Scientific and Technical Uses.* Eastman Kodak Co. 91 pp.

Kraus, G. 1872. *Zur Kenntnis der Chlorophyllfarbstoffe und ihre Verwandten.* Stuttgart.

Kugler, H. 1930–32. Blütenokologie Untersuchungen mit Hummeln. I, III, IV. *Planta* 10:229–280, 16:227–276, 16:534–553.

———. 1936. Die Ausnutzung der Saftmalsumfärbung bei den Roszkastanienblüten durch Bienen und Hummeln. *Ber. Dt. Bot. Ges.* 60:128–134.

———. 1943. Hummeln als Blütenbesucher. *Ergebn. Biol.* 19:143–323.

———. 1947. Hummeln und die UV-Reflexion an Kronblättern. *Naturwissenschaften* 34:315–316.

———. 1950. Der Blütenbesuch der Schammfliege *(Eristalomyia tenax). Zeit. Vergl. Physiol.* 32:328–347.

———. 1951. Blütenokologische Untersuchungen mit Goldfliegen (Lucilien). *Ber. Dt. Bot. Ges.* 64:327–341.

———. 1956. Uber die optische Wirkung von Fliegenblumen auf Fliegen. *Ber. Dt. Bot. Ges.* 69:387–398.

———. 1963. UV-Musterungen auf Blüten und ihr Zustandekommen. *Planta* 59:296–329.

———. 1971. UV-Musterung bei Alpenblumen. *Jahrb. Ver Schultze Alpenfl. Tiere* 36:61–65.

Kühn, A. 1924. Zum Nachweiss des Farbuterscheidungsvermögens der Bienen. *Naturwissenschaften* 12:116.

———. 1927. Uber den Farbensinn der Bienen. *Zeit. Vergl. Physiol.* 5:762–800.

Kullenberg, B. 1956. On Scents and colours of *Ophrys* flowers and their specific pollinators among the aculeate Hymenoptera. Suensk. *Bot. Tidskr.* 50:25–46.

———. 1961. Studies in *Ophrys* pollination. *Zoologiska bidrag fran Uppsala* 34:1–340.

Küppers, H. 1973. *Color: Origin, Systems, Uses.* Trans. by F. Bradly. Van Nostrand Reinhold Ltd., London.

Kyhos, D. W. 1971. Evidence of different adaptations of flower color variants of *Encelia farinosa. Madroño* 21:49–61.

Labhart, T. 1974. Behavioral analysis of light intensity discrimination and spectral sensitivity in the honeybee, *Apis mellifera. J. Compl. Physiol.* 95:230–216.

Land, E. H.1977. The retinex theory of color vision. *Sci. Am.* 237(6):108–128.

Lane, M. A. 1979. Ligule microcharacters and flavonoid chemistry in the study of the *Xanthocephalum* complex. *Bot. Soc. Am. Misc. Publ.* 157:61.

———. 1980. Systematics of *Amphiachysis, Greenella, Gutierrezia, Gymnosperma, Thurovia* and *Xanthocephalum* (Compositae: Astereae). Ph.D. dissertation, University of Texas, Austin.

Laverty, T. 1980. The flower visiting behaviour of bumblebees; learning and flower complexity. *Can. J. Zool.* 58:1324–1335.

Leleji, O. I. 1973. Apparent preference by bees for different flower color cowpeas [*Vigna sinensis* L. (Savi ex Hassk.)]. *Euphytica* 22:150–153.

Levin, D. A. 1969. The effect of corolla color and outline on interspecific pollen flow in *Phlox*. *Evolution* 23:444–455.

———. 1972a. The adaptedness of corolla-color variants in experimental and natural populations of *Phlox drummondii*. *Am. Natur.* 106:57–70.

———. 1972b. Low frequency disadvantage in the exploitation of pollinators by color variants in *Phlox*. *Am. Natur.* 106:453–460.

Levin, D. A. and H. W. Kerster. 1970. Phenotypic dimorphism and population fitness in *Phlox*. *Evolution* 24:128–134.

Levin, D. A. and B. A. Schaal. 1970. Corolla color as an inhibitor of interspecific hybridization in *Phlox*. *Am. Natur.* 104:273–283.

Levy, M. 1978. Flavonoids and pollination ecology: pigments of systematists' imagination? *Phytochem. Bull.* 11:1–16.

Levy, M. and D. A. Levin. 1974. Novel flavonoids and reticulate evolution in the *Phlox pilosa–P. drummondii* complex. *Am. J. Bot.* 61:156–167.

———. 1975. The novel flavonoid chemistry and phylogenetic origin or *Phlox floridana*. *Evolution* 29:487–499.

Lewis, H. 1953. The mechanism of evolution in the genus *Clarkia*. *Evolution* 7:1–20.

Lloyd. D. G. 1969. Petal color polymorphism in *Leavenworthia* (Cruciferae). *Contr. Gray Herb. Harvard Univ.* 198:9–40.

Lothmar, R. 1933. Neue Untersuchungen über den Farbensinn der Bienen, mit besonderer Berücksichtigung des Ultravioletts. *Zeit. Vergl. Physiol.* 19:673–723.

Lubbock, J. 1881. Observations on ants, bees, and wasps. Part IX. Colors of flowers as an attraction to bees: experiments and considerations thereon. *J. Linn. Soc. London Zool.* 16:110–112.

Ludwig, F. 1885. Die biologische Bedeutung des Farbenwechsels mancher Blumen. *Biol. Centralbl.* 4:196.

———. 1887. Einige neue Falle von Farbenwechsel in verbluhenden Blüthenstanden. *Biol. Centralbl.* 6:1–3.

Lutz, F. E. 1924. Apparently non-selective characters and combinations of characters, including a study of ultraviolet in relation to the flower-visiting habits of insects. *Ann. N.Y. Acad. Sci.* 29:181–283.

———. 1933. "Invisible" colors of flowers and butterflies. *Nat. Hist., N.Y.* 33:565–576.

MacNichol, E. F. 1964. Three-pigment color vision. *Sci. Am.* 197(12):1–10.

Manning, A. 1956a. The effect of honey guides. *Behaviour* 9:114–139.

———. 1956b. Some aspects of the foraging behaviour of bumblebees. *Behaviour* 9:164–201.

Maxwell, J. C. 1855. Experiments on colour, as perceived by the eye, with remarks on colorblindness. *Trans. Roy. Soc. Edinburgh* 21(2). *In* W. D. Niven. 1890. *The Scientific Papers of James Clerk Maxwell.* Cambridge University Press, Vol. I, Ch. VII, pp. 126–154

and plate. Reprinted 1965 by Dover Publ. Inc., New York.

Mazokhin-Porshnyakov, G. A. 1959. Otraheniye ultravioletovikh luchei tsvetkami rastenii i zreniye nasekomikh. *Ent. Obozr.* 38:312–325.

———. 1962. Kolorimetricheskoye dokazetel'stvo trikhromazii tsvetovogo zreniya pchelinikh (na primere schmelei). *Biofizika* 7:211–217.

———. 1969. *Insect Vision.* Trans. by R. and L. Masironi; ed. by T. H. Goldsmith. Plenum Press, New York.

Meeuse, B. J. D. 1966. The voodoo lily. *Sci. Am.* 215:80–88.

Menzel, R. 1971. Uber den Farbensinn von *Paravespula germanica* F. (Hymenoptera). ERG and selektive Adaptation. *Zeit. Vergl. Physiol.* 75:86–104.

Menzel, R. and J. Erber. 1978. Learning and memory in bees. *Sci. Am.* 239(1):102–108, 110.

Mogford, D. J. 1974. Flower color polymorphism in *Cirsium palustre.* 2. Pollination. *Heredity* 33:257–263.

———. 1978. Pollination and flower color polymorphism, with special reference to *Cirsium plaustre. In* A. J. Richards (ed.), *The Pollination of Flowers by Insects* (Linnean Soc. Symp. No. 6). Academic Press, London, pp. 191–200.

Möller, F. 1957. Strahlung in der unteren Atmosphare. *In* S. Flugge (ed.), *Handbuch der Physik.* Springer Verlag, Berlin and Heidelberg, Vol. 48, pp. 155–253.

Moller-Räcke, I. 1953. Farbensinn und Farbenblindheit bei Insekten. *Zool. Jahrb. Abt. Allg. Zool. Physiol. Tiere* 63:238–274.

Mosquin, T. 1969. The spectral qualities of flowers in relation to photoreception in pollinating insects. *Abs. 11th Int. Bot. Congr.,* Seattle, p. 153.

Müller, F. 1877. Flowers and insects. *Nature (London)* 17:78–79.

Müller, H. 1876. On the relations between flowers and insects. *Nature (London)* 15:178–180.

———. 1881. *Die Alpenblumen ihre Befruchtung durch Insekten und ihre an passungen an dieselben.* Englemann, Leipzig.

———. 1883a. *The Fertilization of Flowers.* Trans. by D'A. W. Thompson. Macmillan, London.

———. 1883b. The effect of the change of colour in the flowers of *Pulmonaria officinalis* upon its fertilizers. *Nature (London)* 28:81.

Mulligan, G. A. 1972. Autogamy, allogamy, and pollination in some Canadian weeds. *Can. J. Bot.* 50:1767–1771.

Mulligan, G. A. and P. G. Kevan. 1973. Color, brightness and other floral characteristics attracting insects to the blossoms of some Canadian weeds. *Can. J. Bot.* 51:1939–1952.

Newton, I. 1704. Optiks: or, a treatise of the reflexions, refractions, inflexions and colors of light. The Royal Society, London.

Ornduff, R. and T. Mosquin. 1970. Variation in the spectral qualities of flowers in the *Nymphoides indica* complex (Menyanthaceae) and its possible adaptive significance. *Can. J. Bot.* 48:603–605.

Ostler, W. K. and K. T. Harper. 1978. Floral ecology in relation to plant species diversity in the Wasatch Mountains of Utah and Idaho. *Ecology* 59:848–861.

Parker, W. H. 1980. Contrasting patterns of ultraviolet absorbance/reflectance in *Limnanthes* flowers: a novel mechanism of elaboration and evolutionary significance. *Abs. II Int. Cong. Syst. Evol. Biol., Vancouver,* p. 304.

Pedersen, M. W. 1967. Cross-pollination studies involving three purple-flowered alfalfa, one white-flowered line and two pollinator species. *Crop Sci.* 7:59–62.

Pedersen, M. W. and F. E. Todd. 1949. Selection and tripping in alfalfa clones by nectar collecting honeybees. *Agron. J.* 41:247–249.

Petersen, B. 1968. Pollination of some tundra plants with miniature flowers on Niwot Ridge in Boulder County, Colorado. Unpub. Ph.D. thesis, University of Colorado. 106 pp.

Plateau, F. 1899. Nouvelles recherches sur leso rapports entre les insectes et les fleurs. Etudes sur le role de quelques organes dits vexillaires. Deuxieme partie. Le choix des couleurs par les insectes. *Mem. So. Zool. Fr.* 12:336–370.

———. 1900a. Ibid. Troisieme partie. Les syrphides admirent-ils les couleurs des fleurs? *Mem. So. Zool. Fr.* 13:266–285.

———. 1900b. Experiences sur l'attraction des insectes par les étoffes colorées et les objets brillants. *Ann. Soc. Ent. Belg.* 44:174–188.

Proctor, M. and P. Yeo. 1973. *The Pollination of Flowers.* Collins, London.

Rafinski, J. N. 1979. Geographic variability of flower colour in *Crocus scepusiensis* (Iridaceae). *Pl. Syst. Evol.* 131:107–125.

Raven, P. 1972. Why are bird-visited flowers predominantly red? *Evolution* 26:674.

Richtmyer, F. K. 1923. The reflection of ultraviolet by flowers. *J. Opt. Soc. Am.* 7:151–168.

Ripps, H. and R. A. Weale. 1969. Color vision. *Annu. Rev. Psych.* 20:193–216.

Rushton, W. A. H. 1975. Visual pigments and color blindness. *Sci. Am.* 232(3):64–74.

Schaal, B. A. and W. J. Leverich. 1980. Pollination and banner markings in *Lupinus texensis* (Leguminosae). *Southwest. Natur.* 25:280–282.

Schemske, D. W. 1976. Pollinator specificity in *Lantana camara* and *L. trifolia* (Verbenaceae). *Biotropica* 8:260–264.

Schlecht, P. 1979. Colour discrimination in dim light: an analysis of the photoreceptor arrangement in the moth *Deilephila. J. Comp. Physiol.* 129:257–267.

Scogin, R. 1976a. Floral UV patterns and anthochlor pigments in the genus *Coreopsis* (Asteraceae). *Aliso* 8:429–431.

———. 1976b. Anthochlor pigments and pollination biology. I. The UV absorption of *Antirrhinum majus* flowers. *Aliso* 8:425–427.

Scogin, R. and K. Zakar. 1976. Anthochlor pigments and floral UV patterns in the genus *Bidens* (Asteraceae). *Biochem. Syst. Ecol.* 4:165–167.

Scogin, R., D. A. Young, and C. E. Jones. 1977. Anthochlor pigments and pollination ecology. II. The ultraviolet pattern of *Coreopsis gigantea. Bull. Torrey Bot. Club* 104:155–159.

Scora, R. W. 1964. Dependency of pollination on patterns in *Monarda* (Labiatae). *Nature (London)* 204:1011–1012.

Seybold, A. and A. Weissweiler. 1944. Spectraphotometrische Messungen an Blumenblättern. *Bot. Arch.* 45:358–386.

Sharp, J. L. and J. James. 1979. Color preference of *Vespula squamosa. Environ. Entomol.* 8:708–710.

Shelford, R. 1903. Observations on some mimetic insects and spiders from Borneo and Singapore. *Proc. Zool. Soc. London* 2:230–284.

Silbergleid, R. E. 1976. Visualization and recording of long-wave ultraviolet reflection from natural objects. Parts 1 and 2. *Funct. Photog.* 11:20–29, 30–33.

———. 1977. Evaluation of ultraviolet imagery. Mimeo sheets. 7 pp. (unpubl).

———. 1979. Communication in the ultraviolet. *Ann. Rev. Ecol. Syst.* 10:373–398.

Smith, A. J. E. 1963. Variation in *Melampyrum pratense* L. *Watsonia* 5:336–367.

Snyder, A. W. and W. H. Miller. 1972. Fly color vision. *Vision Res.* 12:1389–1396.

Sprengel, C. K. 1793. *Das endeckte Geheimniss des Natur in Bau und in der Befurchtung der Blumen.* F. Vieweg aelt., Berlin. (Facsimile Drucken, Wissenschaftliche Classiker, Vol. III. 1893, Mayer and Muller, Berlin.)

Stace, H. M. and Y. J. Fripp. 1977. Raciation in *Epacris impressa.* II. Habitat differences and flowering times. *Aust. J. Bot.* 25:315–323.

Stebbins, G. L. 1974. *Flowering Plants. Evolution above the Species Level.* The Belknap Press of Harvard University Press, Cambridge, Mass.

Strüwe, G. 1972a. Spectral sensitivity of the compound eye in butterflies. *(Heliconius). J. Comp. Physiol.* 79:191–196.

———. 1972b. Spectral sensitivity of single photoreceptors in the compound eye of a tropical butterfly *(Heliconius numata). J. Comp. Physiol.* 79:197–201.

———. 1973. Spectral sensitivity of the compound eye in a moth. Intra- and extracellular recordings. *Acta Physiol.* 87:63–68.

Sturgeon, K. B. and J. B. Mitton. 1980. Cone color polymorphism associated with elevation in white fir, *Abies concolor,* in southern Colorado. *Am. J. Bot.* 67:1040–1045.

Swihart, S. L. 1972. The neural basis of colour vision in the butterfly, *Heliconius erato. J. Insect Physiol.* 18:1015–1025.

Thien, L. B. 1971. Orchids viewed with ultraviolet light. *Bull. Am. Orchid Soc.* 10:877–880.

Thien, L. B. and B. G. Marcks. 1972. The floral biology of *Arethusa bulbosa, Calopogon tuberosus,* and *Pogonia ophioglossoides* (Orchidaceae). *Can. J. Bot.* 50:2219–2235.

Thomas, I. and H. Autrum. 1965. Die Empfindlichkeit

der dunkel- und helladaptierten Biene *(Apis mellifica)* fur spektrale Farben: zum Purkinje-Phänomen der Insekten. *Zeit. Vergl. Physiol.* 51:204–218.

Thompson, W. R., J. Meinwald, D. Aneshansley, and T. Eisner. 1972. Flavonols: pigments responsible for ultraviolet absorption in nectar guides of flowers. *Science (Washington)* 177:528–530.

Thorp, R. W. 1979. Honeybee foraging behavior in California almond orchards. *Proc. IV Int. Symp. Pollination.* Sp. Misc. Pub. Mc Agric. Expl. Stn. No. 1, pp. 385–392.

Thorp, R. W., D. L. Briggs, J. R. Estes, and E. H. Erickson. 1975. Nectar fluorescence under ultraviolet irradiation. *Science (Washington)* 189:476–478.

Tikhomirov, B. A. 1959. *Relationships of the Animal World and the Plant Cover of the Tundra.* Trans. by E. Issakoff and T. W. Barry; ed. by W. A. Fuller. Boreal Institute, University of Alberta, Edmonton.

Tinbergen, N., B. J. D. Meeuse, L. K. Boerema, and W. W. Varossieau. 1942. Die Balz des Samfalters *Eumenis (Satyrus)* semele L. *Z. Tierpsychol.* 5:182–226.

Tutin, T. G., V. H. Heywood, N. A. Burgess, D. H. Valentine, S. M. Walters, and D. A. Webb (eds.). 1964. *Flora Europaea.* Cambridge University Press.

Utech, F. H. and S. Kawano. 1975. Spectral polymorphisms in angiosperm flowers determined by differential ultraviolet reflectance. *Bot. Mag. Tokyo* 88:9–30.

Valentine, D. H. 1975. The taxonomic treatment of polymorphic variation. *Watsonia* 10:385–390.

Verity, E. 1980. *Color Observed.* Van Nostrand Reinhold Co., New York.

Vogel, S. 1978. Pilzmuckenblumen als Pilzmimeten. Pt. I. *Flora* 167:329–366; Pt. II. *Flora* 167:367–398.

Waser, N. M. and M. V. Price. 1981. Pollinator choice and stabilizing selection for flower color in *Delphinium nelsonii. Evolution* 35:376–390.

Waser, N. M. and L. A. Real. 1979. Effective mutualism between sequentially flowering plant species. *Nature (London)* 281:670–672.

Weevers, T. 1952. Flower colours and their frequency. *Acta. Bot Neerl.* 1:81–92.

Weindorfer, G. 1903. Some comparisons of the alpine flora of Australia and Europe. *Vict. Natur.* 20:64–70.

Wickler, W. 1968. *Mimicry in Plants and Animals.* Trans. by R. D. Martin. Weidenfeld Nicholson, London.

Wigglesworth, V. B. 1972. *The Principles of Insect Physiology,* 7th ed. Wiley, New York.

Williams, C. E. 1904. Notes on the life history of *Gongylus gongylodes,* a mantis of the tribe Empusidea and a floral simulator. *Trans. Ent. Soc. London* 1904:125–133.

Woodson, R. E. 1964. The geography of flower color in butterfly weed. *Evolution* 18:143–163.

Wright, W. D. 1969. *The Measurement of Colour,* 4th ed. Adam Hilger Ltd., London.

Wyszecki, G. and W. S. Stiles. 1967. *Color Science: Concepts and Methods, Quantitative Data and Formulas.* John Wiley and Sons, New York.

Young, T. 1807. *A Course of Lectures on Natural Philosophy and the Mechanical Arts.* J. Johnson, St. Paul's Churchyard, London, Vol. I. Lecture XXXIX, On the nature of light and colours, pp. 457–471 + plates XXIX and XXX. Reprinted 1971 by Johnson Reprint Corp., New York and London.

Zeigenspeck, H. 1955. Die Farben- und UV-Photographie und ihre Bedeutung für die Blütenbiologie. *Mikroskopie* 10:323–328.

2
POST-POLLINATION PHENOMENA AND ADAPTIVE FLORAL CHANGES

David F. Gori

Department of Ecology and Evolutionary Biology
University of Arizona
Tucson, Arizona

ABSTRACT

Certain flowers, either in response to pollination or with age, rapidly change their characteristics and become inconspicuous, unattractive, or inaccessible to pollinators. The temporal pattern of these changes is quite distinct from simple floral senescence and suggests a selective advantage to species possessing this trait. Three hypotheses for the evolution of floral change are presented, and critical predictions are outlined. The hypotheses are then tested using data for a lupine, *Lupinus argenteus*. Floral change in this species appears to have evolved to maximize the time pollinators spend on plants. Ultimately the existence of floral changes may depend on the relative advantages of retaining pollinated/inviable flowers on plants.

KEY WORDS: Floral change, senescence, floral retention, postpollination.

CONTENTS

INTRODUCTION

Flowers of many plant species do not remain constant in form or color throughout their tenure on a plant. Instead, following pollination or simply as a result of age, they become inconspicuous, unattractive, or inaccessible to pollinating agents. The general significance and prevalence of this phenomenon is not known, as detailed descriptions of flower changes are rare (see Sussenguth, 1936; Faegri and van der Pijl, 1972). In this chapter, I argue that the pattern of these changes is quite distinct from simple floral senescence and suggest that these changes may have considerable adaptive significance for species possessing them. I then briefly review existing hypotheses and propose two additional ones to account for the evolution of these behaviors. Critical predictions for these hypotheses are subsequently outlined and tested with data for a lupine, *Lupinus argenteus*. Next, I consider under what conditions floral changes might evolve. To a large extent, this depends on the advantages of retaining pollinated or inviable flowers on plants. Lastly, I speculate that various types of floral change may represent different solutions to the same evolutionary problem, each having evolved under a different set of constraints imposed by the perceptual system of specific pollinators. While interesting in their own right, floral changes also may be of considerable importance in understanding how pollinator–flower relationships evolve, especially in the context of optimal foraging theory.

PATTERNS OF CHANGE IN FLORAL CHARACTERS: POST-POLLINATION PHENOMENA AND ADAPTIVE NONINDUCED CHANGES

There are three basic temporal patterns of floral change (Fig. 2-1). In the first, simple senescence, all floral characters change synchronously and rapidly at the end of a flower's lifespan. These events involve changes in color, cessation of nectar and odor production, and collapse and abscission of the corolla. The physiological mechanisms involved in these changes are reviewed by Mayak and Halevy (1980) for the first time. In the second, changes are induced by pollination at variable floral ages and result in a rapid change in the state of the flower. The change may be limited to only a single character (e.g., a change in banner color spot) or to several (e.g., color, nectar, odor). If no pollination occurs, flowers remain constant in their characteristics until

simple senescence occurs. I have termed these changes pollination-induced. In broader view, the proximate stimuli for the initiation of these changes may include events associated with insect visitations such as petal depression or stigmatic contact instead of pollination per se. In the third pattern, the floral change again involves a limited number of floral characters, but occurs after some fixed interval in the lifespan of the flower. These noninduced changes appear to correspond temporally with the termination of stigma receptivity and pollen viability (Schaal and Leverich, 1980; Brenda Caspar, pers. comm.). Despite their inviability, however, flowers are retained on plants for a period of time before the onset of simple senescence.

At least five basic types of change can occur in floral characteristics following pollination or at termination of floral viability (ie. stigma receptivity and pollen viability). They are color change, termination of odor and/or nectar production, change in flower orientation, collapse of flower parts, and abscission of the corolla. Although it is possible that floral changes have no general significance and merely represent the standard aging process of flowers, several repeated patterns strongly suggest otherwise.

First, floral changes that are induced by pollination occur rapidly relative to the lifespan of unpollinated flowers (Table 2-1). Furthermore, flowers that undergo change, whether in response to pollination or floral inviability, are characteristically retained on plants for extended periods prior to floral senescence, suggesting a relationship between retention and floral change. For example, many species in the genus *Lupinus* undergo a floral change that is restricted to a small spot on the banner petal. In *L. argenteus* and *L. texensis* change occurs at the termination of floral viability, five to six days following flower opening. These flowers, however, are retained on inflorescences an additional five to six days before wilting (Gori, unpub. data; Barbara Schaal, per. comm.).

Second, many floral changes involve only selective parts of the flower while the rest remains unchanged until senescence occurs (Kugler, 1936, 1950; Dunn, 1956; Jones and

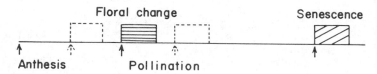

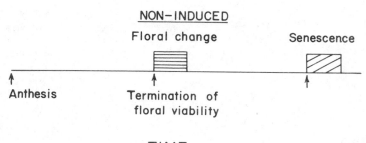

Figure 2-1. Three temporal patterns of floral change. In the first, simple senescence, changes in all floral characteristics occur synchronously at the end of floral lifespan. In the second, pollinator-induced change, change in a limited number of floral characters occurs rapidly following pollination. The timing of this event and therefore the ensuing change is variable. Following change, flowers are retained on plants until simple senescence occurs. In the third, noninduced change, change in a limited number of flower characters occurs after a fixed period of time, independent of the timing of pollination. Changed flowers are retained on plants until simple senescence.

Buchmann, 1977; Wainwright, 1978; Cruden and Hermann-Parker, 1979; Schaal and Leverich, 1980; Appendix to this chapter).

Third, pollinators are clearly able to recognize floral changes and visit only viable or unpollinated flowers. In making this distinction, pollinators increase their foraging efficiency, since flowers that have undergone change invariably offer no food rewards (Table 2-2).

Fourth, floral changes are often highly nonrandom and are easily understood in terms of how different pollinating agents perceive flowers. Such changes include:

A. Color Changes. Color change may occur over the entire flower or may be restricted to selected parts, such as spots and nectar guides, that are critical to pollinator orientation. Flowers may change to colors that are outside the perceptual range of specific pollinators or that match the background color of flowers or the environment. For example, many insect-pollinated species change from a light color to red, which is visually imperceptible to many insects (Raven, 1972). Species pollinated by nocturnal moths commonly change from white to a dark color, presumably reducing contrast with the dark background as, for example, in *Cobaea scandens* (Proctor and Proctor, 1978). In lupines, a yellow spot on the banner petal changes to blue, thus matching the color of the

Table 2-1. A summary, for species showing floral changes, of their temporal pattern, the lifespan of unpollinated flowers, and the length of time changed flowers are retained on plants prior to floral senescence. Values in parentheses refer to the period over which changes occur. For species showing noninducible changes, the time difference between the retention period following change and the lifespan of unpollinated flowers represents the approximate timing post-anthesis of programmed floral change.

Species	Change	Lifespan of unpollinated flowers	Period retained following change	Reference
Lupinus argenteus	Noninduced	11–12 days	5–6 days	Gori (unpub. data)
L. nanus	?	5–6 days	3 days	Dunn (1956)
L. bicolor	?	2 days	1 day	Dunn (1956)
L. sparsiflorus	Induced (4 hr)	4–5 days	3–4 days	Wainwright (1978)
L. arizonicus	Induced (4 hr)	4–5 days	3–4 days	Wainwright (1978)
L. blumeri	Induced (24 hr)	10 days	8 days	Gibson (unpub. data)
L. texensis	Noninduced	11 days	5 days	Schaal and Leverich (1980)
Caesalpinia eriostachys	Induced (4 hr)	3–4 days	2–3 days	Jones and Buchmann (1974)
C. pulcherrima	?	2 days	1–1.25 days	Cruden and Hermann-Parker (1979)
Phyla incisa	Induced (24 hr)	2 days	at least 1 day	Estes and Brown (1973)
Combretum farinosum	?	5 days	3 days	Schemske (1980)
Cymbidium spp.	Induced (4 hr)	5–6 weeks	3–4 days	Arditti and Flick (1976), Harrison and Arditti (1976)
Lantana camara	?	3 days	2.4 days	Barrows (1976)
L. trifolia	?	2–3 days	1–2 days	Schemske (1976)

surrounding petal. The result of these changes is to make flowers inconspicuous or unattractive, either at a distance (presumably the case when change is over the entire flower) or at close range (when changes involve selected floral parts involved in pollinator orientation).

B. *Termination of Odor and/or Nectar Production.* Certain flowers often stop their odor production, and, in response, pollinators will not visit them (e.g., moths, Brantjes, 1976). While this may be the only post-pollination change, it often accompanies other changes (e.g., color changes in *Parkinsonia aculeata,* Jones and Buchmann, 1974). Since some nectar reflects UV radiation, it is also possible that certain pollinators can visually assess the amount of nectar present in open flowers and thereby avoid nectarless, inviable flowers (Thorp et al., 1975).

C. *Changes in Flower Orientation.* For certain flower species, pollinators must be properly oriented with respect to the flower in order to effect pollination (Faegri and van der Pijl, 1972). Changes in flower position may make additional visits difficult and may also serve as a visual cue, signaling the absence of rewards. For example, moths in the genus *Macrosila,* which pollinate *Angraecum* spp., possess downward-curved proboscises that fit into similarly curved flowers. Following pollination, flowers reorient upward, making contact with the nectaries and reward removal by moths virtually impossible (Strauss and Koopowitz, 1973).

D. *Collapsed Flower Parts.* In some species the entire flower rapidly collapses following pollination and becomes inaccessible to pollinators seeking additional visits (*Petunia hybrida,* Gilissen, 1977). In other species these changes are remarkably selective, involving specific petals and sepals as in *Bulbophyllum macranthum* (Millar, 1978), *Cypripedium arietinum* (Case, 1964), and *Caesalpinia eriostachys* Jones and Buchmann, 1974). Selective collapse of floral parts may protect the stigma from damage or pollen loss as well as prevent additional pollinator visits.

E. *Corolla Abscission.* In certain flower spe-

Table 2-2. Summary of reported reward availabilities in changed flowers (negligible rewards/rewardless) and pollinator behavior toward them (avoidance).

Species	Negligible rewards/ rewardless	Pollinator avoidance	Reference
Phyla incisa	?	Yes	Estes and Brown (1973)
Phyrrhopappus carolinianus	Yes	Yes	Estes and Thorp (1975)
Lantana camara	Yes	Yes	Barrows (1976), Dronamraju (1958)
Lupinus blumeri	Yes	Yes	Gibson and Gori (unpub. data)
L. argenteus	Yes	Yes	Gori (unpub. data)
L. texensis	Yes	Yes	Schaal and Leverich (1980)
L. nanus	Yes	Yes	Dunn (1956)
L. arizonicus	Yes	Yes	Wainwright (1978)
L. sparsiflorus	Yes	Yes	Wainwright (1978)
Malvaviscus arboreus	Yes	?	Gottsberger (1971)
Caesalpinia pulcherrima	Yes	Yes	Cruden and Hermann-Parker (1979)
Caesalpina eriostachys	Yes	Yes	Jones and Buchmann (1974)
Parkinsonia aculeata	Yes	Yes	Jones and Buchmann (1974)
Combretum farinosum	Yes	Yes	Schemske (1980)
Bulbophyllum macranthum	?	Yes	Millar (1978)
Cymbidium hybrids	?	Yes	Arditti and Flick (1976)
Aesculus hippocastanum	Yes	Yes	Faegri and van der Pijl (1972)
Cryptantha breviflora	Yes	Yes	B. Caspar (pers. comm.)
C. nana	Yes	Yes	B. Caspar (pers. comm.)
Aesculus hippocastanum	?	Yes	Kugler (1936)
Senecio jacobaea	?	Yes	Kugler (1950)

cies, for example in *Rhododendron,* the entire corolla falls off shortly after pollination. Corollas persist if flowers remain unpollinated.

Finally, floral changes may arise independently several times within genera or may be retained or lost among closely related species, depending on specific ecological circumstances. For example, 26 out of 48 lupine species that have been examined undergo floral color change. Such changes occur sporadically in several different sections of the genus. If these taxonomic divisions reflect phylogenetic relationships, then this pattern suggests either independent evolution or selective loss of floral color change within the genus *Lupinus* (Thomas C. Gibson, unpub. data).

PREVALENCE IN NATURE

A systematic survey of plant species for the presence of floral changes has not been at-tempted, and therefore their prevalence is currently unknown. A brief survey, using photographs in regional floras, revealed that over 6% of 621 species examined showed a distinct change in the color of selective floral parts, mostly in internal design such as nectar guides (Thomas C. Gibson, unpub. data). Color changes over the entire flower were excluded because of their ambiguity. Since this survey did not include whole flower color changes, color change in the UV spectrum, nectar and/ or odor changes, position changes, and collapsed or abscissed corollas, this figure is likely to be a conservative one.

GENERAL HYPOTHESES AND PREDICTIONS

At present only a very limited theoretical framework exists to explain the significance of post-pollination phenomena and adaptive noninduced changes. Several advantages have been suggested for floral changes, including a putative role in reducing nectar thievery (Barrows, 1976), increasing pollination efficiency (Barrows, 1976; Proctor and Proctor, 1978; Schaal and Leverich, 1980), and maintaining pollinator constancy (Heinrich, 1975). With the exception of Schaal and Leverich (1980), however, these hypotheses are vague and fail to explicitly consider the selective pressures operating on both plants and pollinators in the evolution of these traits. In light of this criticism, I now present three hypotheses for the evolution of floral change, and then discuss predictions that will allow their discrimination.

Hypothesis I. Pollination Interference: Increased seed set due to the reduction of superfluous pollinator visitation.

Consider a hypothetical curve that describes how the percentage of ovules fertilized in a flower varies with the number of visits to that flower. While such curves have not yet been constructed for any species, two types are perhaps most likely. First, the percentage fertilized may initially increase at a constant rate with the number of pollinator visits, since with additional visits more pollen reaches the stigma. Due to the finite number of ovules/flower or limited stigmatic surface, the curve will eventually level off (Fig. 2-2A). When seed set shows a monotonic increase with increasing visits/flower, natural selection should favor those individuals whose flowers remain attractive after partial pollination. For certain species a second curve is possible. This curve increases, peaks, and then declines with additional pollinator visits. There may be a number of reasons for the reduction in seed set. Additional visits may dislodge more ungerminated pollen from stigmas than is deposited, substitute self-pollen for cross-pollen, or cause physical damage to reproductive structures during legitimate or illegitimate flower visits (McDade and Kinsman, 1980). In these cases, natural selection should favor individual plants that make their pollinated flowers inconspicuous, unattractive, or inaccessible to pollinators (or robbers).

The following circumstantial evidence supports this hypothesis. Many flower species that

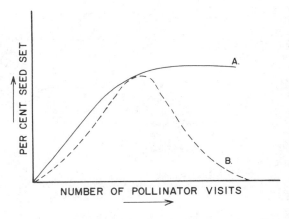

Figure 2-2. Two hypothetical relationships between percent seed set and the number of pollinator visitations to a flower. (A) Percent seed set initially increases at a constant rate with additional visits as more pollen reaches the stigma, and then levels off. The asymptote results from the finite number of ovules per flower, or limited stigmatic surface. (B) Percent seed set initially increases, peaks, and then declines with additional pollinator visitation. Factors potentially responsible for this final decline include the dislodgement of more ungerminated pollen from stigmas than is deposited, substitution of self-pollen for cross-pollen, and physical damage to reproductive structures.

exhibit floral changes have delicate pollination mechanisms that may be easily damaged by superfluous visits (e.g., the thin styles of *Lupinus* species; see also Jones and Buchmann, 1974, and Faegri and van der Pijl, 1972, for descriptions of similar flowers) or mechanisms that require precise placement of pollen on stigmas where additional visits may easily dislodge ungerminated pollen (e.g., *Cypripedium arientinum* trap flowers). The consequences of pollen dislodgement are more serious in flower species that have pollinia or packets of pollen where large reductions in seed set can result from additional visits.

Hypothesis II. Pollinator Foraging Efficiency: Increase in plant fitness due to increased pollinator residence times (the number of flowers visited) on plants.

Pollinators regularly forage in environments in which the resources vary in space and time. Natural selection should favor pollinators capable of assessing local resource harvesting rates and making foraging decisions that maximize their rates of energy intake (Pyke et al.,

1977). Several empirical studies demonstrate that pollinators do behave in general accordance with this prediction (Gill and Wolf, 1975; Witham, 1977; Hartling and Plowright, 1978; Pyke, 1978).

This hypothesis views floral changes as signals that have evolved to advertise the location of nonrewarding flowers and thereby maximize the time pollinators spend on plants. Plants possessing these signals allow pollinators to forage exclusively on flowers potentially containing rewards, resulting in a high resource-harvesting rate. If plants retain nonrewarding flowers but provide no signal to their location, pollinators will visit both types indiscriminately, resulting in a relatively lower harvesting rate on these plants. As a consequence of these different harvesting rates, pollinators will tend to spend more time on plants that signal. Increasing pollinator residence times on plants will result in greater seed set, pollen donation, or both. This hypothesis assumes that there is a fitness advantage to plants that retain inviable or pollinated flowers. This advantage may be through preferential pollinator visitation resulting from the enhanced attractiveness of the inflorescence or the entire plant.

Hypothesis III. Pollination Efficiency: Increase in plant fitness due to the restriction of pollinator visitations to viable, receptive flowers.

This hypothesis was first proposed by Schaal and Leverich (1980). According to their arguments, floral changes ensure that pollinator visitations are restricted to receptive flowers, thus preventing the wastage of incoming pollen on pollinated or unreceptive flowers. In cases where changes signal floral inviability, the collection of inviable pollen by pollinators is also prevented. Inviable pollen can potentially compete with viable pollen collected on the same plant for the stigmatic space of neighboring plants (Waser, this book). This hypothesis also assumes that there is a fitness advantage to plants retaining pollinated or inviable flowers.

The three hypotheses clearly are not mutually exclusive, yet it should be possible to discriminate between the alternatives. Hypothesis I predicts a specific relationship between the number of insect visits and seed set. Supple-

mentally increasing the number of visits by hand-pollinating flowers or restricting pollinators to foraging in cages should, at some point, decrease seed set if damage to reproductive structures occurs. If potential removal of ungerminated pollen is important, brushing stigmas or artificially selfing flowers in a self-incompatible species at the onset of floral change should result in a reduction in seed set.

Hypothesis II predicts that a reduction in signal reliability will lead to a reduction in pollinator residence times on plants. Reducing signal reliability can be accomplished by removing rewards from a portion of the flowers that have not yet undergone change. Observations of pollinator behavior on manipulated plants will indicate whether residence times are affected. A reduction in seed set or in the amount of pollen carried off the plant should result if Hypothesis II is correct. If individuals that do not undergo floral changes are found in the population, pollinator behavior on them may be observed directly and fitness consequences ascertained. In this latter experiment, the fitness effects due to Hypotheses II and III are confounded, since pollinators presumably would regularly visit inviable flowers.

Discriminating between Hypotheses II and III will be difficult, primarily because both may have operated simultaneously in the evolution of floral change. The most useful information for this discrimination is obtained if the signal can be manipulated so that rewardless, inviable flowers are visited, or if individuals not showing color change are found in the population. In both of these cases, the possibility of pollen wastage and the collection of inviable pollen exists. Reward harvesting rates also should be reduced, resulting, potentially, in shorter pollinator residence times. Any decrement in fitness observed, then, is attributable to the combined effects of Hypotheses II and III. Comparing the results of these experiments with those obtained when rewards are removed from normally rewarding flowers will indicate the effect due solely to Hypothesis III. In this latter experiment, typically only viable flowers will be visited; hence there is little potential for pollen wastage and the collection of inviable pollen. Reward-harvesting rates by pollinators, however, would be reduced, and any observed fitness effects therefore would solely be due to the reduction in pollinator residence times (Hypothesis II). Information on the dynamics of pollen loading and carryover, the number of visits required for maximum seed set, and the availability of viable and inviable pollen in the different flower types will also be important in providing supportive evidence. That is, if pollen carryover is long, the number of insect visits is in excess of that required for maximum seed set, and the availability of inviable pollen in flowers is negligible, then the selective pressures described in Hypothesis III should be relatively unimportant in the evolution of floral changes.

LUPINES: A TEST OF HYPOTHESES

The most extensive work to date on floral changes has been done on species in the genus *Lupinus*. Many lupines undergo a floral color change that is restricted to a small spot on the banner petal. When flowers first open, the spot is yellow, and, either in response to pollination or with age, it changes to blue or red-purple, closely matching the surrounding coloration of the banner (Table 2-1). It is not completely clear what role the contrasting spot plays, although it probably functions in pollinator orientation. Foraging insects place their heads at the base of the spot and by pushing down on the keel with their legs, cause stigma exertion and the release of pollen. Flowers are arranged either spirally or in distinct whorls on inflorescences, and flower maturation proceeds from the bottom upward. Following color change, flowers are retained on inflorescences for extended periods of time. Pollen appears to be the only reward used by lupines to attract pollinators, and quantitative counts indicate

greatest availability in flowers with a yellow spot (Table 2-3).

In only one study has there been a systematic attempt to test hypotheses relating to the selective advantage of color change (Gori, unpub. ms.). A brief summary of the results follows. The Pollination Interference Hypothesis (I) was examined for *L. argenteus* by brushing the stigmas of experimental plants with a clean paintbrush at the onset of color change and twice daily for two additional days following this change. All flowers produced by experimental plants were exposed to the preceding treatment. At the end of the season, data on seed set and plant dry weight were collected for the 10 experimental plants and for 39 control plants randomly selected from the population. The regression of seed set and plant dry weight for control plants is shown in Fig. 2-3. The regression is statistically significant, with dry weight explaining almost 50% of the variation in seed set ($y = .83x + 115.6$, $r^2 = .49$, $p < .01$). The 10 experimental plants are also plotted in the figure. All fall close to the line within the 95% confidence intervals calculated for the regression. There is no significant tendency for experimental points to fall consistently below the line. Therefore, it appears, first, that additional visits do not result in the predicted reduction in seed set, and, second, that the onset of color change occurs at a time when pollination interference is no longer possible. The selective pressures considered in Hypothesis I, therefore, have not contributed significantly to the evolution of floral color changes in *L. argenteus*.

The Pollinator Foraging Efficiency Hypothesis (II) was investigated by experimentally reducing signal reliability. This was accomplished by removing all pollen from half of the yellow flowers. Pollen removal was performed

Table 2-3. Pollen availability in lupine flowers of different colors/ages. Numbers represent the mean ($\pm$ SE) number of pollen grains present in individual flowers.

| | Yellow banner spot | | Purple banner spot |
	(1–2 days)	(3 days)	(5–6 days)
Lupinus argenteus	18,876 $\pm$ 2,734	9,716 $\pm$ 1,201	471 $\pm$ 301
	$n = 15$	$n = 15$	$n = 15$
L. texensis (Schaal and	33,019 $\pm$ 891		209 $\pm$ 69
Leverich, 1980)	$n = 10$		$n = 10$

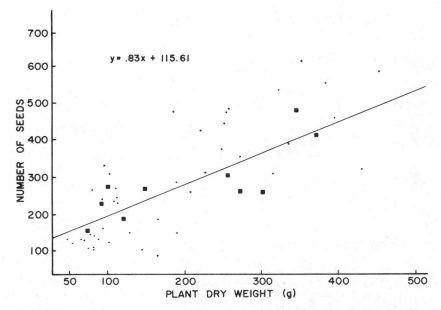

Figure 2-3. Number of seeds produced per plant as a function of plant dry weight (grams) for randomly selected individuals of *Lupinus argenteus* (closed circles, $y = .83x + 115.6$, $r^2 = .495$). Seed sets for the 10 experimental plants receiving stigma brushing at the onset of color change are also plotted (squares). All experimental points are within the 95% confidence intervals calculated for the regression and therefore statistically indistinguishable from control plants.

only on the lower flowers in order to produce a spatially realistic analogue to plants retaining nonrewarding flowers in the absence of floral signals. The manipulation was performed daily on experimental plants for the entire period in which they were in bloom. At the end of the season, seed set and plant dry weights were measured. Residence times of pollinators on plants including the number of flowers visited per inflorescence and the number of inflorescences per plant were observed, both on experimental plants and on controls. Mean number of visits per stalk as a function of the number of flowers actually containing rewards is summarized in Fig. 2-4 for experimental and control plants. Reducing signal reliability resulted in a significant reduction in the mean number of visits per inflorescence (multiple *t*-tests: $t > 2.86$, $p < .01$ in all cases). In addition, pollinators visited fewer inflorescences before leaving plants in the experimental group compared to the controls ($3.03 \pm .24$, $n = 60$, vs. $7.59 \pm .75$ inflorescences/plant, $n = 56$; $t_{63df} = 5.81$, $p < .001$).

Reduced pollinator residence times had no

apparent effect on female fitness, since all seed sets for experimental plants fell within the calculated 95% confidence intervals and show no consistent tendency to lie above or below the regression line (Fig. 2-5). Reductions in residence times, however, should strongly affect the male component of plant fitness. Although there are no published data indicating a positive relationship between time pollinators spend on plants (number of flowers visited) and the amount of pollen carried off the plant, studies on pollen carryover strongly suggest that on the average more pollen is picked up on a flower than is deposited (Thomson and Plowright, 1980; M. Price and N. Waser, unpub. data). Therefore, increases in the number of flowers visited should increase the absolute amount of pollen leaving the plant. These results suggest that the selective pressures described by the Pollinator Efficiency Hypothesis (II) are sufficient to account for the evolution of floral color change.

What about the Pollination Efficiency Hypothesis (III)? Quantitative determination of the number of pollen grains present in flowers

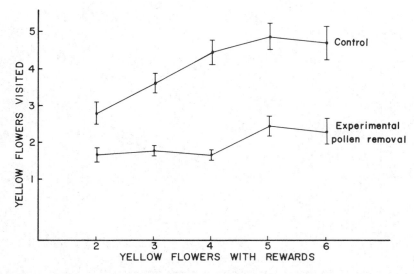

Figure 2-4. The mean number (± SE) of yellow flowers visited per stalk as a function of the number actually containing rewards for control and experimental plants. Pollen was removed from the lower one-half yellow flowers on experimental plants by repeated depression of the keel petals. Pollinators (*Bombus* spp.) visit significantly fewer flowers per stalk on pollen removal plants than on controls (multiple *t*-tests: $t > 2.86$, $p < .01$ in all cases).

of different ages and color indicates that changed flowers have very little pollen (Table 2-3). In addition, the pollen that is available is qualitatively different from that of yellow flowers, being less sticky and cohesive and there-

fore probably less likely to adhere to pollinators. These qualitative and quantitative differences suggest that even if inviable flowers were visited, their contribution to an insect's pollen load would be insignificant. Further,

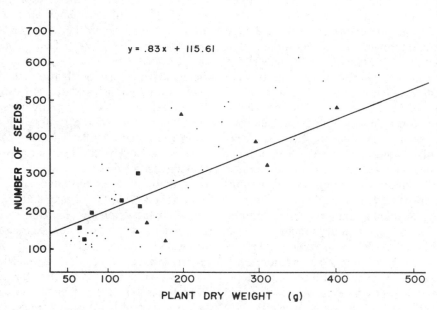

Figure 2-5. Number of seeds produced per plant as a function of plant dry weight (grams) for randomly selected individuals of *Lupinus argenteus* (closed circles, $y = .83x + 115.6$, $r^2 = .495$). Seed sets for experimental pollen removal plants (triangles) and plants not undergoing floral color change (squares) are also plotted. All of these latter points are within the 95% confidence intervals calculated for the regression.

seed set data collected for individuals in the population not showing color change, but retaining nonrewarding flowers, indicate no apparent effect of pollen wastage on seed set (Fig. 2-5). This and other supporting evidence for *L. argenteus* suggest that the number of visitations received by plants exceeds that required for seed set. Thus Hypothesis III appears to be relatively unimportant in the evolution of floral color change in *L. argenteus*. Signal reliability appears to function solely in increasing pollinator residence times on plants, and, further, it is selection on the male component of plant fitness that has been primarily responsible for the evolution of floral color change in *L. argenteus*.

Why do plants retain inviable, nonrewarding flowers on inflorescences? Experiments on *L. argenteus* show that the retention of such flowers enhances overall plant attractiveness, resulting in a higher visitation rate by pollinators (Gori, unpub. data; a similar result was also obtained for *Lotus scoparius*, C. E. Jones, unpub. data). Further, the differential selection of plants depends on differences in the total number of flowers on inflorescences rather than in the number actually providing rewards. Experimental reductions in flower number consistently result in lower visitation rates irrespective of changes in the numbers of flowers with rewards. On the basis of this evidence, I conclude that banner color is indiscernible to pollinators at a distance but functions as a short-distance signal, operative when pollinators are foraging on plants. This may also be true of many other nonlupine species, and suggests that the evolution of floral changes has accompanied that of floral retention. Retention serves as an efficient means for individual plants to increase their apparent attractiveness to pollinators. For this advantage to be fully realized, however, a signal indicating the location of nonrewarding flowers is required so that reward-harvesting rates and pollinator residence times on plants are maximized.

GENERAL DISCUSSION

I have cited a number of examples of floral change, some that have been demonstrated to be pollination- or visitation-induced and others that appear to be noninducible. In the majority of cases, the experiments that would permit a determination of induced vs. noninduced change have not been performed. In any case, insects clearly can distinguish the change and visit only unpollinated or viable flowers. This discrimination is advantageous, since flowers that have undergone changes either are nonrewarding or contain only negligible rewards. I have then asked whether these floral changes have any adaptive significance for species showing them, and have concluded on the basis of their temporal pattern, nonrandom nature, and variable occurrence among closely related species, that they do. Three hypotheses were presented to account for the evolution of these floral changes, and critical predictions were outlined. The hypotheses were explicitly tested on a lupine, *Lupinus argenteus*. The results suggest that color change functions in this species solely to increase pollinator residence times on plants, and, further, that selection has acted primarily on the male component of plant fitness in the evolution of this trait.

Why are floral changes inducible in some species while noninducible in others? Let us first consider the costs and benefits of inducibility. The benefits include a reduction in energy expenditures associated with the termination of rewards prior to the time of programmed floral inviability or senescence, and fitness gains associated with preventing additional visits during this period. There is also presumably a cost to plants in producing and maintaining enzyme systems capable of responding to the variable timing of pollination and pollen depletion.

Inducibility may depend on between-year or within-season variation in the rates of pollen depletion and in the timing of pollination. When the variation is small, the timing of these events is highly predictable, and plants that program floral changes to coincide temporally with them will minimize the potential benefits associated with inducible change. The costs, however, will be unaffected, so that noninducible changes are favored. As the variation increases, the timing of pollen depletion and pollination becomes less predictable, and benefits to inducible change increase. When benefits exceed costs, inducible change becomes advantageous.

At present there are no available data to test these predictions adequately. However, information is at least consistent with them for *L. argenteus*, a perennial that grows in alpine meadows in the Rocky Mountains. Annual variation in pollinator densities, at least in the area in which this species was studied, is small (David Inouye, pers. comm.). This should result in a low between-year variation in the timing of pollen depletion and pollination, and a noninducible change for *L. argenteus* is predicted. Stigmas are receptive for a period of two days, commencing three days after flower opening. The extremely high rate of insect visitation (18 visits/flower/day) suggest that pollination occurs within the first day of receptivity. However, pollen is normally available until just prior to floral color change. Experiments indicate that seed set in *L. argenteus* is not pollinator-limited but is determined by resources allocated to reproduction (Gori, unpub. ms.). Male fitness, however, appears to be limited by the availability of pollinators. Thus the fitness gain through male function resulting from additional visitations would be greater than the fitness loss through female function. (Additional visitations produce no significant effect on seed set.) As predicted, *L. argenteus* possesses a noninduced color change that occurs temporally with pollen depletion and the onset of floral inviability.

With the exception of corolla abscission, floral changes involve the retention of inviable and nonrewarding flowers on plants. This suggests a causal relationship between the two. Floral retention can serve two functions: (1) it allows the remobilization and recycling of structural proteins and nutrients from flowers back to the plant; (2) it serves as an energy-efficient means of enhancing the overall attractiveness of the inflorescence or plant. The potential advantages to floral retention are clearly not mutually exclusive, and both may operate. Flower retention also has potential costs, including metabolic expenditures associated with floral maintenance and fitness costs resulting from pollination interference and reduced pollinator residence times if harvesting rates on plants that retain flowers are less than

on "nonretainers." Whether or not retention evolves in a population will depend on the relative magnitude of these costs and benefits. Clearly, if accompanied by floral changes, the costs associated with retention will be reduced and the conditions for its evolution relaxed. Further, because of this potential for cost reduction, there will always be a selective advantage to the evolution of floral changes, especially if they are sufficiently subtle so that pollinators cannot distinguish between flowers at a distance but can when actually foraging on the plant.

In the case of floral abscission in response to pollination, plants actually reduce their overall attractiveness. This type of floral change is expected to evolve when the retention of flowers is not advantageous for the reasons discussed above, but the energy saving associated with pollination-induced abscission and the disadvantages associated with additional visitation following pollination and reward cessation are significant.

Retention in the absence of floral change is expected when there is no fitness cost due to pollination interference, pollen wastage, or reduced pollinator residence times. In certain situations, plants may actually benefit from reduced residence times. The amount of pollen carried off a plant should be positively related to the amount of time pollinators spend on plants. However, as residence time increases, the potential for considerable intra-plant transfer of pollen also increases, and in a self-incompatible species this may reduce female fitness through pollen competition on stigmatic surfaces. Interestingly, many species of lupines do not undergo floral color change, although it is unknown whether they also retain inviable, nonrewarding flowers. If so, this genus will provide an opportunity to test some of the hypotheses presented here.

Why have some flower species evolved different types of changes to make themselves inconspicuous, unattractive, or inaccessible? While it is possible that a random event in the plant's past may determine which type evolves, the relationship between the plant and its particular pollinators may also be important. Because pollinators perceive flowers in different ways, a plant becomes constrained in the number of ways it can make itself inconspicuous, unattractive, or inaccessible. For instance, changes to red coloration after pollination are effective only for insects that do not perceive red, but ineffective for bird-pollinated flowers. It is interesting that floral changes in bird-pollinated species consist of a distinct lightening of the red coloration as in *Malvaviscus arboreus* (Gottsberger, 1971) and *Bouvardia glaberrima* (Gori, pers. obs.) or abscission of the entire corolla as in certain bird-pollinated *Rhododendron*. Termination of nectar and/or odor production for these flowers would be ineffective as signals. In general, birds have a poorly developed sense of smell, and nectar is contained in tubular nectaries, making direct visual assessment difficult. Similarly, moth-pollinated flowers are white, generally lack color details such as nectar guides, bloom at night, and have strong odors (Faegri and van der Pijl, 1972). The structure of the moth's eye prevents it from seeing detail in dark environments (Dethier, 1963). Therefore, color changes over the entire flower, termination of odor production, and flower reorientation would be effective as signals, while color change over select floral parts as, for example, in internal designs, would not. Nectar is contained in tubular nectaries, so visual assessment is not possible.

Many interesting theoretical and empirical questions relating to the evolution of floral changes remain to be answered. Experimental tests of the hypotheses presented here are almost completely lacking. Empirical work on many species is critically needed in order to evaluate their importance and to establish the relative contribution of male vs. female components of plant fitness in the evolution of these changes. Until only recently plant ecologists have assumed that selection on female fitness (seed set) has been the primary force in the evolution of floral displays, plant mating systems, and reproductive strategies (Cruden, 1976, 1977; Faegri and van der Pijl, 1972, Schaffer and Schaffer, 1977). Recent theoretical work, however, suggests that perhaps male function has been relatively more important (Willson and Price, 1977; Charnov, 1979; Willson, 1979), and several empirical studies have supported this view at least for characters

of floral display (Willson and Price, 1977; Sutherland, ms.; Gori, ms.). It is also important to establish whether observed floral changes are inducible, and, if not, whether they coincide temporally with the termination of floral viability. As a larger sample of species is obtained, hypotheses relating to the ecological determinants of inducible vs. noninducible floral changes can be explicitly tested. A further study of lupines, particularly those that do not undergo a color change, as well as other nonlupine species, will be valuable in understanding the relationship between flower retention and pollinator behavior that obviates the need for floral changes in some species and not in others. Ultimately, we are interested in determining how selective pressures acting on floral changes vary with the identity of pollinators, plant form and life habit (annual vs. perennial), mating system, population density, investment in reproduction, and numerous other features of a species' ecology. Further research on floral changes promises a better understanding of plant–pollinator coevolution, particularly the influence of pollinator energetics on the evolution of floral characters and the relative importance of male and female fitness in the evolution of plant traits.

ACKNOWLEDGMENTS

I am particularly grateful to Thomas C. Gibson for initial inspiration and for generously sharing ideas, unpublished data, and many of the examples of floral changes that appear in this chapter. I would also like to acknowledge James H. Brown, Richard Inouye, James Munger, and Nickolas Waser for reading and commenting on an earlier draft of this manuscript and Brenda Caspar, Steven Buchmann, David Inouye, Donald Koehler, and Barbara Schaal for allowing me to cite their unpublished data. I am also grateful to Susan Anderson, James H. Brown, Nancy Huntly, Richard Inouye, James Munger, Mary Price, Steven Sutherland, and Nickolas Waser for numerous discussions. This research was generously supported by grants from the University of Arizona Graduate College, Dr. Cynthia Carey, and Angelo P. Gori.

LITERATURE CITED

Arditti, J. and H. Flick. 1976. Post-pollination phenomena in orchid flowers, IV. Excised floral segments of *Cymbidium. Am. J. Bot.* 63:201–211.

Arditti, J. and R. L. Knauft. 1969. The effects of auxin, actinomycin D, ethionine, and puromycin on post-pollination behavior by *Cymbidium* (Orchidaceae) flowers. *Am. J. Bot.* 56(6):620–628.

Barrows, E. M. 1976. Nectar robbing and pollination of *Lantana camara* (Verbenaceae). *Biotropica* 8:132–135.

Benseler, R. W. 1975. Floral biology of California. *Madrono* 23:41–53.

Brantjes, N. B. M. 1976. Riddles around the pollination of *Melandrium album* (Caryophyllaceae) during the oviposition by *Hadena bicruris* (Noctuidae, Lepidoptera). *Proc. K. Ned. Akad. Wet. Ser. C Biol. Med. Sci.* 79(1):1–12.

Brooks, A. E. 1978. *Australian Native Plants for Home Gardens,* 6th ed. Lothian, Melbourne. 162 pp.

Case, F. 1964. *Orchids of the Western Great Lakes Region.* Cranbrook Institute of Science, Bloomfield Hills. 147 pp.

Charnov. E. L. 1979. Simultaneous hermaphroditism and sexual selection. *Proc. Natl. Acad. Sci. U.S.A.* 76(5):2480–2484.

Corner, E. J. H. 1964. *The Life of Plants.* World Publ. Co., Cleveland. 314 pp.

Cruden, R. W. 1976. Fecundity as a function of nectar production and pollen-ovule ratios. *Linn. Soc. Symp. Ser.* 2:171–178.

———. 1977. Pollen-ovule ratios: a conservative indicator of breeding systems in plants. *Evolution* 31:32–46.

Cruden, R. W. and S. M. Hermann-Parker. 1979. Butterfly pollination of *Caesalpinia pulcherrima,* with observation on a pschycophilous syndrome. *J. Ecol.* 67:155–168.

Darwin, C. R. 1897. *The Different Forms of Flowers on Plants of the Same Species.* Appleton, New York. 352 pp.

Dethier, V. G. 1963. *The Physiology of Insect Senses.* Methuen, London. 266 pp.

Dronamraju, K. R. 1958. The visits of insects to different coloured flowers of *Lantana camara* L. *Curr. Sci.* 27:452–453.

Dunn, D. B. 1956. The breeding systems of *Lupinus,* group Micranthi. *Am. Midl. Natur.* 55:443–472.

Erickson, R., A. S. George, N. G. Marchant and M. K. Morcambe. 1973. *Flowers and Plants of Western Australia.* A. H. and A. W. Reed, Sidney. 260 pp.

Estes, J. R. and L. S. Brown. 1973. Entomophilous, intrafloral pollination in *Phyla incisa. Am. J. Bot.* 60:228–230.

Estes, J. R. and R. W. Thorp. 1975. Pollination ecology of *Pyrrhopappus carolinianus* (Compositae). *Am. J. Bot.* 62:148–159.

Everett, T. H. 1980. *New York Botanical Garden Encyclopedia of Horticulture,* Vol. 2. Garland Publ. Co., New York. 247 pp.

Faegri, K. and L. van der Pijl. 1972. *The Principles of Pollination Ecology.* Pergamon Press, Oxford.

Galbraith, J. 1967. *Wildflowers of Victoria,* 3rd ed. Longmans, Railway Crescent. 163 pp.

Gilissen, L. J. W. 1977. Style-controlled wilting in the flower. *Planta* 133:275–280.

Gill, F. B. and L. L. Wolf. 1975. Foraging strategies and energetics of east African sunbirds at mistletoe flowers. *Am. Nat.* 109:491–510.

Gottsberger, G. 1971. Colour change of petals in *Malvaviscus arboreus* flowers. *Acta Bot. Neerl.* 20:381–388.

Graf, A. B. 1978. *Tropica.* Roehrs Co., East Rutherford, N.J. 1120 pp.

Harrison, C. R. and J. Arditti. 1976. Post-pollination phenomena in orchid flowers. VII. Phosphate movement among floral segments. *Am. J. Bot.* 63:911–918.

Hartling, L. K. and R. C. Plowright. 1978. An investigation of inter- and intra-inflorescence visitation rates by bumblebees on red clover with special reference to seed set. *Proc. IVth Int. Symp on Pollination.* Md. Agric. Exp. Sta. Spec. Misc. Publ. 1, pp. 457–460.

Heinrich, B. 1975. Energetics of pollination. *Annu. Rev. Ecol. Syst.* 6:139–170.

Ingram, C. 1967. The phenomenal behavior of a South African *Gladiolus. J. Roy. Hort. Soc.* 92:396–398.

Jones, C. E. and S. L. Buchmann. 1974. Ultraviolet floral patterns as functional orientation cues in hymenopterous pollination systems. *Anim. Behav.* 22:481–485.

Kugler, H. 1936. Die Ausnutzung der Saftmalsumfärbung bie den Roszkastanienblüten durch Bienen und Hummeln. *Ber. Dt. Bot. Ges.* 60:128–134.

———. 1950. Der Blťenbesuch der Schammfliege *(Eristalomyia tenax). Zeit. Vergl. Physiol.* 32:328–347.

Mayak, S. and A. H. Halevy. 1980. Flower senescence. In K. V. Thimann (ed.), *Senescence in Plants.* CRC Press, Inc., Boca Raton, Fla., pp. 132–156. 276 pp.

McDade, L. and S. Kinsman. 1980. The impact of floral parasitism in two neotropical hummingbird pollinated plant species. *Evolution* 34(5):944–958.

Metcalf, L. J. 1972. *The Cultivation of New Zealand Trees and Shrubs.* A. H. and A. W. Reed, Wellington. 292 pp.

Millar, A. 1978. *Orchids of Papua New Guinea.* University of Washington Press, Seattle. 101 pp.

Prance, G. T. and A. E. Prance. 1976. The beetle and the water lily. *Garden J. N.Y. Bot. Garden* 26:118–121.

Proctor, J. and S. Proctor. 1978. *Nature's Use of Color in Plants and their Flowers.* Peter Lowe, London. 116 pp.

Pyke, G. H. 1978. Optimal foraging in bumblebees and coevolution with their plants. *Oecologia* 36:281–294.

Pyke, G. H., H. R. Pulliam, and E. L. Charnov. 1977. Optimal foraging: a selective review of theory and tests. *Q. Rev. Biol.* 52:137–154.

Raven, P. H. 1972. Why are bird-visited flowers predominantly red? *Evolution* 26:674.

Rickett, H. W. 1966. *Wildflowers of America,* Vol. 4, *The Southwestern States.* McGraw Hill, New York.

Schaal, B. A. and W. J. Leverich. 1980. Pollination and banner markings in *Lupinus texensis* (Leguminosae). *Southwest. Natur.* 25:280–282.

Schaffer, W. M. and M. V. Schaffer. 1977. The adaptive significance of variation in reproductive habit in Agavaceae. *In* B. Stonehouse and C. M. Perrins (eds.),

Evolutionary Ecology. Macmillan, London, pp. 261–276.

Schemske, D. W. 1976. Pollinator specificity in *Lantana camara* and *L. trifolia* (Verbenaceae). *Biotropica* 8:260–264.

———. 1980. Floral ecology and hummingbird pollination of *Combretum farinosum* in Costa Rica. *Biotropica* 12(3):169–181.

Strauss, M. and H. Koopowitz. 1973. Floral physiology of *Agraecum* I: Inheritance of post-pollinator phenomena. *Am. Orchid Soc. Bull.* 42(6):495.

Sussenguth, K. 1936. Über den Farbwechsel von Blüten. *Dt. Bot. Ges.* 54:409–417.

Thorp, R. W., D. L. Briggs, J. R. Estes and E. H. Erickson. 1975. Nectar florescence under ultraviolet irradiation. *Science* 189:476–478.

Thomson, J. D. and R. C. Plowright. 1980. Pollen carry-over, nectar rewards, and pollinator behavior with special reference to *Dievilla lonicera*. *Oecologia* 26:68–74.

Trauseld, W. R. 1969. *Wildflowers of the Natal Drahensberg*. Purnell, Cape Town. 220 pp.

Verdoorn, I. C. 1973. The genus *Crinum* in southern Africa. *Bothalia* 11:27–52.

Vincett, B. A. 1977. *Wildflowers of Central Saudi Arabia*. Pl. M.E., Milano. 114 pp.

Wainwright, C. M. 1978. The floral biology and pollination ecology of two desert lupines. *Bull. Torrey Bot. Club.* 105:24–38.

White, W. C. 1951. *Flowering Trees of the Caribbean*. Rinehart and Co., New York. 125 pp.

Willson, M. F. 1979. Sexual selection in plants. *Am. Nat.* 113(6):777–790.

Willson, M. F. and P. W. Price. 1977. The evolution of inflorescence size in *Asclepias* (Asclepiadaceae). *Evolution* 31:495–511.

Witham, T. G. 1977. Coevolution of foraging in *Bombus* and nectar dispensing in *Chilopsis:* a last dreg theory. *Science* 197:593–595.

APPENDIX. EXAMPLES OF FLORAL CHANGES IN PLANTS

AMARYLLIDACEAE			
Brodiaea versicolor	California	Yellow flowers turn white or purple	Everett (1980)
Crinum euchrophyllum	South Africa	White flowers turn deep rose	Verdoorn (1973)
ASTERACEAE			
Angianthus preissianus	Australia	Yellow center turns red	Galbraith (1967)
Senecio jacobaea	Europe	Yellow florets turn brown	Kugler (1950)
BORAGINACEAE			
Cryptantha breviflora	Utah	Yellow flowers turn white; odor change	B. Casper (pers. comm.)
Cryptantha nana	Utah	Yellow flowers turn white; odor change	B. Casper (pers. comm.)
Heliotropium curassavicum	California	Yellow center changes to bluish purple	N. McCarten (Pers. comm.)
Moltkopisis ciliata	Saudi Arabia	White flowers turn red-purple	Vincett (1977)
Myosotis australis	Australia	Yellow center turns red	Galbraith (1967)
Myosotis discolor	United States	Yellow flowers turn red-blue	Faegri and van der Pijl (1972)
Myosotis sauveolens	Australia	Yellow center turns white on white background	Galbraith (1967)
CACTACEAE			
Pelycyphora pseudopectinata	Mexico	Pink flowers turn white	T. Gibson (pers. comm.)
COMBRETACEAE			
Combretum farinosum	Trop. Africa	Flowers turn from green to orange to red	Schemske (1980)
CRASSULACEAE			
Crassula recurva	South Africa	White flowers turn red	Trauseld (1969)
Crassula rubicunda	South Africa	White flowers turn red	Trauseld (1969)
ERICACEAE			
Rhododendron species	North America	Rapid corolla abscission following pollination	T. Gibson (pers. comm.)
FABACEAE			
Argyrolobium rupestre	South Africa	Yellow flowers turn orange	Trauseld (1969)
Bauhina flammifera	—	Yellow flowers turn red	Corner (1964)

Species	Change	Location	Reference
Caesalpinia eriostachys	Banner petal folds down over stamens	C. America	Jones and Buchmann (1974)
Caesalpinia pulcherrima	Yellow petal margins turn red	Mexico	Cruden and Hermann-Parker (1979)
Cercidium floridum	Banner petal folds over stigma and style	SW. U.S.	D. Cornejo (pers. comm.)
Cercidium microphyllum	White spot turns yellow on a yellow background	SW. U.S.	D. Cornejo (pers. comm.)
Lupinus argenteus	Yellow spot on banner petal turns purple on a blue background	SW. U.S.	Gori (unpub. data)
Lupinus arizonicus	Yellow spot on banner petal turns blue on a blue background	SW. U.S.	Wainwright (1978)
Lupinus bicolor	White spot on the banner turns purple on a blue background	SW. U.S.	Dunn (1956)
Lupinus blumeri	Yellow spot on the banner turns purple on a blue-purple background	SW. U.S.	T. Gibson and D. Gori (pers. obs.)
Lupinus nanus	White spot turns blue on a blue background	SW. U.S.	Dunn (1956)
Lupinus sparsiflorus	Yellow spot turns blue on a blue background	SW. U.S.	Wainwright (1978)
Lupinus texensis	White spot turns red-purple on a blue background	SW. U.S.	Schaal and Leverich (1980)
Parkinsonia aculeata	Banner petal becomes darker orange	C. America	Jones and Buchmann (1974)
GOODENIACEAE			
Dampiera eriocephala	White spot on petal turns purple on a blue background	W. Australia	Erickson et al. (1973)
Dampiera linearis	White spot on petal turns blue on a blue background	W. Australia	Erickson et al. (1973)
HIPPOCASTANACEAE			
Aesculus hippocastanum	Nectar guides turn from yellow to red	Europe	Faegri and van der Pijl (1972), Kugler (1936)
Aesculus californica	Yellow nectar guides turn red	California	Benseler (1975)
IRIDACEAE			
Gladiolus grandis	Flowers brownish during day, blue at night	South Africa	Ingram (1967)

APPENDIX. EXAMPLES OF FLORAL CHANGES IN PLANTS

LILIACEAE

Trillium grandiflorum	E. N. America	White flowers turn deep red	T. Gibson (pers. comm.)
Trillium ovatum	E. N. America	White flowers turn deep red	T. Gibson (pers. comm.)
Trillium pusillum	E. N. America	White flowers turn deep red	T. Gibson (pers. comm.)

LOGANIACEAE

Gelsemium sempervirens	S. US.	Change in the UV pattern	D. Koehler (pers. comm.)

MALVACEAE

Malvaviscus arboreus	Mexico	Red flowers become progressively lighter	Gottsberger (1971)
Tibouchina grandifolia	Brazil	Flowers purple with white inside; central portion turns red-orange	Graf (1978)

MYRTACEAE

Baeckea gunniana	Australia	Yellow center turns red	Galbraith (1967)
Chamelaucium megalopetalum	Australia	White flowers turn red	Galbraith (1967)
Leptospermum myrsinoides	Australia	White flowers turn pink-red	Galbraith (1967)
Micromyrtus ciliatus	Australia	White flowers turn red	Brooks (1978)
Thryptomene calycina	Australia	Yellow center turns red	Galbraith (1967)
Verticordia grandiflora	W. Australia	Yellow center turns red	Erickson et al. (1973)
Verticordia huegelii	W. Australia	White flowers turn red	Erickson et al. (1973)
Verticordia plumosa	W. Australia	White-pink center turns red	Erickson et al. (1973)

NYMPHACEAE

Victoria amozonica	S. America	White flowers turn red	Prance and Prance (1976)

ONAGRACEAE

Fuchsia excorticata	New Zealand	Calyx tube green-purple, turns red-purple	Metcalf (1972)

ORCHIDACEAE

Angraecum eburneum	Madagascar	Flower rotates 180 degrees	Darwin (1897), Strauss and Koopowitz (1973)
Angraecum sesquipetale	Madagascar	Flower rotates 180 degrees	Strauss and Koopowitz (1973)
Bulbophyllum macranthum	New Guinea	Two lateral petals fold over column	Millar (1978)
Calypso bulbosa	N. America	Rapid collapse of flower parts	T. Gibson (pers. obs.)
Catasetum species	C. and S. America	Flower rotates 180 degrees	Darwin (1897)
Crytochus arcuata	S. Africa	White flowers turn apricot	T. Gibson (pers. comm.)

Species	Location	Change	Reference
Cymbidium species	India	Swelling and straightening of the column; lips turn from some color to solid red	Harrison and Arditti (1976), Arditti and Knauft (1969)
Cypripedium arientinum	N. America	Dorsal sepal collapses over pouch opening	Case (1964)
PAPILIONACEAE			
Daviesia pectinata	W. Australia	Yellow central spot to deep red	Erickson et al. (1973)
Hovea trisperma	W. Australia	White spot on the banner petal turns blue on blue background	Erickson et al. (1973)
POLEMONIACEAE			
Cobaea scandens	Mexico	White flowers turn dingy purple	Proctor and Proctor (1978)
Linanthus androsaceus	U.S.	White flowers turn pink	Rickett (1966)
Linanthus montanus	U.S.	White flowers turn pink	Rickett (1966)
ROSACEAE			
Raphiolepis umbellata	Asia	White center of the flower turns purple	T. Gibson (pers. comm.)
RUBIACEAE			
Bouvardia glaberrina	SW. U.S.	Red flowers become progressively lighter	G. Byers (pers. comm.)
SCROPHULARIACEAE			
Euphrasia glacialis	Australia	White center of the flower turns red	Galbraith (1967)
SOLANACEAE			
Brunfelsia americana	Caribbean	White flowers turn yellow	White (1951)
Brunfelsia latifolia	Caribbean	Pale violet flowers turn white	White (1951)
Brunfelsia paucifolia	Caribbean	Violet flowers turn white	White (1951)
Petunia axillaris	Temp. S. Amer.	Rapid collapse of the corolla	Gilissen (1977)
Solanum macranthum	Caribbean	Blue-purple flowers turn white	White (1951)
VERBENACEAE			
Lantana camara	C. America	Yellow flowers turn red	Barrows (1976), Schemske (1976)
Lantana trifolia	C. America	Yellow circle inside flower fades to white	Schemske (1976)
Phyla incisa	U.S.	Yellow nectar guides turn rose-purple	Estes and Brown (1973)

3
FLORAL FRAGRANCES AS CUES IN ANIMAL BEHAVIOR

Norris H. Williams

Department of Biological Science
Florida State University
Tallahassee, Florida[1]

ABSTRACT

Floral fragrances are often a major means of attracting insects and other pollinators to flowers. In a number of instances the floral fragrances serve as cues that orient, or otherwise influence the behavior of pollinators. A great deal of the recent work on the identification of floral fragrance components has been conducted with members of the Orchidaceae. In particular, studies on the genera *Ophrys, Cypripedium,* and others in Europe, and the male-euglossine-bee-pollinated orchids of Central and South America have yielded the identity of a number of floral fragrance compounds. The floral fragrances of *Ophrys* often elicit pseudocopulatory behavior by their pollinators. The fragrances of a number of neotropical orchids are collected by male euglossine bees with the tarsal brushes and are then stored in the hind tibiae. It is postulated that the male bees use these compounds as precursors of their own sex pheromones. The floral fragrance compounds identified to date include monoterpenoids, hydrocarbons, aromatics, and some nitrogenous compounds. Several methods have been developed for collecting floral fragrance compounds for study, the most promising of which include the use of adsorbents such as Tenax and the enfleurage technique with oil-impregnated paper strips. Identification of floral fragrance compounds by gas chromatography should be supplemented by confirmation with mass spectrometry, nuclear magnetic resonance spectrometry, field trials, and other bioassays.

KEY WORDS: Fragrances, orchids, *Ophrys,* euglossine bees, fragrance collection, fragrance analysis.

CONTENTS

[1]Present address: Department of Natural Sciences, The Florida State Museum, University of Florida, Gainesville, Florida.

INTRODUCTION

Floral fragrances have been suspected to be of some importance in attracting pollinators to plants for a number of years. For the most part, attention has been paid to insect pollinators, although some attention has been paid to the attraction of mammals (especially bats) to floral fragrances. The work with insects has been mainly concerned with bees (and to a much lesser extent flies and beetles). Although there had been some preliminary evidence earlier that insects were attracted to floral odors, it was not until the mid-1950s that real progress began to be made on the study of floral odors and their perception by animals. The main reasons for the progress made in the past few decades are the rapid technological advances that have been made in gas chromatography and gas chromatography/mass spectrometry. Before the advent of easily available gas chromatographs, there were no simple, reliable methods to study floral fragrances other than the human nose.

Floral fragrances, along with locale odor, odors from the Nasonov gland, etc., had been suspected of having some role in the recruitment of worker honeybees to food plants. The early work of von Frisch suggested that floral fragrances were very important, but this theory was rapidly overtaken by von Frisch's own theory on the dance language of bees. There is still controversy about the relative and absolute importance of olfaction and dance in the recruitment of honeybees to a food source. A number of papers have criticized the experimental methods, results, or conclusions used to support the dance language hypothesis (Johnson, 1967; Wenner and Johnson, 1967; Wenner et al., 1967, 1969; Johnson and Wenner, 1970; Wells and Wenner, 1971). Rebuttals followed some of these papers, by von Frisch (1967, 1974), Gould et al. (1970), and Gould

(1975, 1976). Rosin (1978, 1980) has lately rebutted some of the rebuttals, and it is obvious that the issue is still unsettled. The above references will give an introduction to the literature, but rather than be drawn into the controversy, I will spend the remainder of this chapter on other work and let those interested in honeybees delve into the specialized honeybee literature and controversy.

The non-insect visitors attracted to flowers by floral fragrances are mainly bats. Certainly bats find some of the flowers they visit for nectar at least in part by floral fragrance. Meeuse (1961) discussed bat pollination, and indicated that at least one chemical compound, diacetyl, had been identified as a bat attractant. Most of the current literature simply refers to the odor of bat flowers as "musky," or some similar nontechnical term. There are several recent studies of bat pollination, but for the most part they are concerned with various aspects of the ecological relationships between bats and flowers, and the floral fragrances have received little attention (Heithaus et al., 1974; Heithaus et al., 1975). Howell (1977) stated that olfaction plays a major role in the orientation of bats as they visit various flowers. Rourke and Wiens (1977) reviewed the literature on pollination by nonflying mammals, and mentioned that the flowers visited by these mammals produce copious amounts of nectar and often have a nutty or yeasty type of odor. They noted that others had reported an odor of sour milk for other geoflorous Proteaceae, as well as an odor of caraway liquor. Rourke and Wiens suggest that the cryptic flowers of a number of geoflorous Proteaceae produce odor, and that the "odor must be the primary attracting mechanism regardless of the pollinator." Lumer (1980) recently presented evidence that *Blakea chlorantha* (Melastomataceae) is pollinated by mice, and odor is certainly implicated in finding the flowers. Steiner (1980) showed that *Mabea occidentalis* (Euphorbiaceae) is visited by bats, the Red Wooly opossum, and noctuid moths. He reports the flowers have a strong musty odor, as is typical of bat-pollinated flowers. Sussman and Raven (1978) reviewed the literature on additional pollination systems by mammals and discussed bat pollination in an evolutionary context. None of these authors has provided an identification of the components of the floral fragrances that attract the bats or other mammals, and this is likely to be a profitable area of research in the future.

Flies are often attracted to flowers, particularly those with the odor and appearance of decaying meat. Most of the compounds are usually of an aminoid nature, but Metcalf et al. (1975) showed that male oriental fruit flies are attracted to methyl eugenol. This compound has been isolated from the flowers of a number of species of plants, and apparently serves as a food lure or aggregation lure. The flies respond to the compound in very low concentrations, and Metcalf et al. (1975) showed that 10^{-3} to 10^{-2} μg was attractive to the fruit flies. They suggested that the olfactory threshold approached that of the silkworm moth, 10^{-4} μg. It is certainly likely that a number of other nonaminoid compounds will prove to be the attractants of flies to flowers. More often, however, pollination by dipterans is associated with foul-smelling flowers. Ackerman and Mesler (1979) discussed the attraction of fungus gnats to *Lister cordata* (Orchidaceae), a plant whose flowers have a "truly repulsive" fragrance, and Mesler et al. (1980) reported that *Scoliopus bigelovii* (Liliaceae) had a very similar fragrance and also was pollinated by fungus gnats. Although the components of the floral fragrances were not identified, Ackerman and Montalvo (pers. comm.) report that the flowers have the aroma of "molluscs beginning to go bad." Faegri and van der Pijl (1966, 1979) discussed fly pollination in some detail, and divided the general class into myophily and sapromyophily. Some myophious flowers lack any perceptible odor, but the sapromyophilous flowers have the definite aroma of decaying

protein, and Faegri and van der Pijl suggest that this aroma, in conjunction with the color of the flowers, is important in attracting pollinators.

Meeuse (1978) discussed the physiology of sapromyophilous flowers, particularly the role of heat in the dispersal of odor and attraction of pollinators. He has worked mainly with members of the Araceae, some of which produce heat as a means of volatilizing the floral fragrance. *Arum, Sauromatum,* and other genera attract both beetles and flies to the inflorescence. Meeuse and his co-workers (Chen and Meeuse, 1971, 1972; Meeuse, 1978) have shown that the inflorescences produce indole or skatole (often species-specific as to which is produced), and that the role of heat produced by the inflorescences is to aid in the volatilization and dispersal of this odor. The beetles and midges are attracted to the odor and fall into the floral chamber. Dormer (1960) and Meeuse (1978) have described the pollination of *Arum* in detail, and Knoll (1926, cited in Meeuse, 1978) convincingly demonstrated that the heat produced by the inflorescence is not the source of attraction, but is only useful in volatilizing the floral fragrances.

Thien (1974) has shown that several species of *Magnolia* are pollinated almost exclusively by beetles, and he described the pollination mechanism in detail. The flowers of the species studied are often strongly protogynous and produce strong floral fragrances to which the beetles appear to be attracted. Thien et al. (1975) identified several of the compounds present in the floral fragrances of eight species of *Magnolia* and in one species of *Liriodendron.* They used gas chromatography (GC) and GC/mass spectrometry to identify the floral fragrance compounds, which were mainly aliphatic hydrocarbons, methyl esters, and terpenes. They were able to use the floral fragrance compounds taxonomically, but apparently did not do any bioassays to determine if the floral fragrances alone would attract the pollinating beetles.

Although the studies cited above have been very important in contributing to our knowledge of the role of floral fragrances in attracting pollinators, the work of Kullenberg (and his later associates), beginning in the early 1950s, was the real start of the study of floral fragrances as important cues in animal behavior. Although Kullenberg studied a number of species of insects and their relationships to the floral fragrances of a wide variety of species of flowering plants, I will limit my discussion to his work with members of the Orchidaceae. In an amazingly simple experiment, Kullenberg (1956) showed that male *Andrena* bees were attracted to the scent of *Ophrys lutea*. He simply covered the flowers with a piece of cloth for a few hours, removed the cloth and put it in the flight area of several male *Andrena* bees. The male bees landed on the cloth and began moving about, with their wings in motion, and later they began to palpate the cloth with their abdomens. The results of experiments and observations up through 1956 have been summarized by Kullenberg (1956). In studying populations of *Ophrys insectifera,* he found that males of *Gorytes* were attracted to the scent of the flowers. Although they were attracted to the floral scent, which Kullenberg believed to be chemically very similar to the scent produced by the abdomens of female *Gorytes,* the male bees did not attempt to copulate with the flowers. In one set of experiments, Kullenberg made dummy flowers and treated them with chemicals suspected as part or all of the floral fragrance of the orchid. Male *Gorytes* approached the dummy scented with nerolidol most frequently, but also approached a farnesol-scented dummy and an unscented control dummy. Kullenberg suggested on the basis of this and other experiments that the males of *G. mystaceus* confused the scent of *O. insectifera* with the odors of farnesol, hydroxycitronellal, and nerolidol, but he found that only the scent of the flower actually "excited" the insects. He concluded that these compounds are similar to the floral fragrance of the orchid, and that the scents do attract males on mating flights. He further concluded that the scents of the flowers do not elicit the complete copulatory behavior of the male bees. Kullenberg's 1956 paper and the works cited therein provide a good introduction to his early work on the attraction of hymenoptera to *Ophrys* flowers.

RECENT WORK ON OPHRYS, CYPRIPEDIUM, PLATANTHERA, AND OTHER ORCHIDS

A number of species of orchids have intricate pollination mechanisms. In many cases, odor or floral fragrance is thought to be of primary importance in attracting pollinators to the flowers. Recently Kullenberg, Bergstrom, Nilsson, and other European scientists have been working on a variety of groups of orchids and have been very successful in identifying the floral fragrance components and the role the fragrance components play in attracting pollinators. Some of the best examples of the role of floral fragrances as cues in animal behavior come from their work. In addition, they have devised a number of techniques for studying the minute amount of odor produced by these flowers and by the insects that pollinate the orchids. Their techniques will be discussed later in this chapter.

The details of pollination in a number of species of *Ophrys* have been summarized by Kullenberg in a series of papers (Kullenberg, 1961, 1973a,b, 1975, and references cited therein) and by Kullenberg and Bergstrom (1973, 1975, 1976a,b) and Bergstrom (1978). There are about 30 species (or taxonomic units) in *Ophrys,* occurring mainly in the Mediterranean region. The flowers produce no nectar, and the pollen is not available as food. The flowers attract only males of several genera of Hymenoptera, the primary distance attractant being the fragrance produced by the labellum of the flower. Females are not attracted to the flowers. The fragrance produced by any one species of *Ophrys* is a complex mixture of chemicals, and the genus as a whole produces (1) terpenoids, (2) fatty acid derivatives, and (3) other compounds, such as aromatics or nitrogen-containing compounds. Bergstrom (1978) suggests that the fatty acid derivatives and terpenoids may be the most important compounds in distance attraction in the range of approximately a meter or so, that sesquiterpenes may also be important attractants at shorter ranges, and that sesquiterpenes may be the chemicals responsible for the actual releasing of the pseudocopulatory behavior. It seems

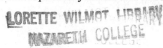

conclusively proved that some compound(s) from the flowers do attract male bees. Once the bee has been attracted to the flower, visual and tactile cues appear to play a role in releasing further copulatory behavior. Bergstrom lists five events (and their stimuli) in the pollination visits to *Ophrys:* (1) distant attraction (chemical), (2) close-range orientation (chemical and visual), (3) landing on the flower (chemical and visual), (4) pseudocopulation (chemical, visual, and tactile), and (5) copulation discontinuance (internal signals). He postulates that there must be a "functional overlap" of the compounds produced by the females bees (which attract the male) and the compounds produced by the flowers (which also attract the male). Kullenberg and Bergstrom (1973) point out that the insects will neither approach the flowers nor land on them without the chemical stimulation. The initiation of copulation, or pseudocopulation, is essential for the pollination of the flower. The tactile and visual stimulation depend to a large extent on the construction of the labellum, the form and distribution of the hairs on the labellum, and the color patterns. However strong the tactile and visual stimuli, Kullenberg and his co-workers believe that the chemical stimulus is continuously necessary to release the tactile responses of the male bee.

Different species, or species groups, of *Ophrys* produce distinctive quantitative and qualitative blends of floral fragrance compounds. The particular blend produced by a given species (or species group) appears to be important in the selective attraction of a specific pollinator (or group of pollinators). In addition to the selective attraction of a group of pollinators, there is a mechanical component to the isolating mechanisms between species. In some species, the male bee lands on the flower and positions himself along the labellum such that the pollinia are deposited on the head; however, in other species the male insect backs into the flower and receives the pollinia on the tip of the abdomen. This differential placement of pollinia is an effective mechanical isolating mechanism between species. Additional phenological isolating mechanisms also exist in the genus.

Kullenberg and Bergstrom (1976a,b) and Bergstrom (1978) have studied not only the floral fragrance compounds produced by *Ophrys,* but also the cephalic compounds produced by some of their pollinators, especially species of *Andrena, Colletes,* and *Argogorytes.* The Dufour glands of the latter genus contained several compounds "which show a remarkable similarity to the pattern of hydrocarbons in *Ophrys insectifera,*" an orchid pollinated by that genus of bee (Bergstrom, 1978).

Although great advances have been made in the past 20 years in the study of the *Ophrys*–pollinator relationship, it is obvious that much more work needs to be done. Bergstrom (1978) presents four hypotheses on the *Orphrys*–pollination/pollinator pseudocopulation syndrome. The first hypothesis stresses the overall morphology of the labellum and would emphasize its visual aspects. The second hypothesis suggests that the *Ophrys* flowers mimic the odor produced by female insects that serve to attract males. They have limited, preliminary evidence to support this hypothesis. The third hypothesis suggests that the flowers produce extranormal stimuli that attract and excite the males. These compounds would not be the same as those produced by the females, but would compete with them in attracting males. A fourth hypothesis is not so clearly stated, but in essence encompasses all aspects of the behavior not covered exclusively by the first three hypotheses. Bergstrom presents four lines of attack on the problem from the chemical point of view. At the time of his writing, the chemical analyses were not complete enough to allow a clear choice between the competing hypotheses. Bergstrom then provides a possible mode of evolution of the *Ophrys*–pollinator relationship and suggests future analytical chemical work to test some of the evolutionary hypotheses. Whatever the evolutionary history of the group, and regardless of whether the floral fragrance compounds are the same as those produced by the females, mimic those produced by the females, or act as extranormal stimuli, it seems clear that the floral fragrance compounds do serve to attract the pollinators from a distance and influence them to land and begin their copulatory behavior.

Nilsson studied the floral biology of *Pla-*

tanthera (1978), *Cypripedium* (1979a), *Herminium* (1979b), and *Dactylorhiza* (1980). *Dactylorhiza sambucina* produces mainly limonene and caryophyllene in its floral fragrance, compounds found to some extent in other nonorchidaceous flowers visited by the same species of *Bombus* that visit *Dactylorhiza*. This species of *Dactylorhiza* is a deceit flower and produces no nectar. Nilsson does not suggest that the orchid is mimicking other flowers, but suggests that pollination by bumblebees in the spring might require the production of strong floral fragrances. In this case, the floral fragrance appears to be simply a means of attracting the bumblebee to the flowers, with visual cues taking over once the bee is close to or on the flower.

Platanthera chlorantha produces fragrances that differ in composition between night and day (Nilsson, 1978). The flowers produce at least 11 compounds in their fragrance, with caryophyllene the major component in the day and methyl benzoate and monoterpenes the major components at night. The greenish-white flowers produce nectar, and Nilsson showed that the moths that visit the flowers take the nectar. He also demonstrated that the floral fragrance, as well as visual stimuli, is important in attracting the moths and in evoking landing on the flower and subsequent feeding. Nilsson states that "The landing response appears then to be stimulated by the combined action of visual cues and higher odour concentration. . . ."

Nilsson (1979a) provided a complete study of the pollination of *Cypripedium calceolus* by solitary bees. Female *Andrena haemorrhoa* were the most common pollinators. The floral fragrance contains nine or more compounds, with octyl and decyl acetate being the prevalent ones. Tengo and Bergstrom (1977) earlier had found that at least two species of *Andrena* that visit *C. calceolus* contain identical compounds in their cephalic secretions. It is known that male *Andrena* use the cephalic secretions to mark objects to which females are attracted. Furthermore, Bergstrom and Tengo (1974) found farnesene in the Dufour gland of female *Andrena*, and farnesene is also present in the floral fragrance of *C. calceolus*. Moreover, female *Andrena* are known to mark the nest with

compounds from the Dufour gland. Nilsson (1979a) makes the intriguing suggestion that the floral fragrance of *Cypripedium* interferes with the "normal pheromone controlled behavior patterns of some of its pollinators." He goes on to suggest that "Inborn alighting reactions might perhaps be released during a near-by flight over its labellum. Bees visiting already occupied flowers suggest that the addition of bee-perfumes to the labella enhances the deceptive ability."

The floral fragrance of *Herminium monorchis* contains a number of monoterpenes and anisaldehyde (Nilsson, 1979b). Nilsson showed by using tape-traps that the floral fragrance ("honey-like") is important in attracting pollinators (mainly Hymenoptera and Diptera). The fragrance is instrumental in attracting a variety of small wasps to the inflorescences, and Nilsson suggests that the floral fragrances are both attractants and releasers of feeding behavior in their pollinators. In addition to the olfactory cues, Nilsson suggests that the green-yellow color is a visual orientation cue at close range for the small wasps.

ATTRACTION OF MALE EUGLOSSINE BEES TO FLORAL FRAGRANCES

In a series of publications Dodson (1962a–1978b; van der Pijl and Dodson, 1966) has described the pollination of a number of genera and species of orchids by male euglossine bees. As the male euglossine bee–orchid pollination relationship has been reviewed in some detail recently (Williams, 1981a), I will give only a brief perspective here of the "euglossine syndrome" pollination system. The Euglossini (Hymenoptera: Apidae) are exclusively neotropical. The flowers that possess the euglossine syndrome lack nectar, and pollen is not available as food. The male bees visit the flowers and collect the chemical constituents of the floral fragrances with specialized tarsal brushes on the front feet (Fig. 3-1). The bees then launch into the air and transfer the collected floral fragrances to the inflated hind tibiae (Figs. 3-2 through 3-6). In the process of brushing to collect the fragrances, or in launching into the air to transfer, the bees often come into contact with the anther and re-

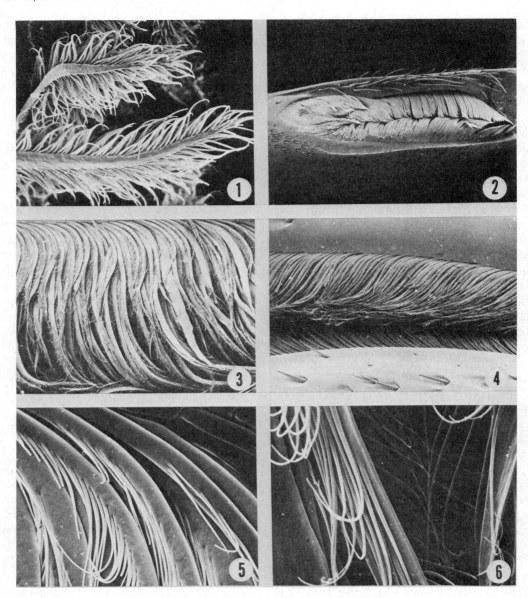

Figures 3-1 through 3-6. Specialized structures on front and hind legs of male euglossine bees for collecting and storing floral fragrance compounds. 1. Detail of collecting setae from front tarsal brush of *Eufriesia ornata*, ×1000. 2. Hind tibia of *Eufriesia pulchra*, ×20. Note the conspicuous slit surrounded by setae. The hind tibia is greatly inflated and capable of storing vast quantities of floral fragrance compounds. 3. Detail of setae of slit on hind tibia, *Eualema nigrita*, ×140. 4. Hind tibial slit of *Exaerete frontalis*, ×125. *Exaerete* is a nest parasite of other euglossine bees, and has a very reduced hind tibial slit. 5. Detail of base of hind tibial setae, *Eulaema nigrita*, ×1000. 6. Detail near apex of hind tibial setae, *Eulaema nigrita*, ×2050. All are scanning electron micrographs.

move the pollinarium. If the stigma is receptive, and if the bee is already carrying a pollinarium, the pollinia are often deposited in the stigma during the brushing or transferring actions. The collection of floral fragrances by the male euglossine bees is not limited to the Orchidaceae, but is also found in some species of *Spathiphyllum* and some species of *Anthurium* (Araceae), and in at least one species of *Gloxinia* (Gesneriaceae), *Cyphomandra* (Solanaceae), and *Dalechampia* (Euphorbiaceae), and in scattered species in other families of angiosperms (Vogel, 1966b; Williams and Dressler, 1976; Armbruster and Webster, 1979).

Dodson and Frymire (1961a,b) first described the attraction of the male euglossine bees to the orchid floral fragrances. Early field experiments involving living flowers indicated that the male bees were attracted mainly or exclusively to the floral fragrances. In the mid-1960s a program was begun to identify the components of the floral fragrances of male-euglossine-bee-pollinated orchids. The early work was first concerned with demonstrating that different species of orchids that had different pollinators also had different floral fragrances (Dodson and Hills, 1966). This was demonstrated for a number of species of *Catasetum, Gongora,* and *Stanhopea* (Fig. 3-7).

The early work (discussed in more detail below) led to the tentative identification (now mainly substantiated by more refined chemical analyses) of 11 compounds in 84 species of orchids (Hills et al., 1968, 1972). Field trials of these and other compounds (another 30 or so compounds are present, and some were tentatively identified with less confidence) showed that some of the floral fragrance compounds were potent attractants for males of a variety of species of euglossine bees. Cineole, methyl salicylate, benzyl acetate, methyl cinnamate, eugenol, and methyl benzoate were all found to be attractants, with cineole being the best

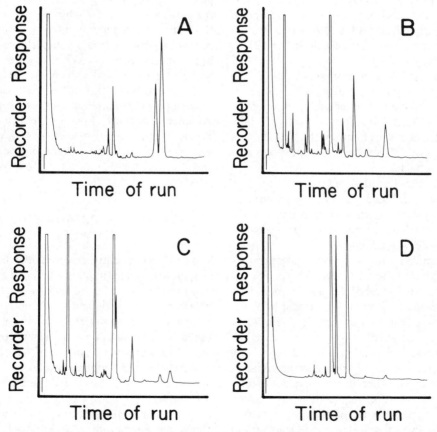

Figure 3-7. Comparison of gas chromatograph profiles of fragrances of four species of male-euglossine-bee-pollinated orchids. A. *Gongora superflua.* B. *Gongora tricolor.* C. *Gongora truncata.* D. *Clowesia warczewitzii.* Fragrances were collected by the Tenax-charcoal method described in the text. Samples were chromatographed on ⅛ in × 6 ft stainless steel columns packed with 3% Carbowax 20M, temperature programmed from 70 to 160°C at 8° per minute and held at 160°C for 12 minutes. Flame ionization detector with nitrogen as a carrier gas.

general attractant by far. Alpha- and beta-pinene, although almost ubiquitous in orchid floral fragrances, proved to be nonattractants, and even seem to be repellents to some species of euglossine bees. The field bioassay consists of tacking a 5 × 5 cm blotter pad to an upright object (tree or pole) and saturating the blotter with the compound to be tested. A compound is considered an attractant if a bee approaches the pad, lands and brushes, launches into the air and transfers the compound to the hind tibiae, and returns to the pad to brush again. The species attracted and their behavior have been described in detail by Dodson et al. (1969), Dodson (1970, 1975d), Evoy and Jones (1971), Williams and Dodson (1972), and Williams (1981a). The composition of floral fragrance has been described for *Catasetum* (Hills et al., 1972) and *Anguloa* (Williams et al., 1981), and reports on floral fragrance compositions for 11 other genera are in preparation (Williams, Whitten, and Dodson, in prep).

Dodson (1970) and Williams and Dodson (1972) discussed the specificity of attraction of various compounds isolated from orchid floral fragrances. There are approximately 60 species of euglossine bees in central Panama, of which at least 35 species are attracted to cineole, 11 species to methyl salicylate, and 6 species to benzyl acetate. If cineole and benzyl acetate are combined (cineole:benzyl acetate, 1:39), only eight species are attracted to the mixture. Only two species were attracted when alpha-pinene was added to the mixture, making a fragrance similar to that produced by *Stanhopea tricornis*. Of the two species attracted to the artificial fragrance, one *(Eulaema meriana)* is a known pollinator of *S. tricornis,* and the other species *(Euglossa dodsoni)* is too small to pollinate the orchid. Williams and Dodson (1972) showed that even though some chemicals are potent attractants, the addition of other compounds to a pure chemical or mixture may completely destroy the attraction potential of the major attractant. They also showed that chemicals that are structurally similar to orchid fragrance compounds (but not known themselves from orchid floral fragrances) may attract a few individuals of a few species, but in general structural analogs are not good attractants.

The many papers of Dodson (1962a–1978b), Dodson et al. (1969), Dressler (1966, 1967, 1968a,b, 1976a,b, 1977a,b, 1978a,b, 1979a,b), and their co-workers (Hills et al., 1972; Williams and Dodson, 1972) have shown rather conclusively that the floral fragrances of a number of species of male-euglossine-bee-pollinated orchids are the primary attractant of the pollinators, that the species-specific pollinators (or groups of pollinators) serve as isolating mechanisms between related species, and that the floral fragrances may serve to attract pollinators over long distances and may be important in long-distance pollen flow. Although we do not yet know just why the male euglossine bees visit orchid flowers and collect floral fragrances, it is clear from the amount of time the male bees spend doing this, and the specificity involved, that it is probably of some major importance to the bees. Dodson et al. (1969) offered several hypotheses to explain why the male bees visit orchid flowers. Williams (1981a) discussed all of the competing hypotheses, and presented some evidence in support of one of them (to be discussed in more detail below). A review of the biological relationships of male euglossine bees and orchid flowers is given by Williams (1981a).

FRAGRANCE PRODUCTION BY FLOWERS

Fahn (1979) recently reviewed the literature on secretory tissues in plants, and it is apparent that much more needs to be known about the structure, physiology, and biochemistry of fragrance production by flowers. Vogel (1961, 1963a,b), who discussed the structure of the osmophores of several species of Araceae, Orchidaceae, and Asclepiadaceae, noted that the subepidermal cells store starch, and postulated that the starch is converted to floral fragrance, since the starch disappears after the floral fragrance has been produced.

Swanson et al. (1980) studied the ultrastructure of glandular ovarian trichomes in two species of *Cypripedium* (Orchidaceae).

They had noted that neutral red stained not only osmophores on the petals, but also the bulbous ovarian trichomes. They found that the tip of the bulbous cells had the morphology often found in secretory cells. Smooth endoplasmic reticulum was abundant, dictysomes were limited in quantity, and lipid droplets were present in the cell near the apex. These and other morphological characters indicate that the bulbous trichomes are secretory, and Holman and co-workers are apparently analyzing the chemical nature of the product of these cells.

In the male-euglossine-bee-pollinated orchids, one might expect that the osmophore region would be developed in such a manner that the surface area would be greatly increased over that found on a typical petal, either by having a great deal of finely divided tissue with many infoldings, or by producing numerous trichomes. In the case of *Clowesia* (Fig. 3-8) the surface area is increased by having numerous mounds of tissues, and in *Polycycnis* the osmophore region has numerous trichomes, each also with increased surface area (Fig. 3-9). In some species (*Stanhopea*, Fig. 3-10) the cuticle is quite thin in the area in which floral fragrances are produced, as was expected; however, in some other species (*Cycnoches*, Fig. 3-11), the surface area is not noticeably increased, and contrary to what was expected, the cuticle is extremely thick. It appears then that the structure of the osmophore region in the male-euglossine-bee-pollinated orchids is quite variable from species to species, or from genus to genus. Apparently the orchids man-

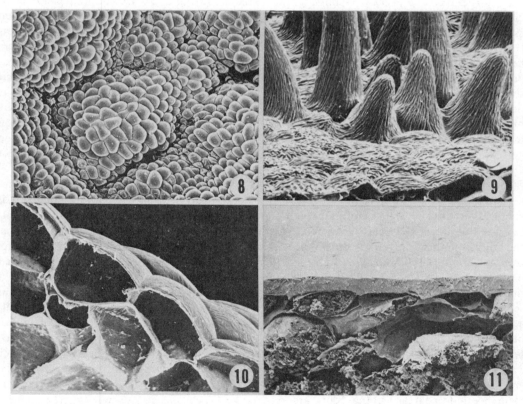

Figures 3-8 through 3-11. Scanning electron micrographs of floral fragrance-producing regions of the labella of four species of male-euglossine-bee-pollinated orchids. 8. *Clowesia russelliana*, rugose mounds of tissue at rear of hypochile, ×100. 9. *Polycycnis gratiosa*, epidermis and trichomes of osmophore, ×450. 10. *Stanhopea pulla*, cross section at rear of hypochile, ×580. Note thin walls of epidermal cells. 11. *Cycnoches egertonianum*, cross section through osmophore, ×500. Note thick walls and cuticle.

age to manufacture and secrete odor via a number of different types of surface configurations. This obviously needs additional investigation, and we are actively studying the structure of osmophores at the light and electron microscope levels.

Although in some flowers the floral fragrances might occur as minute droplets of monoterpenes in the cytoplasm, which diffuse through the epidermal layer and cuticle to the outside (Fahn, 1979), in the flowers that attract the male euglossine bees we have not been able to demonstrate any accumulation of volatile compounds in the epidermis or subepidermal layers prior to the release of the odor. We have tried solvent extraction, steam distillation, and crushing the portion of the flower where the odor is produced, but in no case were we able to detect floral fragrances. It appears that, in the male-euglossine-bee-pollinated orchids, the floral fragrance is actively produced as it is released from the surface of the flower (Hills, Williams, and Dodson, unpub.). We believe that Vogel's (1963a, 1966a) suggestion is probably correct, that the plant accumulates starch reserves that are actively metabolized to floral fragrance compounds in the tissues immediately below the epidermis and are then released through the cuticle. Vogel (1966a) suggested that the fragrance of the flower accumulates in the cavities on the rugose surface of the osmophore region in *Stanhopea* as a microliquid layer. We have not been able to substantiate this suggestion, but that could be because we used different techniques. Although it is not clear just how the fragrance is released from the flower (our transmission electron micrographs do not show any definite microchannels through the cuticle), it does seem to be clear that the fragrance does not accumulate to any degree in the osmophore, but is actively produced and released from the flower.

In the male-euglossine-bee-pollinated flowers, which depend on the brushing action of the male bees for pollination, the odors are secreted in a very localized area. For the bee to get to the source of the odor, it must orient itself on the flower in such a manner that it receives the pollinarium on the body or head in the proper place for later pollination of the stigma. It is in the actual brushing, approach to the source of odor, or launching into the air for transferring that the male bee receives the pollinarium or the stigma strips the pollinia from the insect. In some other flowers, such as several species of the Pleurothallidinae (Orchidaceae), the source of the floral fragrance is not near the reproductive structures, but rather is located on the tips of the sepals or petals (Vogel, 1966a). In these cases, it is assumed that the pollinating insect is attracted from a distance by the floral fragrances, and that close-range orientation is cued by visual aspects of the flower.

In the male-euglossine-bee-pollinated orchids, aroids, gesneriads, and other groups, there is no evidence that there is any significant production of heat to aid in the volatilization of the floral fragrance. It is known that the floral fragrances are not produced in large quantities on cool, overcast days, and that on sunny, warm days a great deal of floral fragrance is produced. It seems that these flowers rely on the ambient temperature to aid not only in the production of floral fragrances, but also in the dispersal of the fragrance. There is an ecological association with this, in that the pollinating insects are not usually active on cool, overcast days, and as a consequence the lack of odor production on such days is not a detriment.

As mentioned above, Meeuse and his coworkers have shown that certain members of the Araceae produce heat that aids in the volatilizion of the floral fragrance compounds. The compounds produced, indole and skatole, are slightly larger than the monoterpenes and simple aromatics produced by flowers of a number of species, and perhaps the additional production of heat is very necessary to volatilize these compounds sufficiently to attract pollinators. However, some orchids do produce skatole and do not produce heat; so heat is not always necessary to volatilize this compound.

COMPOSITION OF FLORAL FRAGRANCES

Most of the floral fragrance compounds seem to be either terpenoids, simple aromatics, aminoid compounds, or hydrocarbons. Of the terpenes, the most common floral fragrance compounds seem to be monoterpenes. These compounds consist of two isoprene units joined by a head-to-tail (usually) condensation to produce a ten-carbon, branched molecule. Monoterpenes may be open-chain, monocyclic, or bicyclic. Open-chain compounds are represented by geraniol, geranial, myrcene, nerol, and ocimene. Monocyclic monoterpenes include alpha-phellandrene, limonene, cineole, pulegone, menthone, menthol, alpha-terpinene, and carvone. Bicyclic monoterpenoids include compounds such as alpha- and beta-pinene, camphor, and thujane, although not all of these have been found in floral fragrances. Caryophyllene, and other sesquiterpenes, also occur in some floral fragrances.

Aromatic consituents of floral fragrances include such compounds as vanillin, methyl salicylate, methyl benzoate, piperonal, benzyl acetate, eugenol, methyl cinnamate, 2-phenyl ethyl acetate, and 2-phenyl ethanol. The aromatic compounds are derived from several different biosynthetic pathways, but very little (or nothing) has been done with biosynthetic pathway analysis in floral fragrances.

Among the nitrogen-containing compounds found in floral fragrances, indole and skatole (3-methyl indole) are among the most common. Skatole also has been shown to be a potent attractant of male euglossine bees. The foul smell of many fly- and beetle-pollinated flowers is usually stated to be due to aminoid compounds or amines, although specific identifications are conspicuously lacking in the literature. Obviously this field is ripe for future investigation.

A number of fatty acid derivatives and hydrocarbons have been reported from floral fragrances (Thien et al., 1975; Bergstrom, 1978), as well as methyl esters. Thien et al. (1975) and Bergstrom (1978) provide long lists of the compounds they have found in *Magnolia* flowers and *Ophrys* flowers respectively.

Undoubtedly, a large number of additional compounds will be identified in the future from floral fragrances. The advent of smaller, more affordable GC/MS systems will facilitate identifications, and one hopes that additional field bioassays will be used to study the biological activity of floral fragrance compounds, their relationships in mixtures, and their importance in attracting pollinators.

FRAGRANCE RECEPTION BY INSECTS

An extremely large literature is available on olfaction in insects, from the reception of pheromones by moths to the aggregation pheromones of bark beetles. A great deal of the literature is concerned with electroantennograms of various insects; however, the literature verifying olfaction and odor reception by pollinators is miniscule. An electroantennogram is obtained by inserting microelectrodes into the end of the antenna and the dorsal ganglion, blowing a fragrance across the antenna, and recording the response on an oscilloscope. One of the major works on olfaction and antennal reception in insects that pollinate flowers (other than honeybees) is that by Priesner (1973), who prepared extracts from the labella of 11 species (18 different forms or subspecies) of *Ophrys* and studied the electroantennogram responses of approximately 50 species of bees (both males and females). In addition, he studied electroantennogram (EAG) responses to a variety of chemicals thought to be present in the floral fragrances of a number of species of *Ophrys*. He found that labellar extracts from 7 species of *Ophrys* elicited very strong EAG responses in certain species of bees, but the remaining 11 forms did not elicit high responses. *Ophrys lutea* elicited very high responses in males of *Andrena pubescens; O. sphecodes* ssp. *sphecodes* and *O. arachnitiformis* elicited responses from males of *A. jacobi;* and *O. apifera, O. bombyliflora, O. scolopax,* and *O. tenthredinifera* elicited responses from males of *Eucera longicornis* and *E. tuberculata.* Males of 22 additional species of *Andrena* did not display EAG responses to the 18 labellar extracts, nor did males of 11 additional genera of apoid bees. The females of all species tested failed to give EAG responses to the labellar extracts. Priesner noted that "The EAG patterns to the flower extracts closely resembled the groups of *Ophrys* species and pollinators established on the base of their attraction specificity by Kullenberg." Very high EAGs in males of *E. longicornis* and *E. tuberculata* were elicited by γ-cadinene, a compound assumed to be the main terpenoid component in the labellar extract of several species of *Ophrys*. Other sesquiterpenes were at least 1000 times less effective in eliciting EAG responses. Priesner concluded that the electroantennogram response of *Eucera* males to γ-cadinene is similar to the EAG response of the same males to labellar extracts of species of *Ophrys* assumed to have this compound in their floral fragrances. The chemical identification of this compound was based on independent work by Bergstrom and E. and S. Stenhagen.

Field experiments with male euglossine bees indicate that the antennae are the site of odor reception. Evoy and Jones (1971) point out that the male bees require the complete sequence of landing, brushing, and launching into the air to complete the next stage, the transfer of the collected material to the inflated hind tibiae. If the chemical is applied directly to the front tarsal brushes, the bee attempts to clean himself rather than trying to transfer the material to the hind tibia. If the chemical (eugenol in this case) was accidentally dropped on the antennae, the bee also attempted to clean himself. In a set of field experiments (Adams, Dodson, Evoy, and Williams, unpub.) the antennae were removed from the head of the bee, and the bee lost the ability to orient toward the source of the odor. Jones (cited in Evoy and Jones, 1971) removed the antennae in a laboratory situation, and the insect lost the ability to orient toward the source of the attractant. Evoy and Jones concluded that the antennae are the primary sense organ in the chemotactic response of the male euglossine bees to the source of the odor, or floral fragrance.

Threshold quantities have not been experimentally determined for male euglossine bees attracted to orchid floral fragrances, nor for that matter have threshold quantities been experimentally determined for most other groups of insects attracted to floral fragrances. Metcalf et al. (1975), however, have determined that approximately 10^{-3} to 10^{-2} μg of methyl eugenol on filter paper will attract the oriental fruit fly, and they suggest that the threshold quantity may approach the 10^{-4} μg threshold of silkworm moths. This threshold quantity represents very few molecules in the air. This area obviously needs additional work.

The distance over which an insect might be attracted to a floral fragrance of course depends to a large extent on the threshold quantity needed to stimulate the insect. For male euglossine bees we do know that the males will fly up to approximately 1 km over water to an odor source located on a buoy. Ackerman took a canoe out to several locations in Lake Gatun at varying distances from Barro Colorado Island (BCI) in Panama. He was able to attract male euglossine bees to an odor source, and noted that the bees appeared to be coming from BCI. They certainly left the odor source and returned in the direction of BCI. He used distances of 50 meters, 200 meters, and 1 km from the shore. The other shore was even farther away than the BCI shore. He did note that predominantly the larger species of *Eulaema*, rather than the smaller species of *Euglossa*, were attracted to the odor source on the lake, although sampling on the island showed that the smaller species were very abundant (Ackerman, 1981).

At least as a working model, Kullenberg and Bergstrom (1976a) postulate that the male bees attracted to *Ophrys* flowers are able to perceive the floral fragrance up to a kilometer from the source. It seems reasonable to suggest that a number of insects are able to perceive floral fragrances over a distance of at least several meters.

UTILIZATION OF FLORAL FRAGRANCE COMPOUNDS

In the case of beetle, bat, rodent, and marsupial pollination, the floral fragrances seem to be mainly distance attractants. In the case of some types of fly-attracting flowers, the floral fragrances might also serve as releasers of feeding behavior, or in some specialized cases as releasers of mating, copulatory, or egg-depositing behavior. Honeybees possibly use floral fragrance compounds as markers of food plants, but because of the controversy concerning anything involving honeybees, I will refrain from discussing this subject (see references cited above).

Bergstrom (1978), Kullenberg and Bergs-trom (1975), and Tengo (1979) discussed the possible relationships between the floral fra-grances of *Ophrys* and the insects that polli-nate the flowers, and Bergstrom (1978) has nicely summarized the hypotheses that have been put forward to explain the relationship between the pollinators and the plants (see above). Determination of which of the hy-potheses will remain viable awaits additional experimental work by the European group.

Nilsson's (1979a) study of *Cypripedium cal-ceolus* provides some of the best evidence of the use of floral fragrances as confuser com-pounds, which deceive the insect and elicit a response normally caused by some other stim-ulus. As far as can be determined, however, the floral fragrances of *C. calceolus* are not collected by the bees or otherwise utilized, but are simply attractants that release a particular type of behavior in the visiting insects.

In 1969 we presented three hypotheses to explain the curious behavior of male euglossine bees and their relationships with orchid flowers (Dodson et al., 1969). I have recently sum-marized the progress that has been made in testing these hypotheses since that time (Wil-liams, 1981a), and will only briefly summarize the work here.

1. Dodson (1966b) had earlier suggested that the male euglossine bees lived longer if they had access to floral fragrance compounds than if they did not. Ackerman, however, has managed to keep bees that were hatched from nests alive in captivity for over two months (Ackerman, pers. comm.). The male bees might indeed need the compounds for some metabolic reason, but no firm data are in hand to support this hypothesis.

2. We suggested that the male bees might use the floral fragrance compounds unmodified to attract other males to a mating site (Dodson et al., 1969), and Dodson (1975d) later elab-orated on this hypothesis by proposing that a particular male bee collects a variety of floral fragrances and becomes an "attractor male." This attractor male establishes a lek, which in-volves attracting a number of males of the same species to the display site. Kimsey (1980) studied the mating behavior of two species of

euglossine bees (*Eulaema meriana* and *Eug-lossa imperialis*), and described their territo-rial display behavior. She defined leks in a very conservative way, and by her conservative def-inition male euglossine bees do form leks; that is, several individuals of a species may estab-lish territories near each other in favorable sites. Kimsey, however, noted that the males only interact in defending their territories from other conspecific males. The matings she ob-served were all in solitary territories, and she observed that the male bees do not appear to use the collected floral fragrance compounds to attract other conspecific males. She concluded that males are attracted to treefalls, areas that provide a number of potentially favorable ter-ritorial sites, and that the males actively de-fend their territories from other males. Dodson (1966b) had observed mating in two different species (*Eulaema cingulata* and *Euglossa ig-nita*), and territorial behavior has been ob-served in these species as well as in *Eulaema polychroma* (Dressler, 1967). The question of leks, their formation, and their composition is still not settled, partly because different species have been studied by different investigators, and partly because different definitions of leks have been used by different investigators. Kim-sey concluded that some male euglossine bees establish leks by her definition, but they were not the same type of lek that Dodson had in mind. She furthermore concluded that the leks were not obligatory for mating, but rather were facultative leks that were in the transi-tional state of development in the evolution of leks. A great deal of additional field work will be necessary to settle the question.

3. A third hypothesis we (Dodson et al., 1969) suggested was that the male bees might convert the floral fragrance compounds to fe-male sex attractants. There is some prelimi-nary evidence to support this hypothesis, as I have summarized (Williams, 1981a) and dis-cussed in detail elsewhere (Williams, 1980, 1981b). The essence of this hypothesis is that the male bees modify the floral fragrance com-pounds, either in the hind tibiae or in the man-dibular glands, and transfer the product to the mandibular glands if the modification took place in the hind legs. I have shown elsewhere

that the mandibular gland of the male of each species of euglossine bee studied to date has a species-specific quantitative and qualitative blend of compounds. The defensive use of these compounds is ruled out because (1) females do not seem to have much of anything in their mandibular glands, (2) each species' composition is unique (quantitatively and qualitatively), and (3) it does not seem parsimonious for each species to have its own unique blend of compounds for defensive use. This hypothesis needs a great deal of additional field and laboratory work either to refute it or to help substantiate it. Such field and laboratory work is currently under way.

TECHNIQUES FOR COLLECTING AND ANALYZING FLORAL FRAGRANCES

The human nose is a fairly good tool for studying floral fragrances, but its ability to identify floral fragrance compounds both quantitatively and qualitatively is definitely limited, and the results obtained from simply smelling a flower do not lend themselves to definitive identification or easy publication. There are about five techniques currently being used to collect floral fragrances for further analytical work.

Hills et al. (1968, 1972) used chambers constructed of ¼-inch clear Plexiglas to concentrate floral fragrances. They designed the boxes so that the entire inflorescence could be placed in the chamber without their having to remove the flowers from the inflorescence or the inflorescence from the plant. In this way the floral fragrance production could be studied over several days. The chambers were not completely airtight, but were glued on all edges. Small holes were drilled in the sides, and Swagelok fittings with gas chromatograph septa were secured in the fittings. The lid, which was slotted to allow easy placement of the inflorescence in the chamber, was lined with self-adhesive weatherstripping foam tape, and the entire lid was secured to the box with masking tape. The hole through which the inflorescence is routed was stuffed with foam plugs. Each chamber and lid was washed with hot soapy water after use and baked in a 70°C oven overnight. No contamination was found when empty chambers were analyzed. The inflorescence was placed in the box between 0730 and 0900 each morning, or when the flowers were producing "good odor," and an equilibration time of at least 30 minutes was allowed before a sample was taken for analysis. Sampling was done by inserting a 10-milliliter gas-tight syringe through the septum and withdrawing a sample of the concentrated floral fragrance, which was then analyzed by gas chromatography.

Holman and Heimermann (1973, 1976) have developed a technique using glass fiber paper that has been impregnated with a thermally stable oil, Convalex 10 pump oil dissolved in ether. The paper is saturated with the mixture, and the ether driven off. The papers are then purged with a flow of helium at 200°C for 18 hours to get rid of any substances that would be volatile under the GC conditions to be used later, and the purified paper strips are then stored in jars or in aluminum foil. Floral fragrances are collected by putting living flowers in jars along with the paper strips, the strips and flowers remaining in the jars for at least one day, often longer. The paper strip is eventually removed from the jar, placed in an aluminum foil wrapper, and returned to the laboratory for GC/MS analysis. Holman and Heimermann have modified the injection port of their GC to accept the paper strip. The port is purged with helium, closed, and heated to 170°C. A valve is opened to move the vapor into the GC column, and then rapidly closed. The sample is then chromatographed isothermally at 160°C. A similar method is used to introduce the floral fragrance sample into the GC/MS.

Bergstrom (1973, 1975, 1978) and Stallberg-Stenhagen et al. (1973) have developed several methods to study floral fragrances. The simplest method they used was to place the labellum of *Ophrys scolopax* in a splitter-free intake system for a few minutes, raise the temperature to 130°C, and then temperature-program the GC column. Bergstrom has also used a modified enfleurage technique in which flower parts are placed in contact for three to four days with Chromosorb G 60/80 mesh coated with silicone high-vacuum grease. The adsorbent is purified before use by purging it with nitrogen at 300°C for several days. The adsorbent is then degassed in a pre-column tube, the tube is removed from the system and replaced by an empty tube, and the sample, which had been driven onto the head of the GC column, is now analyzed by temperature-programming the GC from ambient up to approximately 170°C. Apparently, the flowers have also been analyzed by placing them in the pre-column tube and flushing them with gas, but the amounts of material collected in this manner are not sufficient for good mass spectra. Bergstrom (1978) has recently been using either Porapak Q or SE-30 on Chromosorb W as an adsorbent in the pre-column tube with good results.

Nilsson (1978) used glass collection tubes packed with Chromosorb G 60/80 mesh coated with 10% silicone grease, or packed with Porapak Q 50/80 mesh adsorbent. The collection tubes, which were conditioned before use, were then attached to larger glass tubes in which flowers had been placed. Air was drawn through the concentration tube into the collection tube for from several hours to several days. When sufficient material had been collected, the collection tube was placed directly into the GC inlet, heated to 150°C for 20 minutes, and then removed. The GC oven was then programmed from ambient to 225°C at 4 or 8°C per minute. Retention times and mass spectra were then obtained for identification.

Armbruster and Webster (1979) and Armbruster (pers. comm.) have tried to collect the floral fragrances of *Dalechampia* by suspending the flowers over dishes filled with cyclohexane, and were able to collect a small amount of floral fragrance in this manner for later GC and GC/MS analyses. A major disadvantage of this method is that the flowers do not remain in very good condition for very long when suspended over cyclohexane vapors.

We have recently developed a modification of the collection tube packed with adsorbent for the collection of floral fragrances (Fig. 3-12). This method was modified from the techniques of Bergstrom and his co-workers, Zlat-

kis et al. (1973, 1974) Bertsch et al. (1974), Nilsson (1978), Brown and Purnell (1979), Tsugita et al. (1979), and Simon et al. (1980). For concentrating the floral fragrances we use the same type of Plexiglas boxes as discussed above. Filtered air is drawn through the chamber by a vacuum pump or, more normally, by an aspirator connected to a water faucet. A collection tube is placed in the air stream. This collection tube, or cartridge, is filled with Tenax GC and activated charcoal separated by a plug of glass wool, with the ends also plugged with glass wool. Collection takes anywhere from an hour to a day. During collection, the sample is drawn through the tube so that it first comes into contact with the Tenax GC. The sample is purged from the tube with a nitrogen stream, and to desorb the sample, the cartridge is placed in a copper or stainless steel tube wrapped with heating tape. The cartridge is reversed, so that the stream of nitrogen first comes into contact with the charcoal. This is necessary to prevent irreversible adsorption of the Tenax held compounds onto the charcoal. A series of Swagelok reduction fittings from ½ inch to ⅟₁₆ inch are used on the exit end of the tube, and both ends of the copper tube are fitted with high-temperature gaskets made from high-temperature GC septa. The sample is heated to 200°C as the stream of nitrogen purges it from the cartridge. The ⅟₁₆-inch Swagelok fitting contains a vespel ferrule to make a gas-tight connection to a glass capillary tube. This capillary tube is inserted through a block of solid aluminum into the exit end of the copper tube, and the aluminum block is equipped with a copper cold finger immersed in liquid nitrogen. After a short period of time, the capillary tube is removed, and the condensed floral fragrance is eluted with 0.5 ml of solvent. The sample is now ready for GC or GC/MS analysis.

All of these methods have advantages and disadvantages. The method of Hills et al. (1968, 1972) requires the use of large-capacity gas-tight syringes and does not produce a liquid sample. The sample is not as concentrated as one desires for additional chemical work, nor is is possible to do much chemistry with the small sample so obtained. An advantage is that

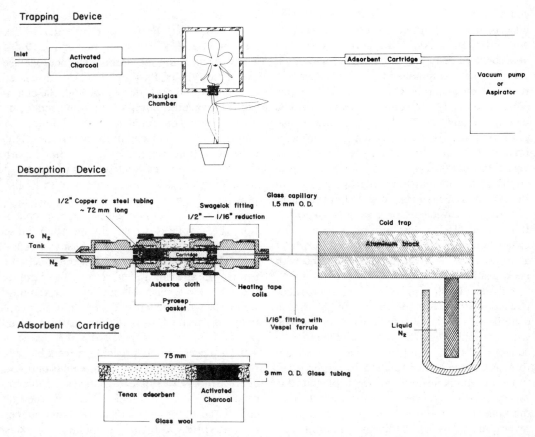

Figure 3-12. Cartridge method of collecting floral fragrances. A. General layout for collecting fragrances. The absorbent cartridge should be placed so that the fragrance passes through the Tenax first. B. Desorption device. The adsorbent cartridge should be placed so that the nitrogen purge gas passes through the activated charcoal first. C. Detail of adsorbent cartridge. The cartridge must be conditioned before use by placing it in the desorption device and purging it with a flow of nitrogen while it is heated to 275°C.

the method does not require any special equipment for collection and desorption of the sample. The Holman oil-paper strip has the advantage of using a material that is easily mailed to workers anywhere, can be used in the field, and requires no special field equipment. It has the disadvantage of requiring fairly elaborate preparation of the papers strips and the necessity of a specialized injection port on the GC or GC/MS. We have, however, obtained good results by placing rolled-up strips in a cartridge and desorbing them by our method.

The pre-column tube of Bergstrom and Nilsson and other workers has the advantage that it does not harm the flowers, can be used easily, and concentrates a good amount of fra-grance. It has the disadvantage of requiring a specially adapted gas chromatograph inlet system. Collecting floral fragrances over cyclohexane (or any other solvent) has the advantage of being inexpensive, simple, and easily done in the field. It has the disadvantage of not giving very much sample and of eventually killing the flowers.

Our cartridge system has the advantage of not requiring a special inlet for the GC and of producing abundant sample as a liquid that is easily stored for additional work, and being fairly simple to build and use. It has the disadvantage of requiring special desorbing equipment and access to liquid nitrogen, and the possibility that some compounds may undergo reactions or rearrangements on the activated charcoal. We have not had any problems to date with reactions or rearrangements, but this is at least a theoretical possibility.

We hope that others will try our cartridge method. We feel that trapping fragrances on only one adsorbent, such as Porapak or Tenax, does not trap all of the compounds present in a floral fragrance. Several monoterpenes are not easily adsorbed (or they rapidly desorb) on Tenax. The additional use of activated charcoal yields some of these monoterpenes. The pieces for our system are easily made by a general metal-working machine shop, and no special glass blowing is required. The cartridges are conditioned before use by simply heating them in the desorption tube with a stream of nitrogen, and are easily cleaned for re-use by purging with nitrogen in the heated desorption tube. The fact that each cartridge can be used over and over again reduces the cost.

Once a floral fragrance has been collected and concentrated by whatever method is used, the sample must be analyzed. Vogel (1966a) used thin-layer chromatography to try to identify the floral fragrance components of *Stanhopea tigrina*. While thin-layer chromatography is useful in a number of analytical situations, it is not the method of choice in studying floral fragrance compounds. The vast majority of current studies on floral fragrances utilize gas chromatography (GC) or gas chromatography connected with mass spectrometry (GC/MS).

Gas chromatography without the use of GC/MS yields a great deal of useful information. GC work can utilize large packed columns (¼ inch O.D.), necessary if one is injecting 10-milliliter gas samples, or smaller (⅛ inch) packed columns for liquid samples and smaller gas samples. Better separations are often obtained using glass capillary columns. Whatever the chromatographic conditions, the result is a chromatogram with (one hopes) a number of peaks. Identification, at least tentative, can be made by chromatographing the sample on several different columns and comparing the retention times of the unknown compounds (or, better, relative retention times) with the retention times of known compounds. When supplemented with co-injection of known and unknown, this method often yields a good tentative identification. If the gas chromatograph is equipped with an effluent splitter, the operator is able to smell a compound as it elutes from the end of the column, since only a portion is routed to the detector, and a large fraction can be diverted through the effluent splitter. This allows the operator to collect a particular peak as it emerges from the splitter so that further analyses can be done on a particular compound. In the past, compounds collected in this manner could be analyzed by mass spectrometry, but with the advent of more convenient GC/MS systems it is usually easier to go directly into the mass spectrometer.

The GC/MS system allows a fairly good identification of a number of floral fragrance compounds. When used in conjunction with known compounds, good GC and GC/MS can give very reliable identifications. It is of course desirable to have additional confirmation of identity. There are a number of current treatments of gas chromatography and combined GC/MS that give details of these types of analyses (e.g., Middleditch, 1979).

High performance liquid chromatography (HPLC) is useful in separating liquid samples of floral fragrances. This is particularly useful when one is attempting to obtain enough material for a structural confirmation with nuclear magnetic resonance spectrometry. While the usefulness of HPLC is almost certain to increase in the study of floral fragrances, it is unlikely to match the usefulness of GC and GC/MS in the near future.

FIELD TRIALS AND BIOASSAYS

Once the major (and one hopes minor) components of a floral fragrance have been identified, it is necessary to determine what, if any, biological activity each component of the floral fragrance has. The most comprehensive studies of biological activity are those by Kullenberg, Berstrom, and co-workers on the floral fragrances of *Ophrys* and its pollinators, and the work of Dodson, Dressler, Williams, and co-workers on the relationships of male euglossine bees to the flowers they pollinate and the relationships with the floral fragrances that have been identified from these flowers.

Although the components of the extracts were not identified, Kullenberg (1973b) showed that different thin-layer chromatography fractions of labella of various species of *Ophrys* had different levels of ability to excite the insects that pollinate them. The entire extract was usually the most potent excitant. The three fractions he used and their compositions were: (1) mainly hydrocarbons, (2) aldehydes, and (3) aliphatic alcohols and sesquiterpene alcohols. In general, Kullenberg showed that the third fraction was more potent than the first two fractions. Tengo (1979) continued this type of experimentation and, based on the earlier work of Kullenberg (1973b) that indicated that the alcohols were the most potent in eliciting behavioral responses from a number of *Ophrys* pollinators, used 1-octanol as a field test substance. This compound was very attractive to males of several species of *Andrena;* it has been identified in the mandibular glands of at least one species of *Andrena,* and it has recently been identified as a major component of the floral fragrance of *Ophrys lutea* (Tengo, 1979, and Borg-Karlsson, cited by Tengo). Interestingly, Priesner (1973) found that γ-cadinene, rather than the alcohols, gave the best electroantennogram responses. There obviously is still need for a great deal of additional work on *Ophrys* and its pollinators.

With the identification of a number of floral fragrance compounds from male-euglossine-bee-pollinated orchids (Hills et al., 1968), the biological activity of a number of compounds has been studied. Dodson et al. (1969), Dodson (1970, 1975d), Hills et al. (1972), and Williams and Dodson (1972) have presented some of the results of the biological activity and specificity of orchid floral fragrance compounds. We have found that cineole, methyl salicylate, benzyl acetate, vanillin, skatole, methyl cinnamate, methyl benzoate, and eugenol are the better attractants, with cineole, methyl salicylate, skatole, eugenol, and benzyl acetate being the best general attractants. Alpha- and beta-pinene are not attractants,

but serve to modify the attraction potential of other compounds. Whereas cineole is an excellent general attractant, other compounds such as 2-phenyl ethanol may attract only one or two species of bees in a given area, but will attract a large number of individuals of the species they do attract.

The field bioassay of suspected floral fragrance compounds has been very useful in aiding in the tentative identification of these compounds. In our earlier work, we tentatively identified floral fragrance compounds on the basis of gas chromatographic retention times on two or more columns, smelling the suspected compound as it eluted from the GC column, and co-injecting known and unknown compounds. A suspected compound tentatively identified in such a manner was then subjected to a bioassay in the field. If a suspected unknown floral fragrance compound chromatographed like a known compound A, if it smelled like known compound A, and if compound A attracted the same species of bee that was attracted to the floral fragrance, we considered our tentative identification fairly good. It is significant that more advanced techniques, such as GC/MS (Holman and Heimermann, 1973, 1976), have confirmed many of our earlier identifications, and in no case have any of our tentative identifications been refuted. We are continuing GC/MS work to substantiate our earlier identifications.

Our use of floral fragrance compounds to attract male euglossine bees has had several very productive side-effects. We have now been able to collect numerous individuals of what at one time were very rare species. Furthermore, we have found a large number of new species of euglossine bees by attracting them to isolated individual floral fragrance compounds. An additional benefit has been the aid of floral fragrance compound specificity in working out species relationships in the euglossine bees. For example, we had known for some time that there were two distinct size groups within what was normally called *Eulaema meriana.* Partly through our use of floral fragrance compounds, we have been able to separate *E. bombiformis* from *E. meriana.* Preliminary results indicate that specificity of attraction to a compound or group of compounds will be of considerable use in delimiting the subgenera of *Euglossa*.

FUTURE TRENDS AND FUTURE NEEDS

There are at least four areas in the study of floral fragrances and their role in animal behavior and pollination biology that need to be addressed in general in the future. The easy access to gas chromatography, GC/MS, and other analytical techniques should mean that rapid advances will be made in these areas in the next few years.

1. There is an urgent need for positive identifications of floral fragrance compounds. Pollination biology is a science with enough history now to go beyond the level of simply describing an odor as being musky, foul, fruity, etc., and we need good identifications of the major and minor components of the floral fragrances. One hopes that a number of the floral fragrance compounds will become commercially available, so that field bioassays of biological activity will be likely to be done.

2. Additional work needs to be done on the reception of individual compounds by the insects that actually pollinate the flowers from which a particular chemical has been identified. We need to know if a particular chemical that is almost ubiquitous in floral fragrances, but which fails to attract insects, such as alpha-pinene, is actually perceived by the bee, or if the bee is not attracted to the compound because it does not perceive it. Likewise, it is desirable to know if some of the repellent compounds in some floral fragrances, especially in some of the male-euglossine-bee-pollinated plants, are truly perceived by the insects, or if they simply block the receptors for other compounds that do attract the insects.

3. A third area needing extended investigation is the study of control mechanisms in fragrance production. We need to know what types of variation occur over the period in which the flower is producing fragrance; we need to know if there is any variation in the compounds produced from one day to the next in those flowers that produce fragrance over a period of several days. We need better information on the stimuli that initiate fragrance production, and on the stimuli that shut off fragrance production. For instance, in male flowers of *Catasetum*, once the pollinarium has been removed from it, the flower ceases to produce fragrance in a matter of hours, and will not produce fragrance the following day.

4. There is a need for a study of intraspecific variation in floral fragrance composition. Along with this, there is the repeated need for good voucher specimens of those plants that have been studied. We must make sure that those plants that are studied in pollination biology are the same ones studied in floral fragrance analyses. If there is intraspecific variation in floral fragrance, and we know this is so in several species, then it is imperative that careful study be done of the pollinators of particular odor forms.

5. Finally, we need to try to determine if all of the information we have gathered and intend to gather can be interpreted in any systematic, taxonomic, or evolutionary manner. There seems to be ample work for all who want to become involved.

ACKNOWLEDGMENTS

A portion of the work reported here was supported by a grant from the National Science Foundation, Washington, D.C. (DEB-7911556). I thank W. M. Whitten, A. M. Pridgeon, and R. Gallup for help with various aspects of the gas chromatography and photography; I thank W. M. Whitten, J. D. Ackerman, J. T. Atwood, and A. M. Montalvo for discussing some of the topics covered here with me. I thank Ron Matusalem for inspiration.

LITERATURE CITED

Ackerman, J. D. 1981. Phenological relationships of male Euglossini (Hymenoptera, Apidae) and their fragrance hosts. Ph.D. dissertation, Florida State University, Tallahassee, Fl.

Ackerman, J. D. and M. R. Mesler. 1979. Pollination biology of *Listera cordata* (Orchidaceae). *Am. J. Bot.* 66:820–824.

Armbruster, W. S. and G. L. Webster. 1979. Pollination of two species of *Dalechampia* (Euphorbiaceae) in Mexico by euglossine bees. *Biotropica* 11:278–283.

Bergstrom, G. 1973. Studies on natural odoriferous compounds. VI. Use of a pre-column tube for the quantitative isolation of natural, volatile compounds for gas chromatography mass spectrometry. *Chem. Scripta* 4:135–138.

————. 1975. Development of an integrated system for the analyses of volatile communication substances in social Hymenoptera. *In* C. Noirot, P. E. Howse, and C. Le Masne (eds.), *Pheromones and Defensive Secretions in Social Insects.* I.U.S.S.I., Dijon, France, pp. 173–187.

————. 1978. Role of volatile chemicals in *Ophrys*-pollinator interactions. *In* G. Harborne (ed.), *Biochemical Aspects of Plant and Animal Coevolution.* Academic Press, New York, pp. 207–230.

Bergstrom, G. and J. Tengo. 1974. Farnesyl and geranyl esters as main volatile constituents of the secretion from Dufour gland in 6 species of *Andrena* (Hymenoptera, Apidae). *Chem. Scripta* 5:23–38.

Bertsch, W., A. Zlatkis, H. M. Liebich, and H. J. Schneider. 1974. Concentration and analysis of organic volatiles in Skylab 4. *J. Chromatog.* 99:673–687.

Brown, R. H. and C. J. Purnell. 1979. Collection and analysis of trace organic vapour pollutants in ambient atmospheres. *J. Chromatog.* 178:79–80.

Chen, J. and B. J. D. Meeuse. 1971. Production of free indole by some aroids. *Acta Bot. Neerl.* 20:627–635.

————. 1972. Induction of indole synthesis in the appendix of *Sauromatum guttatum* Schott. *Plant Cell Physiol.* 13:831–841.

Dodson, C. H. 1962a. The importance of pollination in the evolution of the orchids of tropical America. *Am. Orchid Soc. Bull.* 31:525–534, 641–649, 731–735.

————. 1962b. Pollination and variation in the subtribe Catasetinae (Orchidaceae). *Ann. Missouri Bot. Gard.* 49:35–56.

————. 1963. The Mexican stanhopeas. *Am. Orchid Soc. Bull.* 32:115–129.

————. 1965a. *Agentes de polenización y su influencia sobre la evolución en la familia Orquidacea.* Univ. Nac. Amaxonia Peruana, Inst. General de Investigaciones. 128 pp.

————. 1965b. Studies in orchid pollination: The genus *Coryanthes. Am. Orchid Soc. Bull.* 34:680–687.

————. 1966a. Studies in orchid pollination: *Cypripedium, Phragmopedium* and allied genera. *Am. Orchid Soc. Bull.* 35:125–128.

————. 1966b. Ethology of some bees of the tribe Euglossini (Hymenoptera: Apidae). *J. Kansas Ent. Soc.* 39:607–629.

————. 1966c. Studies in orchid pollination: The genus *Anguloa. Am. Orchid Soc. Bull.* 35:624–627.

————. 1967a. Studies in orchid pollination. The genus *Notylia. Am. Orchid Soc. Bull.* 36:209–214.

————. 1967b. Relationships between pollinators and orchid flowers. *Atas do Simpósio sôbre a Biota Amazônica* 5:1–72.

————. 1967c. El género *Stanhopea* en Colombia. *Orquideología* 2:7–27.

————. 1970. The role of chemical attractants in orchid pollination. *In* K. L. Chambers (ed.), *Biochemical Coevolution.* Oregon State University Press, Corvallis, pp. 83–107.

————. 1975a. Clarification of some nomenclature in the genus *Stanhopea* (Orchidaceae). *Selbyana* 1:46–55.

————. 1975b. Orchids of Ecuador: *Stanhopea. Selbyana* 1:114–129.

————. 1975c. *Dressleria* and *Clowesia:* A new genus and an old one revived in the Catasetinae (Orchidaceae). *Selbyana* 1:130–137.

————. 1975d. Coevolution of orchids and bees. *In* L. E. Gilbert and P. H. Raven (eds.), *Coevolution of Animals and Plants.* University of Texas Press, Austin, pp. 91–99.

————. 1978a. Three new South American species of *Catasetum* (Orchidaceae). *Selbyana* 2:156–158.

————. 1978b. The catasetums (Orchidaceae) of Tapakuma, Guyana. *Selbyana* 2:159–168.

Dodson, C. H. and G. P. Frymire. 1961a. Preliminary studies in the genus *Stanhopea* (Orchidaceae). *Ann. Missouri Bot. Gard.* 48:137–172.

————. 1961b. Natural pollination of orchids. *Missouri Bot. Gard. Bull.* 49:133–152.

Dodson, C. H. and H. G. Hills. 1966. Gas chromatography of orchid fragrances. *Am. Orchid Soc. Bull.* 35:720–725.

Dodson, C. H., R. L. Dressler, H. G. Hills, R. M. Adams, and N. H. Williams. 1969. Biologically active compounds in orchid fragrances. *Science* 164:1243–1249.

Dormer, K. J. 1960. The truth about pollination in *Arum. New Phytol.* 59:298–301.

Dressler, R. L. 1966. Some observations on *Gongora. Orchid Dig.* 30:220–223.

————. 1967. Why do euglossine bees visit orchid flowers? *Atas do Simpósio sôbre a Biota Amazônica* 5:171–180.

————. 1968a. Pollination by euglossine bees. *Evolution* 22:202–210.

————. 1968b. Observations on orchids and euglossine bees in Panama and Costa Rica. *Rev. Biol. Trop.* 15:143–183.

————. 1976a. How to study orchid pollination without any orchids. *In* K. Senghas (ed.), *Proc. 8th World Orchid Conf.,* Frankfurt, Germany, pp. 534–537.

————. 1976b. Una *Sievekingia* nueva de Colombia. *Orquideología* 11:215–221.

————. 1977a. Dos *Polycycnis* nuevas de Sur America. *Orquideología* 12:3–13.

————. 1977b. El género *Polycycnis* en Panama y Costa Rica. *Orquideología* 12:117–133.

————. 1978a. New species of *Euglossa* from Mexico and Central America. *Rev. Biol. Trop.* 26:167–185.

————. 1978b. An infrageneric classification of *Euglossa,* with notes on some features of special taxonomic importance (Hymenoptera: Apidae). *Rev. Biol. Trop.* 26:187–198.

————. 1979a. Una *Sievekingia* llamativa de Panama. *Orquideología* 13:221–227.

————. 1979b. *Eulaema bombiformis, E. meriana,* and

Mullerian mimicry in related species (Hymenoptera: Apidae). *Biotropica* 11:144–151.

Evoy, W. H. and B. P. Jones. 1971. Motor patterns of male euglossine bees evoked by floral fragrances. *Anim. Behav.* 19:583–588.

Faegri, K. and L. van der Pijl. 1966. *The Principles of Pollination Ecology*. Pergamon Press, London. 248 pp.

———. 1979. *The Principles of Pollination Ecology*, 3rd ed. Pergamon Press, Oxford. 244 pp.

Fahn, A. 1979. *Secretory Tissues in Plants*. Academic Press, London. 302pp.

Frisch, K. von. 1967. *The Dance Language and Orientation of Bees*. Harvard University Press, Cambridge, Mass. 566 pp.

———. 1974. Decoding the language of the bee. *Science* 185:663–668.

Gould, J. L. 1975. Honeybee recruitment: the dance-language controversy. *Science* 189:685–693.

———. 1976. The dance-language controversy. *Q. Rev. Biol.* 51:211–244.

Gould, J. L., M. Henerey, and M. C. MacLeod. 1970. Communication of direction by the honeybee. *Science* 169:544–554.

Heithaus, E. R., P. A. Opler, and H. G. Baker. 1974. Bat activity and pollination of *Bauhinia pauletia:* plant–pollinator coevolution. *Ecology* 55:412–419.

Heithaus, E. R., T. H. Fleming, and P. A. Opler. 1975. Foraging patterns and resource utilization in seven species of bats in a seasonal tropical forest. *Ecology* 56:841–854.

Hills, H. G., N. H. Williams, and C. H. Dodson. 1968. Identification of some orchid fragrance components. *Am. Orchid Soc. Bull.* 37:967–971.

———. 1972. Floral fragrances and isolating mechanisms in the genus *Catasetum* (Orchidaceae). *Biotropica* 4:61–76.

Holman, R. T. and W. H. Heimermann. 1973. Identification of components of orchid fragrances by gas chromatography–mass spectrometry. *Am. Orchid Soc. Bull.* 42:678–682.

———. 1976. The chemical composition of fragrances of some orchids. *In* H. H. Szmant and J. Wemple (eds.), *First Symposium on the Scientific Aspects of Orchids*. Chemistry Department, University of Detroit, Detroit, Mich., pp. 75–89.

Howell, D. J. 1977. Time sharing and body partitioning in bat–plant pollination systems. *Nature* 270:509–510.

Johnson, D. L. 1967. Honeybees: do they use the direction information contained in their dance movements? *Science* 155:844–847.

Johnson, D. L. and A. M. Wenner. 1970. Recruitment efficiency in honeybees: studies on the role of olfaction. *J. Apic. Res.* 9:13–18.

Kimsey, L. S. 1980. The behaviour of male orchid bees (Apidae, Hymenoptera, Insecta) and the question of leks. *Anim. Behav.* 28:996–1004.

Kullenberg, B. 1956. Field experiments with chemical sexual attractants on aculeate hymenopteran males. I. *Zool. Bidrag Uppsala* 31:253–352, 5 plates.

———. 1961. Studies on *Ophrys* L. pollination. *Zool. Bidrag Uppsala* 34:1–340.

———. 1973a. New observations on the pollination of *Ophrys* L. (Orchidaceae). *Zoon Suppl.* 1:9–14, 2 plates.

———. 1973b. Field experiments with chemical sexual attractants on aculeate hymenopteran males. II. *Zoon Suppl.* 1:31–43, 2 plates.

———. 1975. Chemical signals in the biocoenosis. *Biological Signals 73–85*. Kungl. Fysiografiska Sallsakapet, Lund, Sweden.

Kullenberg, B. and G. Bergstrom. 1973. The pollination of *Ophrys* orchids. *In* G. Bendz and J. Santesson (eds.), *Chemistry in Botanical Classification. Nobel Symp. 25.* Nobel Foundation, Stockholm, pp. 253–258.

———. 1975. Chemical communication between living organisms. *Endeavour* 34:59–66.

———. 1976a. The pollination of *Ophrys* orchids. *Bot. Notiser* 129:11–20.

———. 1976b. Hymenoptera aculeata males as pollinators of *Ophrys* orchids. *Zool. Scripta* 5:13–23.

Lumer, C. 1980. *Blakea chlorantha* (Melastomaceae): a rodent pollinated plant in the neotropics. *Bot. Soc. Am. Misc. Ser. Publ.* 158:68. (Abstract)

Meeuse, B. J. D. 1961. *The Story of Pollination*. Ronald Press, New York. 243 pp.

———. 1978. The physiology of some sapromyophilous flowers. *In* A. J. Richards (ed.), *The Pollination of Flowers by Insects* (Linnean Soc. Symp. No. 6). Academic Press, London, pp. 97–104.

Mesler, M. R., J. D. Ackerman, and K. L. Lu. 1980. The effectiveness of fungus gnats as pollinators. *Am. J. Bot.* 67:564–567.

Metcalf, R. L., W. C. Mitchell, T. R. Fukuto, and E. R. Metcalf. 1975. Attraction of the oriental fruit fly, *Dacus dorsalis,* to methyl eugenol and related olfactory stimulants. *Proc. Natl. Acad. Sci. U.S.A.* 72:2501–2505.

Middleditch, B. S. (ed.). 1979. *Practical Mass Spectrometry*. Plenum Press, New York. 387 pp.

Nilsson, L. A. 1978. Pollination ecology and adaptation in *Platanthera chlorantha* (Orchidaceae). *Bot. Notiser* 131:35–51.

———. 1979a. Anthecological studies on the lady's slipper, *Cypripedium calceolus* (Orchidaceae). *Bot. Notiser* 132:329–347.

———. 1979b. The pollination ecology of *Herminium monorchis* (Orchidaceae). *Bot. Notiser* 132:537–549.

———. 1980. The pollination ecology of *Dactylorhiza sambucina* (Orchidaceae). *Bot. Notiser* 133:367–385.

Pijl, L. van der and C. H. Dodson. 1966. *Orchid Flowers: Their Pollination and Evolution*. University of Miami Press, Coral Gables, Fla. 214 pp.

Priesner, E. 1973. Reaktionen von Riechreszeptoren männlicher Solitärbienen (Hymenoptera, Apoidea) auf Inhaltsstoffe von *Ophrys*-Blüten. *Zoon Suppl.* 1:43–54, 1 plate.

Rosin, R. 1978. The honeybee "language" controversy. *J. Theor. Biol.* 72:589–602.

———. 1980. Paradoxes of the honeybee "dance language" hypothesis. *J. Theor. Biol.* 84: 775–800.

Rourke, J. and D. Wiens. 1977. Convergent floral evolution in South African and Australian Proteaceae and its possible bearing on pollination by nonflying mammals. *Ann. Missouri Bot. Gard.* 64:1–17.

Simon, P. W., R. C. Lindsay, and C. E. Peterson. 1980. Analysis of carrot volatiles collected on porous polymer traps. *Agric. Food Chem.* 28:549–552.

Stallberg-Stenhagen, S., E. Stenhagen, and G. Bergstrom. 1973. Analytical techniques in pheromone studies. *Zoon Suppl.* 1:77–82.

Steiner, K. E. 1980. The reproductive biology of *Mabea occidentalis* (Euphorbiaceae). *Botany and Natural History. A Symposium Signalling the Completion of the "Flora of Panama,"* Panama, Panama. pp. 125–127. (Abstract)

Sussman, R. W. and P. H. Raven. 1978. Pollination by lemurs and marsupials: an archaic coevolutionary system. *Science* 200:731–736.

Swanson, E. S., W. P. Cunningham, and R. T. Holman. 1980. Ultrastructure of glandular ovarian trichomes of *Cypripedium calceolus* and *C. reginae* (Orchidaceae). *Am. J. Bot.* 67:784–789.

Tengo, J. 1979. Odour-released behavior in *Andrena* male bees (Apoidea, Hymenoptera). *Zoon* 7:15–48.

Tengo, J. and G. Bergstrom. 1977. Comparative analyses of complex secretions from heads of *Andrena* bees (Hym., Apoidea). *Comp. Biochem. Physiol.* 57B:197–202.

Thien, L. B. 1974. Floral biology of *Magnolia*. *Am. J. Bot.* 61:1037–1045.

Thien, L. B., W. H. Heimermann, and R. T. Holman. 1975. Floral odors and quantitative taxonomy of *Magnolia* and *Liriodendron*. *Taxon* 24: 557–568.

Tsugita, T., T. Imai, Y. Doi, T. Jurata, and H. Kato. 1979. GC and GC-MS analysis of headspace volatiles by Tenax GC trapping techniques. *Agric. Biol. Chem.* 43:1351–1354.

Vogel, S. 1961. Die Bestäubung der Kesselfallen-Blüten von *Ceropegia*. *Beitr. Biol. Pfl.* 36:159–237.

———. 1963a. Duftdrüsen im Dienste der Bestäubung: über Bau und Funktion der Osmophoren. *Akad. Wiss. Lit. (Mainz), Abh. Math.-Naturwiss. Kl., Jahrgang* 1962:599–763.

———. 1963b. Das sexuelle Anlockungsprinzip der Catasetinen- und Stanhopeen-Blüten und die wahre Funktion ihres sogenannten Futtergewebes. *Oesterr. Bot. Z.* 100:308–337.

———. 1966a. Scent organs of orchid flowers and their relation to insect pollination. *In* L. R. DeFarmo (ed.), *Proc. 5th World Orchid Conf.,* Long Beach, Calif., pp. 253–259.

———. 1966b. Parfümsammelnde Bienen als Bestäuber von Orchidaceen und *Gloxinia*. *Oesterr. Bot. Z.* 113:302–361.

Wells, P. H. and A. M. Wenner. 1971. The influence of food scent on behavior of foraging honeybees. *Physiol. Zool.* 44:191–209.

Wenner, A. M. and D. L. Johnson. 1967. Honeybees: do they use direction and distance information provided by their dances? *Science* 158:1072–1077.

Wenner, A. M., P. H. Wells, and F. J. Rohlf. 1967. An analysis of the waggle dance and recruitment in honeybees. *Physiol. Zool.* 40:317–344.

Wenner, A. M., P. H. Wells, and D. L. Johnson. 1969. Honeybee recruitment to food sources: olfaction or language? *Science* 164:84–86.

Williams, N. H. 1980. Why male euglossine bees visit orchid flowers. *Botany and Natural History. A Symposium Signalling the Completion of the "Flora of Panama,"* pp. 139–140. (Abstract)

———. 1981a. The biology of orchids and euglossine bees. *In* J. Arditti (ed.), *Orchid Biology: Reviews and Perspectives,* II. Cornell University Press, Ithaca, N.Y. In press.

———. 1981b. Why male euglossine bees visit orchid flowers: Another hypothesis. *In* W. G. D'Arcy and M. D. Correa (eds.), *Botany and Natural History. A Symposium Signalling the Completion of the "Flora of Panama."* Submitted for publication.

Williams, N. H. and C. H. Dodson. 1972. Selective attraction of male euglossine bees to orchid floral fragrances and its importance in long distance pollen flow. *Evolution* 26:84–95.

Williams, N. H. and R. L. Dressler. 1976. Euglossine pollination of *Spathiphyllum* (Araceae). *Selbyana* 1:349–356.

Williams, N. H., J. T. Atwood, and C. H. Dodson. 1981. Floral fragrance analysis in *Anguloa, Lycaste,* and *Mendoncella* (Orchidaceae). *Selbyana* 5. 5:291–295.

Zlatkis, A., H. A. Lichenstein, and A. Tishbee. 1973. Concentration and analysis of trace volatile organics in gases and biological fluids with a new solid adsorbent. *Chromatographia* 6:67–70.

Zlatkis, A., W. Bertsch, D. A. Bafus, and H. M. Lieblich. 1974. Analysis of trace volatile metabolites in serum and plasma. *J. Chromatog.* 91:379–383.

4
BUZZ POLLINATION IN ANGIOSPERMS

Stephen L. Buchmann

U. S. Department of Agriculture
Agricultural Research Service
Carl Hayden Bee Research Center
2000 East Allen Road
Tucson, Arizona
and
Department of Ecology and Evolutionary Biology
University of Arizona
Tucson, Arizona

ABSTRACT

Vibrational pollination in angiosperms, its characteristics (specific plant and bee co-adaptations), and selected case histories in the historical development of this branch of anthecology are summarized and reviewed. Presently known information on the distribution of taxa (orders) that exhibit anther dehiscence by apical slits, pores, or valves within the Angiospermae (Anthophyta) is presented in figure form. Within the subclass Dicotyledoneae the following orders contain at least some poricidal members: Laurales, Ranunculales, Dilleniales, Theales, Lecythidales, Ebenales, Malvales, Capparales, Ericales, Primulales, Rosales, Fabales, Myrtales, Rafflesiales, Santalales, Euphorbiales, Sapindales, Geraniales, Polygalales, Polemoniales, Scrophulariales, Gentianales, Lamiales, and Rubiales. In the subclass Monocotyledoneae, the following orders contain at least some apically dehiscent taxa: Arales, Commelinales, and Liliales. In all, the dicots account for a total of 24 orders, 64 families, 473 genera, and about 19,000 species. The monocots are represented by 3 orders, 8 families, 71 genera, and about 1,150 species. Thus, poricidal dehiscence of some sort is found in 27 angiosperm orders, 72 families, 544 genera, and at least 15,000 to 20,000 species. This represents about 6 to 8% of the 225,000 presently known species of flowering plants. Since access to the pollen is severely limited by the small terminal pores, or apical slits, bees evolved a behavioral repertoire (shivering their indirect flight muscles) to effectively release the pollen. Whether this behavior is stereotyped (instinctual) or learned during the first floral encounters is discussed. Exceptions to this "buzz pollination rule," in poricidal taxa that are not known to be vibrated, include the 12 families dehiscent by valves, and a few other forms with very sticky pollen. Convergent and parallel staminal evolution in porose flowers and the specialized cases of heteranthery, enantiostyly, pollen dimorphism, floral deception, "milking anthers," gleaning, and biting bees are also discussed.

KEY WORDS: Buzz pollination, vibration, vibratile, poricidal anthers, pollen, heteranthery, enantiostyly, bees, apoidea, floral deception.

CONTENTS

INTRODUCTION

The stamens of the majority of the world's angiosperms are longitudinally dehiscent, releasing their pollen passively, whereupon bee pollinators may collect pollen mixed with nectar for use in larval cell provisions. The anthers of most of the 225,000-plus described species of flowering plants are dehiscent, releasing their pollen grains typically by complete longitudinal stomial slits (i.e., releasing pollen all at once along the entire length of the anthers, see Fig. 4-1), by valves or true apical pores (Halsted, 1890; van Tieghem, 1891; Kerner von Marilaun, 1895; Harris, 1903, 1905a,b; Burck, 1891, 1906a,b, 1907; Staedtler, 1923; Shoenichen, 1924; Matthews and MacLachlan, 1929; Richter, 1929; Maheswari, 1950; Canright, 1952; Aleksandrov and Dobrotvorskaya, 1960; Buchmann and Hurley, 1978; Fahn, 1974; Schmid, 1976; Schmid and Alpert, 1977). Schmid (1976) should be consulted for a more complete review of these and related references concerning anther dehiscence.

For this chapter, I consider apically (= poricidally) dehiscent flowers to be those species whose anthers have partial longitudinal loculicidal slits, e.g., extending ⅒ to ⅛ from apex to base or rarely transversely, and those that have true rounded apical pores, either sessile or on stalks (see Fig. 4-1). In the broadest sense, "poricidally dehiscent taxa" include genera and species that dehisce by means of short apical slits, short transverse slits, valves, or true apical pores (usually rounded, or elevated on a rostrum of connective tissue, or with very narrow sessile pores). This inclusive usage of "poricidal" has been used in the construction of a table (see Table 5 in Buchmann, 1978) presenting the known cases of poricidally dehiscent angiosperms, the nature of their dehiscence, the number of taxa involved, and a brief description of their biogeography. No attempt has been made to classify or categorize the various types of porose dehiscence modes in angiosperm anthers. This task is presently a controversial and complex subject in staminal evolution, and a full account treating the dehiscence of anthers will be presented elsewhere (Buchmann and Schmid, in prep.). The list of poricidal forms as presented earlier

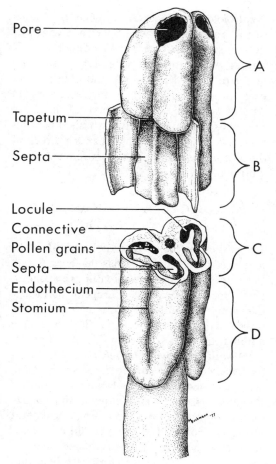

Pore

Tapetum

Septa

Locule
Connective
Pollen grains
Septa
Endothecium
Stomium

A

B

C

D

Figure 4-1. The real system. The intact anther apex with terminal pores is shown in (A). In (B) the anther walls have been pulled back to reveal the inner septum and tapetum. A cross section (C) of the anther reveals the hollow locules with their septa and associated structures. The base of the intact anther is shown in (D) along with the unopened stomium (slit where dehiscence occurs in longitudinally dehiscent anthers).

dehiscent upon desiccation, thus resulting in an underestimate of the number of pored taxa), and from living flowers interacting in natural populations with their coevolved native pollinators. The last method is obviously the best one for delimiting the presence and extent of localized apical dehiscence in a particular taxon. Even with these qualifiers, it is suggested that approximately 95% of all true poricidally dehiscent angiosperm families and genera have been accurately represented by my methods. Information contrary to the information presented in that earlier table, such as additions or deletions to my list, would be welcomed by the author, especially from botanists and entomologists with first-hand knowledge and field experience with apically pored taxa.

This unique form of pollination has been termed "buzz pollination" by one worker (Buchmann, 1974), since it recalls the audible buzz component of intrafloral bee behavior during the rapid floral visitations. The terms vibratile and vibrational pollination have also been used to emphasize the vibratory character of this means of pollen expulsion and efficient harvesting by bees (Buchmann et al., 1977).

Bees that are able to buzz, first alight on the corolla (or directly on the anther cone or individual anthers for smaller bees), curl around the anther cone or directly over the pores, and then grasp the stamens tightly with wings held stationary over the thorax and abdomen. By rapidly contracting and relaxing their large indirect flight muscles, uncoupled from the axillary sclerites and the regular flight musculature, they transmit strong vibrations throughout their bodies. Body parts such as the thorax, abdomen, or legs that are in direct contact with the anthers or staminal cone strike the anthers rapidly (about 50–2,000 Hz), immediately causing the anthers, locules, and sometimes the entire flower (as in the case of *Solanum* species) to vibrate with the same frequency as the vibrating thorax of the buzzing bee. The strong, rapid vibrations of the bees' integument and the flower striking the boundary layer of air are responsible for the characteristic audible (often up to 5 meters

(Buchmann, 1978) should be considered as a "state-of-the-art" working list to which some taxa will have to be added in the future, and a few removed, based upon more detailed anatomical studies and observations of their pollination made in the field. In its present state, this information must be used cautiously, since these data were gathered widely from the literature, from floras and other systematic works, by studying herbarium sheets (not a very satisfactory method, since apically dehiscent forms may appear later as longitudinally away) component of this behavior, and are the reason for naming this buzz pollination. The bee-induced floral vibrations function to loosen the locule-contained pollen, producing strong pollen grain/locule wall interactions that result in rapid expulsion of most pollen from the anther apices within a few seconds, or a fraction of a second (see Buchmann and Hurley, 1978, for a more complete discussion). Floral position is almost immaterial during floral vibration, since efficient pollen removal can be effected by most bees (or artificially with a tuning fork or "vibrator" as a surrogate bee), vibrating apically dehiscent anthers presented at any angle. The pollen is not simply shaken out of a pendant flower, but is forcibly expelled, owing to the biophysical interactions brought about by the bees rapidly vibrating the androecium. Such shivering of the flight muscles is used for rapid pollen harvesting from porose flowers by many female bees (and even one syrphid fly, *Volucella mexicana*) on nearly every continent. Buzz pollination and hence apical dehiscence are likely quite ancient and represent millions of years of adaptations and counter-adaptations by plants and their buzz pollinating bees, culminating in the vast array (at least 72 families of plants and 544 genera) of poricidal flowers and the use of this pollen-collection behavior by bees in nearly every family within the bee superfamily Apoidea.

Statements in the literature such as "bumblebees have been observed foraging for it [pollen] by vibrating their wings at a high frequency accompanied by a high-pitched buzzing sound" (e.g., Macior, 1974) abound, yet are somewhat inexact, since the buzzing one hears is not produced by movements of the wings, as during flight, but rather is due to the interaction of the bees' cuticle with the boundary layer of air adjacent to the bees. This acoustical component of the buzzing behavior is really just a by-product of the actual interaction, and doesn't have much biological meaning. What makes buzzing work as a pollen release/harvesting method are the high frequency vibrations transmitted to the floral substrate by the bees. The audible component is a serendipitous bonus, however, for it allows floral biologists to locate these bees on flowers in the dim morning light when one can hear but not see them. One can also time the floral buzzes with a stopwatch throughout the day, or for different bee species, and obtain accurate information on the time necessary to empty a flower of its pollen (i.e., handling time) or the number of flowers visited per unit time, etc., since the audible buzz is a direct indication of the actual time spent for each floral vibration.

HISTORICAL DEVELOPMENT OF BUZZ POLLINATION

The earliest reference to actual vibratile pollination of poricidal plants has not been found. It will probably be located within the voluminous nineteenth century European literature. Christian Konrad Sprengel, often cited as the father of pollination biology, did illustrate a number of porose taxa in his monumental 1793 work *(Das endeckte Geheimnis der Natur im Bau in des Befruchtung der Blumen)*. Genera illustrated by Sprengel include: *Arbutus, Berberis, Erica, Kalmia, Solanum,* and *Vaccinium*. However, as keen an observer of natural history as Sprengel apparently did not make, or never witnessed, the connection between porose anthers and bee vibratory behavior. Knuth and Loew, in their classic four-volume *Handbuch der Blutenbiologie* (1895–1905), also talked about pollination in some poricidal forms, but did not discuss true buzz pollination.

The floral biology of one American species of *Cassia (C. fasciculata)* has been examined repeatedly since the late 1800s (Leggett, 1875; Todd, 1882; Robertson, 1890; Harris and Kuchs, 1902; Thorp and Estes, 1975). Other species of *Cassia* have also been studied since the 1800s (Leggett, 1881; Meehan, 1886; Burck, 1887a,b; F. and H. Muller in Knuth and Loew, 1895–1905; Kirchner, 1911; van der Pijl, 1954; Michener, 1962; Wille, 1963; Linsley and Cazier, 1972; Buchmann, 1974; Silander, 1978). The reader should consult Thorp and Estes (1975) for a good account of *Cassia* anthecology. Leggett (1875) posed the question early for *Cassia:* How is the abundant pollen discharged from apical pores? Burck (1887a,b) may have been the first to publish observations of anther vibration and subsequent pollen discharge by release of a small cloud of pollen. Floral vibration of *Cassia* has been observed many times (Burck, 1887a,b; Knuth and Loew, 1895–1905; Kirchner, 1911; van der Pijl, 1954; Michener, 1962; Wille, 1963; Linsley, 1962b, 1970; Linsley and Cazier, 1972; Hardin et al., 1972; Buchmann, 1974; Thorp and Estes, 1975; Salinas and Sánchez, 1977).

Vibratile behavior also has been noted for bees collecting pollen from other plants with anthers opening via apical pores, for example: *Amoreuxia* (Buchmann, unpub.); *Centradenia* (Almeda, 1977); *Cochlosperum* (Roubik et al., 1981); *Cyanella* (Dulberger and Ornduff, 1980); *Dodecatheon* (Macior, 1964, 1970a); *Lycopersicon* (Rick, 1950; Rick et al., 1978); *Melampyrum* (Meidell, 1944); *Melastoma* (van der Pijl, 1954); *Monochaetum* (Almeda, 1978); Monotropoideae (Wallace, 1975); *Mouriri* (Buchmann and Buchmann, 1981); *Pedicularis* (Macior, 1968a,b, 1969, 1970b, 1973, 1977); *Rhynchanthera* (Laroca, 1970); *Solanum* (Herbst, 1918; Linsley, 1962a,b, 1970; Michener, 1962; Linsley and Cazier, 1963; Macior, 1964, 1970a; Hardin et al., 1972; Bowers, 1975; Buchmann et al., 1977; Symon, 1979); *Tibouchina* (Laroca, 1970); *Tococa* (Laroca, 1970); *Vaccinium* (Roberts, 1978; Rajotte and Roberts, 1978a,b; Cane et al., unpub.); *Wachendorfia* (Ornduff and Dulberger, 1978); and *Xiphidium* (Buchmann, 1980).

Charles Darwin in the months just prior to his death had become interested in heteranthery in poricidal flowers. A letter written by Darwin just nine days before his death on April 19, 1882, was a request to Professor J. E. Todd of Iowa for more information and seeds of *Solanum rostratum* (see Fig. 4-2; Bell et al., 1959). Darwin had been interested in experimenting with them in his greenhouse. It was probably his last scientific inquiry, and it is tempting to think of him making turns, and kicking pebbles to count the rounds on his beloved sandwalk while pondering *Solanum rostratum*.

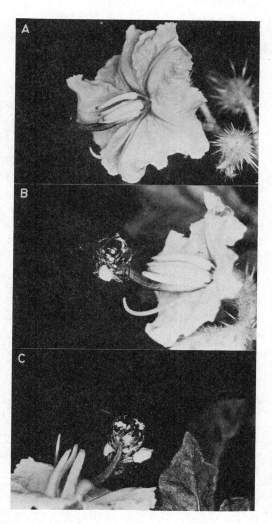

Figure 4-2. A, Flower of *Solanum rostratum* (Solanaceae), demonstrating poricidal anthers, a dimorphic androecium (heteranthery), and the deflection of the style opposite that of the large stamen (enantiostyly). B, A female *Protandrena* (formerly *Psaenythia mexicanorum*) (Anthophoridae) in the position in which floral vibration occurs. In this case the bee is vibrating the single large anther. C, A female *Protandrena mexicanorum* grasping the dimorphic anther loosely while grooming and packing a large load of *Solanum* pollen into her scopae.

One person stands out in the early history of the study of poricidal plants. The pioneering American botanist James Arthur Harris spent a large part of his varied and productive career cataloging poricidal dehiscence in angiosperm stamens. This author rediscovered these marvelous early contributions of Harris during his early search for examples of porose anthers. Harris and Kuchs (1902) originally made some observations on the pollination of *S. rostratum* and *C. chamaecrista* (= *C. fasciculata*). They detailed the prominent enantiostyly and heteranthery, and milking behavior, but remarkably did not recount floral vibration by the bees (*Agapostemon, Bombus,* and others known to buzz) they observed. This led Harris (1903) to explore polygamy and floral abnormalities in *Solanum,* and to conduct his Ph.D. dissertation research (a catalog of poricidal dehiscence in angiosperms), at Washington University on this subject (1905a). This was immediately followed (Harris, 1905b) by the publication of "The dehiscence of anthers by apical pores" in the 16th annual report of the Missouri Botanical Gardens. Harris (1905c) also tried to correlate the influence of the Apoidea on the distribution (biogeography) of floral types. Harris (1906) explored the "anomalous" anthers of *Dicorynia, Duparquetia,* and *Strumpfia,* and later (1909, 1914) examined heteranthery and physiological dimorphism in the anthers of *Lagerstroemia.* Harris was truly a pioneer in poricidal dehiscence mechanisms, but it is remarkable that he was apparently unaware of the necessity of pollen release by bee vibration in the plants he studied. The reader is directed to Fig. 4-3 for the modern (*sensu* Buchmann) placement of poricidal taxa among the angiosperm orders.

Another author to treat briefly poricidal flowers as a group was Kugler (1955). He described these flowers as "Glockenblumen" (campanulas or bell flowers), and described the blossoms as hanging bells with dissemination devices (i.e., anther cone). The poricidal morphology was not associated with active bee vibration by Kugler. Flowers treated were: Amaryllidaceae *(Galanthus, Leucojum),* Boraginaceae *(Borago, Cerinthe, Symphytum),* Ericaceae *(Andromeda, Calluna, Erica, Rhododendron, Vaccinium),* Primulaceae *(Cyclamen, Solandella),* Pyrolaceae *(Pyrola),* and Solanaceae *(Solanum).*

After a long period of neglect, two seminal papers appeared (Michener, 1962; Wille, 1963) that once again brought buzz pollination to the attention of the floral biology audience. This can be seen by the number of citations these two papers have received since they were published. Buzzing and "gleaning" pollen collection were mentioned by both, but Wille (1963) added the category of "biting" bees. Wille observed *Trigona* biting small holes in the anther through which they extracted pollen. This has also been recorded by F. and H. Muller (cited in Knuth and Loew, 1895–1905), and for other bees (*e.g., Chilicola* and *Xylocopa). Xylocopa* commonly pierces corollas to rob nectar and may extensively chew porose anthers in addition to buzzing to effect pollen release. "Milking" of anthers by bees, i.e., grasping the anther base with mandibles and squeezing pollen toward and out the pores, with or without vibration, has been reported several times (Leggett, 1881; Todd, 1882; Robertson, 1890; Harris and Kuchs, 1902; Bowers, 1975; Thorp and Estes, 1975). The combination of milking anthers and buzzing simultaneously was observed by Bowers (1975) for *Bombus* on *S. rostratum* and by Thorp and Estes (1975) for bumblebees on *Cassia fasciculata.*

The papers of Michener and Wille properly lead us into the modern period, with several workers making important contributions up to the present time. This period is exemplified by the excellent field observations on *Cassia* by several authors (Thorp and Estes, 1975; Salinas and Sánchez, 1977). Detailed descriptions of floral anatomy and pollinator intrafloral behavior on many species of *Cassia* were made. Other details of floral morphology in relation to buzz pollination were given, such as the hooded, stiff, cucullus petal (Thorp and Estes, 1975).

The expertise and ingenuity of this era is seen in the many excellent contributions on *Dodecatheon* (Macior, 1964, 1970a), *Solanum* (Macior, 1964, 1970a), and *Pedicularis* (Macior, 1968a,b, 1969, 1970b, 1973, 1977). These observations on floral vibration by different bees are especially important because of the macrocinematography of bees buzzing flowers in the field. This has allowed critical behavioral observations to be made that would not have been possible with the unaided eye. We also see clear floral modification experiments (usually mutilation type) to test the responses of bees to modified flowers lacking specific stimuli (Macior, 1964). These are reminiscent of the pioneering work of Clements and Long (1923) in this area. Macior (1968a) also tape-recorded the buzzes of various bees working porose flowers along with their flight sounds, at different ambient temperatures. These were then analyzed with regard to frequency on an oscilloscope. This has since been done for bees on *Solanum* (Buch-

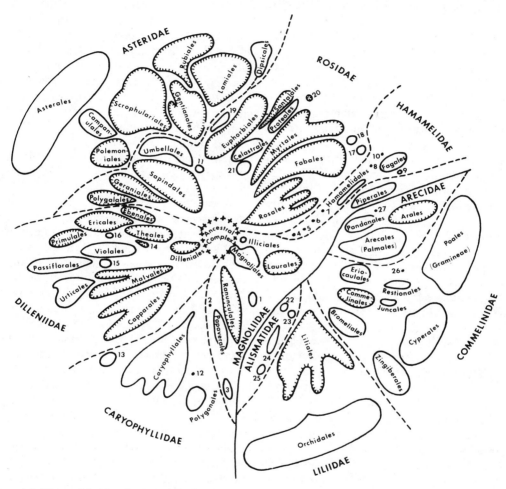

Figure 4-3. Diagram showing the suspected affinities and relative degree of specialization of the orders of angiosperms. The orders are represented as a series of cross sections of the branches of the phylogenetic tree as it arises from the largely unknown angiosperm ancestral complex. Orders in which at least some species dehisce by means of apical slits, pores, or valves are delimited by interior hatched outlines. Redrawn after Stebbins, 1974.

DICOTYLEDONS

1. Nympheales
2. Sarraceniales
3. Aristolochiales
4. Trochodendrales
5. Cercidphyllales
6. Didymeleales
7. Eupteleales
8. Eucommiales
9. Casuarinales
10. Leitneriales
11. Juglandales
12. Batales
13. Plumbaginales
14. Lecythidales
15. Salicales
16. Diapensiales
17. Podostemales
18. Haloragales
19. Cornales
20. Rafflesiales
21. Rhamnales

MONOCOTYLEDONS

22. Alismatales
23. Triuridales
24. Najadales
25. Hydrocharitales
26. Typhales
27. Cyclanthales

mann, 1974), *Vaccinium* (Cane et al., unpub.), and other plants using airborne and substrate vibration analysis (Buchmann, unpub.). Several biophysical models have been proposed to account for vibrationally induced pollen expulsion (see Buchmann and Hurley, 1978; DeTar et al., 1968) in these flowers.

The important contribution to evolutionary shifts from reward to deception in pollen flowers by Vogel (1978) must be included here. This is a thought-provoking study that should stimulate evolutionary ecologists into rigorous field experiments on the function(s) of deception (pseudopollen, anther dummies, etc.) in pollen flowers.

Important recent contributions have been made with respect to the number of pollen grains per anther in heterantherous forms, and more recently for enantiostyly in *Solanum* (Bowers, 1975), and for *Cyanella* (Dulberger and Ornduff, 1980) and *Wachendorfia* (Ornduff and Dulberger, 1978). The density-dependent control of reproductive success was studied recently in *Cassia biflora* by Silander (1978). Outcrossing and the reproductive biology of *Cochlosperum* was examined recently (Roubik et al., 1982).

A recent shift includes the chemical ecology of these plants, especially in relation to pollen chemistry (caloric values, total protein, amino acids, lipids, carbohydrates including starch, etc.) as in the work by Baker and Baker (1979) and Buchmann (unpub.). The genus *Mouriri* (Melastomataceae) was demonstrated to be buzz pollinated, and also to offer bees complex lipoidal mixtures in elaiophors (Buchmann and Buchmann, 1981). This was the first observation of floral lipid (via elaiophors) production in the Melastomataceae, and by a poricidal flower. The entire tribe Memecyleae (including *Memecylon, Mouriri,* and *Votomita*) includes about 300 species, all of which are apparently buzz pollinated and are oil flowers in the sense of Vogel (1974).

ANTHER DEHISCENCE

In its most primitive form, as in *Degeneria* (Degeneriaceae), the laminar stamen resembles the (micro-) sporophyll of a cryptogam or cycad, since the thecae (anther halves containing two pollen sacs or male sporangia) are present as individual sporangia, or pollen sacs, on a small sporophyll. In more derived taxa, the laminar portion becomes reduced, but the thecae retain the nature of sporangia with an opening mechanism. The most common method of anther opening or rupture (equals dehiscence) is a longitudinal slit along the stomium along the entire length of the thecae (see Fig. 4-1 for details). Often the locule walls will turn back, further exposing the pollen grains within. As a matter of terminology, the rupture of the stomium accounts for the so-called anther dehiscence, while pollen is "shed" from the dehisced anthers for wind transport or for biotic transfer by animal pollination vectors.

A complete classification including figures of dehiscence modes in angiosperm anthers is in progress (Buchmann and Schmid, in prep.), so it will not be presented here. Suffice it to say that there exists a bewildering array of anther types, involving poricidal dehiscence by apical slits, valves, pores, and intermediates between pores and normal longitudinal stomial dehiscence (Figs. 4-4, 4-5, 4-6). The simplest form of porose dehiscence occurs when partial longitudinal breaks in the stomium start at the apex and continue about one-sixth to one-third of the way down the length of the thecae (Fig. 4-1). This results in a functionally poricidal condition in which the anthers shed their pollen through apical slits (e.g., the teardrop-shaped slits in many *Solanum* species). An apical pore is usually small, round, and located at the anther apex, although such pores may be of almost any shape, may have jagged edges, or can be located at the tips of long tubes of anther connective tissue, making them rostrate (Fig. 4-4D,J). Valvular dehiscence is a specialized type in which the pores, two to four per anther, may be located anywhere along the thecae, and are covered with small tissue flaps known as lids or valves. Examples of valvular anther dehiscence are found in many genera in the following families: Atherospermataceae, Berberidaceae, Canellaceae, Gomortegaceae, Gyrocarpaceae, Hamamelidaceae, Hernandiaceae, Lauraceae, Monimi-

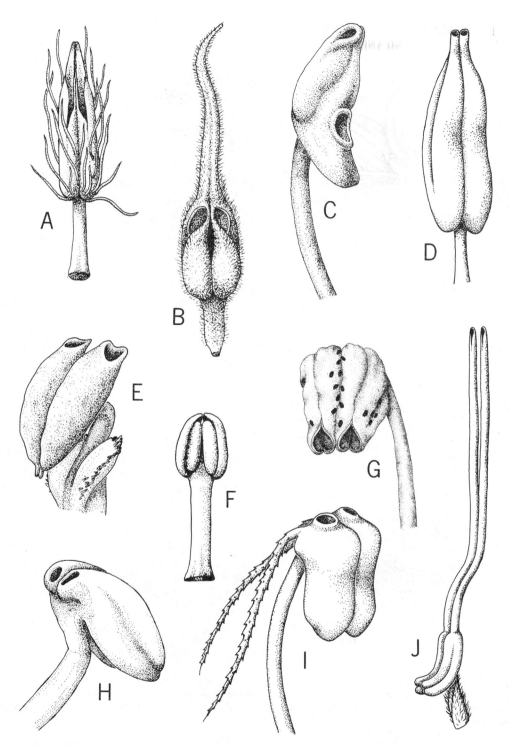

Figure 4-4. Selected stamens showing poricidal dehiscence (by apical slits or true pores) of anthers, and other features including rostrate pores (D, J), elongation of anther connectives (A, B, D, J), and oil glands or elaiophors on the dorsal aspect of the connective in the Memecyleae (C). A, *Apeiba tibourbou* (Tiliaceae); B, *Sloanea terniflora* (Elaeocarpaceae); C, *Mouriri myrtilloides* subsp. *parvifolia* (Melastomataceae); D, *Solanum elaeagnifolium* (Solanaceae); E, *Chimaphila umbellata* (Pyrolaceae); F, *Xiphidium caeruleum* (Haemodoraceae); G, *Bixa orellana* (Bixaceae); H, *Pyrola elliptica* (Pyrolaceae); I, *Cassiope mertensiana* (Ericaceae); J, *Demosthenesia cordifolia* (Vacciniaceae).

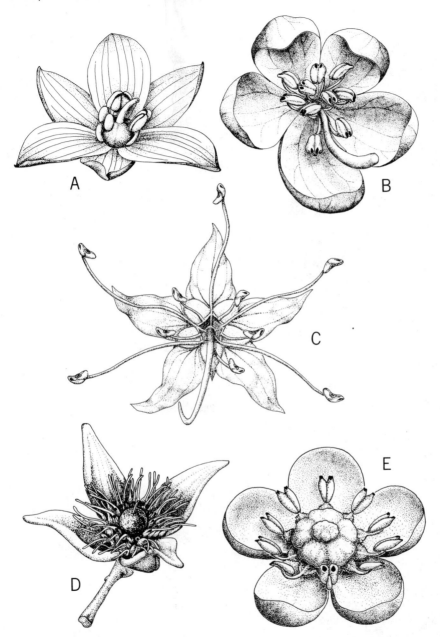

Figure 4-5. Poricidal flowers demonstrating variation in stamen length, number, placement, pore types, and modifications of the anther connective. A, *Xiphidium caeruleum* (Haemodoraceae); B, *Pyrola elliptica* (Pyrolaceae); C, *Mouriri mytilloides* subsp. *parvifolia* (Melastomataceae); D, *Sloanea terniflora* (Elaeocarpaceae); E, *Chimaphila umbellata* (Pyrolaceae).

aceae, Myrtaceae, Olacaceae, and the Podphyllaceae. A table of these families and the associated 69 genera will be presented elsewhere (Buchmann, 1982). Although porose, valvularly dehiscent taxa are generally not buzz pollinated, owing to sticky pollen in clumps, and therefore remain as a small number of exceptions to the general rule of vibratory pollen release in poricidal forms.

Poricidal anthers are considered to be any

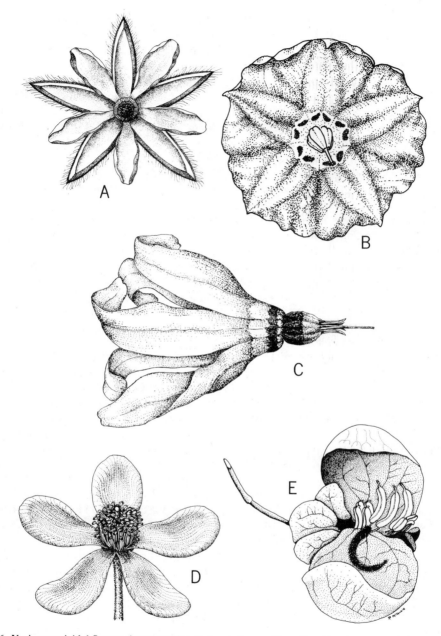

Figure 4-6. Various poricidal flowers demonstrating *Solanum*-type, with anthers forming a cone connivent about the pistil (A, B, C), "shaving-brush" presentation (D), and heteranthery and enantiostyly in a *Cassia* (E). A, *Apeiba tibourbou* (Tiliaceae); B, *Solanum parishii* (Solanaceae); C, *Dodecatheon clevelandii* (Primulaceae); D, *Bixa orellana;* E, *Cassia quiedondilla* (Fabaceae).

that dehisce through a pore (apical slit, round, jagged, rostrate, or valvular) at the apex of the theca. The pollen grains in such forms usually occur singly, as monad pollen, although tetrads, groups of four united grains, are known in some Ericaceae and Pyrolaceae, which are still pollinated in a vibratile manner.

Sometimes longitudinally dehiscent and poricidal anthers also bear elongated processes arising from the anther connective, which extend well beyond the anther pores. These may be quite long, even several times the total length of the anther (see Fig. 4-4J). Such an elongate connective, or awn, is commonly found in anthers of the Melastomataceae, and the Elaeocarpaceae (as in *Sloanea,* Fig. 4-4B). These prolonged connectives are thought to act as "triggers" that, when moved by a pollinator during a floral visit, probably function to bump the anthers and release additional pollen onto these insects. Thus, these flowers are likely adapted for legitimate buzz pollination, but also for pollen release onto smaller nonhymenopterous insect visitors that are incapable of floral vibration. In the Ericaceae especially (as in *Arctostaphylos* and *Cassiope;* see Fig. 4-4I), there are long pendulant anther filaments that may also serve to loosen additional pollen onto visiting insects. These processes are always paired and usually pendant, arising from connective tissue. Until direct observations and experimental manipulation, such as mutilations, are made, we can only speculate upon their true biological function(s).

In most flowering plants, the pollen sacs open by means of complete longitudinal breaks along the stomium (e.g., as in Asteraceae, Magnoliaceae, and Papaveraceae). A classical fibrous layer (endothecium = fibrous, contractile or mechanical layer), which is a subepidermal layer with secondary wall thickenings, or sclereids, has been routinely implicated in the rupture of anther locules during dehiscence. An endothecium is not present in many species, including some aquatic forms such as the Zosteraceae, and in some cleistogamous flowers. The absence of an endothecial layer from some species, however, indicates that an alternate mechanism for dehiscence and subsequently pollen release will be found. In taxa with poricidal anthers (Bixaceae, Cochlospermaceae, Dilleniaceae, Fabaceae, Melastomataceae and Solanaceae, etc.), dehiscence is accomplished largely by the disintegration of the locule wall.

Although most botanists probably believe that the physical forces resulting in stomial disruption and therefore anther dehiscence must in some way be related to the endothecium, when one is present, the precise nature

of the biophysical mechanisms for anther dehiscence remains to be elucidated (Schmid, 1976, 1982; Schmid and Alpert, 1977). With this emphasis placed on the inferred role of the endothecium, most plant anatomists have simply disregarded the histology of the proximal portion of the stamen and the supporting filament. Only recently has one worker (Schmid, 1976, 1982; Schmid and Alpert, 1977) emphasized that some of the controlling factors in dehiscence might be mediated by the structural nature of the filament, since the anther is totally dependent upon the rodlike filament for nutrient and water transport. Since mechanisms of anther dehiscence in longitudinal forms are so poorly understood, the nature of dehiscence in truly poricidal forms is certainly a fertile field for experimentation and ecological plant anatomy.

PORICIDAL FLOWERS

The primary purpose of an earlier study (Buchmann, 1978) was to document the known cases of poricidal dehiscence (in the all-inclusive sense including apical slits and valvular dehiscence) in angiosperms, along with data on the numbers within taxa, and some information on their biogeography. I have made every attempt to locate, by literature searches and personal investigations, all plants known to dehisce by means of pores, including forms that are not buzz pollinated (see below, section on "Poricidal Plants That Are Not Buzz Pollinated").

During the early phases of this study, I discovered the work of the late James Arthur Harris (1903, 1905a,b,c, 1906, 1909, 1914; Harris and Kuchs, 1902) on the subject of the distribution of apical dehiscence in angiosperm stamens. Harris recognized a total of 31 families (24 dicot and 7 monocot; see Table 4-1) and 313 genera as being poricidal. To this list, I have added many other poricidal taxa (Table 4-2), while deleting a few that Harris treated. Harris's goals were the following, and they have largely been mine: (1) the compilation of a list of taxa showing apical dehiscence by pores, for use in future biological investigations; (2) the determination of the distribution

Table 4-1. Families containing at least some or all poricidal taxa, as recognized by
J. A. Harris (M = monocot, D = dicot).

Acanthaceae	(D)	Fabaceae	(D)	Amaryllidaceae	(M)
Balanophoraceae	(D)	Loganaceae	(D)	Araceae	(M)
Begoniaceae	(D)	Melastomataceae	(D)	Commelinaceae	(M)
Bixaceae	(D)	Myrsinaceae	(D)	Liliaceae	(M)
Cyrillaceae	(D)	Ochnaceae	(D)	Mayacaceae	(M)
Dilleniaceae	(D)	Pittosporaceae	(D)	Pontederiaceae	(M)
Ebenaceae	(D)	Polygalaceae	(D)	Rapateaceae	(M)
Elaeocarpaceae	(D)	Pyrolaceae	(D)		
Ericaceae	(D)	Rubiaceae	(D)		
Euphorbiaceae	(D)	Solanaceae	(D)		
Flacourtiaceae	(D)	Theaceae	(D)		
Gentianaceae	(D)	Tremandraceae	(D)		

Totals = 31 families (24 dicot and 7 monocot).

of apically dehiscent taxa; (3) the demonstration of any morphological and anatomical similarities in floral structure between different families and genera (floral convergence and parallelism); and (4) testing the hypothesis that the floras of certain regions are far richer in forms exhibiting apical dehiscence than in other phytogeographic regions. All of this raw biogeographical information is presented in an earlier paper (Buchmann, 1978), and an updated version is in progress (Buchmann, 1982).

One of the most striking features of the "syndrome" of vibratile pollination is the independent evolution in floral morphology (pores, modified connectives, anthers, etc.) and resultant buzz pollination in so many phylogenetically unrelated and diverse families among angiospermous plants (see Vogel, 1978). Presumably these floral convergences, or parallelisms, began in a tight coevolutionary relationship between early angiosperms that dehisced by pores, and primitive bees. The plant family Melastomataceae appears by the Eocene and was probably poricidal then, although this is not apparent in the fossils which are lacking flowers. The oldest bees also appear in the Eocene (F. M. Carpenter, pers. comm.), but these are eusocial apids such as *Trigona,* indicating that bees surely had a longer fossil history. This author, at least, believes that the bees departed from their Sphecid ancestors early in the Cretaceous. The evolutionary origin of the angiosperms is equally obscure, with one author (Axelrod, 1952) proposing an even earlier origin (Jurassic) than the commonly held Cretaceous origin.

A total of 27 orders within the class Angiospermae (or division Anthophyta) were found to contain at least some apically dehiscent members. Within the subclass Dicotyledoneae, a total of 24 poricidal orders were found to be distributed in the following manner: superorder Magnoliidae (Laurales, Ranunculales); superorder Dilleniidae (Capparales, Dilleniales, Ebenales, Ericales, Lecythidales, Malvales, Primulales, Theales); superorder Rosidae (Euphorbiales, Fabales, Geraniales, Myrtales, Polygales, Santalales, Sapindales); superorder Asteridae (Gentianales, Lamiales, Polemoniales, Rubiales, Scrophulariales). Within the subclass Monocotyledoneae, 3 orders were found: superorder Lilidae (Liliales); superorder Commelinidae (Commelinales); superorder Arecidae (Arales, poricidal but not buzzed). In all, the dicots accounted for a total of 24 orders, 64 families, 473 genera, and about 19,000 species worldwide. The monocots were represented by 3 orders, 8 families, 71 genera, and about 1,150 species. Thus, poricidal dehiscence of some sort (true or valvular, etc.) was found in 27 angiosperm orders, 72 families, 544 genera, and some 15,000 to 20,000 species on a worldwide basis (see Fig. 4-3 for the orders that contain at least some porose members). These figures again represent the total number of taxa presently known to be poricidal, and not the number of plant families, genera, and species that are buzz pollinated. I assume that only 5 to 10% of the por-

Table 4-2. List of poridically dehiscent angiosperm families, including families with at least one taxon dehiscing by pores, apical slits, or valves (M = monocot, D = dicot), as recognized by Buchmann (1978).

Acanthaceae	(D)	Lythraceae	(D)
Actinidiaceae	(D)	Malpighiaceae	(D)
Aextonicaceae	(D)	Mayacaceae	(M)
Araceae	(M)	Melastomataceae	(D)
Atherospermataceae	(D)	Mendonciaceae	(D)
Balanophoraceae	(D)	Monimiaceae	(D)
Begoniaceae	(D)	Myoporaceae	(D)
Berberidaceae	(D)	Myrsinaceae	(D)
Bixaceae	(D)	Myrtaceae	(D)
Boraginaceae	(D)	Myzodendraceae	(D)
Byblidaceae	(D)	Olacaceae	(D)
Canellaceae	(D)	Ochnaceae	(D)
Clethraceae	(D)	Pentaphylacaceae	(D)
Cochlospermaceae	(D)	Pittosporaceae	(D)
Commelinaceae	(M)	Podophyllaceae	(D)
Cyrillaceae	(D)	Polygalaceae	(D)
Dilleniaceae	(D)	Pontederiaceae	(M)
Dipterocarpaceae	(D)	Primulaceae	(D)
Ebenaceae	(D)	Pyrolaceae	(D)
Ehretiaceae	(D)	Rafflesiaceae	(D)
Elaeocarpaceae	(D)	Rapateaceae	(M)
Epacridaceae	(D)	Resedaceae	(D)
Ericaceae	(D)	Roridulaceae	(D)
Euphorbiaceae	(D)	Rubiaceae	(D)
Fabaceae	(D)	Rutaceae	(D)
Flacourtiaceae	(D)	Saurauiaceae	(D)
Gentianaceae	(D)	Scrophulariaceae	(D)
Gomortegaceae	(D)	Sladeniaceae	(D)
Gyrocarpaceae	(D)	Solanaceae	(D)
Haemodoraceae	(M)	Sterculiaceae	(D)
Hernandiaceae	(D)	Strychnaceae	(D)
Iridaceae	(M)	Theaceae	(D)
Krameriaceae	(D)	Thunbergiaceae	(D)
Lauraceae	(D)	Tiliaceae	(D)
Lecythidaceae	(D)	Tremandraceae	(D)
Liliaceae	(M)	Viscaceae	(D)

Totals = 64 dicot families, 8 monocot families, 72 families in all.

icidal plants are exceptions to the rule and are not vibrated by bees.

A biogeographic analysis of the co-distribution, if one exists, for bee taxa and poricidal plants, will be difficult to document. Areas with high bee and porose plant species diversity and richness may be fortuitous, or may represent real coevolved associations. It will not be easy to differentiate plant associations with numerous buzz flowers from areas that have inherent plant species richness.

A recent review (Michener, 1979) showed that bees realize their greatest abundance and diversity in warm temperate, xeric regions, especially the Mediterranean basin, the Californian region, and adjacent desert areas. Some dry areas, like central Chile or the western part of southern Africa, have less rich faunas. Arid tropical regions, as well as tropical savannas, have relatively depauperate bee faunas. Warm temperate, mesic areas, such as found in eastern North America and Europe, also have rich faunas, although not as rich as the southwestern United States. The moist lowland tropics (both neo- and paleotropics) vary considerably in bee abundance and diversity.

Thus, unlike many taxa that abound in the tropics, bees attain their greatest species diversity and richness in warm temperate areas.

PORICIDAL PLANTS THAT ARE NOT BUZZ POLLINATED

If a flower has poricidal stamens and light, dry pollen, the probability is very great that it will be buzzed by bees in its native habitat. However, there are exceptions to buzz pollination in a few families and genera with apical pores.

Not all plants with poricidal dehiscence are pollinated in a vibratory manner by bees. Usually these taxa dehisce by means of apical slits, pores, or valves, but some features (usually pollen made sticky by the presence of excess pollenkitt on the exine surface) preclude them from being vibrated by pollinating bees. The number of these poricidal but nonvibratile plants is small (probably only 5–10% of the total number of porose species).

The largest taxon that is an exception is the monocotyledonous family Araceae, with about 115 genera and 2,000 species. Dehiscence is by means of longitudinal slits or by poricidal means in an undetermined number of genera. In these forms, each locule dehisces by an irregular apical slit, or in a few cases, through an elongate tubular connective process. The pollen is free or conglomerated in a vermiform column (Harris, 1905a). Thus, the pollen is made sticky by pollenkitt, and would not be released by vibrating the anthers. In addition, where the pollen vectors are known for aroids, they are usually small flies or beetles. The only known records of bee visitations to araceous genera are by males of some euglossine bees (Apidae: Euglossini), which visit them to collect floral volatiles.

The next taxa that must be considered poricidal but non–buzz pollinated are the 12 families and approximately 69 genera that are valvularly dehiscent by means of valves or lids. In most of these families, the flowers are not showy (except for the Berberidaceae) and do not possess the necessary floral characteristics usually associated with the syndrome of melittophily. Their pollen grains are normally sticky, presented in masses, and usually attached to the undersides of the valves, and thus available for transport by any generalized floral visitor.

In the Ericaceae, the genus *Rhododendron* has two, large, rounded apical pores per anther but is not buzz pollinated, or only rarely so. The pollen is present in tetrads and is attached in long sticky strings known as viscin threads, again making vibration an inefficient mode of pollen extraction for bees. The approximately 600 species in this genus are largely visited and pollinated by bees, moths, and birds, especially in the Australasian region of the paleotropics. The author has, however, observed bumblebees (*Bombus* spp.) buzzing flowers of *Rhododendron* in California. This vibratory technique was a seemingly inefficient method of pollen harvesting on these flowers, as judged by the long time spent working individual flowers and the small pollen loads collected by bees.

In the genus *Krameria* (Krameriaceae), the anthers are dehiscent by means of jagged pores. However, the pollen is extruded in large sticky masses, and the flowers are not usually visited by bees actively seeking pollen (are visited by oil-collecting anthophorid bees instead).

The parasitic flowering plants (families: Balanophoraceae, Myzodendraceae, Rafflesiaceae, Rapateaceae, and Viscaceae) also possess some poricidal members, but their minute flowers and sticky pollen would likely preclude vibration by bees. Pollination information is scanty or lacking for most of these groups.

Lastly, the small aquatic family Mayacaceae should be discounted as a vibratile taxon. Even removing these families from the vibratile taxa in the poricidal table (Table 4-2) still leaves 52 families as probably having at least some buzz pollinated taxa, on the basis of their floral morphologies and associated features. It remains, however, for many of these poricidal familes and genera to be documented as such by critical field observations.

POLLEN AS AN ATTRACTANT AND POLLEN ODOR

Early in the evolution of the angiosperms, pollen certainly must have been the original attractant and reward for anthophilous insects

(Faegri and van der Pijl, 1971), Even prior to the evolution of the flower, plant microspores, or fungal spores, etc. were probably exploited by early insects as high-protein, energy-rich food. Today it is likely that the pollen of many tropical cycads is readily eaten by beetles, considered by some floral biologists to be the legitimate cycad pollinators (most are cited as wind-pollinated), as well as the most likely candidates, along with flies (see Thien, 1980), as the earliest pollinators of flowering plants.

Pollen, as a floral "reward," is apparently more selective than nectar in attracting insects, especially among the Apoidea. Female bees usually exhibit a high degree of fidelity (termed oligolecty or -trophy for utilization of a few related species, or monolecty in the extreme case of one floral source) to one pollen source, at least on a given trip. Temporally, as plants come into bloom or fade out, long-lived bee species or those that are multivoltine must track the floral resources and will shift to other species that satisfy their nutritional requirements. The honeybee is a good example of a polylectic species, and undoubtedly has the greatest diet breadth of any bee. Shorter-lived oligoletic bees have specific pollen "preferences," possibly based on chemical phagostimulants including pollen volatiles, and/or upon floral odor adsorbed by the pollenkitt, or upon some type of larval conditioning to the proper pollen during larval growth and feeding upon the pollen provisions (sometimes termed the "Hopkins Host Selection Principle"). One author (Schmalzel, 1980, unpub. ms.) has hypothesized that amino acids in pollen may act as phagostimulants alone or in concert with other substances for the prey recognition of pollen by *Apis*. Other authors (Lepage and Boch, 1968; Hopkins et al., 1969) feel that some types of volatile lipids are the key phagostimulants for pollen recognition as food by bees. However, Loper (pers. comm.) bioassayed synthetic octadecatrienoic acid (*cis*-9, 12, *trans*-2) as identified by Hopkins et al., (1969) and found no honeybee phago- or olfactory stimulation.

Whatever the key phagostimulants for pollen recognition are, bees must obtain some chemosensory and mechanical feedback from setae and sensillae on their appendages. This information would then initiate a behavioral sequence for grooming the pollen from body areas and packing it into appropriate scopal transport devices, rather than grooming it off the body.

Nectar, however, is widely exploited by many insect orders, and both male and female bees usually collect nectar (for maintenance and an additional carbohydrate source in pollen provisions) from a wider array of plant species than they do for pollen. Pollen, when functioning as an attractant/arrestant is generally well exposed and available for rapid collection by generalized and specialized insects. This is in contrast to most nectar, which is often secreted in concealed nectaries, and as such is accessible to specialized visitors only. An analogous situation is seen in flowers with poricidal anthers, where the abundant pollen (except where pollen grain number has undergone reduction) is usually well hidden and available only to specialized bees that have learned to vibrate the flowers, or to bees that glean spent grains after vibration by other bees, or bite at the anther apices to release pollen. Thus, the pollen is not seen by the bees unless some is exposed at the pores. Most, if not all, porose flowers bear anthers that are large, turgid, often bright yellow and appear as if they contain their full complement of pollen, even if they have been visited and are empty. Thus, deceit (the appearance of plump apparently pollen-filled anther locules) in pored anthers likely plays an important role in attracting pollinating bees and directing insect movements. Such "dummy anthers" would result in more geitonogamy or xenogamy (Frankel and Galun, 1977) by deceiving incoming pollen-laden bees into buzzing spent flowers, and thereby moving more pollen from flower to flower (see Vogel, 1978).

Many workers have observed that certain pollen has a specific odor different from, or the same as, the overall floral scent (due to osmophores on the perianth; see Vogel, 1963). That such a specific pollen odor in pollen-only flowers might lead bees at close range to the pollen, much as visible nectar guides function, is a possibility. Some workers (von Frisch, 1923; von Aufsess, 1960) have demonstrated experimentally that honeybees are able to differentiate between pollen odor and the general floral scent, and could be conditioned to each separately. It is very likely that other species of bees possess similar discriminatory abilities. Other cases of (undoubtable) pollen odor have been described by Porsch (1954), and the author has noted strong pollen odor in species of *Dodecatheon* (Primulaceae), and some *Solanum* spp. (Solanaceae), which was similar to the overall floral odors. At least one melastrome, *Bellucia imperialis,* has pollen with a very distinct odor (S. Renner, pers. comm.). Clearly, the odors from isolated pollen grains (easy to isolate and usually abundant in most buzz flowers) should be compared by headspace capillary gas chromatography to the volatiles produced by the osmophore areas of the floral perianth. Only then can their volatile constituents be analyzed quantitatively. Louveaux (1960) further suggested that pollen contains phytosterols that may "attract" insects by virtue of the odor that they may impart to pollen grains. In the final stage of pollen wall development, lipoidal and other substances may be accumulated on the exine surface. This material, which has been collectively called pollenkitt by Pankow (1957) and Knoll (1930) imparts color, odor, and varying degrees of stickiness due to the relative abundance of lipids present. The pollenkitt would also be the first layer on the pollen that a pollinator would contact (except for components such as amino acids that can rapidly diffuse out of pollen grains), and thus should contain at least some phagostimulants.

Finally floral odor, often very strong (as in some *Solanum* spp. and *Cassia* spp.), might serve to reinforce the pollen deceit strategy employed by poricidal flowers. Thus, a brightly colored, fully turgid buzz flower with plump yellow anthers may contain no pollen (previously harvested by earlier bees), and yet succeed in acquiring the services of bees as pollen vectors.

FLORAL AND BEE COADAPTATIONS FOR BUZZ POLLINATION
Plant Adaptations

The floral adaptations for pollination really delimit vibratile pollination. Poricidal flowers are generally small or average-size flowers that are usually produced near stem tips on shrubs, or on the outer canopy edge in trees. Some tropical forms are true giants, however; flowers such as *Tibouchina* (Melastomataceae), *Amoreuxia,* and *Cochlospermum* (Cochlospermaceae) may be several inches across. The plants may occur in a clumped distribution, or may be at infrequent intervals (a "trapline distribution"). Floral anthesis usually occurs before or at sunrise, and the flowers are freshest (pollen available and stigma receptive) during the early morning hours. Toward the end of their floral life history the blossoms begin to desiccate, wilt, fade in color, and drop from their pedicels. Dramatic post-pollination or induced floral color changes, as in caesalpinaceous legumes (banners changing from yellow to red) are rare here. *Borago* (Boragainaceae) does have blue flowers when fresh that turn pink as they age. The floral ontogeny is short, generally limited to one to two days. In some genera, notably *Solanum,* a few new flowers may be produced by the plant during the day instead of all being produced early in the morning. Most buzz plants produce few to many flowers per plant each day over a long blooming season that may last weeks to several months. Most of these plants are very dependable sources of pollen for bees, since they present many flowers over an extended period of time. The reproductive biology of poricidal plants is very diverse and complicated, and no generalizations can be made at this time. For an excellent discussion of the reproductive biology of a very large genus, the reader is directed to Symon (1979), who discusses sexual forms (e.g., hermaphrodite, monoecious, dioecious, andromonoecious, and androdioecious) in Australian species of *Solanum.* This is an excellent genus for the study of reproductive biology, since it has evolved innumerable sexual forms and is evolutionarily very successful (the second largest genus of flowering plants with over 2,000 described species). Floral symmetry is usually radial (actinomorphic), especially in the "*Solanum*-type" flowers as in *Solanum* and many Melastomataceae (*Melastoma* and *Miconia*). The petals are free or partially united, and rarely fused into a floral tube, as typified by the urceolate corolla of *Arctosta-*

phylos in the Ericaceae. Here the ovary is superior and the anthers are firmly or loosely united, connivent, around the slender central style. The characteristics of the androecium really set these flowers apart from all others. Occasionally zygomorphic, bilateral flowers are found, as in some species of *Cassia*. Even more highly derived flowers have evolved from simpler bilateral forms (e.g., forms exhibiting some degree of torsion as in some large melastome flowers and *Amoreuxia* in the Cochlospermaceae). If the stamens range from five to ten they will be in the form of an anther cone, or else there may be hundreds of stamens grouped in the form of a "shaving brush." Some tropical forms (*Bixa* in the Bixaceae) exhibit this androecial type. Bees working these flowers grasp groups of stamens rather than individual anthers, buzz, then move on to grasp another bunch intraflorally, or fly on to another flower. Stamens within the androecium may be grouped into functionally different sets. These sets can be nearly equal (as in *Amoreuxia* with one-half long and one-half short stamens in two groups), or can involve drastically different anatomical differences and two to three subsets within the androecium (e.g., some *Solanum* spp. *Cassia,* and many melastomes). This androecial dimorphism is termed heteranthery and along with enantiostyly (in which the style is deflected to the right or left of the floral axis opposing the stamens, as in some *Solanum, Cassia, Wachendorfia*) will be discussed in a later section. The pollen itself may be dimorphic, as will be discussed later.

Generally, the flowers are showy with "lively," highly saturated contrasting colors. The most prominent feature of the floral color in the *Solanum*-type (*sensu* Vogel, 1978) is a whitish, blue, or purple corolla contrasting dramatically with the central bright yellow anther cone. This is also common in many melastomes. Even in white flowers, the anthers are usually a vivid yellow. Among some *Cassias* a petal (the "cucullus") has become modified into a toughened and strongly curled petal that is thought to further direct bees into the androecium (Thorp and Estes, 1975). Another feature of floral color, especially in *Solanum,*

are the pseudonectaries. These are modified areas of perianth tissue at the very base of the petals surrounding the anther cone. They are usually bright green with a shiny surface (flattened epidermal cells and lack of papillae), and have a thin whitish ring of color before the overall corolla color takes over. These may function as false nectaries stimulating bees to probe for nectar that is not present (insects are rarely observed to probe at these), or they may function as pollen guides directing bees to the central stamens.

Yellow or yellow-orange is the predominant color for pollen, owing to carotenoids and lipids within the grains and dissolved in the pollenkitt (Knoll, 1930; Pankow, 1957). Very rarely is angiosperm pollen bright blue, pink, red, or green. Pollen is often whitish or cream-colored. This is the case in poricidal flowers, where the pollen is usually a light cream color and the external locule walls are a bright yellow. Thus the anthers are mimicking abundant pollen even while empty, functionally acting as "pollen dummies" (see discussion on floral deceit by buzz flowers). The pollen is hidden from view for legitimate buzzing bees in addition to being relatively safe from discovery by pollen thieves. Pollen is very difficult to remove from porose flowers unless they are vibrated. It has been suggested that since pollen can be exposed for long periods in flowers with longitudinal dehiscence, the yellow color due to carotenoids evolved as shielding pigments to protect the tube and generative nuclei from the damaging effects of ultraviolet radiation (Stanley and Linskens, 1974). It is suggested here that most pollen in poricidal forms is whitish because it remains hidden and protected from UV radiation inside the anther locules, thus not needing the protective shield of carotenoid pigments.

The flowers are usually distinct enough, with contrasting colors in the visible spectrum (400 to 700 nm), that no reinforcing ultraviolet floral patterns are needed to direct insect visits. Fairly strong ultraviolet floral patterns have been found, however, in forms like *Solanum* with green and white pseudonectaries against a blue or purple corolla (Buchmann, 1978).

Floral scents produced by volatiles (usually terpenes, sesquiterpenes, etc.) in specialized osmophores (Vogel, 1963) are usually present in these flowers. Floral aromas have been stated to be absent, weak, and strong in buzz flowers. Volatiles produced by flowers such as *Amoreuxia, Cassia, Solanum,* and many melastomes can be quite strong and sweet-smelling, as typical for most melittophilous blossoms. To date, none of these scents has been characterized chemically. To determine whether a particular species has a floral scent, the blossoms may be placed in a clean, closed glass jar in the sun. The headspace odor is allowed to develop and is sniffed later. This technique works elegantly, and the author has determined that many "odorless" flower claims in the literature are not substantiated.

As mentioned, *ad infinitum,* these are pollen-only flowers with few to many stamens that dehisce in a poricidal fashion by apical slits or true pores. The anthers are usually basifixed, not versatile, and the thecae may be elongated. Floral nectar is almost never produced. In some ericads (e.g., *Arctostaphylos* and *Vaccinium*) floral nectar is produced as an exception to the general rule of producing only pollen. Extrafloral nectar is a common feature in some tropical species of *Cassia.* When present, floral nectar conforms to expected qualitative and quantitative values for sugars, amino acids, etc., that have been determined for other bee-pollinated plants.

Pollen grains may be more abundant per flower or per anther than in nonporose relatives, as in some *Solanum* spp. and melastomes, or may have undergone a drastic reduction in number per locule (as in those flowers practicing heteranthery and floral deceit—again *Solanum* spp., *Cassia* spp., etc.). In cases of actual pollen "superabundance," this may just reflect the physical ability to pack more small pollen grains into a given volume than would be possible with larger-diameter grains. Until actual pollen counts per anther and per flower, along with pollen:ovule ratios, are made, the distribution of pollen abundance or reduction will remain unknown.

Monad pollen (usually single tricolporate grains) is the usual type produced by poricidal plants, although tetrads, even with sticky viscin threads, occur in the Ericaceae and Pyrolaceae. A remarkable convergence in the form and size of pollen in vibratile pollinated plants has occurred among phylogenetically distantly related genera and families across nearly all the angiosperm order. Pollen from vibratile pollinated plants is characterized by its small size (range about $5-40\ \mu$, mean about $25\ \mu$ in diameter as opposed to the $5-210\ \mu$ range and mean of $34\ \mu$ as given by Roberts and Vallespir, 1978, for angiosperm pollen in general), relatively dry surfaces (due to small amounts of pollenkitt), and smooth exine with little sculpturing (Buchmann, unpub.). Pollen grains $\geq 100\ \mu$ and $\leq 10\ \mu$ constitute less than 4% of the total size range for angiosperm pollen (Roberts and Vallespir, 1978). This is especially noticeable in families not usually characterized by small dry grains with featureless exines. Thus the pollen is small and dry and doesn't stick to itself readily or clump, but it can be tightly bound by electrostatic interactions. Dry grains are a necessity for pollen expulsion from the apical anther pores by bee vibrations. Sticky entomophilous pollen forms clumps, and thus cannot be removed effectively by vibration. The small size of these "pollen-only" buzz grains is probably related, in those where they are superabundant, to the need of producing enough pollen grains that some fraction of them escape ingestion or collection and grooming into scopae by the pollinating bees. The largest proportion of pollen grains groomed from the bees' body and packed into scopal transport devices (see Michener et al., 1978; Roberts and Vallespir, 1978; Thorp, 1979) are utilized by the bees for imaginal and larval food, but for the plant are lost to potential pollination. A few viable grains remaining after bee grooming are the pool of grains available for pollen transfer. The necessity to produce an abundance of pollen is obviously offset by those with reduced pollen (those practicing deceit), and by the fact that buzz pollination is a very effective mode of pollination, wasting little pollen. This efficiency is largely due to the filiform nature of the style, and the fact that the stigma is normally crammed into areas on the bees that are not easily groomed by the bees, i.e., "safe sites" for pollen. These areas occur in the proboscideal fossa, between the coxal bases, and on the thoracic dorsum in nototribic forms. Pollen delivery is usually sternotribic or pleurotribic in poricidal plants.

The pollen chemistry in these flowers has just begun to be examined. Since these grains are usually small, the energy reserve product within them is lipid, and they are lipid-rich (internally, but not with surface pollenkitt). They also contain little or no starch and appear to represent selection in favor of starchless (oil-rich) pollen grains. Baker and Baker (1979) stated: "The tendency to starchlessness and small pollen grain diameter is at an extreme in those species where the flowers are buzzed through small pores by visiting bees." Studies of other aspects of buzz pollen chemistry including sugar and starch analyses, lipids, amino acids, protein content, and calorific content are under way (Buchmann, unpub.) and promise to be just as exciting as nectar chemistry. Buzz grains appear to be high in calories and protein especially. At least several species of *Solanum* have total crude protein levels of about 60%. This is extraordinary because, to date, the highest crude protein value for any pollen was ca. 40% in a paniculate *Agave* (Agavaceae).

Recently (Buchmann and Buchmann, 1981), a new floral reward for poricidal flowers was discovered in the tropical tribe Memecyleae (Melastomataceae). Flowers that produce lipids in specialized glands (elaiophors) are well documented (Vogel, 1974), but were unknown in the melastome family. These authors studied *Mouriri myrtilloides* in Panama. Buzzing bees were represented by *Euglossa* (Apidae), *Paratetrapedia* (Anthophoridae), and *Trigona* (Apidae) which were pollen scavengers. A deeply concave elaiophor was discovered on the dorsal aspect of the anther connective (see Fig. 4-4C). A viscous yellow oil was produced that contained a rich mixture of components: at least 13 fatty acids (including methyl stearate and 12-methyl tetradecanoate), glucose, amino acids (alanine, glycine, proline), carotenoids, an auto-fluorescent green phenolic, and possibly saponins. This complex floral oil was collected by *Paratetrapedia* and *Trigona* (*T. pallens*) females. On the basis of the presence of elaiophors in the tribe (*Memecylon, Mouriri,* and *Votomita*), this specialized melastome tribe should now be considered true oil-flowers in addition to being buzz pollinated. To date, this is the only known case of floral lipid production (via elaiophors) by a poricidal flower.

Bee Adaptations

Bees using floral vibration can be small to large in size. They occur in 6 of the 11 bee families within the superfamily Apoidea. The major family not involved is the leafcutter bee family (Megachilidae). Buzzing is poorly developed in the Andrenidae (found only in *Protandrena,* see Fig. 4-2), and information is lacking on the Stenotritidae (Stenotritinae), Ctenoplectridae (Ctenoplectrinae), and the Fideliidae. Thus, the only feature that universally characterizes all of these bees is the possession of well-developed indirect flight muscles and the physiological/behavioral ability of individuals to learn to buzz flowers. Only female bees visit the flowers, since pollen is almost always the only reward, and male bees do not seek out pollen. These females are relatively constant to one pollen source on a given foraging bout, but may be quite catholic in their overall diet breadth (see Schmalzel, 1980, for a discussion of diet breadth).

These bees are usually active in the early morning (matinal) when anthesis occurs in most buzz flowers and may be active again in the late afternoon toward dusk (crepuscular). This is especially true in xeric regions in the American Southwest, where, in the hot midday, bees could easily become stressed for water or undergo problems with thermoregulation.

Intrafloral buzzing behavior changes during the course of the day. When the first bees arrive at the pollen-rich flowers at sunrise, they land on individual flowers and buzz briefly (about 0.1–1 second). These short buzzes are sufficient to allow the bees to work fewer flowers to obtain a full load of pollen. As the standing crop of pollen declines toward midday, owing to frequent bee visits, the buzzes gradually lengthen (to about 1–8 seconds), and the bees have to work many more flowers to obtain the same amount of pollen (per unit time) that they collected during the few hours following sunrise. Early in the morning the floral vibrations are typically single buzzes, and the bees usually stay at one position on the flowers. Later in the day, with less available pollen, bees will use long trains of multiple buzzes, while at the same time rotating on the flower. Presumably this results in the most efficient extraction of the last few remaining pollen grains within each locule. This behavior is enhanced by the floral deceit (i.e., plump yellow anthers) played by the plants, resulting in bees vibrating full or empty anthers. Clearly this results in a greater amount of pollinator movement, and thus higher levels of pollen transfer (geitonogamy and xenogamy).

The pollen is usually transported to the nest dry in the scopae, its adhesion aided by electrostatic forces (Erickson, 1975; Erickson and Buchmann, this book). Some bees, especially *Bombus* species, will admix nectar with the buzzed pollen and carry it doughlike back to the nest. Since the nectar is not derived from the same flowers as the pollen, bees must apportion their foraging bouts to include interspersed trips for nectar, or a terminal nectar trip just before the trip home (also see a later discussion of "carryover phenomenon"). The plants chosen as nectar sources vary widely for both male and female bees.

Female bees in many diverse families and genera all have evolved similar morphological adaptations of the scopae which allow them to collect and transport the small, dry pollen grains these plants offer. Their scopal devices, usually on the metathoracic legs, are clothed with fine, dense plumose setae with interstices about 10–20 μ in size, closely matching the diameter of vibratile-released grains that are carried (Linsley and Cazier, 1972; Roberts and Vallespir, 1978; Thorp, 1979). The fine setae are also a favorable environment enhancing the electrostatic hold on the pollen grains.

During a foraging bout, or at its end, a female may elect to groom the pollen from the body setae and pack it into the scopa. This may be done while hovering near the flowers,

as in some *Anthophora,* or more often while the bee hangs from the flowers by the mandibles and a prothoracic leg. Not all areas on the bees are within reach of grooming movements. These may be termed "safe sites" for pollen residence. They occur on the occiput of the head, in the proboscideal fossa, on the centerline of the thoracic dorsum, and between the coxal bases. Pollen lodged in these safe sites represents the pool of pollen available for pollination, and therefore ultimate fertilization. Most poricidal plants have exserted slender styles with small capitate stigmas. On subsequent floral visits, the stigma is often pushed into the coxal safe site, thus effecting efficient sternotribic pollination. Pleurotribic pollination is the most common form of pollen delivery, but it can also occur pleurotribically, as in enantiostylic forms, or nototribically in other pored flowers.

Three categories of visitation types for bees that visit buzz flowers were given earlier by Wille (1963). These categories are: (1) the buzzing bees, (2) the biting bees, and (3) the gleaning bees. Wille was studying the anthecology of a tropical senna, *Cassia biflora,* in Costa Rica when he categorized the bees in this way. The breakdown is still a valid one. Legitimate buzz pollinators using vibration and producing audible buzzes fall into the buzzing bee class. Gleaning bees are usually smaller bees, oftentimes halictids and honeybees, that are incapable of buzzing or don't buzz the flowers. They simply scrabble around the anthers and corolla, picking up pollen grains that fell to the floral surface during previous buzzes by other bees. Sometimes a further distinction is made, and certain gleaning bees are said to be "milking" the anthers. In the neotropics the gleaning bees using milking are often social bees of the genus *Trigona,* the omnipresent pollen scavengers in tropical habitats. These bees will grasp the base of the anther with their mandibles and try to squeeze pollen out of the pores, or bite directly into the anthers. Both techniques are slow and sloppy but do work to some extent. Biting bees may include *Trigona* and *Xylocopa,* which can do great damage to anthers, literally chewing them off in their search for the hidden pollen.

BEHAVIORAL USES OF VIBRATION BY ACULEATE HYMENOPTERA

The Hymenoptera originated at some time in the Mesozoic Era during the late Triassic about 180 million years ago. These earliest Hymenoptera, suborder Symphyta, were ancestors of the modern sawflies and related groups. They possessed a broad attachment of the abdomen to the thorax, and were already plant feeders and pterygote (winged) insects with complete metamorphosis.

The large indirect flight muscles of many bees and wasps have, in the course of their evolution, acquired special social significance (Michener, 1974; Buchmann, unpub.). Vibrations in Hymenoptera can be produced directly (i.e., under the insects' control), or as a by-product of some behavior. Vibrations resulting as a by-product of other behaviors would include those from beating wings in flight, while shivering, or from tapping body appendages. Strong vibrations result from the contraction and relaxation of the large indirect flight musculature in the pterothorax. When the large muscles function, they contract many times per second, slamming the cuticle into the boundary layer of air, or against any substrate contacting the thorax. These strong thoracic vibrations produce loud audible buzzing (due to slapping of cuticle against air, not motion of attached wings), as a by-product of their contraction. Direct vibrations can be produced by "intentional" tapping of body parts against substrates, rubbing a file across a scraper (stridulation), or shivering the flight muscles. These vibrations from thoracic muscular contraction can serve many functions, such as producing heat for thermoregulation (Heinrich, 1974); incubating brood (e.g., *Bombus*); producing "alarm sounds or buzzes" to frighten potential predators or copulation sounds during mating and courtship; aiding in nest or burrow construction; thermoregulatory fanning (e.g., *Apis*); communication (acoustical component of waggle dance in *Apis*); or vibrating flowers.

Vibration serves Hymenoptera as an aid in loosening and manipulating dry soil particles during nest construction (Spangler, 1973).

Ants, bees, and wasps produce mechanical vibrations by manipulating body parts (stridulation), tapping the substrate, or contracting the flight muscles to produce strong vibrations and audible buzzes. It is usually these vibrations passing to or through a solid substrate contacting the insect, and not their airborne audible (to humans) component, that play a role in insect behavior. It is believed that most Hymenoptera are deaf to airborne sound, but are especially sensitive to substrate-transmitted vibrations.

The most common vibrations produced by ants, velvet ants (Multillidae), and certain wasps are those produced by stridulation. Stridulation is the act of rubbing a hardened scraper over a filelike surface. The frequency of the vibration depends on the rate of the scraper hitting the file and the resonant frequency of the specific insect part, or of the entire insect. These are loud stridulations, presumably produced to deter predators (Eisner, 1980). For an extended discussion of stridulation in ants see Spangler (1967). Audible "alarm sounds" are also produced by the flight muscles in many bees and wasps when they are pinched or otherwise restrained. They can be loud, and would transmit strong vibrations to any predator handling them. Experiments should confirm the selective value of these adaptations (combination of loud sound and strong pulsed vibrations to the sensitive hands or mouths of predators) to frighten predators into quickly releasing the offending insects. Interestingly, these sounds are produced readily, along with stinging motions, by male bees and wasps in addition to females. This behavior has resulted in the author dropping harmless male bees on many occasions.

Vibration is probably used more for the manipulation (loosening and settling/tamping) of soil particles by bees and wasps during burrow and nest construction than for any other purpose. Here, again the vibrations are produced as a by-product of the operation of the flight muscles. Ground-nesting wasps (e.g., *Ammophila* spp.) commonly use vibrations when closing their nest holes (Powell, 1964; Menke, 1965). They can also be heard buzzing to manipulate the soil while deep in their burrows.

These strong vibrations pass from the pterothorax to the whole body, including the mandibles. While holding a pebble or soil clod, the *Ammophila* mandibles function as a miniature "air hammer" to work the soil.

Mud dauber wasps (e.g., *Sceliphron*) produce audible sounds and vibrations while applying fresh mud to their nests (Evans and Eberhard, 1970). The function is unknown but it probably has the effect of rapidly smoothing out the mud ball over the fresh nest surface, and settling out the soil particles by size (Buchmann, unpub.). This presumably makes the nest stronger and less susceptible to crushing or break-in by predators than if no vibration were used during mud application.

Audible sounds are sometimes produced by bees during courtships or copulation. These relatively loud, interesting sounds appear to be produced by flight muscle buzzes in some species and stridulation in others. This was first observed in an anthophorid bee, *Centris palida* (Buchmann and Alcock, unpub.). The first published account of courtship stridulation sounds was by Rozen (1977), who described a South African bee (*Meganomia binghami*, Melittidae) that produces a rasping squeak during copulation. The pregradular stridulatory area is located on the side of male and female metasomal terga IV, V, and VI. Copulatory stridulation has also been documented in another anthophorid bee *(Diadasia rinconis rinconis)* by Buchmann (unpub.). This form of communication among mating bees may be found to be widespread, as their mating strategies become better known. The airborne component of these copulatory buzzes or squeaks probably doesn't have much biological meaning. The strong vibrations passed to the female are probably the communicative mode between the two sexes, if such a communication channel exists. The vibrations would likely have a quiescent effect on the female, thus allowing the male a successful intromission.

Thus the powerplant, indirect flight muscles, to drive the wings during flight was already present in the earliest Hymenoptera 180 million years ago. This musculature was a preadaptation for floral buzzing and all other behavioral uses for buzzing, and only needed to

be acted upon by natural selection during the long evolutionary history of the Hymenoptera. Many species may have a genetic component to buzzing under certain conditions (i.e., during alarm sounds or copulatory buzzes), while it is likely that the novel behavioral shift to the use of buzzing for pollen collection has been learned (in hose species physiologically capable of buzzing) during their first few floral encounters with poricidal blooms, and this must be learned anew each generation by every individual. There is, however, some controversy on this point.

In an early paper on buzz pollination in *Cassia* and *Solanum* (Michener, 1962), this subject was treated briefly. Michener states:

> It is not obvious what characteristics the flowers of *Cassia* and *Solanum wendlandii* might have that may assist in the acquisition of buzzing behavior by bees of different systematic groups and in different parts of the world. It does seem safe to say that only a small percentage of the individuals of these species of bees ever come in contact with flowers requiring this behavior for successful pollen collecting, and it is therefore evident that the buzzing behavior must be inherited merely as a potential activity, first stimulated, possibly, for each individual by the difficulty of obtaining pollen from the flowers concerned or by some other stimulus.

The floral cues that elicit vibratory behavior in *Solanum* types are probably the connivent arrangement of the anthers around the pistil (as in Fig. 4-6B). Since there are many porocidal flowers with anthers arranged in loose groups, the bees must be able instantly to recognize that the anthers are pored, requiring the bees to grasp the stamens firmly and buzz. To settle this larger question of "nature vs. nurture" of the buzzing itself would require detailed observations and natural experiments with just-emerged naive bees on buzz flowers in a large enclosure. Some data of this sort exist. During a study of the biology of the bee genus *Agapostemon* (Halictidae), Roberts (1969) made some observations of bees in an insectary as they worked *Cassia* and *Solanum* flowers. Roberts believed that the bees took a firm hold of the anther apices so as not to be dislodged by their own vibrations. He noted that bees often lose their hold and fall off the flowers. He also stated that the buzzing behavior is inherited rather than learned because inexperienced bees reared in the insectary went to the flowers of *Solanum carolinense, S. rostratum,* and *Cassia fasciculata* and began buzzing without hesitation. It is difficult to know if he observed the actual first few floral encounters with these poricidal flowers. Results of other observations on naive bees experiencing poricidal forms for the first time are not so conclusive. Bumblebees *(Bombus edwardsii)* were reared in a large outdoor insectary, and fed on sugar syrup and given longitudinally dehiscent flowers *(Senecio)* to feed upon (Buchmann, unpub.). When offered pored flowers of *Solanum parishii,* the bees readily landed on the flowers, and probed for nectar on these nectarless flowers time after time without using vibration. These observations were repeated on numerous individuals during hundreds of floral encounters, and buzzing was never seen. To complicate matters, flowers of *Senecio,* which were placed within the cage as a natural nectar source, were vibrated! The nearly closed flowers of this composite were repeatedly vibrated by the bees, though no pollen reward was forthcoming. The factors involved here are unknown, although this does indicate that naive bees have the capacity to learn the buzzing behavior during the first few floral encounters. Additional experimental observations on this topic are desperately needed before any firm conclusions can be drawn.

Honeybees, whether for some unknown physiological cause or a behavioral reason, have not learned to buzz flowers for pollen. They instead resort to the less time- and energy-efficient method of "gleaning" spent grains from the floral corollas that are left by legitimate buzz pollinators. This is difficult to understand, since *Apis* routinely utilizes vibration of the indirect flight muscles for communication (component of straight run of waggle dance), other substrate-carried sounds, endogenous individual and colonial thermoregulation, and communication between gynes in their cells ("queen piping"). However, during thousands of hours of observation by this au-

thor and others, *Apis* has never been observed to vibrate a flower to release pollen. Within the Apini, the genus *Melipona,* a relative of *Apis,* routinely uses vibration to harvest pollen. Other nonbuzzing bees include all of the Megachilidae, and most if not all of the Andrenidae (except *Protandrena = Psaenythia* which vibrates *Solanum* in Arizona; see Fig. 4-2B,C). Why these two large families have not utilized floral vibration as a pollen-harvesting technique is unknown. They will produce alarm sounds when held, and have large indirect flight muscles. The machinery is there, but the learned behavior is not developed.

Sometimes bees are observed to vibrate (producing an audible buzz or multiple buzzes) flowers that dehisce by longitudinal means. This is not very common, but has been recorded for several bee taxa working a few species of flowers. It was first noticed by Meidell (1944) that bees (females of *Megachile willoughbiella*) would vibrate the flowers of the nonporose *Melampyrum pratense* (Scrophulariaceae) while working the flowers for both pollen and nectar. Since no other megachilid is known to use vibration, this observation should be repeated. The author has seen species of bumblebees buzzing flowers of wild rose *(Rosa californica)* by grasping groups of hundreds of stamens which this flower offers in a dense mass resembling a shaving brush. The exact floral cues that function as stimuli eliciting buzzing behavior on these otherwise normal longitudinally dehiscent flowers are unknown. That the bees display such behavior suggests that it is an individually learned response early in the foraging ontogeny of those species, and also can be an efficient mechanism of pollen collection on nonporose forms where the stamens are tightly clumped. It would be interesting to determine if these *Bombus* could harvest more pollen per unit time or on fewer flowers than other bees not buzzing the wild roses.

One fascinating aspect observed in several bee species is the "carryover phenomenon" that occurs when females are buzzing for pollen on porose flowers, then abruptly shift, usually to begin a nectar-foraging bout on a nearby plant with normal longitudinal stomial dehiscence. At these times, the bees continue to buzz these nonporose anthers for a short time before the behavior ceases. This unusual behavior was observed in an earlier study of the floral biology of species of *Solanum* by Linsley and Cazier (1963). They stated it most succinctly: "The behavior pattern which results in vibration of the anthers of *Solanum* carries over to *Mentzelia,* where it is not obviously needed." Another case of carryover buzzing was observed by the author during the course of studies on the anthecology of *Solanum douglasii* and *S. xanti* (Buchmann et al., 1977). The bees were females of *Anthophora urbana* that were vibrating flowers of *Solanum* just after sunrise. After the females had buzzed from 50 to 100 of the nightshade flowers, they would abruptly fly to flowers of *Phacelia tanacetifolia,* which were growing intertwined among the *Solanums. Phacelia* flowers were utilized as one of the primary nectar sources, and a very minor pollen source for the *Anthophora* females. The bees would vigorously buzz the last few *Solanum* flowers, then continue to vibrate the first five to 10 *Phacelia* flowers they encountered before the buzzing damped off, and they began to probe for nectar and scrabble normally for the *Phacelia* pollen. Macior (1968a, 1974) also describes this carryover phenomenon for bumblebees buzzing flowers of *Dodecatheon,* and then also vibrating flowers of *Pedicularis groenlandica,* a nonporicidal species in Colorado, where the two plants are sympatric. To the best of my knowledge, these are the only known accounts of this interesting behavior in the literature. This carryover buzzing behavior should be looked for during all pollination studies of buzz plants, since it may help to explain the key stimuli that initiate and terminate the vibration of flowers for pollen.

DISTRIBUTION OF BUZZING BEES AMONG THE APOIDEA

Bees that routinely vibrate flowers are found in most of the major families, with the exception of the leafcutter bees (Megachilidae; one *Megachile* sp. may possibly vibrate). Why leafcutter bees do not buzz flowers is unknown.

There is presently no information on buzzing by bees in the small families Fideliidae, Stenotritidae (Stenotritinae), or Ctenoplectridae (Ctenoplectrinae). Some observations on Melittidae (*Macropis* sp.; Cane et al., unpub. ms.) buzzing deerberry *(Vaccinium stamineum)* flowers have recently been made.

Within the primitive Colletidae (mud or membrane bees), taxa that vibrate flowers can be found among the Colletini, Chilicolinae, Diphaglossinae. Specifically, the following genera are known to buzz poricidal flowers: *Caupolicana, Colletes,* and *Ptiloglossa.* The small family Oxaeidae, comprising about 20 species in 4 genera *(Mesoxaea, Notoxaea, Oxaea, Protoxaea),* is largely confined to the neotropics, especially in Mexico and South America. All of the known genera have been observed at poricidal flowers, and were vibrating them to release pollen (Hurd and Linsley, 1976). This is an especially interesting family from the standpoint of buzz pollination. The oxaeids, insofar as known, are narrowly polylectic, utilizing pollen from related plants in the Northern and Southern Hemispheres (amphitropical disjuncts), especially species of *Cassia* (Fabaceae), and *Solanum* (Solanaceae), and the caltrop family (Zygophyllaceae genera related to *Larrea).* Female oxaeids also posses especially dense scopae with finely branched plumose setae that have interstices about 10 μ apart (Roberts and Vallespir, 1978) with which to transport the fine-grained pollen they encounter while working species of senna or nightshades. Within the Andrenidae, the subfamily Adreninae contains the genus *Protandrena,* which has been observed taking pollen from *Solanum.* It is the only genus in this large cosmopolitan family known to engage in vibratile manipulation of apically dehiscent flowers. The sweat bees, or Halictidae, represent a large family in which buzzing is well developed in the following taxa: Halictinae (Halictini and Augochlorini), Nomiiae, and Dufoureinae. The genera known to buzz flowers include the following: *Agapostemon, Augochlora, Augochlorella, Augochloropsis, Nomia, Lasioglossum, Pseudoaugochlora,* and *Pseudoaugochloropsis.* Within the large family Anthophoridae, several tribes (Xylocopini,

Exomalopsini, Anthophorini, Centridini) contain genera and species that regularly buzz pollinate flowers. The following genera are known to vibrate flowers: *Amegilla, Anthophora, Centris, Epicharis, Exomalopsis, Svastra, Thygater,* and *Xylocopa.* More genera are likely to be added to this list. In the family Apidae, the tribes Bombini, Euglossini, and Meliponini all contain genera that regularly buzz and exploit poricidal angiosperms. The following genera are known to do so: *Bombus, Eufriesia, Euglossa, Eulaema, Euplusia,* and *Melipona.* Undoubtedly, other species and genera will be added to the known list of buzz pollinators. These records will have to come from new pollination field observations, since many of the older observations were not carefully made to ascertain whether the bees were vibrating the flowers.

VIBRATILE POLLEN HARVESTING BY SYRPHID FLIES

To date, flowers with poricidal dehiscence are known to be exploited only by female bees visiting them to collect pollen, with one notable exception. During the course of previous studies (Buchmann et al., 1977), it became apparent that something quite different was happening. While watching *Xylocopa tabaniformis orpifex,* a small black carpenter bee, vibrating the anther cones of *Solanum douglasii,* I realized that some of the visitors were not xylocopids at all, but instead were flies mimicking the carpenter bees. The small xylocopid is a legitimate buzz pollinator and is the model for its Batesian mimic, *Volucella mexicana* (Syrphidae). *Volucella* is a large, black sryphid fly that closely resembles *Xylocopa* in size, shape, color of body and wings, and in its nondipteran behavior. *Volucella* would visit hundreds of *Solanum* blossoms by grasping them tightly and vibrating the flight muscles to produce single or multiple buzzes that were about the same frequency as the carpenter bee buzzes. The floral buzzes produced by these flies efficiently released large amounts of *Solanum* pollen, which was readily seen on the venter of captured specimens. After buzzing the nightshade flowers, the flies moved to nearby mus-

tard inflorescences (*Brassica nigra* and *B. campestris*) where they probed for nectar, ingesting both *Brassica* nectar and *Solanum* pollen. Gut analyses confirmed the presence of small, smooth *Solanum* grains eaten by the flies. The ingestion of high-protein pollen by syrphid flies undoubtedly contributes greatly to their nutritional status and the ability of female flies to oviposit frequently.

Volucella mexicana may be considered a good Batesian mimic of *Xylocopa,* exploiting a novel food source while also receiving some protection from predation by looking, sounding, and otherwise behaving like a robust carpenter bee, equipped with a potent venom and long sting. Many genera and species of flower flies, Syrphidae, are found on flowers where they consume nectar and pollen. It would not seem too surprising, then, to find other examples of mimetic buzzing syrphids like *Volucella* on other poricidal plants. It would seem easy for these insects to utilize this preadapted buzzing behavior, enabling them to exploit a rich source of pollen previously unavailable to them.

MODIFICATIONS OF THE ANDROECIUM: HETERANTHERY, ENANTIOSTYLY, AND DECEPTION

The deceptive attraction of insects by flowers is found widely among angiosperms. In the past most of these examples have not come from pollen flowers, although deceit plays a large role here. A brief review of pollen flowers is in order. These are flowers that offer a "surplus" of pollen as the only reward to their pollinators, instead of nectar or other substances. They are mostly cup-shaped flowers that present their pollen in a showy androecium (collection of stamens) with the anthers and pollen advertising themselves optically at close range.

Recently Vogel (1978) proposed three broad categories for the classification of pollen flowers. It should be mentioned that these are of independent but parallel derivation in unrelated families, and of different evolutionary states and ages. Vogel's classification follows:

1. The Magnolian type is probably the most ancient type of pollen flower. The very abun-

dant pollen production is the result of primary polyandry. The pollen is usually shed onto the receptacle in "mess and soil" pollination (Faegri and van der Pijl, 1979), where it is eaten and the flowers are pollinated by beetles. This flower type prevails in the Magnoliales and Nymphaeales and includes some of the Araceae.

2. The *Papaver* type also produces copious pollen as the result of polyandry. The pollen is not shed but remains exposed in the anthers. Although this group exhibits many primitive traits, it is primarily a melittophilous specialization. The pollen is used not so much for imaginal as for larval food. Bees gather the pollen by swallowing, brushing or scraping, vibration, or scrambling around the flower accomplishing mess and soil pollination. Minor pollinators may also include beetles and flies. This is also a heterogeneous assemblage. The Papaveraceae and Ranunculaceae *(Anemone, Ranunculus, Thalictrum)* belong here, as do more primitive families (Actinidiaceae, Begoniaceae, Cistaceae, Cochlospermaceae, Dilleniaceae, Guttiferae, Ochnaceae, Paeoniaceae, Tiliaceae). Others such as *Rosa,* some *Rubus, Anoplobatus,* some Aizoaceae and Portulacaceae *(Glottiphyllum, Portulaca),* Fabaceae (*Acacia* and *Mimosa*), and Cactaceae *(Opuntia)* members of mostly nectariferous families appear to have been secondarily derived from early polyandrous or oligandrous nectar flowers.

3. The *Solanum* type comprises all of the oligandrous pollen flowers, i.e., those with few stamens. This is a modern, strictly bee-pollinated specialization. The flowers are radial (actinomorphic) or bilateral (zygomorphic) in symmetry. The general trend here is for the few anthers to become enlarged, able to produce more pollen, and usually showy with bright yellow anthers contrasting with the corolla colors (see Fig. 4-6B). The few anthers also mimic pollen copiousness by exhibiting a yellow and swollen appearance. The pollen is powdery, dry, and released as a "cloud" by floral vibrations of bees working the anthers. The position of the style is usually independent of the position of the anthers, owing to the overall dusting of pollinators in poricidal forms. Some

of these oligandrous pollen flowers, in the Dilleniaceae *(Hibbertia, Schumacheria),* Ochnaceae *(Luxemburgia, Ouratea),* Primulales *(Ardisia, Cyclamen, Dodecatheon, Lysimachia),* Melastomataceae, Malpighiaceae, and Commelinales, probably never had a nectary, and likely evolved from polymerous ancestors. Some, however, were derived from oligandrous nectar flowers and retain a rudimentary nonfunctional nectary. Pollen flowers in this group with radial symmetry occur in *Aristea, Barbacenia, Calectasia, Cucurligo, Cyphomandra, Dichorisandra, Galanthus, Hypoxis, Leucojum, Luzuriaga, Lycianthes, Ramonda, Sabatia, Solanum, Sowerbia,* and the Cyanastraceae, Mayacaceae, and Rapateaceae (see Figs. 4-5 and 4-6). Forms with bilateral symmetry occur in *Alonsoa, Chironia, Exacum, Henriettea, Orphium, Pyrola,* the Byblidaceae, and the Roridulaceae. A somewhat aberrant form in this class is represented by some papilionaceous pollen flowers (Genisteae) and by louseworts *(Pedicularis)* with concealed pollen but adapted for bee pollination (Vogel, 1978).

Among nectar flowers, the evolutionary trend has been for the reduction of anthers leading to less pollen wastage and more precise pollen transfer. Oligandrous pollen flowers, however, rely on increased anther size and more visibility, but the pollen quantity decreases (see discussion in Vogel, 1978). Most poricidal anthers have persistent yet paperlike locule walls (endothecium may be reduced), and keep their turgid appearance before dehiscence and during pollen presentation, until long after the anthers are devoid of pollen. The advantages of this superb deceit in stimulating floral visitation and greater levels of outcrossing are obvious. Deceptive pollen mimics or anther dummies are a feature of many other flowers (e.g., stylodia in *Begonia*), or the enlarged connectives (as in *Axinaea, Blakea, Loreya*), by trichome tufts (e.g., *Anagallis, Arthropodium, Bulbine, Narthecium, Stypandra, Tradescantia, Xyris*) or swollen yellow filaments (in *Cleome, Conandron, Dianella, Keraudrenia*). All of these serve to increase pollinator activity and attraction and may even distract pollinators from the real anthers so the pollen supply is somewhat protected from predation (by pollen thieves or legitimate bee pollinators). This could maximize the pollen pool allocated for pollination.

Within this oligandrous trend pollen flowers use other strategies to maintain their showiness and yet hold back some pollen as food (lost to pollination) for bees, while preserving enough pollen for fertilization. The powerful combination of heteranthery and deceit is just such a method and is widespread among poricidal forms. Heteranthery, or heterandry, (unlike heterostyly) is confined to pollen flowers and is the dimorphism within the androecium of an individual flower (as in Fig. 4-6E). The critical feature is that heteranthery represents a *functional* division of labor within the androecium, which separates the showier feeding anthers from the fertilizing anthers (pollen reward from pollination). Thus, the floral advertisement function is delegated to the "fodder" androecium, to attract pollinators, while the real pollination anthers are usually inconspicuous and the same background color as the rest of the flower. These pollinating anthers are exposed to be touched (therefore transferring pollen to bees) by unsuspecting bees manipulating the showier fodder or feeding stamens. Therefore small amounts of viable pollen are precisely placed in areas where bees cannot groom ("pollen safe sites"), and in turn are in the correct position for the stigma to pick up pollen on the bees' next visit. The difference between the strikingly colored feeding anthers and the dull-colored pollinating anthers is very noticeable in these heterantherous genera.

Heteranthery is known to most pollination ecologists who are familiar with senna (*Cassia* in the Fabaceae) or the tropical Melastomataceae (Fig. 4-6E). Again there are two (or even three in some *Cassia*) types of stamens, one of which provides pollen to bees for larval provisions while the other type dusts the bee with pollen during floral vibration. Pollen not groomed into scopae or ingested is then available for pollination. In the Melastomataceae many of the flowers are pink or purple, and the pollination anthers are the color of the corolla while the food anthers are yellow. The pollinating anthers are supported on jointed fila-

ments in the lower half of the flower, and serve as a support platform for a bee collecting pollen from the showy food-stamens which are themselves rather inaccessible. Small bees usually curl around individual anthers to buzz without a "platform." *Cassia,* bearing five to ten anthers, provides pollination, fodder, and sometimes a third rudimentary set of anthers. The dry pollen is ejected by vibration of the pollination and feeding anthers (may produce infertile or otherwise less nutritious degenerative pollen). The feeding anthers of *Commelina coelistris* no longer produce pollen, but a milky fluid instead (Faegri and van der Pijl, 1979). These are "advertising" connectives that are not eaten as claimed in the literature (Vogel, 1978). Whether the pollinators utilize the milky fluid in these anthers, or whether these staminodes have also lost the feeding function, acting only as pure advertising organs, is not clear; only field observations will clarify this unusual case.

Heteranthery is also well developed in nonporicidal polyandrous forms such as the oft-cited case of myrtle (*Lagerstroemia indica),* or less developed as in *Verbascum thapsus* (Faegri and van der Pijl, 1979). *Lagerstroemia* is a rare case of actinomorphic heteranthery.

Bilateral polyandrous flowers exhibiting heteranthery include the following: *Amoreuxia* (Cochlospermaceae), *Hibbertia* sect. *Hemipleurandra* (Dilleniaceae), *Swartzia* (Fabaceae), and *Cleome hirta* (Capparidaceae). All of these have a distinct bouquet of feeding stamina, but very few fertilizing stamina. The partial or full replacement of real anthers and pollen by "dummies" is confined to the fodder stamens. Vogel (1978) terms this "haptonasty" and cites the case of *Sparmannia africana* (Tiliaceae) as an example. The staminodia are peripheral with yellow filament knobs acting as feeding anthers. While working the flowers, visitors "fondle" the dummies while the fertile stamens and stigma touch the venter of the bees' abdomen. Heteranthery probably arose independently many times. In oligandrous pollen flowers, such as *Solanum,* the dimorphism (if present) starts in a mutualistic way in which the feeding anthers offer real but often sterile fodder pollen (collapsed or with little cytoplasm), and then goes toward partial deceit. The gradual emergence of heteranthery from ancestors with monomorphic androecia can be seen in many groups (*Cassia, Solanum,* and melastomes). These trends are briefly addressed by Vogel (1978).

Heteranthery is mutualistic, and pollen is offered as a true reward in most cases. Thus the deception is only partially developed. A small amount of pollen will be offered and collected, from either monomorphic anthers in isotemonic flowers (terminology of Vogel, 1978) or actual "fodder pollen" in heterantherous flowers. Even flowers with drastic reductions in the number of pollen grains in the thecae offer enough pollen that can be gathered by bees with some left over for pollination/fertilization. Obviously in nectarless pollen flowers, at least a minimal pollen reward must be provided, or bees would soon cease to visit the flowers and no cross pollination would occur.

Orchids have taken the deception game to the extreme. Some genera (e.g., *Arethusa, Calopogon, Eria, Maxillaria, Onicidium, Polystachya, Pogonia, Thelymitra*) make a gamble toward complete deceit (i.e., only offer floral volatiles, but no nectar or pollen utilizable to bees) and approach floral parasitism. No true loose pollen is offered, since the pollen in orchids is sequestered away in pollinia that bees do not exploit for food. Some do have a sort of pollen imitation, such as labellar trichomes to function as pollen dummies. In two cases (*Maxillaria* and *Polystachya*) the trichomes produce a sort of granular pseudopollen said to contain starch. This case should be examined again. Pseudopollen in *Polystachya,* however, is empty and inedible (Vogel, 1978). This may be a case of the orchids in a community floral mimicry association where they can get by with no bee-exploitable pollen, when all other non-orchids nearby provide a pollen reward. They would be rare mimics of widespread models, so pollinators would continue to visit them for no reward. One orchid genus of western Australia *(Elythranthera)* has even mimicked poricidal buzz pollinated species, and may get pollinated by deceit (S. D. Hopper, pers. comm.)!

One feature that often accompanies heteranthery in pored plants is enantiostyly (or enantiomorphy). This is the deflection of the style either to the left or right of the floral axis (dextrostylous or sinstrostylous orientation). Often in oligandrous flowers one stamen will be elongated, possibly a different color from the rest of the androecium, and will be deflected in the opposite direction of the style (as in *Solanum rostratum* in Fig. 4-2A). These flowers can be thought of as being left- and righ-handed ("enantiomers") and occur on the same plant with equal frequency.

Enantiostyly is found in a few poricidal plants but is widely distributed across unrelated families. It is present in the following: *Solanum* subgenus *Androcera* which includes *S. rostratum* and *S. citrullifolium*), *Cassia* (Fabaceae), Haemodoraceae *(Barberetta, Bilatris, Wachendorfia),* and the Tecophilaeaceae *(Cyanella).* It will undoubtedly be found in other plants as they are studied in detail.

Enantiostyly has hardly been noticed by floral biologists, so it is not surprising that few detailed studies have been made. Recently, *Cyanella* species (Tecophilaeaceae) and *Wachendorfia* (Haemodoraceae) have been examined with regard to enantiostyly (Ornduff, 1974; Ornduff and Dulberger, 1978; Dulberger and Ornduff, 1980). *Cyanella* has well-developed enantiomorphs with one larger stamen opposing the style as in *S. rostratum.* This stylar deflection left (L) or right (R) must result in pollen deposition from that anther to one side of the bee's abdomen. Therefore, the stigma, pointing in the opposite direction, can only receive pollen that fell on the same side of the insect during a previous visit to the opposite morph. Enantiostyly would then represent a mechanical system to promote pollinations between left- and righ-handed morphs, presumably increasing geitonogamy. To what extent this floral dimorphism may increase xenogamy is unknown. Functionally *Cyanella* is unique, since usually one flower per plant is produced daily. Effectively this makes every plant either right- or left-handed, which would increase the level of outcrossing. A 1:1 population ratio of R or L plants was also observed.

Bowers (1975) examined enantiostyly in several populations of *Solanum rostratum*

(Fig. 4-2) in Oklahoma. She found that the style and aberrant large lower anther reverse positions in successive flowers on a raceme. Right- and left-handed flowers are produced in a 1:1 ratio within a plant (many flowers produced per plant per day), and also for a population ratio. The large anther is the pollinating anther, but no differences in pollen fertility (as judged by staining and hand pollinations) between the two anther types were found. Bowers (1975) also failed to find inviable pollen as had Linsley and Cazier (1963). Geitonogamy was found, but xenogamy was more frequent in the largest populations. In the case of both *Cyanella* and *S. rostratum,* the enantiomorphy is not well developed enough to result in pollen deposition in two discrete patches on opposite sides of the bees. Thus, the stigmatic surface probably receives pollen from the long anther of an alternate morph, but also a fraction of pollen grains from the short upper staminal set of the same flower. There are no significant pollen stainability differences in either case. For *Cyanella,* even if some degree of selfing occurs, the self-incompatibility mechanisms would prevent self-fertilization. Dulberger and Ornduff (1980) suggest that the enantiomorphic strategy may have the dual effect of promoting outbreeding and of reducing total pollen production to the amounts required by pollinators between different types of flowers. Only field experiments with radioactively labeled pollen and careful experiments with pollen flow and pollinators can provide the answers to questions about the biological functions of heteranthery and enantiomorphy.

POLLEN DIMORPHISM

Pollen can be dimorphic intraspecifically or intraflorally in several important ways. The most common dimorphism is probably the variable pollen viability found between plants or within different anthers, in heteranthic forms, in a flower. Dimorphism also occurs in size, shape, and equatorial diameter, features of the exine, and the chemical composition of the pollen itself.

Pollen dimorphism in pollen viability, usually judged by staining with aniline blue in lactophenol, has been investigated for several poricidal plants. One of the first studies was performed by Linsley and Cazier (1963) on bee pollen loads from *Solanum elaeagnifolium.* They found that 64% of the grains were inviable (did not stain blue), and suggested that the high proportion of inviability was due to an inherent genetic factor, and not the result of high external temperatures reported earlier by Stow (1927) for another *Solanum.* When mixed populations of *S. rostratum* and *S. elaeagnifolium* were tested, Linsley and Cazier (1963, 1972) reported the percentage inviable as only 15.6% of the total. Bowers (1975) found no differences in pollen viability between the large and small anthers within a flower. In *S. douglasii,* Jones and Buchmann (unpub.) found only 5% inviable grains. One species of *Cassia (C. quiedondilla)* produces an extremely high percentage of inviable pollen. The normal "pollinating anthers" produced 95% inviable pollen, while the "fodder" anthers produced 96% inviabile. The smallest "rudimentary" anthers produced the highest pollen viability, i.e., only 87% inviable (Buchmann, 1974). In *Cassia bacillaris* both types of anthers have viable pollen, while in *Melastoma malabathricum* one set of anthers produces nonviable pollen (Percival, 1965). In four species of *Cyanella* (Tecophilaeceae), Dulberger and Ornduff (1980) found no significant differences in pollen staining ability between dimorphic stamen types. They also found that the number of pollen grains in the lower anthers exceeded that in the upper anthers. The ratio of pollen grains produced in the pollinating part of the androecium to the number of grains in the feeding part varied from 1.33 to 4.25 in the form species of *Cyanella.* This is an unusual case and the reasons for such high inviability are unknown. Other species should be checked using chemical tests and also hand pollinations with pollen from each anther set in heteranthic forms to see if the production of inviable pollen is widespread among poricidal plants.

Prominent size dimorphisms, in volume, equatorial diameter, and overall shape, occur in many taxa of poricidal plants. Such variability seems to be especially widespread in *Cassia, Solanum,* and many genera within the Haemodoraceae and Melastomataceae. If heteranthery is well developed, there is a good chance of finding at least a slight size dimorphism in pollen from the different anther subsets within the androecium.

Several species of *Cassia* have been examined in fair detail. Linsley and Cazier (1972) examined *Cassia bauhinoides* and found two distinct size classes of pollen. The larger grains were 80 μ in diameter and comprised 21% of the total grains, whereas the smaller grains were only 27 μ and made up 79% of the total number. These abundant, small, and apparently viable grains have been interpreted by Linsley and Cazier (1972) as a food "bonus" for pollinating bees. *Cassia quiedondilla* was examined briefly by Buchmann (1974), who found two size classes of pollen present. The larger grains averaged 32.5 μ in diameter, while the small grains were only 20 μ in diameter, and represented 43.9% of the total. What we desperately need for species with size-dimorphic pollen are accurate counts of total grains per anther in each set and the distribution of pollen size-classes among these different anther sets.

The genus *Tripogandra* in the Commelinaceae offers a good example of both heteranthic development and concomitant pollen dimorphism. These features were noted earlier by Lex (1961) and later by Mattsson (1976), and again mentioned in Faegri and van der Pijl (1979). *Tripogandra* produces two distinct sets of stamens, and each whorl of stamens develops its own characteristic type of pollen. One type of pollen is spheroidal and fertile, and the other very elongated and sterile. It has been suggested that the elongated pollen is offered to the insect as food, since it is a more "economic" way of producing pollen with smaller plant energy reserves, especially for nitrogenous compounds. The amount of pollen produced in the long-filamented anthers of *T. grandiflora* is, however, about half of that produced in the shorter anthers. Mattsson (1976) prefers to think of the production of carotenoids as a most important feature in the floral biology of *T. grandiflora*. He considers the structural and chemical changes of the pollen as accidental consequences of the shift in metabolism toward increased synthesis of these compounds ontogenetically, and believes they have little value in the process of selection. Many other examples of size dimorphisms occur but will not be treated here. It is tempting to speculate on the possible adaptive significance of dimorphic pollen with respect to the energy reserves they contain for development of the rapidly growing pollen tube penetrating the stylar tissue on its way to the ovules. Surely intraspecific pollen competition occurs down the "stylar racetrack," but it has not been documented. The reader is further directed to an excellent recent review, by Vogel (1978), on the evolutionary shifts from reward to deception that have occurred in pollen flowers, especially porose ones.

Differences in chemical composition, chemical dimorphism, between pollen grains within the same flower but from heterantherous subsets probably occur, but they have not been documented. There must be differences between viable and empty shell-like "sham pollen" in addition to finer chemical differences. One of the problems with this line of research has been to obtain enough pollen (usually several hundred milligrams is required) for chemical analyses. A floral "vibrator" has recently been developed that will allow large amounts of buzz pollen to be collected for chemical analysis (Buchmann and Spangler, unpub.).

BIOPHYSICAL MODELS FOR BUZZ POLLINATION IN ANGIOSPERMS

A biophysical model for the pollen/locule wall interactions resulting in pollen expulsion upon bee or artificial (e.g., a tuning fork) vibration was developed earlier (Buchmann and Hurley, 1978). This model is presented in an abbreviated verbal form in this section, so that it may be compared with the model presented by DeTar et al. (1968) for acoustically forced vibration of tomato blossoms. The governing differential equations and supporting equations for these two models will not be given here to save space, and the interested reader should consult the original papers for the mathematical details.

The model (Buchmann and Hurley, 1978) was created with the morphology of *Solanum* (Solanaceae) anthers in mind, but the results obtained are generally applicable to any apically dehiscent flower that is vibrated by bees to release pollen. The anthers (Fig. 4-1) were modeled as a rectangular parallelepiped (see Fig. 4-7) with an apical pore, and containing numerous small particles (= pollen grains). As the box vibrates, particles striking the walls are assumed to rebound elastically. If a pollen grain strikes a receding wall, it loses energy. If

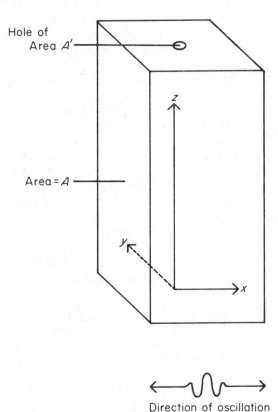

Hole of Area A'

Area $= A$

Direction of oscillation

Figure 4-7. The geometry of the model system. The poricidal anther is modeled as a rectangular box (retangular parallelepiped) that oscillates in the direction shown. The axial coordinates (x, y, z) are given along with the area of the wall (A). The apical pore shown at the top of the box has an area of A'.

a grain strikes an advancing wall, it gains energy in the collision. In each oscillation, there is a net gain in the energy of the particles. As the anther (rectangular box) is shaken, vibrational energy is transmitted from the pterothorax of the bee to the flower, the pollen grains gaining significant energy. As the energy increases, and the particles move about more and more vigorously, they will begin to escape through the hole in the box (or the anther pore; see Fig. 4-1). The rate at which particles leave the box is calculated as a function of the geometry of the model system and the frequency of the floral vibration.

Buchmann and Hurley (1978) divided their model into three parts. In the first section they determined the rate at which particles are escaping from the hole as a function of the energy of the particles within the box. Second, they determined the rate at which the particle energy is changing with time. There will be an increase in energy due to the rapid vibration of the box and a decrease due to the particulate loss from the apical pore. Lastly, they put these two results together and obtained a pair of coupled, first-order, ordinary differential equations whose solution determines both the energy and the particle number as a function of time.

In order to demonstrate the feasibility of the mechanism proposed, they (Buchmann and Hurley, 1978) analyzed quantitatively the following model. A rectangular box (the anther; see Fig. 4-7) contains identical particles (= pollen grains) of mass m. The box oscillates along the x axis with a period τ. For simplicity, we shall let the box move in the positive x direction with a constant velocity V for a time $\tau/2$. Particles striking the walls rebound elastically.

We know that, on very general grounds (by the second law of thermodynamics), in one complete oscillation of either wall there must be a net gain in energy of the particles inside the box. It is impossible to create a device, operating in a cycle, whose sole effect is to remove heat energy from a system and convert this energy into work. There is a small hole (in real flowers the terminal anther pores or slits;

see Figs. 4-1 and 4-7, respectively) in the box of area A'. Particles striking this hole escape from the box in the form of a cloud when all the indentical particles are considered.

Initially, Buchmann and Hurley assumed the particles to be at rest (i.e., pollen in the flower is loosely packed or settled near the basal half of the anther). As the box is shaken, vibrational energy is transmitted from the thorax of the bee to the flower, the particles gaining significant energy. They assumed that through interparticle collisions the velocity distribution is the Maxwell-Boltzmann distribution. They also neglected the viscous effects of air on the particle motion. It is not difficult to show that this is a good approximation for the system under study. As the energy increases, the particles begin to move about more and more vigorously, they will begin to escape through the apical pore or hole in the model box (Fig. 4-3). The objective is to determine the rate at which particles leave the box; in particular, how long it takes to empty the box. This is, of course, a very brief verbal version of the more detailed quantitative model presented by these authors.

The quantitative and predictive results of the Buchmann and Hurley model can be summarized. These equations were integrated numerically for several values of e (i.e., the coefficient of restitution of the pollen grain colliding with the anther wall), and the results are tabulated in Table 4-3. The value taken for A'/A (i.e., the ratio of the area of the hole to the area of the oscillating surfaces) was 0.044. This is an approximate estimate for *Solanum xanti.* Buchmann and Hurley (1978) have computed the elapsed time for 90% of the pol-

Table 4-3. Elapsed time for 90% of the particles (pollen grains) to emerge from the anther pores.

e*	ϵ'	t'
1.0	14000.0	7.0
0.5	670.0	20.0
0.2	8.5	80.0
0.1	4.6	105.0

*The coefficient of restitution of the pollen grains colliding with the inner locule walls is given by e. When $e = 1$, there is an elastic collision. ϵ' is the energy of the emerging particles at the given time t'. All quantities are dimensionless.

len grains to emerge from the anther. The energy of particles emerging at the given time is ϵ'. It is not the average energy of the emerging particles (the energy increases monotonically).

For *Solanum xanti,* γ, the volume of one locule, is about 6.8×10^4 cm^3, and A, the mean area striking the pollen grains, is 1.8×10^{-2} cm^2. The mean velocity with which the bees vibrate the anthers is difficult to estimate. It is proportional to the product of the amplitude times the frequency. If we choose a lower bound for the amplitude of 0.5 mm and a frequency of 500 Hz, then V is approximately 25 cm/sec^{-1}. Thus, $\gamma/AV = 1.5 \times 10^{-3}$ sec. To calculate the evacuation time then, we must multiply t' in Table 4-3 by 1.5×10^{-3} sec. For example, if $e = 0.1$, then $t = (1.5 \times 10^{-3}$ sec$)$ $(105) = 0.16$ sec, or 90% of the pollen grains would be vibrated from the anther in 0.16 sec. This corresponds to field observations of vibrating bees that take approximately 1 to 5 sec to buzz a flower with five anthers, such as *Solanum.* The calculated height to which the pollen cloud rises above the terminal pores (about 1.4 cm) also corresponds to observations made in the field.

To calculate the energy (ϵ) of the emerging particles, we must return to the scaling equation for the energy, namely, $\epsilon' = \epsilon/\frac{1}{2}mV^2$ (Reif, 1965). Now the energy itself is not directly observable. We can, however, observe the height to which the pollen grains rise above the anther pores. This height is related to the energy by the relation $\epsilon = mgh$ or, solving for h, $h = \epsilon/mg$. In terms of ϵ' we have $h = (V^2/2g)\epsilon'$. If we take V as before to be 25 cm/sec^{-1} and let $e = 0.1$ so that from Table 4-3, $\epsilon' = 4.6$, then $h = 1.4$ cm above the anther pores.

In our necessarily simplified model for vibrational pollen release, we have obviously not considered, or mathematically derived, every minute detail of the buzz syndrome in nature. Variables that were difficult to measure, or to deal with mathematically, and thus are not included in the model are: (1) possible non-unidirectional plane of anther vibration; (2) features of the roughened locule walls (termed Ubisch bodies); (3) nature of electrostatic forces among pollen grains and between pollen/locule walls; (4) two size classes of pollen grains in some anthers [e.g., *Cassia* and some *Solanum* spp.; see Linsley (1962a), Linsley and Cazier (1963), and Buchmann (1974)], which would have no effect on the evacuation times for the pollen, but would affect the height above the terminal pores to which the pollen rises; (5) viscosity effects of pollen/air interactions—from a vacuum experiment (flowers vibrated in vacuum behaved normally with the pollen cloud rising higher than in air), and calculation of a small Reynold's number, these effects have been discounted; (6) some amount of lipoidal pollenkitt existing on the surface of even relatively dry pollen grains (as in *Solanum*), which may account for some pollen sticking together; (7) different sizes and shapes of flowers, stamens, pores, locules, and pollen (these were too complex to be treated in the model, but we feel that even the most complicated vibratile flower might behave as predicted if the stamens vibrate as one unit); (8) bee production of floral vibrations differing in frequency and amplitude, which as such are not treated here; (9) exact amplitude of the floral oscillation, which is difficult to determine empirically, but must be small (we used an approximate figure in our calculations); (10) exact figures for pollen velocity, height distribution to which pollen grains rise, and escape velocity, which are not available; (11) duration of the floral vibration, which is highly variable (from 0.1 to 8 or more seconds), depending upon the time of day (i.e., depending on standing crop of pollen within the flowers), or bee species doing the buzzing, but presumably makes no difference in our treatment, and (12) effects of relative humidity and high ambient temperatures on anther dehiscence and pollen shedding in poricidal forms, which are not known.

In addition to the biophysical model presented earlier (Buchmann and Hurley, 1978), and reviewed herein, there have been few attempts to examine quantitatively or model the complex vibrations responsible for pollen discharge in buzz flowers. One other model was constructed for the acoustically forced vibration of greenhouse tomato (*Lycopersicon* spp.) blossoms by agricultural engineers (DeTar et

al., 1968, 1971). The latter studies were unknown to Buchmann and Hurley (1978) and represent two independent approaches and theoretical models to attempt to explain floral vibration resulting in pollen expulsion in porose flowers. These workers (DeTar et al., 1968, 1971) develop a rigorous analytical computer program for finding the natural frequencies of a nonuniform, viscoelastic shell beam, which provides the mode shapes and the shear, and moment values at any cross section of the tomato anther cone.

A detailed mathematical examination of their model will not be presented here, but, rather, general biological principles and applications resulting from their analyses will be summarized. The DeTar model (DeTar 1968; DeTar et al., 1968, 1971) is an adaptation of a work originally presented by Alley (1962, 1965) for NASA. The Alley method, which is in turn similar to the methods presented by Pestel and Leckie (1963) and Thomson (1965), is one of the most comprehensive methods of beam analysis yet developed. The procedure also incorporates the effects of shear deformation and rotary inertia for a fibrating nonuniform beam, two components that are usually considered negligible in beam analyses. However, in the case of the tomato anther cone, at the higher harmonics, the effects of shear deformation were found by DeTar (1968) to be important. The method requires the simultaneous solution of four first-order differential equations of motion of a beam, using a numerical technique involving a truncated Taylor series (a transfer matrix method).

Physical properties of the tomato anther cone including: length, diameter, and weight per unit length, cross-sectional area, moment of inertia, tension, compression, shear, and relaxation terms (using an Istrom* testing machine), were measured, and average values were used numerically in the model. These values were used as parameters for predicting the natural frequency of vibration of the anther

---

*Mention of a trademark, proprietary product, or vendor does not constitute a guarantee or warranty by the USDA and does not imply its approval to the exclusion of other products or vendors that may also be suitable.

cone. The cone was idealized as a nonuniform, internally damped cantilever beam. A transfer matrix method was used to determine the natural frequencies of several modes, and the state vector at all points in the cone was determined. DeTar et al. (1968) felt that the fundamental frequency of the floral vibration would be affected by the stiffness of the clamping (i.e., the flexible pedicel connecting flower to stem), which would drop off to zero as the stiffness is reduced to zero. They also measured the vibration frequency of the tomato flower to be about 2–30 Hz (the fundamental frequency), and determined that the blossom itself would vibrate at the frequency of an equivalent "free–free" beam.

Their (DeTar et al., 1968) predicted and empirical experimental results and the model's predictions were found to be in close agreement. The model predicted the lowest mode to occur at 4.5 kHz, and the best greenhouse results occurred at 4.5 kHz. A 40-watt sonic transducer, capable of sound pressure levels of 155 db at close range, was used to provide peak vibrational intensities at 1.2, 4.5, 6.3, 9.9, and 15.1 kHz. The predicted frequencies were 1.2, 4.5, 9.6, 14.8, and 20.9 kHz. Results on greenhouse tomato blossoms were measured in three different ways: a rating of visible pollen scatter, number of fruits set, and the number of seeds per fruit. The dynamic response of the flower was predicted successfully for the second through the fifth modes. The optimum sound frequency as measured by seed counts in tomatoes "set by sound" was 4.5 kHz (the predicted second mode frequency). Also the optimum frequency, as measured by the first visible pollen rain, was 5.5 kHz. Their use of 155 db at 4.5 kHz produced a 75% set of tomato fruit (even self-compatible tomato cultivars usually realize only a 30% fruit set).

Although using high sound pressure levels to vibrate flowers is not completely analogous to bees buzzing flowers by direct contact, much can be learned from the DeTar model. Work is now in progress (Buchmann, in prep.) to measure the airborne and floral vibrations (with a transducer and accelerometer), while small and large bees are actually vibrating flowers (*Solanum* spp.). Thus measurements

of floral amplitude, the range of bee frequencies and bee thoracic temperatures (by the "grab/stab" method), and the length of floral buzzes (equals handling times to collect pollen) should give an idea of the amount of energy expended by bees to harvest pollen by intrafloral vibration.

ELECTROSTATICS AND BUZZ POLLINATION

The background biophysical data for the electrostatic interactions between insects and flowers is given elsewhere in this book (see Erickson and Buchmann). The suite of morphological adaptions plants have evolved for vibratile pollination (e.g., poricidally dehiscent anthers, abundant pollen, small smooth powdery-dry pollen, with little surface sculpturing, and almost on oily pollenkitt) are especially important when electrostatic parameters are considered. Since most of these buzz flowers are pendant during vibration, one might imagine the pollen simply falling out of the anther pores with a large percentage of it missing the bee, as if the flower were a salt shaker turned upside down. Thus, in buzz plants anther vibrations would tend to scatter the pollen randomly, making this a somewhat inefficient pollen-gathering technique. If this were true, much of the pollen would be lost and unavailable for pollination or larval brood provisions. Such an inefficient system would tend to be selected against. However, in the presence of an electrostatic field around a visiting bee, most of the pollen would actively precipitate onto the bee, thereby greatly increasing the efficiency of pollen collection. The probable important role of electrostatics in buzz pollination was recognized earlier (Buchmann, 1978; Buchmann and Hurley, 1978).

Prior to floral visitation by bees, the small, light dry pollen will be held in place largely by electrostatic interactions (and gravity) between the pollen and interior locule walls. Release of pollen and accessibility to it are severely limited by the small apical anther pores. Upon the approach and landing of a bee, the electrical microenvironment will begin to change, via induction, as the flower and bee dissipate each other's opposite charge. This suspected electrostatic shift might have the effect of weakening the attraction of the grains to the anther walls. The pollen would now be ready for active vibrationally induced discharge which usually requires less than one second of bee vibration. A biophysical mechanism for this vibratory pollen-shedding has been proposed (Buchmann and Hurley, 1978, summarized earlier). During this short interval the bee still carries a net positive charge, and conditions are set for the electrostatic attachment of the negatively charged pollen onto the bee. Thus, even if some pollen is discharged to the sides or up toward the bee's dorsum, it should be held strongly and pulled onto the bee's integument. Very little pollen should be lost to air currents near the bee. The pollen can then be easily recovered by the female bee during grooming movements, except on the occipital region of the head, proboscideal fossa, or centerline area of the thoracic dorsum where most solitary and social bee species cannot reach during grooming movements (Buchmann, unpub. obs.).

When a bee visits succeeding flowers, any pollen not packed into scopal transport devices will be available for pollination (i.e., deposition onto stigmatic surfaces). As mentioned previously, these flowers are characterized by filiform, dry stigmas with little or no sticky stigmatic exudates. The transfer of pollen from bee to stigma and the adhesion of relatively dry buzz pollen onto a dry stigmatic surface is also likely mediated electrostatically. Once the pollen grain has landed on such a dry stigma, it will very soon be firmly anchored by the rapidly growing pollen tube penetrating the stigma.

It is presumed, based on certain physical laws, that a foraging bee is gradually discharged by repeated contact with flowers of the opposite charge, or by the pollen precipitating onto its body. Thus many questions remain about, for example, the rate of discharge of the bee over time, the bee's ability to recharge on short flights between flowers, and the efficacy level of an electrostatic system with repeated floral contact during a single foraging bout.

ENERGETIC ASPECTS OF BUZZ POLLINATION

The physical and energetic aspects of pollen collection by bees have hardly been treated in any of the recent literature on optimal foraging theory. This is unfortunate, since floral nectar is largely used for flight fuel and maintenance, whereas pollen protein is essential for the production of bee larvae. It is suggested that pollen harvesting by floral buzzing is an important behavioral method for the rapid and energetically efficient collection of pollen in an optimal foraging sense. The audible floral buzzes that accompany vibration are a good measure of the handling time needed to harvest a poricidal flower for pollen. The duration of floral buzzing per unit time is very short (about 0.1 second) at sunrise on *Solanum* ssp. when pollen is most abundant, and much longer (up to 10 seconds or more) later in the day when the standing crop of pollen per flower is very low (Buchmann, unpub.). The novel behavioral use of flight muscle contraction by some bees to aid in pollen harvesting, in addition to possible electrostatic precipitation, is seen as a precise method for the rapid collection of large amounts of pollen (*Apis* does not buzz). Data will be presented elsewhere to demonstrate that bees that harvest pollen by floral vibration are able to collect far more pollen (in milligrams/unit time) on buzz-adapted flowers than on either flowers that offer both nectar and pollen, or nonvibratile pollinated flowers.

Heinrich (1972) provides data that *Bombus* buzzing of flowers of *Solanum dulcamara* results in the elevation of the temperature of the thoracic flight muscles. This has been verified in *Bombus* and other bees (e.g., *Anthophora, Centris*) by "grabbing and stabbing" them with a thermocouple thermometer immediately after the floral buzzes (Buchmann, unpub.). This thermoregulatory ability (a byproduct of muscular contraction and vibration) would allow females possessing it a headstart in collecting pollen from poricidal flowers with anthesis before sunrise, the coolest part of the day, perhaps giving them a competitive advantage over other bees (e.g., small nonbuzzing gleaning bees). This was proposed by Thorp and Estes (1975). *Bombus* is in an especially favorable position, since they have a "fossil fuel" (= stored honey) supply within the nest that they can use as a flight fuel before leaving the nest on foraging bouts.

CONCLUSIONS AND FUTURE DIRECTIONS

Buzz pollination is a unique method of pollen harvesting from poricidal anthers, and is used by hundreds of bee species worldwide. Numerous phylogenetically diverse plants (at least 54 families and 357 genera) have evolved this form of concealed pollen presentation that requires bee vibration. Many of these plants have also developed an elaborate division of labor within the androecium (i.e., heteranthery), or stylar enantiomorphy producing left- and right-handed flowers that may result in greater levels of outcrossing. Many bright yellow pored anthers are now seen in their true role of shamming pollen copiousness as a form of deception to assure the services of pollen vectors for pollination, and ultimately fertilization and seed production. The production of anther dummies, pseudopollen, or some pollen with reduced viability is part of this elaborate plant deception strategy. A large proportion of pollen is offered to pollinators for imaginal or larval food, and therefore lost to pollination, while a small fraction is lodged in "pollen safe sites" on bees, inaccessible to grooming movements, and is available for pollination.

Future directions for experimental buzz pollination research might include the following: (1) mutiliation experiments to determine the floral cues that elicit floral vibration; (2) experiments and hand pollinations to test function(s) of enantiostyly and heteranthery; (3) counts of pollen grains per locule, per anther and flower in dimorphic androecia, along with pollen viability and pollen grain:ovule ratios; (4) reproductive biology of pollen flowers; (5) pollen chemistry (total protein. lipids, carbohydrates, amino acids, and calorific content); (6) headspace collection and gas chromatographic/mass spectroscopic isolation and identification of floral volatiles in pollen flowers;

(7) sound analyses (airborne and substrate recordings) of floral buzzes in comparison to alarm buzzes and flight sounds for the same bee species at different temperatures and times of day; (8) competition for pollen among buzzing, biting, and gleaning bees on a given poricidal plant, or between nonporicidal and porose flowers within a given plant community; (9) energetics of optimal foraging on pollen flowers, handling times for pollen extraction on different species and through the day, ratio of pollen to nectar flowers on a given trip, and "grab/stab" thoracic temperatures of bees during floral buzzes; (10) limitation of fruit set due to available plant energy or pollination limitations (Bateman Principle); (11) allocation of resources to male and female function in pollen flowers.

ACKNOWLEDGMENTS

I wish to thank the following for their helpful comments on this or an earlier draft of the manuscript, and for many stimulating discussions regarding vibratile pollination: Herbert and Irene Baker, Marlo D. Buchmann, C. Eugene Jones, E. G. Linsley, Gerald M. Loper, David W. Roubik, Robert J. Schmalzel, Rudolf Schmid, Justin O. Schmidt, Hayward G. Spangler, Robbin W. Thorp, Stefan Vogel, and Grady L. Webster. Very special thanks go to Paula McKenzie (Allen) (for Fig. 4-6E) and to Marlo D. Buchmann (for all others) for preparation of the illustrations, and to Elizabeth L. Sager and Deana L. Cassa for typing the manuscript. I also thank Renaté Kempf for translation of the Kugler paper.

LITERATURE CITED

Aleksandrov, V. G. and A. V. Dobrotvorskaya. 1960. Obrazovanie tychinok i formirovanie fibroznogo sloya v ply'nikakh tsvetkov nekotorykh rastenif. *Botanichniĭ zhurnal, Kyyiv* 45:823–831.

Alley, V. L. 1962. A matrix model for the determination of the natural vibrations of free–free unsymmetrical beams with application to launch vehicles. NASA TN D-1247.

———. 1965. A method of determining modal data of a non-uniform beam with effects of shear deformation and rotary inertia. NASA TN D-2930.

Almeda, F. 1977. Systematics of the neotropical genus *Centradenia* (Melastomataceae). *J. Arnold Arboretum* 58(2):73–108.

———. 1978. Systematics of the genus *Monochaetum* (Melastomataceae) in Mexico and Central America. U.C. Publ. Bot. 75:1–134.

Aufsess, A. von. 1960. Geruchliche Nahorientierung der

Biene bei entomophilen und ornothophilen Bluten. *Zeit. Vergl. Physiol.* 43:469–498.

Axelrod, D. T. 1952. A theory of angiosperm evolution. *Evolution* 6:29–60.

Baker, H. G. and I. Baker. 1979. Starch in angiosperm pollen grains and its evolutionary significance. *Am. J. Bot.* 66(5):591–600.

Bell, P. R., J. Challinor, J. B. S. Haldane, P. Marler, and H. L. K. Whitehouse. 1959. *Darwin's Biological Work: Some Aspects Reconsidered.* John Wiley and Sons, New York.

Bowers, K. A. W. 1975. The pollination ecology of *Solanum rostratum* (Solanaceae). *Am. J. Bot.* 62(6):633–638.

Buchmann, S. L. 1974. Buzz pollination of *Cassia quiedondilla* (Leguminosae) by bees of the genera *Centris* and *Melipona. Bull. So. Calif. Acad. Sci.* 73(3):171–173.

———. 1978. Vibratile "buzz" pollination in angiosperms with poricidally dehiscent anthers. Unpub. Ph.D. dissertation, entomology. University of California, Davis. 238 pp.

———. 1980. Preliminary anthecological observations in *Xiphidium caeruleum* Aubl. (Monocotyledoneae: Haemodoraceae) in Panama. *J. Kansas Ent. Soc.* 53(4):685–699.

———. 1982. Poricidal dehiscence of angiosperms and vibratile pollination. Submitted to U.C. Publ. in Bot.

Buchmann, S. L. and M. D. Buchmann. 1981. Anthecology of *Mouriri myrtilloides* (Melastomataceae: Memecyleae), an oil flower in Panama. *Biotropica* 13(2):7–24. Supplement on reproductive botany.

Buchmann, S. L. and J. P. Hurley. 1978. A biophysical model for buzz pollination in angiosperms. *J. Theor. Biol.* 72:639–657.

Buchmann, S. L., C. E. Jones, and L. J. Colin. 1977. Vibratile pollination of *Solanum douglassii* and *S. xanti* (Solanaceae) in southern California. *Wasmann J. Biol.* 35(1):1–25.

Burck, W. 1887a. Notes biologiques: relation entre l'hétérostylie dimorphe et l'heterostylie trimorphe. *Ann. d. Jard. Bot. de Buitenzorg* 6:251–267.

———. 1887b. Notes biologiques. 2. dispositions des organes dans les fleurs dans les but de favoriser, l'autofécondation. *Bull. d. Jard. Bot. de Buitenzorg* 6:354–365.

———. 1891. Beitr. z. Kenntnis d. myrmekoph. Pflanzen. *Ann. d. Jard. Bot. de Buitenzorg.* 10:119–127.

———. 1906a. Over den inrloed der nectariën en andere suikerhoudende weefsels in de bloem op het openspringen der helmknoppen. *Versl. Gewone Vergad. Wisen Natuurk. Afd., K. Akad. Wet. Amst.* 15:278.

———. 1906b. On the influence of the nectaries and other sugarcontaining tissues in the flower on the opening of the anthers. *Proc. Sect K Akad. Wet. Amst.* 9:390.

———. 1907. On the influence of the nectaries and other sugarcontaining tissues in the flower on the opening of the anthers. *Recl. Trav. Bot. Néerl.* 3:163 [issued 1906].

Canright, J. E. 1952. The comparative morphology and relationships of the Magnoliaceae. I. Trends of specialization in the stamens. *Am. J. Bot.* 39:484–497.

Clements, F. E. and F. L. Long. 1923. *Experimental Pollination: An Outline of the Ecology of Flowers and Insects.* Carnegie Institute of Washington, Washington, D.C., No. 336.

DeTar, W. R. 1968. Acoustically forced vibration of the tomato blossom. Unpub. Ph.D. thesis, Purdue University, Lafayette, Ind.

DeTar, W. R., C. G. Haugh, and J. F. Hamilton. 1968. Acoustically forced vibration of greenhouse tomato blossoms to induce pollination. *Trans. ASAE* 11(5):731–735, 738.

———. 1971. Vibrational analysis of a non-uniform, viscoelastic beam, for an agricultural application. Unpub. paper, No. 71–691. American Society of Agricultural Engineers.

Dulberger, R. and R. Ornduff. 1980. Floral morphology and reproductive biology of four species of *Cyanella* (Tecophilaeaceae). *New Phytol.* 86:45–56.

Eisner, T. 1980. Chemistry, defense, and survival: case studies and selected topics. *In* M. Locke and David S. Smith (eds.), *Insect Biology in the Future.* "VBW 80." Academic Press, New York, pp. 847–878.

Erickson, E. H. 1975. Surface electric potentials on worker honeybees leaving and entering the hive. *J. Apic. Res.* 14(3/4):141–147.

Evans, H. E. and M. J. W. Eberhard. 1970. *The Wasps.* University of Michigan Press, Ann Arbor, p. 102.

Faegri, K. and L. van der Pijl. 1971. *The Principles of Pollination Ecology,* 2nd ed. Pergamon Press, London, 291 pp.

———. 1979. *The Principles of Pollination Ecology,* 3rd rev. ed. Pergamon Press, London. 244 pp.

Fahn, A. 1974. *Plant Anatomy,* 2nd ed. Pergamon Press, Oxford.

Frankel, R. and E. Galun (eds.). 1977. *Pollination Mechanisms, Reproduction and Plant Breeding. Monographs on Theoretical and Applied Genetics 2.* Springer-Verlag, New York. 281 pp.

Frisch, K. von. 1923. Über die "Sprache" der Bienen eine tierpsychologische Untersuchung. *Zool. Jahrb. (Physiol.)* 40:1–86.

Halsted, B. D. 1890. Notes upon the stamens of Solanaceae. *Bot. Gaz.* 15:103–106.

Hardin, J. W., G. Doerksen, D. Herndon, M. Hobson, and F. Thomas. 1972. Pollination ecology and floral biology of four weedy genera in southern Oklahoma. *Southwest. Natur.* 16(3/4):403–412.

Harris, J. A. 1903. Polygamy and certain floral abnormalities in *Solanum. Trans. Acad. Sci. St. Louis* 13:185–202.

———. 1905a. The dehiscence of anthers by apical pores. Ph.D. dissertation, Washington University. 146 pp.

———. 1905b. The dehiscence of anthers by apical pores. *Missouri Bot. Gard., 16th Ann. Rept.,* 167–257.

———. 1905c. The influence of the Apidae upon the geographical distribution of certain floral types. *Can. Entomol.* 37:353–356, 373–380, 393–398.

———. 1906. The anomalous anther structure of *Dico-*

rynia, Duparquetia, and *Strumpfia. Bull. Torrey Bot. Club* 33:223–228.

————. 1909. Variation and correlation in the flowers of *Lagerstroemia indica. Rept. Missouri Bot. Gard.* 20:97–104.

————. 1914. On a chemical pecularity of the dimorphic anthers of *Lagerstroemia indica,* with a suggestion as to its ecological significance. *Ann. Bot.* 28:499–507.

Harris, J. A. and O. M. Kuchs. 1902. Observations on the pollination of *Solanum rostratum* Dunal and *Cassia chamaecrista* L. *Kansas Univ. Sci. Bull.* 1:15–41.

Heinrich, B. 1972. Energetics of temperature regulation and foraging in a bumblebee, *Bombus terricola* Kirby. *J. Comp. Physiol.* 77:49–64.

————. 1974. Thermoregulation in bumblebees. I. Brood incubation by *Bombus vosnessenskii* Queens. *J. Comp. Physiol.* 88:129–140.

Herbst, P. 1918. Über Läutäusserungen einiger chilenischer Blumenwespen (Apidae). *Dt. Ent. Z.,* 93–96.

Hopkins, C. Y., A. W. Javans, and R. Boch. 1969. Occurrence of octadeca-*trans*-2, *cis*-9, *cis*-12-trienoic acid in pollen attractive to the honeybee. *Can. J. Biochem.* 47:433–436.

Hurd, P. and E. G. Linsley. 1976. The bee family Oxaeidae with a revision of the North American species (Hymenoptera: Apoidea). *Smith. Contr. Zool.* 220:1–75.

Kerner von Marilaun, A. 1895. *The Natural History of Plants.* Translated from the German and edited by F. W. Oliver. London.

Kirchner, O. 1911. *Blumen und Insekten, ihre Anpassungen aneinander und ihre gegenseitige Abhägigkeit.* B. G. Teubner, Leipzig and Berlin, 436 pp.

Knoll, F. 1930. Über Pollenkitt und Bëstaubungsart. *Z. Bot.* 23:609–675.

Knuth, P. and E. Loew. 1895–1905. *Handbuch der Blutenbiologie.* I–III, 2. Engelmann, Leipzig. 2973 pp.

Kugler, H. 1955. *Einführung in die Blütenokologie,* 1st ed. Gustav Fsicher, Stuttgart, pp. 188–193. 345 pp.

Laroca, S. 1970. Contribuicão para o conhecimento das relacões entre abelhas e flôres: coleta de Pólen das anteras tubulares des certas Melastomataceae. *Revista Floresta* 2(2):69–74.

Leggett, W. H. 1875. 178. *Cassia. Bull. Torrey Bot. Club* 6:171.

————. 1881. 96. Fertilization of *Rhexia virginica* L. *Bull. Torrey Bot. Club* 8:102–104.

Lepage, M. and R. Boch. 1968. Pollen lipids attractive to honey bees. *Lipids* 3:530–534.

Lex, R. E. 1961. Pollen dimorphism in *Tripogandra. Baileya* 9:53–56.

Linsley, E. G. 1962a. The colletid *Ptiloglossa arizonensis* Timberlake, a matinal pollinator of *Solanum. Pan-Pacific Entomol.* 38(2):75–82.

————. 1962b. Ethological adaptations of solitary bees for the pollination of desert plants. *Sartryck ur Meddelande nr 7 Sveriges Froodlkareforbund,* 189–197.

————. 1970. Some competitive relationships among matinal and late afternoon foraging activities of caupolicanine bees in southeastern Arizona (Hymenoptera: Colletidae). *J. Kansas Ent. Soc.* 43(3):251–261.

Linsley, E. G. and M. A. Cazier. 1963. Further observations on bees which take pollen from plants of the genus *Solanum. Pan-Pacific Entomol.* 30:1–18.

————. 1972. Diurnal and seasonal behavior patterns among adults of *Protoxaea gloriosa* (Hymenoptera: Oxaeidae). *Am. Mus. Novitat.* 2509:1–25.

Louveaux, J. 1960. Recherches sur las récolte du pollen par les abeilles. Ann. l'abeille 3.

Macior, L. W. 1964. An experimental study of the floral ecology of *Dodecatheon meadia. Am. J. Bot.* 51:96–108.

————. 1968a. Pollination adaptation in *Pedicularis groenlandica. Am. J. Bot.* 55:927–932.

————. 1968b. Pollination adaptations in *Pedicularis canadensis. Am. J. Bot.* 55:1031–1035.

————. 1969. Pollination adaptation in *Pedicularis lanceolata. Am. J. Bot.* 56:853–859.

————. 1970a. The pollination ecology of *Dodecatheon amethystinum* (Primulaceae). *Bull. Torrey Bot. Club* 97:150–153.

————. 1970b. The pollination ecology of *Pedicularis* in Colorado. *Am. J. Bot.* 57:716–728.

————. 1973. The pollination ecology of *Pedicularis* on Mount Rainier. *Am. J. Bot.* 60:863–871.

————. 1974. Behavioral aspects of coadaptations between flowers and insect pollinators. *Ann. Missouri Bot. Gard.* 61(3):760–769.

————. 1977. The pollination ecology of *Pedicularis* (Scrophulariaceae) in the Sierra Nevada of California. *Bull. Torrey Bot. Club* 104(2):148–154.

Maheswari, P. 1950. *An Introduction to the Embryology of Angiosperms.* McGraw-Hill, New York.

Matthews, J. R. and C. M. MacLachlan. 1929. The structure of certain poricidal anthers. *Trans. Proc. Bot. Soc. Edinburgh* 30:104–122.

Mattsson, O. 1976. The development of dimorphic pollen in *Tripogandra* (Commelinaceae). *In* I. K. Ferguson and J. Muller (eds.), *The Evolutionary Significance of the Exine.* (Linnean Soc. Symp. Series No. 1). Academic Press, New York, pp. 163–183.

Meehan, T. 1886. On the fertilization of *Cassia marilandica. Proc. Acad. Nat. Sci. Phila.* 38:314–318.

Meidell, O. 1944. Notes on the pollination of *Melampyrum pratense* and the "honeystealing" of bumblebees and bees. Bergens. Museum. Aarbok Naturwitenskayselig rekke, Norway. 11, 12 pp.

Menke, A. S. E. 1965. A revision of the North American *Ammophila* (Hymenoptera: Sphecidae). Ph.D. dissertation, University of California, Davis, p. 27.

Michener, C. D. 1962. An interesting method of pollen collecting by bees from flowers with tubular anthers. *Rev. Biol. Trop.* 10(2):167–175.

————. 1974. *The Social Behavior of the Bees: A Comparative Study.* The Belknap Press of Harvard University Press, Cambridge, Mass. 404 pp.

————. 1979. Biogeography of the bees. *Ann. Missouri Bot. Gard.* 66:277–347.

Michener, C. D., M. L. Winston, and R. Jander. 1978. Pollen manipulation and related activities and structures in bees of the family Apidae. *Univ. Kansas Sci. Bull.* 51(19):575–601.

Ornduff, R. 1974. Heterostyly in South African flowering plants: a conspectus. *J. S. Afr. Bot.* 40:169–187.

Ornduff, R. and R. Dulberger. 1978. Floral enantiomorphy and the reproductive system of *Wachendorfia paniculata* (Haemodoraceae). *New Phytol.* 80:427–434.

Pankow, H. 1957. Über Pollenkitt bei *Galanthus nivalis* L. *Flora* 146:240–253.

Percival, M. S. 1965. *Floral Biology*. Pergamon Press, Oxford. 239 pp.

Pestel, E. C. and F. A. Leckie. 1963. *Matrix Methods in Elastomechanics*. McGraw-Hill, New York.

Pijl, L. van der. 1954. *Xylocopa* and flowers in the tropics. I–III. *Proc. Kon. Ned. Ak. Wet. ser.* C 57:413–423, 514–562.

Porsch, O. 1954. Geschlechtagebundener Blütenduft. *Öst. Bot. Z.* 101:359–372.

Powell, J. A. 1964. Additions to the knowledge of the nesting behavior of North American *Ammophila* (Hymenoptera: Sphecidae). *J. Kansas Ent. Soc.* 37:240–258.

Rajotte, E. G. and R. B. Roberts. 1978a. Nectar sugar dynamics of highbush blueberry cultivars (*Vaccinium corymbosum* L.). *Proc. IVth Int. Symp. on Pollination*. Md. Agric. Sta. Spec. Misc. Publ. 1:157–164.

———. 1978b. Complementary foraging of bee species in blueberries. *J. N.Y. Ent. Soc.* 84:315.

Reif, F. 1965. *Fundamentals of Statistical and Thermal Physics*. McGraw-Hill, New York, pp. 268, 293.

Richter, S. 1929. Über den öffnungsmechanismus der Antheren bei einigen Vertretern der Angiospermen. *Planta* 8:154–184.

Rick, C. M. 1950. Pollination relations of *Lycopersicon esculentum* in native and foreign regions. *Evolution* 4:110–122.

Rick, C. M., M. Holle, and R. W. Thorp. 1978. Rates of cross-pollination in *Lycopersicon pimpinellifolium;* Impact of genetic variation in floral characters. *Pl. Syst. Evol.* 129:31–44.

Roberts, R. B. 1969. Biology of the bee genus *Agapostemon* (Hymenoptera: Halictidae). *Univ. Kansas Sci. Bull.* 48(16):689–719.

———. 1978. Energetics of cranberry pollination. *Proc. IVth Int. Symp. on Pollination*. Md. Agric. Exp. Sta. Spec. Misc. Publ. 1:431–440.

Roberts, R. B. and S. R. Vallespir. 1978. Specialization of hairs bearing pollen and oil on the legs of bees (Apoidea: Hymenoptera). *Ann. Ent. Soc. Am.* 71(4):619–627.

Robertson, C. 1890. Flowers and insects. V. *Bot. Gaz.* 15:199–204.

Roubik, D. W., J. D. Ackerman, C. Copenhaver, and B. Smith. 1982. Stratum, tree and flower selection by tropical bees: implications for the reproductive biology of outcrossing *Cochlospermum vitifolium* in Panama. *Ecology* 63:712–720.

Rozen, J. G. 1977. Biology and immature stages of the bee genus *Meganomia* (Hymenoptera: Melittidae). *Am. Mus. Novitat.* No. 2630, pp. 1–14.

Salinas, A. O. D. and M. S. Sánchez. 1977. Biologia floral del genero *Cassia* en la region de los Tuxtlas, Veracruz. *Boletin de la Sociedad Botanica de Mexico* No. 37:5–45.

Schmalzel, R. J. 1980. The diet breadth of *Apis* (Hymenoptera: Apoidea). Unpub. M.S. thesis, University of Arizona. 79 pp.

Schmid, R. 1976. Filament histology and anther dehiscence. *Bot. J. Linnean Soc.* 75:303–315.

———. 1982. Functional and ecological interpretations of floral and fruit anatomy. *Bot. Rev.* 43. In press.

Schmid. R. and P. H. Alpert. 1977. A test of Burck's hypothesis relating anther dehiscence to nectar secretion. *New Phytol.* 78:487–498.

Schoenichen, W. 1924. *Biologie der Blütenpflanzen: eine Einführung an der hand mikroshopischer Übungen*. Freiburg i. Br.

Silander, J. A. 1978. Density-dependent control of reproductive success in *Cassia biflora*. *Biotropica* 10(4):292–296.

Spangler, H. G. 1967. Stridulation and related behavior in harvester ants. Unpub. Ph.D. dissertation. Kansas State University, Manhattan. 152 pp.

———. 1973. Vibration aids soil manipulation in Hymenoptera. *J. Kansas Ent. Soc.* 46(2):157–160.

Sprengel, C. K. 1793. *Das entdeckte Geheimnifs der Natur im Bau und in der Befruchtung der Blumen*. Facsimilie-Druck, Berlin: Mayer and Muller (1893).

Staedtler, G. 1923. Über Reduktionserscheinungen im bau der Antherenwand von Angiospermen-Blüten. *Flora, Jena* 116:85–108.

Stanley, R. G. and H. F. Linskens. 1974. *Pollen. Biology, Biochemistry, Management*. Springer-Verlag, New York. 307 pp.

Stebbins, G. L. 1974. *Flowering Plants: Evolution above the Species Level*. Belknap Press of Harvard University Press, Cambridge, Mass. 399 pp.

Stow, I. 1927. A cytological study on pollen sterility in *Solanum tuberosum*. *Jap. J. Bot.* 3(3):217–237.

Symon, D. E. 1979. Sex forms in *Solanum* (Solanaceae) and the role of pollen collecting insects. In *The Biology and Taxonomy of the Solanaceae* (Linnean Soc. Symp. Series No. 7). Academic Press, New York.

Thien, L. B. 1980. Patterns of pollination in the primitive angiosperms. *Biotropica* 12(1):1–13.

Thomson, W. T. 1965. *Vibration Theory and Application*. Prentice-Hall, Englewood Cliffs, N.J. pp. 243–246, 278–281.

Thorp, R. W. 1979. Structural, behavioral, and physiological adaptations of bees (Apoidea) for collecting pollen. *Ann. Missouri Bot. Gard.* 66:788–812.

Thorp, R. W. and J. R. Estes. 1975. Intrafloral behavior of bees on flowers of *Cassia fasciculata*. *J. Kansas Ent. Soc.* 48:175–184.

Tieghem, P. van. 1891. *Traité de botanique*, 2nd ed. Paris.

Todd, J. E. 1882. On the flowers of *Solanum rostratum* and *Cassia chamaecrista*. *Am. Nat.* 16:281–287.

Vogel, S. 1963. Duftdrusen im Dienste der Bestaubung

over Bau und Funktion der Osmophoren. *Abhandlungen Mathematischnatur-wissenschaft-liche Klasse Akademie Wissenschaften, Jahrg. Nr., Wisebaden* 10:601–763.

über Bau und Funktion der Osmophoren. *Abhandlungen Mathematischnatur-wissenschaft-liche Klasse Akademie Wissenschaften, Jahrg. Nr., Wisebaden* 10:601–763.

Vogel, S. 1974. *Ölblumen und ölsammelnde Bienen.* Steiner, Wisebaden. 267 pp.

Vogel, S. 1978. Evolutionary shifts from reward to deception in pollen flowers. *In The Pollination of Flowers by Insects* (Linnean Soc. Symp. Series No. 6) pp. 89–96.

Wallace, G. D. 1975. Studies of the Monotropoideae (Ericaceae): taxonomy and distribution. *Wasmann J. Biol.* 31(1/2):1–88.

Wille, A. 1963. Behavioral adaptations of bees for pollen collecting from *Cassia* flowers. *Rev. Biol. Trop.* 11(2):205–210.

Section II
CHEMICAL AND PHYSICAL ASPECTS OF POLLINATION

5
FLORAL NECTAR SUGAR CONSTITUENTS IN RELATION TO POLLINATOR TYPE

Herbert G. Baker and Irene Baker

Botany Department
University of California
Berkeley, California

ABSTRACT

The literature on the sugars that can be found in floral nectar is reviewed. A summary of the analyses we have made of the sugar compositions of floral nectar from 765 species of plants from a wide range of habitats and climatic zones is presented.

Evidence is presented that:

1. There is general constancy within a species in the placing of nectar in one of four classes, *viz.* sucrose/hexose ratios (by weight) of: < 0.1 (hexose-dominant), 0.1–0.499 (hexose-rich), 0.5–0.99 (sucrose-rich), and > 0.999 (sucrose-dominant). This constancy survives considerable environmental variation.

2. Most nectars contain sucrose, glucose, and fructose (often together with traces of other sugars). It is rare to find a nectar in which only one sugar is detectable, and none of our nectars contained only fructose, or even sucrose and fructose in the absence of glucose. The other combinations were all found.

3. Certain families are characterized by sucrose-rich (or dominant) nectars, e.g., Lamiaceae and Ranunculaceae. Other families are characteristically hexose-rich (or dominant), e.g., Brassicaceae and Asteraceae. Still other families show diversity, even within genera, e.g., Scrophulariaceae.

4. There are similarities to be seen in sugar-ratios between plants with the same pollinator type, even if the plants are taxonomically unrelated. Thus hummingbird-flowers have high sucrose/hexose ratios, while passerine (perching) bird-flowers have low ratios. The possible evolutionary background to the distinction between hummingbird and perching bird flower nectars is discussed. Nectars of flowers pollinated by bats of the suborder Microchiroptera, in the American tropics, are hexose-dominated or hexose-rich, but those visited by Megachiroptera in the Old World tropics may be somewhat richer in sucrose. Nectars from lepidopteran-pollinated flowers (visited by hawkmoths, or by "settling" moths, or by butterflies) are characteristically sucrose-rich or sucrose-dominant, in strong contrast to the bat-flower nectars. Nectars of flowers visited by long-tongued bees are usually sucrose-rich, while those pollinated by short-tongued bees or flies are generally hexose-rich or dominant. Short-tubed "bee- and butterfly-" flowers are hexose-rich or dominant. Wasp-flower nectars seem to be rather rich in sucrose.

These ratios presumably reflect pollinator preferences in each case, but physiological explanation of the differences is needed. Explanations will also have to take into account other chemicals in nectar (amino acids, lipids, etc.) and morphological and behavioral flower features, as well as factors external to the plant itself (but part of the ecosystem in which the plant grows).

In *Penstemon (sensu lato)*, in the Scrophulariaceae, 7 species that are pollinated primarily by hummingbirds have sucrose-rich or sucrose-dominant nectars, while 17 species primarily pollinated by insects are hexose-rich or hexose-dominant (with only one exception that is sucrose-rich). This distinction occurs even when species in the two groups are so closely related that they form fertile hybrids in nature, providing evidence of recent change in nectar-sugar chemistry.

In the tree genus *Erythrina* (Fabaceae) there is striking difference between the sucrose-rich (or dominant) condition of the hummingbird-pollinated species and the hexose-rich or hexose-dominant condition of all the species that are primarily pollinated by perching birds (whether in the Old World or the New). A prediction is made that hexose-dominant *Erythrina crista-galli*, which is widely cultivated, will prove to be primarily pollinated by perching birds in its South American home. A hybrid between this species and the sucrose-rich *E. herbacea* from Florida shows an intermediate sugar ratio.

The trumpet-vine genus *Campsis* (Bignoniaceae) contains two species: *C. radicans* in North America and *C. grandiflora* in eastern Asia. The former has sucrose-dominant nectar, the latter hexose-dominant nectar, corresponding to pollination by hummingbirds and perching birds, respectively. Their hybrid (*C. × tagliabuana*) is a common horticultural item; it has hexose-dominance comparable to that of its *grandiflora* parent.

Inga vera var. *spuria* (Fabaceae—Mimosoideae), in Costa Rica, is a tree that produces nectar in the late afternoon that is sucrose-dominated, and it is visited then by hummingbirds, various bees, some wasps, and moths. But about 4 hours later, the nectar, now sourly odoriferous, is taken by bats. A change has taken place in the nectar, possibly under the influence of microorganisms, but with an increase in hexose proportion that is favorable to the bats.

In the display of close coordination of nectar-sugar ratios with pollination types, occasional exceptions must be expected. One is provided by *Mutisia viciaefolia* from the Altiplano of Peru, which is unusual for a member of the Asteraceae in being adapted morphologically to hummingbird pollination. Adaptation appears to be complete except for hexose-richness of the nectar. Although the high-altitude occurrence of the plant may, in itself, favor hexose-richness, this might also represent "phylogenetic constraint"—sucrose-richness being very rare in the Asteraceae—and the appropriate genetic mechanism for sucrose-richness may not be available in the tribe Mutiseae. However, even if phylogenetic constraint may be acting, the appropriateness of the rest of the syndrome can produce success even if one character is out of keeping.

Further advances in studies of nectar chemistry in relation to pollination biology will come about if investigations are made on an *ecosystem* basis, and suggestions of necessities for such research are made.

KEY WORDS: Nectar, sugars, sugar-ratios, pollinators, phylogenetic constraint.

CONTENTS

INTRODUCTION

For several years, we have been making chemical analyses of nectars from as wide a range of floral sources as possible. Our published results (Baker and Baker, 1973a,b, 1975, 1976a,b, 1977, 1979a,b; Baker, 1977, 1978; Baker et al., 1978) have mostly been concerned with the amino acids that are present in floral nectar, but we have also given attention to other constituents.

Most important, in terms of concentration, among the nectar constituents are the sugars, to which much attention has been given because they are the basis of the energy reward that flower-visitors receive when they take nectar from flowers. Many studies have been made of the caloric content of this nectar reward in relation to the energetics of the flower-visitors—reviewed by Heinrich (1975, 1979)—but see Bolten et al. (1979) and Inouye et al. (1980). Such studies measure the sugars present as "sucrose equivalents" and generally do not take account of whether the sugar is in the form of glucose, fructose, sucrose, or some other sugar. It is our belief that, with their different sweetnesses (at least to the human palate) and evidence of differential attractiveness to bees and birds (von Frisch, 1950; Hainsworth and Wolf, 1976; Stiles, 1976), the different sugars that are to be found in different nectars, and the different proportions in which they occur, are worthy of careful examination.

It is the purpose of this chapter to report on the analyses of nectars from 765 species of flowering plants and to draw conclusions from the analyses. We shall show that there is an interplay between the constraints of phylogenetic relationship, on the one hand, and positive pollinator-adaptation, on the other, in determining the chemical composition of nectar.

Thus, there were three questions to be asked: First, to what extent does nectar tend to have a relatively constant composition within a taxon (species, genus, and family)? Second, how much modification in its composition takes place when an evolutionary relationship is struck up with a different pollinator class? Third, are certain classes of pollinators ruled out by inability of the plant to produce an appropriate nectar reward? We cannot answer all these questions firmly, for we shall only deal with a restricted number of sugars among the many chemicals that can be found in nectar, and our sizable sample of flower species is small compared with the multiplicity of species in existence, but we believe that our data point the way toward the answers.

METHODS OF ANALYSIS

Previous studies of the sugar composition of floral nectars have been numerous (including those reported in the following papers: Bonnier, 1879; von Planta, 1886; Beutler, 1930; Vansell, 1944a,b; Wykes, 1952a,b, 1953; Zimmermann, 1953, 1954; Bailey et al., 1954; Shuel, 1956; Furgala et al., 1958; Mauritzio, 1959, 1962; Percival, 1961; Lüttge, 1961; Kartashova and Novidova, 1964; Kleinschmidt et al., 1968; Baskin and Bliss, 1969; Yakovleva,

1969; Jeffrey et al., 1970; Butler et al., 1972; van Handel et al., 1972; Gottsberger, et al., 1973; Zauralov and Yakovleva, 1973; Brewer et al., 1974; Finch, 1974; Watt et al., 1974; Elias and Gelband, 1975; Elias et al., 1975; Stockhouse, 1975; Hainsworth and Wolf, 1976; Stiles, 1976; Baker et al., 1978; Baker, 1978; Käpylä, 1978; Macior, 1978; Baker and Baker, 1979a,b; Scogin, 1979; Steiner, 1979; Meeuse and Schneider, 1980; and others). However, comparisons of the results obtained in these studies are made only with difficulty because (a) many of these analyses were qualitative only, (b) they were made by a great variety of methods, of differing sensitivity, and (c) even when quantitative, the results are expressed in many different ways. Also, very few of these workers have attempted to relate their analyses to the kinds of pollinators that the plants reward with these nectars. Consequently, it is not surprising that there has been no consensus of opinion regarding the questions that we posed earlier in this paper.

Our own method of analysis involves the sampling of *freshly* produced nectar with finely-drawn-out micropipettes, spotting the nectar onto a strip of chromatography paper immediately, and drying it to await analysis. Subsequently, the spot is eluted with distilled water and the solution concentrated in a vacuum desiccator.

Respotting on Whatman No. 1 chromatography paper is followed by separation of the sugars by single-direction descending paper chromatography. The solvent system used is *n*-propanol : ethyl acetate : water (14 : 4 : 2 v/v/ v/), a slight modification of that given in Smith (1969, p. 314). The increased proportion of ethyl acetate serves to compact the individual sugar spots considerably. The running time is usually 65–72 hours, and several nectars can be run at a time. A standard mixture of known sugars is run on each paper as a control.

After air-drying, the papers are dipped in the staining reagent (Gal, 1968), which consists of two solutions mixed just prior to use. Solution A is 75 mg oxalic acid dissolved in 15 ml ethanol; solution B contains 150 mg of *p*-aminobenzoic acid in 25 ml chloroform and 2 ml acetic acid. After air-drying again, the papers are held at 110°C for 20 minutes, by which time pentoses become red, and all other sugars produce a brown color. Under ultraviolet (UV) illumination, all sugars fluoresce, and each spot can be outlined in pencil. In this manner, spots can be detected even when the amount of the sugar in question is too small to produce detectable color in visible light.

After cutting out the individual sugar spots, the stained sugars are eluted into 4 ml of 50% methanol, and their fluorescence is measured in a filter fluorometer. (We have used Turner model 111, with primary filter Corning 7-60, and secondary filter Kodak 12). Calibration curves correlating fluorescence values with amounts of sugar have been obtained for each sugar by chromatographing known quantities and obtaining their fluorescence values in the same manner. As the correlations are linear only in dilute solutions, it is necessary, on occasion, to further dilute the stained sugar with 50% methanol. However, by this method, sugar amounts from 0.5 μg upward may be estimated quantitatively.

This method is more sensitive than some other chromatographic methods that have been used (e.g., Percival, 1961; Hainsworth and Wolf, 1976; Stiles, 1976); so we have picked up sugar occurrences that were not recorded by other investigators. Even so, in our reports we do not say that a sugar is "absent" but simply that it was "not detectable" (N.D.).

The results obtained by this method of nectar analysis are comparable, quantitatively, with data obtained by technologically more complicated methods. Table 5-1 shows an example of such correspondence in analyses of floral and sub-bracteal nectars from *Gossypium hirsutum*. The analysis reported by Vansell (1944a) was made by traditional direct analytical chemical methods; the analysis by Butler et al. (1972) was of silylated derivatives of sugars by gas–liquid chromatography. Our analyses also agree well with paper chromatographic analyses of nectars of comparable species by other investigators (e.g., Mauritzio, 1959; Watt et al., 1974).

A most important technical point is that the nectar should *not* be allowed to stand in the

Table 5-1. Analyses of the proportions of the sugars present in floral and sub-bracteal nectar of *Gossypium hirsutum*.

FLORAL NECTAR	Sucrose	Glucose	Fructose
Vansell, 1944a	.022	.465	.514
Butler et al., 1972	.027	.542	.431
Baker and Baker	.022	.549	.429
SUB-BRACTEAL NECTAR	Sucrose	Glucose	Fructose
Butler et al., 1972	.135	.468	.396
Baker and Baker	.124	.456	.439

liquid condition. If it does, significant amounts of sucrose may be broken down, ultimately to increase the glucose and fructose concentrations; this may be caused by natural acidity of the nectar, or it may take place under the influence of enzymes occurring naturally in the nectar or brought in by microorganisms (especially yeasts). For the record, it may be pointed out that most nectars are on the acid side of neutrality (the lowest in our experience being pH 2.8 in *Strelitzia reginae*), although alkaline nectars occur [up to pH 8.2 in *Rhododendron ponticum* and pH 10 in *Viburnum costaricanum* (Baker and Baker, unpub.)]. Further work on the significance of pH variation in nectars is being carried out.

THE SUGARS IN NECTAR

All nectar analysts agree that the most common sugars in nectar are the disaccharide sucrose and the hexose monosaccharides glucose and fructose. These are the "big three." However, quite often other sugars are present in small amounts. Thus, the monosaccharides galactose and arabinose have been recorded by Watt et al. (1974) and Gottsberger et al. (1973), respectively, and the presence of mannose in nectar of *Tilia* spp. under drought stress has been implied by Crane (1977). The disaccharides maltose and melibiose occur occasionally (also see Percival, 1961, and later authors), whereas the trisaccharides melezitose and raffinose occur more frequently (according to Mauritzio, 1959, 1962; Percival 1961). We have found all of these sugars occasionally in nectars with the exception of arabinose and mannose, and we agree that melezitose is not uncommon. Occurring rarely (so far only in the Orchidaceae) are the disaccha-

ride cellobiose (Baskin and Bliss, 1969) and the tetrasaccharide stachyose (Jeffrey et al., 1970). More doubtful are the suggested occurrences of gentiobiose and lactose (Baskin and Bliss, 1969), manninotriose (Jeffrey et al., 1970), and the "insect sugar" trehalose (Percival, 1965, p. 85).

Although the unusual oligosaccharides are present in some completely fresh nectars (as we can testify), they may increase in amount in nectar that has been allowed to stand as liquid, for some of them are products of the breakdown of sucrose (Mauritzio, 1959). If fresh nectar is analyzed, it is our experience that the unusual sugars of all kinds are present only in very small quantities. Consequently, in our statistical adventures, we have confined ourselves, for the present, to the relative proportions by weight of the three common sugars, sucrose, glucose, and fructose, and the ratio of sucrose to glucose plus fructose in particular.

An explanation of our terminology is in order. A perfectly "balanced" nectar, containing equal quantities by weight of sucrose, glucose, and fructose, would have a sucrose/hexose ratio of 0.333/0.667 = 0.5. If there is as much sucrose (by weight) as the combined hexoses, the ratio will be 1.0. Consequently, we have designated nectars with a sucrose/hexose ratio of more than 0.999 "sucrose-dominated," while those with ratios between 0.5 and 0.999 are "sucrose-rich." Nectars with ratios between 0.1 and 0.499 are "hexose-rich," and those with ratios of less than 0.1 are "hexose-dominated." In some previous publications (Baker, 1978; Baker and Baker, 1979a,b) any ratio below 0.5 was called "sucrose-poor," but we believe it is preferable to use "hexose-rich" and "hexose-dominant" as the natural counterparts to "sucrose-rich" and "sucrose-dominant."

CONSISTENCY IN DETERMINATION

Furgala et al. (1958) studied nectar-sugar composition in several leguminous crops. They found no variation in nectar "type" within species. In 1961, Percival, in her semiquantitative study of nectars from 899 species of flowering plants, found significant variation from sample to sample of the same species in only 61 cases, that is, in less than 7% of the total. To be sure, for ourselves, of reasonable responsibility in reporting nectar type for a species, we have made multiple determinations wherever possible, and, in addition, we have looked particularly carefully at three species, *Gelsemium sempervirens* (Loganiaceae), *Agoseris glauca* (Asteraceae), and *Trifolium repens* (Fabaceae). These are all outcrossing species (*Gelsemium* is heterostylous; the *Agoseris* is homostylous, and may be self-incompatible—in any case it has showy capitula); the *Trifolium* is also homostylous but self-incompatible. In addition, the three color forms of *Kentranthus ruber* (Valerianaceae) were also examined. Tables 5-2 through 5-5 contain the data. In these species there is a high degree of consistency in the analyses, not only in the sucrose/hexose ratios but also in the glucose/fructose ratio. Cultivated material from several sources of *Trifolium repens* gave some suggestion of interpopulation differentiation (Table 5-5),

but the distinction is only between sucrose-rich and sucrose-dominant.

It is known that variations in environmental conditions can influence the volume of nectar that is produced by a flower and the overall concentration of solutes in the nectar (Shuel, 1955, 1957; Percival, 1965), but do these variations influence the relative proportions of the constituent sugars? There is no good evidence on this matter. Consequently, we experimented with material of the garden weed *Cymbalaria muralis (Linaria cymbalaria)*, a native of Europe that can flower throughout the year in Berkeley in a variety of temperatures and photoperiods. That the species is affected physiologically by this variation in environmental conditions is shown by the fact that, at the height of the summer, its corollas are almost white, whereas in cooler seasons, they are lilac-colored (Baker and Baker, 1977).

We grew ramets derived from the division of a single plant and put them in a series of six growth chambers (with three temperature regimes and two photoperiods) (Table 5-6). At low temperature ($10°C$) and short photoperiod (8 hours), no nectar was produced. At low temperature and long photoperiod (16 hours), very little nectar was produced (too little to establish the presence of hexoses). At higher temperatures ($26.7°/10°C$, and $26.7°C$ constantly), with short and long photoperiods, sucrose-dominant nectar was consistently produced. It may be noted that the amino acid

Table 5-2. Proportions of the sugars present in nectars from seven plants of *Gelsemium sempervirens* in cultivation in Berkeley.

	1	2	3	4	5	6	7
Melezitose	.036	.053	.050	.018	.023	.039	.029
Maltose	N.D.	N.D.	N.D.	N.D.	N.D.	N.D.	N.D.
Sucrose	.683	.692	.541	.617	.717	.652	.590
Glucose	.157	.129	.173	.235	.153	.168	.213
Fructose	.124	.125	.235	.129	.107	.145	.168
$\dfrac{S}{G+F}$	2.434	2.724	1.326	1.693	2.751	2.086	1.547
$\dfrac{G}{F}$	1.266	1.032	0.736	1.822	1.430	1.159	1.268

N.D. = not detectable.

Table 5-3. Proportions of the sugars present in nectars from seven plants of *Agoseris glauca* from Pennsylvania Mountain, Colorado.

	1	2	3	4	5	6	7
Melezitose	N.D.	N.D.	N.D.	N.D.	N.D.	N.D.	.030
Maltose	N.D.	.023	N.D.	N.D.	trace	N.D.	N.D.
Sucrose	N.D.	.032	N.D.	N.D.	.012	.047	.127
Glucose	.638	.621	.650	.667	.639	.542	.501
Fructose	.362	.323	.349	.332	.348	.412	.342
$\frac{S}{G + F}$	0.021	0.034	0.036	0.008	0.013	0.049	0.150
$\frac{G}{F}$	1.762	1.923	1.862	2.009	1.836	1.316	1.465

Table 5-4. Proportions of the sugars present in plants of three color varieties of *Kentranthus ruber*.

	White	Pink	Red
Melezitose	.018	.021	.018
Maltose	N.D.	N.D.	N.D.
Sucrose	.476	.490	.542
Glucose	.332	.320	.303
Fructose	.173	.169	.137
$\frac{S}{G + F}$	0.942	1.003	1.230
$\frac{G}{F}$	1.919	1.893	2.212

Table 5-5. Sucrose/hexose ratios for separate plants of *Trifolium repens* from five populations cultivated in Berkeley.

	$\frac{S}{G + F}$					$\bar{x}$
Cultivar Aber. 100	0.714	0.929	0.792	0.723		0.790
Cultivar S184	0.877	0.680	0.912	1.096	1.461	1.005
Berkeley—Campus lawn	2.323	2.207				2.265
Berkeley—Residence lawn	0.618					0.618
Tredegar, Wales	1.066	1.252				1.159

Table 5-6. Sucrose/hexose ratios of nectars produced by ramets of a clone of *Cymbalaria muralis* grown in different controlled environments.

Light/dark temperature	10°C	10°C	26.7/10°C	26.7/10°C	26.7°C	26.7°C
Photoperiod	8 hr	16 hr	8 hr	16 hr	8 hr	16 hr
Sucrose		+	.547	.619	.536	.571
Glucose		?	.217	.181	.214	.202
Fructose		?	.236	.200	.250	.227
$\frac{S}{G + F}$	—	—	1.220	1.625	1.154	1.329
$\frac{G}{F}$	—	—	0.919	0.905	0.856	0.890

Table 5-7. Numbers of nectars with detectable sugar combinations.

Sucrose	7
Glucose	2
Fructose	0
Sucrose + glucose	29
Sucrose + fructose	0
Glucose + fructose	78
Sucrose + glucose + fructose	649
	765

complements of these nectars also remained constant in the different environments (Baker and Baker, 1977). Such consistency in nectar constitution may be important in giving its nectar a uniform "taste," recognizable by a visitor to the flowers (Baker, 1978; Baker and Baker, 1977).

So with some degree of confidence, we may now look at the data from the 765 species sampled and draw some conclusions.

GENERAL CONSIDERATION OF NECTAR SAMPLES

A qualitative summary of the data for 765 species is given in Table 5-7. The first conclusion is that a great majority of nectars contain some proportion of all three major sugars, sucrose, glucose, and fructose. Second, although a few nectars appear to contain (in detectable quantity) only sucrose and very few only glucose, no nectars contain only fructose. Other investigators (e.g., Percival, 1961; Hainsworth and Wolf, 1976; Macior, 1978) have found nectars containing sucrose and fructose in which they have failed to detect glucose, but no such nectar has appeared in our samples. By contrast, sucrose and glucose (with an apparent absence

of fructose) and glucose and fructose (with sucrose undetectable) occur moderately frequently.

The very frequent occurrence together of sucrose, glucose, and fructose by no means implies that most nectars are "balanced," with equal quantities by weight of the three sugars. In fact we find such "balanced" nectars to be rare. Sucrose is often present in "overbalancing" amounts ("sucrose-rich" and "sucrose-dominant" nectars), but slightly more often "hexose-rich" or "hexose-dominant" is the appropriate description (Table 5-8) (see also Zauralov and Yakovleva, 1973).

Not even the two hexoses, glucose and fructose, need be "in balance"; in fact, they usually are not. In our sample of nectars, glucose more frequently outweighs fructose, although an overbalance in the other direction has been reported by Mauritzio (1959) from a smaller sample of "honeybee" flowers. Transglucosidases and transfructosidases have been shown to be present in some nectars (Zimmermann, 1953, 1954). Also glucose and fructose are surely secreted from the nectary independently and are not necessarily products of the hydrolysis of sucrose in the nectar, so there is no need to expect equality of representation of glucose and fructose. However, invertase was reported in the nectar of *Euphorbia pulcherrima* (Frey-Wyssling et al., 1954). The subject of glucose/fructose ratios is worthy of separate investigation, but, for the present, we shall confine our attention to sucrose/hexose ratios.

WITHIN-FAMILY RESEMBLANCES

In 1961, Percival showed that some families of flowering plants may be characterized by su-

Table 5-8. Numbers and relative proportions of the four categories of sugar ratios (hexose-dominant, hexose-rich, sucrose-rich, and sucrose-dominant) in the 765 species analyzed.*

N	< 0.1	0.1 to 0.499	0.5 to 0.999	> 0.999
765	196	231	149	190
	(.25)	(.30)	(.19)	(.25)
		(.55)		(.44)	

*Ratio calculated as S/(G + F) where S = sucrose, G = glucose, F = fructose, N = number of species.

crose-rich nectars, while others usually show hexose-rich nectars. Similar remarks were made about smaller samples by Zauralov and Yakovleva (1973) and Watt et al. (1974). With our samples, with more quantitative analyses, we can confirm Percival's findings generally. Thus, the Brassicaceae and Asteraceae are generally hexose-rich, while the Ranunculaceae and Lamiaceae exemplify sucrose-rich families (Table 5-9).

Each of the G-values in the table measures the significance of the deviation of the data for a particular family from the overall picture. G-values measure the goodness of fit of two spectra that are being compared. They have a similar statistical basis to chi-square values but are more accurate for the present purpose (see Sokal and Rohlf, 1969, Ch. 16).

However, the Scrophulariaceae show that uniformity is not necessarily the picture in all large families. *Pedicularis, Castilleja, Mimulus, Scrophularia, Bacopa, Phygelius,* and *Cymbalaria* show sucrose-richness or even dominance. *Penstemon,* however, spans a range from sucrose-dominance to hexose-dominance, while *Veronica* appears to be regularly hexose-dominant.

Such data as these, when they can be presented in detail, should be interesting to taxonomists and phylogenists, but the probability remains that the sucrose/hexose ratio of a nectar is *also* affected by the pollination biology of the species concerned. After all, the biological function of floral nectar is the reward of appropriate flower-visitors; presumably, different kinds of flower-visitors have different "taste" preferences and different nutrient needs that they will attempt to satisfy. How much does restriction in the potentiality of a plant species for meeting those preferences and needs influence the development of a working relationship with a particular kind of pollinator?

NECTAR SUGAR RATIOS AND POLLINATION BIOLOGY

Percival (1961) hinted at a relationship between pollination biology and nectar constitution when she pointed out that families with deep-tubed flowers tend to produce sucrose-rich nectar, whereas those with open or shallow-tubed flowers tend to be hexose-rich. Presumably, these states are related to the preferences and needs of the pollinators of these kinds of flowers.

But only for two classes of visitors have these preferences been investigated experimentally: honeybees and hummingbirds.

Honeybees

Pioneer studies were carried out by von Frisch (summarized in von Frisch, 1950) in which he demonstrated that, out of 34 sugars and related chemicals, only 9 "tasted sweet" to bees. These included sucrose, maltose, glucose, fructose, trehalose, and melezitose, all of which have been claimed to be present in some nectars. Apparently tasteless to bees were lactose,

Table 5-9. Porportions of species in the four sugar-ratio categories in some of the larger taxonomic families investigated.

N		$\dfrac{S}{G + F}$				G^*	P^*
		< 0.1	0.1 to 0.499	0.5 to 0.999	> 0.999		
10	Brassicaceae	.90	.10	0	0	20.21	< .001
52	Asteraceae	.51	.42	.06	.02	43.76	< .001
765	OVERALL	.25	.30	.19	.25	—	—
53	Scrophulariaceae	.09	.43	.21	.26	9.31	.025
21	Lamiaceae	0	.33	.24	.43	12.99	.005
21	Ranunculaceae	0	.10	.19	.71	26.19	< .001

*Tested against OVERALL (actual numbers utilized). For G-statistic see text. P = probability of difference from OVERALL.

melibiose, raffinose, xylose, and arabinose, all of which have also been claimed as nectar constituents. Among those sugars somewhat repellent to bees are cellobiose and gentiobiose (which have been claimed for some orchid nectars). Geissler and Steche (1962) suggested, and Barker and Lehner (1974b) confirmed, that galactose is toxic to honeybees. Crane (1977) has discussed the toxicity to honeybees of the hexose sugar, mannose, with the implication that it occurs in the nectar of some species of *Tilia* when these trees are subjected to drought stress. We have not identified it in nectar from unstressed trees of *T. platyphyllos* growing in Tredegar (Wales) and Berkeley.

So, if sucrose, glucose, and fructose are all acceptable to bees, the question is which, if any, do they prefer? From experiments carried out in England, Wykes (1952a,b) concluded that honeybees prefer a "balanced" nectar, with roughly equal representation of sucrose, glucose, and fructose. That such balanced nectars are uncommon, however, was suggested by the nectar survey by Percival (1961), and, later, Waller (1972) disputed Wykes's claim directly. His experiments showed that honeybees prefer a sucrose-rich liquid. Barker and Lehner (1974a,b) showed that starved honeybees would load their crops with sucrose, glucose, or fructose in quantity when each was offered singly, but that a mixture of sucrose with a hexose sugar caused even greater loading. These authors apparently did not try a combination of glucose and fructose, nor did they try all three sugars together.

From this blurred picture, the impression that one is left with is that honeybees are able to make good use of a sucrose-rich nectar, but that they do not demand it. And it must be remembered that honeybees (native to the Old World), although tremendously important as pollinators of contemporary plants, may not be representative of bees as a whole, so that generalizations about bee preferences and needs should not be based on honeybees alone.

Hummingbirds and Perching Birds

More experimental work has been done with the nectar-sugar preferences of hummingbirds.

From experiments with hummingbird "feeders," Hainsworth and Wolf (1976) concluded that the composition of artificial nectars had only a small effect on food choice "but the preferred sugar compositions appear to be the most common in nectars of plants visited by hummingbirds" (Hainsworth and Wolf, 1976, p. 101). This order of preference appeared to be SFG = SF > S > FG > SG > F > G (where S = sucrose, G = glucose, and F = fructose). Of 123 nectars of hummingbird flowers from North America, Costa Rica, Peru, and Ecuador that they analyzed qualitatively (Hainsworth and Wolf, 1976, Table 2), 90 were SFG, 13 SG, 7 SF, 8 GF, 5 S, and none was G or F. Thirty-one of these nectars also contained what they called "polysaccharides" (presumably oligosaccharides other than sucrose). In their paper, SFG merely indicates the presence of all three sugars; it does not imply that they are necessarily "in balance."

Using a different constellation of hummingbird species to provide experimental animals, Stiles (1976) found a somewhat different order of preference for nectar sugars (S > G > F, "with an equal parts mixture of the three sugars falling somewhere in the middle"). Nearly all birds rejected fructose in any comparison test. Van Riper (1960) also found the broadtailed hummingbird *(Selasphorus platycercus)* to prefer sucrose and glucose.

One feature remains clear from these three studies, and that is that hummingbirds prefer a sucrose-rich nectar (whatever other sugars may be present), which was confirmed by Stiles in field studies with Anna hummingbirds *(Calypte anna)* in southern California (Stiles, 1976, Table 15).

Stiles (1976) hinted that other kinds of flower-visiting birds may not have this same preference for sucrose over other sugars and, consequently, may not be rewarded with sucrose-rich nectar by the flowers they visit. Similarly, Cruden and Toledo (1977), using nectar analyses supplied by us, pointed out that *Erythrina breviflora*, which is pollinated in Mexico chiefly by orioles and tanagers (Cruden and Hermann-Parker, 1977), produces nectar that is markedly less rich in sucrose than the nectar of the predominantly hummingbird-pollinated *E. coralloides*. Subsequent chemical analyses by us have shown comparable results from nectars of other hummingbird-pollinated and perching bird–pollinated *Erythrina* species (Baker and Baker, 1979a,b; Feinsinger et al., 1979; Morton, 1979; Steiner, 1979). (See also Sherbrooke and Scheerens, 1979.)

Consequently, we decided to analyze our nectar data on the basis of principal pollinator classes, paying particular attention to the sucrose/hexose ratios. The attribution of each species to a particular pollinator class is made either on the basis of personal experience of it in nature and statements about its pollinators in the literature or on the basis of flower morphology, flower color, scent, and phenology (for syndromes see Knuth, 1906–9; van der Pijl, 1960–61; Baker and Hurd, 1968; Faegri and van der Pijl, 1971; Proctor and Yeo, 1972; Armstrong, 1979; Ford et al., 1970; etc.). Also some species have more than one important pollinator type; in these cases the data were used in each connection. The results of the analyses are shown in Table 5-10.

The first item to point out is that hummingbird-flower nectars *are* prevailingly sucrose-rich or sucrose-dominant. This is in striking contrast to the nectars of those species that are visited predominantly or only by perching birds (of the order Passeriformes), where the nectars are, almost without exception, hexose-rich or hexose-dominated. The difference between the two sets of data is very highly significant statistically (Table 5-11).

The perching birds represented by this sample of nectars belong to a mixture of families within which adaptation to nectar drinking may have evolved independently. Table 5-11 also shows that the selection of a hexose-rich or hexose-dominated nectar is characteristic of the flowers utilized by sunbirds (Nectariniidae) and white-eyes (Zosteropidae), which tend to visit the same kinds of flowers mostly in Africa and Asia, and also the honeyeaters (Meliphagidae), which are particularly characteristic of Australia, as well as the honeycreepers (Drepaninidae) of the Hawaiian Islands. The flowers visited by lorikeets (Psittacidae, of the order Psittaciformes) (only

Table 5-10. Numbers of species in each of the four sugar-ratio categories arranged by predominant pollinators.

	$\frac{S}{G+F}$						
	< 0.1	0.1 to 0.499	0.5 to 0.999	> 0.999	N	G*	P**
OVERALL	195	231	149	190	765	—	—
Hummingbirds	0	18	45	77	140	119.52	< .001
New World passerines	11	1	0	0	12	25.16	< .001
Sunbirds, etc.	24	9	2	0	35	28.07	< .001
Honeyeaters	18	4	0	0	22	36.87	< .001
Honeycreepers	5	1	0	0	6	10.57	< .02
Lorikeets, etc.	1	2	0	0	3	3.69	.30
Hawkmoths	2	8	19	32	61	41.16	< .001
Settling moths	3	14	11	15	43	70.07	< .001
Butterflies and skippers	5	17	24	29	75	24.23	< .001
Short-tongued bees and butterflies	23	21	3	0	47	38.07	< .001
Short-tongued bees	115	103	28	17	263	75.47	< .001
Long-tongued bees	13	75	49	66	203	42.40	< .001
New World bats	9	18	0	0	27	32.51	< .001
Old World bats	1	3	2	1	7	1.36	.90
Nonvolant mammals	0	2	2	1	5	13.44	< .01
Wasps	2	7	4	5	18	1.24	.75
Beetles	1	3	2	3	9	1.22	.75
Flies	29	27	7	9	72	14.82	< .001

*G = G-statistic (see text).
**P = probability of difference from OVERALL.

Table 5-11. Proportions of the species in each of the four sugar-ratio categories arranged according to types of birds that are their primary pollinators.

N		$\dfrac{S}{G+F}$				$G*$	$P*$
		< 0.1	0.1 to 0.499	0.5 to 0.999	> 0.999		
140	HUMMINGBIRDS	0	.13	.32	.55	—	—
12	New World passerines	.92	.08	0	0	76.05	$< .001$
35	Sunbirds, etc.	.69	.26	.06	0	124.05	$< .001$
22	Honeyeaters	.82	.18	0	0	107.85	$< .001$
6	Honeycreepers	.83	.17	0	0	42.22	$< .001$
3	Lorikeets	.33	.67	0	0	16.19	.001
78	All perching birds	.76	.22	.03	0	99.32	$< .001$

*Tested against HUMMINGBIRDS (actual numbers used).

represented by three Australian species in our study) also seem likely to be hexose-rich. But the most striking evidence of a difference from the hummingbird/sucrose relationship is shown by the New World passerine-bird flowers, which may be sympatric with the hummingbird flowers. These flowers are visited by a variety of passerine birds, as different as orioles (Icteridae), tanagers (Thraupidae), and bananaquits (Parulidae), yet they have nectar that is overwhelmingly hexose-dominated.

In Michigan, Southwick and Southwick (1980) observed ruby-throated hummingbirds taking phloem sap from holes in the trunks of paper birch trees *(Betula papyrifera)* made by yellow-bellied sapsuckers. During the nesting season, these hummingbirds neglected flower nectar and concentrated upon this source of liquid nourishment which these authors believe to be rich in sucrose.

A physiological explanation of the contrast between hummingbirds and perching birds in relation to the nectars they imbibe is required, and we hope that animal physiologists will be inspired to provide it.

Possible factors may include differences in taste of the sugars. Many hummingbird flowers are believed derived from bee-flower ancestry, and, as we shall show later, flowers polinated by long-tongued bees also tend to show sucrose-richness or dominance. The hummingbirds subsist on these nectars together with insects that they catch. By contrast, many of the passerine birds, or their relatives, will take

fruit juices as well, and these we find (Baker and Baker, unpub.) to be consistently hexose-rich or hexose-dominated. Perhaps the fruits have provided a conditioning of the taste sense of the perching birds. Other suggestions will be made later.

It should be noted that hummingbirds, in the right circumstances of hunger, will not spurn Old World bird flowers when these are provided for them in cultivation in California. In Berkeley, Anna and Allen hummingbirds freely visit *Eucalyptus ficifolia* from Australia (ratio 0.023), *Phormium tenax* from New Zealand (ratio 0.006), and *Kniphofia uvaria* from South Africa (ratio 0.038)—all species with hexose-dominant nectar.

Bat Flowers

Another group of nectars that is strikingly coherent and distinct is that of the flowers in the New World tropics that are visited by bats (all of which belong to the suborder Microchiroptera; Baker, 1973). These nectars, with a single exception (to be examined later in this paper), are consistently hexose-rich or hexose-dominant (Table 5-12). Hexose dominance was also remarked upon in a previous publication (Baker, 1978), for which a smaller sample of species was available. For example, our analysis of nectar collected from the tropical epiphyte *Markea neurantha* (Solanaceae) by Voss et al. (1980) showed such strong hexose dominance that sucrose was not detectable.

Table 5-12. Proportions of the species in each of the four sugar-ratio categories arranged according to their mammalian pollinators.

N			$\dfrac{S}{G + F}$				G*	P*
			< 0.1	0.1 to 0.499	0.5 to 0.999	> 0.999		
27	NEW WORLD BATS		.33	.67	0	0	—	—
7	Old World bats		.14	.43	.29	.14	10.85	< .02
5	Nonvolant mammals		0	.40	.40	.20	14.73	< .005

*Tested against NEW WORLD BATS (actual numbers used).

The difference between these bat flower nectars and those of the hummingbird flowers is highly significant ($G = 98.87$; df 4; $P < .001$).

In the Old World tropics, flower visitation is carried out by the other suborder of bats, the Megachiroptera (Baker, 1973). We have not yet been able to obtain many nectar samples from flowers visited by these bats, but the data available suggest that they are less heavily on the hexose side. E. Gould (1978) provided analyses that suggest at least the presence of sucrose in nectar from *Oroxylum indicum* (Bignoniaceae), while Lüttge (1961) found sucrose, glucose, and fructose in nectar of *Musa sapientum,* the cultivated banana plant, which, in its native Southeast Asia, is visited by nectar-lapping bats (van der Pijl, 1956).

Once again, a physiological explanation of why bats should appear to select for hexose-richness in nectar is required, but the fact that many nectar-drinking bats also chew fleshy or juicy fruits (and swallow the liquid contents) may have significance. By this practice, they may have developed a preference for the hexose sugar-richness that is common in fruits (Baker and Baker, unpub.) and have selected for this among the flowers.

There are very few analyses yet available of nectars that cater to the needs of nonvolant mammals that visit flowers (Table 5-12), but a trend toward sucrose-richness is worth following up.

Moth Flowers

Contrasting strongly with the hexose-richness of the bat-flower nectars is the prevailing sucrose-richness or dominance of the nectars of the equally nocturnal moth flowers (Table 5-13). A distinction has been made between those flowers whose nectar is taken by hawkmoths (Sphingidae) and those visited by "settling moths" (Noctuidae, Geometridae, Pyralidae, etc.), but the settling moth flowers are also predominantly sucrose-rich or sucrose-dominant.

Is the significance of this that it makes

Table 5-13. Proportions of the species in each of the four sugar-ratio categories arranged according to bat- or lepidopteran-pollination.

N			$\dfrac{S}{G + F}$				G*	P*
			< 0.1	0.1 to 0.499	0.5 to 0.999	> 0.999		
27	New World bats		.33	.67	0	0	65.93	< .001
7	Old World bats		.14	.43	.29	.14	6.21	.1
61	HAWKMOTHS		.03	.13	.31	.52	—	—
43	Settling moths		.07	.33	.26	.35	7.18	.05
75	Butterflies		.07	.23	.32	.39	3.93	> .25

*Tested against HAWKMOTHS (actual numbers used).

moth-flower nectar easily distinguishable from bat-flower nectar? This is possible because the noticed cases of hawkmoth and settling moths visiting "bat flowers" both involve Old World tree species. Thus, hawkmoths have been photographed visiting a bat flower (*Kigelia africana,* Bignoniaceae; Harris and Baker, 1958) in Africa, and noctuid moths have been photographed visiting planted *Durio zibethinus* (Bombacaceae) in Honduras (Baker, 1970). However, the evolutionary significance of the difference between moth-flower nectars and bat-flower nectars must take into account the fact that sucrose-richness also appears to characterize Temperate-Zone moth flowers as well as those from the tropics, and, it will be seen later, butterfly flowers as well.

Sucrose-richness may be a feature of Lepidoptera-adapted flowers as a whole, as suggested by Percival's (1961) observation of a nonquantified association between deep-tubed flowers and sucrose-richness of the nectar.

Butterfly Flowers

With regard to flowers visited by butterflies, a distinction must be made between two subgroups. On the one hand, there are those flowers that are primarily butterfly flowers, with deep, narrow corolla-tubes and relatively copious nectar production, and, on the other hand, those smaller flowers (often grouped in conspicuous inflorescences) that are visited almost equally by butterflies and "short-tongued" bees.

Taking the "true" butterfly flowers first (Table 5-13), they, like the moth flowers, are seen to be sucrose-rich. This group contains such notable "butterfly bushes" as *Buddleja*

davidii (ratio = 0.737), *Caesalpinia pulcherrima* (ratio = 1.237), and *Antigonon leptopus* (ratio = 0.667). In strong contrast, the relatively short-tubed "bee-and-butterfly" flowers, which provide good foraging for both bees and butterflies (often providing an excellent standing platform for prolonged visits by the insects), show a picture of hexose-rich nectars in our samples (Table 5-14). In comparison with "true" butterfly flowers, the difference is highly significant ($G = 65.27$; $d.f.$ 4; $P < .001$). Many of the "bee-and-butterfly" flowers are members of the Asteraceae, and, as that family is characterized by hexose-richness of the nectar, that may be a partial explanation of the picture we obtain. However, as will be seen shortly, hexose-richness seems to prevail among short-tubed and open bee-flowers, also.

In this connection, the results of an investigation of nectar feeding by *Colias* butterflies at high elevations in the Rocky Mountains of Colorado (Watt et al., 1974) are of interest. They found low sucrose/hexose ratios in the nectars of the species that were visited. However, in the subalpine zone, all but one [*Gilia (Ipomopsis) aggregata*] of the species on their list would be classified by us as "bee-and-butterfly" flowers, and 9 out of the 17 species are members of the Asteraceae. In the alpine zone, *Colias meadii* was seen to visit 8 flower species, 7 of which are Asteraceae, and all of which are "bee-and-butterfly" flowers. Our own continuing studies of nectars from flowers in the alpine zone at Jones Pass and Pennsylvania Mountain, in the Colorado Rockies, suggest that the great majority of plants growing in the tundra, whatever their visitors, have low sucrose/hexose ratios (Baker and Baker, unpub.); so the limitation may be in what the

Table 5-14. Proportions of the species in each of the four sugar-ratio categories arranged according to pollination by butterflies or by short-tongued bees and butterflies.

N		$\dfrac{S}{G+F}$				G^*	P^*
		< 0.1	0.1 to 0.499	0.5 to 0.999	> 0.999		
75	BUTTERFLIES	.07	.23	.32	.39	—	—
47	Short-tongued bees and butterflies	.49	.45	.06	0	65.27	< .001

*Tested against BUTTERFLIES (actual numbers used).

plants can produce in an alpine climate rather than what the flower visitors might prefer. In fact, Watt et al. (1974) point out that *Colias* butterflies have the enzymatic capability of interconverting sucrose, glucose, and fructose, and so can make use of nectars with any combination of these sugars that is provided to them.

Bee-Flowers

We have divided "bee-flowers" into two subgroups according to whether their nectar would be available only to "long-tongued" bees or whether it would be available to "short-tongued" bees (and to flies as well). This distinction follows one made by Robertson (1928); the threshold for distinction is at a tongue length of about 6 mm. Honeybees just about reach the threshold, and we have counted them with the short-tongued bees.

The difference in the results for the two kinds of bee-flower (Table 5-15) is statistically significant ($G = 126.72$; *d.f.* 4; $P < .001$). The long-tongued bees tend to be rewarded with sucrose-rich nectar, whereas the short-tongued bees and flies *may* get a sucrose-rich reward but most often do not. However, we should sound a word of caution about this contrast. In characterizing long-tongued bee-flowers as offering a sucrose-rich reward, we could fall into the trap set for us by phylogenetic constraint. Many of the long-tongued bee-flowers in our sample come from the Lamiaceae and Ranunculaceae, families characterized by sucrose-rich nectar, as well as from the genera of the Scrophulariaceae in which sucrose-richness is common. By contrast, many of the short-tongued bee-flowers are members of the Brassicaceae and Asteraceae which have the opposite characterization. It is difficult to see how one can resolve which is the cart and which is the horse in these relationships.

Other Insects

Some wasp flowers have been sampled, and the results from 18 species (Table 5-16) suggest that the proportion of sucrose to hexose is greater in them than for short-tongued bee-flowers ($G = 13.70$; *d.f.* 4; $P < .005$).

With beetles, the number of flower species (Table 5-16) is too small for conclusions to be drawn. With flies, the numbers are large enough for it to be possible to claim hexose-richness or dominance as a characteristic (Table 5-16). It should be noted, however, that most of these "fly-flowers" are also "short-tongued bee-flowers," so the lack of a significant difference between them is hardly surprising.

GENERAL CONSIDERATIONS OF CORRELATIONS

Within our sample of floral nectars we have demonstrated prevailingly high sucrose/hexose ratios in species pollinated by hummingbirds, Lepidoptera (especially sphingids), long-tongued bees, and possibly wasps, while the opposite condition prevails in nectars from flowers pollinated by perching birds and by bats (at least the Microchiroptera). On the whole, the flowers that are visited by short-tongued bees and flies also tend to be hexose-rich.

Any explanation of these ratios must await correlations that we have not yet completed,

Table 5-15. Proportions of the species in each of the sugar-ratio categories arranged according to bee tongue length.

N		$\dfrac{S}{G + F}$				G^*	P^*
		< 0.1	0.1 to 0.499	0.5 to 0.999	> 0.999		
263	SHORT-TONGUED BEES	.44	.39	.11	.07	—	—
203	Long-tongued Bees	.06	.37	.24	.33	126.72	< .001

*Tested against SHORT-TONGUED BEES (actual numbers used).

Table 5-16. Proportions of the species in each of the four sugar-ratio categories arranged according to pollination by various insects.

| N | | | $\dfrac{S}{G+F}$ | | | | |
		< 0.1	0.1 to 0.499	0.5 to 0.999	> 0.999	G^*	P^*
263	SHORT-TONGUED BEES	.44	.39	.11	.07	—	—
18	Wasps	.11	.39	.22	.28	13.70	< .005
9	Beetles	.11	.33	.22	.33	8.65	< .05
72	Flies	.40	.38	.10	.13	2.61	.5

*Tested against SHORT-TONGUED BEES (actual numbers used).

with such parameters as the amount of nectar produced and its total sugar content taken into account in each case. At first sight it does not look as if these correlations will be high. Large amounts of nectar are produced by sucrose-rich hummingbird flowers and by hexose-rich bat flowers, for example, and the overall concentrations of sugars in bat and hawkmoth flowers (with contrasting sucrose/hexose ratios) tend to be low in both cases (Baker, 1978).

The work of Wolf (1975) with sunbird foraging in Kenya suggests that the nectars taken by these birds may have about the same overall sugar concentrations (measured as "sucrose equivalents") as those of hummingbird flowers. This is also true of the Old World perching bird flowers in our sample. If this be true, the easy explanation that hummingbird-flower nectar is adapted by its high sucrose/hexose ratio to the high energy needs of these darting and hovering birds does not have the ring of truth to it, for equal weights of sucrose and hexose sugars (giving equal "sucrose equivalents") have roughly equal caloric contents.

The possibility cannot be ignored that the "taste" of these nectars (which will vary with the proportions of the constituent sugars) may be important in determining the acceptability of a particular nectar to a particular kind of flower-visitor. It has already been suggested in this chapter that the hexose-richness of juicy fruits may have conditioned perching birds and bats to a "taste" for a hexose-rich nectar. It is also true that nectars containing different mixtures of sugars will have different viscosities (I. Baker, in prep.), and this may influence the rate of feeding of a flower-visitor.

Also, Watt et al. (1974) point out that if some osmotic phenomenon limits sugar uptake by the flower visitor, the lower osmotic pressure exerted by a solution of oligosaccharides compared to a solution containing the same weight of monosaccharides (but a larger number of molecules) per unit volume will be advantageous. This could be an explanation of why some nectars are sucrose-rich or dominant, but, in view of the fact that these are in a minority in our sample (and the same is true for the sample reported on by Zauralov and Yakovleva, 1973), there must also be an explanation of why the majority of nectars are hexose-rich rather than sucrose-rich.

Again, Watt et al. (1974, p. 369) have a suggestion. They point out that "each formation of a saccharide bond costs the plant three high-energy phosphate bonds, from ATP or its equivalent, since phosphorylations of monosaccharides are essential steps in their assembly to . . . oligosaccharides." Thus, it might be advantageous to keep the excretion of oligosaccharides in nectar to the minimum that will still attract an appropriate flower-visitor. This would fit with the suggestion made elsewhere (Baker, 1978) that, to a bat, a primary attribute of floral nectar is that it is a refreshing drink rather than an important source of nutrients. Similarly, Watt et al. (1974) have suggested that the dilute nectars taken by *Colias* butterflies in the Colorado mountains may be important in maintaining the water balance of these insects.

However, obviously, neither sucrose/hexose ratios nor any other single feature of nectar sugar chemistry is the only factor involved in determining the foraging behavior of flower-visitors. Adaptation of nectar to a particular kind of flower-visitor may involve other chemicals in the nectar, such as lipids, amino acids, and proteins (Baker and Baker, 1973a,b, 1975, 1977, 1979b; Baker, 1977, 1978). Thus, members of the Asteraceae usually produce nectars relatively strong in amino acids (Baker and Baker, 1973b, 1975) which may compensate for relatively low sucrose/hexose ratios in some flowers that attract Lepidoptera.

Nectars may even contain toxic alkaloids, nonprotein amino acids, or other unpleasant substances (Baker and Baker, 1975; Baker, 1977, 1978), as well as phenolic substances. Even a suitable nectar may be unavailable to a particular visitor if the flowers are unattractive by reason of shape, color, scent, or time of opening, of if they are borne in an inappropriate position or are impossible for that visitor to enter. Clearly what is needed is a series of studies of pollinator rewards on an ecosystem basis, rather than just taking each flower species and each visitor species in isolation. One such study has recently been completed (Yorks, 1979). We are pressing forward in such studies in tundra and in tropical forests.

Also, success in attracting pollinators is a *relative* matter. The presence in the vicinity of other flowers with more desirable nectar will result in the neglect of the less desirable one

(Vansell et al., 1942, etc.), although this is a "two-sided coin" because *absence* of highly attractive competitors may mean that only partially desirable nectar will be taken by "desperate" visitors (von Frisch, 1950, etc.). Thus, Anna hummingbirds resident in Northern California during the winter months will visit any available garden flower.

However, it does seem that, within certain constraints imposed by membership of a particular family, coevolution with flower-visitors has gone on and has modified nectar composition. This can be further exemplified by consideration of individual plant genera (see also, Harborne, 1977, p. 52).

PENSTEMON AND ITS RELATIVES

A natural group of species for examination is provided by the large genus *Penstemon* (Scrophulariaceae) and its close relatives *Keckiella* and *Chionophila*. The genus *Keckiella* was the subgenus *Hesperothamnus* of *Penstemon* until removed by Straw (1966, 1967). Table 5-17 shows the sucrose/hexose ratios of those species whose nectars we have examined. It can be seen that, in accordance with the general pictures already drawn earlier in this chapter, the hummingbird-pollinated species are sucrose-rich or sucrose-dominant, whereas, on

Table 5-17. Sugar ratios shown by species of *Penstemon, Keckiella*, and *Chionophila* (Scrophulariaceae).

HUMMINGBIRDS		INSECTS (MOSTLY BEES)	
P. barbatus	0.901	P. heterodoxus	0.452
P. centranthifolius	1.626	P. oreocharis	0.359
P. kunthii	1.068	P. procerus	0.226
P. bridgesii	1.130	P. rydbergii	0.297
P. newberryi	1.012	P. virens	0.143
K. cordifolia	0.502	P. whippleanus	0.210
K. ternata	1.275	P. deustus	0.577
$\bar{x}$ = 1.075		P. eriantherus	0.853
Range 0.502–1.626		P. alpinus	0.261
?BEES OR HUMMERS		P. speciosus	0.329
P. campanulatus	0.413	P. secundiflorus	0.206
P. hartwegii	0.677	P. spectabilis	0.205
K. antirrhinoides	0.603	P. azureus	0.404
K. breviflora	0.727	P. heterophyllus	0.143
$\bar{x}$ = 0.605		P. laetus	0.242
Range 0.413–0.727		P. davidsonii	0.087
		C. jamesii	0.045
		$\bar{x}$ = 0.296	
		Range 0.045–0.853	

the whole, the insect-pollinated species (mostly bee-pollinated) have lesser sucrose/hexose ratios.

These data are the more impressive because the hummingbird-pollinated species belong to four different subgenera of *Penstemon* (Bennett, 1960) and to *Keckiella,* so that adaptation to these birds has apparently evolved separately at least five times. In the cases of *P. spectabilis* (which Straw, 1956, has shown is pollinated primarily by pseudomasarid wasps) and *P. centranthifolius* (which is hummingbird-pollinated) the species are closely enough related that they hybridize in nature, and the hybrids are fertile. Nevertheless, their sucrose/hexose ratios are very different—evidence of recent change in nectar-sugar chemistry, one way or the other. The same is true of *Penstemon davidsonii* (bumblebees) and *P. newberryi* (hummingbirds). The miscellaneous group in the bottom left-hand column of Table 5-17 seem poised to go either way.

ERYTHRINA

For the genus *Erythrina* (Fabaceae—Papilionoideae), Table 5-18 shows a striking picture. The hummingbird-pollinated species have moderately high sucrose/hexose ratios, whereas the species that are primarily pollinated by perching birds have very low sucrose/hexose ratios. This distinction will be the subject of detailed treatment in another publication (Baker and Baker, in prep.). Of interest here is the extremely low ratio obtained from the nectar of a tree of *E. crista-galli* growing in the University of California Botanical Garden, Berkeley (Baker and Baker, 1979b).

Erythrina crista-galli is native from eastern Brazil to Argentina and Uruguay, but is com-

Table 5-18. Sucrose/hexose ratios of *Erythrina* species pollinated by hummingbirds and perching (passerine) birds, respectively.

HUMMINGBIRDS	$\frac{S}{G+F}$	PERCHING BIRDS	$\frac{S}{G+F}$	
E. amazonica	0.557	E. abyssinica	0.072	O.W.
E. americana	0.845	E. acanthocarpa	0.067	O.W.
E. atitlanensis	1.198	E. breviflora	0.049	N.W.
E. berteroana	0.767	E. burano	0.021	O.W.
E. chiapasana	2.030	E. caffra	0.023	O.W.
E. chiriquensis	0.612	E. dominguezii	0.032	N.W.
E. cobanensis	2.069	E. falcata	0.039	N.W.
E. corallodendrum	0.735	E. fusca**	0.039	N.W. + O.W.
E. coralloides	0.921	E. humeana	0.033	O.W.
E. costaricensis	0.811	E. latissima	0.074	O.W.
E. flabelliformis	1.470*	E. lysistemon	0.047	O.W.
E. folkersii	2.187	E. megistophylla	0.015	N.W.
E. guatemalensis	0.888	E. perrieri	0.071	O.W.
E. herbacea	1.048	E. poeppigiana	0.029	N.W.
E. lanceolata	0.609	E. resupinata	0.030	O.W.
E. macrophylla	1.268	E. tahitensis	0.056	O.W.
E. pallida	1.652	E. variegata	0.017	O.W.
E. rubrinervia	1.076	E. verna	0.032	N.W.
E. salviiflora	1.301	(E. crista-galli	0.028)	N.W.
E. speciosa	1.046			
E. tajumulcensis	0.755			
E. sp. (Opler 1680)	0.575			
E. sp. (Cruden 2286)	0.675			

Range = 0.557–2.187
All New World

Range = 0.015–0.074
N.W. = New World
O.W. = Old World

*Includes data from Sherbrooke and Scheerens (1979).
**Also visited by bats.

monly cultivated in warmer parts of the world (Krukoff and Barneby, 1974). However, despite its abundance, there is only one record of hummingbird visits to this species (Pickens, 1931, from the literature survey by Toledo, 1974). Even this record is presented without details and was probably made in California on cultivated material and with hummingbirds that differ from those in its native area. It is our experience with the tree growing in the University Botanical Garden, Berkeley (Fig. 5-1) that Anna hummingbirds may visit it, but that they frequently take nectar "illegitimately" from above the flower, poking their beaks into a fold in the standard, making no contact with anthers or stigma. They make approaches from in front of the flower (legiti-

mately), in addition. We believe that in the inverted flower position, so characteristic of this species, the standard provides a standing platform for small perching birds (which could also stand on an adjacent flower) and would contact anthers and stigma with the tops of their heads as they probe for nectar. Thus, it is interesting to find that Ali (1932) saw cultivated trees in India pollinated successfully by sunbirds, which must perch while they feed. Similar observations were made in South Africa by Guillarmod et al. (1979).

Van der Pijl (1937) recorded that, in Java, *E. crista-galli* flowers are "robbed" by perching birds that take nectar from the side of the flower. (Van der Pijl was under the impression that, because this species is native to South

Figure 5-1. Flowers of *Erythrina crista-galli* on a tree growing in the University of California Botanical Garden, Berkeley.

Table 5-19. Nectar-sugar composition in two *Erythrina* species and the hybrid between them.

	Erythrina crista-galli	*Erythrina × bidwillii*	*Erythrina herbacea*
Melezitose	.009	.008	.017
Maltose	.015	N.D.	N.D.
Sucrose	.032	.177	.394
Glucose	.417	.434	.318
Fructose	.527	.381	.272
Ratio $\frac{S}{G+F}$	0.034	0.217	0.668

America and has flowers that face outward from the inflorescence, it is usually hummingbird-pollinated.) Werth (1915) recorded the visits of bumblebees to the flower. Careful observation of this species in its South American native area should establish whether or not *E. crista-galli* is another passerine (perching) bird species or not, but the sucrose/hexose ratio of its nectar points that way.

What about the inheritance of sugar characteristics in *Erythrina*? With woody plants it is hard to make experimental genetical studies. However, horticulturists have helped by making hybrids. One such is *Erythrina × bidwillii* (Table 5-19), the hybrid between *E. crista-galli* (low sucrose/hexose ratio) and *E. herbacea* (high ratio). The sucrose/hexose ratio of the hybrid is intermediate between the ratios of the parents (Baker and Baker, 1979b).

CAMPSIS

Another interesting question is raised by the genus *Campsis* in the Bignoniaceae. This genus of vines contains only two species: *C.*

radicans, which is native to the southern and midwestern states of the United States, and *C. grandiflora* (= *C. chinensis*), which is native to China (and possibly native to or introduced at an early date to Japan). Although they are so disjunct geographically, and probably have been so since some time in the Tertiary Period, they can be crossed, and rather fertile material of hybrid origin (*C. × tagliabuana*) is commonly cultivated (under a variety of names, including those of the two species). Probably, as a result of the raising of further generations from seed, and of back-crossing to the parents, *C. × tagliabuana* is now a "hybrid" swarm.

The Old World *C. grandiflora* appears to be adapted to pollination by passerine birds, and the New World *C. radicans* is an acknowledged hummingbird flower. In cultivation, the hybrids are visited frequently by bees as well.

Table 5-20 contains data derived from the analyses of nectars carried out on cultivated material of the two species and of hybrids by Colliva and Giulini (1970) as well as our own analyses of nectar from four clones of *C. radicans* and from ten clones of hybrid material. The high sucrose/hexose ratio in *C. radicans* is seen to be appropriate to the hummingbirds who service it, whereas the very low ratio for *C. grandiflora* fits with its probable pollination by passerine birds in China. The many determinations of a low sucrose/hexose ratio in the hybrid material appear to demonstrate that this condition is genetically partially or completely dominant over sucrose-richness, a finding that is being tested in other cases where closely related species with differing sucrose/hexose ratios can be hybridized.

Table 5-20. Sugar complements of *Campsis* and *Distictis* (Bignoniaceae).

	Melez.	Malt.	Sucr.	Gluc.	Fruct.	$\frac{S}{G+F}$	$\frac{G}{F}$
Campsis radicans (4 collections)	.015	.012	.616	.152	.206	1.721	0.733
Campsis × tagliabuana (10 collections)	.006	.011	.054	.578	.351	0.058	1.647
Campsis grandiflora (Colliva and Giulini, 1970)	N.D.	N.D.	trace	.600	.400	near 0	1.500
Distictis buccinatoria (4 collections)	.053	N.D.	.390	.265	.292	0.703	0.908

There is one other analysis of floral nectar of *Campsis radicans* in the literature. This was made by Elias and Gelband (1975), on material collected in Illinois, and showed an absence of sucrose (with glucose and fructose represented about equally). Our best explanation of this discordant result is that introgression of *C. grandiflora* genes may have taken place (via hybridization between *C. radicans* and cultivated *C.* × *tagliabuana*) in the history of this midwestern material.

Another hummingbird-pollinated "trumpet" vine in the Bignoniaceae, *Distictis buccinatoria (Phaedranthus buccinatorius)*, is in common cultivation in the San Francisco Bay Region. A native of Mexico, whose congeneric relatives are confined to the New World tropics (Gentry, 1973, 1974), it has a thoroughly appropriate sucrose-rich constitution (Table 5-20).

INGA VERA VAR. SPURIA

We have already reported that New World bat-pollinated species in our sample are characterized by hexose-dominant or hexose-rich nectars. We have only encountered one apparent exception that showed sucrose-domination in freshly produced nectar (Table 5-21). It is *Inga vera* var. *spuria,* a tropical forest tree (of the Fabaceae—Mimosoideae) from Central America where the samples that we analyzed were collected from freshly opened flowers by Dr. Paul A. Opler and by Ms. Suzanne Koptur, in Costa Rica. The pollination of this species by bats has been studied by Salas (1974).

Salas reports that the flowers are visited, after their late-afternoon opening, by moths,

Table 5-21. Sugar complements of *Inga vera* var. *spuria* sampled in the afternoon and morning.

	4 P.M.	6 P.M.	Morn.
Melezitose	.053	.019	.014
Maltose	N.D.	.015	.021
Sucrose	.597	.321	.147
Glucose	.160	.414	.461
Fructose	.190	.231	.357
$\frac{S}{G + F}$	1.781	0.498	0.180

hummingbirds, various bees, and some wasps, to all of whom a high sucrose content would be appropriate. But Salas (1974) also reports that a sour odor, which makes the flower attractive to bats, is not developed immediately when the flowers open, but becomes apparent about four hours later, and then the flowers are visited by bats. He suggests that this phenomenon is due to the fermentative actions of microorganisms, particularly yeasts. If this is true, we can look to some hydrolysis of the sucrose to glucose and fructose as part of the fermentation process. Thus, the sucrose/hexose ratio of the nectar could be lowered to a level in accord with that found in other bat-flowers. Although not made on the same tree in the same night, the data (Table 5-21) show a diminution of the sucrose/hexose ratio as the night wears on.

MUTISIA VICIAEFOLIA

Although, in general, there is a good correlation between nectar sugar ratios and pollinators, it is not to be expected that there will be a complete absence of exceptions. We realize that adaptation to a particular kind of pollinator involves a syndrome of characters, and one of these characters may be at variance with the overall pattern and still allow good adaptation.

Thus, hummingbird pollination is rare in the Asteraceae, and this could be linked to the prevailing hexose-richness of the nectar in this family. Hexose-richness is still found in the one hummingbird-pollinated member of this family that we have studied, *Mutisia viciaefolia* from the Altiplano of Peru (Fig. 5-2 and Table 5-22).

However, this species has a vine habit that brings its flowers up into the flying range of the hummingbirds. Its inflorescence is columnar, standing clear of the foliage, with long florets facing upward where they can be easily probed by a hovering bird. The flower-heads are of an appropriate orange color and are odorless. The pollen is prominently displayed on the well-exserted stigmas. Everything fits with hummingbird pollination except for the hexose-rich nectar—and this may be where phylogenetic constraint is felt. The Asteraceae are charac-

Figure 5-2. Flower-heads of *Mutisia viciaefolia* on a vine growing in the University of California Botanical Garden, Berkeley.

terized by hexose-rich (or dominant) nectar, although a very small number of sucrose-rich nectars have been found in the family (Percival, 1961; Lüttge, 1961; Baker and Baker, unpub.), and it may be that the tribe Mutisiae lacks the genetic apparatus for production of sucrose-rich nectar. It is also possible that, in the severe climatic conditions of the Altiplano there may be energetic reasons for the production of hexose-rich rather than sucrose-rich nectar (as suggested earlier in this chapter). Nevertheless, the syndrome of characters is sufficiently complete to be workable.

CONCLUSION

We are now in a position to give partial answers to the three questions posed at the beginning of this chapter.

First, there *is* a tendency for intrafamilial resemblances in the ratio between sucrose and the hexose sugars in nectar. There is also a general agreement between sugar ratios and the nature of the principal pollinators.

Second, some modification of nectar-sugar proportions can take place when a new flower/pollinator relationship is set up.

Third, there may be some tendency to predispose members of a particular family to pollination by particular classes of pollinators by reason of nectar chemistry, but this can be outweighed by adaptations in the morphological (and phenological) features of flowers and inflorescences.

FUTURE STUDIES

Future studies of floral nectar sugars should be coordinated with studies of other chemicals present in the nectar, whether these be attractive or repellent to visitors. Then, the program should include investigations that have an ecosystem basis, with numbers and densities of plants and potential pollinators elucidated. The range of the potential pollinators available in the ecosystem will be related to the range of rewards profferred by the plants. The phenology of the plants and of the flower-visitors will help to indicate when and to what degree there is likely to be competition between plants for pollinators (or the converse), which, again, is likely to influence the closeness with which the sugar reward is matched to the needs of the pollinators. The picture will also be influenced by the breeding system of the plants (whether they require outcrossing, are capable of selfing, or are even autogamous, or are given to apomixis or vegetative reproduction), determining how much the plants have need for the services of pollinators.

For any one ecosystem these desiderata will require that a major effort be made, but it is only with this information in hand that comparisons between ecosystems and studies of an

Table 5-22. Sugar complement of nectar of *Mutisia viciaefolia*.

	Sucrose	Glucose	Fructose	$\dfrac{S}{G + F}$
Mutisia viciaefolia	.175	.486	.339	0.224

evolutionary nature can be made with confidence.

ACKNOWLEDGMENTS

We are grateful to many persons who have collected nectar samples that have been incorporated into our analyses, including R. W. Cruden, G. W. Frankie, E. O. Guerrant, W. A. Haber, S. Koptur, W. Lucas, E. McClintock, P. A. Opler, R. A. Robichaux, C. E. Turner, and P. F. Yorks. F. Bowcutt, J. Cronhart, E. O. Guerrant, D. McElfresh, D. Schoen, and C. E. Turner assisted in making the analyses.

Valuable financial assistance was received from the National Science Foundation (grants BMS 73 01619 A01, BEB 76-19919, and DEB 78 11728).

LITERATURE CITED

Ali, S. A. 1932. Flower birds and bird flowers in India. *J. Bombay Nat. Hist. Soc.* 35:573–605.

Armstrong, J. A. 1979. Biotic pollination mechanisms in the Australian flora—a review. *N. Z. J. Bot.* 17:467–508.

Bailey, M. E., E. A. Fieger, and E. Oertel. 1954. Paper chromatographic analyses of some southern nectars. *Gleanings Bee Cult.* 82:401–403, 472–474.

Baker, H. G. 1970. Two cases of bat pollination in Central America. *Rev. Biol. Trop.* 17:187–197.

———. 1973. Evolutionary relationships between flowering plants and animals in American and African tropical forests. *In* B. J. Meggers, E. S. Ayensu, and W. D. Duckworth (eds.), *Tropical Forest Ecosystems in Africa and South America: A Comparative Review.* Smithsonian Institution Press, Washington, D.C., pp. 145–159.

———. 1977. Non-sugar components of nectar. *Apidologie* 8:349–356.

———. 1978. Chemical aspects of the pollination biology of woody plants in the tropics. *In* P. B. Tomlinson and M. Zimmermann (eds.), *Tropical Trees as Living Systems.* Cambridge University Press, New York, pp. 57–82.

Baker, H. G. and I. Baker. 1973a. Amino acids in nectar and their evolutionary significance. *Nature, Lond.* 241:543–545.

———. 1973b. Some anthecological aspects of the evolution of nectar-producing flowers, particularly amino acid production in nectar. *In* V. H. Heywood (ed.), *Taxonomy and Ecology.* Academic Press, London, Ch. 12, pp. 243–264.

———. 1975. Nectar constitution and pollinator-plant coevolution. *In* L. E. Gilbert and P. H. Raven (eds.), *Animal and Plant Coevolution.* University of Texas Press, Austin, pp. 100–140.

———. 1977. Intraspecific constancy of floral nectar amino acid complements. *Bot. Gaz.* 138:183–191.

———. 1979a. Sugar ratios in nectars. *Phytochem. Bull.* 12:43–45.

Baker, H. G. and P. D. Hurd. 1968. Intrafloral ecology. *Annu. Rev. Ent.* 13:385–414.

Baker, H. G., P. A. Opler, and I. Baker. 1978. A comparison of the amino acid complements of floral and extrafloral nectars. *Bot. Gaz.* 139:322–332.

Baker, I. and H. G. Baker. 1976a. Analyses of amino acids in flower nectars of hybrids and their parents, with phylogenetic implications. *New Phytol.* 76:87–98.

———. 1976b. Analyses of amino acids in nectar. *Phytochem. Bull.* 9:4–7.

———. 1979b. Chemical constituents of the nectars of two *Erythrina* species and their hybrid. *Ann. Missouri Bot. Gard.* 66:446–450.

Barker, R. J. and Y. Lehner. 1974a. Acceptance and sustenance value of naturally occurring sugars fed to newly emerged adult workers of honeybees (*Apis mellifera* L.). *J. Exp. Zool.* 187:277–286.

———. 1974b. Influence of diet on sugars found by thin-layer chromatography in thoraces of honeybees, *Apis mellifera. J. Exp. Zool.* 188:157–164.

Baskin, S. I. and C. A. Bliss. 1969. Sugar occurring in the extrafloral exudates of the Orchidaceae. *Phytochemistry* 8:1139–1145.

Bennett, R. W. 1960. *Penstemon Nomenclature.* Supplement to Bulletin XIX. American Penstemon Society, Arlington, Va.

Beutler, R. 1930. Biologisch-chemische Untersuchungen am Nektar von Immenblumen. *Zeit. Vergl. Physiol.* 12:72–176.

Bolten, A. B., P. Feinsinger, H. G. Baker, and I. Baker. 1979. On the calculation of sugar concentration in flower nectar. *Oecologia (Berl.)* 41:301–304.

Bonnier, G. 1879. Les nectaires. *Annales des Sciences Naturelles (Botanique)* 8:1–213.

Brewer, J. W., K. J. Colyard, and C. E. Lott, Jr. 1974. Analysis of sugars in dwarf mistletoe nectar. *Can. J. Bot.* 52:2533–2538.

Butler, G. D., G. M. Loper, S. E. McGregor, J. L. Webster, and H. Margolis. 1972. Amounts and kinds of sugars in the nectars of cotton (*Gossypium* spp.) and the time of their secretion. *Agron. J.* 64:364–368.

Colliva, V. and P. Giulini. 1970. Osservazioni sull'ibrido *Campsis* × *tagliabuana* (Vis.) Massalongo e sulle sue specie genitrice. *Giorn. Bot. Ital.* 104:469–482.

Crane, E. 1977. Dead bees under lime trees. *Bee World* 58:129–130.

Cruden, R. W. and S. Hermann-Parker, 1977. Defense of feeding sites by orioles and hepatic tanagers in Mexico. *Auk* 94:594–596.

Cruden, R. W. and V. M. Toledo. 1977. Oriole pollination of *Erythrina brevifolia* (Leguminosae): evidence for a polytypic view of ornithophily. *Pl. Syst. Evol.* 126:393–403.

Elias, T. S. and H. Gelband. 1975. Nectar: its production and functions in trumpet creeper. *Science, N.Y.* 189:289–291.

Elias, T. S., W. R. Rozich, and L. Newcombe. 1975. The foliar and floral nectaries of *Turnera ulmifolia* L. *Am. J. Bot.* 62:570–576.

Faegri, K. and L. van der Pijl. 1971. *The Principles of*

Pollination Ecology, 2nd ed. Oxford, Pergamon Press.

Feinsinger, P., Y. B. Linhart, L. E. Swarm, and J. A. Wolfe. 1979. Aspects of the pollination biology of three *Erythrina* species on Trinidad and Tobago. *Ann. Missouri Bot. Gard.* 66:451–471.

Finch, S. 1974. Sugars available from flowers visited by the adult cabbage root fly, *Erioischia brassicae* (Bch.) (Diptera, Anthomyiidae). *Bull. Ent. Res.* 64:257–263.

Ford, H. A., D. C. Paton, and N. Forde. 1979. Birds as pollinators of Australian plants. *N.Z. J. Bot.* 17:509–520.

Frey-Wyssling, A., M. Zimmermann, and A. Mauritzio. 1954. Über den enzymatischen Zuckerumbau in Nektarien. *Experientia* 10:490–491.

Frisch, K. von 1950. *Bees—Their Vision, Chemical Senses and Language.* Cornell University Press, Ithaca, N.Y.

Furgala, B., T. A. Gochnauer, and F. G. Holdaway. 1958. Constituent sugars of some northern legume nectars. *Bee World* 39:203–205.

Gal, A. E. 1968. An improved method for the detection and radioassay of monosaccharides on thin-layer or paper chromatograms. *J. Chromatog.* 34:266–268.

Geissler, G. and W. Steche. 1962. Natürliche Trachten als Ursache für Vergiftungen bei Beinen und Hummeln. *Z. Bienenforsch.* 6:77–92.

Gentry, A. H. 1973. Generic delimitations of Central American Bignoniaceae. *Brittonia* 25:226–242.

———. 1974. Studies of Bignoniaceae 11: a synopsis of the genus *Distictis. Ann. Missouri Bot. Gard.* 61:494–501.

Gottsberger, G., J. Schrauwen, and H. F. Linskens. 1973. Die Zucker-bestandteil des Nektars einiger tropischer Blüten. *Portug. Acta. Biol. Ser. A* 13:1–8.

Gould, E. 1978. Foraging behavior of Malaysian nectar-feeding bats. *Biotropica* 10:184–193.

Guillarmod, A. J., R. A. Jubb, and C. J. Skead. 1979. Field studies of six southern African species of *Erythrina. Ann. Missouri Bot. Gard.* 66:521–527.

Hainsworth, F. R. and L. L. Wolf. 1976. Nectar characteristics and food selection by hummingbirds. *Oecologia (Berl.)* 25:101–113.

Harborne, J. B. 1977. *Introduction to Ecological Biochemistry.* Academic Press, London.

van Handel, E., J. S. Haeger and C. W. Hansen. 1972. The sugars of some Florida nectars. *Am. J. Bot.* 59:1030–1032.

Harris, B. J. and H. G. Baker. 1958. Pollination in *Kigelia africana* Benth. *J. W. Afr. Sci. Assn.* 4:28–30.

Heinrich, B. 1975. Energetics of pollination. *Annu. Rev. Ecol. Syst.* 6:139–170.

———. 1979. *Bumblebee Economics.* Harvard University Press, Cambridge, Mass.

Inouye, D. W., N. D. Favre, J. A. Lanum, D. M. Levine, J. B. Myers, M. S. Roberts, F. C. Tsao, and Y. Y. Wang. 1980. The effects of nonsugar nectar constituents on estimates of nectar energy content. *Ecology* 61:992–996.

Jeffrey, D. C., J. Arditti, and N. Koopowitz. 1970. Sugar content in floral and extrafloral exudates of orchids: pollination, myrmecology and chemotaxonomy implication. *New Phytol.* 69:187–195.

Käpylä, M. 1978. Amount and type of nectar sugar in some wild flowers in Finland. *Ann. Bot. Fennici* 15:85–88.

Kartashova, N. N. and T. N. Novidova. 1964. A chromatographic study of the chemical composition of nectar. *Isv. Tomskogo Otd. Vses, Botan. Obschestva* 5:111–119 (in Russian). (*Chem. Abstr.* 64: no. 8641a, 1966.)

Kleinschmidt, M. G., A. K. Dobrenz, and F. G. McMahon. 1968. Gas chromatography of carbohydrates in alfalfa nectar. *Plant Physiol.* 43:665–667.

Knuth, P. 1906–9. *Handbook of Flower Pollination.* Trans. by P. A. Davis. 3 vols. (I, 1906; II, 1908; III, 1909). Clarendon Press, Oxford.

Krukoff, B. A. and R. C. Barneby. 1974. Conspectus of species of the genus *Erythrina. Lloydia* 37(3):332–459.

Lüttge, U. 1961. Über die Zusammensetzung des Nektars und der Mechanismus seiner Sekretion. I. *Planta* 56:189–212.

Macior, L. W. 1978. Pollination interactions in sympatric *Dicentra* species. *Am. J. Bot.* 65:57–62.

Mauritzio, A. 1959. Papierchromatographische Untersuchungen an Blütenhonigen und Nektar. *Ann. de l'Abeille* 4:291–341.

———. 1962. From the raw material to the finished product: honey. *Bee World* 43:66–81.

Meeuse, B. J. D. and E. L. Schneider. 1980. *Nymphaea* revisited: a preliminary communication. *Israel J. Bot.* 28:57–79.

Morton, E. S. 1979. Effective pollination of *Erythrina fusca* by the orchard oriole *(Icterus spurius):* coevolved behavioral manipulation. *Ann. Missouri Bot. Gard.* 66:482–489.

Percival, M. S. 1961. Types of nectar in angiosperms. *New Phytol.* 60:235–281.

———. 1965. *Floral Biology.* Pergamon Press, Oxford.

Pickens, A. L. 1931. Some flowers visited by birds. *Condor* 33:22–38.

Pijl, L. van der. 1937. Disharmony between Asiatic flower-birds and American bird-flowers. *Ann. Jard. Bot. Buitenz.* 48:17–26.

———. 1956. Remarks on pollination by bats in the genera *Freycinetia, Duabanga* and *Haplophragma,* and on chiropterophily in general. *Acta Bot. Neerl.* 5:135–144.

———. 1960–61. Ecological aspects of flower evolution. Parts I and II. *Evolution* 14:403–416; 15:44–59.

von Planta, A. 1886. Über die Zusammensetzung einiger Nektar-Arten. *Hoppe-Seyl. Z. (Z. Physiol. Chem.)* 10:227–247.

Proctor, M. C. F. and P. R. Yeo. 1972. *The Pollination of Flowers.* Collins, London.

van Riper, W. 1960. Does a hummingbird find nectar through its sense of smell? *Sci. Am.* 202:157 ff.

Robertson, C. 1928. *Flowers and Insects*. Publ. by the author, Carlinville, Ill.

Salas Duran, S. 1974. Analisis del sistema de polinizacion de *Inga vera* subespecie *spuria*. Tesis para Licenciado en Biología, Universidad de Costa Rica.

Scogin, R. 1979. Nectar constituents in the genus *Fremontia* (Sterculiaceae): sugars, flavonoids and proteins. *Bot. Gaz.* 140:29–31.

Sherbrooke, W. C. and J. C. Scheerens. 1979. Ant-visited extrafloral (calyx and foliar) nectaries and nectar sugars of *Erythrina flabelliformis* Kearney in Arizona. *Ann. Missouri Bot. Gard.* 66:472–481.

Shuel, R. W. 1955. Nectar secretion in relation to nitrogen supply, nutritional status, and growth of the plant. *Can. J. Agric. Sci.* 35:124–138.

———. 1956. Studies of nectar secretion in excised flowers. I. The influence of cultural conditions on quantity and composition of nectar. *Can. J. Bot.* 34:142–153.

———. 1957. Some aspects of the relation between nectar secretion and nitrogen, phosphorus and potassium nutrition. *Can. J. Pl. Sci.* 37:220–236.

Smith, I. 1969. *Chromatographic and Electrophoretic Techniques,* Vol. I, *Chromatography,* 3rd ed. W. Heinemann, London.

Sokal, R. R. and F. J. Rohlf. 1969. *Biometry*. W. H. Freeman, San Francisco.

Southwick, E. A. and A. K. Southwick. 1980. Energetics of feeding on tree sap by ruby-throated hummingbirds in Michigan. *Am. Midl. Natur.* 104:328–334.

Steiner, K. E. 1979. Passerine pollination of *Erythrina megistophylla* Diels (Fabaceae). *Ann. Missouri Bot. Gard.* 66:490–502.

Stiles, F. G. 1976. Taste preferences, color preferences, and flower choice in hummingbirds. *Condor* 78:10–26.

Stockhouse, R. E. 1975. Nectar composition of hawkmoth-visited species of *Oenothera* (Onagraceae). *Gr. Basin Natur.* 35:273–274.

Straw, R. M. 1956. Floral isolation in *Penstemon*. *Am. Nat.* 90:47–53.

———. 1966. A redefinition of *Penstemon* (Scrophulariaceae). *Brittonia* 18:80–95.

———. 1967. *Keckiella:* new name for *Keckia* Straw (Scrophulariaceae). *Brittonia* 19:203–204.

Toledo, V. M. 1974. Observations on the relationship between hummingbirds and *Erythrina* species. *Lloydia* 37:482–487.

Vansell, G. H. 1944a. Cotton nectar in relation to bee activity and honey production. *J. Econ. Ent.* 37:528–530.

———. 1944b. Some western nectars and their corresponding honeys. *J. Econ. Ent.* 37:530–533.

Vansell, G. H., W. G. Watkins, and R. K. Bishop. 1942. Orange nectar and pollen in relation to bee activity. *J. Econ. Ent.* 35:321–323.

Voss, R., M. Turner, R. Inouye, M. Fisher, and R. Cort. 1980. Floral biology of *Markea neurantha* Hemsley (Solanaceae), a bat-pollinated epiphyte. *Am. Midl. Natur.* 103:262–268.

Waller, G. D. 1972. Evaluating responses of honeybees to sugar solutions using an artificial-flower feeder. *Ann. Ent. Soc. Am.* 65:857–862.

Watt, W. B., P. C. Hoch, and S. G. Mills. 1974. Nectar resource use by *Colias* butterflies. *Oecologia (Berl.)* 14:353–374.

Werth, E. 1915. Kurzer Überblick über die Gesamtfrage der Ornithophilie. *Bot. Jahrb.* 53:313–378.

Wolf, L. L. 1975. Energy intake and expenditures in a nectar-feeding sunbird. *Ecology* 56:92–104.

Wykes, G. R. 1952a. An investigation of the sugars present in the nectar of flowers of various species. *New Phytol.* 51:210–215.

———. 1952b. The preferences of honeybees for solutions of various sugars which occur in nectar. *J. Exp. Biol.* 29:511–518.

———. 1953. The sugar content of nectars. *Biochem. J.* 53:294–296.

Yakovleva, L. P. 1969. The qualitative sugar composition of the nectar of some honey plants. *Tr. Nauchno-issled. Inst. Pchel.,* 236–244 (in Russian). (*Apic. Abstr.* 22:91.)

Yorks, P. F. 1979. Chemical constitution of nectar and pollen in relation to pollinator behavior in a California ecosystem. Ph.D. thesis (Botany), University of California, Berkeley.

Zauralov, O. A. and L. P. Yakovleva. 1973. Sugar composition of the nectars of some nectariferous plants. *Rast. Resur.* 9:444–451 (in Russian). (*Chem. Abstr.* 80:12451 n. 1974.)

Zimmermann, M. 1953. Papierchromatographische Untersuchungen über die pflanzliche Zuckersekretion. *Ber. Schweiz. bot. Ges.* 63:402–429.

———. 1954. Über die Sekretion saccharosespaltender Transglukosidasen in pflanzlichen Nektar. *Experientia (Basel)* 10:145–149.

6
EVOLUTION AND DIVERSITY OF FLORAL REWARDS

Beryl B. Simpson
Department of Botany
University of Texas
Austin, Texas
and
John L. Neff
Austin, Texas

ABSTRACT

The suite of substances used by plants as floral rewards for pollinating animals is quite restricted. Among these, pollen and nectar are by far the most widely used. Numerous factors seem to have influenced the evolutionary "choice" of certain substances as floral rewards. These include the availability in a plant of a substance, or a precursor of it, that could be switched into use as a floral reward; the cost of the production of the substance; and the flexibility of the reward system in terms of further evolution. Animal visitors have tended to select for substances that are easily handled and/or consumed. In the case of food rewards, those substances that contain appreciable calories or limiting nutrients per unit volume relative to other reward substances have been selectively favored. Rewards requiring great specializations for use on the part of pollinating animals tend to be rarer that those that require fewer such specializations. An examination of the substances used as modern rewards in the light of these selective constraints highlights the reasons for the widespread use of pollen and nectar.

KEY WORDS: Floral rewards, pollination, nectar, pollen, floral oils, sexual attractants, brood places.

CONTENTS

INTRODUCTION

The evolution of the diversity of flower form and color and its obvious relationship to pollination is widely appreciated and has been the subject of many studies for over a hundred years. However, the selective forces and constraints that have led to the modern array of floral rewards have received comparatively little discussion, in part because of the dominance of nectar and pollen as pollinator rewards. Our purpose here is to attempt to determine the origin and development of the present spectrum of floral rewards, and to learn why some are so successful and why others are restricted to one or a few angiosperm groups. The basis for much of the following discussion rests on our determination of the initial use for products that have become floral rewards, and an assessment of the behaviors and morphological structures of animals that would have fostered the production of certain types of rewards. We consider two kinds of substances to be preadapted for use as floral rewards: (1) integral parts of flowers; and (2) substances present in nonfloral tissues that can be produced in concentrated amounts in flowers and thus serve as rewards, or that can be easily chemically modified into suitable rewards. Superimposed upon availability of a reward substance or precursor of it within a plant is, however, the cost of its production relative to the gain it yields in terms of securing pollinator services.

We further assume that floral rewards that require the fewest specializations on the part of floral visitors will be the most widely used. Accordingly, we examine how reward substances might have been initially gathered, and how various modifications on the part of the pollinators for their collection and/or use might have been evolved. In this case, the ease of use of a florally produced substance has superimposed upon it its quality for the user. By quality, we mean its character in comparison with similar substances produced by plants in contexts other than as floral rewards, or over other kinds of rewards. A particular reward might have *superior* quality for one of several reasons. For example, it might contain a higher energy content or amount of a limiting nutrient per unit volume than another reward. It could also be more easily handled than similar substances produced elsewhere in a plant, or than other types of rewards.

POLLEN

The production of pollen is a characteristic of all angiosperms and is, by definition, an integral part of the pollination process. Pollen, in fact, predates angiospermy (Pettit and Beck, 1967) and would have been available as a food for flower-visiting animals from the inception of the angiosperms. Ovules, also a requisite part of angiosperm sexual systems, may have played a role as a pollinator reward in the earliest pollination systems (Crepet, 1979), but they are no longer significant in modern systems (yet see "Brood Places," below). The advantages of using small, dispersible cells instead of large, sessile ovules should be obvious. Nevertheless, the common statement that pollen is an "easy" substance for a plant to use as a floral reward because it has to be produced "anyway," is much too simplistic. Its use as a major floral reward is due not only to its presence, but also to its nature. However, none of the characteristics of pollen (e.g., shape, size, or even chemistry) has been conclusively demonstrated to have resulted from selection by pollinators. Rather, most characteristics can be explained in terms of reproductive functions.

Pollen exine sculpturing has proved to be a valuable taxonomic tool, but the reasons for the intricate patterning have yet to be explained. Attempts to relate exine patterning to pollinator type or behavior have, in general, failed (Taylor and Levin, 1975). Intuitively, exine sculpturing, in terms of the number and positioning of the copli, pores, reticula, striae, verrucae, etc., should have little correlation with types of pollinating agents because the two differ so greatly in scale. Sculpturing patterns are on the order of a few microns at most, whereas insect pollen-collecting structures are more typically on the order of tenths of millimeters. The most convincing hypotheses about the adaptive function of various pollen wall features are those that relate them to recognition and germination on stigmatic surfaces and/or to water absorption and retention (Heslop-Harrison, 1976, 1979). Exceptions appear to lie in the correlation between relatively smooth exine surfaces and vibratile pollination systems (see Buchmann, this book) and wind-pollinated systems. Pollen of most entomophilous species is spiny or coated with pollenkitt, which serves to promote clumping and thus transport of several grains at one time.

Pollen size has similarly been linked in various contexts to diverse groups of pollinators or modes of pollination. Pollen diameters (excluding water-pollinated systems) range from about 5 μm to over 210 μm (Erdtman, 1966; Roberts and Vallespir, 1978; Baker and Baker, 1979), with a corresponding 74,000-fold difference in the ranges of volume per grain. Most pollen diameters, however, fall in a range between 15 and 60 μm. Although the correlation is far from perfect, the most obvious relationship is between pollen diameter and the length of the style as originally pointed out by Darwin (1877) in reference to heterostyly and later explored more fully by Covas and Schnack (1945), Taylor and Levin (1975), Lee (1978), and Baker and Baker (1979). Lee's (1978) analysis also suggested that large flowers with large styles and large pollen are pollinated by large animals. Although the matter apparently has not been investigated, there is reason to suspect that large pollen grains may be superior rewards for pollen-feeding insects. The sporopollenin content of the exine includes a high proportion of the total energy content of pollen (as measured by techniques of bomb calorimetry; Colin and Jones, 1980), but it is indigestible to most groups of pollinating animals (McLellan, 1977). Although partially offset by increases in exine thickness (Lee, 1978), increases in pollen volume should increase the proportion of the grain that is available as nutrients for pollen consumers. Yet, this hypothetical increase in nutrient availability to pollen consumers may be offset by shifts in cytoplasmic energy reserves (e.g., oil vs. starch; Baker and Baker, 1979) or increased

difficulty on the part of an insect in handling large grains.

One correlation between a pollen characteristic and its use as a reward that appears to be substantiated is that between the amount of pollen produced and the extent to which it serves as the floral reward. The most impressive series of examples are flowers that use pollen as the only reward for bees (e.g., Papaveraceae; *Solanum,* Solanaceae; *Cassia,* Fabaceae). In such groups, there appears to be an increase in the amount of pollen produced over related taxa that use other rewards or a combination of rewards (e.g., Papaveraceae vs. the Fumariaceae). In some cases in which reduction of pollen production has been ascribed to an increased "efficiency" of pollen transfer, the actual causal factor may have been decreased use of pollen as a floral reward.

Chemically, pollen has often been described as an excellent food source (hence reward) for flower-visiting animals because of its high nitrogen content and its complex array of chemicals relative to generalized plant cells. As in the case of morphological features, however, almost none of the chemical characteristics of pollen can be credited with certainty to direct selection by pollinating agents. High nitrogen levels, cited as the primary reason for the widespread collection and metabolism of pollen by various animals, occur in wind-dispersed pollen as well (Todd and Bretherick, 1942), and are undoubtedly related to the fact that pollen in both cases is the male gametophyte and must contain sufficient nutrients for its germination and growth (Colin and Jones, 1980). Still, anemophilous pollen has been shown to be of low nutritive value for *Apis mellifera* (Maurizio, 1960), although the reasons for these findings are not known. It should be noted that even within entomophilous groups of plants that present pollen as a reward, pollens differ with respect to nutritive value for specific taxa of bees (Levin and Haydak, 1957; Campana and Moeller, 1977). Maurizio's (1960) results might, therefore, be a species-specific phenomenon. In cases in which an insect can metabolize only particular kinds of pollen, it is impossible to determine whether the chemistry of only that particular pollen was suitable for the animal, or whether selective pressures on the insect led to restriction in the repertoire of pollen used and a subsequent loss in the ability to use other types of pollen.

The relative amounts of starch or oil in the cytoplasm of pollen and the degree to which the pollen is used as a primary reward were investigated by Calvino (1952), who pointed out that starch was particularly prevalent in the pollen of anemophilous species. Recently, Baker and Baker (1979) have pursued this idea in greater detail. According to the hypotheses of these workers, pollen with abundant starch (and apparent low lipid content) is of lesser value as a food source for insects, primarily bees, than pollen with low starch and comparatively high lipid levels. Consequently, groups of plants that use pollen as one of the major, or the only, floral reward should have oily pollen. Plants offering other rewards in addition to, or instead of, pollen should have starchy pollen. This shift is supposed to occur because starch is less expensive for the sporophyte or gametophyte to make than oil, and should therefore be favored as an energy-conserving mechanism. An example with which we are familiar exists in the Malpighiaceae. Malpigh species we have tested that offer floral oils as virtually their only reward, have, as postulated, starchy pollen. Species of this family (e.g., members of the Galphimieae such as *Thyrallis angustifolia*) that do not secrete oils, and that now rely on pollen as the sole reward, have nonstarchy pollen. Yet, the oil–starch dichotomy is not a simple one, as all pollen contains some lipid, and sugars may be the major energy reserve in nonstarchy pollen. Moreover, truly confirmatory evidence for the selective pressures leading to the presence or absence of starch in pollen is still lacking. No one has tested whether starchy pollen is, in fact, an inferior food source compared with nonstarchy pollen. Studies on honey-bee metabolism (Chauvin, 1968) have shown that these bees cannot digest amylopectin, but the fact that the starch in pollen was absorbed indicated that amylopectin was not the primary kind of starch in the grains. Chauvin suggested that

pollen starch is of a different, digestible type (α-amlylose?), or that enzymes produced by the pollen itself or by the microflora of the bees' intestinal tracts bring about the starch breakdown. Our studies on the solitary bee *Diadasia afflicta*, which provisions its nests exclusively with the starchy pollen of *Callirhoe* spp. (Malvaceae), show that these bees can remove all of the starch from the grains. The large number of other solitary bee taxa that specialize on groups of the Malvaceae and Onagraceae (Linsley, 1958; Moldenke, 1979) with starchy pollen, also indicates an ability to digest efficiently pollen high in starch. While the presence of starch in the pollens of these groups may represent "taxonomic constraints" in the sense of Baker and Baker (1980), it argues against strong selective pressures for alteration of this component of pollen chemistry in response to pollinator preferences. It would, moreover, be instructive to investigate the relationship between the starch content of a pollen grain and the utilization of it, or its replacement by the growing gametophyte.

In some cases, particular compounds present in pollen have been singled out as important factors in specific plant species–animal vector relationships. One of these is the proposed correlation between high proline and tyrosine levels in the pollen of saguaro (*Carnegiea gigantea*, Cactaceae) and *Agave* spp. (Agavaceae) and their being pollinated by bats (Howell, 1974). Although Howell stated that these amino acids have no particular role in plant reproductive biology, it is now clear that proline and hydroxyproline play an important role in cell wall formation during pollen tube growth (Dashek and Harwood, 1974). High proline contents are known to be present in large anemophilous pollen grains with fast-growing pollen tubes, such as *Zea mays* (Poaceae) (Anderson and Kulp, 1922); and thus high proline content, and high protein content in general, may be a common feaure of pollen that must produce tubes that can rapidly traverse long, ephemeral styles (Britikov et al., 1964; Stanley, 1971).

Early data also suggested that the high proline content of pollen was an important factor influencing choice of pollen by bees. More recent evidence tends to refute this hypothesis (Barker, 1972), although there may have been problems with experimental controls of pollen consumption.

Howell (1974) also emphasized the high protein content of pollen of flowers pollinated by bats, but the figures on which she based her belief that the protein levels of these pollens were comparatively high bear scrutiny. Most studies of chemical composition of pollen of entomophilous plants have used honeybee-collected pollen (Todd and Bretherick, 1942; Vivino and Palmer, 1944; Weaver and Kuiken, 1951; Levin and Haydak, 1957; McLellan, 1977). The use of bee-collected pollen is obviously advantageous when large amounts of pollen are needed for analysis, but this procedure can lead to erroneous results because bees add significant amounts of sugar (as regurgitated nectar) to the pollen pellets (Todd and Bretherick, 1942). In one study in which both hand- and bee-collected pollen of the same species were analyzed, the figures suggest that over 40% (by weight) of the pollen pellets consisted of sugars added by the bees (Todd and Bretherick, 1942). This example is probably extreme, but the data do suggest that the appreciable amounts of bee-added sugars can seriously distort calculations of percent composition of substances in pollen when bee pellets are used. An additional complication is added to the calculation of chemical consituents because pollen is quite hygroscopic, and the moisture content can vary widely depending on the degree of hydration. Both of these factors will tend to lead to an underestimation in the percentage of protein in bee-collected pollen. Howell (1974) considered the protein values of 22.9% for the paniculate agaves and 43.7% for *Carnegiea gigantea* to be unusually high, but we suggest that an application of a correction factor to the bee-pellet determined values with which she compared her hand-collected values shows that the *Carnegiea* value is high, but the agave figure is below average. Curiously, her values for *Opuntia versicolor* (8.93% protein) and *Agave schottii* would appear to be among the lowest for all published protein values. Clearly, in future studies, there is a need for more analyses of hand- (or machine-) collected pollens and presentation of percentage values in terms of known moisture levels and/or dry weight.

In terms of compounds other than amino acids, honey-bees are known to be attracted to specific components of some pollen, particularly octadeca-*trans*-2, *cis*-9, *cis*-12-trienoic acid (Lepage and Boch, 1968; Hopkins et al., 1969). In this case, the chemical involved does appear to be a good candidate for a specific pollen substance that is produced in direct response to selection for pollinator attraction, but the taxonomic distribution or possible plant physiological functions of this and similar compounds are as yet unexplored.

Obviously, the complexity of pollen chemistry provides numerous opportunities for the development of coevolutionary interactions. What is difficult to unravel, and more difficult to prove, is the cause-and-effect relationships between various characteristics of pollen and its use as a reward by specific pollinators.

A special case of pollen modification and the role it plays as a reward does appear in food pollen (Vogel, 1978). In every case in which food pollen is present, there is a morphological differentiation between the anthers that produce the food pollen and those that produce the functional pollen. Likewise, all cases appear to involve bees as the primary pollinating agents. Usually, anther differentiation takes the form of disparate filament lengths, resulting in a placement of the functional pollen on a part of the insect where it will not be groomed off or placed in the scopae. In contrast, food anthers are positioned in such a way that pollen is easily gathered. Food pollen has been reported in some cases to be starchy (e.g., Commelinaceae; Baker and Baker, 1979, and our data), an apparent contradiction to the hypothesis that pollen that serves primarily as a food should be oily. In several cases, food pollen has been reported to be sterile. While this seems to be a logical consequence of noninvolvement in fertilization, there is a problem depending upon the definition of sterile. Sterility defined as nongerminability or loss of fertilization capacity is one thing; sterility defined by absence of staining with aniline blue is another. This stain indicates the presence of callose, a saccharide found in the intine of pollen grains and along the inside walls of pollen tubes. Grains that stain may be functionally sterile or fertile. While it is possible that the lack of an intine alone could make a grain inviable, it is probable that inviability is due to the lack of cytoplasm that usually accompanies the absence of an intine. The lack of cytoplasm would obviously make a pollen grain useless as a food source. The reports of various authors (Linsley and Cazier, 1963, 1970; Buchmann, 1974) using cotton blue for staining and showing up to 85 or 95% sterility in some food anthers, thus seem strange. It is possible that deceit is actually involved, but it has not been suggested.

From an animal's point of view, as pointed out above, pollen represents a concentrated nitrogen source combined with numerous other compounds of potential nutritive value. What has seldom been pointed out is that pollen is a convenient food source because it requires a minimum of adaptations on the part of an animal for its use. Virtually any mandibulate insect and many nonmandibulate ones as well (many adult Diptera and even some Lepidoptera such as *Heliconius*—Gilbert, 1972; DeVries, 1979) can feed on pollen, although various modifications of the mandibles, maxillae, or other portions of the mouthparts enhance the efficiency of pollen feeding.

A more pressing problem for animals is how to extract the nutritional components of pollen from within the indigestible (for all animals except some Collembola; Scott and Stojanovich, 1963) exine. Grinfel'd (1975) has shown that in various pollen-feeding beetles and some primitive pollen-feeding Lepidoptera *(Micropteryx)* the mandibles may be modified for grinding pollen grains to extract their contents. The asymmetrical condition of the mandibles in various Thysanoptera may have evolved as an adaptation to piercing individual grains in order to extract the contents (Grinfel'd, 1959).

The more common patterns of pollen digestion, however, do not require mechanical destruction or damage to the exine. These other methods are not all equally suited to all types of pollen, and the extent of their usage is not yet well established (Stanley and Linskens, 1974). In bees and other organisms equipped with a similar proventricular valve, simple osmotic shock may frequently be sufficient to lyse or burst the pollen cell membranes at the pores. The pollen is transferred from the crop in which relatively low osmotic differentials exist between pollen cell contents and the crop fluids, to the ventriculus in which there is a higher difference between the two (Kroon et al., 1974). In aperturate pollen grains (Linskens and Mulleneers, 1967), a pollen tube may be readily produced in the presence of acid (Johri and Vasil, 1961). This pseudogermination may be an important mechanism of pollen digestion by animals with low gut pH values such as pollen-feeding bats (Howell, 1974). Other specific compounds that elicit germination of pollen tubes of various plant groups, or enzymes that allow digestion within the intact, ungerminated pollen grain may be present in some animals, but apparently they have not been investigated in any detail. In general, however, pollen feeding seems to require little or no modification in gut morphology, as implied by the fact that bees have relatively simple intestinal tracts (Snodgrass, 1956).

While pollen may be a high-quality food source, the relative slowness of the digestive process for the large quantities needed for nutrition, and the fact that more than half of the caloric value of the grain is tied up in the indigestible exine (McLellan, 1977), means that pollen alone is unlikely to be an adequate sole energy source for anthophilous animals such as bees with high energy demands.

One drawback to pollen as a reward is that it is nonrenewable within a flower. Pollen production of an individual flower is fixed well before anthesis, and thus pollen presentation is a matter of doling out fixed quantities of rewards. Plants can alter the presentation schedule by sequential dehiscence of the anthers or, in the case of poricidal and some other anthers, by a within-anther time-release mechanism (Buchmann et al., 1977; pers. obs.). In contrast to nectar, a plant has little leeway to alter or replenish pollen rewards in response to varying visitation schedules. Moreover, there is no evidence that pollen can be resorbed by the parent plant.

All of the preceding has discussed pollen as a food source for the individual collecting it. In the case of bees, the foremost modern pollen consumers, there is a high pollen demand for the provisioning of the nests, or in the case of honeybees, the production of glandular larval food. Because of this demand, selection has led to a wide array of structural modifications for collection and transportation of pollen (Thorp, 1979). These are not, however, modifications of the mandibles. Almost all of the mandibular modifications of bees can usually be related to nest construction in females and mating in males (Michener and Fraser, 1978). The most striking modifications for collection and transport are those involving external body features. Besides bees, only a few insects use pollen and nectar as the primary larval provisions. In the most diverse of these groups, the Masaridae (Vespoidea), pollen transport has remained strictly internal. Some masarid wasps, such as *Trimeria buyssoni,* known to collect pollen from narrow tubular flowers such as those of *Glandularia* (Verbenaceae) or *Heliotropium* (Boraginaceae), do possess special arrays of setae on the forelegs that facilitate pollen extraction just as similar organs do among some bees. However, the range of variations on this theme is small compared to the spectrum of collecting structures in bees.

External pollen transport has also evolved in the fig wasps (Agaonidae), but in this case the pollen is not used as a larval provision (see "Brood Places," below). One of the few characters separating bees from the rest of the Sphecoidea is the presence of branched body hairs (Michener, 1944; Brothers, 1975). Although there is some controversy on this point due to the presence of strict internal pollen transport in some putatively primitive bees (euryglossines and hylaeines), we suggest that the evolution of branched body hairs was the feature that facilitated external transport of pollen and led to the initial radiation of the bees. Reduction of the extent of branched hairs in the pollen transport organs has evolved repeatedly among bees associated with unusually large pollen grains (over 60 μm), in bees employing agglutination to hold pollen in a cohesive mass, and in the Megachilidae that use abdominal scopae. Increased branching of the body hairs has occurred in groups associated with very small pollen grains (less than 15μm) and in many oil-collecting groups (Roberts and Vallespir, 1978). Overall reduction of external pollen transport is characteristic of twig nesting bees because of selection for a reduction in the cross-sectional body area.

Cantharophily has received much publicity (from Diels, 1916, to Baker and Hurd, 1968) as *the* ancestral mode of angiosperm pollination. The hypothesized primitiveness of this association stems, in part, from the alliance of beetles with the so-called primitive angiosperms (Magnoliales; Grant, 1950). Beetles are often said to have been practically the only insects available as pollinators at the time of angiosperm emergence. It has also long been thought that the beetle–magnolialian pollination system is unmodified or unspecialized. Recently, several authors (Gottsberger, 1974; Thien, 1974; Meeuse, 1978; Schneider, 1979) have pointed out the weaknesses in an hypothesis that compares these modern pollination systems to primitive angiosperm pollination syndromes. Both Gottsberger (1974) and Thien (1974) have shown the specialized nature of many of these systems, and Grinfel'd (1975) has pointed out that many of the presumed archaic beetle pollination systems involve modern beetle groups. Moreover, contrary to popular opinion, there were many insect groups available as potential pollinators during the middle Cretaceous when angiosperms arose and/or began to diversify (Doyle and Hickey, 1976). Among them were Hymenoptera, Orthoptera, Thysanoptera, Neuroptera, and, possibly, Lepidoptera (Grinfel'd, 1975). Spore feeding was presumably well established before angiosperms arose (Crepet, 1979). Collembola, now a minor group in terms of pollination, probably played a role in these early systems (Kevan et al., 1975), as may have the now extinct Paleodictyoptera and Megasecoptera, groups commonly found in Carboniferous and Permian deposits (Smart and Hughes, 1973).

NECTAR

Nectar is now the single most important floral reward. From the plant's point of view, it is extremely easy to produce and, compared to various other rewards, relatively inexpensive (Table 6-1). Nectar is generally derived from phloem sap (Gunning and Steer, 1975) that has undergone any of a number of modifications along the route from the sieve elements to the secretory cells. The phloem sap consists primarily of sucrose mixed with smaller quantities of polysaccharides, amino acids, vitamins, lipids, and inorganic ions (Ziegler and Ziegler, 1962; Zimmermann and Brown, 1977; Fahn, 1979a); so one would expect to find some quantities of all of these kinds of chemicals in nectar. Most physiologists conclude that during the process of nectar secretion, varying amounts of these compounds and ions probably are removed from the phloem solution, leaving the amounts found in nectar, rather than being specifically secreted into the solution (Gunning and Steer, 1975). One principal group of compounds removed, at least in part, along the route to secretion is amino acids (Gunning and Steer, 1975). Physiological studies also clearly show that glucose and fructose, two of the very common sugars found in nectar, are never found in phloem sap and must be produced from the breakdown of sucrose in the nectary cells (Zimmermann and Brown, 1977).

Not only is nectar the most common reward offered by animal-pollinated hermaphroditic flowers, it is also offered by most zoophilous dioecious species (excluding those involving deceit) and many monoecious species as well. It has the distinct advantage, from the animal's point of view, of being very simple to use. Any animal can metabolize a simple sugar solution.

The precise origin of nectar as a reward is impossible to determine. In fact, the anatomical simplicity of most nectaries and their appearance on numerous, distinct plant organs (Fahn, 1979b) indicate that nectaries have developed several times independently throughout the angiosperms. Potential pollinators could have had the behavioral trait of searching for sweet secretions before floral nectaries were developed. Such a behavioral pattern could have been evolved by proto-pollinators already accustomed to feeding on honeydew (the excretion of sweet liquids by insects such as aphids probably predated angiosperms); extrafloral nectar; ripe, sweet, animal-dispersed fruits; stigmatic secretions (insects commonly collect stigmatic secretions of some gymnosperms); or even fungal exudates. In many rusts, the transfer of spermatia to receptive hyphae (most rusts are heterothallic) is mediated

Table 6-1. Energy values of floral rewards.

Reward		Heat of combustion[1] joules/gm	Digestible energy[2] per gram	Oil Equivalents[3]
Oil:	Stearic Acid[a]	40.17	40.17	1.00
Sugar:	Glucose[b]	15.69	15.69	2.56
	Sucrose[b]	16.53	16.53	2.43
	nectar—20% sucrose	3.31	3.31	12.14
	nectar—40% sucrose	6.61	6.61	6.08
	nectar—60% sucrose	9.92	9.92	4.05
Starch		17.57	17.57	2.29
Pollen:	Wind-pollinated monocots[c]	21.76		
	Wind-pollinated dicots[c]	24.60		
	Animal-pollinated dicots[c]	24.14		
	Animal-pollinated dicots[d]	23.84	11.25[4]	3.57

1. Energy values determined by bomb calorimetry. Energy values for floral oils unavailable but stearic acid should be a reasonable approximation: a—from Swern, 1979; b—from Merrill and Watt, 1955; c—from Colin and Jones, 1980; d—from McLellan, 1977. (Values from a, b, c converted from calories/gm to joules/gm by using a factor of 4.18.)
2. Assumes 100% digestibility for sugar, starch, lipid, and protein; adjustments for protein metabolism not included.
3. Calculated grams of reward to equal estimated digestible energy of one gram oil.
4. May overestimate digestible energy owing to inclusion of sugars added by bees.

by insects that visit the spermagonia to feed on a sugary solution exuded from the spermagonial opening that protrudes from the host plant (Craigie, 1927; Ingold, 1971; Alexopoulous and Mims, 1979). In gymnosperms such as *Ephedra,* some partially insect-pollinated species secrete sugary solutions from the integumentary or perianth parts (Bino and Meeuse, 1981). These secretions are gathered by numerous groups of insects that potentially carry pollen from male to female flowers. Angiosperms could improve upon the prefloral nectar sources by producing a large volume of nectar accessible in a floral cup. Once attracted, an animal could easily learn to recognize a flower and reduce its searching time by repeatedly visiting a uniform search image.

Nectar as a reward can be modified in a number of ways to increase constancy or specificity, including subtle alterations in nectar chemistry, variations in volumes or concentrations of the solution, or variations in the time of its production. Almost all of the variations in nectar production can be correlated with different parameters of the pollination system. Since Baker and Baker (this book) summarize these relationships in detail, we will not discuss each of them. However, we do wish to point out that caution must be exercised in interpreting nectar characteristics strictly in terms of pollinator selection. For example, Percival (1961) noted the dominance of hexose sugars in shallow, exposed nectaries, whereas sucrose tended to dominate in deep hidden nectaries. Recent work (Corbet et al., 1979; Plowright, 1981) has focused on the importance of humidity in controlling nectar concentrations. At a given relative humidity, a nectar containing only sucrose must have a higher concentration than a nectar consisting of hexoses in order to maintain an equilibrium with the air (Corbet et al., 1979). This simple physical relationship is probably a principal factor influencing the dominance of hexose sugars in shallow, exposed nectaries. As also noted by Corbet et al. (1979), the presence of other compounds in nectar may be in part related to their influence on vapor pressure biochemistry. While high inorganic cation concentrations have been shown to have deterrent effects on honeybees (Waller

et al., 1972), an additional or perhaps initial function for these ions may have been to alter vapor pressure relationships.

Various animals now display adaptations that can be related to specific nectar-feeding behaviors. In particular, elongations of mouthparts or beaks for narrow and/or deep corollas have evolved repeatedly in various animal groups (see Proctor and Yeo, 1972, or Daly et al., 1978). A special case can be seen in carpenter bees (*Xylocopa* spp., Anthophoridae), where thickened galae serve as piercing organs for robbing tubular corollas (Schremmer, 1972). In some cases such as hummingbirds (Colubridae) or horseflies (Tabanidae), elongate mouthparts originally may have been evolved for some other aspect of feeding, but now facilitate the use of narrow, tubular corollas.

STIGMATIC EXUDATES

Stigmatic exudates have received little attention as rewards for pollinators although, in some cases, they may be the primary reward. However. liquid secretions on the stigmatic surface are absent in plants with sporophytic incompatability (Heslop-Harrison et al., 1975) and are usually present in minor quantities in other taxa. Generally, the exudates are composed of lipids, amino acids, phenolics, alkaloids, and antioxidants (Martin, 1969; Baker et al., 1973). Free sugars are occasionally present (Portnoi and Horovitz, 1977). Major functions for the components of stigmatic exudates include roles in pollen capture and germination as well as stigma protection and resistance to dessication (Baker et al., 1973). Various uses of stigmatic exudates as pollinator rewards have been postulated (Baker et al., 1973; Lord and Webster, 1979), but in general the evidence is weak. Significant exceptions appear to be species of *Aristolochia* (Aristolochiaceae), which have stigmatic secretions that are extremely rich in amino acids and are fed upon by pollinating flies trapped in the blossoms (Baker et al., 1973), and species of *Anthurium* (Araceae). Croat (1980) has reported sugar levels of up to 8% in the copious exudates of *Anthurium seibertii.* Some palms (see Simpson and Neff, 1981) also produce abundant secretions from the stigmatic surfaces. While not truly homologous, the sugary pollination drop of the gymnosperm *Ephedra campylopoda* (Gnetaceae) may also serve as a pollinator reward (Porsch, 1910, 1916).

Thus, stigmatic exudates undoubtedly do serve as reward substances in some plants, but primarily in those that have a very short or rudimentary style and a relatively broad stigma. In other flowers, in which these morphological features do not occur, potential damage to the style and disruption of pollen tube growth appear to outweigh any selection for use of these substances as rewards. Nectar can contain all of the constituents found in stigmatic exudates and can be produced in a place that minimizes damage to the gynoecium.

OILS

Floral oils, a relatively recently discovered reward (Vogel, 1969, 1974; Simpson et al., 1977; Simpson and Neff, 1981), differ from pollen, nectar, and stigmatic exudates in that there does not appear to be any plant substance from which they are derived although there are numerous cases in which similar oils are incorporated into various common plant lipids. In all species investigated, floral oils—identified as saturated, β-acetoxy free fatty acids or diacylglycerols (diglycerides) derived from them—are secreted in appreciable amounts from structures termed elaiophores (Vogel, 1974). The lipids are collected from all of the New World tropical oil-producing groups (the most numerous assemblage) by female bees of the Anthophorinae (Hymenoptera: Anthophoridae), which have arrays of modified collecting setae on the basitarsi of the fore- and/ or midlegs or even the abdomen (Neff and Simpson, 1981). The chemical structures of the lipids of Old World taxa have not been determined, and the mixture of compounds reported to be in the anther secretions of *Mouriri* (Melastomataceae; Buchmann and Buchmann, 1981) differs from those found in all of the other New World oil flower species analyzed to date (Simpson and Neff, 1981).

While free fatty acids are ubiquitous in almost all living organisms, large concentrations of them are apparently rare. Appreciable quantities of lipids can be present in nectar, but at least some of these seem to be unsaturated lipids (Baker, 1978), probably phenolics, terpenes, or glycerides containing unsaturated fatty acids. Lipids are often found on pollen surfaces, sometimes in such large amounts that they may constitute a major source of oils for pollen-collecting bees. Interesting in this context is the fact that oil-collecting anthophorine bees all use, although not exclusively, "dry" (non-lipid-coated) pollen from buzz flowers (Buchmann, this book), whereas non-oil-collecting congeners do not. It seems most likely that this difference is related primarily to the mechanics of pollen transportation rather than to any inherent difference in nutritional factors. To our knowledge, however, the lipids on pollen wall surfaces have not been identified, and the tacit assumption has been that they are primarily glycerides, the most abundant kind of neutral lipids found in plants.

High lipid contents are now known to occur in some kinds of food bodies such as Beltian bodies (Rickson, 1975) and Beccariian bodies (Rickson, 1980) and in the seeds of numerous angiosperms (e.g., all of the oil seed crops; Hilditch and Williams, 1964). None of these structures has any association with the pollination system per se and, in all cases, they contain oils in the form of triacylglycerols, not free fatty acids (triacylglycerols being composed of fatty acids joined to glycerol). The three most common fatty acids (oleic, linoleic, and palmitic) found in triacylglycerols have chain lengths of C 18, C 18, and C 16 respectively; the last is saturated. The three most common saturated β-acetoxy free fatty acids of the floral oils that we have examined have chain lengths of C 16, C 18, and C 20. Noteworthy in this context is Vogel's report (1974) of an acetic acid–containing diacylglycerol as the principal component of the floral lipids of *Calceolaria pavonii* (Schrophulariaceae). In commercially important triacylglycerols, acetic acid is the only straight, short-chain acid never found (Institute of Shortening and Edible oils, Inc., 1974). Moreover, the presence of the β-

position substitution of the acetate group in the saturated floral fatty acids is apparently unique (Simpson et al., 1977).

It has been suggested that prefloral location of lipids similar to those now produced by elaiophores may have been present in vegetative glandular trichomes (Vogel, 1974), but it is difficult to see what function they would have served. Since the floral oils are apparently edible and are metabolized even by fungi (Simpson and Neff, 1981), an anti-predator function seems unlikely. In cases in which the chemistry of lipid secretory products from glandular vegetative trichomes has been determined, the oils have been found to be predominantly terpenes (Fahn, 1979b).

Nevertheless, despite their differences in chemical structure from common fatty acids and their apparent absence in nonfloral plant structures, floral oils are assuredly synthesized via the pathways used to make fatty acids for triacylglycerols, phospholipids, etc. Differing biosynthetic steps would include the absence of a series of desaturations and the attachment of an acetate group in the β position or, if a diacylglycerol is produced, the addition of acetic acid to the glycerol backbone.

While it is thus not difficult to postulate the probable biosynthetic pathway that was switched into floral oil production, it is difficult at this time to see how the oil-collecting habit originated in bees, especially since the full role of the oils in the life history of bees is still not completely understood (Simpson and Neff, 1981). It is certain that they function, at least in part, as a metabolite for the larvae. Evidence from different studies suggests that, in some cases, lipids are the only liquid in the provisions (Vogel, 1974), but that, in others, variable amounts of nectar can be used with the oils in the larval food mass (Neff and Simpson, 1981). Oil has an advantage over nectar in that it contains more energy per unit volume than even pure sugar (9 cal/gram vs. 4 cal/gram for sugars, Table 6-1). However, oils must be more energetically expensive for a plant to make than nectar.

Baker (1978) has suggested that the oil flower–oil collection syndrome arose from a nectar production–nectar feeding system. This suggestion stemmed from findings that lipids are found, sometimes in large quantities, in many plant nectars (Baker and Baker, 1982). There are several problems, however, with the assumption that this sort of nectar-lipid occurrence was the precursor of oil flowers. First, the lipids in nectar seem to be of a different type from those secreted by elaiophores. Second, no appreciable amounts of sugar have ever been found in elaiophore secretions. Third, and most important, floral oils are not collected in the same way as nectar. If oil flowers arose simply as the end of a nectar continuum, one would expect that the oils would be collected with the mouthparts, or modifications of them. Instead, oil collection requires the use of specialized structures formed by modified setae (Neff and Simpson, 1981). The oils are transported on modified scopal hairs (Roberts and Vallespir, 1978) once they they are collected.

Oil production–oil collection is, therefore, a fairly tight system requiring specializations on the part of both the plant and the pollinator. A switch to floral oil production by a plant species is apparently much easier, in an evolutionary sense, than a switch to oil collecting. In the New World tropics, many genera of six unrelated families (Iridaceae, Orchidaceae, Malpighiaceae, Krameriaceae, Scrophulariaceae, Solanaceae, and, possibly, a seventh, Melastomataceae) have switched into an oil flower syndrome. Only two (or three) tribes of one family of bees have the morphological specializations necessary for collection of these oils. If the temperate (*Lysimachia,* Primulaceae) and Old World tropical groups (*Momordica* and *Thladiantha,* Cucurbitaceae) do, in fact, have a system analogous or identical to the New World tropical systems involving anthophorine bees, they have not undergone similar coevolutionary developments. Within the New World tropics, specificity between plant species and anthophorine species appears to be slight except on the distributional fringes of both participants, and there is no known correlation between the geographical range of a given oil plant and any particular oil-collecting visitor(s). However, oil bees are completely bounded by the distributional limits of oil flowers as a group, and the converse is also generally true.

We have suggested (Neff and Simpson, 1981) that oil collection began as a secondary association between flowers that had lipid-secreting patches and bees that initially visited for a different reward (pollen?). Presumably, these bees had unspecialized tarsal setae, but a large enough concentration of them on the tarsi to absorb liquids from diffuse trichome patches. Once the process of tarsal absorption was established, selection would have led to the striking array of setal combs and brushes now found in the Anthophorinae (Neff and Simpson, 1981). The setal patterns of bees in general seem to respond readily to natural selection and show a great diversity within the Apoidea.

SEXUAL ATTRACTANTS

The use of aromatic oils as floral rewards has received much publicity since its description in 1961 (Dodson and Frymire, 1961a,b), but the evolution of the system and its advantages to the pollinating euglossine bees is still not known (Williams, 1982). The most convincing theory to us concerning the advantage to the bees is that the compounds they collect serve as pheromone precursors and thus eliminate some steps of chemical synthesis on the part of the male euglossines. The origin of the production of these scent-producing compounds within flowers does not seem hard to explain, since the same or similar compounds (terpenes, aldehydes, ketones) produced via the cineole–indole–terpene pathway are common throughout various groups of plants, both vascular and nonvascular, and were presumably present in the progenitors of the angiosperms.

The paradox with the euglossine–volatile oil system lies not so much in explaining how natural selection could have led to an abundant floral production of these compounds, but rather how the entire system got started in the first place. Outside of the cases in which pseudocopulation is involved, male bees are generally not primary pollinators of any species. In the euglossine system, males are very often the only pollinators. Specificity is related to particular scents or mixtures of scents to which only one or a few species of bees are attracted. However, the specific scent or scent mixture will attract males regardless of the emission source (e.g., filter paper, unrelated plant species), showing that scent is not only the reward but also the attractant.

In the volatile-oil reward system, even more than in the neutral-oil system, there have been extensive morphological, and presumably physiological (Williams, this book), modifications of the bees for collection and utilization of the compounds. These include modifications of the tarsal setae (different, however, from those of the anthophorines) and bizarre changes in the hindlegs for absorption of the fragrances. While it would probably be advantageous for any bee that uses male pheromones in mating (assuming this is the use of the compounds) to collect fragrances that would bypass some biosynthetic steps of pheromone synthesis, it is possible that the use of such rewards is restricted because the kind of morphological variation necessary to permit selection to lead to floral scent utilization was not present in other groups of bees.

Although lacking the specificity of the male euglossine system, acquisition of chemicals used in sexual attractants from floral rewards has also been reported in danaid butterflies. Adult males ingest pyrrolizidine alkaloids from various plant parts (Boppré, 1978) including, at least in some cases, nectar (Deinzer et al., 1977), which they convert to dihydropyrrolizine pheromones for use in mating. Rothschild and Marsh (1978) note that nectar alkaloids occasionally have a direct aphrodisiac effect. Indirect evidence suggests that males of other lepidopteran families (Ithomiidae, Arctiidae, Ctenuchidae) may be involved in similar systems (Pliske, 1975).

RESINS AND GUMS

Of even more restricted use than, but evolutionarily related to, the use of volatile oils, is the presentation of floral resin or gums as a reward for female bees. Resin or gum production as "floral" rewards have thus far been reported in *Dalechampia* (Euphorbiaceae; Armbruster and Webster, 1979) and *Clusia* (Clusiaceae; Skutch, 1971). In *Dalechampia* at least, different species of the genus exude either volatile oils or resin from glands on the pseudanthium. If scents are produced, male euglossines visit the flowers and are the pollinators. If resin is produced, female euglossine, meliponine, and anthidiine bees visit the flowers to gather it for use as nest lining material. Chemically, resins are composed primarily of terpenes; so the switch between scent and resin production is not difficult to imagine.

Resin is, however, primarily a vegetative material held in canals or pockets from which it oozes if the plant is pierced. As in the case of volatile terpenes, resins appear to serve an anti-predator function in vegetative tissues (Berryman, 1972). Many bees will collect resin exuding from vegetative tissues and carry it to the nest where it is worked into a coating for the nest walls. As humans early discovered (e.g., resinous coatings to seal Egyptian coffins and pitch inside of Greek wine vessels), resin is a good water repellent. In small quantities, resins are not toxic.

Given, then, that bees will search out and use resinous material, one must ask as Armbruster did (manuscript), why bees would preferentially use *Dalechampia* resin and thereby become effective pollinators. Armbruster's present hypothesis is that *Dalechampia* resin is more easily worked than vegetative resins. In addition, we might point out that sources of vegetative resin are unpredictable, but bees can learn the image pattern of a flower and would become predictable visitors if the resin source were guaranteed.

There are no special structural modifications needed on the part of the bees for the collection of floral resins beyond those already present for collection of vegetative resins. Therefore it seems that any bee that uses resin might become a part of this reward system. In fact, the most common tropical bees that use resins for nest linings are euglossines and meliponines, both of which visit *Dalechampia*.

The situations involving *Clusia* and perhaps *Mouriri* (Melastomataceae—Buchmann and Buchmann, 1981; Simpson and Neff, 1981) are less clear, since the complete chemistry of the collected secretions and their uses are unknown. Skutch (1971) implies that in addition to using *Clusia* resin (or gum) as a material for nest construction, female bees throw it onto invading predators.

FOOD TISSUES

Various floral tissues (see Simpson and Neff, 1981) of diverse plant species have been modified over time into food bodies that serve as rewards for visiting animals. Pollen, of course, is not included in the category of food tissues even though it is sometimes specifically produced as a food source (see above). Food tissues are generally of epidermal or parenchymal origin and contain greater than normal quantities of sugar, starch, protein, or lipid (or various combinations). The patches of these cells are chewed by flower visitors, usually beetles, but occasionally by fruit bats and/or other mammals that ultimately serve as pollen vectors.

In general, the substances used as "food" in these tissues are compounds found in all plant cells, and are commonly accumulated in certain organs such as rhizomes, tubers, storage roots, phloem, and seeds. No novel biosynthetic pathways are thus necessary for any of the substances that provision food tissues. All that is needed is translocation to the vacuoles or in situ production. One could envision increased specialization of tissues (removed from the sexual parts) that were initially gnawed by unspecific flower visitors that successfully carried pollen from flower to conspecific flower during feeding forays. Starch, a product found in some food tissues (Porsch, 1906), should be a relatively inexpensive reward for the plant to provide. However, the provisioning of food tissues with protein as in the case of *Calycanthus* (Rickson, 1979) would be relatively expensive for a plant, particularly if nitrogen were limited.

Modifications on the part of the animal for feeding on food tissues are mimimal, since simple chewing apparati are all that are required. Restriction of floral visitors to those that consume only the food bodies and do not damage the ovary or anthers, and that will have a high probability of successful pollen transfer, seems to be accomplished by the use of fragrances in *Calycanthus* (Calycanthaceae) (McCormack, 1975). The attraction of animals to most species with food bodies or food hairs (e.g., *Vanilla planifolia,* Orchidaceae) has not been determined but presumably also involves odor cues.

While food tissues can be inexpensive rewards and secure the effective services of pollinators, they are rarely employed in the angiosperms. It appears that selection for this type of reward system is counterbalanced by the high mortality that often accompanies such a system. Cox (1980) has postulated that dioecy in *Freycinetia* (Pandanaceae) has evolved in response to selection for the protection of the megagametophyte from the bats that serve as pollinators while consuming nutritive inflorescences and subtending bracts. Moreover, in beetle and mammal systems that use food tissues as floral rewards, there are constraints imposed on floral morphology. Flowers must be stout or large and last several days. The cost of producing the flower becomes comparatively large. In cases in which there are food scales or "pseudopollen" (*sensu* van der Pijl and Dodson, 1966), floral forms are much more intricate, and the reward systems essentially mimic pollen reward systems.

BROOD PLACES

The provision of an insect with a brood place has, on occasion, become the reward offered by flowers to pollinating insects. It seems most probable that such a system evolved from an initial parasitic relationship involving, in different cases, gall-forming, seed-feeding, or phytophagous insects. Correspondingly, there must have been a stage during which the negative effects outweighed the good, if any. While it might appear that there would be a large pool of potential insects that might become incorporated into such a system, very few are actually involved, and the use of a flower as a brood place as a reward is very rare in the angiosperms. The line between true parasitism and providing enough pollinator service to counterbalance the deleterious effects of ovule

and seed destruction is a fine one. In at least one case, *Melandrium album* (Caryophyllaceae) and its pollinator *Hadena bicuris* (Noctuidae) (Brantjes, 1976), calculations indicate that the negative effects of the larvae will usually exactly counterbalance the pollination services of the adults, leading to an effective net loss for the plants. Since most floral parasites are larvae of insects that never or rarely carry pollen (usually adults oviposit into the side of a flower), there generally would be selection only to prevent parasitism. Even in cases in which the adults of parasitic larvae do carry pollen and effect pollination, selection would normally have favored a nonparasitic floral visitor. Utilization of a parasite as a pollinator would be likely if a plant became dependent on it when other pollinators disappeared or were never present, as might occur in a population established by long-distance dispersal. In general, however, the cost outweighs the gain, and alternatives are too readily available to favor such a system.

In the few cases where a brood place is the floral reward, there is generally great fidelity between the partners. The most advanced systems involve morphological and behavioral adaptations on the part of both the plants and the insects. These include the famous *Yucca* (Agavaceae)–*Tegiticula* (Incurvariidae) and the *Ficus* (Moraceae)–agaonid wasp pollination systems. Figs are an extremely specialized case that seems continuously to reveal more intricacies as investigations continue (Ramirez, 1969; Galil and Eiskowitch, 1973; Wiebes, 1979). The plants now produce, in some cases, a variety of temporal fruits crops and various flower types within a synconium. The wasps are also modified (e.g., wingless males, pockets on the thorax and legs of the females for pollen transfer) compared to related groups, and show high specificity to particular figs. The *Yucca* system is less specialized and in some cases even employs nectar as part of the total reward system. Compared to figs and fig wasps, morphological specializations of both the plants and the moths are slight, but the presence of specialized pollen-gathering maxillary palps and the behavioral adaptation of the female moths of gathering a pollen ball for the sole purpose of placing it in the stigmatic cavity is an obvious adaptation for pollination. In both of these cases, selection has led to some sort of mechanism that prevents total destruction of the seed crop. In the case of the figs, only certain flowers can be used for egg laying. In *Yucca,* the larvae are small relative to the entire seed contents of a capsule, and the fruits with "excess" ovipositions can be aborted.

Some method of control on the part of the plant for limiting the extent of parasitism by internal feeding larvae is probably a prerequisite for the success of a brood flower system. In *Melandrium–Hadena* and similar systems, the larvae are largely external feeders, and it seems unlikely that they can achieve the intimacy present in the yucca or fig systems without changes such as radical increase in fruit size, reductions in the sizes of the moths, or some kind of behavioral "restraint" on the part of the ovipositing moths. All of these changes are likely, however, to be initially disadvantageous for one or both of the partners.

FUTURE DIRECTIONS

Several things have become apparent in our analysis of the spectrum of floral rewards and our attempt to assess their evolutionary development. First, the chemistry of substances used as rewards needs much more work. Even nectar, which is relatively well known, contains components such as lipids that have not yet been precisely characterized. The role of physical factors in determining or influencing nectar composition also needs further investigation. A greater understanding of biosynthetic pathways will help elucidate how some rewards such as floral oils or terpenes might be produced by altering only the final steps of common pathways. We obviously also need a much better understanding of the nutritional requirements of animal vectors in order to say with certainty what substances are superior to others as food. For nonnutritive rewards, we are still ignorant of the use of most collected substances (volatile oils, some floral exudates). In order justifiably to speak of energetic advantages to either the plant or the pollinator, we should have some method of determining

the cost of synthesis to the plant relative to gains in effective pollinator service. Table 6-1 provides a very preliminary attempt to assess the total energy and the digestible energy of various rewards and to give an indication of interchangeable quantities of these substances in terms of energy, using oil as a standard. Obviously, such a scheme is grossly simplified because it ignores the fact that some substances contain compounds composed, at least in part, of limited nutrients. Finally, we need data on the foraging costs of most animal vectors and their relationships to the energetic gains received from particular rewards.

Naturally, the questions that now need answers are the difficult ones, and that explains in large measure why they remain unresolved. However, with modern technology and carefully designed experiments, all are tractable.

ACKNOWLEDGMENTS

We thank Kalen Jacobson for helping with the manuscript and NSF for the support of our *Krameria* work.

LITERATURE CITED

Alexopoulous, C. J. and C. W. Mims. 1979. *Introductory Mycology,* 3rd ed. John Wiley and Sons, New York.

Anderson, R. J. and W. L. Kulp, 1922. Analysis and composition of corn pollen, Preliminary report. *J. Biol. Chem.* 50:433–453.

Armbruster, W. S. and G. L. Webster. 1979. Pollination of two species of *Dalechampia* (Euphorbiaceae) in Mexico by euglossine bees. *Biotropica* 11:278–283.

Baker, H. G. 1978. Chemical aspects of the pollination biology of woody plants in the tropics. B. Tomlinson and M. H. Zimmermann (eds.), *Tropical Trees as Living Systems.* Cambridge University Press, New York, pp. 57–82.

Baker, H. G. and I. Baker. 1979. Starch in angiosperm pollen grains and its evolutionary significance. *Am. J. Bot.* 66:591–600.

———. 1980. Starch and starchless pollen in a taxonomic context. *Bot. Soc. Am. Misc. Publ.* 158:9 (Abstract).

———. 1982. Chemical constituents of nectar in relation to pollination mechanisms and phylogeny. *In* M. H. Nitecki (ed.) *Biochemical Aspects of Evolutionary Biology.* 131–171.

Baker, H. G. and P. D. Hurd. 1968. Intrafloral ecology. *Annu. Rev. Ent.* 13:385–414.

Baker, H. G., I. Baker, and P. A. Opler. 1973. Stigmatic exudates and pollination. *In* N. B. M. Brantjes (ed.),

Pollination and Dispersal. Dept. of Botany, University of Nijmegen, Nijmegen, Netherlands, pp. 47–80.

Barker, R. J. 1972. Whether the superiority of pollen in the diet of honeybees is attributable to its high content of free proline. *Ann. Ent. Soc. Am.* 65:270–271.

Berryman, A. A. 1972. Resistance of conifers to invasion by bark beetle–fungus associations. *BioScience* 22:598–602.

Bino, R. J. and A. D. J. Meeuse. 1981. Entomophily in dioecious species of *Ephedra:* a preliminary report. *Acta Bot. Neerl.* 30:151–153.

Boppré, M. 1978. Chemical communication, plant relationships and mimicry in the evolution of danaid butterflies. *Ent. exp. app.* 24:264–277.

Brantjes, N. B. M. 1976. Riddles around the pollination of *Melandrium album* (Mill.) Garcke (Caryophyllaceae) during the oviposition by *Hadena bicuris* Hufn. (Noctuidae, Lepidoptera), II. *Proc. K. Ned. Akad. Wet. Ser. C Biol. Med. Sci.* 79:127–141.

Britikov, E. A., N. A. Musatova, S. V. Vladimertseva, and M. A. Protsenko. 1964. Proline in the reproductive system of plants. *In* H. F. Linskens (ed.), *Pollen Physiology and Fertilization.* North-Holland, Amsterdam, pp. 77–85.

Brothers, D. J. 1975. Phylogeny and classification of the aculeate Hymenoptera, with special reference to Mutillidae. *Univ. Kansas Sci. Bull.* 50:483–648.

Buchmann, S. L. 1974. Buzz pollination of *Cassia quiedondilla* (Leguminosae) by bees of the genera *Centris* and *Melipona. Bull. S. Calif. Acad. Sci.* 73:171–173.

Buchmann, S. L. and M. D. Buchmann. 1981. Anthecology of *Mouriri myrtilloides* (Melastomataceae: Memecyleae), an oil flower in Panama. *Biotropica* 13 (supplement): 7–24.

Buchmann, S. L., C. E. Jones, and L. J. Colin. 1977. Vibratile pollination of *Solanum douglasii* and *S. xanti* (Solanaceae) in southern California. *Wasmann J. Biol.* 35:1–25.

Calvino, E. M. 1952. Le sostanze di riserva dei pollini e il loro significato, filogenetico, ecologico, embriologico. *Nuovo G. Bot. Ital. N.S.* 49:1–26.

Campana, B. J. and F. E. Moeller. 1977. Honeybees: preference for and nutritional value of pollen from five plant sources. *J. Econ. Ent.* 70:39–41.

Chauvin, R. 1968. Digestion et nutrition des adultes. *In* R. Chauvin (ed.), *Traite de Biologie de l'Abeille,* Tome 1. Masson et Cie, Paris, pp. 347–377.

Colin, L. J. and C. E. Jones. 1980. Pollen energetics and pollination modes. *Am. J. Bot.* 67:210–215.

Corbet, S. A., D. M. Unwin, and O. E. Prys-Jones. 1979. Humidity, nectar and insect visits to flowers, with special reference to *Crataegus, Tilia* and *Echium. Ecol. Ent.* 4:9–22.

Covas, G. and B. Schnack. 1945. El valor taxonómico de la relacion longitud del pistilo: volumen del grano de polen. *Darwiniana* 7:80–90.

Cox, P. A. 1980. Flying fox pollination and the maintenance of unisexuality in populations of *Freycinetia reineckei* (Pandanaceae). *Abstr., Second Int. Congr.*

Syst. Evol. Biol., University of British Columbia, Vancouver, Canada, 17–24 July 1980, p. 169.

Craigie, J. H. 1927. Discovery of the function of the pycnia of the rust fungi. *Nature* 120:765–767.

Crepet, W. L. 1979. Insect pollination: a paleontological perspective. *BioScience* 29:102–108.

Croat, T. B. 1980. Flowering behavior of the neotropical genus *Anthurium* (Araceae). *Am. J. Bot.* 67:888–904.

Daly, H. V., J. T. Doyen, and P. R. Ehrlich. 1978. *Introduction to Insect Biology and Diversity.* McGraw-Hill, New York.

Darwin, C. 1877. *The Different Forms of Flowers on Plants of the Same Species.* John Murray, London.

Dashek, W. V. and H. I. Harwood. 1974. Proline, hydroxyproline and lily tube elongation. *Ann. Bot.* 38:947–959.

Deinzer, M. L., P. A. Thompson, D. M. Burgett, and D. L. Isaacson. 1977. Pyrrolizidine alkaloids: their occurrence in honey from tansy ragwort (*Senecio jacobaea* L.). *Science* 195:497–499.

DeVries, P. J. 1979. Pollen-feeding rainforest *Parides* and *Battus* butterflies in Costa Rica. *Biotropica* 11:237–238.

Diels, L. 1916. Kaferblumen bei den Ranales und ihre Bedeutung für die Phylogenese der Angiospermen. *Ber. Dt. Bot. Ges.* 34:758–774.

Dodson, C. H. and G. P. Frymire. 1961a. Preliminary studies in the genus *Stanhopea* (Orchidaceae). *Ann. St. Louis Bot. Gard.* 48:137–173.

———. 1961b. Natural pollination of orchids. *Missouri Bot. Gard. Bull.* 49:133–152.

Doyle, J. A. and L. J. Hickey, 1976. Pollen and leaves from the mid-Cretaceous Potomac group and their bearing on early angiosperm evolution. *In* C. B. Beck (ed.), *Origin and Early Evolution of Angiosperms.* Columbia University Press, New York, pp. 139–206.

Erdtman, G. 1966. *Pollen Morphology and Plant Taxonomy: Angiosperms.* Almquist, Stockholm.

Fahn, A. 1979a. Ultrastructure of nectaries in relation to secretion. *Am. J. Bot.* 66:977–985.

———. 1979b. *Secretory Tissues in Plants.* Academic Press, New York.

Galil., J. and D. Eiskowitch. 1973. Further studies on pollination ecology in *Ficus sycomorus* II: pocket filling and emptying by *Ceratosolen arabicus* Mayr. *New Phytol.* 73:515–520.

Gilbert, L. E. 1972. Pollen feeding and reproductive biology of *Heliconius* butterflies. *Proc. Natl. Acad. Sci.* 69:1403–1407.

Gottsberger, G. 1974. The structure and function of the primitive angiosperm flower—a discussion. *Acta Bot. Neerl.* 23:461–471.

Grant, V. 1950. The pollination of *Calycanthus occidentalis. Am. J. Bot.* 37:294–297.

Grinfel'd, E. K. 1959. The feeding of thrips (Thysanoptera) on pollen of flowers and the origin of asymmetry in the mouthparts. *Ent. Rev.* 38:798–804.

———. 1975. Anthophily in beetles (Coleoptera) and a critical evaluation of the cantharophilous hypothesis. *Ent. Rev.* 54:18–22.

Gunning, B. E. S., and M. W. Steer. 1975. *Ultrastructure and the Biology of Plant Cells*. Arnold, London.

Heslop-Harrison, J. 1976. The adaptive significance of the exine. *In* K. Ferguson and J. Muller (eds.), *The Evolutionary Significance of the Exine*. Academic Press, New York, pp. 27–37.

————. 1979. An interpretation of the hydrodynamics of pollen. *Am. J. Bot.* 66:737–743.

Heslop-Harrison, J., I. Heslop-Harrison, and J. Barber. 1975. The stigma surface in incompatibility responses. *Proc. Roy. Soc. London, Ser. B* 188:287–297.

Hilditch, T. P. and P. N. Williams. 1964. *The Chemical Constitution of Natural Fats*. Chapman and Hall, London.

Hopkins, C. Y., A. W. Jevans, and R. Boch. 1969. Occurrence of octadeca-*trans*-2, *cis*-9, *cis*-12-trienoic acid in pollen attractive to the honeybee. *Can. J. Bot.* 47:433–436.

Howell, D. J. 1974. Bats and pollen: physiological aspects of the syndrome of chiropterophily. *Comp. Biochem. Physiol.* 48A:263–276.

Ingold, C. T. 1971. *Fungal Spores: Their Liberation and Dispersal*. Clarendon Press, Oxford.

Institute of Shortening and Edible Oils, Inc. 1974. *Food Fats and Oils*. Institute of Shortening and Edible Oils, Inc., Washington D.C..

Johri, B. M. and I. K. Vasil. 1961. Physiology of pollen. *Bot. Rev.* 27:325–281.

Kevan, P. G., W. G. Chaloner, and D. B. O. Saville. 1975. Interrelationships of early terrestrial arthropods and plants. *Paleontology* 18:391–417.

Kroon, G. H., J. P. Praagh, and H. H. van Velthius. 1974. Osmotic shock as a prerequisite to pollen digestion in the alimentary tract of the worker honeybee. *J. Apic. Res.* 13:177–181.

Lee, S. 1978. A factor analysis study of the functional significance of angiosperm pollen. *Syst. Bot.* 3:1–19.

Lepage, M. and R. Boch. 1968. Pollen lipids attractive to honeybees. *Lipids* 3:530–534.

Levin, M. D. and M. H. Haydak. 1957. Comparative value of different pollens in the nutrition of *Osmia lignaria* Say. *Bee World* 38:221–227.

Linskens, H. F. and J. M. L. Mulleneers. 1967. Formation of "instant pollen tubes." *Acta Bot. Neerl.* 16:132–142.

Linsley, E. G. 1958. The ecology of solitary bees. *Hilgardia* 27:540–599.

Linsley, E. G. and M. A. Cazier. 1963. Further observations on bees which take pollen from plants of the genus *Solanum*. *Pan-Pacific Entomol.* 39:1–18.

————. 1970. Some competitive relationships among matinal and late afternoon foraging activities of caupolicanine bees in southeastern Arizona (Hymenoptera, Colletidae). *J. Kansas Ent. Soc.* 43:251–261.

Lord, E. M. and B. D. Webster. 1979. The stigmatic exudate of *Phaseolus vulgaris* L. *Bot. Gaz.* 140:266–271.

Martin, F. W. 1969. Compounds from the stigmas of ten species. *Am. J. Bot.* 56:1023–1027.

Maurizio, A. 1960. Bienenbotanik. *In* A. Budel and E. Herold (eds.), *Biene und Bienenzucht*. Ehrenwirth, Munchen, pp. 68–104.

McLellan, A. R. 1977. Minerals, carbohydrates and amino acids of pollens from some woody and herbaceous plants. *Ann. Bot.* 41:1225–1232.

McCormack, J. H. 1975. Beetle pollination of *Calycanthus floridus* L.: pollinator behavior as as function of volatile oils. Ph.D. thesis, University of Connecticut, Storrs.

Meeuse, A. D. J. 1978. Nectar secretion, floral evolution and the pollination syndrome in early angiosperms I–II. *Proc. K. Ned. Akad. Wet. Ser. C Biol. Med. Sci.* 81:300–312, 313–326.

Merrill, A. L. and B. K. Watt. 1955. Energy value of foods—basis and derivation. *U.S. Dept. Agric. Handbook* 74:105.

Michener, C. D. 1944. Comparative external morphology, phylogeny, and a classification of the bees (Hymenoptera). *Bull. Am. Mus. Nat. Hist.* 82:157–326.

Michener, C. D. and A. Fraser. 1978. A comparative anatomical study in bees (Hymenoptera: Apoidea). *Univ. Kansas Sci. Bull.* 51:463–487.

Moldenke, A. R. 1979. Host–plant coevolution and the diversity of bees in relation to the flora of North America. *Phytologia* 43:357–419.

Neff, J. L. and B. B. Simpson. 1981. Oil-collecting structures in the Anthophoridae (Hymenoptera): morphology, function and use in systematics. *J. Kansas Ent. Soc.* 54:95–123.

Percival, M. S. 1961. Types of nectar in angiosperms. *New Phytol.* 46:142–173.

Pettit, J. M. and C. B. Beck. 1967. Seed from the Upper Devonian. *Science* 156:1727–1729.

Pijl, L. van der and C. H. Dodson. 1966. *Orchid Flowers: Their Pollination and Evolution*. University of Miami Press, Coral Gables, Fla.

Pliske, T. E. 1975. Pollination of pyrrolizidine alkaloid-containing plants by male Lepidoptera. *Environ. Ent.* 4:474–479.

Plowright, R. C. 1981. Nectar production in the boreal forest lily *Clintonia borealis*. *Can. J. Bot.* 59:156–160.

Portnoi, L. and A. Horovitz. 1977. Sugars in natural and artificial pollen germination substrates. *Ann. Bot.* 41:15–20.

Porsch, O. 1906. Beiträge zur "histologischen Blutenbiologie." *Öst. Bot. Z.* 56:135–143.

————. 1910. *Ephedra campylopoda* C. A. Mey. eine entomophile Gymnosperme. *Ber. Dt. Bot. Ges.* 28:404–412.

————. 1916. Die Nektartropfen von *Ephedra campylopoda Ber. Dt. Bot. Ges.* 34:202–212.

Proctor, M. and P. Yeo. 1972. *The Pollination of Flowers*. Taplinger, New York.

Ramirez, B. W. 1969. Fig wasps: mechanism of pollen transfer. *Science* 163:580–581.

Rickson, F. R. 1975. The ultrastructure of *Acacia cornigera* L. Beltian body tissue. *Am. J. Bot.* 62:913–922.

————. 1979. Ultrastructural development of the beetle

food tissue of *Calycanthus* flowers. *Am. J. Bot.* 66:80–86.

―――. 1980. Developmental anatomy and ultrastructure of the ant-food bodies (Beccariian bodies) of *Macaranga triloba* and *M. hypoleuca* (Euphorbiaceae), *Am. J. Bot.* 67:285–292.

Roberts, R. B. and S. R. Vallespir. 1978. Specialization of hairs bearing pollen and oil on the legs of bees (Apoidea: Hymenoptera). *Ann. Ent. Soc. Am.* 71:619–627.

Rothschild, M. and N. Marsh. 1978. Some aspects of danaid/plant relationships. *Ent. exp. appl.* 24:437–450.

Schneider, E. L. 1979. Pollination biology of the Nymphaeaceae. *In: Proc. Fourth Intl. Symp. on Pollination,* 11–13 Oct. 1978, Beltsville, Md., pp. 419–429.

Schremmer, F. 1972. Der Stechsaugrüssel, der Nektarraub, das Pollensammeln und der Blütenbesuch der Holzbienen *(Xylocopa)* (Hymenoptera, Apidae). *Zeit. Morph. Tiere* 72:263–294.

Scott, H. G. and C. L. Stojanovich. 1963. Digestion of juniper pollen by Collembola. *Florida Entomol.* 46:189–191.

Simpson, B. B. and J. L. Neff. 1981. Floral rewards: alternatives to pollen and nectar. *Ann. Missouri Bot. Gard.* 68:301–322.

Simpson, B. B., J. L. Neff, and D. Seigler. 1977. *Krameria,* free fatty acids and oil-collecting bees. *Nature* 267:150–151.

Skutch, A. F. 1971. *A Naturalist in Costa Rica.* University of Florida Press, Gainesville.

Smart, J. and N. F. Hughes. 1973. The insect and the plant: progressive palaeoecological integration. *In* H. F. Emden (ed.), *Insect Plant Relationships.* John Wiley and Sons, New York, pp. 143–155.

Snodgrass, R. E. 1956. *Anatomy of the Honeybee.* Comstock, Ithaca, N.Y.

Stanley, R. G. 1971. Pollen chemistry and tube growth. *In* J. Heslop-Harrison (ed.), *Pollen: Development and Physiology.* Appleton-Century-Crofts, New York, pp. 131–155.

Stanley, R. G. and H. F. Linskens. 1974. *Pollen: Biology, Biochemistry, Management.* Springer-Verlag, New York.

Swern, D. (ed.). 1979. *Bailey's Industrial Fat and Oil Products,* 4th ed. John Wiley & Sons, New York.

Taylor, T. N. and D. A. Levin. 1975. Pollen morphology of the Polemoniaceae in relation to systematics and pollination systems: scanning electron microscopy. *Grana* 15:91–112.

Thien, L. B. 1974. Floral biology of *Magnolia. Am. J. Bot.* 61:1037–1045.

Thorp, R. W. 1979. Structural, behavioral, and physiological adaptations of bees (Apoidea) for collecting pollen. *Ann. Missouri Bot. Gard.* 66:788–812.

Todd, F. E. and O. Bretherick. 1942. The composition of pollens. *J. Econ. Ent.* 35:312–317.

Vivino, E. A. and L. S. Palmer. 1944. The chemical composition and nutritional value of pollens collected by bees. *Arch. Biochem.* 4:129–136.

Vogel, St. 1969. Flowers offering fatty oil instead of nectar. *Abstr. XI Int. Botanical Congr.,* p. 229.

―――. 1974. Ölblumen und ölsammelnde Bienen. *Akad. Wissen. Literatur Math.-Natürwiss. Klasse. Tropische und subtropische Pflanzenwelt* 7:1–267.

―――. 1978. Evolutionary shifts from reward to deception in pollen flowers. *In* A. J. Richards (ed.), *The Pollination of Flowers by Insects.* Academic Press, New York, pp. 89–96.

Waller, G. D., E. W. Carpenter, and O. A. Ziehl. 1972. Potassium in onion nectar and its probable effect on the attractiveness of onion flowers to honeybees. *Am. Soc. Hort. Sci. Proc.* 97:535–539.

Weaver, N. and K. A. Kuiken. 1951. Quantitative analysis of the essential acids of royal jelly and some pollens. *J. Econ. Ent.* 44:635–638.

Wiebes, J. T. 1979. Co-evolution of figs and their insect pollinators. *Annu. Rev. Ecol. Syst.* 10:1–12.

Williams, N. H. 1982. The biology of orchids and euglossine bees. *In: Orchid Biology: Reviews and Perspectives. II.* Cornell University Press, Ithaca, N.Y. In Press.

Ziegler, H. and I. Ziegler. 1962. Die wasserlöslichen Vitamine in den Siebröhrensaften einiger Baume. *Flora* 152:257–278.

Zimmermann, M. H. and C. L. Brown. 1977. *Trees. Structure and Function.* Springer-Verlag, New York.

7

VISIBLE FLORAL PIGMENTS AND POLLINATORS

Ron Scogin

Rancho Santa Ana Botanic Garden
Claremont, California

ABSTRACT

The behavior of potential pollinators is strongly influenced by the color of a flower. Each pollinator class exhibits innate or entrained preferences among floral colors. The pigments responsible for imparting floral colors include three distinct chemical classes: anthocyanins, carotinoids, and betalains. The presence of these light-absorbing pigments, either singly or in concert, will determine flower color. Additional factors that determine flower color include the amount of pigments present, the presence of additional phenolic compounds (co-pigments), and the degree of metal chelation to yield supramolecular complexes. Floral colors and the pigments responsible for them occur nonuniformly on a geographical basis. Blue floral colors are more common in temperate regions, whereas tropical floras are enriched with red flowers. This appears to be correlated with the color preferences of principal pollinators in different areas.

KEY WORDS: Anthocyanin, anthocyanidin, carotinoid, betalain, co-pigment, metal chelation, anthochlor pigment, flavonol.

CONTENTS

INTRODUCTION

The phenomenon of color vision has a sporatic distribution within the animal kingdom. Organisms capable of color vision include the mammalian primates, most birds and insects, many fish and reptiles, and a few molluscs (cephalopods) and crustaceans. Among these organisms the most important agents of plant pollination are the birds and insects. Nonprimate mammals such as bats and rodents may also pollinate flowers, but they are essentially colorblind and are attracted by other signals.

A number of features influence the attractiveness of a flower to a potential pollinator. Among these features are the color, surface texture, shape, degree of dissection, and scent of the flower (Proctor and Yeo, 1973). Attractants can be usefully divided into those operating over long distances (arbitrarily defined as greater than one meter) and those operating over short distances, depending upon the distance from the flower at which the pollinator's sensory system becomes aware of the attractant. In general, visual attractants act over much longer distances than olfactory ones. The maximum distance at which a flower (or inflorescence) can be perceived by a pollinator is determined by the flower (or inflorescence) size and the visual contrast between the flower and its surrounding background. It is in providing enhanced contrast with the vegetative background that floral coloration plays its premier role in long-distance, pollinator attraction.

One must constantly bear in mind that pollinators are not enticed to visit a flower by color or any other single attractant acting in isolation, but rather by a complex syndrome of characters to which the pollinator is innately attracted or behaviorally entrained. Isolating color as an attractive character is somewhat artificial and is made more so by restricting our consideration to visible colors, to the exclusion of ultraviolet (UV) wavelengths. Several authors have recently pointed out the hazards of considering colors, whether visible or ultraviolet, independently, rather than as a comprehensive visual stimulus to the pollinator (Kevan, 1972; Levy, 1978). Mindful of these caveats, the present review will treat only visible colors and the pigments responsible for them, noting that an important and integral constituent of floral coloration is the contribution of near-ultraviolet wavelengths.

Natural selection has acted to control the elaboration of numerous floral pigments in several chemical classes, with absorption–reflection properties over wavelengths to which pollinator visual systems are sensitive and which provide striking visual contrast between floral parts and surrounding background vegetation or between different portions of a single flower. Many pollinator classes find this visual contrast inherently attractive.

A substantial body of literature treating pollinator visual physiology and color-elicited behavioral responses demonstrates that particular pollinators or classes of pollinators have a characteristic predilection in the color of the flowers they visit. The literature establishing these patterns is reviewed in Procter and Yeo (1973) and Faegri and van der Pijl (1979) and is summarized in Table 7-1. The visual sensitivity spectrum of some pollinators, such as birds, is nearly identical to that for humans; shifted, in the case of hummingbirds, about 30 nm toward shorter wavelengths (Huth and Burkhardt, 1972; Goldsmith, 1980). In insects, however, the visual spectrum is shifted substantially to shorter wavelengths, making insects sensitive to near-ultraviolet wavelengths but often insensitive to red coloration.

Whether pollinator color preferences are innate or are learned is a difficult question to answer unequivocally, and the answer most likely varies among groups of organisms. Ilse (1928) demonstrated that in six species of newly emerged, laboratory-reared butterflies, statistically significant, inherited color preferences existed. Visits to a color other than the innately preferred color could be entrained with

Table 7-1. Color preferences and qualities of pollinator classes.

Pollinator	Preferences
Bats	Cream, purple, greenish (drab)
Bees	Yellow, blue (bright)
Beetles	Cream, greenish (drab)
Birds	Scarlet (bright)
Butterflies	Red, yellow, blue, pink (bright)
Moths	
Nocturnal	White
Diurnal	Red, purple, pink (bright)
Flies	
Carrion, dung	Brown, purple (dull)
Others	Cream, white
Wasps	Brown (dull)

food rewards, but the preferred colors were never completely abandoned, and entrained butterflies, when given a choice, frequently would merely shift to visiting that tint of the preferred color most closely resembling the training color. In contrast, Bene (1941) and Grant (1966) have argued that the red coloration of flowers so frequently visited by hummingbirds does not reflect an innate preference for red among these pollinators. Grant (1966) postulates that the prominence of red among hummingbird flowers is a learned preference, most frequently associated with migratory hummingbird taxa in which new food resources must regularly be discovered. The red coloration is "chosen" to minimize competition for food resources with red-blind insect taxa.

Experimental studies have demonstrated that pollinators express preferences for particular color morphs when confronted by a color choice under field conditions. Such preferences were exhibited by lepidopterans presented with both intraspecific (Levin, 1972) and interspecific (Levin, 1969) choices in the genus *Phlox* and by bees confronted with polymorphic populations (with respect to corolla color) of *Levenworthia* (Lloyd, 1969) and *Cirsium* (Mogford, 1974). Hummingbirds and bumblebees also preferentially chose pigmented over albino flowers for visitation in *Delphinium* (Waser and Price, 1981). (See review by Kay, 1978.)

One may conclude from this information that different pollinator classes have different floral color preferences, whether innate or learned, and that these color preferences operate under field conditions in a biologically significant way. The forces of natural selection, then, must operate strongly to regulate the pigments responsible for imparting color to flowers.

THE PIGMENTS IMPARTING FLORAL COLORATION

The chemistry, biosynthesis, and systematic distribution of floral pigments have been extensively reviewed elsewhere (Harborne et al., 1975; Goodwin, 1976). Only those aspects of floral pigments relating directly to interaction with pollinators will be treated here.

The major pigments imparting coloration to flowers are divided among three chemical classes: flavonoids, carotinoids, and betalains. Frequently, representatives of two (and rarely all three) of these classes co-occur within the same flower. Some biological redundancy appears to exist in that the same floral color (to human vision) can be produced by representatives of different chemical classes; e.g., the same bright yellow coloration can be due to flavonoids (anthochlor pigments or 8-OH flavonols), carotinoids (xanthophylls), or betalains (betaxanthins). This apparent redundancy most likely reflects our imperfect understanding of pollinator perception of color or of alternative, additional biological roles for these compounds in the plants elaborating them.

Flavonoids

Anthocyanins. Anthocyanins, a subgroup of the plant phenolic pigments termed flavonoids, are the most widely distributed floral pigments in nature, accounting for the cyanic coloration (orange-red-blue-purple) in most flowers. They are water-soluble pigments that occur in the vacuole of plant cells, frequently concentrated at the tissue surface in epidermal cell layers. Anthocyanins are conveniently extracted into acidified methanol and easily identified by paper chromatographic techniques (Harborne, 1967). These pigments are naturally occurring pH indicators: their presence in alcoholic solution can be confirmed quickly by the presence of a reversible color change from red under acidic conditions to blue in an alkaline environment.

The structures of 164 different, naturally occurring anthocyanins have been determined (Gibbs, 1974), their structural diversity being derived largely from the number, identity, and position of attachment of sugar moieties to the anthocyanidin structure. (The acid-stable, skeletal pigment structure, with sugars removed, is termed an anthocyanidin.) All anthocyanins occur naturally with from one to several sugars attached. The nature and number of sugar substitutions influences only minimally the light absorption properties of an anthocyanin, so that the nature of anthocyanin sugar substituents has little bearing on floral coloration. By contrast, the structure of the skeletal anthocyanidin is paramount in color determination. Of the 15 naturally occurring anthocyanidins (Timberlake and Bridle, 1975), only three are of major importance. Pelargonidin, cyanidin, and delphinidin differ only in having one, two, and three hydroxyl groups, respectively, on the anthocyanidin's phenyl, "B" ring (see Fig. 7-la). The number of "B"-ring hydroxyl groups strongly affects the light absorption properties and therefore the color of the compounds, with the result that pelargonidin is orange-red, cyanidin is cherry-red (magenta), and delphinidin is blue-purple (mauve). As a first visual approximation, scarlet and orange-red flowers contain pelargonidin; crimson and magenta flowers contain cyanidin; and mauve, purple, and blue flowers contain delphinidin.

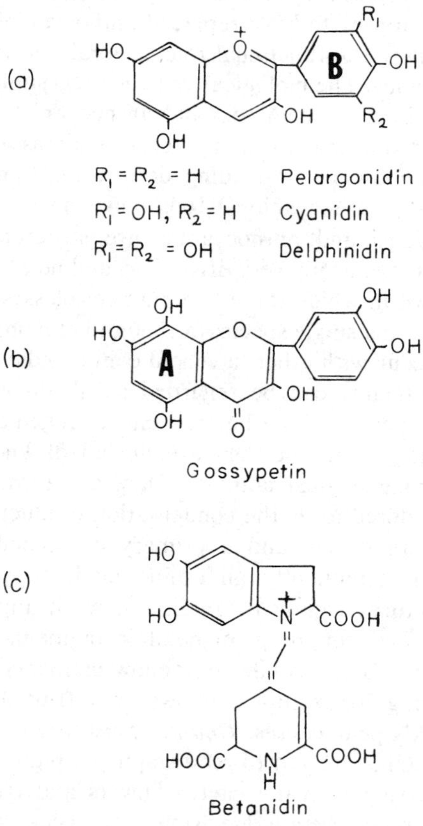

It is a common occurrence for native plants to contain a mixture of several anthocyanins that can be based on one or, less frequently, several anthocyanidins. This tendency to contain a mixture of pigments is amplified in garden plants selected for striking appearance and variety in color morphs. Among ornamental horticultural plants as many as four aglycones (anthocyanidins) can occur in a single flower of a selected cultivar (e.g., *Gladiolus*). Similarly, 18 anthocyanins based on six aglycones have been identified in various combinations in selected color forms of *Pisum sativum*.

Yellow Flavonoid Pigments. While the commonly occurring flavones and flavonols are frequently referred to as pigments, the majority of them, in fact, contribute little to light absorption in flowers over the human visible spectrum. These compounds often occur (as glycosides) in perfectly white flowers (Roller, 1956). Their only contributions from the viewpoint of human vision are to add "body" to the white flower and often to impart a cream or ivory coloration. The presence of these compounds in flowers may, however, be much more significant when considered from the viewpoint of the insect visual spectrum and taking into account their intense near-ultraviolet wavelength absorption. There are two flavonoid pigment groups whose light absorption properties cause them to impart a strong yellow coloration to flowers in which they occur: (1) the anthochlor pigments and (2) 6-OH and 8-OH substituted flavonols.

1. Anthochlor Pigments. The anthochlor pigments are comprised of two classes of flavonoids, chalcones and aurones, that historically have been considered together because of a shared color-change reaction. They change visibly from yellow to orange-red in the presence of a base such as ammonia vapor. Later studies have revealed that chalcones and aurones are structurally distinct, but biogenetically related flavonoid classes.

Anthochlor pigments alone can produce intense yellow floral coloration, as in the yellow snapdragon *(Antirrhinum majus)*, or, more

(a)

R₁ = R₂ = H   Pelargonidin
R₁ = OH, R₂ = H   Cyanidin
R₁ = R₂ = OH   Delphinidin

(b)

Gossypetin

(c)

Betanidin

Figure 7-1. Representative structural formulae for (a) anthocyanidins, (b) an 8-OH flavonol, and (c) a betalain.

frequently, they can co-occur with carotinoids, as in the genera *Coreopsis* and *Bidens* (Harborne, 1976).

2. 6- and 8-Substituted Flavonols. Flavonols, which are substituted in the 6 and/or 8 position of the aromatic "A" ring with a hydroxyl or methoxyl group, provide strong yellow coloration to flowers in which they occur (see Fig. 7-1b). Gossypetin (8-hydroxyquercetin) is the principal yellow pigment in the cotton flower *(Gossypium herbaceum),* in the California flannel bush *(Fremontia californica),* and elsewhere (Harborne, 1976). Hydroxylation of flavonols at the 6 position is less common, but has been reported from the Asteraceae and Fabaceae (Harborne, 1967).

Carotinoids

Carotinoids are lipid-soluble pigments that occur in chromoplasts in the cytoplasm of plant cells. They function as accessory pigments to chlorophyll in photosynthesis in all plants. In addition, most yellow-colored flowers owe their coloration to high concentrations of carotinoid in the corolla. Carotinoids are divided into two subclasses, xanthophylls and carotenes, depending on whether oxygen is present or absent, respectively, in the molecular structure. These pigments can be distinguished from less commonly occurring yellow pigments, such as certain flavonoids, by their high solubility in such lipid solvents as petroleum ether and diethyl ether, in which flavonoids are only slightly soluble.

Carotinoid pigments can impart colors to flowers ranging from light yellow through deep orange, depending upon the quantity of the carotinoid present and its molecular structure. Flowers containing carotinoid pigments can be divided into three main groups:

1. Light yellow and lemon yellow colored flowers that usually have xanthophylls as their major carotinoid constituents, frequently as epoxides (compounds with a three-membered, oxygen-containing bridge); e.g., yellow *Calendula officinalis.*

2. Orange-yellow colored flowers having carotenes such as β-carotene (in *Narcissus majalis*) or lycopene (in orange *Calendula officinalis*) as major carotinoid constituents.

3. A few taxa that have taxon-specific pigments, such as eschscholtzxanthin which imparts the deep orange color to petals of the California poppy *(Eschscholtzia californica).*

A listing of plant species and their floral carotinoid constitution may be found in Goodwin (1974).

Betalains

The betalains, a class of water-soluble, vacuolar plant pigments, are completely restricted in occurrence to a single taxonomic order, the Centrospermae (or Caryophyllales), in which they appear to have replaced anthocyanins in fulfilling the functional role of floral and fruit pigments. The biological and chemical similarities between betalains and anthocyanins are so striking that for many years betalains were known by the misleading designation "nitrogenous anthocyanins." It is now known that betalains and anthocyanins are structurally and biogenically very dissimilar, and no case is known in which these two pigment classes co-occur in a single species (although betalains do co-occur with other flavonoid compounds).

Betalains can be regarded as alkaloids in that they possess a heterocyclic, nitrogen-containing ring, but they are devoid of known pharmacological activity. They are formally considered to be the condensation products of betalamic acid and a primary or secondary amine (frequently an amino acid) to yield structures such as betanidin (shown in Fig. 7-1c). Two subgroups of betalain pigments are known. Betaxanthins are yellow pigments, occurring, for example, in flowers and fruit of the prickly-pear cactus, *Opuntia ficus-indica.* Betacyanins are red to reddish-purple pigments, occurring in many cactus flowers and fruits, beet roots *(Beta vulgaris),* and the cockscomb flower *(Celosia cristata).*

Like anthocyanins, the betalains occur naturally glycosylated with one or several sugars (Piattelli, 1976). Additional similarities to anthocyanins include absorption spectra maxima and some paper chromatographic properties. The most reliable way to distinguish betalains from anthocyanins is by paper electrophoresis at pH 2.4, under which conditions betalains migrate as anions, but anthocyanins are immobile (Harborne, 1973).

The betalains appear to be an example of functional convergence at the biochemical level in that they function to replace anthocyanins as floral and fruit pigmentation agents in all families of the Centrospermae except the Molluginaceae and Caryophyllaceae (which produce anthocyanins). It should be emphasized that the absence of anthocyanins in most families of the Centrospermae does not preclude the co-occurrence of betalains with other flavonoid classes (e.g., betalains co-occur with isorhamnetin glycosides in the flowers of *Opuntia lindheimeri;* Rosler et al., 1966).

METHODS OF PIGMENT ANALYSIS

In order to obtain meaningful spectroscopic results or accurate chromatographic mobilities, floral pigments first must be purified to separate them from the numerous other plant constituents that are equally soluble in commonly used extracting solvents.

Extraction Procedures and Precautions

Anthocyanins. Anthocyanins (both aglycones and glycosides), like all other flavonoid classes, are highly soluble in methanol, which is generally the solvent of choice. In contrast to other flavonoids, anthocyanins are very unstable under neutral and alkaline pH conditions. Under these conditions the anthocyanin quickly converts to a colorless, leuco-base form and subsequently degrades to forms that are not easily analyzed. As a result, anthocyanin solutions should always be maintained at an acidic pH by the addition of small amounts of organic (e.g., acetic) or mineral (e.g., hydro-

chloric) acid. A suitable extracting solvent is methanol containing 1% concentrated HCl. If the methanolic solution is observed to become colorless during extraction, small amounts of acid should be added to restore coloration. Anthocyanins also decay slowly at room temperature and when exposed to light; so they are best stored under refrigeration and in the dark.

Carotinoids. Carotinoid pigments are quite unstable when removed from their natural milieu; they are subject to atmospheric oxidation and to *cis-trans* isomerization. Atmospheric oxidation and subsequent decoloration are especially rapid when the pigment is exposed to air on a large-surface-area matrix as is encountered when performing thin-layer chromatography (TLC). Carotinoids can be extracted into a variety of solvents such as benzene, ether, petroleum ether, methanol, carbon disulfide, or chloroform. Extraction (using peroxide-free solvents) and purification are best performed as quickly as is feasible, and the resulting carotinoid solutions should be stored in the dark, refrigerated, and under nitrogen gas.

Betalains. Betalains are stable under usual extraction conditions, although they may undergo irreversible color changes under strongly alkaline conditions. They are usually extracted into distilled water (which is then acidified to prevent color-change reactions), frequently concentrated by adsorption onto a strongly acidic resin column, and eluted using acidified methanol.

Purification and Identification

The following descriptions of techniques are abbreviated. For a more complete and detailed treatment readers should consult the following: for flavonoids, Harborne (1967, 1973); for carotinoids, Davies (1976) and Moss and Weedon (1976); for betalains, Piatteli (1976).

Anthocyanins. Anthocyanins are usually purified by one-dimensional paper chromatography (PC) in a suitable solvent such as butanol–acetic acid–water (BAW, 4:1:5). The resolved, brightly colored bands are cut out, and the pigment is eluted into acidified methanol. One chromatographic separation is usually adequate to purify an anthocyanin sufficiently for chromatographic analysis. If complete purity is required, as for spectroscopic analysis, subsequent one-dimensional PC separation in an alternative solvent, such as 15% acetic acid with 1% concentrated HCl added, should be performed.

In order to identify the anthocyanidin (aglycone), an aliquot of the purified pigment is removed for hydrolysis. This aliquot is taken to dryness by evaporating off the methanolic solvent, and the sample is then dissolved in a small volume (5–10 ml) of 2 N HCl. This aqueous solution is heated to 100°C for 45 min. to 1 hr, allowed to cool, and then shaken with a very small volume (about 0.5 ml) of amyl alcohol. The anthocyanidin pigment will concentrate in the amyl alcohol layer, which can be removed, spotted on paper, and chromatographed in a suitable solvent such as Forestal's solvent, preferably along with known standards for comparison. The spot color and mobility will identify the anthocyanidin. See Harborne (1967) for useful solvents and R_f values.

A generally accepted proof of identity of a known anthocyanin (glycoside) is agreement of the pigment's R_f values in four solvent systems with those R_f values reported in the literature. An indispensable table of anthocyanin R_f values in various solvents can be found in Harborne (1967). Of course, the spot color and anthocyanidin identification must be consonant with this identity. It must be borne in mind that R_f values can vary $\pm 5\%$, depending on the conditions of chromatography, type of paper, laboratory temperature, presence of impurities, and other imponderables. The structure determination of a novel anthocyanin is not difficult for a trained natural products chemist, but it is generally beyond the interests of a pollination biologist.

Carotinoids. The crude carotinoid extract is frequently subjected to an initial fractionation before final purification is begun. The extract is shaken in a mixture of equal volumes of petroleum ether (or heptane) and 90% aqueous methanol. Upon sitting, two immiscible layers will form. The nonhydroxylated carotenes will be in the upper, petroleum layer, and the xanthophylls with two or more hydroxyl groups will be in the lower, methanolic layer. Monohydroxyxanthophylls tend to be distributed in both phases (Moss and Weedon, 1976). Individual carotinoids can then be purified by preparative TLC or, in larger quantities, by column chromatography. A variety of solvent systems and support matrices are available for such chromatography. The choice of the most suitable system depends upon the compounds to be separated and must be determined by a preliminary survey. A very useful tabulation of solvent systems, chromatographic supports, and carotinoid mobilities and spectra is presented by Davies (1976).

Once purified, a carotinoid can be provisionally identified by an agreement between its chromatographic mobility and visible-wavelength spectroscopic maxima and those of a compound reported in the literature. Chromatography (usually TLC) should be performed in several solvent systems, and spectroscopic maxima should be determined in several (at least four) solvents. The structure determination of a novel carotinoid is best left to the carotinoid chemist.

Betalains. Betalains can be purified by preparative paper electrophoresis (at pH 2.4 in a formate buffer) or by chromatography on a polyamide column. Betalains are the least-well-characterized class of floral pigments, with the result that comprehensive catalogs of their characteristics are lacking. A preliminary identification can be made by determining a betalain's electrophoretic mobility under defined conditions relative to a known standard betalain, usually betanin. A useful table of betacyanin electrophoretic mobilities is given by Mabry and Dreiding (1968).

THE DETERMINATION OF FLOWER COLOR

Among the chemical classes of floral pigments, the anthocyanins are the best studied, both intensively and extensively, with regard to biosynthesis, regulation, distribution, and ecological function. As a result, most of the subsequent discussion will deal with anthocyanin pigments.

The fundamental determinant defining which pigments will be present in the flowers of a particular taxon (and, therefore, its floral color) is the biosynthetic potential inherent in the genome of the taxon. Among the three primary anthocyanidins, cyanidin is considered to be the basic or primitive type (Harborne, 1977). Pelargonidin is considered to be derived in nature from cyanidin by means of a "loss" mutation in the biogenic pathway. The change from a cyanidin-producing taxon to a pelargonidin-producing mutant results from the mutational loss of enzymatic introduction of the second (3') hydroxyl group onto the "B" ring. Delphinidin, by contrast, is considered to be derived from cyanidin biosynthesis by a "gain" mutation. In the cyanidin-producing to delphinidin-producing transition, a new biosynthetic potential is gained by the mutational acquisition of an enzyme capable of attaching a third hydroxyl group to the "B" ring. The proviso "in nature" in the previous statement must be emphasized because among horticultural taxa where cultivars are artifically selected, the mutants developed are always loss mutations proceeding from delphinidin-containing to cyanidin-containing or from cyanidin-containing to pelargonidin-containing (Harborne, 1967).

Generally the expression of the biosynthetic potential of a genome is a conservative, slowly changing character, especially if the chemical products are not under strong selective pressure. It is this rule that has allowed profitable use of comparative biochemistry in taxonomic studies—the assumption that shared chemical constituents imply a common phylogeny. The result is that, to a first approximation, the suite of pigments in any taxon will reflect its phylogenetic history. The validity of this is borne out by the successes of studies correlating phenolic chemistry with taxonomic arrangement (Harborne, 1975).

Subject only to constraints placed on it by genome limitations on biosynthetic capacity, a plant can modify its pigments in response to adaptive pressures through natural selection. Among the secondary, plant phenolic compounds, the anthocyanins are unique in having at least one well-understood biological function (perhaps *inter alia*), namely, visitor attraction for pollination and fruit dispersal. The biological roles of other phenolic compounds (flavones, flavonols, quinones, coumarins, cinnamic acids, etc.) are generally unknown or unproven. Anthocyanins provide biologists a unique opportunity to study the forces of natural selection operating upon the production and regulation of a secondary plant compound under known selective pressures.

As noted earlier, the nature and positioning of sugar constituents on an anthocyanidin molecule have little effect on its color. Hence biosynthetic potential and enzymatic specificity with regard to sugar substituents have little bearing upon floral interaction with animal visitors. The anthocyanidin aglycones alone define the color of a particular flower. With respect to the major anthocyanidins present in a flower, selective pressures can shift floral coloration from red (cyanidin-based) to red-orange (pelargonidin-based), or from red to blue or purple (delphinidin-based) in order to attract the most profitable pollinator. Methylation of the hydroxyl groups of an anthocyanidin occasionally occurs, especially in certain plant groups, and has the effect of shifting the absorption maximum by 5–10 nm toward shorter wavelengths. Such an absorption shift has only a slight effect on floral coloration, at least as perceived by the human eye.

Factors Modifying Pigment Expression

Within the constraints established by biosynthetic potential (i.e., with a given suite of anthocyanidins), selective fine tuning for flower color can be accomplished by means of several processes involving modification of pigment-color expression. Variable factors consist of pigment intensity, color hue, and pigment segregation within the flower—all under both genetic and environmental control.

The intensity of coloration is greatly influenced simply by the quantity of pigment present in floral tissues. The amount of pigment can vary from 0.01% to 15% of dry weight in flowers (Harborne, 1977). Within a single taxon the differences in pigment amount can range over a factor of 10 in different lines of selected cultivars between pale and highly colored forms. The regulation of pigment amount is under the control of relatively few genes; a single gene controls pigment intensity in *Primula sinensis* (Harborne and Sherratt, 1961) and in *Solanum iopetalum* (Harborne, 1976), whereas two genes control the color forms of *Torenia fournieri* (Endo, 1962). The mechanism by which these genes regulate pigment amount is unknown.

Co-pigmentation. The presence of additional phenolic compounds, especially flavones, flavonols, and hydrolyzable tannins, can significantly influence the absorption properties of anthocyanin pigments in vivo and thus alter floral coloration. Under physiological conditions (in weakly acidic, aqueous solution), the absorption maxima of anthocyanins are shifted toward longer wavelengths by 5 to 30 nm with the addition of flavonol glycosides to the solution (Asen et al, 1972). Co-pigmentation of anthocyanins with flavonols has the effect of increasing the blueness of floral coloration to human vision and is the result of hydrogen bonding between the anthocyanin and its co-pigment to form highly labile, molecular complexes. Selection for particular color hues that have as their basis co-pigmented complexes may explain why flavone/flavonol glycosides are almost universal constituents of flowers even though they contribute little to direct visual absorption.

The molar ratio of the anthocyanin to its co-pigment affects the degree of spectral shift in absorption peaks (Asen et al., 1972). As a result, not only the presence but also the amount of flavone/flavonol can influence color expression in a flower. Therefore, not only the anthocyanin but also the presence and amount of co-pigment become important factors in fine-tuning flower color to pollinator preference or entrainment. Similarly the genes controlling flavone/flavonol biosythesis and regulation become significant in influencing pollinator attraction.

Metal Chelation.   Metal ion chelation seems to be important in the blue coloration of some flowers (e.g., the blue corn flower, *Centaurea cyanus,* and *Commelina communis*) (Asen and Jurd, 1967). Iron and magnesium have been identified in pigment complexes consisting of an anthocyanin, a metal ion, and a flavone glycoside. The relative roles of metal complexing and co-pigmentation in the bluing effect of these pigment complexes is uncertain at present. The effectiveness of color modification due to metal ion chelation is strongly influenced both by the edaphic environment of the plant and by the plant's nutrient status with respect to nitrogen, phosphorus, and potassium, which influence the uptake and accumulation of soil metals. Since so many environmental factors are involved coordinately in color modification by metal chelation, it seems unlikely that it would play a major role in pigment modification for pollinator attraction.

Differential Pigmentation.   Numerous examples are known of differential cyanic coloration within a single flower. A common expression of this phenomenon is the occurrence of floral nectar guides, which frequently consist of a series of lines or spots in which pigmentation is much more intense than in the background. The occurrence of color bloches at the base of petals is another example of differential coloration. The mechanism by which this differential pigment synthesis or accumulation occurs is unknown, but several possibilities exist. One would be differential biosynthesis of pigments in certain cells but not in the surrounding ones, to yield, say, a nectar guide spot. Another possibility would be general biosynthesis followed by transport (by an unspecified physiological process) and sequestering of pigments in particular cells. A third possibility could be differential biosynthesis or sequestering of a co-pigment that greatly enhances anthocyanin color expression. The relative importance of these possible mechanisms it totally unknown and represents fertile ground for future research.

Anthocyanins are among the end products of a lengthy biosynthetic pathway (the shikimic acid pathway) that produces as intermediate products flavones and flavonols (see Fig. 7-2), and any factors that influence fla-

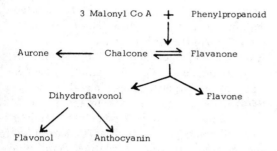

Figure 7-2. Proposed biogenic relationships among classes of flavonoids mentioned in text. (Redrawn from Grisebach, 1973.)

vonoid production will also influence anthocyanin production. A number of different factors have been demonstrated to influence the biosynthesis of flavonoids, including anthocyanins. These factors include light intensity, temperature extremes, water stress, atmospheric composition, and mineral nutrient status (McClure, 1975). Unfortunately, none of the physiological factors examined can be envisioned as acting differentially within a single flower to produce nectar guides, etc.

Cell coloration, as in nectar guides, may be developmentally determined very early in bud development. Single isolated protoplasts from *Nemesia strumosa* buds proliferate into cell populations of either red or reddish-blue coloration (reflecting different amounts of contained anthocyanin), depending on the position of the original protoplast in the floral bud (Hess and Endress, 1972).

Mechanistically, the very meager results available suggest that flavonoid accumulation is responsive to both substrate induction and end-product inhibition in determining the amount of flavonoid present within a plant tissue.

PIGMENTS, PHYTOSOCIOLOGY, AND POLLINATORS

It has frequently been observed that the range of floral colors present in a local flora is not uniform among all geographical regions or phytosociological communities. High-altitude, alpine floras are rich in white and yellow flowers (Kevan, 1972), whereas the tropics are enriched in brightly colored, red and orange flow-

ers. Temperate regions, which are climatically intermediate, show a wider range and more uniform balance in floral colorations than the other regions. A dissenting view to these observations has been presented by Weevers (1952), who argues that floral colors are harmonically distributed geographically, and that the evolution of flower color must, therefore, be independent of environment, including pollinators. The single exception noted by Weevers is the prevalence of blue flowers at high elevation and the depauperization of blue flowers in tropical regions.

The traditional explanation for a geographically nonharmonious distribution of floral colors has been that pollinators are also nonuniformly distributed. The anthophilous fauna of higher elevations (and latitudes) are primarily insects, especially flies; temperate regions have a balanced insect fauna enriched in bees; and the red flowers of the tropics are attributed to a high frequency of anthophilous birds and mammals. This model posits that natural selection has operated to increase the frequency of floral coloration (and the pigments responsible) to match the color preferences of the local pollinator spectrum. If these explanations of floral color ranges, geography, and pollinators are accurate, then a geographically nonharmonious distribution of floral pigments should also be found. Most comparative studies of floral plant pigments have been based upon taxonomic surveys rather than geographical surveys. In the relatively few geographically based surveys conducted, such a pigment disharmony is suggested. A survey of Himalayan plants (Acheson, 1956) revealed a high proportion of delphinidin-containing plants (61%) and few containing pelargonidin (2%). By contrast, in a survey of tropical taxa in Trinidad (Forsyth and Simmonds, 1954), 17% of the taxa surveyed contained pelargonidin, and 41% contained delphinidin. In a general survey of the Australian flora Gascoigne et al. (1948) reported little pelargonidin (2% of the taxa), cyanidin as commonly occurring (47%), and an unusually high proportion of delphinidin-containing plants (63%). In a worldwide floral survey Beale et al. (1941) also found a high frequency of occurrence of pelargonidin in plants native to tropical and subtropical regions (38% and 31%, respectively). By contrast, pelargonidin occurred in only 6% of temperate plants, in which delphinidin was the most common anthocyanidin (59%), and cyanidin frequently occurred (34%) (Beale et al., 1941).

At a higher geographical resolution, Baker and Hurd (1968) suggest that the dominant flower color is a function of the ecological community in which a plant occurs. They observe that the dominant color among herbaceous plants in California grasslands, open woodlands, and deserts is yellow, whereas the most common floral colors of the fir and redwood forests are white and pink. These patterns are explained by the predominance of bees in the former open habitats and the dominance of moths and flies on the forest floor. Investigations studying the correlation (if any) between floral pigments and plant community are completely absent.

The single report explicitly addressing the correlation of floral pigments and pollinator class was that of Harborne and Smith (1978), in which floral pigments and reported pollinators were compared for selected taxa in the plant family Polemoniaceae. This family was chosen for examination because adaptive radiation within it has led to servicing by many pollinator classes, and detailed information is available on specific pollinators for many of its taxa (Grant and Grant, 1965). Harborne and Smith (1978) reported an "obvious correlation . . . between aglycone type, flower color and pollinating vector." They noted that hummingbird flowers were red and contained pelargonidin, bee flowers were blue or purple and contained delphinidin, and lepidopteran flowers were generally pink and contained cyanidin/delphinidin mixtures.

A cautionary note on the generalizations of Harborne and Smith's report must be raised in that, like most pollination studies to date, this was performed in a temperate region on a family with its distribution centered in the American West. The only tropical member examined was the hummingbird-pollinated genus *Cantua*. Generalizations generated from temperate-region studies may not be universally transferable to tropical or high-latitude regions. For example, Scogin (1980) reported that in the largely tropical family Bignoniaceae, 89% of the anthocyanin-containing taxa examined contained only cyanidin glycosides as floral pigments. Yet this family is considered to be largely bee-pollinated (Gentry, 1974). The anomalous taxa in the Bignoniaceae were mostly pelargonidin-containing, bird-pollinated genera such as *Phaedranthus* and *Spathodea*. Studies relating floral pigments and pollinators are in their infancy. In all global regions, and especially in the tropics, many more studies must be performed before valid generalizations regarding plant pigments and pollinators can be constructed.

Pigment–pollinator correlation studies to date have used a taxonomic base to define the survey range, usually the plant family. The rationale for this approach has been to examine pigment–pollinator adaptive co-radiation against the common genetic background of a taxonomic unit (the plant family). A more profitable way to examine this problem in the future may be to use the ecological community or habitat as a survey base for floral pigments and pollinators and to determine if pigment classes correlate with the pollinator spectrum commonly cocurring in specific plant associations. This has the effect of greatly increasing the resolution of earlier surveys of pigments as a function of geography. Unfortunately, floral coloration, pigment occurrence, and plant pollinator data for ecosystems and communities are usually gathered independently in different types of research and are seldom collected coordinately. Such data are sorely needed to confirm evolutionary postulates regarding the selective pressures for particular pigment classes. Work in progress in the author's laboratory is seeking to gather these three data sets for the diverse plant communities occurring in the state of California.

FUTURE RESEARCH

The anthocyanin floral pigments present a unique test system among secondary plant constituents for constructing a comprehensive evolutionary model for the forces of natural selection operating on a secondary plant product.

Anthocyanin evolution can be examined from the level of genetic-biosynthetic control through physiological regulation to selective adaptation of floral pigmentation under field conditions. Substantial data exist already in these areas, and large pieces of the puzzle are presently falling into place.

The biosynthetic pathways and enzymology of flavonoids (including anthocyanins) are largely known (Hahlbrock and Grisebach, 1975) and are under continuing investigation. Regulation of anthocyanin biosynthesis under laboratory conditions is well studied (Mc-Clure, 1975), although data regarding regulation under field conditions are absent. Field studies of pollinator responses to pigment alternatives and concomitant implications for plant reproductive success have been performed (Levin, 1978), but need to be greatly extended. A pressing need at present is for studies of pollinator response to variations in the complete petal reflectance spectrum (including both UV and visible wavelengths), which will aid in an understanding of the role and distribution of co-pigments in pollination ecology.

LITERATURE CITED

Acheson, R. M. 1956. The anthocyanins of some Himalayan flowers. *Proc. Roy. Soc. London B* 145:549–554.

Asen, S. and L. Jurd. 1967. The constitution of a crystalline, blue cornflower pigment. *Phytochemistry* 6:577–584.

Asen, S., R. N. Stewart, and K. H. Norris. 1972. Copigmentation of anthocyanins in plant tissues and its effect on color. *Phytochemistry* 11:1139–1144.

Baker, H. G. and P. D. Hurd, Jr. 1968. Intrafloral ecology. *Annu. Rev. Ent.* 13:385–414.

Beale, G. H., J. R. Price, and V. C. Sturgess. 1941. A survey of anthocyanins. VII. The natural selection of flower colour. *Proc. Roy. Soc. London B* 130:113–126.

Bene, F. 1941. Experiments on the colour preference of black-chinned hummingbirds. *Condor* 43:237–242.

Davies, B. H. 1976. Carotinoids. *In* T. W. Goodwin (ed.), *Chemistry and Biochemistry of Plant Pigments*, Vol. II. Academic Press, New York, pp. 38–165.

Endo, T. 1962. Inheritance of anthocyanin concentration in flowers of *Torenia fournieri. Jap. J. Genet.* 37:284–290.

Faegri, K., and L. van der Pijl. 1979. *The Principles of Pollination Ecology.* Pergamon Press, New York. 244 pp.

Forsyth, W. G. C. and N. W. Simmonds. 1954. A survey of the anthocyanins of some tropical plants. *Proc. Roy. Soc. London B* 142:549–564.

Gascoigne, R. M., E. Ritchie, and D. R. White. 1948. A survey of anthocyanins in the Australian flora. *J. Proc. Roy. Soc. N.S.W.* 82:44–70.

Gentry, A. H. 1974. Flowering phenology and diversity in tropical Bignoniaceae. *Biotropica* 6:64–68.

Gibbs, R. D. 1974. *Chemotaxonomy of Flowering Plants.* McGill-Queens University Press, Montreal. 2372 pp.

Goldsmith, T. H. 1980. Hummingbirds see near ultraviolet light. *Science* 207:786–788.

Goodwin, T. W. 1974. *Comparative Biochemistry of Carotinoids.* Chapman and Hall, London. 356 pp.

———— (ed.). 1976. *Chemistry and Biochemistry of Plant Pigments.* Academic Press, New York. 870 pp.

Grant, K. A. 1966. A hypothesis concerning the prevalence of red coloration in California hummingbird flowers. *Am. Nat.* 100:85–97.

Grant, V. and K. Grant. 1965. *Flower Pollination in the Phlox Family.* Columbia University Press, New York. 180 pp.

Grisebach, H. 1973. Comparative biosynthetic pathways in higher plants. *In* T. Swain (ed.), *Chemistry in Evolution and Systematics.* Butterworths, London, pp. 487–513.

Hahlbrock, K. and H. Grisebach. 1975. Biosynthesis of flavonoids. *In* J. B. Harborne, T. J. Mabry, and H. Mabry (eds.), *The Flavonoids.* Academic Press, New York, pp. 866–915.

Harborne, J. B. 1967. *Comparative Biochemistry of the Flavonoids.* Academic Press, New York. 383 pp.

————. 1973. *Phytochemical Methods.* Chapman and Hall, London. 278 pp.

————. 1975. The biochemical systematics of flavonoids. *In* J. B. Harborne, T. J. Mabry, and H. Mabry (eds.), *The Flavonoids.* Academic Press, New York, pp. 1056–1095.

————. 1976. Functions of flavonoids in plants. *In* T. W. Goodwin (ed.), *Chemistry and Biochemistry of Plant Pigments,* Vol. I. Academic Press, New York, pp. 736–778.

————. 1977. *Introduction to Ecological Biochemistry.* Academic Press, New York. 243 pp.

Harborne, J. B. and H. S. A. Sherratt. 1961. Plant polyphenols. 3. Flavonoids in genotypes of *Primula sinensis. Biochem. J.* 78:298–306.

Harborne, J. B. and D. M. Smith. 1978. Correlations between anthocyanin chemistry and pollinator ecology in the Polemoniaceae. *Biochem. Syst. Ecol.* 6:127–130.

Harborne, J. B., T. J. Mabry, and H. Mabry. 1975. *The Flavonoids.* Academic Press, New York. 1204 pp.

Hess, D. and R. Endress. 1972. Anthocyansynthese in isolierten Protoplasten von *Nemesia strumosa* var. Feuerkoenig. *Z. Pflanzenphysiol.* 68:441–449.

Huth, H. H. and D. Burkhardt. 1972. Der spektrale Sehbereich eines Violettohr-Kolibris. *Naturwissenschaften* 59:650.

Ilse, D. 1928. Uber den Farbensinn der Tagfalter. *Zeit. Vergl. Physiol.* 8:658–692.

Kay, Q. O. N. 1978. The role of preferential and assortative pollination in the maintenance of flower colour polymorphisms. *In* A. J. Richards (ed.), *The Pollination of Flowers by Insects.* Academic Press, New York, pp. 175–190.

Kevan, P. G. 1972. Floral colors in the high arctic with reference to insect–flower relations and pollination. *Can. J.Bot.* 50:2289–2316.

Levin, D. A. 1969. The effect of corolla color and outline on interspecific pollen flow in *Phlox. Evolution* 23:444–455.

———. 1972. The adaptedness of corolla-color variants in experimental and natural populations of *Phlox drumondii. Am. Nat.* 106:57–70.

———. 1978. The origin of isolating mechanisms in flowering plants. *Evol. Biol.* 11:185–317.

Levy, M. 1978. Flavonoids and pollination ecology: pigments of systematist's imagination? *Phytochem. Bull.* 2:35–42.

Lloyd, D. G. 1969. Petal color polymorphism in *Leavenworthia. Contr. Gray Herb.* 198:9–40.

Mabry, T. J. and A. S. Dreiding. 1968. The betalains. *In* T. J. Mabry (ed.), *Recent Advances in Phytochemistry.* Appleton-Century-Crofts, New York, pp. 145–160.

McClure, J. W. 1975. Physiology and function of flavonoids. *In* J. B. Harborne, T. J. Mabry, and H. Mabry (eds.), *The Flavonoids.* Academic Press, New York, pp. 970–1055.

Mogford, D. J. 1974. Flower color polymorphism in *Cirsium palustre.* 2. Pollination. *Heredity* 33:257–263.

Moss, G. P. and B. C. L. Weedon. 1976. Chemistry of the carotinoids. *In* T. W. Goodwin (ed.), *Chemistry and Biochemistry of Plant Pigments,* Vol. I. Academic Press, New York, pp. 149–224.

Piattelli, M. 1976. Betalains. *In* T. W. Goodwin (ed.), *Chemistry and Biochemistry of Plant Pigments,* Vol. I. Academic Press, New York, pp. 560–596.

Proctor, M. and P. Yeo. 1973. *The Pollination of Flowers.* William Collins, Glasgow. 418 pp.

Roller, K. 1956. Uber Flavonoide in weissen Blumenblattern. *Z. Bot.* 44:477–500.

Rosler, H., U. Rosler, T. J. Mabry, and J. Kagan. 1966. The flavonoid pigments of *Opuntia lindheimeri. Phytochemistry* 5:189–192.

Scogin, R. 1980. Anthocyanins of the Bignoniaceae. *Biochem. Syst. Ecol.* 8:273–276.

Timberlake, C. F. and P. Bridle. 1975. The anthocyanins. *In* J. B. Harborne, T. J. Mabry, and H. Mabry (eds.), *The Flavonoids.* Academic Press, New York, pp. 214–266.

Waser, N. M. and M. V. Price. 1981. Pollinator choice and stabilizing selection for flower color in *Delphinium nelsonii. Evolution* 35:376–390.

Weevers, T. 1952. Flower colors and their frequency. *Acta Bot. Neerl.* 1:81–92.

8
ELECTROSTATICS AND POLLINATION

Eric H. Erickson

U.S. Department of Agriculture
Agricultural Research Service
University of Wisconsin
Madison, Wisconsin
and
Stephen L. Buchmann

U.S. Department of Agriculture
Agricultural Research Service
Carl Hayden Bee Research Center
Tucson, Arizona
and
Department of Ecology and Evolutionary Biology
University of Arizona
Tucson, Arizona

Living nature is based on a relatively small
number of basic principles which are clearly
adapted to the most varied ends.
Albert Szent-Gyorgyi

ABSTRACT

This chapter is intended as a review of the electric relationships between plants and their pollinators. It also presents data on weather parameters and acquired surface potentials on honey bees arriving and departing from a hive. Plants possess negative surface electric potentials under fair-day conditions that are subject to diurnal and annual fluctuations. Insects acquire electric potentials via their contact with substrates or the atmosphere. We hypothesize that electrostatics plays an important role in pollen transfer to pollinating insects, and in the subsequent attachment of pollen to the stigma. Electrostatics is also discussed in relation to wind and buzz pollination.

KEYWORDS: Electric fields, electrostatics, pollinator behavior, pollen collection, honeybees, buzz pollination.

CONTENTS

PLANTS

Plants possess negative surface charge densities under fair-day conditions and are surrounded by electrostatic fields that are subject to diurnal and annual fluctuations (Khvedelidze, 1958; Maw, 1962b, 1963). These observations are in accord with the concept of continuous electron flow from the earth to the atmosphere. The magnitude of the fields depends in part on the chemical composition of the plant and its cuticle, height, and environment, including surrounding plants. The distribution of the electric field around the plant varies with its shape. The laws of physics suggest that plant electric fields (gradient of the potentials) should be greatest near sharp points where plant terminals are most numerous (e.g., flowers) and least on broad flat surfaces (e.g., leaves) (Maw, 1962a; Chalmers, 1967). The very highest potentials will probably be on the terminal plant parts.

Since plants are covered with cuticular waxes possessing excellent dielectric properties, it follows that if plants are emitting a continuous low-level electron radiation, surface emissions would vary in density depending on wax thickness, trichome densities, and the diurnal rhythm of the plant. Differences in charge distribution would be expected among different surfaces; for example, glabrous surfaces should differ from hirsute surfaces. Wet surfaces reduce or eliminate charges, whereas drying under certain conditions may restore such charges (Edwards, 1960).

Marty (1943) alluded to the possible influence of atmospheric electricity on plant fruit set and nectar flow, and Khvedelidze (1958) first recognized the relationship between the life processes of the plant and recurring electric potentials. More recent investigations have provided little to correlate these associations with the plant, the insects associated with the plant, or coevolutionary interactions between the two.

It has been shown that spores and pollen grains can be readily charged, and that these charges affect aerial dispersal, distribution, and ultimately germination upon the stigma (Rack, 1959; Potamina and Shmigel, 1960; Baba, 1961; Swinbank et al., 1964). Although these tests were carried out using artificially high potentials for charging, it is possible that pollen and spores normally possess low-level charges characteristic of the species. Leach (1976) demonstrated that the explosive liberation of spores by certain epiphyllous fungi is largely a hygroscopic/electrostatic event; perhaps periods of pollen shedding after anther dehiscence, and insect pollination also coincide with periodic peaks in the electric fields exhibited by plants. Peak times of electric fields are on clear, warm, dry, bright days, which are also conducive to good pollinator flight activity.

INSECTS

Insects possess surface electrical charges acquired by virtue of their contact with the earth or their movement through the atmosphere (Maw 1963; Chalmers, 1967; Carlton, 1971, 1976; Erickson, 1975; Warnke, 1979). Correlations between electric fields, magnetic fields, or atmospheric ionization and the activities, communication, or metabolism of insects have been described by various authors. There is evidence that electric stimuli can elicit in insects behavioral responses that are in many ways similar to those elicited by changes in light, temperature, and relative humidity (Maw, 1961a,b, 1962a; Altmann, 1963; Callahan, 1967).

The behavioral responses of meloid (blister beetle) larvae are examples of the effects of electrical stimuli, etc. Eggs of the phoretic species (Nemognathinae) are deposited on or near the flowers of the host plant. Upon hatching, the first-instar larvae (triungulins) reach a flower and await the arrival of an acceptable host (a solitary female bee). These larvae perform a "grasping stance" when stimulated by the touch of a camel's hair brush, air currents, or musical note (about 160 Hz) that corresponds to the wingbeat frequency of an approaching host bee. These larvae are also able to spin a silk strand from the anal region (Hocking, 1949). The triungulins can be made to spin silk when disturbed by substrate vibration or when a negative electrostatic field is produced near them (Erickson and Werner, 1974). This field can be produced with an ebonite rod that had been rubbed against wool. If the negatively charged rod is held near a larva and then withdrawn, the larva produces an attachment to the substrate, releases its hold, and begins spinning silk until the larva reaches the rod. While suspended it assumes the characteristic grasping stance that happens naturally upon the approach of a bee host. Larvae could be induced to spin an unbroken strand of silk up to 12 cm long.

Erickson and Werner (1974) assumed that the silk was not used for ballooning but rather served to ensure larval floral attachment until transfer to a host was completed, or that the silk strand aided the transfer of the larva to the bee. Since bees just prior to floral landings may have acquired strong positive surface charge densities (data will be presented elsewhere), the dramatic behavioral responses of the larvae to induced electrostatic fields suggest that silk production by an "electrically alerted" larva may be an evolved mechanism for facilitating host attachment. This example remains a well-documented behavioral adaptation to, and use of, differences in naturally occurring electrostatic fields.

That atmospheric electricity (ion density) affects flight activity is suggested by increased light trap catches at the approach of a storm. Flying insects are repulsed by a positively charged screen; negative fields attract or stimulate certain insects (Johnson, 1969). Similarly, electrical conditions have been shown to affect both speed and duration of flight activity (Edwards, 1960). Controversial Russian experiments with a variety of animals deprived of air ions may indicate that animals cannot survive without them (see Hansell, 1961).

POLLINATORS

Flying insects are found to be variously charged (Edwards, 1962). The quantity and life of the charge vary, depending on atmospheric conditions. In still air the charge persists for several minutes. From this it is assumed that insects in flight acquire a net positive charge due to their passage through electrified air (presumably due to the stripping away of electrons via friction). Studies (Erickson, 1975) have shown that under fair-day conditions honeybees return to the hive with significant positive electric potentials (1.8 V maximum) on their bodies (Fig. 8-1). These charges were continuously monitored as the bee walked through an electrically shielded glass tube and across a (9 mm I.D.) copper detector ring at the hive entrance (Erickson, 1975; Erickson et al., 1975).

Electric charges on honeybees are associated only with periods of foraging activity (extended flight at a constant altitude). Surface charges on bees that are flying but not foraging are small to nonexistent, perhaps because bees in "play-flight" are constantly changing altitude. Bees leaving the hive have only slight charges, often of negative sign (Fig. 8-1). Thus, after foraging the hive appears to act as an electrical sink. Under dull-day conditions (high humidity, overcast skies) the magnitude of charges on incoming bees is much reduced, and in advance of electrical storms the character of recorded surface potential differences is aberrant. Diurnal and annual rhythms in the magnitude of the charges on bees returning to the hive are evident (Figs. 8-1 – 8-3). The level

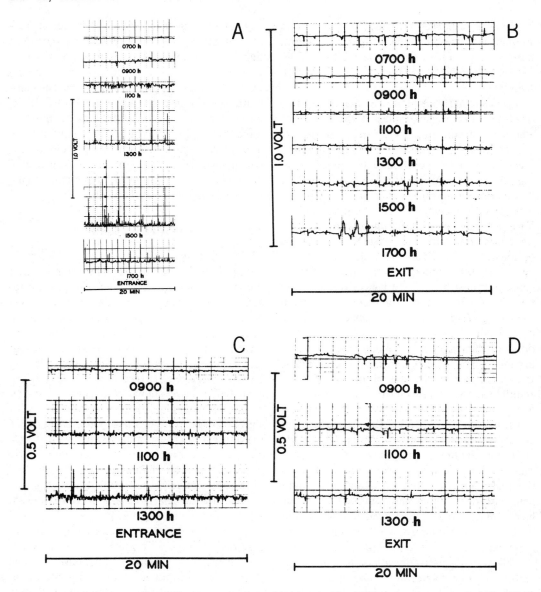

Figure 8-1. Surface electric potentials on bees entering and leaving the hive. Each peak represents a single bee. Up is positive, down is negative. A and B, fair-day conditions; C and D, inclement weather.

of foraging/flight activity increases during periods of atmospheric electrical activity preceding a storm (Schua, 1951, 1952).

Information transfer between terrestrial life forms via induced and biologically generated electric fields now seems likely. This communication mode has been shown to exist in aquatic organisms (Warnke, 1979) and may

be important for honeybees as well. Altmann (1959) suggested that electrical communication may occur among bees. Warnke (1973) has shown that the sensitivity of the pulvilli of bees to electrical stimulation is very high, exceeding even that of the antennae by a wide margin. Electric "loading" of honeybees varies according to their environment and results in

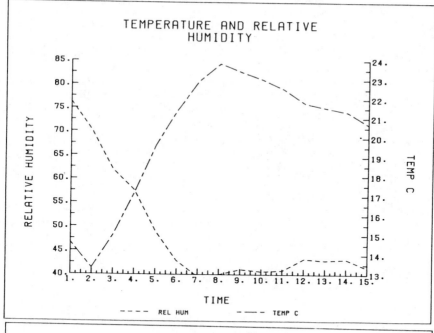

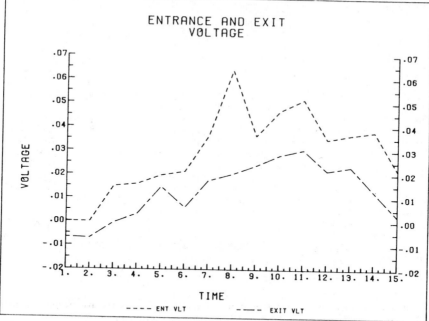

Figure 8-2. Mean hourly surface electric potentials on bees entering and leaving the hive for the period 5 to 25 June 1972. Graphed above are corresponding means for temperature and humidity. (Hour 1 is 0800, hour 15 is 2100 CST.)

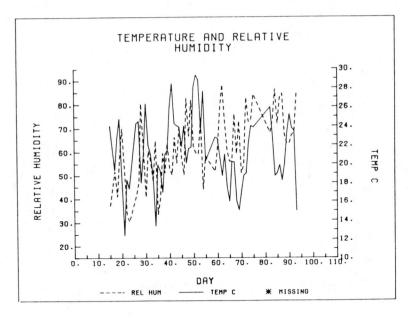

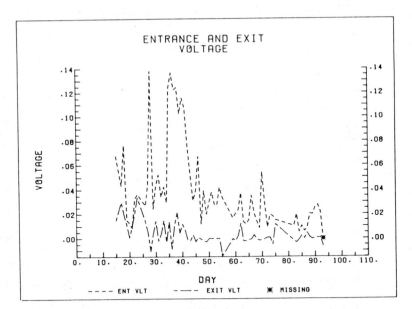

Figure 8-3. Mean daily surface electric potentials on bees entering and leaving the hive for the period 15 June to 1 September 1972. Graphed above are corresponding means for temperature and humidity.

differences in motor activity (Altmann and Warnke, 1971). Bees respond variously to electric stimulation depending upon atmospheric conditions (Erickson, unpub. data; Warnke, 1973).

POLLINATION

Since plants are negatively charged and bees after flight are positively charged, it must be assumed that as a foraging bee approaches the flower, an electrical field is established. This field will grow in strength as the gap narrows. It must be further presumed that once the bee contacts the flower, both are discharged. When a bee lands on a flower or inflorescence, some of its positive charge will dissipate (i.e., charge is neutralized) to the substrate and temporarily modify the floral charge. This presents the possibility that the bee leaves behind an "electrical footprint" that subsequent flower visitors could detect. After departure of the bee, a period of 5 to 40 minutes (given other variables) may be required for the flower (dielectric) to regain its normal charge density, depending on local atmospheric conditions (Grman et al., unpub. data). This "footprint" might be detected via either antennae or pulvilli, by subsequent bee visitors (both conspecific and interspecific) hovering near or contacting previously visited flowers. This signal would thus identify a recently visited spent flower (i.e., one with a reduced standing crop of nectar or pollen), making that flower less worthwhile to visit. Many bees perform such "approach then avoid" behavior when examining flowers while hovering next to them, especially late in the afternoon (Jones and Buchmann, 1974; Jones, 1978; and Buchmann, unpub. obs.). The time needed for a flower to regain normal charge could well coincide with the time interval needed by the flower to secrete additional nectar. Hence, optimal pollinator energetics would be served by the bees' ability to detect and respond to electric parameters. Other post-pollination/fertilization or "induced changes" include floral color and odor changes (see Gori, this book), and even odorous allomone markers, as in Nasanov scent-marking by *Apis* workers or by female solitary bees (Frankie and Vinson, 1977; Vinson et al., 1978), are well documented. Induced electrostatic floral charges could be one of a multitude of undiscovered biological and biophysical cues to which bees are able to respond physiologically.

There is some evidence that electric charges on flowers may also play a passive role in pollination; namely, effecting pollen transfer from anther to bee and subsequently onto the stigma of another flower via electrostatic precipitation (Erickson, 1975). Thus, an electrified bee, or other insect, may be an effective pollinator without having to contact anthers or stigmas directly. Electrostatic precipitation could be especially effective in pollen flowers, which are flowers that offer a surplus of pollen as the only reward to their insect visitors, in lieu of other substances (nectar or lipids). Pollen flowers have recently been divided into three major types (Vogel, 1978): the magnolian-type (Magnoliaceae and others), the *Papaver*-type (Papaveraceae and Ranunculaceae), and the *Solanum*-type (covered in Chapter 4, this book). In these latter flowers the pollen grains are powdery and could be effectively harvested by a charged pollination vector. When bees are working certain flowers, and with backlighting to the observer, a "pollen cloud or rain" can often be seen settling out onto the bees. This attraction is presumably electrostatic in nature, and has been noted on a thistle (*Cirsium* sp.) (Erickson, unpub. obs.) and on nightshades (*Solanum* spp.) (Buchmann, unpub. obs.). The plumose setae of bees, a characteristic of the superfamily, undoubtedly contribute to the notable efficiency of bees as pollinators by enhancing the potential gradients near the bees under fair-day conditions. The relationships between atmospheric conditions, surface electric charges on bees and plants, and optimal conditions for anther dehiscence, pollen shedding, and stigmatic receptivity need to be investigated.

ELECTROSTATICS AND BUZZ POLLINATION

An extended account of buzz pollination is given elsewhere (see Buchmann, this book). This unique form of pollination is exemplified by such genera as *Cassia* (senna), *Dodecatheon* (shooting star), *Lycopersicon* (tomato), *Solanum* (deadly nightshade), and *Vaccinium* (blueberry and cranberry), along with many others. Most of these flowers produce no floral nectar, but do offer a large amount of pollen as a reward for visiting female bees. This pollen is invariably small unless in a tetrad (usually 5–30 μm in equatorial diameter), light, and powdery-dry in texture (Buchmann, unpub.). The powdery texture is a consequence of the minute amount of oily surface pollenkitt on these buzz grains as compared to normally sticky entomophilous grains. The exine also tends to be abnormally smooth even if the family to which the vibratile genus belongs is usually characterized by other palynological features. The anthers in these forms dehisce by small apical pores or slits, and the androecium may form a cone, connivent about the pistil. The style is usually filiform with a relatively small dry stigmatic surface with short epidermal papillae.

When a pollen-collecting female bee alights on one of these flowers, it collects pollen by rapidly contracting the indirect flight muscles, thus producing strong vibrations that are transmitted directly to the anthers. This vibration rapidly produces a directed stream or pollen cloud from the anther pores that primarily strikes the venter of the bee, and sometimes the pleural and dorsal areas also. Since most of these buzz flowers are pendant when vibrated, one might imagine the pollen simply falling out of the anther pores, with a large proportion of it missing the bee, as if a salt shaker had been turned upside down. Thus, in buzz plants anther vibrations would tend to randomly scatter pollen, making this a somewhat inefficient pollen-gathering technique. If this were true, much of the pollen would be lost and unavailable for pollination or for brood provisions. Such a system would indeed be inefficient and would tend to be selected against. However, in the presence of an electrostatic field, most pollen would precipitate onto the bee, which would greatly increase the efficiency of this system. The probable important role of electrostatics in buzz pollination was .. recognized earlier (Buchmann, 1978, unpub. data).

We feel that prior to floral visitation by bees, the small, light, dry pollen will be held in place largely through electrostatic interactions between the pollen and the interior locule walls. Release of the pollen and accessibility to it are also limited by the small apical anther pores. Upon the approach and landing of a bee, the electrostatic microenvironment will begin to rapidly change (via induction) as the flower and bee dissipate each other's opposite charge. This suspected electrostatic shift might have the effect of weakening the attraction of the grains to the locule walls. The pollen would now be ready for active vibrationally induced discharge (usually requiring less than one second). A biophysical model for this phenomenon has been proposed (Buchmann and Hurley, 1978). During this short interval the bee still carries a net positive charge, and conditions are set for the electrostatic precipitation of the negatively charged pollen onto the bee. Thus, even if some pollen is discharged to the sides or up toward the bee's dorsum, it should be held strongly and pulled onto the bee's integument. Very little pollen should be lost to air currents near the bee. The pollen can then be easily recovered by the bee during grooming movements except on the occipital head region, proboscideal fossa, or centerline area of the thoracic dorsum where most bees cannot reach (Buchmann, unpub. obs.).

When the bee visits succeeding flowers, any pollen not packed into scopal transport devices will be available for pollination (i.e., deposition on the stigmatic surface). As mentioned previously, these flowers are characterized by filiform, dry stigmas with little or no sticky stigmatic exudates. The transfer of pollen from bee to stigma and adhesion of the relatively dry pollen to a dry stigmatic surface is also likely mediated electrostatically. It is not known if a bee rubs pollen directly off onto the stigma, or whether it is "pulled off" of the bee by electrostatic forces; but once the pollen is in place on the small epidermal papillae, it will be held tightly until it is further anchored by the penetration of the growing pollen tube.

We presume, based on certain physical laws, that a foraging bee is gradually discharged by repeated contact with flowers (of the opposite sign), or by the pollen precipitating onto its body. Thus many questions remain such as the rate of discharge of the bee over time, the bee's ability to recharge on short flights between flowers, and the efficacy level of an electrostatic syndrome with repeated floral contact during a single foraging bout.

WIND POLLINATION

Abiotic pollination by wind (anemophily) is widespread among most gymnosperms and many angiosperms but found predominantly within the Poaceae, Cyperaceae, and Juncaceae. With very few exceptions, however, abiotic pollination is thought to be wasteful of pollen, in that the transfer process is nondirectional. In these plants, flowers are typically diclinous, and anthers are exposed near the stem tips. Anther dehiscence and pollen shedding from inflorescences are related to daily patterns of temperature, relative humidity, and probably changing microenvironmental electrostatic conditions (see Schmid, 1976; Schmid and Alpert, 1977).

The pollen grains have a small negative charge when shed from the plant, but we assume they would soon acquire a strong positive charge as they are passively carried by the wind. As the air mass carrying the pollen grains approaches a female plant, the airspeed will drop, and the grains will begin to settle out of the air toward the plant and ground. The grains should be attracted to oppositely charged male and female plants. Adding to this effect is the fact that many wind-pollinated trees are leafless at pollination, so that there would be less interference from broad leaves blocking the incoming pollen rain. Grains will impact upon stems and leaves, but since the female flowers are usually located on the outer periphery of the plant surface (i.e., terminals), and because these terminal projections have the strongest negative electric charges (Chalmers, 1967; Maw, 1962a), the incoming "pollen rain" should be selectively attracted to the female flowers (or any flowers near terminals). This may help to offset the pollen wastage due to the small point targets relative to the larger background of nontarget soil, stems, and leaves. Most wind-pollinated gymnosperms trap airborne pollen with a sticky pollination droplet that is drawn into the micropylar area, carrying viable pollen that will ultimately fertilize the ovules. This pollen "droplet capture" may therefore have a large electrostatic component. In some gymnosperms (notably *Ephedra, Pinus,* and *Welwit-*

schia) the pollination droplet is greatly extruded and probably captures more pollen than a deeply seated droplet.

A biophysical examination of stigmas of wind-pollinated plants (which are often plumose and thereby increase surface area) might reveal that in addition to passively filtering pollen from the air, they function as miniature electrostatic capture devices for pollen. However, until further measurements of plant surface potential differences are made, this must remain a speculative but tantalizing hypothesis. Recently, Niklas (1981) constructed models of lobed ovules of fossil seed plants and bombarded them with pseudopollen in wind tunnel experiments. He concluded that the laminar airflow becomes turbulent near the ovules (lobed salpinx), and that maximum impact of pseudopollen was in regions of high airflow disturbances. Thus most pollen was discharged on the downwind surfaces as a result of a sharp drop in airflow rate. This "snow fence" scenario proposed by Niklas does much to explain pollen deposition in early seed plants, and is probably applicable to extant wind-pollinated species. We would only add that electrostatic attraction should be considered in addition to airflow, and that together they could form an efficient mechanism for pollen capture by anemophiles.

SUMMARY

Electric charges occur almost ubiquitously on substrates, and in the atmosphere, and constitute an ever present feature of the environment within which all terrestrial organisms have evolved. Atmospheric fields are positive during fair weather and negative during stormy weather. These potential gradients are also characterized by circadian and circannual rhythms.

Plants are grounded and bear negative surface charge densities that depend upon the size, shape, and surface characteristics of the plant and the microenvironment. Within a plant canopy there are differences in potentials with the lowest on stems and leaves, and the highest charges on stem terminals where insect- and wind-pollinated flowers usually occur (data to be presented elsewhere). Periods of pollen shedding and insect pollination frequently coincide with periodic peaks in the electric potential differences exhibited by plants and the atmosphere.

Insects possess diverse surface electric charges acquired by virtue of their contact with biotic and abiotic substrates, and their passage through the atmosphere. These electrical stimuli are believed to elicit behavioral responses in many insects similar to those elicited by changes in light, temperature, and relative humidity.

Flying insects have been found to possess variable charges, with the quantity and duration of the charge depending on atmospheric conditions. In this chapter we review the evidence that honeybees under fair-day conditions return to the hive with significant positive electric potentials (1.8 V maximum) on their bodies (see Erickson, 1975). These charges are acquired during flight and foraging and may play an important role in aiding pollen collection by bees, especially in buzz-pollinated taxa. When a bee lands on a flower, their charges may be mutually dissipated. The possibility that bees leave behind "electrical footprints" that subsequent visitors could detect was also presented. These signals might delimit a spent flower (i.e., one without nectar or pollen rewards) from a good flower that had not been previously visited.

There have been suggestions that surface charges on flowers may also play a passive role in pollination, just by affecting pollen transfer from anther to bee and then by transferring pollen from the bee onto the stigma of another flower by electrostatic precipitation. This may be an important feature of the buzz pollination syndrome.

Within wind-pollinated plants, pollen shedding is related to diurnal patterns of temperature, relative humidity, and probably microenvironmental electrostatic conditions. It was further suggested that owing to the greater surface charge density on plant terminals, where the female flowers are normally located, wind-borne pollen should be selectively attracted to the female flowers.

By applying biophysical principles, in this case electrostatics, to pollination ecology, we may be able to more clearly understand and explain such transfer processes as pollen shedding and aerial distribution, electrostatic precipitation of pollen from anther to bee, transfer of pollen from bee to stigma, and the possibility of electrical charges left by bees on flowers as post-pollination cues to other bees. We hope that the application of electrostatics to pollination studies will grow rapidly and provide insights into unsolved or little-understood problems that relate to pollen transfer phenomena in floral biology. Atmospheric potential gradients can be measured relatively easily with a portable field mill (Grman, Erickson, and Whitefoot, unpub. ms.), and atmospheric ionization conditions can also be quantified. Technically, the most difficult to measure, but most revealing, electrical measurements will be the surface potential differences on different plant parts. Instrumentation (such as piezoelectric devices, etc.) is now being constructed by one of us (Erickson) that will enable us to make accurate and precise measurements of plant surface charge densities. Floral measurements prior to, during, and immediately following bee visitations should be especially informative.

Obviously, little is yet known regarding the importance of electric phenomena in bee–flower interactions. Exciting new research now under way will add to this body of knowledge. Still, definitive results may not be achieved easily. The celerity with which these mysteries are uncovered will depend upon the number of properly trained scientists (teams of biophysicists and entomologists) who are challenged to investigate them.

ACKNOWLEDGMENTS

The authors wish to thank the following for their helpful criticism of this manuscript: Michael E. Grman, Hayward G. Spangler, Justin O. Schmidt, Robert J. Schmalzel, Paul Cooper, C. Eugene Jones, M. Shapiro, W. E. Evenson, R. Nanes, Y. I. Lehner, and B. W. Ross. We also thank P. M. Crump and K. V. Boggs, who assisted in preparing the figures, and Elizabeth L. Sager, for typing of the manuscript.

LITERATURE CITED

Altmann, G. 1959. Der Einfluse statischer elektrischer Felder auf den Stoffwechsel der Insekten. *Z. Bienenforsch.* 4:199–201.

———. 1963. Die physiologische Wirkung elektrischer Felder auf Tiere. [The physiological effect of electric fields on animals.] *Det. Zool. Ges., Verhandl.* 26:360–366.

Altmann, G. and U. Warnke. 1971. Einfluss unipolar geladener Luftionen auf die motorische Aktivität der Honigbienen. *Apidologie* 2:309–317.

Baba, S. 1961. Isoelectric zones of vegetative and generative nuclei in pollen grains of *Tradescantia*. *Kyoto U. Col. Sci. Mem. Ser. B* 28(3):359–363.

Buchmann, S. L. 1978. Vibratile "buzz" pollination in angiosperms with poricidally dehiscent anthers. Entomology Ph.D. dissertation, University of California, Davis. 238 pp.

Buchmann, S. L. and J. P. Hurley. 1978. A biophysical model for buzz pollination in angiosperms. *J. Theor. Biol.* 72:639–657.

Callahan, P. S. 1967. Insect molecular bioelectronics: a theoretical and experimental study of insect sensillae as tubular waveguides, with particular emphasis on their dielectric and thermoelectric properties. *Ent. Soc. Am. Misc. Publ.* 5(7):315–347.

Carlton, J. B. 1971. The quantitative measurement of an electrostatic charge on a housefly by capacitor techniques. Ph.D. Dissertation, Texas A. & M. University, College Station. 113 pp.

———. 1976. Electronic tracking system for studying the "oscillatory effect" in insects. USDA ARS-S-96. 7 pp.

Chalmers, J. A. 1967. *Atmospheric Electricity*. Pergamon Press, New York.

Edwards, D. K. 1960. Effects of experimentally altered unipolar air ion density upon amount of activity of the blowfly *Calliphora vicina*. *Can. J. Zool.* 38:1079–1091.

———. 1962. Electrostatic charges on insects due to contact with different substrates. *Can. J. Zool.* 40(4):479–584.

Erickson, E. H. 1975. Surface electric potentials on worker honeybees leaving and entering the hive. *J. Apic. Res.* 14(3/4):141–147.

Erickson, E. H. and F. G. Werner. 1974. Bionomics of nearctic bee-associated Meloidae (Coleoptera). A comparative analysis of larval host-seeking behavior among the meloinae and nemognathinae. *Ann. Ent. Soc. Am.* 67(6):903–908.

Erickson, E. H., H. H. Miller, and D. J. Sikkema. 1975. A method of separating and monitoring honeybee flight activity at the hive entrance. *J. Apic. Res.* 14:119–125.

Frankie, G. W. and S. B. Vinson. 1977. Scent marking of passion flowers in Texas by females of *Xylocopa virginica texana* (Hymenoptera: Anthophoridae). *J. Kansas Ent. Soc.* 50:613–625.

Hansell, C. W. 1961. An attempt to define "Ionization of the Air". *Proc. Intl. Conf. on Ionization of the Air.* 1:1–10.

Hocking, B. 1949. *Hornia minutipennis* Riley: A new record and some notes on behavior (Coleoptera, Meloidae). *Can. Entomol.* 81:61–66.

Johnson, C. G. 1969. *Migration and Dispersal of Insects by Flight*. Methuen & Co., Ltd., London. 763 pp.

Jones, C. E. 1978. Pollination constancy as a pre-pollination isolating mechanism between sympatric species of *Cercidium*. *Evolution* 32(1):189–198.

Jones, C. E. and S. L. Buchmann. 1974. Ultraviolet floral patterns as functional orientation cues in hymenopterous pollination systems. *Anim. Behav.* 22:481–485.

Khvedelidze, M. A. 1958. In regard to bio-electric potentials of plants. *Uspekhi Sovrem Biol.* 46(1):33–47.

Leach, C. M. 1976. An electrostatic theory to explain violent spore liberation by *Drechslera turcica* and other fungi. *Mycologia* 68(1):63–86.

Marty, J. 1943. Letter. Electric atmospheric effect upon bee activity and vitality. Silverton, Oregon. To Pacific Coast Bee Lab., Davis, California.

Maw, M. G. 1961a. Behaviour of an insect on an electrically charged surface. *Can. Entomol.* 93:391–393.

———. 1961b. Suppression of oviposition rate of *Scambus buolinane* (Htg.) (Hymenoptera: Ichneumonidae) in fluctuating electrical fields. *Can. Entomol.* 93:603–604.

———. 1962a. Behaviour of insects in electrostatic fields. *Proc. Ent. Soc. Manitoba* 18:1–7.

———. 1962b. Some biological effects of atmospheric electricity. *Proc. Ent. Soc. Ont.* 92:33–37.

———. 1963. Physics in entomology: sound and electricity in insect behaviour. *Proc. N.C. Br. Ent. Soc. Am.* 18:6–10.

Niklas, K. J. 1981. Simulated wind pollination and airflow around ovules of some early seed plants. *Science* 211(4479):275–277.

Potamina, N. D. and V. N. Shmigel. 1960. The effect of a high potential electrostatic field on the pollen of some fruit trees. *Bot. Zhur. (Moskva)* 45(2):266–272.

Rack. K. 1959. Untersuchungen über der elektrostatische Ladung der *Lophodermium*-sporen. *Phytopath. Z.* 35(4):439–444.

Schmid. R. 1976. Filament histology and anther dehiscence. *Bot. J. Linn. Soc.* 73:303–315.

Schmid, R. and P. H. Alpert. 1977. A test of Burck's hypothesis relating anther dehiscence to nectar secretion. *New Phytol.* 78:487–498.

Schua, L. 1951. Der Einfluss des Wetters auf das Verhalten der Honigbienen [The influence of weather on the behavior of honey bees.] *Dt. Zool. Ges., Verhandl.* 1950:183–186.

———. 1952. Untersuchungen über den Einfluss meteorologischer Elemente auf das Verhalten der Honigbienen *(Apis mellifica)*. *Zeit. Vergl. Physiol.* 34:258–277.

Swinbank, P., J. Taggart, and S. A. Hutchinson. 1964. The measurement of electrostatic charges on spores of

Merulius lacrymans (Wulf.). *Fr. Ann. Bot.* 28(110):239–249.

Vinson, S. B., G. W. Frankie, M. S. Blum, and J. W. Wheeler. 1978. Isolation, identification, and function of the Dufour gland secretion of *Zylocopa virginica texana* (Hymenoptera: Anthophoridae). *J. Chem. Ecol.* 4(3):315–323.

Vogel, S. 1978. Evolutionary shifts from reward to deception in pollen flowers. *In* A. J. Richards, (ed.) *The Pollination of Flowers by Insects* (Linnean Soc. Symp. No. 6), Academic Press, London, pp. 89–96.

Warnke, U. 1973. Physikalisch-physiologische Grundlagen zur luftelektrisch bedingten "Wetterfühligkeit" der Honigbiene *(Apis mellifica)*. Dissertation, Universität des Saarlandes.

———. 1979. Information transmission by means of electric biofields. *Proc. Symp. on Electromagnetic Bio-Information,* Marburg, pp. 55–79.

Section III
POLLINATOR ENERGETICS

9
INSECT FORAGING ENERGETICS

Bernd Heinrich

Department of Zoology
University of Vermont
Burlington, Vermont

ABSTRACT

The energetics of pollination concerns pollination biology from the perspective of the energy investment for payoffs of pollinators, as well as the energy investment for pollination payoffs of plants. This perspective on pollination biology, though only one of several possible perspectives, leads to a set of predictions on the reproductive biology of plants. However, neither the foragers nor the plants are governed solely by energetics. The forager's behavior is often constrained by factors other than those relating to immediate energy economy, and the plants may have other options available for pollination besides those dictated by the forager's energetics. The different potential strategies of both the pollinators and the plants are presumably influenced by quantitative energetics. Some quantitative aspects of the different strategies (as well as potential methods of determining them) are here discussed.

KEY WORDS: Foraging behavior, optimal foraging, food rewards, mimicry, wind pollination, flower specialization, constancy, sphinx moths, bumblebees, solitary bees.

CONTENTS

INTRODUCTION

The nectar from flowers is a product adapted specifically to attract pollinators. The evolutionary specifications on the plant have been to make this product highly attractive. At the same time, however, the plant must be penurious in dispensing it.

There are at least two reasons why plants should limit nectar production. First, it costs them energy resources that could be incorporated into seeds. Second, the larger the quantity of the reward (relative to the needs of the pollinators) above the minimum required to ensure pollination of the flowers, the more it would be counterproductive for cross-pollinations. It would restrict the spread of pollen to other plants for which it is "designed."

It should be of considerable economy for a plant to restrict forager access to the food rewards, thus sequestering them for a select clientele of pollinators. Common features used to exclude nonpollinating foragers from nectar include long tubular corollas, complex floral morphologies, hidden food rewards, poisons, and unique signaling cues (Heinrich, 1975a).

It sometimes may even be advantageous for the plant to exclude some individuals of the insect species adapted as its primary pollinators. If all individuals of a common pollinator species (which can potentially use any of a number of different kinds of flowers) have easy access to the food rewards of a highly dispersed flower, then there might not be enough reward left over to make it economically worthwhile for all individual pollinators of that species to specialize as is required for interplant pollen transfer. However, if only a few individuals learn to gain access to the food rewards, then the larger food rewards remaining can be sufficient for the specialized individuals of that species to visit and pollinate other flowers of the plant species that may be widely dispersed.

In summary, even though evolution has ensured that flowers provide worthwhile resources, the plant's strategy has been to be selective in the dispensation of rewards. The forager's strategy, in turn, has been to circumvent the plant's mechanisms that limit access. The plant has evolved flowers with design and reward complexities to use the forager to the plant's best advantage. The ensuing coevolutionary relationships, with food energy as the primary payoff to pollinators for cross-pollination service, has evolved as an intricate game of strategies and counterstrategies. The study of these strategies, particularly those of plants, dates back at least to Conrad Sprengel (1793).

The study of animal foraging strategies in the context of the plant's flowering strategies is relatively recent. Two main approaches have been followed. On the one hand, measurements or deductions of energy expenditure of foraging have been examined as adaptive responses to the caloric rewards provided by flowers (Heinrich, 1972; Heinrich and Raven, 1972). Other studies have been concerned with determinations of "optimum" choice of different flowers, patches of flowers, and/or patterns of movement between them with the basic assumptions of maximizing energy returns or minimizing energy costs (Pyke, et al., 1977).

My aim here is to provide a synthesis, combining recent work on insect foraging energetics with models of foraging "optimization." I hope to provide a broad picture of foraging economics, to give insights on where we are and where we might proceed.

I will consider primarily factors that affect immediate cost–benefit functions of foraging. How these may translate to "profit" beyond immediate returns in the overall strategy of the nectivore awaits future studies.

Immediate aspects of the foraging behavior that affect foraging returns can be analyzed at several levels, each with its own options and possibilities. These levels include flower choice, flower handling, and movement pattern between flowers, plants, and patches. Using available data on the foraging behavior of different insect nectivores, I will examine the caloric costs and payoffs of alternative foraging behaviors. Some of the potential implications of these results to the breeding biology of the plants are considered, particularly as they relate to pollination of flowers that the pollinators do not identify as individuals.

METHODS

The following methods are provided as a guide to aid potential investigators interested in insect foraging energetics. The methods described here are not necessarily meant to refer exclusively to the research reported in this article, nor do they represent an exhaustive survey. They are, however, meant to provide a springboard to a variety of potential research.

Nectar Volume. The most commonly used and convenient method of measuring nectar volume of flowers is to take up the nectar in capillary tubes thar are commercially available in 1, 2, 5, 10, and 20 μl and larger sizes. Using clean, dry tubes the nectar will be taken up by capillary action. A millimeter rule can be used to measure the recovered volume of any one tube. Thus, given a 32 mm length of the 1 μl tubes, nectar samples of approximately 0.03 μl can be measured by this method. However, the florets of many flowers, such as those of composites, provide considerably less nectar than this. Very small nectar volumes can be recovered using fine-bore capillary tubing made by drawing out melted glass tubing over a Bunsen burner. Approximate nectar volumes recovered can be determined from the size of spots created on filter paper.

To facilitate withdrawing very small volumes of nectar, it is often necessary to cut off the flower above the nectary in order to contact the nectar droplet with the capillary tube. It may also be necessary to squeeze the base of the flower to bring the nectar within reach of the capillary tube. The nectar can also be removed gravimetrically by placing the flowers in small tubes and centrifuging them.

Sugar Concentration. The most convenient method of measuring the sugar concentration of nectar is from the refractive index by refractometer. Commercially available hand-held pocket refractometers can be used to measure nectar samples near 1 μl. Sugar concentrations are measured in "sucrose equivalents." Many nectars also contain the monosaccharides glucose and fructose (Percival, 1961), the hydrolysates of sucrose. The refractive indexes of equimolar solutions of glucose and fructose are about half that of equimolar sucrose. In terms of caloric equivalents, however, a given weight or percent of sucrose is equivalent to a given weight or percent of glucose and fructose. One milligram of sugar represents 3.7 calories. Although it makes little difference in terms of caloric equivalents whether the nectar contains sucrose, glucose, or fructose, some nectars also contain small quantities of other sugars (Wykes, 1953) that may not be equally utilizable by a nectivore (Loh and Heran, 1970). In addition, the other inclusions of nectar, such as amino acids (Baker, 1977), could also affect the refractometer readings, so that for precise caloric bookkeeping it may not only be necessary to identify the different nectar constituents, but also to analyze the metabolic efficiencies of their conversions in the animal under consideration.

So far, questions that would make such precision necessary have not been asked. As far as we know, the common nectivores utilize sucrose, glucose, and fructose equally well, and the other nectar inclusions, although they are important, do not significantly affect the energy budgets.

Energy Budgets. The rate of energy intake of nectivores foraging from flowers can be estimated knowing their rates of flower visitation and the caloric equivalents recovered on the average per flower. Usually insects take most of the nectar pool out of each flower they visit, except in flowers like jewelweed *(Impatiens)* with a recurved nectar spur where tongue length can limit the distance that the insect can reach into the flower. When the nectar is not all pooled, as in desert willow *(Chilopsis)*, bees may take all of the pooled nectar, leaving the small unpooled remnants as long as the other available flowers continue to yield the pooled nectar (Witham, 1977).

The energy investment of foraging can be estimated from observed activity schedules, knowing the energy costs of the different activities. Percent of foraging time devoted to flight and perching can be determined from timed durations with a stopwatch. The cost of flight of various insects has been measured in the laboratory in terms of ml O_2/gram body wt/hr. Hovering costs for sphinx moths are given in this chapter (see p. 205). The cost of free flight for honeybees is 80–85 ml O_2/gram/hr (Heinrich, 1980), and that for bumblebees is similar, depending on the nectar load carried aloft in flight (Heinrich, 1975d). Each milliliter of oxygen consumed corresponds to the expenditure of approximately 5.0 calories on a carbohydrate diet (Kleiber, 1961).

When the insect is perched on the flower, its metabolism may approach that at rest; or if it regulates its thoracic temperature by shivering, it may at lower temperature expend nearly as much energy as during flight. Measurements and methods of estimation of metabolic rate of bumblebees during rest (Kammer and Heinrich, 1974) and during shivering on flowers (Heinrich, 1972) have been described.

Mass Marking Bees. Different methods of marking have been devised to help answer different research problems. Each method has unique advantages and limitations. The following describes some of the various possibilities (see also Smith, 1972).

Large numbers of bees can be marked at any one time by spraying the foragers with a hand atomizer (Smith, 1972). The paints used have included titanium oxide in alcohol (which colors them white), basic fuchsin, or fluorescent powders that can be detected in captured bees under ultraviolet (UV) light (Musgrave, 1949).

Frankie (1973) has marked large numbers of wild solitary bees in order to determine general bee mobility between different crowns of a tropical forest. Large numbers of bees were captured by net, placed into a bag and dusted with fluorescent powder, and then released. Subsequent samples of bees from neighboring areas were subjected to UV radiation for detection of possible marked bees.

In social bees it is possible to mark all the bees of a hive by having them pass over a special hive entrance block coated with fluorescent powder (Smith and Townsend, 1951). In this way, the entire foraging populations of a hive of honeybees can be marked in 90 minutes. In addition, bees from a hive may be marked by feeding them radioactive isotopes (Levin, 1960; Lecompte, 1964).

Individual Identification. In order to observe specific individual animals over consecutive foraging trips it will be necessary to mark them. There are at least three potential methods, each suited for specific animals and specific purposes. In insects with either little or no insulating pile on the thorax (such as wasps) one can hold the insect temporarily fast without anesthetizing it, apply a layer of liquid paper (typing correction fluid) to the thorax, and then write a mark or number on it with water-insoluble drawing ink. It is also relatively easy to put a layer of the liquid paper on the thoraces of furry insects, such as bumblebees. However, the surface will not be smooth; so it is difficult to write legible numbers on it. On the other hand, the fuzz provides a good substrate for holding glue to attach prefabricated markers. The plastic markers (2.3 mm diameter) designed by C. H. R. Graze K. G. (Fabrik für Bienengeräte, 7057 Endersbach

bei Stuttgart, Postfach 7, Würtemberg, W. Germany) for honeybee queens are numbered 1–99, with a package of 500 coming in five colors: green, blue, white, yellow, and red. A bottle of glue for the markers comes with the package.

Large numbers of similar tags can also be made from paper, which is then sprayed several times with a plasticizing compound to make it waterproof. Smith (1972) describes a somewhat similar method. He types a series of numbers from 0 to 99 on white paper, which is photographed and reduced in size, in a photographic reproduction process. The numbers are then punched out and fastened onto the bees with Duco cement. By using a series of dots and bars with the numbers, and by using paint of different colors, it is possible to give individual identification to many hundreds of bees.

One disadvantage of this tag system is that the glue does not dry immediately, and the insects have to be anesthetized while the glue hardens so that they do not scrape the markers off. Usually the insect will remain still long enough for the glue to harden after ether narcosis applied only until the animal first ceases movement. Many of the marked individuals probably will not be seen again, but those that remain in the area can be easily identified by their color and number.

As long as bees have already established a foraging area, it is relatively easy to follow selected marked bees in this area by eye without engaging in extensive physical chasing. In open areas, however, bees may wander, and the animals must be physically pursued. It is sometimes useful to follow them with an armful of flags, and to stick flags into the ground after successively visited flowers in order to determine both foraging route and flower constancy.

A third method of marking, which is much less disruptive to the insects, is to daub them with quick-drying enamel paint (such as Testor's Butyrate Dope or other model airplane paint). Bees can be marked with spots on the wings, abdomen, or thorax while they are foraging at flowers, and their foraging activity may be minimally or not at all disrupted.

However, some of the paint wears off in a few days, and it is difficult to accurately identify many individuals with distinctive marks.

Two other, more elaborate methods of marking bees have been developed. Both methods rely on detection at the hive, rather than being adapted for following bees in the field. In one method, devised by Gary (1971), bees are marked wtih a ferrous metal tag in the field. Each tag is associated with a tiny numbered plastic disc (as described above). The bees are released after capture and marking in the field, and the tags with the discs are recovered with magnets installed at hive entrances. This method allows one to map the foraging areas and the different types of flowers utilized by hives, provided the hives are located and equipped with tag-retrieval magnets.

Allen's (1981) system is based on an electronic tag, where each tag used has a specific resonating "signature" that is detected with an electronic scanner located at the hive entrance. The method, when combined with an electronic scale, allows for the measurement of the activity of individual bees at the hive entrance. It is possible to automatically record the times spent in and out of the hive for each individually marked bee, and to record the weight differences or foraging loads brought back with each foraging trip.

ENERGY BALANCE

An insect forager collecting nectar must harvest an amount of food energy that is some minimum threshold above foraging cost, foraging investment for pollen, caloric investment for predator avoidance, and all of a number of other costs related to reproduction. If it were possible to define the broad outline of the energy budget from the standpoint of calories in vs. calories out, one would still be left with determining the biological meaning of such economic bookkeeping.

Beyond the obvious need to have a net positive energy balance is the problem of determining the minimum *rate* of this energy inflow. Many insects have stringent time constraints within which a minimum energy

income must be amassed in order for them to complete their colony cycle. The minimum rate of profit making might determine the biologically relevant amount of profit of individual foraging trips, and the number of trips per unit time. For example, the "strategy" of the honeybee colony in stockpiling food resources is necessary to make full use of the massive and temporarily patchy resources for which the bees are specialized. The degree of spatial patchiness of high-energy resources will determine the need to enhance the communciation capabilities, while the degree of temporal patchiness will determine the need for stockpiling and the necessity to make huge profits on the individual foraging trips that are taken. The food stockpiled in the late spring of one year may be the major resource used in the early spring of the next year for brood rearing. In contrast, "profit" and energy balance can be assessed on a much shorter time scale in bumblebee workers that almost immediately convert their food resources into offspring. On the other hand, the bumblebee drones that leave the colony and forage primarily from low-energy food resources such as goldenrod and other composite flowers in the fall, have only their own energy balance to consider, without the need to amass profit. Thus, the energy balance equations should vary considerably among the different castes and should be independent of foraging cost. But these aspects of foraging energetics and what they might mean to the plants' investment to attract pollinators have so far been almost totally ignored.

Another variable that should affect the acceptable foraging profits concerns division of labor. In social bees, only a given percentage of the bees do the foraging. It would be expected that each of the colony's foragers must bring in more resources than the proportional drain on resources by those individuals that do little or no foraging. Bumblebee colonies are initiated in the spring by the overwintered queen, which initially performs all of the colony's tasks. She must be able to bring in a large foraging profit with each foraging trip in order to devote the majority of her time to brood incubation and other hive duties. Thus, she should make considerably greater energy profits per trip in comparison to drones and workers.

A solitary female bee, on the other hand, does not need to incubate her brood or support nestmates (although high offspring mortality may select for females that make large foraging profits per trip). For a solitary insect forager, it would also be essential for immediate foraging profit to exceed some minimum threshold, but at the present time not enough comparative data on total life strategies are available to determine what this minimum is. So far, we have not been able to specify the minimum timespan needed for foraging, or the number of offspring that must be produced within this timespan.

FLOWER SPECIALIZATION AND FLOWER CHOICE

The flowering plants are noted for their great variety of shapes, colors, scents, and sizes. Moreover, they present a wide range of amounts of food resources in the form of nectar or pollen, or both, in often relatively hidden places within a complex floral morphology. Many of the flower's features are apparently designed to limit (by exclusion) forager access to the food rewards, thereby placing a selective pressure on the foragers to break the defenses of the plants in order to reach the rewards.

In the competition for the often limited food rewards of flowers, foragers can be either generalists and forage from many different kinds of flowers, or they can be specialists and forage from few kinds. The relative advantages of specialization and generalization have long been of interest to ecologists. As MacArthur (1972) has suggested, it makes sense theoretically that a harvester cannot simultaneously be perfect at several jobs; perfection in one is achieved at the cost of reduced efficiency in another.

Relatively few data have been available to test the above assumption that specialization is more efficient (in terms of increasing either resources harvested or reproductive output) than is foraging generalization. However, a number of recent papers on bee foraging provide quan-

titative data relative to specialist–generalist strategies.

One such paper (Strickler, 1979) examines the foraging efficiency of the specialist bee *Hoplitis anthocopoides* (Megachilidae), and four generalist bee species that forage for pollen from *Echium vulgare* (Fig. 9-1). All of the five bee species were solitary bees. For each offspring the female solitary bee constructs a separate cell that she provisions with a mass of pollen. The pollen of each cell is entirely eaten by one larva, providing a measure of the amount of pollen required to produce one offspring. Strickler was able to relate the amount of pollen collected per unit time at *Echium* flowers to the reproductive output in the various bee species. Foraging "efficiency" was calculated on the potential number of offspring produced from the pollen collected per unit "handling time," or per unit of time involved in removing pollen from flowers. Although the results exclude nectar foraging, they nevertheless provide the first comparative data on the selective merits of handling strategies at a specific flower.

By all measures of foraging proficiency examined—search times between flowers, number of flowers visited per stalk, amount of pollen removed per flower, handling time per flower, amount of pollen collected per unit time, and units of foraging time required to collect enough pollen to produce one offspring—the specialist on *Echium vulgare* was superior to all of the four generalists using the same flowers (Table 9-1). The actual foraging advantage of the specialist may even be greater than these results might suggest, since the specialist bee collects nectar simultaneously with pollen, whereas the smaller generalist species, *H. producta* and *C. calcarata,* do not. The specialist can satisfy her metabolic requirements without greatly reducing her pollen foraging time, whereas the generalists visit other flowers to collect nectar or else interrupt pollen foraging on the same flower to probe for nectar.

It might be predicted that the specialist would handle other flowers less efficiently than the generalists. However, Strickler could not test this hypothesis because *H. anthocopoides*

GENERALISTS

SPECIALIST

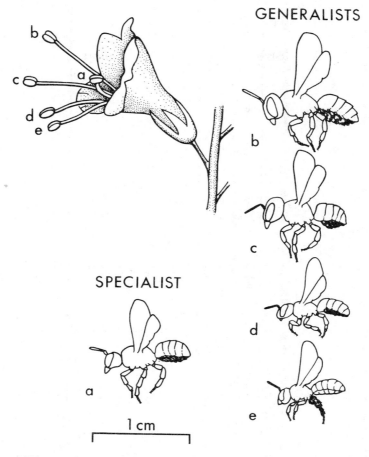

1 cm

Figure 9-1. Flower of *Echium vulgare* and relative sizes of the specialist and four generalist bees (females) that collected pollen from it. Dark areas indicate pollen-collecting hairs. a = *Hoplitis anthocopoides*, b = *Megachile relativa*, c = *Osmia coerulescens*, d = *Hoplitis producta*, and e = *Ceratina calcarata*. Mean dry weight of the females of the five species was 16.6, 16.8, 14.3, 4.4, and 3.6 mg, respectively. (From Strickler, 1979.)

would not collect pollen from flowers other than *Echium vulgare*.

The long-term adaptive advantage of specialization rests on predictability of resources. The host plants of many specialist solitary bees of deserts (MacSwain et al., 1973), for example, are available after rains (Linsley and MacSwain, 1958), which presumably trigger both the emergence of the bees and the germination of the plants. The bees must quickly make use of these ephemeral resources. They do not necessarily need to make use of a temporal progression of plants in flower, if the production of offspring can be ensured in a few days of foraging on a single species. Although the temporal predictability of *Echium* bloom

may be high, its spatial predictability can be low; the bee possibly compensates by increasing its foraging over a wider range (Eickwort, 1973, 1977).

Unlike solitary bees, social bees require a long time to build up a colony before reproductive adults are produced. The colony cycle requires a few months, or a whole growing season, and there is seldom, if ever, any one plant species in bloom throughout the whole tenure of the colony that can satisfy the colony's needs. Commonly, throughout a season, each new cohort of bees faces a different set of plants that must serve as nectar and pollen sources. Morphological and behavioral adaptations of the bee to specialize on a specific

Table 9-1. Comparison of foraging performance of a solitary bee foraging specialist, *Hoplitis anthocopoides*, on its host flower, *Echium vulgare*, with the performance of generalist solitary bees at the same flowers. (From Strickler, 1979.)

	Within stalks interflower flights(s)	Flights between stalks(s)	No. flowers visited per stalk	No. anthers from which pollen removed	Handling time per flower(s)	Pollen removed per flower (mg)	Pollen collection rate (mg/min)	Pollen required per offspring (mg)	Flower handling time to produce offspring (min.)
Specialist:									
Hoplitis anthocopoides	1.0	2.0	2.5	4.0	5.7	.196	2.129	17.8	8.36
Generalists									
1. *Megachile relativa*	—	—	—	3.9	10.4	.145	0.929	35.2	37.8
2. *Megachile* sp.	1.3	3.1	2.1	—	—	—	—	—	—
3. *Osmia coerulescens*	—	—	—	3.7	15.7	.192	0.828	14.4	17.4
4. *Hoplitis producta*	1.7	2.8	2.2	3.1	29.3	.137	0.324	6.4	19.8
5. *Ceratina calcarata*	—	—	—	2.8	31.7	.122	0.308	4.1	13.3
6. *Ceratina* sp.	2.2	4.5	1.9	—	—	—	—	—	—

kind of flower are clearly nonadaptive in such bees which must be able to utilize many kinds of flowers to satisfy the colony economy.

Perhaps the most important group of native bees in terms of pollination ecology in a variety of arctic, north-temperate, and montane habitats are bumblebees. Their foraging behavior, evolved with strong selective pressure to enhance energy profits, is in turn a selective pressure shaping the evolution of flowering strategies. Most aspects of the bees' foraging behavior relate to the enhancement of foraging returns, or energetics. Aside from wide-ranging potential implications in ecology and evolution, there are also practical reasons for studying the bees' mechanisms of foraging optimization. A major consideration is convenience. Bumblebees are ideal because they are common, are easily marked for individual identification for observation in the field, adapt easily to captivity, and occur in pleasing habitats for field biologists. For these and other reasons, we now have a resonably comprehensive picture of their foraging behavior.

Bumblebees are limited in the kinds of flowers that they visit primarily by the minimum rewards the flowers offer, and by the bees' tongue-length, which determines the speed with which the rewards can be harvested. In general, short-tongued bees can harvest faster in short-corolla flowers, and long-tongued bees have a competitive edge in long-corolla flowers (Inouye, 1980).

Bumblebees are possibly greater generalists and opportunists than most other bees. Bumblebees of any one species forage from any of several kinds of floral and nonfloral resources. They may, for example, even forage in large numbers from honeydew produced by scale insects hidden among spruce needles (Heinrich, pers. obs.). In addition, they visit many morphologically complex flowers that cannot be "handled" by any other bees.

Analysis of pollen loads collected by bumblebees indicated that the individuals sometimes visit numerous species of flowers on any one foraging trip (Brian, 1951), so that they were considered to be "inconstant" foragers. However, this interpretation is an oversimplification that largely ignores the sophistication of bumblebee foraging energetics.

Whether a bee is a specialist or a generalist must be assessed, not only by the kinds of flowers it visits, but also by the number and sequence of different flower visits relative to flower availability. Bees could collect relatively pure pollen loads, not because they are flower-constant, but because they have site-fidelity to an area where, there are, at that time, no other acceptable plants in bloom. On the other hand, pollen loads from many species of plants could reflect a sampling behavior of bees that had not yet specialized. Both of the above possibilities have now been experimentally examined (Heinrich, 1976a), and data on the relative costs and benefits of different potential behavioral strategies are available.

Site-fidelity does not necessarily produce flower-fidelity. Many bees may share the same foraging area containing a variety of different plants in bloom, but individual bees of one or more species may utilize the flowers of different plant species in different proportions while foraging at a particular site (Heinrich, 1976b, 1979a). In other words, the bumblebees are generalists as species, but as individuals they specialize. It would appear that being both generalists and specialists, they would have the best (or the worst) of both worlds. I shall provide a model, with quantitative tests, to evaluate the energetic costs and payoffs of the different potential strategies.

One of the first problems that a forager faces in its first exposure to the field is: Which flowers should it visit? Unlike solitary bees such as *Hoplitis anthocopoides*, bumblebees are not strongly preprogrammed genetically to search for a specific flower that is likely to be rewarding, and unlike honeybees, the new forager is not directed by the dances of scouts within the hive to rewarding flowers. Each individual must independently determine which flowers to utilize, and since the foragers returning to a colony at any one time may have a great variety of different pollen loads, it appears that the individuals come to independent decisions.

The flowers of different species of concurrently blooming plants often differ manyfold in their rate of nectar and amount of pollen production. If each bee is attempting to maximize its profits, why are not all bees specializing on

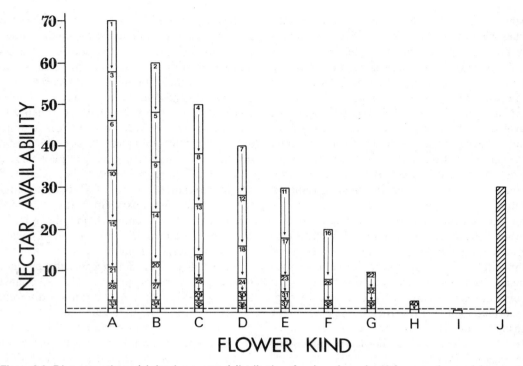

Figure 9-2. Diagrammatic model showing expected distribution of majors chosen by 40 foragers given serial access to a mixture of 9 flower-kinds (A→I), here ranked in order of decreasing nectar availability. (Flower J is a potential new source with a relatively large but, to the bees, unknown food reward.) It is assumed that the bees can rank the relative nectar rewards and utilize first the most rewarding. The numbers 1–40 specify the different bees, and they indicate the order of occupancy at the different flowers. Each bee in this model utilizes a maximum of 11 units of nectar, with 1 unit being the lower threshold of profitability. After the 20th bee has had access to the array all flowers are reduced to 7 nectar units or less, and subsequent foragers receive fewer rewards per unit time. They are then less "satisfied" and become willing to "sample" new flowers, gambling on flower J.

the same flowers? I shall attempt to answer this question by presenting a model (Fig. 9-2) in which bees sample the available reward spectrum, rank it from "richest" (A) to "poorest" (I), and specialize on the most rewarding flower-kind they find (Heinrich, 1976b, 1979a). As soon as bee #1 specializes on flower A, however, it changes the reward spectrum for the other bees. The bees have no way of assessing rate of nectar production; they can only gauge nectar availability. Thus, after bee #1 has removed a certain percentage of the rewards provided by a flower from population A, then the new ranking of nectar rewards for bee #2 may favor visiting flower population B. As more and more foragers are fielded, gradually "filling up" the foraging niches on the most rewarding flowers, the net effect will be that the nectar rewards of all the plants will tend to be-

come equivalent. Furthermore, the number of individual bees specializing on any one flower-kind will be directly related to the food rewards it produces; high-nectar flowers will have many specialists, while low-nectar flowers will have few. The above predictions of the model were generally supported by field observations (Heinrich, 1976b).

In an experimental test of the above model, inexperienced bees *(Bombus vagans)* were released, one at a time, into an exclosure (and enclosure) containing *Impatiens biflora,* the most highly rewarding flower (A), and *Aster novae-angliae* and *Solidago graminifolia* (Fig. 9-3), relatively low-reward flowers (G and H) (Fig. 9-2). As predicted, each bee visited numerous kinds (A to I in Fig. 9-2) of flowers in its first foraging trip out of the hive. By the sixth or seventh foraging trip, almost all bees

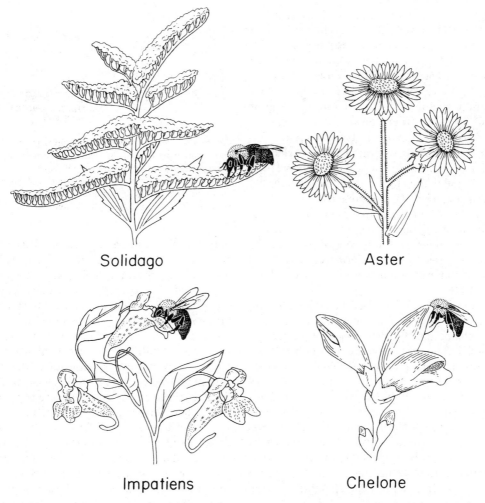

Solidago

Aster

Impatiens

Chelone

Figure 9-3. Four different kinds of flowers used in the learning experiments of bumblebees (Heinrich, 1979a). The top two were handled accurately from the start. In the bottom two, handling accuracy increased with experience.

had specialized on the most rewarding flower, A. (Since bees were only allowed into the foraging area one at a time, the rewards of A remained virtually unchanged, so that the reward ranking remained constant.) However, the rewards of A were then experimentally reduced by allowing many bees to have simultaneous access to these flowers. Again, as predicted by the model, the bees apparently reordered the flower spectrum and utilized flowers, B, C, and D, that had not previously been included in their foraging repertoires (Heinrich, 1979a).

When the bees are abundant, the amount of sugar (in the nectar) available per flower per unit foraging time becomes reduced until it is similar for a broad range of flowers that may have very different rates of nectar production. What determines optimal flower choice when the ranking between different species becomes similar? According to optimality criteria (MacArthur and Pianka, 1966), the bees should become less selective. They should take flowers as they are encountered, if they are to be included in the diet at all.

Only part of the above prediction is observed. When the high-reward flowers became partially depleted, the bees visited on the average three different kinds of flowers, rather than one (Heinrich, 1979a). However, these flowers were not visited on the basis of the frequency with which they were encountered. One individual might, for example, be "majoring" on A and "minoring" on B, while a second one might be majoring on B and minoring on A, C, or some other flower. Similar observations were made in a meadow outside the foraging enclosure. In a meadow with a relatively uniform distribution of fall dandelions, wild carrot, and white clover, one B. vagans was observed majoring on the dandelion and minoring on clover, another was majoring on clover while minoring on dandelion, a third was majoring on clover and minoring on wild carrot, whereas a fourth was majoring on carrot and minoring on dandelion as well as clover (Heinrich, 1976b).

If the bees have evolved behavior that enhances foraging intake, then there are at least two explanations for the "majoring–minoring" behavior. The first concerns the temporal variability of the food resources. Flower availability is constantly changing as species are coming in and out of bloom (Heinrich, 1976c). Even though food rewards may be driven very low owing to competition, there is always the likelihood that a new flower (J, Fig. 9-2) with superior rewards may suddenly appear that provides more food than any of those available at the time. In order to take advantage of such new food sources (as well as diurnal changes of food availability in those already available), it is imperative for the bees to sample. A strict major (on the best flower) is always the best strategy in a constant environment at any *one* time, whereas minoring is a necessary compromise required to track resources changing over time (Oster and Heinrich, 1976). Individual bees *(B. fervidus)* have been observed to maintain the same major and minor specializations for one month in the field (Heinrich, 1976b), but when a minor species is suddenly enriched (by artificially fortifying the flowers with sugar syrup), the bees can change their minor to a major within one foraging trip (Fig. 9-4).

There are several potential options whereby the bees could establish, maintain, and change their foraging specializations. They could start

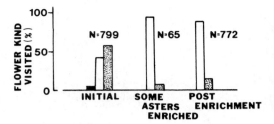

Figure 9-4. Three consecutive foraging trips of a *B. fervidus* worker before, during, and after enrichment (with sugar syrup) of *Aster,* the bee's minor flowers. N = total number of flower visits. Solid bar: *Impatiens biflora;* stippled bars: *Solidago graminifolia;* open bars: *Aster novae-angliae.* (From Heinrich, 1976b.)

their foraging careers by visiting all available flowers, and progressively drop either the least profitable or the morphologically most complex or difficult to handle from their foraging repertoire. Alternately, they could start with the most morphologically "simple" and easy-to-handle flowers and keep adding progressively more "difficult" flowers. I calculated the potential caloric profits from each of the above potential strategies (Fig. 9-5), using data derived from the bees in the enclosure that established their foraging preferences from among a series of different wild flowers. The calculations (based on observed foraging rates of experienced bees, nectar availability, and the assumption made here, for comparative purposes only, that flowers are visited in direct proportion to their abundance) indicate that the energetically most profitable strategy (for inexperienced individuals) is to sample all available flowers and to keep dropping the least profitable flowers. This is, indeed, the strategy that is observed (Fig. 2, Heinrich, 1979a). Correct choice of flower, and hence "sampling," is of overwhelming importance in enhancing foraging profits.

What immediate energetic "cost" do the bees incur for sampling and learning to forage on the most rewarding flower? One way of assessing this cost is to compare the caloric intake derived from strict majoring on a proven reward source, to the caloric intake that is derived by the addition of visits to empty flowers of another kind, the minor. Such a comparison would be difficult to make in the field, since it

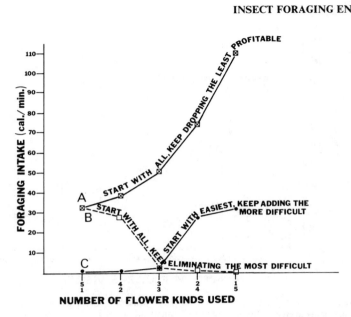

Figure 9-5. Potential foraging intakes in cal/min by *B. vagans,* in a field enclosure starting their foraging career along three potential strategies, A, B, or C. Calculations are based on observed foraging rates and nectar amounts available from the five most common species of flowers in the following order of "difficulty": *Chelone glabra, Impatiens biflora, Trifolium pratense, Aster novae-angliae,* and *Solidago canadensis.* It has been assumed for the purposes of this calculation that all flowers are handled in the manner of experienced bees, and that they are visited in direct proportion to their abundance. The actual strategy corresponds closely to A (see Fig. 2, Heinrich, 1979a).

is not possible to assess the relative reward that the bee might have gained from two different populations of flowers, after they have been visited. In addition, the economics of learning to handle different flowers may be difficult to segregate from the total economics of flower choice.

It was possible in a laboratory study to isolate the energy cost the bees incurred for appropriate flower choice alone. The bees were given a choice between two populations of artifical flowers that differed only in color and amount of food reward provided; visitation to a "white flower" provided the bee with a 1 μl 50% sugar reward, whereas visitation to a "blue flower" of identical morphology provided no reward. Counting the consecutive white and blue flowers visited at different times during the foraging bout gave an accurate cost of the actual caloric intake, relative to the *potential* intake if all visits were restricted to the rewarding flowers. Under the above experimental conditions, *B. terricola* workers had a learning or sampling cost of 30 calories for every 30 calories intake for the first

50 flower visits (Fig. 9-6). For 150 to 200 flower visits, the sampling cost was reduced to 12 calories for every 100 calories and after 300 or more flower visits, the sampling cost stabilized near 4 calories for every 180 calories intake (Fig. 9-6).

In the above experiment, the bees retained blue flowers as their minor, even though these flowers were totally nonrewarding. In addition, they learned to narrow their foraging to blue flowers (when blue flowers rather than white flowers were rewarding) much more quickly than to white flowers (Heinrich et al., 1977). In addition, when both blue and white flowers were rewarding, but with blue flowers having six times more reward than white, they consistently foraged "unoptimally" in the sense that their foraging resulted in only about half the caloric intake that they could have made had they bypassed the white flowers entirely (Fig. 9-7).

A possible explanation for the above, apparently less than fully optimal foraging behavior could be that the bees are not entirely without genetic biases. In general, "fly flowers" tend to

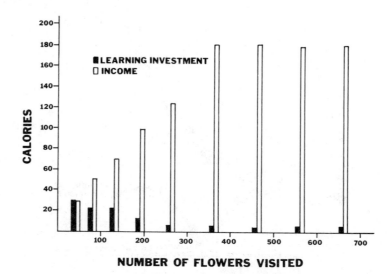

Figure 9-6. Caloric rewards at white (artificial) flowers (each rewarded with 1 μl 50% sucrose solution) and cost of learning by sampling incurred by visiting nonrewarding (blue) flowers over successive sequences of flower visits by a "naive" bumblebee, *Bombus terricola,* visiting a total of 650 flowers (derived from data in Figs. 2 and 3, Heinrich et al., 1977). Each set of two bars represents cumulative calories from 0 (first set) or from previous sets (all beyond first set).

be white (Müller, 1881) and have low food rewards, whereas "bee flowers," many of which are blue, provide large rewards (Heinrich and Raven, 1972). Perhaps it is energetically more advantageous to sample and learn to utilize

blue (and presumably yellow and ultraviolet) rather than white flowers. However, no systematic study has yet been made relating flower color to potential foraging profits.

At the present time, it is not known how

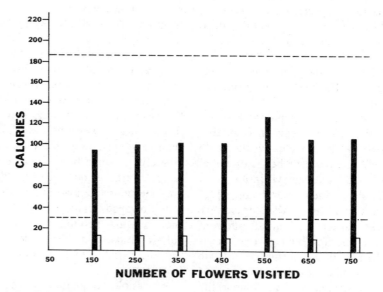

Figure 9-7. "Unoptimal" foraging of *Bombus terricola* workers given a choice between white artificial flowers (1.85 cal/flower) and blue artificial flowers (with an average of 0.31 cal/flower). Data indicate rewards collected at the blue (closed bars) and the white (open bars) flowers in 100-flower intervals, following the first 50 flower visits (derived from data in Fig. 7, Heinrich et al., 1977). Dashed lines indicate potential rewards if only white (upper) or blue (lower) flowers were visited.

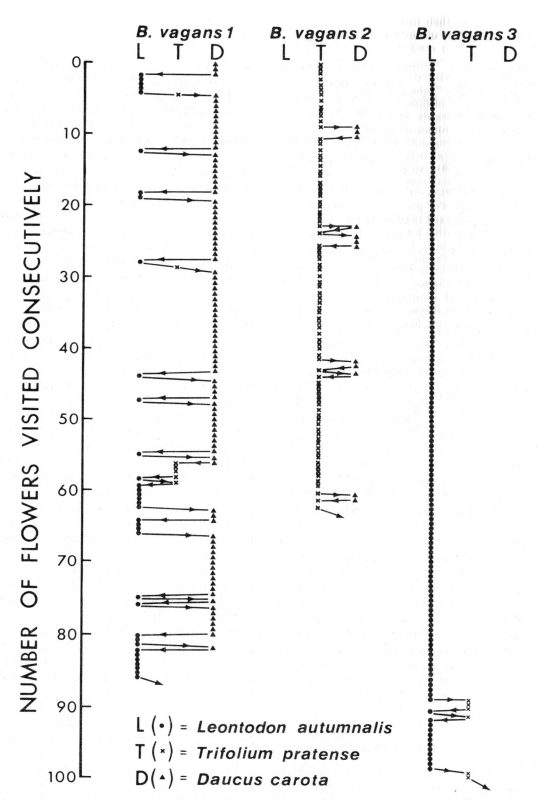

Figure 9-8. Sequence of kinds and numbers of flowers consecutively visited by three different selected *Bombus vagans* workers. Bees generally switched back and forth between major and minor flowers at frequent intervals. The flowers were blooming intermingled in the same field. The duration of observations were: bee 1, 11.0 min.; bee 2, 14.1 min.; bee 3, 14.8 min. (From Heinrich, 1976b.)

bees measure their foraging returns. Many individual flowers of any one kind used by a bee are empty. Net rewards are thus presumably measured by "averaging" the returns from many flowers visited. If so, then the bees may have to make "runs," periodically switching from one kind of flower to another in order to assess the rewards in any one kind. Such "runs," or temporarily "pure" majors, in an overall majoring–minoring strategy, are indeed the general behavior pattern that is observed (Fig. 9-8). All bees can be considered to have "pure" majoring strategies, provided the time intervals over which they are observed are short enough. The greater the difference in the amount of reward between the different available flowers, the longer the "runs" are restricted to the same (the most rewarding) kind of flower. For example, in the enclosure with a half dozen or more kinds of different flowers, the bees *(B. vagans)* that had specialized on jewelweed had unbroken runs on these flowers of at least 30 flowers. When the rewards in these flowers were low, the bees visited, on the average, fewer than 10 flowers in a row before "sampling" one or more of another kind (Fig. 9-5, Heinrich, 1979a).

From the plant's perspective, it is advantageous to make the pollinators' runs as long as possible so that the pollen is deposited on receptive stigmas of the same rather than different species. If "sampling" is a major criterion in making runs of varying lengths, then the predictability of the rewards should determine the length of the runs. In other words, the more variable the rewards, the more sampling is necessary to arrive at the true average. However, in laboratory studies of foraging in *B. edwardsii*, the bees preferred a population with predictable rewards between individual flowers to one providing the same net rewards but with individuals providing unpredictable rewards (Waddington et al., 1981).

FLOWER HANDLING

In order for generalist bee species to enhance their foraging intake at a variety of different kinds of flowers, they must not only correctly evaluate and rank the available reward spectrum, but they must also learn to handle the high-ranking flowers. For example, to collect nectar from *Impatiens biflora,* the long-tongued bumblebees can reach in by bracing themselves with their legs outside the flower entrance, rapidly pushing themselves part-way in and thrusting with the extended proboscis. Short-tongued bumblebees at the same flowers collect nectar by approaching the flower at the other end, and piercing the nectar spur. To collect either nectar or pollen from *Chelone glabra* flowers, the bees have to approach the front of the flower and pry apart the partially closed corolla tube before crawling inside. To collect pollen from *Solanum dulcamara,* the bees have to dangle from the flower, grasp it firmly with their mandibles, and shake it violently by buzzing to dislodge the pollen from the tubular anthers, which is then shed onto the ventor of the abdomen (Heinrich, 1976b). Not all bees in the field are equally skilled. For example, in a field of *Potentilla recta,* a plant whose flowers provide pollen but no nectar, some bees (visiting an average of 11.2 flowers/min.) tongued the flowers for nectar but did not buzz them for pollen. At the other extreme, the highly skilled bees never tongued the flowers, but always buzzed them instead, and visited nearly three times as many individual blossoms per unit time (Table 4, Heinrich, 1976b). Individual variation in foraging skill has been shown in honeybees foraging from vetch, *Vicia cracca* (Weaver, 1957, 1965).

Handling accuracy of morphologically complex flowers is a function of experience. Flowers in open inflorescences (Asteraceae) are handled appropriately by bees with no previous foraging experience (Heinrich, 1979a; Laverty, 1980). However, handling accuracy at zygomorphic flowers such as *Aconitum napellus* (Heinrich, 1976b), *Impatiens biflora,* and *Chelone glabra* (Heinrich, 1979a) and in *Aconitum columbianum, Delphinium barbeyi, Mertensia ciliata, Oxytropic splendens,* and *Pedicularis groenlandica* was low (<50% accurate) in the first 10 flowers encountered, increasing to >90% accuracy (depending on the species) after more than 20 to 100 flower visits (Heinrich, 1979a; Laverty, 1980).

Flowers that are handled inaccurately are

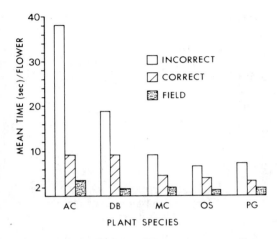

Figure 9-9. Mean duration of correct and incorrect flower visits to five plant species by inexperienced *B. flavifrons* workers and mean duration of flower visits by *B. flavifrons* workers in the field. FL, flower. Plant species: *A. columbianum* (AC), *D. barbeyi* (DB), *M. ciliata* (MD), *O. splendens* (OS), and *P. groenlandica* (PG). (Redrawn from Laverty, 1980 with permission of the publisher.)

energetically expensive because they cause the bee to slow down. For example, Laverty (1980) found that in five different kinds of flowers, experienced bees under experimental conditions required 3–9 sec handling time, whereas inexperienced bees required 6–18 sec handling time (Fig. 9-9). In addition, subsequent improvement in flower handling was apparent even after flowers were handled "correctly," since individuals in the field, which presumably had the longest foraging experience, extracted the pollen and nectar from the same flowers in only 1–3 sec. Even after flowers are handled accurately, the speed of handling is still considerably less than that seen in most experienced foragers (Fig. 9-10).

What is the energetic cost of handling inaccuracy? First, it should be noted that the potential profits available from "difficult" flowers are often enormous in comparison to those that require no skill to manipulate, since fewer bees are capable of handling them. Thus, even if a

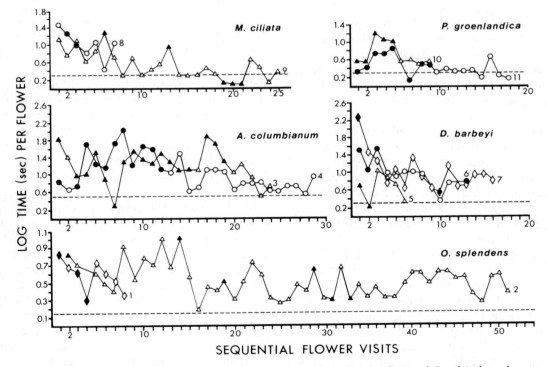

Figure 9-10. Duration of flower visits over foraging trips of inexperienced *B. flavifrons* and *B. sylvicola* workers to five plant species. Each sequence (1–11) represents an individual bee. Open and closed symbols indicate correct and incorrect flower visits, respectively. Dotted line shows mean duration of flower visits by *Bombus* spp. foragers in the field. (Redrawn from Laverty, 1980 with permission of the publisher.)

bee is slowed down at a complex flower, so that it extracts nectar from 5 to 10 times fewer flowers than an experienced forager would, it is still energetically advantageous to have chosen the "difficult" flower. If it had chosen instead to forage from an "easy" flower, it might have made a smaller potential income. Furthermore, not only would the immediate income be much less, there is also little room for future improvement of income.

The caloric cost of learning to manipulate can be estimated from data on the improvement of handling accuracy and rate of flower visitation. For example, in the experiments with *B. vagans* workers in an enclosure (Heinrich, 1979a), the maximum reward available to experienced bees was 110 cal/min. (Fig. 9-11). Assuming that a flower handled "inaccurately" slows the bee down enough so that it does not visit five other flowers, then the "reward" of that flower is six times less than it might be if it were handled accurately. When the potential rewards of the accurately and inaccurately handled flowers are added over the foraging career of the bee, then the difference between the maximum potential and the "actual" rewards provides a measure of the caloric cost of handling inaccuracy. At least in *Impatiens biflora,* the cost of handling inaccuracy diminishes rapidly as 70 flower visits are approached (Fig. 9-11). However, this calculation is undoubtedly an underestimation of the cost, since it does not take improvements of foraging speed into account. Foraging speed may continue to improve long after the flowers are handled accurately (Laverty, 1980). On the other hand, the overall cost of flower handling on the "correct" (most rewarding) flowers may not be great when seen in the context of the total profits that can be made. For example, experienced bees foraging exclusively from *Solidago canadensis* can collect aprroximately 0.01 mg sugar/min. If they forage for one week, at 8 hr/day, they can collect a total of 33.6 mg sugar during this time. On the other hand, if the bee had chosen *Impatiens biflora* instead, initially visiting only one flower per minute (rather than 11/min., as in experienced bees), it still could have collected the 2.7 mg sugar from that flower to make the po-

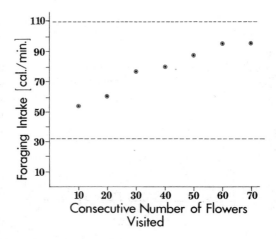

Figure 9-11. Calculated improvement of foraging intakes of bees beginning to forage from jewelweed, *Impatiens biflora.* The calculation is based on observed improvement of handling accuracy (Fig. 8, Heinrich, 1979a) of *B. vagans* workers initiating their foraging careers. It is here assumed for the purposes of the calculation that a flower handled inaccurately slows the bee down enough to make the flower worth six times less than one handled by an experienced bee. Upper dashed line indicates maximum rate of intake by experienced bees specializing on jewelweed. Lower dashed line indicates expected foraging intake if the bees did not exert flower choice and indiscriminately used all flowers in the enclosure (see Heinrich, 1979a) but handled them proficiently.

tential income of 33.6 mg sugar in 12 minutes rather than in one week. Thus, although the cost of learning to manipulate complex flowers can result in a manyfold reduction in potential profit, it is obviously not as important as making the proper flower choice in view of the often substantial difference in nectar rewards between these two flowers.

In the larger, evolutionary sense, foraging profits during flower handling can be increased in two opposite ways. The forager can make a large energetic investment in order to visit and nearly empty many flowers per unit time, or it can reduce energetic investment by visiting fewer flowers and draining nectar fully. Although both strategies can yield the same "profits" in the immediate sense of efficiency (ratio of energy investment and energy yield), the *rates* of profit making (which may often be of greater importance) can vary tremendously.

The relative advantage of one strategy over the other, in enhancing the rate of foraging returns, should depend on the quantity of nectar per flower, and the "expected" reward of other, similar flowers. The feasibility of each strategy, in turn, depends on the flower morphology and the behavior and morphology of the forager.

Possibly the largest energetic investment for flower handling that a forager can probably make is to hover, but hovering is also the mechanism of visiting the most flowers per unit time. Relatively few comparative data are available that directly compare different modes of foraging at given flowers under the same set of conditions such as flower density and temperature. As an approximation, however, hoverers visit three to four times more flowers than the fastest nonhoverers. For example, in a patch of jewelweed *(Impatiens biflora),* the ruby-throated hummingbird visited 37 flowers per minute, whereas bumblebees visited 10 to 12 flowers per minute. Similarly, in a patch of lambkill *(Kalmia angustifolia),* a clear-wing hawkmoth (*Hemaris* sp.) visited 50 flowers per minute, while bumblebee workers *(Bombus ternarius)* visited 20 per minute (Heinrich, 1975b,c).

Considerable data are available on the energetic cost of hovering, and these data indicate that hovering during nectar extraction from flowers is the most energetically costly mode of flower handling that has evolved. The total cost is especially great in larger animals. One of the premier groups of insects that hover during nectar foraging is sphinx moths. They vary some 60-fold in weight, from 0.1 to 6.0 grams. Over this size range, there is a tendency for the rate of energy expenditure during hovering to increase as a function of mass according to the equation $Y = 59.4X^{0.81}$, where $Y = cm^3 \, O_2/hr$ and $X =$ body mass in grams (Bartholomew and Casey, 1978). Extrapolation from this formula (which is based on 28 species from Central America) indicates that a moth weighing 0.1 gram consumes approximately 9.2 $cm^3 \, O_2/hr$, whereas one weighing 6 grams has a 27.6 times greater energy expenditure.

In order to determine how these energy expenditures compare to the energy rewards of flowers, and the strategies of foraging for flowers, it is necessary to convert both oxygen consumption and nectar rewards to the same units.

Moths convert carbohydrates from nectar to lipids before use by the flight muscles (Beenakkers, 1969). Because of the additional metabolism required for this conversion, the number of calories corresponding to 1 $cm^3 \, O_2$ consumed is probably less than the 5.0 that is normally used to convert oxygen consumption to caloric equivalents in a carbohydrate diet. Generally, 3.7 cal/mg is used as the value to convert the caloric expenditure in terms of sugar intake, whether di- or monosaccharides, such as glucose or fructose, are used.

Using the above conversion factors to examine flower handling costs, in terms of sugar available in nectar, delineates some of the energetic constraints of hovering as a flower handling strategy. A sphinx moth weighing 6.0 grams, for example, must take in 5.7 mg sugar/min. in comparison to 0.21 mg for a 0.1 gram moth (Fig. 9-12) just to cover the ener-

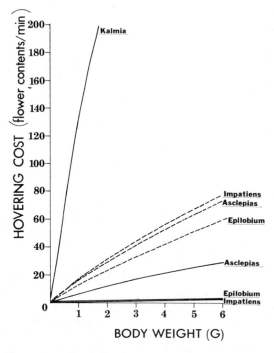

Figure 9-12. Flower handling (while hovering) costs of sphinx moths as a function of body mass in terms of rate of sugar utilization. Based on metabolic data by Bartholomew and Casey (1978).

getic cost of flower handling. Thus, within the sphinx moths, flower handling costs vary at least 27-fold because of differences in size alone. The energetically expensive strategy of hovering, however, must ultimately be assessed in terms of the potential payoff, and not on cost alone.

A comparison of the caloric rewards available from several kinds of flowers used by sphinx moths in North America, relative to flower handling cost (Fig. 9-13), indicates that the caloric income should be well above handling cost, as long as they contain their full complement of nectar. For example, a moth such as *Manduca sexta,* weighing 2 to 3 grams, needs to visit only one jewelweed flower per minute to pay off the energetic cost of one minute's flower handling (hovering), but when visiting a patch of blossoms that has been open to bumblebees and other foragers, it must visit 30 to 45 flowers per minute to pay off the same

hovering cost. The results strongly suggest that competition from other nectivores (in this case bumblebees) causes a reduction of the standing crop of nectar, which would seriously affect the ability of large moths to make an energetic profit. If the large hovering moths used flowers with relatively small nectar rewards (which are adequate for bees), they would incur an energy debt. It is assumed here that the moths could handle individual flowers very rapidly— at rates of 30 to 50 flowers per minute—but more data are needed before more valid comparisons can be made.

Hovering moths have at least one option of reducing flower handling cost below that which might be predicted from the above measurements made in the laboratory—an option not available to hummingbirds. They can rest their front legs on the flower (Fig. 9-14), thus giving themselves a mechanical support that should reduce wing loading and the metabolic cost of flight (Bartholomew and Casey, 1978). Clear-winged sphinx moths (*Hemaris* sp.) routinely support themselves in part with their front legs while hovering at the flowers of *Asclepias syriaca, Pontedaria cordata* (Fig. 9-14), *Prunus virginiana, Kalmia angustifolia,* and presumably other flowers (pers. obs.). It is not known under what conditions this strategy might be used by other sphinx moths.

It is of interest that the substantial energetic cost of hovering greatly restricts the ability of large moths to make a profit from perhaps the majority of flowers. In addition, it puts them at a great competitive disadvantage in terms of flower handling efficiency at many of the flowers that do not produce large food rewards; competitors can thrive while driving the food rewards substantially below the minimum required merely to pay off the flower handling cost. For example, a moth weighing 0.1 gram that requires only 0.21 mg sugar/min. can visit at least 50 flowers/min., but needs to visit only 13 to pay off the handling cost. On the other hand, a 6.0 gram moth, which is not likely to have a much faster flower handling time, would have to visit at least 580 of these flowers per minute just to pay off the flower handling cost. The large moth would appear to be under a distinct handicap, since it would be energet-

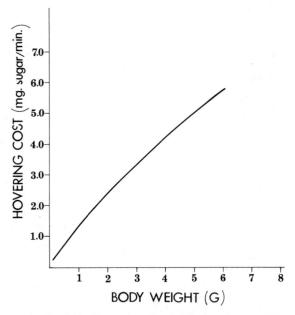

Figure 9-13. Flower handling (while hovering) costs of sphinx moths as a function of body mass in terms of caloric contents of flowers (*Kalmia angustifolia, Impatiens biflora,* and *Asclepias syriaca*). Solid line: flowers with full nectar complement. Dashed lines: based on nectar contents of flowers in the field that had been open to foragers. The calculations for the graph are derived from metabolic data by Bartholomew and Casey (1978), and nectar by Heinrich (1975b, 1976a).

Figure 9-14. Photograph of diurnal clear-winged sphinx moth, *Hemeris* sp., foraging from *Pontederia cordata*. Note use of forelegs for bracing.

ically excluded from many flowers. Why then have moths not evolved ever greater miniaturization, rather than larger size, which can reach twice that of the smallest hummingbirds? At the present time, this is still a mystery. It is possible, however, that other contingencies, such as the ability to fly long distances to find mates, oviposition sites, and isolated food patches, may have favored larger size, necessitating severe compromises in immediate foraging efficiency.

PATTERNS OF MOVEMENT AND "OPTIMAL" FORAGING STRATEGIES

As Horn (1979) has indicated, adaptive or optimality arguments can easily be scorned as "tautological, panglossean, teleological speculations about artifacts of history." Nevertheless, they can serve as references for gathering and putting data into useful contexts. I shall not discuss optimal foraging per se because I do not want to speculate on the ultimate constraints that may or may not have been operating to produce any one specific pattern. Rather, I shall be concerned with observed behavior in terms of patterns of movement and how they relate to the enhancement of foraging intake in regard to variously distributed food rewards.

In some habitats, plants stand out visually. Such conspicuous plants can attract foragers from a distance because of their display. In general, the foragers presumably can enhance foraging intake by making direct visual comparisons between competing individual plants, by flying not only to the closest neighbor but also to the ones with the largest floral display. That is, increased interplant distance is compensated for by larger display. The tendency for pollinators to choose the larger display may be a selective pressure for the production of even larger displays, in a "big bang" reproduction, at the expense of the plant's energy and other reserves (Schaffer and Schaffer, 1979).

In a complex environment, specific plants that are widely separated are often hidden from the view of foragers. Nevertheless, the foragers may identify them by the location where they grow. For example, an understory bush in Costa Rica, *Pentagonia wendlandii* (Rubiaceae), was visited during one blooming period by at least 113 "traplining" euglossine bees (Janzen, 1971). The bees continued to arrive 1 to 2 days after the flowers were removed; they were not attracted to flowers as such, but to the location where they expected flowers.

Although circumstantial evidence for traplining is extensive, proof of its existence and energy economy is sparse. It is generally not physically possible either to follow or to predict the foraging paths of a fast-flying animal moving between widely scattered plants. However, in one study of individually marked bumblebees that were followed on successive foraging trips (in some instances ranging over one month), the bees were shown to revisit specific sequences of flower clumps on successive foraging trips (Heinrich, 1976b).

Traplining can best be examined as a modification of site specificty. After a bee has found large concentrations of food at one site, it will return again and again to that site until the food is exhausted. For example, honeybees restrict visits to specific foraging areas, often several miles from the nest (Beutler, 1951). Individual bees, active at one site some distance from the hive, normally do not switch immediately to new, superior food rewards as they come into bloom closer to the hive, but the hive constantly tracks new food rewards by way of the scouts that direct new recruits to them (von Frisch, 1967).

When the food rewards at a specific site are renewable, as well as limited with regard to the transport capacity of the forager, then the traplining forager must visit several sites in succession on any one trip. Visiting the different sites in a specific sequence may be the best strategy for remembering all of the sites. In addition, it may be the best strategy for enhancing foraging returns at specific flowers; there is little payoff in revisiting flowers before they have had time to refill. At the present time, no attempt has been made to distinguish between the two above strategies which could both act to enhance foraging returns. It is not known what roles, if any, rates of nectar refilling and size of nectar patches play in the establishment and maintenance of foraging traplines.

Another important aspect for the efficient harvesting of the food rewards, besides that of the spatial and temporal distribution of visitations to food patches, concerns the distribution of visits to the food within the patches. A forager may visit either few or many flowers in a patch.

In most situations, the forager does not know whether or not a given flower in a patch has recently been visited and emptied. However, since foragers tend to favor near flowers with visits, and since nectar availability may vary from one patch to another owing to many variables such as genetic differences, edaphic conditions, and previous forager activity, it is generally true that if *some* flowers of a patch are rewarding, then the neighboring flowers of the same kind would tend to offer, on the average, similar rewards. Therefore, it should be advantageous for the forager to remain in the same area after being rewarded, rather than moving quickly to a new area. Conversely, if the forager encounters few or no rewards after sampling a few flowers of a patch, then it may be advantageous to move on and examine flowers of some other patch.

Two "rules" of movement are generally cited that tend to accomplish both of the above. One concerns the distance and the second concerns the angle of interflower flights, the "turning angles." When a poor or unrewarding patch is encountered, bees tend to make relatively long interflower flights, and they tend to have low turning angles. As a consequence, they tend to rapidly move or forage through the patch. For example, bumblebee workers moved on the average <75 mm on 67% of their interflower flights in a highly rewarding patch of white clover, *Trifolium repens,* but only 30% of the flights were this short after the nectar had been depleted. The bees tended to move forward (relative to the last floral move) in only 53% of the interfloral moves in the nectar-rich patch. But when the flowers held little nectar, the bees moved in the forward direction in 82% of their successive interflower moves (Heinrich, 1979b). The logical extreme of having large turning angles and visiting the closest flower would be for the forager to fly back and forth between the same two flowers. Apparently other factors are also involved in enhancing foraging returns.

The results of Pyke (1978) are in agreement with the hypothesis that bumblebees (*B. flavifrons* and *B. appositus*) when departing from an inflorescence tend to maintain a similar direction to that followed when arriving. Perhaps the bees have a memory of the direction taken during the last move (Pyke, 1978) by orientating to polarized light or landmarks. Preliminary observations are in agreement with the hypothesis that the bees do not move out of an area, or through it, simply on the basis of turning angles and interflower flight distance alone. When a barrier was placed in front of them, they moved *around* it, rather than visiting nearby flowers (Heinrich, unpub. obs.)

At least part of the reason why bees turn more on interflower moves after visiting highly rewarding flowers could be that they walk and turn more on these flowers after landing. Presumably the longer they forage and turn on a given (rich) inflorescence, the more likely it is that they will be facing in a different direction when they are ready to depart. The different possible behavioral mechanisms still remain to be elucidated.

The decision of whether or not to visit more nearby flowers or to move on and search for flowers in another patch should be similar whether the "patch" consists of a group of plants, many flowers on a single plant, or a multiflowered inflorescence. By staying to probe the nearest flower or floret, however, the forager might soon start reprobing already emptied florets.

The revisitation of already emptied flowers can be reduced by systematic foraging in one direction. While foraging from globular inflorescences, such as those of clover, for example, a bee may consistently turn either right or left, visiting the florets in a circular sequence as they are encountered. Nevertheless, this potential response, if present, may still not be the most efficient way to accommodate a three-dimensional flower arrangement.

It has been suggested that the behavior of bees to almost invariably move upward on successive flower visits on vertical (hence two-dimensional) inflorescences is a preprogrammed behavior that is adaptive because it results in systematic foraging that minimizes repeat visits to already visited flowers (Heinrich, 1975c, 1979b). However, since vertical inflorescences commonly provide more nectar in the bottom than in the top flowers, it is also possible to make another optimality argument that would account for the same behavior. Pyke (1978), for example, maintains that moving upward is adaptive because the bees first visit the most rewarding flowers, leaving the inflorescence at upper flowers when food rewards fall below a threshold.

A recent experimental laboratory study was designed to examine the two (not necessarily exclusive) hypotheses. In this study, using artifical inflorescences, where the amount of nectar in the different flowers could be strictly controlled, bumblebees *(B. edwardsii)* that had no previous foraging experience at inflorescences always moved upward while visiting con-

secutive flowers on artificial "inflorescences," regardless of whether the rewards were greatest in the bottommost or topmost flowers (Waddington and Heinrich, 1979). However, when rewards were made to be greatest in the bottommost flowers, the bees learned to start lower and to depart before reaching the topmost (empty) flowers. Conversely, when rewards were always greatest in the top of the inflorescence, the bees learned to start in the middle and to depart from the top. These patterns of interflower movement on vertical inflorescences tend to maximize visits to rewarding flowers while minimizing visits to nonrewarding flowers.

In a "featureless" environment, it is probable that the foragers would not be able to maintain specific traplines. Foraging paths would wander, producing a good deal of overlap in flower visitation, at least if the individuals foraged independently of one another. As shown in flock foraging of birds, however, one individual may orient on another, and the group moves forward as an organized "wave," staying in rewarding areas and bypassing picked-over ones (Cody, 1974). The same principle may apply to male *Centris* bees that forage in groups, passing as waves over tightly clumped floral resources, such as the crowns of large tropical trees (Frankie, et al., 1976). Almost no quantitative data are available to evaluate this potential foraging mechanism and its biological significance.

The avoidance of already visited flowers may, at least in some instances, be due to flower-marking by bees. Frankie and Vinson (1977) observed that female *Xylocopa virginiana* avoided flowers of *Passiflora incarnata* that were treated with extracts of female DuFour's glands. The DuFour's gland pheromones are effective in repelling bees for about 10 minutes. The female *Xylocopa*, unlike *Apis* and *Bombus* workers, are territorial. Possibly by excluding other foragers these bees can increase their foraging efficiency by restricting visits to filled flowers, or to those given time enough to recharge some of their nectar.

Bumblebee workers in the laboratory may also mark flowers they visit with scent (Cameron, 1979). The chemical mark appears to *stimulate* probing by other bees rather than discourage it. The significance of this behavior is not clear. One possibility is that these social nonterritorial bees are stimulated to attempt foraging where experienced individuals have already been successful.

The above interpretation is by no means universally applicable to bumblebee foraging as it is observed in the field. For example, in a patch of white clover *(Trifolium repens)* that had low standing crops of nectar and pollen owing to the previous foraging activity of numerous other bumblebees, *B. terricola* workers consistently veered away from many inflorescences after approaching and almost touching them. Up to 31% of approached flower heads were rejected in dense (200–290 heads/m²) stands. The rejection rate was slightly lower (21%) in sparse (20–60 heads/m²) stands. However, none of the heads were rejected in a clover patch that had received no bumblebee foragers for two days.

The basis on which the bumblebees discriminated visited from unvisited heads is unknown. However, since humans could discriminate nectar-rich from nectar-poor heads of white clover on the basis of the intensity of the flower scent, it is possible that the bees could also use the same cue. On the other hand, other high-nectar flowers such as turtlehead *(Chelone glabra)* and monkshood *(Aconitum napellus)* are not rejected by bumblebees, and the same flowers may be visited in succession, within several seconds, by different bees.

As in a larger flower patch, bees allocate their foraging effort on an inflorescence with many florets in relation to the food rewards provided. *Bombus terricola* workers visited an average of only 2.26 florets per clover inflorescence in nectar-poor areas. On the other hand, in areas where the nectar had been allowed to accumulate by excluding bees for two days, they probed 11.6 florets per each inflorescence visited (Heinrich, 1979b). These data suggest that there is (in flower patches) a positive feedback between the amount of food provided and the pollination service received. However, the relationship is not a simple one. If a bee stayed too long at one plant with many flowers, or at one inflorescence, it would eventually spread more of the plant's (or flower's) own pollen on itself, tending to decrease rather than increase cross-pollination service. More data are needed to clarify the relationship between the conflicting demands for pollination.

The problem of whether to stay or to leave extends also to the individual flower. The bees can stay at each flower they probe until they have picked up every last scrap of nectar and pollen, or they can stay only long enough to pick up what is most easily and quickly taken. It may require considerable time and effort to pick up the "dregs." The problem was investigated by Witham (1977) in bumblebees *(B. sonorus)* foraging from desert willow *(Chilopsis)*.

The *Chilopsis* flowers offer only nectar to its large bee pollinators. Approximately 8 μl nectar can accumulate in a pool at the base of the corolla tube. The five grooves that radiate up from this pool can contain an additional total of 1.1 μl. Bumblebees can remove the pooled nectar at 2.0 μl/sec, but they extract the nectar from the grooves at only 0.3 μl/sec.

In order to take the most nectar per unit time the bees should specialize to take the pool nectar, leaving the "dregs." However, when nectar abundance declines, they have nothing to lose by staying and collecting the groove nectar (Witham, 1977).

The nectar-dispensing pattern of *Chilopsis* allows these flowers to be visited at least twice, first by high-energy-demanding bees, and second by food-stressed bumblebees or any of a number of small, energetically less demanding, solitary bees. The plant is, in effect, increasing the number of pollinator visits per given nectar reward by partitioning the reward.

UNOPTIMAL FORAGING?

In a world of all possibilities the bees are unoptimal foragers. But optimality is always determined within a set of constraints. The bees, or other pollinators, are only "optimal" in the sense that we arbitrarily specify what it is that they "should" be able to do or not do; we set limits on their performance and then say they are optimal because they live up to these often arbitrarily low standards that we perceive them to have.

Having thus far played the devil's advocate, I do not, however, propose that we throw the baby out with the bathwater. Nevertheless, I propose that we could learn as much or more about foraging behavior by not making prior assumptions about what should be optimal. Presumably the overall life strategy is, by definition (according to the laws of natural selection) optimal. But it is optimal only in terms of a balance of compromises that sometimes have to be made in the face of conflicting selective pressures. Why not recognize the magnitude of these compromises and point them out quantitatively? This could be done by showing how much the actual behavior deviates from the predicted, if no compromises were made at all. Such calculations also would not be necessary without arbitrary considerations, but at least "optimality" would be falsifiable, and meaningful cross-taxonomic comparisons could be made between different life strategies. Some animals may be able to *forage* more optimally than others because they have fewer conflicting constraints.

The sterile workers of social bees do not "need" to make at least two compromises that some foragers of some other species may have to make. They do not seek mates, and they do not concern themselves with oviposition. However, female butterflies and moths must concern themselves with both, and possibly also with predator-avoidance behavior. They could potentially accomplish at least foraging, mate seeking, and search for oviposition sites in a serial manner, and thus be highly "optimal" in each in turn, or they could engage in all three simultaneously, being "unoptimal" in each because of the necessary compromises that doing all three at the same time entails. Ultimately, the question is whether or why one or the other strategy (which in each case is a part in a different mosaic of at least three strategies) is the best in enhancing reproductive output.

A recent study (Schmidt, 1980) suggests that *Colias* butterflies foraging from a *Senecio* employ the mixed strategy of compromises rather than the serial strategy with optimal foraging. Unlike bumblebees, *Colias* butterflies do not visit nearest flowers on successive flower visits (Schmidt, 1980). Schmidt suggests that this apparently unoptimal foraging

behavior is adaptive, since the butterflies, while ranging widely, may increase their chances of finding both mate and oviposition sites. Unfortunately, however, no data are available on their "simultaneous" success in these two other potential activities that could occur during foraging.

The individual *Colias* butterflies collect food resources only for their own use. If their food requirements can be met incidentally during other pursuits, then the selective pressure to enhance foraging returns relative to other selective pressures could have been relatively weak. Relatively few data are available on "unoptimal" foraging, but it is probable that such behavior could play a significant role in pollination service of plants.

Perhaps one of the clearest examples of unoptimal foraging is that of insects visiting flowers for food rewards even though they offer no food rewards. The behavior is common enough for some plants to rely on it totally for their pollination service.

Knowledge of the fallibility of insect foragers dates back to Sprengel (1793), who discovered that some species of *Orchis* lack nectar secretion and are pollinated by deceiving their pollinators. Seventy-one years later, Charles Darwin (1878) rejected this idea, saying "we can hardly believe in so gigantic an imposture." We know now, however, that the majority of *Orchis* species act by deceit, and one of the mechanisms of deceit is food source mimicry where flowers of a nonrewarding plant mimic those that are rewarding (Pijl and Dodson, 1966).

There are two potential advantages of mimicry that are not mutually exclusive. One, summarized by Wiens (1978), involves the energetic advantage to the plant of reducing the nectar reward: "There is a clear energetic benefit to the plant mimic since no nutritive rewards whatever are given by the operator. It seems difficult to believe that a few microliters of nectar or a few thousand pollen grains would be critical in the overall energy budget of the individual. Nevertheless, this seems to be the best explanation." In order to gain better insights, it may be necessary in the future to look more closely at the role that the production of food rewards to feed pollinators plays in the plant's energy budget, and how these rewards relate to seed production.

At a given amount of reward, rare plants might "need" to mimic a common model in order to be pollinated. However, if the mimicry is effective, and if the model is common enough or provides enough nectar, then the mimic might get pollinated without producing any reward. Almost no data are available on foraging behavior of pollinators in the field relative to their responses to reward amounts and sequences of model and mimics encountered in order to gain insight into the conditions where mimicry might succeed or fail.

Pollinators visit (and pollinate) flowers on the basis of the *average* reward they get from a group of individuals of similar appearance. Therefore, if reduction in nectar production is possible interspecifically in individual plants whose flowers mimic the flowers of other species, then it should be even easier intraspecifically. This might be expected most commonly in plants gowing under crowded conditions, where individuals are not identified by the pollinators.

If individual flowers are not identified, they are visited, and pollinated, regardless of their nectar content. For example, in one experiment, a few muskmelons, *Cucumis melo* (Cucurbitaceae), of a nectarless strain were mixed with normal plants and then made available to honeybees. Both nectar-bearing and nectarless flowers were visited and had good fruit set (Bohn and Mann, 1960). However, when more nectarless than normal flowers were together, the bee visits dropped off, and fruit set was erratic.

In a resource-limited wild population of plants, the individual plants, having flowers with little nectar, should be at an advantage, provided they get pollinated because their neighbors produce nectar. With sustained selection for low nectar reward, the high-energy pollinators should eventually be excluded from the population, visiting competing plant species instead. If the nectar "cheaters" continue to be able to put more calories into seed production than their nectar-producing competitors, then the plants would eventually be without even the low-energy pollinators. They could then avoid extinction by either becoming selfers, adopting some other agent besides insects for pollen transfer, or mimicry (if they are rare). However, if they are common (so that mimicry is precluded), they would, by being in crowds in open places, already be preadapted for wind pollination. As discussed further elsewhere (Heinrich, unpub. ms.), cheating in nectar production is one hypothesis for the evolution of wind pollination. This strategy may, however, ultimately be at least as costly to the plants in terms of resources (Colin and Jones, 1980) as nectar production. At the present time, however, there are few data available to evaluate the above hypothesis on the evolution of wind pollination based on pollination energetics.

SUMMARY AND CONCLUSIONS

Relatively accurate predictions about the foraging behavior of insect pollinators can be made on the basis of the net food rewards that are available from flowers. In this chapter I examine caloric costs and payoffs of different potential foraging strategies, as well as some of the methods of study. These indicate, for example, how much nectar some flowers must "guard" or retain, to be pollinated by insects of given size and energy expenditure. They also show the great potential advantage of "sampling" by individuals of generalist species that are otherwise behaviorally specialized.

Patterns of movement between flowers and plants determine, in part, whether or not individual plants receive direct feedback between food rewards provided and cross-pollination service. The forager's limitations in identifying the most rewarding flowers can affect the plant's options for reducing energy expenditure for pollination service by mimicry (when rare) and possibly by wind pollination (when abundant).

Much needs to be known about the energetics of nectar production in plants, the variability of nectar production between plants, and the correlation of seed production to nectar and pollen production in order accurately to assess the plant's options. In particular, more attention will need to be paid to the individual plant in its success in pollination relative to the forager's behavior. The "unoptimal" foraging behavior of pollinators that visit empty flowers could, if examined in more detail, shed light on some important aspects of floral biology.

LITERATURE CITED

Allen, T. 1981. Ph.D. dissertation, University of California, Berkeley.

Baker, H. G. 1977. Non-sugar chemical constituents of nectar. *Apidologie* 8:349–356.

Bartholomew, G. A. and T. M. Casey. 1978. Oxygen consumption of moths during rest, preflight warm-up and flight in relation to body size and wing morphology. *J. Exp. Biol.* 76:11–25.

Beenakkers, A. M. T. 1969. Carbohydrate and fat as a fuel for insect flight. A comparative study. *J. Ins. Phys.* 14:353–361.

Beutler, R. 1951. Time and distance in the life of the foraging bee. *Bee World* 32:25–27.

Bohn, G. W. and K. K. Mann. 1960. Nectarless, a yield-reducing mutant character in muskmelon. *Proc. Am. Soc. Hort. Sci.* 76:455–459.

Brian, A. D. 1951. The pollen collected by bumblebees. *J. Anim. Ecol.* 20:191–194.

Cameron, S. A. 1979. To probe or not to probe; some signals important in the utilization of flowers by bumblebees. M. S. thesis, Division of Entomology and Parasitology, University of California, Berkeley.

Cody, M. L. 1974. Optimization in ecology. *Science* 183:1156–1163.

Colin, L. J. and C. E. Jones. 1980. Pollen energetics and pollination modes. *Am. J. Bot.* 67:210–215.

Darwin, C. 1878. *The Effects of Cross and Self Fertilization in the Vegetable Kingdom.* John Murray, London.

Eickwort, G. C. 1973. Biology of the European mason bee, *Hoplitis anthocopoides* (Hymenoptera: Megachilidae) in New York State. *Search: Agriculture* 3:1–32.

———. 1977. Male territorial behavior in the mason bee, *Hoplitis anthocopoides* (Hymenoptera: Megachilidae). *Anim. Behav.* 25:542–554.

Frankie, G. W. 1973. A simple field technique for marking bees with fluorescent powders. *Ann. Ent. Soc. Am.* 66:690–691.

Frankie, G. W. and S. B. Vinson. 1977. Scent marking of passion flowers in Texas by females of *Xylocopa virginica texana. J. Kansas Ent. Soc.* 50:613–625.

Frankie, G. W., P. A. Opler, and K. S. Bawa. 1976. Foraging behavior of solitary bees: implications for outcrossing of a neotropical forest species. *J. Ecol.* 64:1049–1057.

Frisch, K. von. 1967. *The Dance Language and Orientation of Bees.* Harvard University Press, Cambridge, Mass.

Gary, N. E. 1971. Magnetic retrieval of ferrous labels in a capture-recapture system for honeybees and other insects. *J. Econ. Ent.* 64:961–965.

Heinrich, B. 1972. Energetics of temperature regulation and foraging in a bumblebee, *Bombus terricola* Kirby. *J. Comp. Physiol.* 77:49–64.

———. 1975a. Bee flowers: a hypothesis on flower variety and blooming times. *Evolution* 29:325–334.

———. 1975b. The role of energetics in bumblebee-flower interrelationships. In L. E. Gilbert and P. H. Raven (eds.), *Coevolution of Animals and Plants.* University of Texas Press, Austin.

———. 1975c. Energetics of pollination. *Ann. Rev. Ecol. Syst.* 6:139–170.

———. 1975d. Thermoregulation in bumblebees. II. Energetics of warmup and free flight. *J. Comp. Physiol.* 96:155–166.

———. 1976a. Resource partitioning among some eusocial insects: bumblebees. *Ecology* 57:874–889.

———. 1976b. The foraging specializations of individual bumblebees. *Ecol. Mon.* 46:105–128.

———. 1976c. Flowering phenologies: Bog, woodland, and disturbed habitats. *Ecology* 57:890–899.

———. 1979a. "Majoring" and "minoring" by foraging bumblebees, *Bombus vagans:* an experimental analysis. *Ecology* 60:245–255.

———. 1979b. Resource heterogeneity and patterns of movement in foraging bumblebees. *Oecologia (Berl.)* 140:235–245.

———. 1980. Mechanisms of body-temperature regulation in honeybees, *Apis mellifera. J. Exp. Biol.* 85:73–87.

Heinrich, B., P. R. Mudge, and P. G. Deringis. 1977. Laboratory analysis of flower constancy in foraging bumblebees: *Bombus ternarius* and *B. terricola. Behav. Ecol. Sociobiol.* 2:247–265.

Heinrich, B. and P. H. Raven. 1972. Energetics and pollination ecology. *Science* 176:597–602.

Horn, H. S. 1979. Adaptation from the perspective of optimality. In O. T. Solbrig, S. Jain, G. B. Johnson, and P. H. Raven (eds.), *Topics in Plant Population Ecology.* Columbia University Press, New York, pp. 48–61.

Inouye, D. W. 1980. The effects of proboscis and corolla tube lengths on patterns and rates of flower visitation by bumblebees. *Oecologia (Berl.)* 45:197–201.

Janzen, D. H. 1971. Euglossine bees as long-distance pollinators of tropical plants. *Science* 171:203–205.

Kammer, A. E. and B. Heinrich. 1974. Metabolic rates related to muscle activity in bumblebees. *J. Exp. Biol.* 61:219–227.

Kleiber, M. 1961. *The Fire of Life: An Introduction to Animal Energetics.* John Wiley & Sons, New York. 454 pp.

Laverty, L. M. 1980. The flower-visiting behavior of bumblebees: floral complexity and learning. *Can. J. Zool.* 58:1324–1335.

Lecompte, J. 1964. Donneés récentes sur la biologie des insectes sociaux acquises par l'utilisation des radioisotopes. *Rev. Zool. Agric.* 63:1–16.

Levin, M. D. 1960. A comparison of two methods of

mass-marking foraging honeybees. *J. Econ. Ent.* 53:696–698.

Linsley, E. G. and J. W. MacSwain. 1958. The significance of floral constancy among bees of the genus *Diadasia* (Hymenoptera, Anthophoridae). *Evolution* 12:219–223.

Loh, W. and H. Heran. 1970. Wie gut können Bienen Saccharose, Glucose, Fructose and Sorbit im Flugstoffwechsel ververten? *Zeit Vergl. Physiol.* 67:436–452.

MacArthur, R. H. 1972. *Geographical Ecology.* Harper & Row, New York.

MacArthur, R. H. and E. R. Pianka. 1966. On the optimal use of a patchy habitat. *Am. Nat.* 100:603–609.

MacSwain, J. W., P. H. Raven, and R. W. Thorpe. 1973. Comparative behavior of bees and Onagraceae IV. *Clarkia* bees of the Western United States. *Univ. Calif. Publ. Ent.* 70:1–80.

Müller, H. 1881. *Die Alpenblumen, ihre Befruchtung durch Insekten und ihre Anpassung an dieselben.* Leipizig.

Musgrave, A. J. 1949. The use of fluorescent material for marking and detecting insects. *Can. Entolmol.* 81:173.

Oster, G. and B. Heinrich. 1976. Why do bumblebees major? A mathematical model. *Ecol. Mon.* 46:129–133.

Percival, M. S. 1961. Types of nectar in angiosperms. *New Phytol.* 60:235–281.

Pijl, L. van der and C. H. Dodson. 1966. *Orchid flowers, their Pollination and Evolution,* Coral Gables, Fla.

Pyke, G. H. 1978. Optimal foraging movement patterns of bumblebees between inflorescences. *Theor. Pop. Biol.* 13:72–98.

Pyke, G. H., H. R. Puliam, and E. L. Charnov. 1977. Optimal foraging theory: a selective review of theory and tests. *Q. Rev. Biol.* 52:137–154.

Schaffer, W. M. and M. V. Schaffer. 1979. The adaptive signficance of variations in reproductive habit in the Agavaceae II. Pollinator foraging behavior and selection for increased reproductive expenditure. *Ecology* 60: 1051–1069.

Schmidt, J. 1980. Pollinator foraging behavior and gene dispersal in *Senecio* (Compositae). *Evolution* 34:934–943.

Smith, M. V. 1972. Marking bees and queens. *Bee World* 53:9–13.

Smith, M. V. and G. F. Townsend. 1951. A technique for mass-marking honeybees. *Can. Entolmol.* 83:346–348.

Sprengel, C. K. 1793. *Das entdeckte Geheimnis der Natur im Bau und in der Befruchtung der Blumen.* Berlin.

Strickler, K. 1979. Specialization and foraging efficiency of solitary bees. *Ecology* 60:998–1009.

Waddington, K. D. and B. Heinrich. 1979. The foraging movements of bumblebees on vertical "inflorescences": an experimental analysis. *J. Comp. Physiol.* 134:113–117.

Waddington, K. D., T. Allen, and B. Heinrich. 1981. Floral preferences of bumblebees *(Bombus edwardsii)* in relation to intermittent versus continuous rewards. *Anim. Behav.* 29:779–784.

Weaver, N. 1957. The foraging behavior of honeybees on hairy vetch. I. Foraging methods and learning to forage. *Insectes Sociaux* 3:537–549.

———. 1965. The foraging behavior of honeybees on hairy vetch. III. Differences in the vetch. *Insectes Sociaux* 12:321–326.

Wiens, D. 1978. Mimicry in plants. In M. K. Hecht, W. C. Steere, and B. Wallace (eds.), *Evolutionary Biology,* Vol. 11. Plenum Press, New York, pp. 365–403.

Witham, T. G. 1977. Coevolution of foraging in *Bombus* and nectar dispensing in *Chilopsis:* a last dreg theory. *Science* 197:593–595.

Wykes, G. R. 1953. An investigation of the sugars present in the nectar of flowers of various species. *New Phytol.* 51:210–215.

10
POLLINATION ENERGETIC AVIAN COMMUNITIES: SIMPLE CONCEPTS AND COMPLEX REALITIES

F. Lynn Carpenter

Department of Ecology and Evolutionary Biology
University of California
Irvine, California

ABSTRACT

I test five hypotheses concerning the effects of avian pollinator energetics on the nectar production and on flower and plant densities in several montane species of hummingbird-visited plants in the California Sierra Nevada. The data in general did not support those hypotheses which assumed that visitation rates limit reproductive output in these plants. The results from this study and others suggest that hummingbirds are so numerous in these areas that visitations are not limiting to the plants.

KEY WORDS: Pollination, energetics, avian communities, nectar, plant reproductive strategies, pollination strategies, flower density, elevation.

CONTENTS

INTRODUCTION

The publication of Heinrich and Raven's feature article in *Science* in 1972 was a turning point in the field of pollination energetics because it stimulated workers in the field to pose testable hypotheses in an effort to discover general trends. Heinrich and Raven's (1972) ideas dealt with the energetic strategy a plant species should evolve to obtain the pollination services of a particular kind of pollinator. Pollinators differ in their energetic requirements, from the low-energy groups such as ants and flies to the high-energy endothermic groups, mammals and birds. A plant must provide enough energetic reward to attract its particular pollinator with its characteristic energetics. But if the plant also is selected to outcross, it must limit the energetic reward so that the pollinator is forced to visit more than one plant. Heinrich and Raven (1972) provide some hypotheses about how factors such as plant density, elevation, and latitude should affect selection for the quantity of energy rewards offered by plants, through their direct effects on pollinator energetics. Their premises assume that pollinator visitations limit the quantity or quality of reproductive output in plants. Since 1972 several studies have emerged with results pertinent to Heinrich and Raven's (1972) ideas. This chapter reviews some of the work done on avian communities and presents my own results from a four-year study on hummingbird-plant communities on the east slope of the central California Sierra Nevada.

THE HYPOTHESES TO BE CONSIDERED

Both birds and insects prefer concentrated sugar solutions over dilute solutions (Percival, 1965; Hainsworth and Wolf, 1976; Stiles, 1976), but bird-pollinated flowers characteristically produce less concentrated nectar than do bee-pollinated flowers (Baker, 1975). Providing dilute nectar may be a mechanism that reduces visits by nonpollinating insects, thereby preventing reduction in the attractiveness of the flower to birds (Bolten and Feinsin-

ger, 1978). Bird-flowers, however, produce large volumes of nectar. Attractiveness of a flower overall is related not just to sugar concentration, but to total energy content (milligrams sugar) determined by volume and concentration. Therefore all the hypotheses and comparisons in this chapter will be made on the basis of amount of sugar provided per flower.

Using data from this review and the Sierra Nevada study, I tested the following energetics hypotheses:

1. Attractant Hypothesis. The amount of floral nectar provided by plants to attract their pollinators correlates positively with the energetic demands of those pollinators (Faegri and van der Pijl, 1966 and many others). Low-energy ectothermic animals, such as ants and flies, pollinate flowers that provide tiny amounts of nectar (Hickman, 1974). I was interested in whether correlations could be found within the high-energy endothermic group of pollinators. Plants pollinated by bees, hummingbirds, and passerine birds should provide amounts of nectar that increase in that order. The order is determined by increasing body size, which brings with it increased total daily energy requirements. Although hovering in hummingbirds is an expensive method of foraging, the larger body size of passerine nectar-feeders results in larger 24-hr energy requirements. The passerine nectar-feeders include Australian honeycreepers, African sunbirds, Hawaiian honeycreepers, and New World passerine pollinators such as tanagers. These birds generally weigh from 10 to 100 grams compared to the usual 2 to 8 grams for hummingbirds. As a result their daily energy requirements range from 10 to 50 kcal/day (MacMillen and Carpenter, 1977; Ford et al., 1979), compared to 6 to 10 kcal/day for hummingbirds (MacMillen and Carpenter, 1977). Therefore, passerine-visited flowers must contain more nectar than hummingbird-visited flowers in order to attract their large pollinators in the numbers and frequencies required for adequate pollination.

2. Cornucopia Hypothesis. Because many plants are not pollinated by a single kind of pollinator, the above hypothesis may be too simplistic (Heinrich, 1977). I expected that within the Sierran community of humming-bird-visited plants, plant species would range from specialists dependent on birds for pollination to generalists that use both birds and insects; I predicted that generalists would be "cornucopia" species, providing sufficient nectar for both pollinator groups, whereas bird-specialized plants would provide only enough for birds. For example, a cornucopia species will remain attractive to birds later in the day even after insects have removed nectar for their requirements. On the other hand, insects may be prevented from visiting specialist flowers by means of morphological features such as long slender corollas, so that specialist flowers need produce less nectar to remain attractive to birds. Whether a plant should evolve to be a pollinator specialist or generalist is a separate question, probably related to the dependability of and constancy of visitation by a particular pollinator to that plant species. I deal here only with the energetics accompanying the strategy once it has evolved. My prediction is related to Heinrich's (1975) observation that weedy, generalized plants that outcross have copious nectar that attracts many pollinators.

3. Nectar-Timing Hypothesis. Observational support for hypotheses 1 and 2 does not necessarily mean that a characteristic nectar reward has evolved in a plant species in response to selection to procure a specific pollinator (hypothesis 1) or to encourage visitation by a diversity of pollinators (hypothesis 2). Instead, physical factors characteristic of the habitat could determine the nectar quantity produced, and the kind of pollinators attracted could then follow secondarily. To see whether pollinator characteristics drive the evolution of pollination energetics, we must ask whether nectar energetics are fine-tuned to pollinator behavior. One such hypothesis is that plants should provide their nectar at the time of day when their pollinators are active. Plants that do not, may lose their nectar to nonpollinating visitors (Smith and Carpenter, 1983), to drying by evaporation, or to dilution and washout by rain, and such waste would be selected against if plants are light- or water-limited. Furthermore, many nectar thieves damage the nectar-

ies or the entire flower (Smith and Carpenter, 1983), and an accumulation of nectar during a time of pollinator inactivity would likely attract such thieves. It is known that plants pollinated by nocturnal hawkmoths, bats, and other mammals often secrete the most nectar at night (e.g., Baker, 1961; Cruden, 1970; Carpenter, 1978a). I predicted that even during daylight hours, plants pollinated by diurnal visitors should be selected to fine-tune secretion. Specifically, many nectar-feeding bird pollinators feed actively in the morning and late afternoon but are relatively inactive in early afternoon (Stiles and Wolf, 1970; Stiles, 1973; Carpenter and MacMillen, 1976; Gass, 1978). This activity pattern is independent of changing patterns in food availability because (1) it occurs at artificial feeders at which food is continuously supplied (Wheeler, 1980); (2) it occurs in natural populations even when natural food availability is greatest in early morning and gradually decreases linearly throughout the day (Gass, 1978; Carpenter, 1979, Figs. 1 and 3); and (3) it occurs probably because the birds are metabolizing quickly and therefore using their food quickly during the cold morning and late afternoon hours, whereas metabolism occurs more slowly during the moderate, warm temperatures of midday (Carpenter and MacMillen, 1976).

Therefore, I predicted that plants specialized for hummingbird pollination should show bimodal nectar secretion patterns. First, secretion should occur predawn or in the early morning so that a large standing crop is available during the morning foraging peak. Second, secretion should slow or stop during midday to prevent the accumulation of large standing crops during a time when many energy-demanding nectar thieves are active. Third, secretion could resume high rates in the middle or late afternoon to increase a plant's attractiveness to birds during their second foraging peak. In contrast, generalized plants pollinated by day-active insects as well as by birds should provide nectar in a way that maintains their standing crops attractive all day. A plant generalist using birds and nocturnal hawkmoths should show nocturnal nectar production and bimodal diurnal production.

4. Flower Density Hypothesis. To be energetically attractive, plant species characterized by high flower density need provide less nectar per flower than species with dispersed flowers because their pollinators expend less foraging energy searching for the next flower and traveling from flower to flower (Heinrich and Raven, 1972). Plants can, in fact, control their density by adopting different seed dispersal strategies, allelopathy, etc.

5. Elevation Hypothesis. To be attractive, plants pollinated by endotherms should produce more energy or be closer together at higher elevations than at lower elevations because temperature generally decreases about 6°C per 1000 meters elevation, causing endothermic energy requirements to increase (Heinrich and Raven, 1972). An elevational difference of 1000 meters theoretically should raise metabolic requirements about 30% in pollinating birds (Calder, 1974). As it is stated by Heinrich and Raven (1972), this hypothesis is simplistic, because local topography and vegetation can affect climate. Some high elevation sites conceivably could be warmer than low elevation sites.

Do we find in fact that the nectar energetics of plants are driven by the energetics of their pollinators? Using data from this review and the Sierran study, I tested the five hypotheses above, and the results conflict. I will deal with the hypotheses, then discuss variability in nectar measurements, and close with a challenge to a basic assumption of the energetics hypotheses. First I will present techniques in some detail because methods must be standardized to reduce variability.

MATERIALS AND METHODS OF THE SIERRAN STUDY

We studied hummingbird-visited plants along an elevational gradient from 1700 to 2800 meters in Mono and Inyo counties, California (Table 10-1). Different species occurred along the gradient; Table 10-1 lists the principal plants studied, all of which are perennials, although *Ipomopsis aggregata* has been reported as biennial (Munz and Keck, 1970). Nectar production in each population of all species except *Castilleja* spp. was measured usually by emptying 5 to 15 flowers on each of 3 to 6 plants at dusk ($N = 30$ to 100 flowers, or spurs for *A. formosa*). We bagged the flowers with fine mesh nylon, and removed the nectar from each flower sequentially at dawn (overnight

Table 10-1. Principal plants in the Sierran study.

Site	Years studied	Habitat	Plant species	Genetic self-compatibility[1]
1700 m	76, 77	Stream bank	*Aquilegia formosa* (Ranunculaceae)	PI
	76, 77, 78, 79	Sagebrush scrub	*Castilleja linariaefolia* (Scrophulariaceae)	I
1850 m	77	Stream side	*Penstemon bridgesii* (Scrophulariaceae)	PI
	78, 79	Rocky hillside	*P. bridgesii*	PI
2300 m	78, 79	Sagebrush scrub on dry hillside	*C. chromosa*	I
2450 m	76, 77, 78, 79	Wet meadow	*C. miniata*	I
	77, 78, 79		*A. formosa*	PI
	76, 78, 79	Forested slope	*Ipomopsis aggregata* (Polemoniaceae)	I
	77, 78	Rocky hillside	*P. bridgesii*	PI
2700 m	77, 78, 79	Rocky slope	*P. bridgesii*	PI
	78		*C. linariaefolia*	I
2800 m	77, 78, 79	Alpine scrub	*C. linariaefolia*	I
	77, 78, 79	Disturbed roadside	*I. aggregata*	I

[1]PI = mechanically self-incompatible and partly genetically self-incompatible; capsule set increases with outcrossing. I = genetically self-incompatible; no capsules result from self-pollination.

production) and at 0930, 1300, 1630, and 2000 hr. We measured the volume of each sample and then measured its sucrose-equivalent concentration (Hainsworth and Wolf, 1972a) with a Bausch and Lomb sugar refractometer. We recorded two categories of flower age, newly opened flowers in anthesis, and mature flowers in stigmatal receptivity. Considering male and female stage nectar production separately does not change my conclusions, so for the purpose of this report the data from both stages are combined.

To avoid damaging nectaries with our repeated sampling, we used a 1 cm length of .61 mm O.D. Intramedic polyethylene tubing glued with Duro superglue to the inside of the tip of a 10 μl glass Wiretrol syringe. The thin flexible tubing is much more gentle on flowers than the glass capillary tubes ordinarily used in nectar sampling, and the Wiretrol has the advantage of syringe action for extraction of difficult-to-remove samples such as the tiny quantities of nectar which are highly concentrated. I express all results in terms of milligrams sugar produced per flower per hour, by multiplying microliters by percent sucrose and by a factor to correct for the fact that my refractometers read in percent by weight rather than percent by volume (Bolten et al., 1979). The correction factor used was derived by D. C. Paton: actual % = 0.820 (read %)$^{1.102}$. Some error is caused by nectar constituents other than sucrose, fructose, and glucose, which contribute to the refractometer reading (Inouye et al., 1980). Nonsugar constituents were not measured or accounted for in my study or in any other studies reviewed herein. Fortunately, such constituents are low or absent from nectar of most hummingbird-pollinated flowers, so that the error is only about 9% (Inouye et al., 1980).

Castilleja flowers are constricted just above the nectary in a way that prevents repeated sampling without destroying the flower. For these species we picked and sampled flowers at dusk to obtain mean standing crop and then bagged comparable flowers. At the end of 24 hr, we measured accumulated nectar, and from this mean I subtracted the mean standing crop from the evening before to give subse-

quent production. The procedure deletes all degrees of freedom so that variation around the mean is not available for *Castilleja* spp. A second difficulty is that some species reduce nectar production or reabsorb nectar when it accumulates in flowers that are not visited (Corbet, 1978). Thus, the amount of nectar obtained from flowers bagged for 24 hr may be less than the total secreted over a comparable period when it is periodically extracted, by visitors or ecologists. To check for this in *Castilleja* spp. on a large sample of flowers, at dawn we picked flowers bagged the previous evening that had accumulated nectar overnight, measured standing crop in unbagged flowers, then bagged a comparable group of flowers presumably containing the same initial mean standing crop. Subtracting mean standing crop at 2000 hr the night before from mean accumulated nectar at dawn yielded overnight production. At 0930 hr we picked the flowers we had bagged at dawn and subtracted mean standing crop at dawn from the amount at 0930 hr to give production over the early morning period. We continued this procedure every 3.5 hr through the day. In this way, we allowed flower visitors to keep the standing crops low before we bagged flowers, and did not allow nectar to accumulate above those standing crop levels for more than 3.5 hr (except for the overnight sample—10 hr). Adding the increments of production over the entire 24-hr period then gave a value that could be compared to the amount accumulated in bags undisturbed for 24 hr to see if the former values are in fact consistently greater. This technique was used also to test the nectar timing hypothesis in *Castilleja* spp.

Most species contain one nectary per flower. *Aquilegia* flowers are divided into five nectar-producing spurs and thus contain five nectaries. Therefore, I express all nectar data on a per-nectary basis.

To determine how specialized a species's population was on hummingbirds for pollination, we marked 10 buds on each of 20 plants in the population, left 15 of these plants open to bird and insect visitors, and caged the other 5 with large cages made of 1.2 cm (½ in.) mesh chickenwire, which prevented bird entry. All flowers were at least 10 cm from the nearest edge of the cage. The cages allowed entry of bumblebees and, surprisingly, even hawkmoths. I compared percent of flowers setting seed capsules by insect pollination alone (caged flowers) to that resulting from pollination by both birds and insects (open controls) to give a measure of how dependent a population was on birds for pollination. The pollens involved are sticky, so that wind was unlikely to affect the results. However, this measure should be considered a rough approximation because several factors affect insect visitation to the cages. The cages themselves reduced visitation frequencies by hawkmoths to *I. aggregata* (Carpenter, unpub. data), and seemed to decrease visitation frequencies by bees to *A. formosa* during one of two years (Lejnieks, pers. comm.). For *I. aggregata,* reproductive output (and therefore pollination success) did not differ between flowers inside or outside of the cages in the year that hawkmoths were most important. Similarly, reproductive output is usually the same in *A. formosa* whether inside or outside of the cages. Thus, reduction in insect visitation frequency caused by cages probably was not important in these cases. Additionally, insect visitors to *P. bridgesii* and *Castilleja* spp. were mostly small, solitary bees whose behavior seemed unaffected by the cages. However, visitation rates also may be affected by the degree of availability of alternative resources for the insects in the vicinity and by the accumulation of nectar in caged flowers, a factor that raised bumblebee visitation to caged *A. formosa* one year (Carpenter, 1979). Thus, several factors act to either underestimate or overestimate the importance of insects in pollination of caged flowers, and the capsule data should be considered only a rough estimate. Capsule-set data were supplemented with data on visitation rates by different pollinator categories (Carpenter, 1979, Table 1 and Fig. 3) and analysis of pollen loads on different pollinators (Carpenter, Lejnieks, and Paton, unpub. data) to give an overall judgment on how dependent a plant was on birds for pollination.

Plant density in each population was measured by counting the flowers within a radius arbitrarily determined by the distance to the fifth nearest neighbor plant around each of 20 plants in the population to give an average value of density that could be compared with values from other populations.

RESULTS FROM THE LITERATURE AND FROM THE SIERRAN STUDY

(1) Attractant Hypothesis

Brown et al. (1978) compiled diverse data on the nectar secretion rates of flowers pollinated by a wide variety of animals studied in many geographic localities. They found a significant positive relationship between nectar production and body size of the pollinator (Fig. 10-1). Considerable variability occurs in the data for large insects, which would include the endothermic bees and hawkmoths, and the authors did not differentiate between passerine and hummingbird data. To be meaningful, differences in nectar production should occur between insect-, hummingbird-, and passerine-pollinated flowers growing in the same region because competition between plants for pollinators is a local phenomenon. In contrast, that an insect-flower produces more nectar in one geographic area than does a bird-flower in a different area is not ecologically meaningful. Variation in data compared geographically results because factors such as climate affect the energetic demands of pollinators and therefore, concomitantly, the hypothesized selection for nectar production in their flowers.

Limited data from two studies comparing the nectar production of plants growing in the same locality but pollinated by representatives of different pollinator categories, are summarized in Table 10-2. Snow and Snow (1980) measured nectar in plants in semi-natural habitats of the Andes of Colombia. These localities had temperate zone climates and were not very different from each other except that one was on the east slope and the other on the west slope of the Eastern Andes. Only plants measured over 24 hr are included here. Hummingbird flowers varied tremendously in mean sugar production between (4-fold) and within (20-fold) sites, but did have higher values in

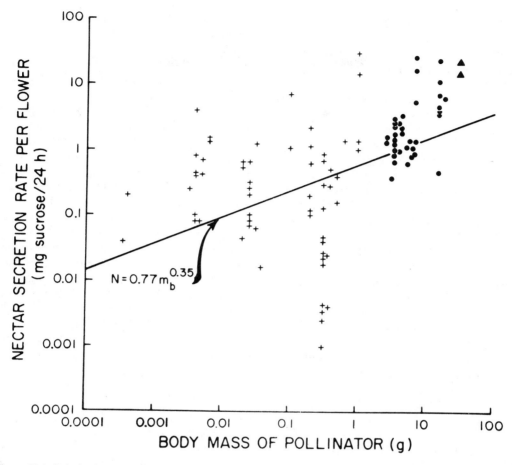

Figure 10-1. Relation between nectar produced over 24 hr in plants, and body mass of their pollinators (from Brown et al., 1978). Relationship is statistically significant but the authors give no *r* value. Circles = birds; triangles = mammals; crosses = insects.

Table 10-2. Nectar production in some bee- and bird-pollinated flowers.

	Mean rate of nectar production (mg sugar per flower per 24 hr)	Range	Number of species or populations sampled
Columbian Andes[1]			
Site 1 2400–2550 m:			
Hummingbird flowers	27.12	3.36–73.44	3 species
Site 2 2300–2500 m:			
Hummingbird flowers	6.24	0.96–17.04	7 species
Bee flowers	0.96	0.72–1.44	3 species
Trinidad and Tobago[2]			
Passerine flowers	634	—	1 species
Hummingbird flowers	80	—	1 species
Sierra Nevada, California[3]			
Hummingbird flowers	1.61	0.22–4.90	10 pops.

[1]Snow and Snow, 1980.
[2]Feinsinger et al., 1979.
[3]Present study, 1978 data only.

general than did the few bee-pollinated plants measured (Table 10-2). On the tropical islands of Trinidad and Tobago, nectar production in tanager- and hummingbird-pollinated species of *Erythrina* was measured, and the former had almost eight times the nectar production of the latter (Feinsinger et al., 1979). Although the data are scant, they are the only ones available for comparing passerine and hummingbird flowers. All of these values are high for insect- and bird-pollinated flowers compared to Fig. 10-1. In contrast, the hummingbird-flower values from the Sierra Nevada (Table 10-2) fall very close to the regression line in Fig. 10-1.

Although we can say that the attractant hypothesis is supported in a coarse way, enormous variations in nectar output are seen in plants with the same pollinators at the same site as well as at different sites (Table 10-2). The challenge to ecologists in the future will be to try to explain this variation. In addition to testing the hypotheses to be discussed below, future workers should look at the role of nectar robbers in plant energetics. The most prolific producer in the Andean study was a *Passiflora* that was robbed heavily by *Diglossa,* a passerine that pierces flowers to steal their nectar without pollinating them. If a plant does not evolve mechanisms to prevent piercing, then it may have to evolve copious nectar production to compensate its pollinators for nectar loss so as to remain attractive to them. Alternatively, piercers may remove nectar thoroughly from only a portion of the flowers on a plant (Colwell *et al.*, 1974) or may damage the nectaries, either of which actions may effectively remove these flowers from the pollinators' itinerary (Colwell *et al.*, 1974). The plant may compensate for this loss in richness by producing flowers with copious nectar, such that the unvisited or undamaged flowers are exceptionally rich. If a relation is found between high incidence of nectar thievery and copious nectar production, it will remain unclear whether the copious production is a strategy to deal with thievery or whether these plants are those selected by nectar thieves because of their copious production, which occurs for some other reason. It will be difficult to differentiate these two possibilities.

In addition, the apparent differences in nectar production between different geographical localities need to be explained. Climate alone seems insufficient because it does not explain the difference between values from the two sites in the Andes or between Sierran and Andean values, both of the latter being areas with temperate climates. Some of the difference may prove to be caused by the prevalence of large nectar robbers like *Diglossa* in the tropical latitudes of Colombia. Nectar robbers and thieves (*sensu* Inouye, 1980) occur at higher latitudes as well, but are more usually insects with smaller appetites and perhaps lower visitation rates. Energetics alone may be insufficient to explain some differences. For example, passerine-pollinated flowers may produce prolific nectar because, unlike hummingbirds, passerines are not dependent on nectar alone for food but also eat insects. Thus, these birds may not be limited by nectar, and the plants they pollinate must be exceptionally rich to be attractive.

(2) Cornucopia Hypothesis

To my knowledge, no comparative data exist that pertain to the hypothesis that generalists should produce more nectar than hummingbird-specialized flowers, except my own on hummingbird-visited flowers.

The percentage by which birds raised capsule set above that set inside cages is assumed to be a measure of the degree of dependency on birds for pollination. The comparison between hummingbird specialization and mean nectar production for each species's population is illustrated in Fig. 10-2. Hawkmoths were important pollinators for one population of *I. aggregata* in one year, bumblebees pollinated *A. formosa,* and small pollen-collecting bees pollinated *C. linariaefolia* in high elevation sites in some years. Thus, sometimes different populations of the same species differed in their degree of dependency on birds for pollination, and the degree to which a given population depended on birds occasionally changed from year to year. These results are consistent with our observations on visitation.

No positive relationship between degree of generalization and nectar production can be

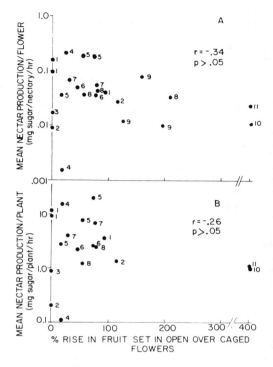

Figure 10-2. Relation between nectar production and dependency of a plant population on hummingbirds for pollination. Abscissa is: (% flowers setting fruit in open treatment) minus (% flowers setting fruit in caged treatment) divided by (% flowers setting fruit in caged treatment), and is assumed to reflect dependency on bird-pollination. A. Ordinate is mean milligrams sugar produced per hour per flower (or per spur for *A. formosa*) averaged over 24 hr. A point is plotted for *C. miniata* in 1979 assuming an abscissal value equal to the mean of the 1977 and 1978 values. Duffield (1972) also experimentally found this species to be highly bird-specialized. B. Sugar production per average plant. Key: 1. *A. formosa* 2450 m; 2. *I. aggregata* 2450 m; 3. *C. linariaefolia* 2700 m; 4. *C. linariaefolia* 2800 m; 5. *P. bridgesii* 2700 m; 6. *P. bridgesii* 2450 m; 7. *P. bridgesii* 1850 m; 8. *C. linariaefolia* 1700 m; 9. *C. miniata* 2450 m; 10. *C. chromosa* 2300 m; 11. *I. aggregata* 2800 m.

detected in Fig. 10-2A because the three most generalized species, *A. formosa, C. linariaefolia* at high elevations, and *I. aggregata* at 2450 meters, produced almost the whole range of nectar values. Generalized species are to the left of an abscissal value of 50%; values to the right are indicative of a statistically significant rise in capsule-set when birds are allowed to pollinate flowers. Two populations, *A. formosa* at 2450 meters and *C. linariaefolia* at 2800 meters, might be considered cornucopia generalists. However, *A. formosa* is robbed (by nonpollinating bumblebees) more than any other species, and its high productivity may have evolved partly to compensate for such loss. Furthermore, the insect pollinator of *C. linariaefolia* at 2800 meters was identified by observation to be a small pollen-collecting megachilid bee (Carpenter, 1978b, Fig. 2); thus, high nectar productivity makes no sense as an attractant strategy for these insects. Nectar values were tremendously reduced in this same population the following year; so these results are ambiguous.

How attractive a given plant is within the population probably depends upon its number of flowers as well as its nectar production per flower; therefore, I calculated total plant attractiveness by multiplying the mean sugar produced per flower by the mean number of flowers per plant (Table 10-3). *Castilleja miniata* was omitted because individual plants cannot be distinguished from intermingling stalks of neighbors. The results in Fig. 10-2B show no difference from the relationship in Fig. 10-2A. Thus, the data did not support the idea that generalized species should be cornucopia nectar producers.

(3) Nectar-Timing Hypothesis

Figures 10-3 through 10-7 show the diurnal patterns of nectar secretion in the Sierran hummingbird flowers. The more specialized species, *Castilleja* and *Penstemon* spp., tended to show bimodal nectar secretion as predicted by this hypothesis. The midday depression of secretion is unlikely to be a physiological response to high ambient temperature because the second spurt of secretion often began between 1300 and 1630 hr, the interval when temperature peaked (see Figs. 10-3 bottom, 10-4 top, and 10-5). This same independence between nectar production and temperature also occurs in a hot-desert hummingbird flower, *Beloperone californica* (Smith and Carpenter, 1983). In some years bimodality in the same population disappeared (Figs. 10-3 top and 10-5 bottom). Because samples were taken at peak flowering that occurred at about the same time each year, the yearly differences

Table 10-3. Number of flowers per plant at peak of flowering.

Species	Site	Year	No. flowers/plant $\bar{x}$	s	N plants	Comparison of mean flowers/plant in successive years 1978/79	1977/78
C. chromosa	2300 m	1979	66.9	23.8	20		
		1978	89.6	33.3	20	1.3	
C. linariaefolia	1700 m[1]	1978	161.6	99.85	20		
		1977	323.6	222.1	13		2.0
	1700 m[2]	1978	339.75	116.0	20		
	2700 m	1978	52.1	16.5	10		
	2800 m	1979	75.9	31.4	20		
		1978	75.5	60.9	20	1.0	
P. bridgesii	1850 m	1979	57.7	39.3	20		
		1978	125.2	54.1	20	2.2	
	2450 m	1978	71.1	34.9	20		
		1977	45.15	31.7	20		0.6
	2700 m	1979	64.75	35.4	20		
		1978	108.9	55.7	12	1.7	
		1977	76.35	56.7	20		0.7
I. aggregata	2450 m	1979	47.6	19.1	19		
		1978	23.1	8.5	19	0.5	
	2800 m	1979	45.0	13.9	20		
A. formosa	2450 m	1979	59.2	25.2	20		
		1978	124.8	39.4	20	2.1	
		1977	89.05	23.3	21		0.7

[1]The principal 1700-meter site (unnamed).
[2]The supplementary 1700-meter site (Wells' meadow).

are not caused by variation in time of sampling.

Measurements of standing crops in open, unbagged flowers showed that the above nectar secretion patterns in conjunction with bird foraging activity usually resulted in a slow decline of nectar available during the day (Carpenter, 1979). A few exceptions occurred, however: in 1978 *A. formosa* and two of three *P. bridgesii* populations exhibited midday accumulations of nectar in unbagged flowers (Carpenter, 1979). The only *P. bridgesii* population in which nectar standing crop declined during the day was also the only one that showed bimodal nectar production that year (Fig. 10-5 bottom). Thus, continual midday production did seem to result in accumulation of large amounts of nectar available in open flowers during midday. These quantities, as suggested above in the section on hypotheses, could act to attract nectar robbers. In fact, they did so in the case of *A. formosa*.

Most populations of *Castilleja* and *Penste-mon* produced little nectar overnight, as indicated by values close to 0 at 0600 hr. Exceptions are: *C. miniata* in one of four years (Fig. 10-3 top), *C. chromosa* (Fig. 10-3 bottom), and *C. linariaefolia* in two of six cases (Fig. 10-4). When nectar is secreted at night, chances are good that it will accumulate and be available to birds at dawn, although hawkmoths conceivably could remove it. Possibly nocturnal nectar is produced only in the hour just before dawn, thereby increasing the chance that it will still be available at dawn. If it is available, then the amount accumulated is the hourly rate graphed at 0600 hr on Figs. 10-3 through 10-7 multiplied by approximately 10 hr, the typical night length. These accumulations will accentuate bimodality of nectar availability to birds. For example, the dawn peak of *C. chromosa* (Fig. 10-3 bottom) will contain about 0.7 mg/flower, greater than the late afternoon peak.

Midday data usually were not obtained from the generalized population of *I. aggre-*

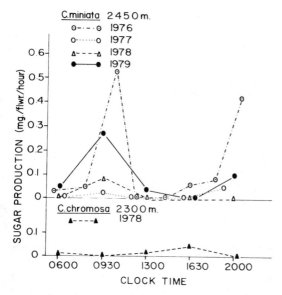

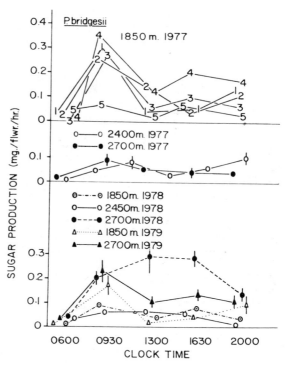

Figure 10-3. Hourly nectar production in milligrams sugar produced per flower in *Castilleja* spp., averaged over 3–3.5 hr sampling intervals during the day and 9.5–10 hr overnight. Values are plotted at the end of the time interval, indicated here and elsewhere in Pacific Daylight Saving Time. Sunrise occurred between 0530 and 0600 hr, sunset between 1930 and 2000 hr. No measure of variability is available because of peculiarities in the sampling technique required for this genus (see methods section). Top: *C. miniata* at 2450 m, production measured in four years. Bottom: *C. chromosa* at 2300 in 1978.

Figure 10-5. Axes same as in Fig. 10-3. *P. bridgesii*. Top: 1977 for 1850 m site. Mean values for each of five plants are plotted separately (solid lines) for 1850 m site to show variation between plants. Middle: 1977 values for 2400 m and 2700 m sites are means of all flowers ± 1 S.E. Bottom: 1978 (three elevations) and 1979 (two elevations). Mean production of all flowers, ± 1 S.E.

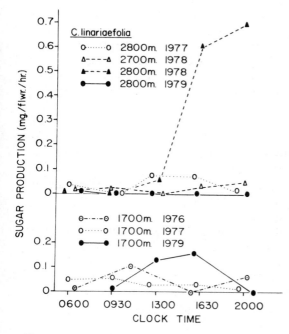

Figure 10-4. Axes same as in Fig. 10-3. *C. linariaefolia* measured in three years at four sites.

gata, because it produced such small amounts that it was difficult to sample effectively after short intervals (note that the ordinate of Fig. 10-6 is an order of magnitude less than that of the other figures). For the year that we collected samples for five intervals (1977 at 2800 meters), the data do suggest a midday depression, which should occur in this species because its insect pollinators are hawkmoths. Hawkmoth activity is either crepuscular or diurnally bimodal in a pattern similar to that of hummingbirds (Carpenter, 1979, Fig. 2). Nocturnal production occurred in three of five cases (Fig. 10-6).

The data for *Aquilegia* suggest that no midday depression of nectar production ever occurs in this generalized species (Fig. 10-7). Midday production would be predicted for *A. formosa,* which attracts day-active bumblebees. In fact, visitation rates by bees to *A. formosa* peak in midday (Carpenter, unpub.

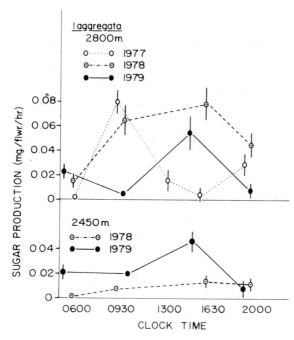

Figure 10-6. Axes same as in Fig. 10-3. *I. aggregata* in 1977 (one elevation), 1978 (two elevations), and 1979 (two elevations). Mean production of all flowers ± 1 S.E.

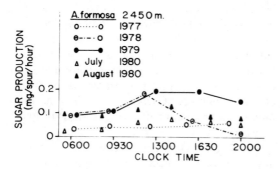

Figure 10-7. Axes same as in Fig. 10-3. *A. formosa* 2450 m. Mean production of all spurs for three of the years of this study; additionally, means are given from measurements taken in July (open circles) and August (closed circles) of 1980. (Lejnieks, Carpenter, and Paton, unpub. data.)

data). Also, substantial nocturnal production occurs in this species, and hawkmoth pollination does occur in years when these insects are abundant (Carpenter, unpub. data).

In summary, the Sierran data tend to support, with some exceptions, the hypothesis that nectar production during the midday period should be conservative in hummingbird-specialized plants because of lowered bird activity, whereas plants pollinated by insects active in midday will produce unimodally.

Several other studies of diurnal patterns of nectar production in bird-pollinated plants have been made. I have selected only those that provide data for a full diurnal period, and involve flowers known to be pollinated by birds. One good study using the accumulation technique (Stiles, 1975) showed morning production but no afternoon peak. However, because large amounts of nectar had accumulated by afternoon, the second peak could be missing because of production shutdown or nectar reabsorption (Corbet, 1978); so unimodality must be questioned. Only repetitive sampling of flowers to prevent nectar accumulation assures detection of both peaks.

In Costa Rica two of four plant species coevolved for hummingbird pollination showed early morning and late afternoon peaks in nectar production, with the largest peak first (Feinsinger, 1976). Confidence limits were not provided; so the significance of the smaller, second peak is uncertain. Many tropical plants have only a morning peak because they drop their flowers in the afternoon, perhaps to deter predation; those with flowers lasting more than one day often do have a late afternoon peak (P. Feinsinger, pers. comm.). A mint, *Leonotis,* pollinated by African sunbirds showed indications of a second peak in production in midafternoon, following a larger early morning peak (Gill and Wolf, 1975). A study of Australian *Eucalyptus* pollinated by honeyeaters showed a slight suggestion of a late afternoon peak in nectar after the first peak in early morning (Bond and Brown, 1979). None of these trends has been analyzed statistically, but together they suggest that bimodal nectar production may occur fairly frequently in plants coevolved for bird-pollination. Bimodality may occur sometimes in insect-pollinated flowers as well (e.g., Frankie and Vinson, 1977; Frankie and Haber, this book). The example given by Frankie (this book) shows the second peak in late morning or early afternoon, and is suggested to occur for reasons other than diurnally varying activity patterns in the pollinators.

In contrast with specialists on birds, *Metrosideros collina* in Hawaii was found to be pollinated by diurnal insects as well as by bimodally active honeycreeper birds, and it produced nectar unimodally (Carpenter, 1976). Thus, the nectar-timing hypothesis, based on an argument of resource conservation by plants, is supported by the Sierran data and data from previously published studies.

(4) Flower Density Hypothesis

The Sierran flora was also used to test the hypothesis that nectar production should be correlated inversely with density. These results are shown in Fig. 10-8. This hypothesis was not supported in spite of significant differences between populations in peak flower density. Flower density at the peak of flowering in the same population was quite comparable from year to year for most of my Sierran popula-

tions. So even though a population did seem to exhibit a characteristic flower density, individuals in sparse populations did not seem to compensate with increased energy provision to remain as attractive as individuals in dense populations.

Pertinent comments about this hypothesis are made in Feinsinger and Colwell's (1978) review paper on tropical hummingbird communities. They state that tropical hummingbird flowers tend to fall into three categories: (1) copious producers, which are sparsely dispersed and traplined by birds (cf. Janzen, 1971); (2) intermediate producers, which are dense in dispersion and defended by territorial individuals; (3) low producers, which are sparse and fed upon by occasional wanderers. That the traplined dispersed species are copious producers compared to the dense, clumped species tends to support the hypothesis, but the sparse, poor producers do not. Perhaps the latter attain reproductive success because the supply of wandering hummingbirds is adequate to fertilize most flowers, a suggestion supported by P. Feinsinger (pers. comm.).

Janzen (1971) also suggests that traplined bee flowers are copious producers. In honeyeater-pollinated *Eucalyptus* in Australia, species with many flowers per tree (= dense clumps) produce less nectar per flower than do species characterized by fewer flowers per tree (Paton and Ford, this book). However, more careful comparisons need to be made between plants whose pollinators, habitat, and other characteristics, except density, are similar, to isolate the relationship between density and selection for nectar production. My data satisfy these requirements and do not support the idea that flower density and nectar production are inversely related.

I point out that this hypothesis predicts an inverse relationship between flower density and nectar secretion per flower based on strategic arguments for optimal pollinator visitation. Such a relationship, if it is observed in the future, does not necessarily support this hypothesis, because nectar secretion rates could be reduced by competition or resource limitation in areas of high flower and/or plant density (see below).

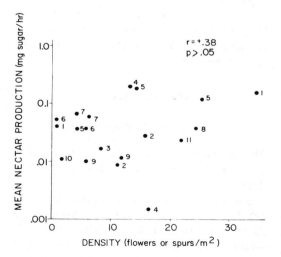

Figure 10-8. Relation between flower density and nectar production. Data points are identified as in the key to Fig. 10-2.

(5) Elevation Hypothesis

Weather stations at three of my major study sites, 1700 meters, 2450 meters, and 2700 meters, showed that temperature was inversely related to elevation. For example, in July and August 1978 the negative correlation between midsummer mean daily maxima and elevation was significant ($r = -.99, p = .05$). Temperature declined $10°C$ with 1000 meters increase in elevation.

Figures 10-4, 10-5, and 10-6 show elevational differences in production in Sierran hummingbird-pollinated species that occur at more than one elevation. The prediction that higher-elevation populations should produce more nectar seemed to be supported by *P. bridgesii* in 1979 (Fig. 10-5) and *I. aggregata* in 1978 (Fig. 10-6). However, the elevational comparisons in these two cases are probably not ecologically valid. First, the mere 335-meter difference in elevation in the two *I. aggregata* sites is not adequate to account for the sixfold difference in production rates seen in 1978, since for a 335-meter increase in elevation, temperatures decline only about $3°C$, given the measured rate above. Such a small decline should relate to only a 15% rise in nectar production according to the energetics hypothesis. Furthermore, although the magnitude of difference in production in *P. bridgesii* populations in 1979 was consistent with the magnitude of the measured decline in temperature with elevation (about 40% increase in avian energy requirements for an 850-meter increase in elevation—Fig. 10-5 bottom), an intermediate population at 2450 meters produced *less* than the 1850-meter population in 1978, and the relationship with elevation was the inverse in 1977 (Fig. 10-5). So none of these cases offers unambiguous support for the hypothesis. Previous studies on hummingbird-flowers in tropical mountains showed that they produce less nectar than do similar hummingbird flowers in the tropical lowlands nearby (Hainsworth and Wolf, 1972b; Stiles, 1978), exactly the opposite of the hypothesized prediction. Stiles (1978) suggests that the low temperatures at high elevations may limit nectar production.

Data conflict with regard to the alternative, that plants at high elevation should have higher flower densities. Table 10-4 shows that some species tended to grow less densely at high elevations. The only supportive result was with *P. bridgesii*. However, although the 2450-meter population is closer in elevation to the 2700-meter population than to the 1850-meter population, its density was indistinguishable from that of the latter. Flower density in this species may relate more to level of habitat disturbance, in that both the 1850-meter and 2450-meter sites are natural rocky slopes, whereas the 2700-meter site is a highly disturbed talus road cut, a situation in which this species thrives.

For the purposes of testing the elevation hypothesis, my most comparable populations are the four *C. linariaefolia* populations, which show a trend opposite to the one hypothesized (Table 10-4). Thus, the results suggest that elevational relationships, if any exist, are more complex than predicted by the energetics hypothesis. It is plausible that local differences in soil, exposure to sun and wind, and humidity exert more control over flower density and nectar secretion rates than do the foraging energetics of pollinators. Mechanistic rather than strategic arguments may be more appropriate, as was suggested also for the previous hypothesis.

To my knowledge, no other data comparing both nectar production and flower density with elevation exist for bird-pollinated plants along an elevational gradient.

DENOUEMENT

Variability in Nectar Data

The complexity of nectar data is illustrated by a summary in Table 10-5 of all the Sierran data.

Samples taken after 24-hr accumulation did not differ from interval samples summed over 24 hr in *P. bridgesii, I. aggregata,* or most *C. linariaefolia* populations. Such data suggest that when nectar accumulated in these flowers, production rates did not slow, nor was nectar reabsorbed as observed in some insect-polli-

Table 10-4. Comparison of flower densities at high and low elevations.

Species	Year	Low elevation	High elevation	Percent difference	Sig. by t-test
		\[a\]Flowers/m^2			
C. linariaefolia					
	1978	16.9 (42.5)[1]	8.4 (5.3)[4]	−50	n.s.
		20.5 (21.7)[2]	13.1 (12.5)[3]	−36	n.s.
P. bridgesii	1979	4.1 (5.5)[1]	25.2(41.1)[3]	+515	$p < .05$
	1978	6.3 (9.2)[1]	5.8(8.6)[2]	−8	n.s.
			14.1 (7.2)[3]	+124	$p < .001$
	1977	0.8 (0.7)[2]	4.2 (3.3)[3]	+425	$p < .001$
I. aggregata	1979	15.7 (20.2)[1]	21.6 (38.6)[2]	+38	n.s.

Sites are indicated by superscripts: *C. linariaefolia*—1 = 1700 m site C, 2 = 1700 m site W, 3 = 2800 m, 4 = 2700 m; *P. bridgesii*—1 = 1850 m, 2 = 2450 m, 3 = 2700 m; *I. aggregata*—1 = 2450 m, 2 = 2800 m.

[a]Mean, with s given in parentheses. N is generally 20 measurements per population.

nated flowers (Corbet, 1978). However, nectar reabsorption or production decline may have occurred in *C. miniata* in those flowers bagged for 24 hr (Table 10-5). Why this should occur in some populations and some species but not in others, is unclear. In contrast, samples taken of *A. formosa* after 24 hr of undisturbed accumulation were higher than the summed samples over five intervals, suggesting that repetitive sampling of the same flowers may have damaged the nectaries in this species. Similarly, in another study (Feinsinger, 1978), some species showed nectar reduction with increased sampling frequency, some showed nectar enhancement, and some showed no effect. No explanation was given in this study either.

Populations in which samples were made in more than one year sometimes showed tremendous yearly variation in mean nectar production. The *C. linariaefolia* population at 2800 meters showed at least an order of magnitude difference in 1978 and 1979 values. Production in *C. miniata* was ten times higher in 1977 than in 1976, and was five times higher in 1978 than in 1977 in *P. bridgesii* at 2700 meters and in *I. aggregata* at 2800 meters.

Nectar data in other studies have proved variable between flowers on the same plant (Feinsinger, 1978), although no other study has followed the same population over a long time to determine yearly variation within a population. My data also are variable between flowers on the same plant (and on different

plants) sampled at the same time, standard deviations generally approaching the means (note values for 24-hr accumulations in Table 10-5). Between-flower variation is thought to increase the number of flowers hummingbirds will probe on a plant before leaving (Feinsinger, 1978).

Between-year variation would seem to relate either to human error if different assistants are involved, or to yearly variation in weather. In my study, 1977 exhibited the most numerous extremes of high or low values (Table 10-5). In this year, two of the three data gatherers were the same as in the moderate production year of 1978. The greatest turnover in assistants occurred in 1979, and nectar values this year were comparable to those of other years. No worker in any year had characteristically high or low values, yet the values gathered by the same worker on the same population often showed great differences between years. For example, my data on *I. aggregata* at 2450 meters varied from .009 mg/flower/hr in 1978 to .027 in 1976. Thus, all things considered, it seems highly unlikely that the large differences between years in some populations were due to human error.

Alternative explanations for great yearly and between-flower variation include possible effects of local physical factors or of competition that supercede selection for pollination energetic strategies. As mentioned previously, plant density could reflect the intensity of com-

Table 10-5. Mean nectar production (averaged over 24 hr).

	Year	mg/fl/hr	S.D.	(N)[1]	Comments
Castilleja linariaefolia					
1700 m	1976	.034	—	(38)	24-hr sample by accumulation
		.034	—	(174)	added sequential samples
	1977	.036	—	(154)	24-hr sample by accumulation
		.037	—	(666)	added sequential samples
	1979	.043	—	(126)	
2700 m	1978	.017	—	(249)	
2800 m	1977	.014	—	(47)	24-hr sample by accumulation
		.053	—	(134)	added sequential samples
	1978	.204	—	(193)	
	1979	.018	—	(56)	24-hr sample by accumulation
		.002	—	(159)	added sequential samples
C. miniata					
2450	1976	.063	—	(19)	24-hr sample by accumulation
		.101	—	(158)	added sequential samples
	1977	.007	—	(78)	24-hr sample by accumulation
		.010	—	(209)	added sequential samples
	1978	.012	—	(158)	
	1979	.079	—	(217)	
[Gass et al., 1976]	1972	.02			24-hr sample by accumulation
	1973	.01			
C. chromosa					
2300 m	1978	.011	—	(709)	
Penstemon bridgesii					
1850 m	1977	.111	.099	(25)	
	1978	.055	.079	(292)	
	1979	.074	.059	(35)	24-hr sample by accumulation
		.068	—	(96)	added sequential samples
2450 m	1977	.050	.097	(292)	
	1978	.036	.045	(243)	
2700 m	1977	.040	.027	(28)	24-hr sample by accumulation
		.036	—	(158)	added sequential samples
	1978	.182	.251	(281)	
	1979	.119	.140	(85)	
Aquilegia formosa[2]					
1700 m	1976	.031	.021	(68)	24-hr sample by accumulation
	1977	.029	.023	(15)	24-hr sample by accumulation
		.013	—	(90)	added sequential samples
2450 m	1977	.061	.038	(13)	24-hr sample by accumulation
		.041	—	(83)	added sequential samples
	1978	.093	.062	(46)	
	1979	.154	.125	(460)	
[Gass et al., 1976]	1973	.05			24-hr sample by accumulation
Ipomopsis aggregata					
2450 m	1976	.029	.020	(18)	24-hr sample by accumulation
		.027	—	(46)	added sequential samples
	1978	.009	.017	(118)	
	1979	.028	.043	(182)	
2800 m	1977	.010	.015	(15)	24-hr sample by accumulation
		.010	—	(150)	added sequential samples
	1978	.051	.056	(101)	
	1979	.024	.057	(181)	

[1]Standard deviations cannot be computed for *Castilleja* species because values result from the subtraction of two means, or for data labeled "added sequential samples" because values result from the addition of the means for all sampling intervals over a 24-hr period.
[2]Values for this species are given in mg/spur/hr.

petition for limited resources and therefore might affect the nectar production rates of individual plants. However, I found that nectar production did not relate to plant density. For example, plant densities were indistinguishable ($p > .9$) in *C. linariaefolia* at 2800 meters in 1978 and 1979, the years nectar production varied by an order of magnitude. Also the number of flowers per plant was not dramatically different during these years, varying at most by a factor of two (Table 10-3). The plants could have attained the same attractiveness by producing fewer flowers but more nectar per flower. Yet this did not occur.

Physical factors may influence nectar energetics. The manner and cost of nectar secretion in plants is poorly understood, but secretion seems to be energy-demanding (Findlay et al., 1971 and Reed et al., 1971, in Stiles, 1978). Also, large amounts of water are provided in the dilute nectar of hummingbird flowers. During my study, 1976 and especially 1977 were drought years, 1978 was exceptionally wet, and 1979 was average. Production did not seem to relate to precipitation, however (Table 10-5). Both high and low nectar values occurred in both 1977 and 1978, the two most extreme years with regard to precipitation. Furthermore, *Castilleja* spp. are root parasites and have been shown to be buffered against environmental fluctuations (Hansen, 1979), yet the 2800-meter population of *C. linariaefolia* showed the largest year-to-year differences in nectar production of any species (Table 10-5). This population showed a strong depression in nectar production from 1978 to 1979 and grew in my highest elevation sites, which could suggest temperature as a variable. However, ten weeks of temperature data gathered at the 2450-meter site showed that maximum and minimum weekly temperatures were not significantly different between 1979 and 1978, so that temperature must be discounted. Thus in my study nectar production was variable and was unrelated to precipitation or temperature. Nectar production was also was unrelated to plant or flower density, which should be measures of competition and of the effort invested by a plant into number of flowers, respectively. However, the Sierran

data were not gathered in a way to specifically test the possibility of a negative relationship between number of flowers per plant and nectar production per flower, and this definitely should be pursued in the future.

The question that is raised by the Sierran data and other studies of pollination energetics in avian communities, is whether energetics patterns exist that reflect adaptive strategies in bird-pollinated plants, or whether nectar production varies randomly. The only one of my initial energetics hypotheses that was supported, besides the coarse support for hypothesis 1 (Fig. 10-1 and Table 10-2), was unimodality of nectar production in pollinator-generalized plants and bimodality of nectar production in plant species specialized on birds for pollination. However, this was the only hypothesis that assumed an energetic strategy should evolve because of a factor (e.g., limited water, loss to nectar thieves) other than competition for limited pollinators. None of the hypotheses predicting energetic fine-tuning as a strategy to attract pollinators was supported. If nectar energetics patterns are random rather than "adaptive" as predicted by optimization hypotheses, what is the cost in fitness? Is it possible that pollinations are rarely limiting, at least in bird-pollinated flowers, so that little cost in fitness is involved? Certainly the optimum energy reward per flower for a species depends on the number of flowers produced per plant, on the importance of outcrossing to the species, and especially on the ratio of pollinators to flowers, which varies greatly in time and space. Thus, in one study (Carpenter, 1976) an optimum nectar production per flower was found in *Metrosideros* that resulted in maximum capsule set, but nectar production in the population of these honeycreeper-pollinated trees averaged higher than that optimum. Perhaps the low optimum measured in one year was caused by the high bird-to-flower ratio that existed that year (Carpenter, 1978b), whereas the population had evolved in response to average conditions of lower bird-to-flower ratios.

In the Sierran study, capsule-set data suggest that in fact pollinators may not limit reproductive output in these plants. Capsule set early and late in the flowering season of four different populations did not differ consistently from capsule set at peak flowering (Carpenter, unpub. data). Thus, overall pollination does not vary significantly over the flora as a whole throughout the season, although it may be less variable between populations at peak bloom (Carpenter, unpub. data). If pollinators were limiting, one would expect that early- and late-blooming flowers should not be adequately pollinated, assuming pollinators are attracted to populations at peak bloom, and this pattern has been found in some insect-pollinated flowers (Zimmerman, 1980). Similarly, capsule set in the Sierra hummingbird flowers was not related to flower density in experiments set up during peak bloom (Carpenter, unpub. data), suggesting that all flowers get pollinated regardless of density. If bird-to-flower ratios were low, birds would be expected to concentrate on denser patches, and capsule set should be low in sparsely flowered areas. Finally, capsule set was not related to differing rates of nectar production between years (Table 10-6). For example, in two of five populations, capsule set was lower in 1979 than in 1978 even though nectar production was higher in 1979 than in 1978. In two of the other populations, capsule set was not significantly greater in 1978 than in 1979 although nectar production was much greater in 1978 than in 1979. Apparently, the hummingbirds tracked resource levels both within and between years and over

all of the flower densities in a manner that rarely allowed bird-to-flower ratios to fall so low as to reduce pollination. In fact, using a different approach, Montgomerie and Gass (1981) showed that generally all nectar was cropped by hummingbirds in their study sites in the mountains of northwestern California and in the Mexican lowlands, as did Carpenter (1978b) for the Sierra sites in this study. This result suggests that bird visitations would not limit reproductive output in these two plant communities. Feinsinger (1978) suggests that hummingbird flowers at his study area in Costa Rica also may not be pollinator-limited. If this conclusion is true, dependable and thorough pollination may be a major advantage of bird-pollination, as suggested by Stiles (1978). In comparison, several studies have shown that insect pollination does limit reproductive output in plants either seasonally (Zimmerman, 1980), in sparsely flowering parts of the population (Platt et al., 1974; Augspurger, 1980), in inflorescences with fewer flowers (Schaffer and Schaffer, 1979; Schemske, 1980), or in flowers providing less nectar (Thomson and Plowright, 1980).

Migratory Rufous hummingbirds are the main pollinators in the Sierran system and breed in the Pacific Northwest in late spring and early summer. Most of the birds in my areas are immatures. Thus, plants in the Cascade-Sierra and Rocky Mountain ranges may be situated in the path of a constant outflow of

Table 10-6. Between-year differences in capsule set in relation to between-year differences in nectar production.

Species' population	Nectar production[1]			% Capsule set		
	1978	1979	Difference (sig.)[2]	1978	1979	Difference (sig.)[2]
C. linariaefolia						
2800 m	.204	.002	+.203[3]	58.7	52.9	+5.8 (n.s.)
P. bridgesii						
1850 m	.055	.068	−.013 (n.s.)	52.5	63.9	−11.4 (n.s.)
2700 m	.182	.119	+.063 ($p < .05$)	60.1	57.6	+ 2.5 (n.s.)
I. aggregata						
2450 m	.009	.031	−.016 ($p < .001$)	74.3	68.9	+5.4 (n.s.)
A. formosa						
2450m	.093	.146	−.053 ($p < .005$)	49.8	38.0	+11.8 (n.s.)

[1]Mean mg sugar produced per flower (or spur) per hour, averaged over 24 hr.
[2]t-test comparing means of 1978 and 1979.
[3]Significance not testable because variance not obtained (see methods section), but magnitude of difference suggests significance.

"newly produced" pollinators that always exceed the numbers required to pollinate all plants. In such a system, plants pay no cost in fitness if they do not increase energetic rewards with decreased plant density, decreased number of flowers per plant, decreased flower density, or increased elevation. The energetics hypotheses fail because their basic assumption that visitations are potentially limiting is not met. Between-community studies should be conducted comparing this system to one in which pollinators may be limiting.

ACKNOWLEDGMENTS

Field assistants for this study were Cameron Cho, Robert Fulton, Mark Hixon, Richard Lejnieks, David Paton, Penelope Paton, and Roberta Soltz. Analysis of data was assisted by Betsy Brenner and Roberta Soltz. This project was funded by N.S.F. grant DEB 77-25124.

LITERATURE CITED

Augspurger, C. K. 1980. Mass-flowering of a tropical shrub *(Hybanthus prunifolius):* influence of pollination attraction and movement. *Evolution* 34:475–488.

Baker, H. G. 1961. The adaptation of flowering plants to nocturnal and crepuscular pollinators. *Q. Rev. Biol.* 36:64–73.

———. 1975. Sugar concentrations in nectars from hummingbird flowers. *Biotropica* 7:37–41.

Bolten, A. B. and P. Feinsinger. 1978. Why do hummingbird flowers secrete dilute nectar? *Biotropica* 10:307–309.

Bolten, A. B., P. Feinsinger, H. G. Baker, and I. Baker. 1979. On the calculation of sugar concentration in flower nectar. *Oecologia* 41:301–304.

Bond, H. W. and W. L. Brown. 1979. The exploitation of floral nectar in *Eucalyptus incrassata* by honeyeaters and honeybees. *Oecologia* 44:105–111.

Brown, J. H., W. A. Calder, and A. Kodric-Brown. 1978. Correlates and consequences of body size in nectar feeding birds. *Am. Zool.* 18:687–700.

Calder, W. A. III. 1974. Consequences of body size for avian energetics. In R. A. Paynter, Jr. (ed.), *Avian Energetics.* Publ. Nuttall Ornith. Club, no. 15. Cambridge, Mass., Ch. 2.

Carpenter, F. L. 1976. Plant–pollinator interactions in Hawaii: pollination energetics of *Metrosideros collina* (Myrtaceae). *Ecology* 57:1125–1144.

———. 1978a. Hooks for mammal pollination? *Oecologia* 35:123–132.

———. 1978b. A spectrum of nectar-eater communities. *Am. Zool.* 18:809–819.

———. 1979. Competition between hummingbirds and insects for nectar. *Am. Zool.* 19:1105–1114.

Carpenter, F. L. and R. E. MacMillen. 1976. Energetic cost of feeding territories in an Hawaiian honeycreeper. *Oecologia* 26:213–223.

Colwell, R. K., B. J. Betts, P. Bunnell, F. L. Carpenter, and P. Feinsinger. 1974. Competition for the nectar of *Centropogon valerii* by the hummingbird *Colibri thalassinus* and the flower piercer *Diglossa plumbea,* and its evolutionary implications. *Condor* 76:447–452.

Corbet, S. A. 1978. Bee visits and the nectar of *Echium vulgare* L. and *Sinapis alba* L. *Ecol. Entomol.* 3:25–38.

Cruden, R. W. 1970. Hawkmoth pollination of *Mirabilis* (Nyctaginaceae). *Bull. Torrey Bot. Club* 97:89–91.

Duffield, W. J. 1972. Pollination ecology of *Castilleja* in Mount Rainier National Park. *Ohio J. Sci.* 72:110–114.

Faegri, K. and L. van der Pijl. 1966. *The Principles of Pollination Ecology.* Pergamon Press, New York.

Feinsinger, P. 1976. Organization of a tropical guild of nectarivorous birds. *Ecol. Mon.* 46:257–291.

———. 1978. Ecological interactions between plants and hummingbirds in a successional tropical community. *Ecol. Mon.* 48:269–287.

Feinsinger, P. and R. K. Colwell. 1978. Community organization among neotropical nectar-feeding birds. *Am. Zool.* 18:779–795.

Feinsinger, P., Y. B. Linhart, L. A. Swarm, and J. A. Wolfe. 1979. Aspects of the pollination biology of three *Erythrina* species on Trinidad and Tobago. *Ann. Missouri Bot. Gard.* 66:451–471.

Ford, H. A., D. C. Paton, and N. Forde. 1979. Birds as pollinators of Australian plants. *N.Z. J. Bot.* 17:509–519.

Frankie, G. W. and S. B. Vinson. 1977. Scent marking of passion flowers in Texas by females of *Xylocopa virginica texana* (Hymenoptera: Anthophoridae). *J. Kansas Ent. Soc.* 50:613–624.

Gass, C. L. 1978. Rufous hummingbird feeding territoriality in a suboptimal habitat. *Can. J. Zool.* 56:1535–1539.

Gass, C. L., G. Angehr, and J. Centa. 1976. Regulation of food supply by feeding territoriality in the Rufous hummingbird. *Can. J. Zool.* 54:2046–2054.

Gill, F. B. and L. L. Wolf. 1975. Economics of feeding territoriality in the golden-winged sunbird. *Ecology* 56:333–346.

Hainsworth, F. R. and L. L. Wolf. 1972a. Crop volume, nectar concentration, and hummingbird energetics. *Comp. Biochem. Physiol.* 42:359–366.

———. 1972b. Energetics of nectar extraction in a small, high altitude, tropical hummingbird. *J. Comp. Physiol.* 80:377–387.

———. 1976. Nectar characteristics and food selection by hummingbirds. *Oecologia* 25:101–113.

Hansen, D. H. 1979. Physiology and microclimate in a hemiparasite *Castilleja chromosa* (Scrophulariaceae). *Am. J. Bot.* 66:477–484.

Heinrich, B. 1975. Energetics of pollination. *Annu. Rev. Ecol. System.* 6:139–170.

———. 1977. Pollination energetics: an ecosystem approach. *In* W. J. Mattson (ed.), *The Role of Arthropods in Forest Ecosystems*. Proc. in Life Sciences. Springer-Verlag, New York, Ch. 5.

Heinrich, B. and P. H. Raven. 1972. Energetics and pollination ecology. *Science* 176:597–602.

Hickman, J. C. 1974. Pollination by ants: a low-energy system. *Science* 184:1290–1292.

Inouye, D. W. 1980. The terminology of floral larceny. *Ecology* 61:1251–1253.

Inouye, D. W., N. D. Favre, J. A. Lanum, D. M. Levine, J. B. Meyers, M. S. Roberts, F. C. Tsao, and Y. Wang. 1980. The effects of nonsugar nectar constituents on estimates of nectar energy. *Ecology* 61:992–996.

Janzen, D. H. 1971. Euglossine bees as long-distance pollinators of tropical plants. *Science* 171:203–205.

MacMillen, R. E. and F. L. Carpenter. 1977. Daily energy costs and body weight in nectarivorous birds. *Comp. Biochem. Physiol.* 56A:439–441.

Montgomerie, R. D. and C. L. Gass. 1981. Energy limitation of hummingbird populations in tropical and temperate communities. *Oecologia.* 50: 162–165.

Munz, P. A. and D. D. Keck. 1970. *A California Flora*. University of California Press, Berkeley.

Percival, M. 1965. *Floral Biology*. Pergamon Press, New York.

Platt, W. J., G. K. Hill, and S. Clark. 1974. Seed production in a prairie legume (*Astragalus canadensis* L.). Interactions between pollination, predispersal seed predation, and plant density. *Oecologia* 17:55–63.

Schaffer, W. M. and M. Schaffer. 1979. The adaptive significance of variations in reproductive habit in the Agavaceae II: Pollinator foraging behavior and selection for increased reproductive expenditure. *Ecology* 60:1051–1069.

Schemske, D. W. 1980. Evolution of floral display in the orchid *Brassavola nodosa*. *Evolution* 34:489–493.

Smith, E. D. and F. L. Carpenter. 1983. Midday nectar depression in *Beloperone californica*. Bull. Southern California Acad. Sci. In press.

Snow, D. W. and B. K. Snow. 1980. Relationships between hummingbirds and flowers in the Andes of Columbia. *Bull. Br. Mus. Nat. Hist. (Zool.)* 38:105–139.

Stiles, F. G. 1973. *Food Supply and the Annual Cycle of the Anna Hummingbird*. Univ. Calif. Publ. Zool. 97. 109 pp.

———. 1975. Ecology, flowering phenology, and hummingbird pollination of some Costa Rican *Heliconia* species. *Ecology* 56:285–301.

———. 1976. Taste preferences, color preferences and flower choice in hummingbirds. *Condor* 78:10–26.

———. 1978. Ecological and evolutionary implications of bird pollination. *Am. Zool.* 18:715–727.

Stiles, F. G. and L. L. Wolf. 1970. Hummingbird territoriality at a tropical flowering tree. *Auk* 87:467–491.

Thomson, J. D. and R. C. Plowright. 1980. Pollen carryover, nectar rewards, and pollination behavior with special reference to *Diervilla lonicera*. *Oecologia* 46:68–74.

Wheeler, T. G. 1980. Experiments in feeding behavior of the Anna hummingbird. *Wilson Bull.* 92:53–62.

Zimmerman, M. 1980. Reproduction in *Polemonium:* competition for pollinators. *Ecology* 61:497–501.

11
THE INFLUENCE OF PLANT CHARACTERISTICS AND HONEYEATER SIZE ON LEVELS OF POLLINATION IN AUSTRALIAN PLANTS

David C. Paton

The Australian Museum
Sydney
NSW Australia

Hugh A. Ford

Department of Zoology
University of New England
Armidale
NSW Australia

ABSTRACT

We show that features of both plants and honeyeaters influence the foraging behavior of honeyeaters in ways that are likely to affect the quantities of pollen arriving at and departing from a plant. Honeyeaters preferred plants with more flowers and plants that produced more nectar. These plants should experience higher levels of pollination. When honeyeaters were defending feeding territories, smaller species were better pollinators than larger species because they visited individual flowers within their territories more frequently. Standing crops of nectar were generally highest and hence visitation rates to flowers and levels of pollination lowest during peak flowering. These seasonal changes in visitation rates were associated with changes in the species of honeyeater exploiting the flowers.

KEY WORDS: Honeyeater, foraging behavior, floral displays, bird pollination, territoriality, nectar.

CONTENTS

INTRODUCTION

A close mutual relationship between Australian honeyeaters (Meliphagidae) and the plants these birds visit has long been recognized, with the birds effecting pollination while collecting nectar (Paton and Ford, 1977; Armstrong, 1979; Ford et al., 1979). Many of the flowers seem structurally well suited for pollination by birds and the birds well adapted with their brush-tipped tongues for collecting nectar, suggesting a long period of association and coadaptation (Sargent, 1928; Keast, 1976, Paton and Ford, 1977). Most of the 67 species of honeyeater that occur in Australia visit flowers to some extent, and many depend almost entirely on nectar, at least for periods of the year (Keast, 1968a; Ford and Paton, 1977; Recher, 1977, 1981; Paton, 1979, 1980; Pyke, 1980). Seasonal movements associated with the flowering of suitable nectar-bearing plants are common, though in areas where there is a continuous cycle of nectar-producing plants or other sugary substances, a proportion of the honeyeater population is resident (Bell, 1966; Keast, 1968b; Ashton, 1975; Ford, 1977, 1979; Paton, 1979, 1980; Recher, 1981).

Much less is known about the plants and whether they depend on birds for pollination. Most plants pollinated by birds in Australia are believed to be outcrossers (Keighery, 1982), so honeyeaters should move between plants to be effective pollinators. However, detailed observations on the foraging behavior of honeyeaters and how they move within and between plants are lacking for all but a few plant species (Hopper and Burbidge, 1978; Pyke, 1981a; Paton, 1982). In this chapter we investigate the influence of plant characteristics and honeyeater size on the behavior of honeyeaters, and discuss how changes in honeyeater behavior affect the quantities of pollen transferred between plants.

For simplicity we shall assume that the plants require or are striving for cross-pollination, and so will restrict ourselves to assessing the behavior of honeyeaters in terms of effecting cross-pollination.

Estimating Levels of Cross-Pollination

Levels of cross-pollination should be related to the frequency and duration of foraging bouts to a plant. The number of visits a plant receives determines the number of times that outcross pollen (i.e., pollen from other conspecific plants) arrives at the plant and also the number of times that its own pollen departs. The length of the visit, coupled with pollen carryover and pollen accumulation on the bird, determines the quantities of pollen involved respectively.

When a honeyeater forages at the flowers of a plant three things should occur: (1) the quantities of outcross pollen that are transferred to stigmas of successive flowers should decrease (see Lertzman and Gass, this book); (2) self pollen should accumulate on the honeyeater; and (3) the quantities of self pollen transferred to stigmas should increase at each successive flower probed. Thus the longer the visit the more outcross pollen (and self pollen) that the plant should receive on its stigmas, and the more self pollen that should depart from the plant. There may be limits to these quantities set by the length of pollen carryover and by the dynamics of pollen loads on honeyeaters. If pollen is carried over to only a few flowers, then visiting more flowers beyond this limit would not increase the amounts of outcross pollen received by the plant. Likewise, if a pollen load on a honeyeater reaches a point of saturation or dynamic equilibrium where there is no net change in the quantity of self pollen carried on the honeyeater (Primack and Silander, 1975), then visiting more flowers beyond this point would not lead to any increase in the quantity of pollen departing on that visit. Pollen carryover and the dynamics of pollen loads on honeyeaters have not been measured for any plant–honeyeater system.

In general, increases in either the frequency or the duration of foraging bouts (within the limits of pollen carryover) should result in increases in the quantities of pollen arriving at and departing from a plant and lead to higher amounts of cross-pollination. We use changes in the frequency and duration of foraging bouts at plants to assess the influence of plant traits and honeyeater size on levels of pollination.

We first generate simple qualitative predictions about changes in the frequency and duration of foraging bouts at plants due to either changes in floral traits or species of honeyeater, and then, whenever possible, support these predictions with field observations. Field observations were collected by watching a series of individually marked plants for one-hour periods throughout the day and recording the frequency and duration of visits to each plant, or by watching color-banded honeyeaters and recording the duration or frequency with which they visited flowers or plants. Further details on field methods and study sites are given in Paton (1980, 1982), Ford (1979), and Ford and Paton (1982).

INFLUENCE OF PLANT CHARACTERISTICS ON THE BEHAVIOR OF HONEYEATERS

Individual plants in a population differ in the number, morphology, and arrangement of their flowers, in nectar production, and in the density of conspecific plants surrounding them. Variations in the number of flowers on a plant and plant density are obvious in most populations, but individual variation in nectar production, floral arrangements, and morphology are not always obvious and rarely have been documented. Each of these characteristics potentially influences the rate at which honeyeaters can harvest nectar, which in turn should influence their foraging behavior and hence

the quantities of pollen arriving at and departing from a plant.

How Should Honeyeaters Respond to Variations in the Floral Displays of Individual Plants?

Some simple predictions concerning the response of honeyeaters to variations in plant characteristics can be made by assuming that honeyeaters forage in ways that maximize their net rate of energy gain while harvesting nectar.

Honeyeaters should prefer those plants that offer them higher than average net rates of energy gain. Plants with more flowers, with simpler floral morphologies, with "simpler" floral arrangements, producing more nectar per flower, or occurring in higher densities supply nectar in ways that allow the nectar to be harvested more rapidly. These plants should receive greater attention from honeyeaters, and according to optimal foraging theory (Charnov, 1976; Pyke et al., 1977), the increase in foraging effort (visits/flower) should be proportional to the increase in the rates at which nectar can be harvested. All other things equal, plants with more flowers should allow honeyeaters to visit more flowers before they have to move to another plant. This enables the birds to visit more flowers per unit time by reducing the frequency of interplant movements, and gives them a higher net rate of energy gain. Furthermore, the frequency of foraging mistakes (i.e., revisits to flowers that have just been probed, from which the bird obtains little or no reward) may be reduced when there are more flowers available on a plant (e.g., Pyke, 1978; Paton, 1979). Therefore plants with more flowers may allow more unvisited flowers to be visited per unit time, further improving the bird's net rate of energy gain. Plants in higher densities also allow more flowers to be visited per unit time by reducing the time (and costs) for interplant movements. Plants with simpler floral morphologies (e.g., shorter corolla tubes) potentially allow honeyeaters to probe flowers and extract nectar more rapidly (e.g., see Hainsworth, 1973; Inouye, 1980). Particular spatial arrangements of flowers on a plant may allow more flowers to be visited per unit time or reduce the frequency of foraging mistakes, enabling more unvisited flowers to be visited per unit time (e.g., see Pyke, 1981b). Finally, plants with higher rates of nectar production offer more nectar per flower.

These plants should be visited more frequently, provided honeyeaters can identify them. Plants in higher densities, with more flowers, with simpler floral morphologies, or with "simpler" arrangements are potentially visible from a distance to foraging honeyeaters and so could be visited more frequently. Honeyeaters, however, may need to learn the locations of plants producing more nectar to be able to visit them more frequently unless other cues, such as flower color or condition, indicate higher levels of nectar (e.g., see Gill and Wolf, 1975a).

Predictions about changes in the duration of foraging bouts with plant characteristics are difficult to generate. Plants with more flowers, however, should receive longer visits than plants with fewer flowers, for reasons outlined above. Predictions about changes in the duration of foraging bouts with other plant characteristics depend on the foraging abilities of honeyeaters within a plant and on the frequency of foraging bouts to the plants. Honeyeaters should stay longer at a plant when they obtain a higher net rate of energy gain. If the increase in the frequency of foraging bouts to the above plants is insufficient to completely compensate for the higher rates at which they offer nectar, then these plants will, on average, give higher net rates of energy gain to honeyeaters. Consequently, they should receive longer visits. For example, if honeyeaters do not visit plants that produce more nectar more frequently, then on average these plants would have higher levels of nectar in their flowers than other plants and should receive longer visits.

Several studies have shown that the probability of honeyeaters making foraging mistakes increases as more probes are made at an inflorescence or plant (Paton, 1979; Pyke, 1981a). Honeyeaters, therefore, should experience a declining net rate of energy gain while foraging at a plant. Under such conditions, honeyeaters should probe fewer flowers at plants in higher densities compared with plants in lower densities. This is so because the costs, in both time and energy, to move to a new plant are less when the plants are closer together. By leaving plants in dense areas earlier, honeyeaters should increase the rates at which they probe unvisited flowers.

Plants with more flowers, with higher rates of nectar production, with simpler flowers, with "simpler" floral arrangements, or occurring in higher densities should be visited more frequently provided honeyeaters can identify them. These plants may also receive longer foraging bouts except for plants occurring in higher densities. As a result, these plants should achieve higher amounts of cross-pollination and contribute more to future generations.

This conclusion should hold provided the availability of honeyeaters does not change with changes in plant characteristics. Such changes could occur when nectar is limiting. Under these conditions honeyeaters are likely to establish feeding territories (Carpenter, 1978; Paton, 1979; Ford and Paton, 1982), with the result that the frequency of visits to individual flowers is reduced (Gill and Wolf, 1975b; Paton, 1979). Studies on nectarivorous birds, including honeyeaters, have found that territories are more likely to be established in areas with dense flowers (or plants) or around plants with many flowers (e.g., Feinsinger, 1978; Paton, 1979; Ford and Paton, 1982). Territories may also be more likely to occur around plants with higher rates of nectar production, or with simple floral morphologies and arrangements. By being included in a territory, plants offering nectar at greater harvestable rates might experience lower rates of visitation than plants offering lower harvestable rewards that are not included in a territory. Within a territory, predictions about changes in the frequency and duration of foraging bouts with plant characteristics should still stand.

How Do Honeyeaters Respond to Variations in the Floral Displays of Individual Plants?

Observations on the behavior of honeyeaters at individual plants differing in plant characteristics are limited to studies on the response of New Holland honeyeaters, *Phylidonyris hovaehollandiae,* to variations in the number of flowers on plants and to variations in nectar production (Paton, 1982).

New Holland honeyeaters responded to variations in the number of flowers on plants of *Correa schlechtendalii* and *Eucalyptus cosmophylla* by increasing both the frequency and the length of foraging bouts at plants with more flowers (Fig. 11-1). Individual flowers on plants with more flowers also received more visits (Fig. 11-1). New Holland honeyeaters responded to variations in nectar production in six similar-sized *Eucalyptus cosmophylla* by increasing the frequency of foraging bouts at those plants producing more nectar (Fig. 11-2). They did not increase the length of their foraging bouts (Fig. 11-2). The same birds re-

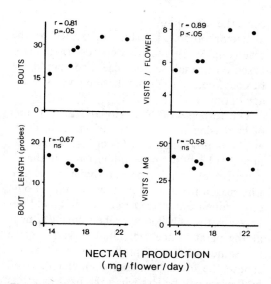

NECTAR   PRODUCTION
(mg / flower / day)

Figure 11-2. Response of New Holland honeyeaters to variations in nectar production of *Eucalyptus cosmophylla.* Correlation coefficients and levels of significance are given on the figure. (From Paton, 1982.)

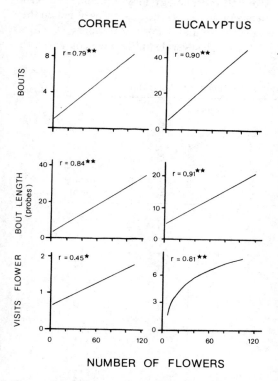

NUMBER OF FLOWERS

Figure 11-1. Response of New Holland honeyeaters to variations in the number of flowers on *Correa schlechtendalii* and *Eucalyptus cosmophylla.* Lines are least squares regression lines. Regression coefficients are given on the figure ($\star$ $p < .05$, $\star\star$ $p < .001$). Twenty-eight *Correa* bushes were watched for 4 hours, and 14 *Eucalyptus* for 5 hours. (From Paton, 1982.)

visited these plants and presumably had learned which plants produced more nectar. Individual flowers on plants producing more nectar also received more visits (Fig. 11-2).

These observations are consistent with the predictions, and plants with more flowers or producing more nectar should experience higher levels of pollination. Many more observations are needed on the responses of honeyeaters to changes in plant traits. These should include manipulative experiments, such as removing flowers from plants, and also detailed observations of the foraging abilities of the birds within plants. Once these detailed observations are available, estimates of the actual changes in the rates of harvestable reward with a plant characteristic will be possible. This should allow more precise predictions about how frequently individual flowers on different plants should be visited. In the case of increasing nectar production, the increase in harvestable rewards should be directly proportional to the increase in nectar production. If honeyeaters forage in ways that maximize their net rate of energy gain, then we would expect the increase in foraging effort to be proportional to the increase in nectar production (i.e., flowers on the different plants should receive the same number of visits per milligram of nectar produced). This appears to be the case (Fig. 11-2), though we stress that this is based on a small sample size, and should not be construed as evidence that honeyeaters forage optimally.

The response of New Holland honeyeaters of increasing the frequency of foraging bouts at plants producing more nectar should result in higher levels of cross-pollination than the alternative of just increasing the bout length. The reason for this is that the amounts of outcross pollen transferred to a flower should decrease with each flower probed during a bout at the plant. Since individual flowers receive the same number of probes under either response, the amounts of outcross pollen received by flowers will be less when the bout length is increased. A similar argument can be developed for the departure of self pollen, since the net amounts of self pollen added to the pollen load of the birds should also decrease with each probe during a bout at the plant. In gen-

eral, increases in the frequency of foraging bouts to a plant will have a greater impact on rates of cross-pollination than increases in the length of a bout.

INFLUENCE OF HONEYEATER SIZE ON THE BEHAVIOR OF HONEYEATERS

Most bird-pollinated plants in Australia are visited by a variety of honeyeaters and other birds that differ in size and morphology (Paton and Ford, 1977). This is illustrated for some common bird-pollinated plants in southeastern Australia (Table 11-1), with bill dimensions and weights of some of the honeyeaters exploiting these flowers given in Table 11-2. Different species of honeyeater potentially provide different service to plants, since differences in size and morphology may influence the bird's behavior as well as the quantities of pollen they collect from and deposit at flowers.

Although several studies have compared the

Table 11-1. Bird visitors to a range of plants in southeastern Australia.

Plant species	Melithreptus	Lichenostomus	Acanthorhynchus	Phylidonyris	Anthochaera	other honeyeaters	other birds	No. species visiting plant
Myrtaceae								
Callistemon macropunctatus	+	+	+	+	+	+	+	19
Eucalyptus cosmophylla	+	+	+	+	+		+	15
Eucalyptus fasciculosa	+	+	+	+		+	+	18
Eucalyptus leucoxylon	+	+	+	+	+	+	+	24
Proteaceae								
Adenanthos terminalis	+	+	+	+	+			8
Grevillea ilicifolia	+	+		+	+	+		10
Grevillea aquifolia	+	+		+	+		+	10
Banksia marginata	+	+	+	+	+		+	22
Banksia ornata	+	+	+	+	+		+	12
Loranthaceae								
Amyema miquelii	+	+	+	+	+		+	22
Amyema pendula	+	+	+	+	+	+	+	13
Lysiana exocarpi		+	+	+	+	+	+	8
Rutaceae								
Correa schlechtendalii	+	+	+	+	+		+	13
Correa reflexa		+	+	+	+			8
Epacridaceae								
Astroloma conostephioides	+	+	+	+	+		+	21
Epacris impressa	+	+	+	+			+	7

Table 11-2. Bill dimensions and weights of some common honeyeaters in southeastern Australia.

Species	Weight (grams)	Bill dimensions (mm)*		
		Length	Depth	Breadth
Eastern spinebill,				
Acanthorhynchus tenuirostris	11.0	22.5	2.7	2.9
White-naped honeyeater,				
Melithreptus lunatus	13.0	12.0	3.9	4.0
Brown-headed honeyeater,				
Melithreptus brevirostris	14.0	12.1	3.9	3.9
Yellow-faced honeyeater,				
Lichenostomus chrysops	17.0	12.5	3.5	3.8
White-plumed honeyeater,				
Lichenostomus penicillata	20.0	11.7	4.0	4.1
Crescent honeyeater,				
Phylidonyris pyrrhoptera	14.5	17.1	3.9	3.5
Tawny-crowned honeyeater,				
Phylidonyris melanops	18.5	18.3	3.8	4.1
New Holland honeyeater,				
Phylidonyris novaehollandiae	20.0	19.0	4.4	4.1
Little wattlebird,				
Anthochaera chrysoptera	65.0	22.8	6.1	5.9
Red wattlebird,				
Anthochaera carunculata	110.0	23.5	6.8	6.7

*Length of exposed culmen; depth and breadth at nares.

quantities and composition of pollen loads on mistnetted honeyeaters (Ford, 1976; Hopper, 1980; Ford and Paton, 1982) and found differences between species, it is not known how pollen loads on mistnetted birds relate to those carried by the birds when foraging. Mistnets dislodge some of the pollen from the birds, and differences in pollen loads between species and even individuals may be due to the length of time the birds were held in the net, to the mistnet affecting the loads of certain species (because of their morphology) more than others, or to differences in the interval between capture and feeding. We show below that species differ in the amount of time they spend feeding, so certain species are more likely to carry heavier pollen loads. What is required is a measurement of the quantities of pollen deposited on stigmas of pollen-free flowers and a measurement of the quantities of pollen lost from or remaining at an intact flower following single visits by different honeyeaters. For the purposes of continuing our discussion we shall assume that the dynamics of pollen loads and pollen carryover are the same for all species, and assess different honeyeaters purely on behavioral attributes.

Larger species of honeyeater have higher energy requirements and can be expected to visit more flowers (of the same standing crop) than a smaller species. As a result, larger species will probably visit more bushes and potentially move pollen farther, provided pollen carryover is long. However, larger species could stay longer at each plant rather than visit more plants. Larger honeyeaters may benefit energetically by remaining longer at plants, since the costs to fly to another plant relative to those of remaining probing flowers are proportionately higher for larger species. Costs for sitting and probing flowers increase with weight at a slower rate ($\propto W^{0.73}$, where W = weight; Wolf et al., 1975) than the costs for flying between plants ($\propto W^{0.97}$; Tucker, 1974). Larger honeyeaters would benefit more than a smaller species by staying longer at a plant. So larger honeyeaters might be expected to probe more flowers at a plant or inflorescence than a smaller species, and this appears to be the case. Larger honeyeaters make more probes at inflorescences of *Amyema pendula* than smaller species (Table 11-3). Differences in size and morphology also influence the speed with which foraging activities are per-

Table 11-3. Feeding rate and probing behavior of eight species of honeyeater feeding on *Amyema pendula* during peak flowering (average of 42 infl./plant, and an average of 8 flowers/infl.) in 1976 near Cranbourne, Victoria.

Bird species	Weight (grams)	Feeding rate (probes/min.)	Probes/inflor. (mean)
Red wattlebird	110	91	7.6
Little wattlebird	65	72	6.8
New Holland	20	67	4.7
White-plumed	20	41	3.2
Yellow-faced	17	39	3.5
Brown-headed	14	34	3.6
White-naped	13	37	3.1
Eastern spinebill	11	36	3.6

formed. Larger honeyeaters probe flowers at faster rates and may fly between plants at faster rates. Changes in speed may then compensate for the increased rates of energy expenditure, and predictions concerning changes in honeyeater behavior at plants with changes in honeyeater size may need to be modified once the speeds for the various components of foraging have been measured.

Feeding rates of large honeyeaters are generally faster than those of smaller species (Table 11-3 and 11-4; Ford, 1979). However, the faster feeding rates are usually insufficient to completely compensate for their higher energy requirements. For example, feeding rates of red wattlebirds are at most three times the feeding rates of eastern spinebills, while the energy requirements are five times greater (Table 11-4). Because feeding rates of large honeyeaters do not completely compensate for their higher energy requirements, larger honeyeaters need more nectar per flower to feed as profitably as a smaller species. They also need a higher minimum nectar level to balance their energy requirements. When nectar levels drop to below this minimum level, larger species must either leave the area or establish feeding territories (Ford, 1979). By establishing feeding territories they can reduce visitation rates to individual flowers and allow nectar levels to build up and remain above the critical level (Paton, 1979). Smaller species are more efficient nectar-feeders and so can potentially visit individual flowers more frequently. Thus smaller species are potentially the better pollinators.

Nectar is often in short supply for much of the year in many areas of southeastern Australia (Ford, 1979; Paton, 1979; Ford and Paton, 1982), and territoriality is common.

Table 11-4. Feeding rates (flowers/minute) for different-sized honeyeaters on a range of plants.

Plant species	Eastern spinebill (40 kJ)*	New Holland honeyeater (75 kJ)	Red wattlebird (200 kJ)
Astroloma conostephioides	26	34	30
Epacris impressa	69	78	—
Eucalyptus cosmophylla	34	35	35
Eucalyptus leucoxylon (S. Aust)	27	43	49
Eucalyptus leucoxylon (Vict.)	—	57	88
Amyema pendula	36	67	91
Banksia marginata	—	1.95	2.01
Banksia integrifolia	1.90	2.87	—

*Estimated daily energy requirement.

Table 11-5. Comparison of the feeding behavior of wattlebirds and New Holland honeyeaters defending territories based on the same plant species.

Plant species	Bird species	No. of flowers on territory	% day feeding	Feeding rate (flowers/min.)	Total visits to flowers in day	Visits/ flower/day
Amyema pendula	Little wattlebird	8,500	63.6	90	49,539	5.8
	New Holland	3,600	67.0	67	39,054	10.9
*Banksia marginata**	Little wattlebird	47	39.0	2.0	527	11.2
	New Holland	32	76.0	2.0	1,029	32.2
Eucalyptus leucoxylon	Red wattlebird	3,000	62.9	88	46,601	15.5
	New Holland	1,460	89.1	57	42,811	29.3

*Inflorescences.

Eastern spinebills, New Holland honeyeaters, and little and red wattlebirds all have been observed defending individual feeding territories (Paton, 1979, unpub.; Ford, 1981; Ford and Paton, 1982). Territories produce sufficient nectar to satisfy the birds' energy requirements but rarely much more (Paton, 1979; Ford, 1981). Wattlebirds often defend feeding territories alongside New Holland honeyeaters. Territories of the larger wattlebirds contained more flowers than those of New Holland honeyeaters, but in proportion to their increased energy requirements. In addition, wattlebirds fed less and visited individual flowers less frequently than New Hollands (Table 11-5). Since smaller honeyeaters are likely to probe fewer flowers during a visit to a plant, the increase in the overall frequency with which New Holland honeyeaters visited flowers should be due to an increase in the number of foraging bouts to a plant. Thus when honey-

eaters are defending territories, smaller species of honeyeater will effect higher amounts of cross-pollination at individual plants. This does not apply for *Eucalyptus leucoxylon,* where territories of both red wattlebirds and New Holland honeyeaters are restricted to parts of a single tree. Cross-pollination in this species will be effected by intruders, or when territory owners return from feeding outside their territories as they do occasionally. However, individual plants of most of the other species of bird-pollinated plants in southeastern Australia produce less nectar than an individual bird requires (Ford et al., 1979), and several to many plants are included in the one territory. Individual plants of *Adenanthos, Grevillea,* and *Banksia* usually produce from 10 to 50 kJ/day and individual *Amyema* and *Lysiana* usually less than 10 kJ/day, while individual *Correa, Astroloma,* and *Epacris* usually produce less than 1 kJ/day. These values compare with daily energy requirements of 40, 75, 130, and 200 kJ for eastern spinebills, New Holland honeyeaters, and little and red wattlebirds respectively.

SEASONAL CHANGES IN VISITATION RATES TO FLOWERS

The floral displays of individual plants pollinated by honeyeaters vary over a flowering season. The most obvious change occurs in the number of flowers in bloom on a plant (e.g., Fig. 11-3), but rates of nectar production and even the arrangement and morphology of flowers may change seasonally as well. The amount of pollination achieved by an individual plant might be expected to vary over a flowering season owing to changes in its floral display. Increases in the floral display (e.g., number of flowers, nectar production) should result in increases in the frequency and possibly the duration of foraging bouts to a plant, and result in higher levels of pollination, provided the availability of honeyeaters does not change.

The availability of honeyeaters, however, may change, owing to either an inability of the birds to keep track of changes in flower numbers or nectar production, changes in their be-

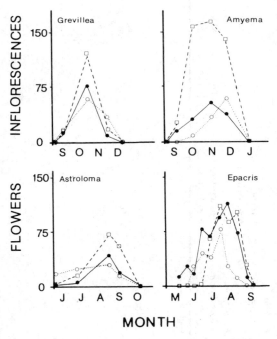

Figure 11-3. Seasonal changes in the number of flowers on three individual plants of *Grevillea aquifolia, Amyema pendula, Astroloma conostephioides,* and *Epacris impressa,* during 1977 flowering season. Data on *Astroloma* and *Grevillea* are from Golton Vale, Victoria, while data on *Epacris* and *Amyema* are from Cranbourne, Victoria.

havior (e.g., territoriality), or changes in the species of honeyeaters exploiting the flowers. Large species of honeyeater, because of their higher energy requirement, tend to prefer areas with dense flowers (Ford, 1979; Ford and Paton, 1982); so they should be more likely to use a plant species during peak flowering rather than at the beginning or end of flowering. Smaller, more efficient honeyeaters can use more sparsely distributed flowers (Ford, 1979). Territoriality may also be more common during peak flowering than at other times, since territories are more likely to be established in areas with high densities of flowers (Paton, 1979; Ford and Paton, 1982). Smaller species of honeyeater may be excluded by larger species at peak flowering. Since increases in the size of the bird and the development of territorial behavior usually result in a reduction in the frequency with which flowers are visited, visitation rates and levels of pollination may be lower during peak flowering.

The number of honeyeaters in an area often follows changes in flower abundance both within a flowering season and between seasons, with numbers increasing as levels of flowering increase, reaching peaks during peak flowering, and then declining as flowering declines (Bell, 1966; Keast, 1968b; Ashton, 1975; Ford, 1979; Paton, 1979). However, levels of flowering and numbers of birds were often scored in qualitative terms, and seasonal changes in the number of birds relative to flowers may have occurred. More accurate assessments are available in some cases. For example, at Cranbourne, near Melbourne, Victoria, honeyeaters tracked changes in flower numbers fairly precisely. Territoriality was common throughout the year in this area, and the birds adjusted the number, size, and location of territories with seasonal changes in flower density and species (Paton, 1979; e.g., Fig. 11-4). At all times there were surplus birds moving through these areas, and responses to increases in nectar availability were rapid.

Changes in the frequency with which flowers were visited still occurred, even when the honeyeaters maintained their numbers, owing to changes in the bird species using the plants. A good example is the sequence of territorial

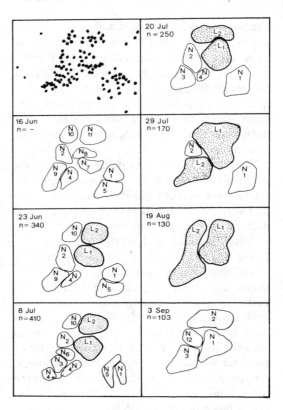

Figure 11-4. Seasonal changes in feeding territories of honeyeaters in a patch of *Banksia marginata* at Cranbourne in 1976. Territories of New Holland honeyeaters (N) and little wattlebirds (L, stippled) are plotted for 7 days between June 16 and Sept. 3. Total number *(n)* of inflorescences in the patch is given under date. Inflorescences were not counted on June 16. The top left-hand figure gives the positions of the major *Banksia* bushes. Numbers within territories identify individual birds.

occupancy in patches of *Banksia marginata,* as illustrated for one patch in Fig. 11-4. Early in the season New Holland honeyeaters controlled all the available flowers. Later little wattlebirds, which are dominant over New Hollands, displaced some of the New Hollands. As flowering levels increased, the number of territories increased. Then, as the number of flowers declined, the wattlebirds enlarged their territories and displaced the New Hollands. Finally the wattlebirds left, and the New Hollands returned. Since little wattlebirds, when defending flowers, visit them less frequently than New Holland honeyeaters (Table 11-5), visitation rates to flowers will change over the season and will be lowest during the middle stages of flowering. The quantities of pollen transferred between flowers, are, then, most likely to be lowest during peak flowering, though in this case the high frequency of visits (see Table 11-5) probably ensures adequate levels of pollen transfer for maximum seed set throughout the flowering season.

Seasonal changes in visitation rates to flowers of other plants have not been assessed directly. However, changes in visitation rates can be inferred from changes in standing crops of nectar. The standing crop or level of nectar in a flower is a function of the time since the flower was last visited. High standing crops indicate lower visitation rates. Standing crops of nectar are usually highest, and we infer that visitation rates to flowers are lowest during peak flowering (Fig. 11-5), Some of these seasonal changes in standing crop may be due to seasonal changes in nectar production. However, for at least *Astroloma conostephioides, Amyema pendula,* and *Epacris impressa* there is no consistent change in nectar production over a flowering season (Paton, 1979, unpub.). These species, then, experience lower visitation rates and lower levels of pollination during peak flowering. At Golton Vale, in central Victoria, increases in nectar standing crops in *Astroloma conostephioides* were associated with changes in bird species and their behavior. Early in the season, only eastern spinebills used the scattered flowers, and nectar levels averaged 10 J/flower by afternoon. As the flowering season progressed, New Holland honeyeaters began using the plants and defending individual nonbreeding feeding territories, and nectar levels averaged 20 J/flower by afternoon. Later in the season and during peak flowering New Holland honeyeaters defended breeding territories, and nectar levels averaged around 50 J/flower. Since nectar production averaged about 48 J/flower/day throughout the season (Paton, 1979), these changes in standing crop indicate about a fivefold change in visitation rates to flowers over the flowering season. In other plant species the higher standing crops of nectar during peak flowering may be due to an inability of the

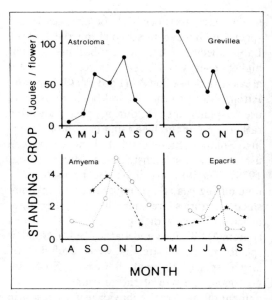

Figure 11-5. Seasonal changes in afternoon standing crops of nectar in *Astroloma conostephioides, Grevillea aquifolia, Amyema pendula,* and *Epacris impressa.* Data on *Astroloma* and *Grevillea* come from South Australia in 1975 (see Ford, 1979). Data on *Amyema* and *Epacris* come from Cranbourne in 1976 (. . . . .) and 1977 (- - - - -). Average standing crop of 30 to 200 flowers is plotted.

birds to keep track of increasing flower numbers (Recher, 1981).

In general, the visitation rates to individual flowers and hence levels of pollination appear to be lowest during peak flowering. This appears to be due, at least in some cases, to larger species exploiting the flowers during peak flowering.

GENERAL COMMENTS

Our assessments of honeyeater behavior suggest that smaller species of honeyeater will affect higher amounts of cross-pollination than larger species, that plants with greater floral displays will achieve higher amounts of cross-pollination, and that cross-pollination may be lowest during peak flowering. These assessments could be confirmed by measuring seed set over a flowering season and on individual plants that differed in their floral displays or in the species of honeyeater visiting their flowers. However, seed set in a plant could reflect the amounts of resources the plant allocates to seeds rather than changes in pollination. Also the frequency of visits to flowers may be more than adequate to guarantee sufficient pollen to maximize seed set on all plants. The advantage to plants that secure greater quantities of outcross pollen may be in having a wider range of pollen types available to fertilize ovules. This advantage might be shown in the quality of the seed crop rather than the quantity, and could be difficult to detect. A third factor might influence seed set as well. So far we have ignored self pollination, since it is unlikely to result in seed set in many of these plants. Self pollination, however, may result in clogging stigmas and preventing eventual cross pollen from germinating. Since the quantities of self pollen deposited on the stigmas of a plant increase as the quantities of outcross pollen increase, potential benefits from increased cross-pollination may not be expressed in seed set.

Measurements of the quantities of pollen on stigmas and the quantities of pollen remaining on anthers of different plants, at different times in the year and under different conditions, are more reliable ways of confirming our assessments. Plants that achieved higher levels of pollination should have more pollen on their stigmas, and less pollen remaining on their anthers. The quantities of pollen remaining in anthers should be a measure of the plant's success in donating pollen. Although some of the pollen on stigmas is likely to be self pollen, differentiating self pollen from outcross pollen may be possible in plants with naturally occurring pollen morphs (e.g., Thomson and Plowright, 1980).

ACKNOWLEDGMENTS

We thank Graham Pyke, Lynn Carpenter, Nick Reid, and Penny Paton for their comments on earlier drafts.

LITERATURE CITED

Armstrong, J. A. 1979. Biotic pollination mechanisms in the Australian flora—a review. *N.Z. J. Bot.* 17:467–508.

Ashton, D. H. 1975. Studies of flowering behaviour in *Eucalyptus regnans* F. Muell. *Aust. J. Bot.* 23:399–412.

Bell, H. L. 1966. A population study of heathland birds. *Emu* 65:295–304.

Carpenter, F. L. 1978. A spectrum of nectar-eating communities. *Am. Zool.* 18:809–819.

Charnov, E. L. 1976. Optimal foraging: the marginal value theorem. *Theor. Pop. Biol.* 9:129–136.

Ford, H. A. 1976. The honeyeaters of Kangaroo Island. *S. Aust. Orn.* 27:134–138.

———. 1977. The ecology of honeyeaters in South Australia. *S. Aust. Orn.* 27:199–203.

———. 1979. Interspecific competition in Australian honeyeaters—depletion of common resources. *Aust. J. Ecol.* 4:145–164.

———. 1981. Territorial behaviour in an Australian nectar-feeding bird. *Aust. J. Ecol.* 6:131–134.

Ford, H. A. and D. C. Paton. 1977. The comparative ecology of ten species of honeyeaters in South Australia. *Aust. J. Ecol.* 3:399–407.

———. 1982. Partitioning of nectar resources in an Australian honeyeater community. *Aust J. Ecol.* 7:149–159.

Ford, H. A., D. C. Paton, and N. Forde. 1979. Birds as pollinators of Australian plants. *N.Z. J. Bot.* 17:509–519.

Feinsinger, P. 1978. Ecological interactions between plants and hummingbirds in a successional tropical community. *Ecol. Mon.* 48:269–287.

Gill, F. B. and L. L. Wolf. 1975a. Foraging strategies and energetics of East African sunbirds at mistletoe flowers. *Am. Nat.* 109:491–510.

———. 1975b. Economics of territoriality in the golden-winged sunbird. *Ecology* 56:333–345.

Hainsworth, F. R. 1973. On the tongue of a hummingbird: its role in the rate and energetics of feeding. *Comp. Biochem. Physiol.* 46A:65–78.

Hopper, S. D. 1980. Bird and mammal pollen vectors in *Banksia* communities at Cheyne Beach, Western Australia. *Aust. J. Bot.* 28:61–75.

Hopper, S. D. and A. H. Burbidge. 1978. Assortative pollination by red wattlebirds in a hybrid population of *Anigozanthos* Labill. (Haemodoraceae). *Aust. J. Bot.* 26:335–350.

Inouye, D. W. 1980. The effect of proboscis length and corolla tube lengths on patterns and rates of flower visitation by bumblebees. *Oecologia* 45:197–201.

Keast, J. A. 1968a. Competitive interactions and the evolution of ecological niches as illustrated by the Australian honeyeater genus *Melithreptus* (Meliphagidae). *Evolution* 22:762–784.

———. 1968b. Seasonal movements in the Australian honeyeaters (Meliphagidae) and their ecological significance. *Emu* 67:159–210.

———. 1976. The origins of adaptive zone utilizations and adaptive radiations, as illustrated by the Australian Meliphagidae. *Proc. 16th Int. Orn. Conf.*, pp. 71–82.

Keighery, G. J. 1982. Bird pollinated plants in Western Australia and their breeding systems. In J. A. Armstrong, J. M. Powell, and A. J. Richards (eds.), *Pollination and Evolution*. Royal Botanic Gardens, Sydney. pp. 77–89.

Paton, D. C. 1979. The behaviour and feeding ecology of

the New Holland honeyeater *Phylidonyris novaehollandiae* in Victoria. Ph.D. thesis, Monash University, Melbourne.

———. 1980. The importance of manna, honeydew and lerp in the diets of honeyeaters. *Emu* 80:213–226.

———. 1982. The influence of honeyeaters on flowering strategies of Australian plants. In J. A. Armstrong, J. M. Powell, and A. J. Richards (eds.), *Pollination and Evolution*. Royal Botanic Gardens, Sydney. pp. 95–108.

Paton, D. C. and H. A. Ford. 1977. Pollination by birds in some native plants in South Australia. *Emu* 77:73–85.

Primack, R. B. and J. A. Silander. 1975. Measuring the relative importance of different pollinators to plants. *Nature* 255:143–144.

Pyke, G. H. 1978. Optimal foraging in hummingbirds: testing the marginal value theorem. *Am. Zool.* 18:739–752.

———. 1980. The foraging behaviour of Australian honeyeaters: a review and some comparisons with hummingbirds. *Aust. J. Ecol.* 5:343–369.

———. 1981a. Honeyeater foraging: a test of optimal foraging theory. *Anim. Behav.* 29:878–888.

———. 1981b. Hummingbird foraging on artificial inflorescences. *Behav. Analysis Lett.* 1:11–15.

Pyke, G. H., H. R. Pulliam, and E. L. Charnov. 1977. Optimal foraging: a selective review of theory and tests. *Qu. Rev. Biol.* 52:137–154.

Recher, H. F. 1977. Ecology of coexisting white-cheeked and New Holland honeyeaters. *Emu* 77:136–142.

———. 1981. Nectar-feeding and its evolution among Australian vertebrates. In J. A. Keast (ed.), *Ecological Biogeography of Australia*. W. Junk, The Hague, pp. 1639–1648.

Sargent, O. H. 1928. Reactions between birds and plants. *Emu* 27:185–192.

Thomson, J. D. and R. C. Plowright. 1980. Pollen carryover, nectar rewards, and pollinator behavior with special reference to *Diervilla lonicera*. *Oecologia* 46:68–74.

Tucker, V. A. 1974. Energetics of natural flight in birds. In R. A. Paynter (ed.), *Avian Energetics*. Publ. Nuttall Ornith. Club No. 15, Cambridge, Mass., pp. 298–333.

Wolf, L. L., F. R. Hainsworth, and F. B. Gill. 1975. Foraging efficiencies and time budgets in nectar-feeding birds. *Ecology* 56:117–128.

12
EXPERIMENTAL STUDIES OF REPRODUCTIVE ENERGY ALLOCATION IN PLANTS

William E. Evenson

Department of Physics and Astronomy
Brigham Young University
Provo, Utah

ABSTRACT

The "Principle of Allocation" is reviewed, and its relevance to plant reproduction is discussed. Experimental studies of energy allocation to reproduction in plants are reviewed. Results of these studies are summarized, and generalizations to be drawn from the studies are proposed and examined. Future trends are briefly projected.

KEY WORDS: Energy allocation, resource allocation, calorimetry.

CONTENTS

AIMS AND LIMITATIONS OF STUDIES OF ENERGY ALLOCATION

Studies of the use of resources for reproduction are directed at increasing our understanding of the operation of natural selection in the evolution of sex and life history strategies, the effects of habitat and competition on resource allocation, and systematic and phylogenetic questions.

Natural selection operates to achieve optimal use of the limited resources available to living organisms. Each individual assimilates a finite amount of available resources that must be allocated to growth, maintenance and defense, and reproduction according to the "Principle of Allocation" (Cody, 1966, following Levins and MacArthur, unpub.).

Interpretation of resource allocation patterns depends critically on the aspect of resource allocation that natural selection tends to optimize. It has been suggested that maximization of expansive energy, i.e., energy used for growth and reproduction, is a basic principle of natural selection (Van Valen, 1976). Indeed, the energy allocated to a particular structure or function, relative to the total energy available to the organism, must give some indication of the relative value placed upon that structure or function by natural selection (Harper, 1967; Gadgil and Bossert, 1970; Harper and Ogden, 1970; Kawano, 1975). However, there is no fundamental reason to grant energy investment a special status relative to the investment of other resources that may be limiting. Limiting resources, be they energy, time, or material, should be the gauge of reproductive investment (Harper and Ogden, 1970; Leigh, 1970; Harper, 1977; Lovett Doust and Harper, 1980; Bierzychudek, 1981).

Limiting resources other than energy may be thought of as providing constraints upon the energy optimization tendency of natural selection. Hence, energy content of tissues may be an appropriate measure of relative resource allocation between plants subject to similar resource regimes. The resource pool may be thought of as determining the possible "fitness sets" (Maynard Smith, 1977, 1978b) within which different strategies of energy allocation can operate. Solbrig and Solbrig (1979) consider, by way of illustration, a population of plants in which individuals develop, by mutation, a chemical in their leaves that is poisonous to a predator. They point out that the energy necessary to produce the compound can no longer be invested in offspring, reducing fecundity. If the mutation is to spread in the population, increased survival due to the new compound must more than compensate for the lowered fecundity. However, energy will be the essential factor reducing fecundity in this evolutionary equation only if the constraints imposed by other resources are relatively constant during many generations of evolutionary change. Bierzychudek (1981) points out that pollinators may be limiting reproductive effort rather than resources. Such possibilities must be considered carefully in the interpretation of resource allocation data.

It must be borne in mind, in energy allocation studies, that caloric energy in tissues is a relatively indiscriminating measure of resource allocation. Conceivably, many different chemical compositions of tissues could result in the same energy content. Furthermore, tissues of different chemistries may be produced from available energy at quite different efficiencies.

Inherent variability of tissues between individuals, which in many cases completely masks differences between energy and biomass distributions (Hickman and Pitelka, 1975), adds further difficulties of interpretation. This within-population variability is doubtless affected by the fact that both absolute and relative energy content of tissues change with time. Whether or not the energy content of standing biomass gives a good estimate of energy assimilation will depend on life history strategy. Successive measurements of the amount of energy within various tissues during the growing season, or an entire lifetime, are always desirable. In practice, measurements of respiration have not been made in plant energy allocation studies; so energy measurements refer to net energy, after respiratory demands have been met, as Ogden (1974) observed.

With full awareness of the problems and potential ambiguities suggested above, measurements of energy allocation among plant tissues, whether by direct calorimetry, dry biomass, or other means of estimation, can give important, if not always precise, information about the strategy by which a plant allocates its resources. Energy allocation studies may be critical to advances in our detailed appreciation of reproductive and life history strategies, since these can be understood, in principle, as adaptations that optimize the use of an organism's resource budget under the environmental conditions in which it grows and propagates (Gadgil and Bossert, 1970; Giesel, 1976).

Plant systematic studies have long stressed

the importance of reproductive characters for understanding both phylogenetic relationships and the processes of plant evolution (Stebbins, 1970, 1971). Principal considerations in any study of higher plant evolution must include pollination mechanisms, seed development and dispersal syndromes, and seedling establishment. Energy studies can help us understand the relative costs of these essential processes; for example, energy allocated to reproduction can be divided between that which is passed on to the next generation ("embryonic and endospermic capital"), and energy necessary for protection and dispersal of this "capital" (Harper and Ogden, 1970). The evolution and ecology of pollination, dispersal, etc., can be studied in this way. For instance, Harper et al. (1970) suggest that large numbers of small seeds or few large seeds represent alternative strategies in the disposition of reproductive resources. While equal biomass alone may not represent equivalent energy investment, relative energies may help to elucidate the ecology of these two alternatives. In fact, these extremes in seed production are often associated with r- and K-strategists, respectively. Gadgil and Solbrig (1972) have shown clearly the importance of measurements of reproductive effort (ratio of energy invested in reproductive tissues, to energy invested in the whole plant), under different densities, to properly assess relative r- vs. K-strategies.

Pollination mechanisms show great diversity (Proctor and Yeo, 1972; Stanley and Linskens 1974; Frankel and Galun, 1977; Faegri and van der Pijl, 1979). The energy requirements for successful pollen dispersal depend on reproductive strategy, pollinator types and their energy requirements, and necessary attracting agents provided by the plant (Heinrich and Raven, 1972; Heinrich, 1975). Hence, selection pressures leading to any particular pollination mechanism are the result of resource allocation to all participants in the pollination process: pollen, anther walls and filaments, nectar, perianth and visual attractants, pollinators, etc. Resource allocation to tissues that participate in the pollination syndrome tends to be correlated between different tissues; for example, energy and other resources allocated to stamens and to petals seem to be related

(Nitsch, 1965; Lovett Doust and Harper, 1980).

The evolution of sex is another topic that has received considerable attention in recent years (Ghiselin, 1974; Williams, 1975; Maynard Smith, 1978a; Lloyd, 1980). Allocation of resources between male and female functions may contribute significantly to the selection forces responsible for the evolution of breeding systems (Charnov et al., 1976; Maynard Smith, 1977). Several studies have been reported that partition reproductive effort between male and female functions with attention to different reproductive strategies (Smith and Evenson, 1978; Lemen, 1980; Lovett Doust and Harper, 1980; Gross and Soule, 1981; D. C. Freeman, pers. comm.).

THE DETERMINATION OF ENERGY ALLOCATION PATTERNS

The most direct method of determining energy allocation to plant tissues is oxygen bomb calorimetry (Paine, 1971). Although this method suffers from the limitations discussed above as well as from some minor experimental subtleties (requiring corrections for burned fuse wire, ash, and side chemical reactions), it is relatively straightforward and reliable. If materials are properly dried, corrections tend to be small for plants, and with reasonable care results can be obtained that are consistent to within 3% (Golley, 1961; Cummins and Wuycheck, 1971; Paine, 1971). The role of the bomb calorimeter is to burn the tissues in an oxygen atmosphere to their original constituents: carbon dioxide, water, and minerals, releasing the binding energy of the tissue molecules in the process. This released energy is measured by a temperature rise in the calorimeter and should be proportional, at the level of efficiency of the production process, to the energy used in producing the tissues.

A still simpler approach experimentally is to estimate energy content from dry biomass. Hickman and Pitelka (1975) found that relative allocations of energy to roots, stems, leaves, flowers, pods, and seeds of five populations of four species in quite different habitats were not significantly different by direct calorimetry and by dry weight, because of typically high within-population variances. They concluded that calorimetry is not necessary for primarily carbohydrate-storing plants. Many recent studies of energy allocation have been based on dry weight, as reviewed below, except for tissues high in lipids (which have an unusually high energy content).

Chemical component analysis of tissues also has been used to estimate energy content. Typical caloric values for tissue components are carbohydrate: 4100 cal/ash-free gram, lipid: 9450 cal/ash-free gram, and protein: 5650 cal/ash-free gram (Paine, 1971). These analyses generally come quite close to calorimeter results. Paine also mentions wet oxidation and thermochemical methods in his review, but these methods have not proved to be important in studies of plant reproductive effort.

Pollen chemistry has been reviewed by Stanley and Linskens (1974), who report that at anthesis grass pollens typically contain more than 50% water, whereas longer-lived pollens may have less than 20% water. Ash content is typically 2.5 to 6.5%, whereas lipids generally constitute something less than 5% of pollen weights. Carbohydrate and protein content vary widely. These figures indicate that dehydration and ash corrections may be important in determinations of caloric content of pollens.

Some workers have used general estimates of resource allocation such as number of flowers, leaf area, number of seeds or fruits, number of pollen grains per ovule, etc. (e.g., Harper, 1977; Law, 1979; Primack, 1979; Lord, 1980; Philipp, 1980; Spira, 1980). These measures give indications of relative resources allocated to reproduction, or to male or female functions. Although it is very difficult to compare most of these measures between studies without knowing more about their energy equivalents, they are so accessible and widely used that their value should not be overlooked in resource allocation studies.

REVIEW OF REPRODUCTIVE ENERGY ALLOCATION STUDIES

Much early work on energetics of plant and animal tissues focused on energy flow through ecosystems, i.e., trophic dynamics, following

Lindeman (1942). Golley (1961) used a Parr oxygen bomb calorimeter and reported average values of caloric content for various plant parts including leaves, stems and branches, roots, litter, and seeds. He emphasized the importance of making fuse, acid, and ash corrections in oxygen bomb calorimetry. Ash corrections made a substantial difference in some of the patterns in his data. He also considered seasonal and habitat variations in caloric content of plant parts. Golley did not report any data specifically on allocation of energy to reproduction in plants.

Cummins and Wuycheck (1971) provided an encyclopedic review of caloric content of plant and animal tissues from the literature as well as their own work. Their review was oriented toward ecosystem energetics, but they reported data for reproductive as well as vegetative organs of some plants.

Studies that have dealt directly with energy allocated to reproduction are reviewed below according to the focus of the research. Much of the work deals with reproductive effort: i.e., the ratio of energy or other resource allocation in reproductive tissues to that in the whole plant. (Table 12-1 summarizes these studies.)

Energetics of Reproductive Strategies

Numerous workers have investigated the relative energetics of sexual and asexual reproduction. Struik (1965) found that asexual reproductive effort, by dry weight, in *Circaea quadrisulcata* (enchanter's nightshade, Onagraceae) was about 12%. In the dry-mesic habitat, where it flowers and fruits abundantly, asexual effort was about twice sexual effort. In the mesic habitat, where it rarely flowers and fruits, asexual effort was about four times sexual effort, sexual effort decreasing by half from dry-mesic to mesic habitat.

Smith (1972) used a Parr semimicro oxygen bomb calorimeter to explore the costs of sexual vs. asexual reproduction in wild strawberries (*Fragaria virginiana,* Rosaceae). He reported caloric values for runners, whole fruits, whole seeds, and the ratio of sexual to asexual reproductive effort under weeded and unweeded conditions in natural habitats. He also used

Table 12-1. Summary of studies of reproductive effort in plants.

Reproductive Strategy	Reference	Taxon	Family
Relative effort to vegetative and sexual reproduction	Struik, 1965	*Circaea quadrisulcata*	Onagraceae
	Smith, 1972	*Fragaria virginiana*	Rosaceae
	Tripathi and Harper, 1973	*Agropyron*	Poaceae
	Ogden, 1974	*Tussilago farfara*	Asteraceae
	Abrahamson, 1975a,b	*Rubus*	Rosaceae
	Kawano and Nagai, 1975	*Allium*	Liliaceae
	Bradbury and Hofstra, 1976	*Solidago canadensis*	Asteraceae
	Sarukhan, 1976	*Ranunculus*	Ranunculaceae
	Turkington and Cavers, 1978	*Medicago* and *Trifolium*	Fabaceae
	Bostock and Benton, 1979	*Achillea millefolium,* *Artemisia vulgaris,* *Cirsium arvense,* *Taraxacum officinale,* *Tussilago farfara*	
	Hancock and Bringhurst, 1980	*Fragaria chiloensis*	Asteraceae
	Pitelka et al., 1980	*Aster acuminatus*	Rosaceae
	Douglas, 1981	*Mimulus primuloides*	Asteraceae
	Grace and Wetzel, 1981	*Typha latifolia*	Scrophulariaceae
			Typhaceae
Trade-off between reproductive effort and vegetative growth	Hickman, 1975	*Polygonum cascadense*	Polygonaceae
	Law, 1979	*Poa annua*	Poaceae
	Hancock and Bringhurst, 1980	*Fragaria chiloensis*	Rosaceae
	Davis, 1981	*Trillium erectum*	Liliaceae
	Primack et al., 1981	*Solidago*	Asteraceae
Relative effort to male and female functions	Putwaine and Harper, 1972	*Rumex*	Polygonaceae
	Smith and Evenson, 1978	*Amaryllis*	Amaryllidaceae
	Wallace and Rundel, 1979	*Simmondsia chinensis*	Buxaceae
	Hancock and Bringhurst, 1980	*Fragaria chiloensis*	Roasceae
	Lemen, 1980	Wind-pollinated plants	
	Lovett Doust and Harper, 1980	*Smyrnium olusatrum*	Apiaceae
	Philipp, 1980	*Stellaria longipes*	Caryophyllaceae
	Gross and Soule, 1981	*Silene alba*	Caryophyllaceae

Table 12-1. Summary of studies of reproductive effort in plants. (*Continued*)

	Reference	Taxon	Family
Cleistogamy vs. chasmogamy; selfing vs. outcrossing	McNamara and Quinn, 1977	*Amphicarpum purshii*	Poaceae
	Schemske, 1978	*Impatiens*	Balsaminaceae
	Waller, 1979, 1980	*Impatiens capensis*	Balsaminaceae
	Lord, 1980	*Lamium amplexicaule*	Lamiaceae
	Spira, 1980	*Trichostema*	Lamiaceae
Energetics of pollens	Colin and Jones, 1980	42 species	
Life History			
Differences between life forms	Struik, 1965	Annual and perennial herbs	
	Harper et al., 1970	Chart showing range of reproductive effort by life form	
	Hickman, 1977	*Polygonum*	Polygonaceae
	Pitelka, 1977	*Lupinus*	Fabaceae
	Abrahamson, 1979	50 species	
	Primack, 1979	*Plantago*	Plantaginaceae
Phenological patterns, young plants, effect of seed size on reproductive strategy of progeny, tropical species, exotic vs. native species, cultivated vs. wild species	Struik, 1965	Annual and perennial herbs	
	Harper and Ogden, 1970	Several species	Asteraceae
	Van Valen, 1975	*Euterpe globosa*	Palmae
	Hawthorn and Cavers, 1978	*Plantago*	Plantaginaceae
	Abrahamson, 1979	50 species	
	Clark and Burk, 1980	*Plantago insularis*	Plantaginaceae
	Sarukhan, 1980	*Astrocaryum mexicanum*	Palmae
	de Ridder et al., 1981	Annual pasture plants	
	Gross and Soule, 1981	*Silene alba*	Caryophyllaceae
Plastic vs. genetically fixed reproductive effort	Abrahamson, 1975a,b	*Rubus*	Rosaceae
	Hickman, 1975	*Polygonum cascadense*	Polygonaceae
	Abrahamson and Hershey, 1977	*Impatiens capensis*	Balsaminaceae
	Hickman, 1977	*Polygonum*	Polygonaceae
	Holler and Abrahamson, 1977	*Fragaria virginiana*	Rosaceae
	Roos and Quinn, 1977	*Andropogon scoparius*	Poaceae
	Grace and Wetzel, 1981	*Typha latifolia*	Typhaceae

	Reference	Species	Family
Succession and Site Disturbance Effects	Gadgil and Solbrig, 1972	*Taraxacum officinale*	Asteraceae
	Abrahamson and Gadgil, 1973	Three herbaceous communities	
	Gaines et al., 1974	*Solidago*	Asteraceae
	Abrahamson, 1975a,b	*Helianthus*	Asteraceae
	Hickman, 1975	*Rubus*	Rosaceae
	Hickman and Pitelka, 1975	*Polygonum cascadense*	Polygonaceae
	Bradbury and Hofstra, 1976	*Lupinus*	Fabaceae
	Werner and Platt, 1976	*Solidago canadensis*	Asteraceae
	Keeley and Keeley, 1977	*Solidago*	Asteraceae
	Roos and Quinn, 1977	*Arctostaphylos*	Ericaceae
	Newell and Tramer, 1978	*Andropogon scoparius*	Poaceae
	Turkington and Cavers, 1978	Three herbaceous communities	
	Abrahamson, 1979	*Medicago* and *Trifolium*	Fabaceae
	Luftensteiner, 1980	52 populations	
	Grace and Wetzel, 1981	Four communities	
	Primack et al., 1981	*Typha latifolia*	Typhaceae
		Solidago	Asteraceae
Habitat Effects			
Density and competition	Harper and Ogden, 1970	*Senecio vulgaris*	Asteraceae
	Smith, 1972	*Fragaria virginiana*	Rosaceae
	Ogden, 1974	*Tussilago farfara*	Asteraceae
	Snell and Burch, 1975	*Chamaesyce hirta*	Euphorbiaceae
	Abrahamson, 1975a,b	*Rubus*	Rosaceae
	Werner and Platt, 1976	*Solidago*	Asteraceae
	Holler and Abrahamson, 1977	*Fragaria virginiana*	Rosaceae
	Hawthorn and Cavers, 1978	*Plantago*	Plantaginaceae
	Luftensteiner, 1980	49 species	
	Pitelka et al., 1980	*Aster acuminatus*	Asteraceae
Nutrients and soil fertility	Ogden, 1974	*Tussilago farfara*	Asteraceae
	Snell and Burch, 1975	*Chamaesyce hirta*	Euphorbiaceae
	van Andel and Vera, 1977	*Senecio sylvaticus*	Asteraceae
	Lovett Doust and Harper, 1980	*Chamaenerion angustifolium*	Onagraceae
		Smyrnium olusatrum	Apiaceae
	Williams and Bell, 1981	Mojave Desert winter annuals	
Moisture	Abrahamson, 1975a,b	*Rubus*	Rosaceae
	Hickman and Pitelka, 1975	*Polygonum cascadense*	Polygonaceae
	Werner and Platt, 1976	*Solidago*	Asteraceae

Table 12-1. Summary of studies of reproductive effort in plants. (*Continued*)

	Reference	Taxon	Family
	Abrahamson and Hershey, 1977	*Impatiens capensis*	Balsaminaceae
	Hickman, 1977	*Polygonum*	Polygonaceae
Elevation, length of growing season, light	Racine and Downhower, 1974	*Opuntia*	Cactaceae
	McNaughton, 1975	*Typha*	Typhaceae
	Jolls, 1980	*Sedum lanceolatum*	Crassulaceae
	Kawano and Masuda, 1980	*Heloniopsis orientalis*	Liliaceae
	Pitelka et al., 1980	*Aster acuminatus*	Asteraceae
	Douglas, 1981	*Mimulus primuloides*	Scrophulariaceae
Insect stem galls	Hartnett and Abrahamson, 1979	*Solidago canadensis*	Asteraceae

numbers of flowers and fruits and numbers and length of runners to help estimate resources allocated to sexual and asexual reproduction. Sexual reproduction cost one to eight times as much as asexual reproduction, depending upon the conditions of the experiment. Weeding did not affect the ratio of sexual to asexual effort because both kinds of reproduction increased at about the same rate.

The extent to which clonal growth (vegetative reproduction) and seed reproduction are alternative processes was investigated in two closely related species of *Agropyron* (wheatgrass, Poaceae), *A. repens* (quackgrass) and *A. caninum* (slender wheatgrass, now *A. trachycaulum*), by Tripathi and Harper (1973). *A. caninum* reproduces only by seeds, whereas *A. repens* reproduces by both seeds and rhizome buds. The total numbers of seeds plus rhizome buds for the two species was comparable, supporting the suggestion that there is a trade-off between the two kinds of reproduction.

Ogden (1974) reported on the reproductive strategy of *Tussilago farfara* (Asteraceae). He measured both dry weight and caloric content using an oxygen bomb calorimeter several times during the growing season. Dead matter was included as well as living matter to estimate both net production and biomass at each of the harvest times. He reported dry weights and caloric values for a fairly detailed division of plant parts under different levels of plant density and soil fertility. Plants growing in poor soils allocated twice as large a fraction of net assimilate to vegetative reproduction as did plants growing in rich soils, although total vegetative reproduction was less in poor soils. The total fraction of resources expended on reproduction was greater in poor soils, most of the reproductive effort going to vegetative reproduction. The fraction allocated to sexual reproduction was not strongly affected by soil fertility. Seed production was little affected by population density, but vegetative reproduction decreased with increased density. His data suggested that seral perennials invest about the same fraction of total net assimilate in reproduction as seral annuals, though less in sexual reproduction.

Reproductive effort by *Rubus hispidus* (northern dewberry or groundberry, Rosaceae) and *R. trivialis* (southern dewberry) in different habitats was reported by Abrahamson (1975a,b). Dry weight was estimated from sizes of organs' using conversion factors developed from measurements of neighboring populations. Sexual reproductive effort (seed production) declined with increased maturity of the site, whereas vegetative reproductive effort was approximately constant across the sites. Likewise, sexual reproductive effort was lower during the drier of the two years of the study, while vegetative reproductive effort was about the same for both years. In stable habitats, low density of vegetation favored vegetative reproduction, whereas high density led to increased seed reproductive effort.

Kawano and Nagai (1975) measured dry weights through the growing season of several organs in three species of *Allium* (onion, Liliaceae) native to Japan. Sexual reproductive effort was low in all three species (means, 1 to 3%), but vegetative reproductive effort, as measured by dry weight of daughter bulbs, was very high in two of the species—averaging about 55% in one of them, and 27% in the other.

Bradbury and Hofstra (1976) studied two populations of *Solidago canadensis* (Canada goldenrod, Asteraceae). Total reproductive effort was estimated by calorimetry (Parr oxygen bomb calorimeter) as well as by seed production, numbers of flowering heads, and rhizome lengths and numbers. Reproductive effort was similar for the two populations (about 12%). Vegetative reproductive effort was small: only about 1.5% of the energy in new tissues was due to new rhizomes.

Reproductive effort in three species of buttercups (*Ranunculus,* Ranunculaceae) was studied by Sarukhan (1976). Measuring dry weights, he found that vegetative reproduction in *R. repens* (creeping buttercup) occurred at the expense of reproduction by seed. Only *R. repens* reproduced vegetatively. All three species had about the same total reproductive effort, with within-species ranges in the neighborhood of 48 to 60%.

Turkington and Cavers (1978) found 20.4%

reproductive effort in *Trifolium repens* (white Dutch clover, Fabaceae). Most of this reproductive effort (18.5%) was to vegetative reproduction. In contrast, *T. pratense* (red clover), which reproduced only sexually, showed 12% reproductive effort.

The reproductive strategies of five perennial composites were studied by Bostock and Benton (1979), in an effort to clarify the strategic significance of vegetative reproduction and relate it to *r*- and *K*-strategies. They studied *Achillea millefolium* (common yarrow or milfoil), *Artemisia vulgaris* (mugwort), *Cirsium arvense* (Canada thistle), *Taraxacum officinale* (dandelion), and *Tussilago farfara,* in the wild as well as in pot-grown plants. They measured dry weights of several organs and reproductive statistics over two years. Seed reproductive effort was measured in three parts: embryo, pericarp and plume, and ancillary reproductive organs (peduncles, capitula, perianth, etc.). Reproductive effort ranged from 11% in *Artemisia* to 46% in *Tussilago.* Sexual reproductive effort was 2–3% in *Achillea* and *Artemisia,* about 7% in *Cirsium,* and 24–26% in *Taraxacum* and *Tussilago.* Vegetative reproductive effort varied from 9% in *Artemisia,* 13% in *Taraxacum,* and 15% in *Cirsium,* to 20% in *Tussilago* and 26% in *Achillea.* The expectations of *r*- and *K*-strategy theory were essentially borne out as reproductive effort was considered in connection with the usual habitat of each species. Bostock and Benton discussed the difficulties of interpreting the role of vegetative reproduction in reproductive effort.

Dry weights of flowers, vegetative tissues, and roots of greenhouse-grown male and female wild strawberries (*Fragaria chiloensis,* Rosaceae) were measured by Hancock and Bringhurst (1980). They found that female plants invested more energy in sexual reproductive effort than males (8.8% vs. 1.5%) at the expense of vegetative growth. Male flowers and peduncles were more costly than female (1.5% vs. 0.8%), but investment by females in fruits more than compensated for this. Male and female plants had about the same vegetative reproductive allocation (about 86%).

Pitelka et al. (1980) considered the effects of light and density on sexual and vegetative reproductive effort in the forest herb *Aster acuminatus* (Asteraceae). They found vegetative reproductive effort to be independent of these factors, while sexual reproductive effort increased with both light level and plant size. No effects of density were observed.

The balance between vegetative and sexual reproduction in *Mimulus primuloides* (monkeyflower, Scrophulariaceae) was studied by Douglas (1981), who measured biomass of various organs from five California populations at different elevations. Vegetative reproductive effort was greatest at a midelevation (2500 meters), ranging from 41% at the lowest elevation, through 56% at midelevation, to 46% at the highest elevation. Sexual reproductive effort was approximately independent of total biomass, varying between 2% and 4% at the four lower elevations. However, a high rate of flower production was apparently genetically fixed in the highest-altitude population, where sexual reproductive effort was 14%, even though seedling establishment was rare. Douglas observed that sexual reproduction may be important only when large areas are opened to invasion. The variation of vegetative reproductive effort with altitude was explained by reduction of plant size at lower elevations due to intraspecific competition and at high elevations by the severe environment. This explanation was supported by greenhouse experiments at three densities. She concluded that reproductive statistics and biomass estimates of reproductive effort are both important for understanding reproductive strategy.

Grace and Wetzel (1981) estimated biomass allocations in *Typha latifolia* (cattail, Typhaceae) from ^{14}C fixation. They found evidence of trade-off between sexual and vegetative reproductive effort. Ramets which flowered showed a marked reduction in vegetative reproductive effort of 10–15%, and flowering correlated with size of ramet, in an open marsh.

The extent to which reproductive effort represents a trade-off with vegetative growth has been addressed by four studies. Hickman (1975) found reproductive allocation by *Polygonum cascadense* (knotweed, Polygonaceae) to be strongly negatively correlated with biomass between populations, but not within populations.

Law (1979) estimated reproductive effort in several *Poa annua* (annual meadow grass or annual bluegrass, Poaceae) populations from the number of inflorescences per plant. High rates of reproduction early in life were associated with lower reproductive rates and smaller plant sizes subsequently, confirming the concept that there is a trade-off "cost" associated with reproduction.

Hancock and Bringhurst (1980) found that female investment in fruits by wild strawberries came at the expense of vegetative growth.

In a four-year study of *Trillium erectum* (wake-robin, Liliaceae), Davis (1981) found that single-flowered individuals expended a significantly smaller ($p < 0.01$) proportion of their resources on sexual reproduction (8.7%) than did double-flowered plants (11.1%). Double-flowered plants were reproductively superior, but 85% of the population were single-flowered. Davis developed evidence that the reason for the predominance of plants with lower sexual reproductive effort was resource limitation, leading to a trade-off between reproductive effort and vegetative and root growth.

Primack et al. (1981) measured dry weight of all leaves, living and dead, stems, and seed heads in five species of goldenrods (*Solidago,* Asteraceae) collected at seed maturation. They also measured numbers of flowering stems, branches, heads, and seeds. They found that individuals within populations increased reproductive effort by increasing reproductive yield, not by decreasing vegetative yield, except *Solidago odora,* which *increased* vegetative yield while increasing reproductive effort. Each species had a distinct pattern of regulation of reproductive effort. These authors considered three habitats: an open, disturbed habitat; a forest edge; and a forest interior. Reproductive effort was least for the forest species and greatest for the open-ground species. These results are in agreement with *r*- vs. *K*-strategy theory and similar to the findings of Abrahamson and Gadgil (1973) on goldenrods.

Studies of relative costs of male and female functions in plants include—in addition to the work of Hancock and Bringhurst (1980) cited above—research by Putwain and Harper (1972). They looked at dry weight allocations in the dioecious sorrel, *Rumex acetosella* (sheep's sorrel, Polygonaceae), and found that male plants invested a greater proportion of resources in roots and vegetative offshoots than did females. The two sexes utilized resources at different times during the growing season.

Smith and Evenson (1978) reported a detailed analysis of floral parts by dry weight and calorimetry (Parr oxygen bomb calorimeter) for some hybrid varieties of *Amaryllis* (Amaryllidaceae). They did not examine total energy assimilate and hence did not discuss relative energy allocated to reproduction, but rather compared the energy allocated to male and female functions in the flower. They reported energy content of pollen; anther walls; ovules; placentae; ovary walls; perianth with filament, style, and stigma; seeds; and capsules. Typical *Amaryllis* flowers invested only 70% as much energy in ovules and placentae as in pollen and only about a third as much energy in ovules as in pollen. Slightly less than 30% of the energy per flower at anthesis was invested in ovary and anthers, and slightly less than 9% in pollen and ovules. When anther walls and ovary walls were included, about 1.5 times more energy was invested in female structures (17.2% of the flower energy in ovary) than in male structures (11.5% of the flower energy in anther). So protective structures for the ovules required considerably greater net investment in female functions than in male functions. About as much energy was invested in each mature seed as in the whole flower at anthesis.

Wallace and Rundel (1979) reported work with male and female shrubs of *Simmondsia chinensis* (jojoba or goat nut, Buxaceae), measuring both biomass and caloric values using a Parr semimicro oxygen bomb calorimeter. They measured energy allocation to leaves, stems, flowers, and fruits, looking at the distribution of living biomass at only a single point in time. They also measured caloric content of leaves, stems, and reproductive structures and estimated reproductive effort. In addition they looked at nitrogen, phosphorus, and glucose equivalents and analyzed the difference in reproductive effort for male and female plants on the basis of all these resources. They found a strong leaf dimorphism between males and females in desert habitats which seemed to be associated with different reproductive energy requirements for the sexes. Female plants allocated relatively more energy to reproduction than did males when seed set was greater than 30%.

Lemen (1980) considered several wind-pollinated species to test models of relative reproductive effort by the sexes. He measured seed weights, anther volumes, and dry weights of above-ground plant parts. In self-compatible, monoecious species about 90% of total reproductive effort was allocated to female flowers, whereas in self-incompatible species, there was about equal allocation of resources to male and female functions. Male plants were generally smaller than female plants in dioecious species.

Lovett Doust and Harper (1980) reported detailed energy allocation to reproductive tissues for an andromonoecious umbellifer, *Smyrnium olusatrum* (Apiaceae). They measured dry weight, nitrogen, phosphorus, and potassium in stamens, pistils, petals, and stylopodia, as well as ratio of male to hermaphrodite flowers and total plant dry weight. These measurements were done for plants grown under six treatment conditions representing different resource regimes. They found that male gametes were more costly than female, but maternal care reversed the total cost.

Philipp (1980) reported a cultivation experiment with *Stellaria longipes* (chickweed, Caryophyllaceae), which has a gynodioecious/gynomonoecious breeding system. He observed number of seeds per capsule, number of flowers per plant and length of reproductive cycle in order to study how female plants are favored over hermaphrodite plants. He found differential resource utilization by sex of plant and by sex of flowers.

Gross and Soule (1981) studied biomass allocation in the dioecious, perennial herb (*Silene alba* (catchfly, Caryophyllaceae). They considered the sex ratio, growth characteristics, and seed size from which the plants were grown. Seed size significantly affected adult plant size and flower production, but these effects were not different by sex of the plants. Reproductive effort differed by sex, consistently across all seed size categories; males allocated relatively more energy to reproduction than females unless seed set was greater than 20%.

D. C. Freeman (pers. comm., 1981) has looked at environmental effects on spinach (*Spinacia,* Chenopodiaceae). He reports a comparison of caloric content of dry fruits vs. dry stamens to estimate relative energy investment in male and female parts. He finds 4163 ± 11 cal/gram for fruits and 3743 ± 41 cal/gram for stamens.

Evenson and co-workers are currently studying energy allocation in castor bean (*Ricinus communis,* Euphorbiaceae) in an attempt to understand reproductive effort, energy requirements of male vs. female functions, and the impact of the high oil content of the seeds on energy allocation patterns.

Reproductive effort in cleistogamous (closed) vs. chasmogamous (open) flowers and the costs of selfing vs. outcrossing have been studied by McNamara and Quinn (1977), Schemske (1978), Waller (1979, 1980), Lord (1980), and Spira (1980). Five New Jersey populations of *Amphicarpum purshii* (peanutgrass, Poaceae) were studied by McNamara and Quinn (1977). Reproductive effort, estimated from dry biomass, was about 29%. This annual, panicoid grass has both subterranean cleistogamous inflorescences and aerial chasmogamous inflorescences. The subterranean inflorescences accounted for 37 to 100% of reproductive effort and were the source of most surviving seedlings.

Schemske (1978) reported on energy allocation to chasmogamous vs. cleistogamous flowers in *Impatiens* (snapweed or jewelweed, Balsaminaceae), using a Phillipson microbomb calorimeter. He reported caloric values and dry weights for peduncles, capsules and placentae, anther caps, and seeds in cleistogamous flowers and peduncles, capsules and placentae, nectar, anthers, and seeds in chasmogamous flowers. He also reported seed and flower production data. He did not find net reproductive energy relative to total energy assimilate. Chasmogamous seeds were two to three times as costly as cleistogamous seeds.

Waller (1979, 1980) examined the relative costs of seeds from cleistogamous flowers and seeds from chasmogamous flowers in *Impatiens* and reported dry weight for plants, flowers, and seeds as well as calories, using caloric values found by Schemske (1978). He assumed that seeds from chasmogamous flowers were always outcrossed and that flowers of a given type were all similar. Unfortunately, these assumptions do not hold uniformly (E. M. Lord, pers. comm.). He found that seeds produced by chasmogamous flowers were about twice as expensive in calories as seeds from cleistogamous flowers and had about 1.5 times more dry weight. The two flower types had quite different flowering phenology.

Lord (1980) considered pollen/ovule ratios in cleistogamous and chasmogamous flowers in *Lamium amplexicaule* (dead nettle, Lamiaceae). She found that the pollen/ovule ratio was much lower in cleistogamous flowers than in chasmogamous flowers. The first-produced cleistogamous flowers seemed to be more efficient seed producers than were chasmogamous flowers, but later ones were often as costly in resources as chasmogamous flowers and had the additional disadvantages of inbreeding. The phenologies of cleistogamy and chasmogamy were quite different.

Spira (1980) used flower diameters, nectar volumes, and pollen/ovule ratios as indicators of energy per flower in *Trichostema* (vinegar weed, Lamiaceae). He found that cross-pollinating species expend more energy per floral unit and are less efficient seed producers than closely related selfing species.

Colin and Jones (1980) reported on pollen energetics as related to pollination modes for 42 plant species, divided into wind-pollinated dicots, wind-pollinated monocots and gymnosperms, and insect-pollinated dicots, and subdivided by life form (trees, shrubs, perennials, annuals) and taxonomic subclass. They found statistically significant differences between the energy content of pollens from wind-pollinated dicots ($\bar{x}$ = 5880 cal/gram) and from wind-pollinated monocots and gymnosperms ($\bar{x}$ =

5220 cal/gram), as well as within these groups. They found no significant difference between pollens from wind-pollinated dicots and those from insect-pollinated dicots ($\bar{x}$ = 5770 cal/gram), although there were significant differences within these groups. There were significant differences in energy content of pollens both between and within taxonomic subclasses of wind-pollinated dicots and monocots.

Life History Effects: Life Forms and Phenologies

Differences in reproductive effort by life form have been noted by several researchers. Harper et al. (1970, following J. Ogden) presented a chart showing expected ranges of reproductive effort for different plant life forms. They expected average reproductive effort to be greatest for maize and barley, and successively smaller for other grain crops, wild annuals, herbaceous perennials, shrubs, and trees.

Struik (1965) measured sexual reproductive effort by dry weight in open and forest habitats for native annual and perennial herbs in southern Wisconsin from May to November on a weekly sampling basis. She found that sexual reproductive effort was greater for annuals than for perennials, and that there was no significant difference in this relationship by habitat. She also found that reproductive effort was greater for annual crop plants than for native annuals. But timothy (*Phleum pratense,* Poaceae), which has been selected for foliage, had low reproductive effort, comparable to lower native perennials.

Energy allocation patterns were estimated by total dry biomass in buds, flowers and seeds, roots, hypocotyls, stems, and leaves in four coexisting annual species of *Polygonum* (knotweed, Polygonaceae) by Hickman (1977). *Polygonum cascadense* (Cascade knotweed) showed an energy allocation pattern that was adaptive along a moisture gradient, whereas the other three species showed patterns that were not affected by the environment. The four species showed niche differentiation based on spatial and temporal displacements and energy allocation differences. The energy allocation patterns could not be explained simply by, for example, r- vs. K-selection theory. *P. kelloggii* showed up to 76% reproductive effort. *P. cascadense* agreed with r- vs. K-selection theory superficially, but closely related species did not. *P. cascadense* is endemic to the area, whereas the other three species are recent migrants. This is an area of highly unpredictable environment. Hickman found that "annuals allocate major growth energy first to primary root, then sequentially to stems, leaves, flowers, and finally to fruits and seeds."

Pitelka (1977) studied three species of lupines (*Lupinus*, Fabaceae), reporting caloric content of various tissues (Phillipson microbomb calorimeter) as well as dry weight. The three species were *L. nanus*, an annual growing in disturbed sites, with 61% reproductive effort, 29% in seeds; *L. variicolor*, a perennial herb growing in stable but harsh environments, with 18% reproductive effort, 5% in seeds; and *L. arboreus*, a shrub growing in successional sites with 20% reproductive effort, 6% in seeds. These species showed a high fraction of their energy budgets in reproductive accessories (other than seeds) compared to other species that have been studied.

Abrahamson (1979) reported studies of 52 populations with 50 species of wildflowers in fields and woods. He found that the proportion of dry mass allocated to seed reproductive organs was greater in fields than in woods. The proportion of dry mass allocated to leaves and below-ground organs was greater in woodland herbs than in field herbs; that allocated to reproduction was greater in field annuals than in field perennials, and greater in exotics than native plants in fields.

The ratio of seed weight to leaf area was used by Primack (1979) as a measure of reproductive effort in annual and perennial species of *Plantago* (plantain, Plantaginaceae). He examined 15 species of *Plantago* in the field and 25 species from herbarium material. Annual species had greater reproductive effort and greater fruit output than perennial species. Spring annuals had higher reproductive effort than summer annuals. Weedy perennials had higher reproductive effort than nonweedy perennials. Primack discussed the adaptive significance of these patterns in light of r- and K-selection theory; his results are in accord with the expectations of theory.

Clark and Burk (1980) considered biweekly resource allocation patterns in two California–Sonoran desert ephemerals, *Plantago insularis* (plantain, Plantaginaceae) and *Camissonia boothii* (Onagraceae). The two species differed considerably in seasonal patterns of growth and reproduction, *Plantago* being earlier than *Camissonia* in reproduction and completion of its life cycle. *Camissonia* stored energy in vegetative tissues early in the season to use for later reproduction.

Two perennial species of *Plantago* were studied by Hawthorn and Cavers (1978) to examine allocation patterns of young plants. *P. major* and *P. rugelii* were grown in a greenhouse with different numbers and kinds of neighbors. Reproductive effort was estimated by dry weight of spikes relative to leaves, caudex, and roots plus spike. Reproductive effort in *P. major* remained about 21% regardless of density, even though the size of the plants was depressed as density increased. About half of this reproductive effort went to seeds and about half to scapes and capsules. The two species have different allocation patterns as young plants and seem to be adapted to different kinds of sites. *P. major*, with early and sustained seed production, seems best adapted to frequently disturbed sites, whereas *P. rugelii*, with greater root allocation and delayed seed production, is better adapted to sites of less frequent disturbance.

Energy allocation in tropical plants has been reported by Van Valen (1975) and Sarukhan (1980). Van Valen inferred reproductive effort and estimated net production for the palm *Euterpe globosa* (Palmae) in Puerto Rico. He found about 17% reproductive effort in individual palms, and about 5% in whole populations of palms. Sarukhan measured total dry matter in palms of *Astrocaryum mexicanum* (Palmae). Total matter processed by a palm to 120 years of age was allocated 65% to vegetative growth, 35% to reproduction, of which 30% was in fruits and 5% in reproductive accessories. Reproductive effort was observed to be essentially the same in the yearly budget of the 120-year-old palm. Sarukhan and coworkers have been able to detect no senescence effects on reproductive activity. These patterns are very different from those proposed by Harper and Ogden (1970) and from most patterns observed in temperate systems.

Reproductive effort of cultivated species as compared to wild species has been reported by Struik (1965), Harper and Ogden (1970), and de Ridder et al. (1981). As mentioned above, Struik found that reproductive effort was greater in annual crop plants than in native annuals. In contrast, reproductive effort in timothy was low compared to native perennials. Harper and Ogden reported crude reproductive efficiencies (harvest indices) from the literature for a few other members of the Asteraceae for comparative purposes, finding about the same reproductive effort (20–30%) in oilseed crops as in two weedy species. de Ridder et al., reviewing dry biomass allocations, found reproductive effort for annual pasture plants to be as high as cultivated species, with mean values of 26–60%.

The extent to which reproductive effort is a plastic rather than genetically fixed feature of the life history pattern has been addressed by Abrahamson (1975a,b), Hickman (1975, 1977), Abrahamson and Hershey (1977), Holler and Abrahamson (1977), Roos and Quinn (1977), and Grace and Wetzel (1981), using greenhouse experiments as well as field observations. Abrahamson reported that two species of dewberries were apparently plastic in the response of their reproductive effort to environmental conditions. Hickman found that *Polygonum cascadense* was plastic in its energy allocation patterns, whereas three other species of *Polygonum* did not respond to environmental differences. Abrahamson and Hershey reported the reproductive effort of *Impatiens capensis* (snapweed or jewelweed, Balsaminaceae) to be primarily fixed genetically. They studied two populations under extremely wet and dry conditions for the species and found no difference in allocation patterns. Plants from these two populations also responded the same under greenhouse conditions. They measured dry biomass, energy content of tissues (Parr

oxygen bomb calorimetry), and reproductive characteristics.

Holler and Abrahamson (1977) reported observations suggestive of some degree of plasticity in reproductive effort of *Fragaria virginiana* (wild strawberry, Rosaceae). Roos and Quinn (1977) found plastic response to the environment in *Andropogon scoparius* (little bluestem, Poaceae). Grace and Wetzel (1981), studying cattails in three habitats as well as under garden conditions, found evidence of both genetically fixed and plastic components of reproductive effort. Several other studies discussed here also suggest great differences from species to species in the degree of plasticity of reproductive effort.

Succession and Reproductive Effort

Much work has been carried out to study the relationship of successional stage to reproductive effort. Most of this work has focused on tests of *r*- vs. *K*-selection theory (MacArthur and Wilson, 1967; Gadgil and Solbrig, 1972).

Gadgil and Solbrig (1972) discussed resource allocation in three dandelion (*Taraxacum officinale,* Asteraceae) populations that contained four biotypes; they estimated energy allocation from dry weights and leaf surface areas. Reproductive effort was also reported for three herbaceous communities that were sampled on a weekly basis. They discussed their data in terms of *r*- vs. *K*-selection. The predominant dandelion biotype in disturbed sites had higher seed output, higher proportion of biomass devoted to reproduction, and lower competitive ability than the biotype that dominated a less disturbed site. In their second study, members of the herbaceous community on the more disturbed site devoted a greater fraction of resources to reproduction, on the average, than members of a community on a less disturbed site.

Reproductive effort in four species of goldenrods (*Solidago,* Asteraceae) was investigated by Abrahamson and Gadgil (1973). They confirmed the hypothesis that reproductive effort declines with increasing successional maturity of the community. In addition to caloric values (Parr oxygen bomb calorimeter), they measured dry weights and sizes of several characters for each of the four species of goldenrods.

Gaines et al. (1974) studied reproductive effort in four species of sunflowers (*Helianthus,* Asteraceae) by measuring dry weight as well as numbers and weights of seeds, numbers of heads, etc. They found that reproductive effort decreased with increasing successional maturity of the habitats of these sunflower species.

Abrahamson (1975a,b) found that sexual reproductive effort (seed production) in dewberries (*Rubus,* Rosaceae) declined with increasing successional maturity of the site, whereas vegetative reproductive effort was approximately constant across the sites. Likewise, sexual reproductive effort was lower during the drier of the two years of the study, while vegetative reproductive effort was about the same for both years. In stable habitats, low density of vegetation favored vegetative reproduction, whereas high density led to increased seed reproductive effort.

Polygonum cascadense (knotweed, Polygonaceae) was studied in several habitats by Hickman (1975). Dry weights of roots, hypocotyls, stems, leaves, buds, flowers, and seeds were measured as well as numbers of buds, flowers, seeds, etc. Greater reproductive effort was expended by this plant in harsh, open habitats than in moderate habitats (up to 71%). Reproductive effort was found to be very plastic in this species, not genetically fixed. This plasticity was apparently an adaptation to short-term environmental unpredictability. Hickman also provided a critical discussion of the measurement of reproductive allocation.

Hickman and Pitelka (1975) measured dry weights and caloric values for three species of *Lupinus* (lupine, Fabaceae) and two populations of *Polygonum cascadense* (knotweed, Polygonaceae) from a dry and a wet meadow. They measured energies and dry weights of roots, stems, leaves, flowers, pods, and seeds. They did not analyze reproductive effort, but the data they report suggest much greater reproductive allocation by *Lupinus* species in disturbed sites than in relatively undisturbed habitats. There was also a greater fraction of energy allocation to seeds at the expense of

roots, stems, and leaves in the *Polygonum* population in the dry habitat compared to the wet meadow.

Bradbury and Hofstra (1976) studied two populations of *Solidago canadensis* (Canada goldenrod, Asteraceae). One population was invading a grass-dominated, recently abandoned pasture. The other was a well-established population in a much older abandoned pasture. Reproductive effort was similar for the two populations (about 12%). Vegetative reproductive effort was small: only about 1.5% of the energy in new tissues was due to new rhizomes; however, rhizome length was greater in the invading population.

Werner and Platt (1976) studied species of co-occurring goldenrods (*Solidago,* Asteraceae). They found that each species produced fewer and heavier propagules with reduced dispersal capacity in prairie compared with old-field populations. So increased competitive pressures apparently favored selection of larger propagules. They also noted an increase in the allocation of biomass to vegetative reproduction with increased soil moisture.

Keeley and Keeley (1977) examined energy allocation patterns of two species of *Arctostaphylos* (manzanita, Ericaceae) to test *r*- vs. *K*-selection theory, studying two stands, one 23 years old, the other 90 years old. They measured dry weight of reproductive parts and dry weight of terminal vegetative growth in *A. glauca* and *A. glandulosa;* they also measured weight and number of fruits. Fruit production suggested greater reproductive effort by *A. glauca* than by *A. glandulosa* in the 90-year-old stand and greater reproductive effort by *A. glauca* in the 90-year-old stand than in the 23-year-old stand. These results are consistent with *r*- vs. *K*-selection theory, but other expected differences did not appear.

Roos and Quinn (1977) studied reproductive effort in six populations of *Andropogon scoparius* (little bluestem, Poaceae). They measured dry weights and seed numbers in this perennial grass in relation to successional stage of the communities. The mean date of first anthesis was progressively later with increasing age of the field. Similarly, the later the date of first anthesis, and the older the field, the lower was the reproductive effort. Reproductive effort ranged from 24% to 54%, which is rather high for a perennial species. Greenhouse experiments indicated that most of the difference in reproductive effort was due to plasticity rather than genetic differences.

Three herbaceous plant communities were studied by Newell and Tramer (1978), who measured dry weights and estimated community-wide reproductive effort at three-week intervals through the growing season. Their study sites included a one-year-old-field, a ten-year-old-field, and a forest. Reproductive effort was highest in the one-year field, with no significant difference between the other two sites.

Turkington and Cavers (1978) estimated reproductive effort by dry weight in four legumes, representing four positions on the r–K continuum. Three of these species, *Medicago lupulina* (black medic, Fabaceae), *M. sativa* (alfalfa), and *Trifolium pratense* (red clover) seemed to follow r- vs. K-selection theory. *T. repens* (white Dutch clover), on the other hand, seemed to behave as an r-strategist in a K-type environment, i.e., an environment with density-dependent regulation. *T. repens* was the only species of the four to reproduce vegetatively. Reproductive effort varied from 31% in *M. lupulina* to 0.4% in *M. sativa*. *T. repens* showed 20.4% reproductive effort, most of which (18.5%) was vegetative reproduction. *T. pratense* showed 12% reproductive effort.

Abrahamson (1979) studied 52 populations with 50 species of wildflowers in fields and woods, as discussed above. He found greater sexual reproductive effort in fields than in woods.

Luftensteiner (1980) studied 49 angiosperms in four plant communities of central Europe, concluding that reproductive effort depends on habitat factors and competition, but not on position in a successional sequence.

Studying three marshes of differing successional maturity, Grace and Wetzel (1981) found greatest reproductive effort by cattails in the open marsh (6% sexual plus 8% vegetative reproductive effort). Cattails in a "cattail marsh" and a "woods marsh" invested only about 2.3% of their resources in vegetative reproductive effort and none in sexual reproductive effort.

Primack et al. (1981) studied five species of goldenrods, as discussed above. They found the least reproductive effort for forest species and the greatest for open-ground species.

Habitat Effects: Biotic and Abiotic Factors

Biotic factors affecting reproductive effort that have been studied include density and competition, insect gall formation, stability of habitat, and successional status. These latter factors were discussed above. Abiotic factors that have been studied include nutrients and soil fertility, moisture, elevation and temperature, length of growing season, and light.

The first calorimetric work focusing on reproductive effort was that of Harper and Ogden (1970), who reported energy allocation to reproductive effort both by dry weight and by calorimetry, using a Gallenkamp adiabatic bomb calorimeter. They gathered *Senecio vulgaris* (common groundsel, Asteraceae) plant material in 25 harvests from seedling to mature. Dry weight of roots, stems, leaves, receptacles, and flowers and leaf area were reported. They also reported dry weight for capitula partitioned into receptacle, floral remains, achenes, and pappus. These measurements were used to estimate net reproductive effort and energy allocation as a function of time through the growing season. Caloric values for hypocotyl, stem, root, leaves, dead leaves, flower buds, perianth, whole flowers, receptacles, and seeds (achenes plus pappus) were measured. The entire experiment was conducted under three stress treatments (different soil volumes).

As discussed above, Smith (1972) studied sexual and asexual reproduction in wild strawberries under weeded and unweeded conditions. He found that weeding did not affect the ratio of sexual to asexual reproduction.

Ogden's work (1974) on *Tussilago farfara* was carried out under various plant densities and soil fertilities, as indicated above. Vegetative reproduction decreased with increased density, but seed production was little affected. Low soil fertility was associated with increased vegetative reproductive effort. Sexual reproductive effort was not strongly affected by soil fertility.

The effects of density on energy and dry-weight allocations in *Chamaesyce hirta* (Euphorbiaceae) were studied by Snell and Burch (1975). They used a Phillipson microbomb calorimeter for energy measurements, although most of the work they reported was based only on dry weight. Snell and Burch found reproductive effort from 7% to 32%, depending on conditions of density and nutrient availability. In agreement with the predictions of *r*- vs. *K*-selection theory, they found that increased density and decreased nutrients each decreased reproductive effort in this annual species.

In Abrahamson's studies (1975a,b) of dewberries, discussed above, higher vegetative reproductive effort was associated with low density of vegetation, whereas higher sexual reproductive effort was associated with high densities. Werner and Platt (1976) found that under competitive pressures goldenrod species produced fewer but larger propagules.

Holler and Abrahamson (1977) studied the effect of density on reproductive effort in wild strawberries (*Fragaria virginiana*, Rosaceae) under laboratory conditions. They measured numbers and sizes of various characters and converted to dry biomass using conversion factors developed in a separate harvesting and weighing experiment. They found that sexual reproductive effort was approximately independent of density, but vegetative reproductive effort declined with increased density. There was no significant difference between plants from habitats of different successional maturity, which suggests some degree of plasticity in reproductive allocation in this species.

In other work already mentioned above, Hawthorn and Cavers (1978) found that reproductive effort in *Plantago major* was about 21% for all densities studied. Luftensteiner (1980) found competition to affect reproductive effort in four plant communities of Central Europe. And Pitelka et al. (1980) found no effect of plant density on reproductive effort in the forest herb *Aster acuminatus*.

The effects of mineral nutrition on reproductive effort in *Senecio sylvaticus* (groundsel, Asteraceae) and *Chamaenerion angustifolium* (fireweed or great willow-herb, Onagraceae) were studied by van Andel and Vera (1977). They performed experiments at three nutrient levels, measuring allocation of dry biomass and of mineral nutrients in plant organs. In *Senecio*, reproductive effort was about 21–24%, independent of nutrient level. Poor nutrition led to a smaller pappus. *Chamaenerion* expended about 11% of its resources in sexual reproduction in plants that flowered. The proportion of individuals that flowered increased with improved mineral nutrient supply. They concluded that *Senecio* is basically an *r*-strategist, whereas *Chamaenerion* is capable of both an *r*- and a *K*-strategy.

Further study of nutrients and reproductive effort was reported by Williams and Bell (1981). They measured allocations of both dry biomass and nitrogen in organs of several Mojave Desert winter annuals. Reproductive effort, as measured by the amount of nitrogen in the tissues, ranged between 43% and 67%, whereas dry biomass reproductive effort was 31–51%. They found that nitrogen-poor plants allocated nitrogen to reproduction at the expense of vegetative organs throughout their life cycle.

The effect of moisture on reproductive effort was noted by Hickman and Pitelka (1975), who found greater sexual reproductive effort in *Polygonum cascadense*, at the expense of vegetative growth, in a dry habitat compared to a wet one. Abrahamson (1975a,b) found lower sexual reproductive effort in dewberries for a dry year compared to a wet one. Vegetative reproductive effort was about the same for both years. Werner and Platt (1976) found increased vegetative reproductive effort in goldenrods with increased soil moisture. As mentioned above, Abrahamson and Hershey (1977) found no effect of moisture on reproductive effort in *Impatiens capensis*, in two habitats as well as greenhouse populations. And Hickman (1977) reported that sexual reproductive effort in *Polygonum cascadense* decreased with increasing moisture, whereas moisture had no impact on reproductive effort in three other coexisting species of *Polygonum*.

McNaughton (1975) studied three species of *Typha* (cattail, Typhaceae) along an environmental gradient from North Dakota to Texas. He measured number of fruits per clone, weight of fruits, number of rhizomes per clone, weight of rhizomes, and foliage heights. He found results consistent with the hypothesis that a shorter growing season leads to a more *r*-selected population.

Populations of *Sedum lanceolatum* (stonecrop, Crassulaceae) at four elevations were studied by Jolls (1980), who measured dry weight of major organs. She found less sexual reproductive effort at higher elevations (about 30%, compared to about 55% for lower elevations).

Kawano and Masuda (1980) reported seasonal and altitudinal effects on reproductive effort in five populations of *Heloniopsis orientalis* (Liliaceae), finding that "energy allocations to total reproductive structures at both the flowering and fruiting stages conspicuously increase in response to an increase in elevation or a decrease in the length of the growing season." The relative amount of energy in a single propagule also increased strongly with elevation. They discussed the concept of *r*- and *K*-selection as related to their data.

The effect of elevation on reproductive effort of *Mimulus primuloides* (monkeyflower, Scrophulariaceae) was reported by Douglas (1981). She measured dry biomass and found maximum vegetative reproductive effort at midelevations, and constant sexual reproductive effort at the four lower elevations, as discussed above. She also found much higher sexual reproductive effort at the highest elevation.

Reproductive effort by *Opuntia* (prickly pear cactus, Cactaceae) in the Galapagos Islands was studied by Racine and Downhower (1974). They measured caloric content of fruits and estimated reproductive effort by fruit energy per unit area of pad surface. Reproductive effort was smaller for cacti on Santa Cruz Island than for those on Santa Fe and Pinzon islands. This difference was related to competition for light with other woody plants and possibly wind intensity: Santa Cruz is a much larger island than the other two, and

Opuntia stands there grow in much greater density and in greater competition with other woody species than on the other two islands. On Santa Cruz, the number of fruits per pad is an increasing function of tree height, whereas on the other two islands no relationship was observed between these two variables.

Pitelka et al. (1980) found that sexual reproductive effort in *Aster acuminatus* increased with both light level and plant size. Vegetative reproductive effort was not affected by these factors.

Hartnett and Abrahamson (1979; see also Stinner and Abrahamson, 1979) studied the effects of stem gall insects on resource allocation in *Solidago canadensis* (Canada goldenrod, Asteraceae). They measured energy content (Parr oxygen bomb calorimetry) and dry weight of stems, leaves, inflorescences, current rhizome, new rhizome, and roots, summing total production through the growing season. They also recorded mean head weight, total number of heads, and mean number of propagules per head. The insect galls were related to a significant reduction in allocation to seed. The changes were greater than those that could simply be accounted for by a proportional change due to increased stem biomass in the gall. Vegetative reproductive allocation was unaffected by stem galls except for some possible reduction with ball galls. They found that a stem gall one season may affect the growth, reproduction, and survivorship of the plant in the following year owing to a decrease in carbohydrate storage in the rhizome.

DISCUSSION

The studies reviewed above explore a wide range of reproductive and life history strategies as well as ecological and habitat effects on reproductive effort. Aspects of reproductive strategies that have been considered include the energetics of sexual vs. asexual reproduction; selfing vs. outcrossing; cleistogamy vs. chasmogamy; and male vs. female costs in dioecy and other breeding systems. Additional topics that have been studied are energy patterns associated with various pollination mechanisms, effects of seed size on reproductive strategy of progeny, and the extent to which reproductive effort represents a trade-off with vegetative growth.

Studies of differences in reproductive effort associated with different life history strategies include research on reproductive effort in various plant life forms; phenological effects; differences between young and mature plants, native and exotic species, cultivated and wild species, tropical and temperate species; and the extent to which reproductive effort is plastic rather than genetically fixed. Some studies have followed resource allocation through several seasons or through an entire life history.

Studies of ecological and habitat influences on reproductive effort have focused principally on differences in reproductive effort associated with successional status. They have also included a few experimental manipulations of resources or plant density, although most have reported observations over a variety of natural habitat conditions.

In much of the work reviewed above, dry weight has proved a sufficiently accurate measure of energy allocation patterns, in accordance with the conclusions of Hickman and Pitelka (1975). The aims of the study as well as the peculiarities of the tissues being examined are of primary importance in determining the necessity of calorimetry (e.g., Colin and Jones, 1980).

These studies confirm the expectation that vegetative and sexual reproduction often compete for the same resources, so that vegetative reproduction is at the expense of sexual reproduction (Harper, 1967; Tripathi and Harper, 1973; Ogden, 1974; Sarukhan, 1976; Grace and Wetzel, 1981). On the other hand, two species of *Trifolium,* one of which reproduced vegetatively, whereas the other did not, showed quite different total reproductive effort, 12% for the sexually reproducing species compared to 20% for the species that reproduced by both mechanisms (Turkington and Cavers, 1978). The species that reproduced vegetatively, *T. repens,* allocated about 18% of its resources to vegetative reproduction and only 2% to sexual reproduction. The other species, *T. pratense,* allocated 12% to sexual reproduction. Similarly, five perennial composites showed quite

different total reproductive effort and very different balances between sexual and vegetative reproduction (Bostock and Benton, 1979).

Vegetative reproductive effort as a percent of total energy assimilate increased under the stress of poor soil fertility in *Tussilago farfara,* whereas sexual reproductive effort remained approximately constant (Ogden, 1974). In the same study, vegetative reproductive effort decreased with increased density. Douglas (1981) found that vegetative reproductive effort decreased with increased intraspecific competition and with increased severity of environment (high altitude). Abrahamson (1975a,b) also found that vegetative reproduction in dewberries decreased with increased density. Ogden (1974) concluded that reproductive effort was about the same for seral perennials as for seral annuals, but perennials invest a greater proportion of their resources in vegetative reproduction and less in sexual reproduction.

Sexual reproductive effort increased with plant size (which increased with light level) in the forest herb *Aster acuminatus,* whereas vegetative reproductive effort remained approximately constant (Pitelka et al., 1980). Abrahamson (1979) found that sexual reproductive effort was greater for herbs in fields than in woods, perhaps also reflecting the effects of light and size, although more clearly related to successional status, as discussed below. The data of Pitelka et al. (1980) suggest that sexual and vegetative reproductive efforts do not compete for resources in the forest; rather, vegetative reproduction seems to have first priority, and sexual reproduction occurs only when there are extra resources. The observations of Struik (1965) on *Circaea quadrisulcata* and Abrahamson (1975a,b) on dewberries also found vegetative reproductive effort to be approximately constant over several habitats, while sexual reproductive effort depended on the favorability of the site. The relative amount of effort expended on vegetative vs. sexual reproduction varies greatly among the species that have been studied, as cited above.

Several studies found that reproductive effort occurs at the expense of vegetative growth (Hickman, 1975; Hickman and Pitelka, 1975;

Law, 1979; Abrahamson, 1980; Hancock and Bringhurst, 1980; Davis, 1981). Lovett Doust and Harper (1980) found that defoliation increased reproductive effort. However, Primack et al. (1981), studying five species of goldenrods, found that individuals within populations increased reproductive effort by increasing reproductive yield, not by decreasing vegetative growth, except *Solidago odora,* whose total dry weight increased with increased reproductive effort. *S. odora* was found in the edge-of-woods habitat. Leonard (1962) pointed out that the effect of developing fruit in limiting vegetative growth is only apparent when in the presence of limiting resources. So under some circumstances, there may not be appreciable costs in reduced vegetative growth due to reproduction.

Studies of selfing vs. outcrossing substantiate the notion that outcrossed seeds are generally more costly than selfed seeds, and indicate that chasmogamy and cleistogamy often follow different phenological patterns (McNamara and Quinn, 1977; Schemske, 1978; Waller, 1979; Lord, 1980; Spira, 1980).

In wind-pollinated species, the theoretical expectation that sex ratios should be female-biased in selfers and about 1:1 in outcrossers has been confirmed (Lemen, 1980).

Studies of dioecy indicate that male and female plants utilize resources differently, by either different phenologies, different habitat selection, or sexual dimorphism (Putwain and Harper, 1972; Wallace and Rundel, 1979; Hancock and Bringhurst, 1980; Lemen, 1980; Philipp, 1980). Furthermore, female plants were found to expend greater sexual reproductive effort than male plants when seed set was greater than 20–30% (Wallace and Rundel, 1979; Gross and Soule, 1981), and vegetative reproductive effort was found in one case to be greater in male plants, whereas sexual reproductive effort was greater in female plants (Putwain and Harper, 1972). In greenhouse-grown wild strawberries, female plants showed greater sexual reproductive effort than male plants, at the expense of vegetative growth, whereas vegetative reproductive effort was about the same for both sexes (Hancock and Bringhurst, 1980). In this same study, how-

ever, while female plants invested a large fraction of their dry biomass (8.0%) in fruits, male plants invested more energy in flowers and peduncles (1.5% vs. 0.8%). Other comparisons of male and female effort have indicated that male gametes are often more costly than female, but protection and nurturing reverse the relative costs (Smith and Evenson, 1978; Lovett Doust and Harper, 1980).

Studies of reproductive effort by species of different life forms confirm the expectation that perennials generally expend smaller proportions of their resources for reproduction than do annuals (Struik, 1965; Harper et al., 1970, following J. Ogden; Ogden, 1974; Pitelka, 1977; Abrahamson, 1979; Primack, 1979). However, Sarukhan (1980) reported unexpectedly large reproductive effort for the tree life form in his study of a tropical palm. Van Valen's (1975) estimate for another species of palm was much closer to the expected range, but it was a much more indirect measurement. Ogden (1974) found that whereas seral perennials invest a smaller proportion of resources in sexual reproduction than do seral annuals, they invest about the same proportion in total reproduction (vegetative and sexual).

The extent to which reproductive effort is plastic or genetically fixed is still an open question. Both patterns have been found for various species (e.g., Abrahamson, 1975a,b; Hickman, 1975, 1977; Abrahamson and Hershey, 1977; Holler and Abrahamson, 1977; Roos and Quinn, 1977; Grace and Wetzel, 1981).

One study (Abrahamson, 1979) found that exotic species in fields expended greater reproductive effort than did natives. Other studies have found cultivated species to have reproductive effort in the same range as wild species (Struik, 1965; Harper and Ogden, 1970; de Ridder et al., 1981).

Several studies have found that the proportion of total assimilated resources devoted to reproduction is greater for colonizing or weedy species and communities on disturbed sites than for less disturbed sites; i.e., reproductive effort decreases with increasing successional maturity of the sites (Harper et al., 1970; Gadgil and Solbrig, 1972; Abrahamson and Gadgil, 1973; Gaines et al., 1974; Abrahamson, 1975a,b, 1979; Hickman, 1975; Hickman and Pitelka, 1975; Bradbury and Hofstra, 1976; Roos and Quinn, 1977; Newell and Tramer, 1978; Primack, 1979; Grace and Wetzel, 1981; Primack et al., 1981). In contrast to all these studies, Luftensteiner (1980) found no effect associated with successional stage.

Low soil fertility led to greater vegetative reproductive effort, with constant sexual effort, in one study (Ogden, 1974). But a study of the annual euphorb, *Chamaesyce hirta,* showed decreased reproductive effort with decreased soil nutrients (Snell and Burch, 1975). Reproductive effort in *Senecio sylvaticus* and flowering individuals of fireweed were independent of nutrient level, although numbers of flowering fireweed plants increased with increased nutrients (van Andel and Vera, 1977). Williams and Bell (1981) reported that desert winter annuals allocated nitrogen to reproduction at the expense of vegetative organs throughout the life cycle when in a nitrogen-limited environment.

Increasing density tends to depress reproductive effort, with greatest effect on vegetative reproduction (Ogden, 1974; Snell and Burch, 1975; Abrahamson, 1975a,b, 1980; Holler and Abrahamson, 1977). Several workers report little or no effect of density on sexual reproductive effort (Ogden, 1974; Abrahamson, 1975a,b; Holler and Abrahamson, 1977; Hawthorn and Cavers, 1978; Pitelka et al., 1980). Density was never great enough in forest populations of *Aster acuminatus* to affect reproductive effort (Pitelka et al., 1980).

Moisture usually has significant impact on reproductive effort, but its precise effects seem to be strongly related to the characteristics of the species being studied (Abrahamson, 1975a,b; Hickman and Pitelka, 1975; Werner and Platt, 1976; Abrahamson and Hershey, 1977; Hickman, 1977). In cases where light may be limiting, increased light increased sexual reproductive effort (Racine and Downhower, 1974; Pitelka et al., 1980). Insect stem galls decreased sexual reproductive effort without affecting vegetative reproductive effort (Harnett and Abrahamson, 1979).

Reproductive effort decreased with increasing length of growing season (or decreasing elevation) in three other studies (McNaughton, 1975; Kawano and Masuda, 1980; Douglas, 1981), although Golley (1961) found higher caloric content (4700–4800 cal/gram dry weight) in plant tissues from alpine tundra and pine communities than in other, lower-elevation or lower-latitude communities (3800–4200 cal/gram dry weight). And Jolls (1980) found less sexual reproductive effort at higher elevations than at lower elevations in *Sedum lanceolatum.*

Dicots included in one study (Colin and Jones, 1980) had significantly higher caloric content in pollen than did wind-pollinated monocots and gymnosperms.

Several studies have been directed explicitly at testing particular ecological or evolutionary theories (Gadgil and Solbrig, 1972; Abrahamson and Gadgil, 1973; Gaines et al., 1974; Abrahamson, 1975a,b, 1979; McNaughton, 1975; Bradbury and Hofstra, 1976; Werner and Platt, 1976; Hickman, 1977; Holler and Abrahamson, 1977; Keeley and Keeley, 1977; McNamara and Quinn, 1977; Newell and Tramer, 1978; Turkington and Cavers, 1978; Bostock and Benton, 1979; Law, 1979; Primack, 1979; Kawano and Masuda, 1980; Lemen, 1980; Luftensteiner, 1980; Pitelka et al., 1980; Davis, 1981; Grace and Wetzel, 1981; Primack et al., 1981). However, the measurement of energy allocation patterns is of relatively recent interest, so many studies have simply explored these patterns in natural or cultivated populations.

Theoretical ideas that need to be examined more closely through resource allocation studies include the concept of *r*- vs. *K*-selection (MacArthur and Wilson, 1967; Pianka, 1970; Gadgil and Solbrig, 1972, and references therein); differential resource utilization by the sexes (Williams, 1975; Maynard Smith, 1978a; Bawa, 1980; Lloyd, 1980) through habitat selection (Freeman et al., 1976; Grant and Mitton, 1979; Wade, 1981; Wade et al., 1981), through sexual dimorphism (Kaul, 1979; Wallace and Rundel, 1979; Hancock and Bringhurst, 1980), or through nonsynchronous resource use in time (Putwain and Harper, 1972; Melampy and Howe, 1977; Opler and Bawa, 1978; Lovett Doust, 1980);

optimal life history strategies (Gadgil and Bossert, 1970; Hirshfield and Tinkle, 1975; Kawano, 1975; Giesel, 1976; Stearns, 1976; Maynard Smith, 1978b, and references therein; Givnish, 1980); coevolution with pollinators and seed dispersal agents (Heinrich and Raven, 1972; Heinrich, 1975; Regal, 1977; Schaffer and Schaffer, 1977, 1979; Levin, 1978; Gilbert, 1980; McKey, 1980), including the effects on energy allocation of mimicry and deception (Heinrich, 1980; McKey, 1980); and the concept of clutch size as "packaging" of energy allocated to sexual reproduction (Levin and Turner, 1977).

FUTURE TRENDS

Future studies of energy allocation to plant reproduction are likely to emphasize comparative allocation patterns associated with a wide range of reproductive and life history strategies. While a few exploratory studies of this type have already been carried out, much more work of this sort is needed, including comparison of plants occupying similar habitats and subject to similar resource regimes, but employing different pollination mechanisms and seed production and dispersal syndromes.

Studies that follow allocation patterns through time and plant development, and hence are sensitive to phenological differences in reproductive effort, should also be important. Phenological separation of male and female functions should prove especially interesting in studies of both reproductive ecology and the evolution of sex.

Work showing ecological and habitat influences on reproductive effort will continue. The most useful future work of this kind will be studies testing particular ecological theories and experimental studies in which ecological variables are carefully manipulated according to sound experimental design. It will be especially interesting to pursue the energetics of successful vs. unsuccessful plant introductions and of competing exotic and native species (e.g., Abrahamson, 1979).

Another area where new insights might be possible is the application of energy allocation studies to systematic questions. Colin and Jones (1980) have found significant differences in caloric content of pollens from taxonomic subclasses of wind-pollinated dicots and monocots. In addition, they found a statistically significant difference in pollen energy content between two tribes in the Asteraceae that have shifted from insect to wind pollination (Ambrosiinae and Anthemidae). This observation lends support to a previous analysis that indicated independent evolution of the two tribes. Much more work needs to be done before the value of caloric studies to plant systematics is fully appreciated.

Reproductive effort may be subject to constraints due to pollinators (Bierzychudek, 1981), developmental pathways, and morphological considerations. Such constraints need to be studied and taken into account by theory.

LITERATURE CITED

Abrahamson, W. G. 1975a. Reproduction of *Rubus hispidus* L. in different habitats. *Am. Midl. Natur.* 93:471–478.

———. 1975b. Reproductive strategies in dewberries. *Ecology* 56:721–726.

———. 1979. Patterns of resource allocation in wildflower populations of fields and woods. *Am. J. Bot.* 66:71–79.

———. 1980. Demography and vegetative reproduction. *In* O. T. Solbrig (ed.), *Demography and Evolution in Plant Populations*. Blackwell Scientific Publications, Oxford, pp. 89–106.

Abrahamson, W. G. and M. Gadgil. 1973. Growth form and reproductive effort in goldenrods (*Solidago*, Compositae). *Am. Nat.* 107:651–661.

Abrahamson, W. G. and B. J. Hershey. 1977. Resource allocation and growth of *Impatiens capensis* (Balsaminaceae) in two habitats. *Bull. Torrey Bot. Club* 104:160–164.

Andel, J. van and F. Vera. 1977. Reproductive allocation in *Senecio sylvaticus* and *Chamaenerion angustifolium* in relation to mineral nutrition. *J. Ecol.* 65:747–758.

Bawa, K. S. 1980. Evolution of dioecy in flowering plants. *Annu. Rev. Ecol. Syst.* 11:15–39.

Bierzychudek, P. 1981. Pollinator limitation of plant reproductive effort. *Am. Nat.* 117:838–840.

Bostock, S. J. and R. A. Benton. 1979. The reproductive strategies of five perennial Compositae. *J. Ecol.* 67:91–107.

Bradbury, I. K. and G. Hofstra. 1976. The partitioning of net energy resources in two populations of *Solidago canadensis* during a single developmental cycle in southern Ontario. *Can. J. Bot.* 54:2449–2456.

Charnov, E. L., J. Maynard Smith, and J. J. Bull. 1976. Why be an hermaphrodite? *Nature* 263:125–126.

Clark, D. D. and J. H. Burk. 1980. Resource allocation patterns of two California–Sonoran desert ephemerals. *Oecologia (Berl.)* 46:86–91.

Cody, M. L. 1966. A general theory of clutch size. *Evolution* 20:174–184.

Colin, L. J. and C. E. Jones. 1980. Pollen energetics and pollination modes. *Am. J. Bot.* 67:210–215.

Cummins, K. W. and J. C. Wuycheck. 1971. Caloric equivalents for investigations in ecological energetics. *Mitt. int. Ver. Limnol.* 18:1–158.

Davis, M. A. 1981. The effect of pollinators, predators, and energy constraints on the floral ecology and evolution of *Trillium erectum*. *Oecologia (Berl.)* 48:400–406.

de Ridder, N., N. G. Seligman, and H. van Keulen. 1981. Analysis of environmental and species effects on the magnitude of biomass investment in the reproductive effort of annual pasture plants. *Oecologia (Berl.)* 49:263–271.

Douglas, D. A. 1981. The balance between vegetative and sexual reproduction of *Mimulus primuloides* (Scrophulariaceae) at different altitudes in California. *J. Ecol.* 69:295–310.

Faegri, K. and L. van der Pijl. 1979. *The Principles of Pollination Ecology*, 3rd rev. ed. Pergamon Press, New York.

Frankel, R. and E. Galun. 1977. *Pollination Mechanisms, Reproduction and Plant Breeding*. Springer-Verlag, Berlin.

Freeman, D. C., L. G. Klikoff, and K. T. Harper. 1976. Differential resource utilization by the sexes of dioecious plants. *Science* 193:597–599.

Gadgil, M. and W. H. Bossert. 1970. Life historical consequences of natural selection. *Am. Nat.* 104:1–24.

Gadgil, M. and O. T. Solbrig. 1972. The concept of *r*- and *K*-selection: evidence from wildflowers and some theoretical considerations. *Am. Nat.* 106:14–31.

Gaines, M. S., K. J. Vogt, J. L. Hamrick, and J. Caldwell. 1974. Reproductive strategies and growth patterns in sunflowers *(Helianthus)*. *Am. Nat.* 108:889–894.

Ghiselin, M. T. 1974. *The Economy of Nature and the Evolution of Sex*. University of California Press, Berkeley.

Giesel, J. T. 1976. Reproductive strategies as adaptations to life in temporally heterogeneous environments. *Annu. Rev. Ecol. Syst.* 7:57–79.

Gilbert, L. E. 1980. Ecological consequences of a coevolved mutualism between butterflies and plants. *In* L. E. Gilbert and P. H. Raven (eds.), *Coevolution of Animals and Plants*, rev. ed. University of Texas Press, Austin, pp. 210–240.

Givnish, T. J. 1980. Ecological constraints on the evolution of breeding systems in seed plants: dioecy and dispersal in gymnosperms. *Evolution* 34:959–972.

Golley, F. B. 1961. Energy values of ecological materials. *Ecology* 42:581–584.

Grace, J. B. and R. G. Wetzel. 1981. Phenotypic and genotypic components of growth and reproduction in *Typha latifolia:* experimental studies in marshes of differing successional maturity. *Ecology* 62:789–801.

Grant, M. C. and J. B. Mitton. 1979. Elevational gradients in adult sex ratios and sexual differentiation in vegetative growth rates of *Populus tremuloides* Michx. *Evolution* 33:914–918.

Gross, K. L. and J. D. Soule. 1981. Differences in biomass allocation to reproductive and vegetative structures of male and female plants of a dioecious, perennial herb, *Silene alba* (Miller) Krause. *Am. J. Bot.* 68:801–807.

Hancock, J. F. and R. S. Bringhurst. 1980. Sexual dimorphism in the strawberry *Fragaria chiloensis*. *Evolution* 34:762–768.

Harper, J. L. 1967. A Darwinian approach to plant ecology. *J. Ecol.* 55:247–270.

———. 1977. *Population Biology of Plants*. Academic Press, London.

Harper, J. L. and J. Ogden. 1970. The reproductive strategy of higher plants. I. The concept of strategy with special reference to *Senecio vulgaris* L. *J. Ecol.* 58:681–698.

Harper, J. L., P. H. Lovell, and K. G. Moore. 1970. The shapes and sizes of seeds. *Annu. Rev. Ecol. Syst.* 1:327–356.

Hartnett, D. C. and W. G. Abrahamson. 1979. The effects of stem gall insects on life history patterns in *Solidago canadensis*. *Ecology* 60:910–917.

Hawthorn, W. R. and P. B. Cavers. 1978. Resource allocation in young plants of two perennial species of *Plantago*. *Can. J. Bot.* 56:2533–2537.

Heinrich, B. 1975. Energetics of pollination. *Annu. Rev. Ecol. Syst.* 6:139–170.

———. 1980. The role of energetics in bumblebee-flower interrelationships. *In* L. E. Gilbert and P. H. Raven (eds.), *Coevolution of Animals and Plants*, rev. ed. University of Texas Press, Austin, pp. 141–158.

Heinrich, B. and P. H. Raven, 1972. Energetics and pollination ecology. *Science* 176:597–602.

Hickman, J. C. 1975. Environmental unpredictability and plastic energy allocation strategies in the annual *Polygonum cascadense* (Polygonaceae). *J. Ecol.* 63:689–701.

———. 1977. Energy allocation and niche differentiation in four co-existing annual species of *Polygonum* in western North America. *J. Ecol.* 65:317–326.

Hickman, J. C. and L. F. Pitelka. 1975. Dry weight indicates energy allocation in ecological strategy analysis of plants. *Oecologia (Berl.)* 21:117–121.

Hirshfield, M. E. and D. W. Tinkle. 1975. Natural selection and the evolution of reproductive effort. *Proc. Natl. Acad. Sci. U.S.A.* 72:2227–2231.

Holler, L. C. and W. G. Abrahamson. 1977. Seed and vegetative reproduction in relation to density in *Fragaria virginiana* (Rosaceae). *Am. J. Bot.* 64:1003–1007.

Jolls, C. L. 1980. Phenotypic patterns of variation in biomass allocation in *Sedum lanceolatum* Torr. at four elevational sites in the Front Range, Rocky Mountains, Colorado. *Bull. Torrey Bot. Club* 107:65–70.

Kaul, R. B. 1979. Inflorescence architecture and flower sex ratios in *Sagittaria brevirostra* (Alismataceae). *Am. J. Bot.* 66:1062–1066.

Kawano, S. 1975. The productive and reproductive biology of flowering plants. II. The concept of life history strategy in plants. *J. Coll. Liberal Arts Toyama Univ.* 8:51–86.

Kawano, S. and J. Masuda. 1980. The productive and reproductive biology of flowering plants. VII. Resource allocation and reproductive capacity in wild populations of *Heloniopsis orientalis* (Thunb.) C. Tanaka (Liliaceae). *Oecologia (Berl.)* 45:307–317.

Kawano, S. and Y. Nagai. 1975. The productive and reproductive biology of flowering plants. I. Life history strategies of three *Allium* species in Japan. *Bot. Mag. (Tokyo)* 88:281–318.

Keeley, J. E. and S. C. Keeley. 1977. Energy allocation patterns of a sprouting and a nonsprouting species of *Arctostaphylos* in the California chaparral. *Am. Midl. Natur.* 98:1–10.

Law, R. 1979. The cost of reproduction in annual meadow grass. *Am. Nat.* 113:3–16.

Leigh, E. G. 1970. Sex ratio and differential mortality between the sexes. *Am. Nat.* 104:205–210.

Lemen, C. 1980. Allocation of reproductive effort to the male and female strategies in wind-pollinated plants. *Oecologia (Berl.)* 45:156–159.

Leonard, E. R. 1962. Inter-relations of vegetative and reproductive growth, with special reference to indeterminate plants. *Bot. Rev.* 28:353–410.

Levin, D. A. 1978. Pollinator behaviour and the breeding structure of plant populations. *In* A. J. Richards (ed.), *The Pollination of Flowers by Insects*. Linnean Soc. Symp. Ser. No. 6. Academic Press, New York, pp. 133–150.

Levin, D. A. and B. L. Turner. 1977. Clutch size in the Compositae. *In* B. Stonehouse and C. Perrins (eds.), *Evolutionary Ecology*. Macmillan Press, London, pp. 215–222.

Lindeman, R. L. 1942. The trophic-dynamic aspect of ecology. *Ecology* 23:399–418.

Lloyd, D. G. 1980. Benefits and handicaps of sexual reproduction. *In* M. Hecht, W. C. Steer, and B. Wallace (eds.), *Evolutionary Biology*. Plenum Press, New York, 13:69–111.

Lord, E. M. 1980. Intra-inflorescence variability in pollen/ovule ratios in the cleistogamous species *Lamium amplexicaule* (Labiatae). *Am. J. Bot.* 67:529–532.

Lovett Doust, J. 1980. Floral sex ratios in andromonoecious Umbelliferae. *New Phytol.* 85:265–273.

Lovett Doust, J. and J. L. Harper. 1980. The resource costs of gender and maternal support in an andromonoecious Umbellifer, *Smyrnium olusatrum* L. *New Phytol.* 85:251–264.

Luftensteiner, H. W. 1980. Der Reproduktionsaufwand in vier mitteleuropaischen Pflanzengemeinschaften

(The reproductive effort in four plant communities of Central Europe). *Pl. Syst. Evol.* 135:235–251.

MacArthur, R. H. and E. O. Wilson. 1967. *The Theory of Island Biogeography*. Princeton University Press, Princeton, N.J.

Maynard Smith, J. 1977. The sex habit in plants and animals. *In* S. Levin (ed.), *Lecture Notes in Biomathematics*. Springer-Verlag, Berlin, 19:315–331.

———. 1978a. *The Evolution of Sex*. Cambridge University Press.

———. 1978b. Optimization theory in evolution. *Annu. Rev. Ecol. Syst.* 9:31–56.

McKey, D. 1980. The ecology of coevolved seed dispersal systems. *In* L. E. Gilbert and P. H. Raven (eds.), *Coevolution of Animals and Plants*, rev. ed. University of Texas Press, Austin, pp. 159–191.

McNamara, J. and J. A. Quinn. 1977. Resource allocation and reproduction in populations of *Amphicarpum purshii* (Gramineae). *Am. J. Bot.* 64:17–23.

McNaughton, S. J. 1975. *r*- and *K*-selection in *Typha*. *Am. Nat.* 109:251–261.

Melampy, M. N. and H. F. Howe. 1977. Sex ratio in the tropical tree *Triplaris americana* (Polygonaceae). *Evolution* 31:867–872.

Newell, S. J. and E. J. Tramer. 1978. Reproductive strategies in herbaceous plant communities during succession. *Ecology* 59:228–234.

Nitsch, J. P. 1965. Physiology of flower and fruit development. *In* W. Ruhland (ed.), *Handbuch der Pflanzenphysiologie*. Springer-Verlag, Berlin, pp. 163–175.

Ogden, J. 1974. The reproductive strategy of higher plants. II. The reproductive strategy of *Tussilago farfara* L. *J. Ecol.* 62:291–324.

Opler, P. A. and K. S. Bawa. 1978. Sex ratios in tropical forest trees. *Evolution* 32:812–821.

Paine, R. T. 1971. The measurement and application of the calorie to ecological problems. *Annu. Rev. Ecol. Syst.* 2:145–164.

Philipp, M. 1980. Reproductive biology of *Stellaria longipes* Goldie as revealed by a cultivation experiment. *New Phytol.* 85:557–569.

Pianka, E. R. 1970. On *r*- and *K*-selection. *Am. Nat.* 104:592–597.

Pitelka, L. F. 1977. Energy allocation in annual and perennial lupines (*Lupinus*: Leguminosae). *Ecology* 58:1055–1065.

Pitelka, L. F., D. S. Stanton, and M. O. Peckenham. 1980. Effects of light and density on resource allocation in a forest herb, *Aster acuminatus* (Compositae). *Am. J. Bot.* 67:942–948.

Primack, R. B. 1979. Reproductive effort in annual and perennial species of *Plantago* (Plantaginaceae). *Am. Nat.* 114:51–62.

Primack, R. B., A. R. Rittenhouse, and P. V. August. 1981. Components of reproductive effort and yield in goldenrods. *Am. J. Bot.* 68:855–858.

Proctor, J. and P. Yeo. 1972. *The Pollination of Flowers*. Taplinger Publishing Co., New York.

Putwain, P. D. and J. L. Harper. 1972. Studies in the

dynamics of plant populations. V. Mechanisms governing sex ratio in *Rumex acetosa* and *Rumex acetosella*. *J. Ecol.* 60:113–129.

Racine, C. H. and J. F. Downhower. 1974. Vegetative and reproductive strategies of *Opuntia* (Cactaceae) in the Galapagos Islands. *Biotropica* 6:175–186.

Regal, P. J. 1977. Ecology and evolution of flowering plant dominance. *Science* 196:622–629.

Roos, F. H. and J. A. Quinn. 1977. Phenology and reproductive allocation in *Andropogon scoparius* (Gramineae) populations in communities of different successional stages. *Am. J. Bot.* 64:535–540.

Sarukhan, J. 1976. On selective pressures and energy allocation in populations of *Ranunculus repens* L., *R. bulbosus* L. and *R. acris* L. *Ann. Missouri Bot. Gard.* 63:290–308.

————. 1980. Demographic problems in tropical systems. *In* O. T. Solbrig (ed.), *Demography and Evolution in Plant Populations*. Blackwell Scientific Publications, Oxford, pp. 161–188.

Schaffer, W. M. and M. V. Schaffer. 1977. The adaptive significance of variations in reproductive habit in the Agavaceae. *In* B. Stonehouse and C. Perrins (eds.), *Evolutionary Ecology*. Macmillan Press, London, pp. 261–276.

————. 1979. The adaptive significance of variations in reproductive habit in the Agavaceae. II. Pollinator foraging and selection for increased reproductive expenditure. *Ecology* 60:1051–1069.

Schemske, D. W. 1978. Evolution of reproductive characteristics in *Impatiens* (Balsaminaceae): the significance of cleistogamy and chasmogamy. *Ecology* 59:596–613.

Smith, C. A., and W. E. Evenson. 1978. Energy distribution in reproductive structures of *Amaryllis*. *Am. J. Bot.* 65:714–716.

Smith, C. C. 1972. The distribution of energy into sexual and asexual reproduction in wild strawberries *(Fragaria virginiana)*. *Third Midwest Prairie Conf. Proc.*, Kansas State University, Manhattan, Sept. 22–23, 1972, pp. 55–60.

Snell, T. W. and D. G. Burch. 1975. The effects of density on resource partitioning in *Chamaesyce hirta* (Euphorbiaceae). *Ecology* 56:742–746.

Solbrig, O. T. and D. J. Solbrig. 1979. *Introduction to Population Biology and Evolution*. Addison-Wesley, Reading, Mass.

Spira, T. P. 1980. Floral parameters, breeding system and pollinator type in *Trichostema* (Labiatae). *Am. J. Bot.* 67:278–284.

Stanley, R. G. and H. F. Linskens. 1974. *Pollen: Biology, Biochemistry and Management*. Springer-Verlag, Berlin.

Stearns, S. C. 1976. Life-history tactics: a review of the ideas. *Q. Rev. Biol.* 51:3–47.

Stebbins, G. L. 1970. Adaptive radiation of reproductive characteristics in angiosperms. I. Pollination mechanisms. *Annu. Rev. Ecol. Syst.* 1:307–326.

————. 1971. Adaptive radiation of reproductive characteristics in angiosperms. II. Seeds and seedlings. *Annu. Rev. Ecol. Syst.* 2:237–260.

Stinner, B. R. and W. G. Abrahamson. 1979. Energetics of the *Solidago canadensis*–stem gall insect–parasitoid guild interaction. *Ecology* 60:918–926.

Struik, G. J. 1965. Growth patterns of some native annual and perennial herbs in southern Wisconsin. *Ecology* 46:401–420.

Tripathi, R. S. and J. L. Harper. 1973. The comparative biology of *Agropyron repens* L. (Beauv.) and *A. caninum* L. (Beauv.) 1. The growth of mixed populations established from tillers and from seeds. *J. Ecol.* 61:353–368.

Turkington, R. A. and P. B. Cavers. 1978. Reproductive strategies and growth patterns in four legumes. *Can. J. Bot.* 56:413–416.

Van Valen, L. 1975. Life, death, and energy of a tree. *Biotropica* 7:259–269.

————. 1976. Energy and evolution. *Evol. Theory* 1:179–229.

Wade, K. M. 1981. Experimental studies on the distribution of the sexes of *Mercurialis perennis* L. III. Transplanted populations under light screens. *New Phytol.* 87:447–455.

Wade, K. M., R. A. Armstrong, and S. R. J. Woodell. 1981. Experimental studies on the distribution of the sexes of *Mercurialis perennis* L. I. Field observations and canopy removal experiments. *New Phytol.* 87:431–438.

Wallace, C. S. and P. W. Rundel. 1979. Sexual dimorphism and resource allocation in male and female shrubs of *Simmondsia chinensis*. *Oecologia (Berl.)* 44:34–39.

Waller, D. M. 1979. The relative costs of self- and cross-fertilized seeds in *Impatiens capensis* (Balsaminaceae). *Am. J. Bot.* 66:313–320.

————. 1980. Environmental determinants of outcrossing in *Impatiens capensis* (Balsaminaceae). *Evolution* 34:747–761.

Werner, P. A. and W. J. Platt. 1976. Ecological relationships of co-occurring goldenrods (*Solidago:* Compositae). *Am. Nat.* 110:959–971.

Williams, G. C. 1975. *Sex and Evolution*. Princeton University Press, Princeton, N.J.

Williams, R. B. and K. L. Bell. 1981. Nitrogen allocation in Mojave Desert winter annuals. *Oecologia (Berl.)* 48:145–150.

Section IV
COMPETITION AND POLLINATION

13
COMPETITION FOR POLLINATION AND FLORAL CHARACTER DIFFERENCES AMONG SYMPATRIC PLANT SPECIES: A REVIEW OF EVIDENCE

Nickolas M. Waser

Department of Biology
University of California
Riverside, California

ABSTRACT

In this chapter I review evidence that competition for pollination has produced or maintains differences in floral characters among sympatric plant species. I begin by proposing a functional definition of competition for pollination and by considering different mechanisms and outcomes subsumed under this definition. With this background, I discuss studies of flower color, flower morphology, and flowering time that involve inspection of single communities, comparison of communities, and experimental manipulation. I conclude that mere inspection of single communities is the weakest approach, and experimentation the strongest approach, for inferring whether and how competition has promoted floral character differences; and that combinations of approaches may be most profitable in improving our scanty understanding of the evolutionary impact of competition for pollination.

KEY WORDS: Arizona, bees, character displacement, Colorado, comparative methods, competition for pollination, *Delphinium nelsonii*, experimental manipulations, flower color, flower constancy, flower morphology, flowering time, hummingbirds, hybridization, interspecific pollen transfer, *Ipomopsis aggregata*, montane islands, null communities, *Penstemon barbatus*, pollen loss, optimal foraging, reproductive success, stigma contamination, Utah.

CONTENTS

INTRODUCTION

The early naturalists found it intuitive that sympatric, ecologically similar species should compete in some fashion for shared necessities of life (e.g., Darwin, 1890, p. 48ff.; Elton, 1927). They were quick to realize that competition could occur not only when aggressive behavioral interaction was observed or when obviously important resources such as food or space were involved, but also in more unorthodox and subtle ways. A good example comes from the writings of the American botanist Charles Robertson. Robertson (1895) apparently was the first to propose that the transfer of pollen by animals is a resource for which plants might compete, and that this competition might be responsible for some differences among plant species in flowering time and other characters.

Competition for pollination has stimulated interest since Robertson's time, especially in the last decade (e.g., Levin and Anderson, 1970; Mosquin, 1971; Reader, 1975; Lack, 1976; Stiles, 1977; Thomson, 1978; Waser, 1978a). There has been some confusion, however, about the exact mechanisms and expected evolutionary outcomes of this interaction. Several things may help to reduce this confusion.

The first step is to adopt a simple functional definition of competition for pollination. Out of necessity, theoretical ecologists did this for competition in general at an early stage. The definition they have incorporated into mathematics is that competition is any interaction involving two or more species (or phenotypes within a species) that causes each competitor to suffer reduced fitness (see MacArthur, 1972, p. 21ff.; Emlen, 1973, p. 306ff.). By extension, I will define competition for pollination as any interaction in which co-occurring plant species (or phenotypes) suffer reduced

reproductive success because they share pollinators.

This definition is sufficiently broad to encompass at least two different sorts of interaction. The first I call *competition through pollinator preference*. This mechanism is implicit or explicit in most discussions of competition for pollination published to date (e.g., Clements and Long, 1923; Free, 1968; Mosquin, 1971; Straw, 1972; Reader, 1975; Carpenter, 1976; Lack, 1976). It occurs when one plant species or phenotype within a species is somehow able to attract pollinators away from others and when this reduction in visitation lowers reproductive success. The second mechanism of interaction is qualitatively different from the first. I call it *competition through interspecific pollen transfer*. It occurs when a pollinator forages without perfect preference in a mixture of plant species (or differently adapted phenotypes of a single species) and causes pollen transfer between them, and when the resulting losses of pollen, receptive stigma surface, and effective pollination movements lower reproductive success (see Lewis, 1961; Levin, 1971; Ganders, 1975; Waser, 1977, 1978a,b; Feinsinger, 1978; Sukhada and Jayachandra, 1980; Thomson et al., 1981). It is important to stress that this second mechanism of competition, unlike the first, can operate even when pollinator numbers are not directly limiting (contrast Straw, 1972; this is the reason I refer to "competition for *pollination*" rather than "competition for *pollinators*" in this review), and in certain cases even when seed set is not reduced (recalling that pollen grains are the vehicle by which half of all nuclear genes reach the next generation, cf. Charnov, 1979). The second mechanism, but not the first, could result in competition even in wind-pollinated systems, so long as the presence of other species causes loss of pollen or stigma surfaces that otherwise would contribute to fertilization events. Finally, one manifestation of competition through interspecific pollen transfer, but not through pollinator preference, is the formation of maladapted hybrids when related varieties or species are involved (cf. McNeilly and Antonovics, 1968; Levin, 1971; Reader, 1975). Recognizing that deleterious hybridization is functionally equivalent to competition through interspecific pollen transfer is important for the discussion that follows. Indeed, if we do *not* equate these two things, we are left in an illogical position because we then have to postulate an abrupt point in a continuum of matings between more and more distantly related sympatric varieties or species beyond which we have "competition" rather than "hybridization."

Although the names I have proposed for the two different mechanisms of competition for pollination described here are awkward, they have the great advantage of being descriptive. I feel that it only reinforces the common confusion about mechanisms to adopt the zoologically derived terms "interference" and "exploitation" instead, as some people have suggested (e.g., Thomson, 1978; Brown and Kodric-Brown, 1979; Pleasants, 1980; Zimmerman, 1980).

A second step toward reducing confusion is to recognize that there are several possible evolutionary outcomes of competition for pollination, and to distinguish these outcomes clearly from the interaction itself. In general, competition is predicted to lead to assemblages of species that differ in characters related to resource use. Two different ways have been proposed by which this could happen, however (cf. Roughgarden, 1976; Thomson, 1981). One view is that competition can cause differential extinction, at least on a local scale, of species too similar to others in their resource requirements (e.g., MacArthur and Levins, 1967; MacArthur, 1972). An alternative view is that competition will constitute a strong selective force promoting evolutionary divergence among co-occuring species in resource use (Lack, 1947; Lawlor and Maynard Smith, 1976; Roughgarden, 1976).

At least in its extreme form, where one "cornucopian" plant species (*sensu* Mosquin, 1971) is supposed to draw pollinators entirely away from others, competition through pollinator preference might indeed cause extinctions. Since this competitive effect is one-sided, however, with the most attractive species suffering little or no fitness loss from the presence of other species, it is hard to see how it alone will lead to stable divergence in flowering time or other attributes related to pollination. Instead, a cornucopian species, in the absence of other restrictions, eventually should drive other species extinct and co-opt their share of the flowering season and of the pollinators (cf. Mosquin, 1971).

In contrast to competition through pollinator preference, competition through interspecific pollen transfer should promote stable divergence in floral traits related to pollination. Careful considerations of published models that implicitly or explicitly deal with effects of interspecific pollen transfer (Levin and Anderson, 1970; Waser, 1977, 1978b) indicates that this type of interaction is unlikely to lead to actual extinction except when plant species share edaphic requirements as well as pollinators (see Waser, 1978b, p. 231). Since all co-occurring species or varieties suffer fitness reductions from the interaction, however, reciprocal divergence is a reasonable expectation, at least when other forces do not counterbalance the fitness advantage of divergence (for example, see Ågren and Fagerström, 1980). For this reason and others, including the fact that competition through interspecific pollen transfer does not require pollinator rarity, I have argued (Waser, 1977, 1978a,b) that this mechanism is more likely than competition through pollinator preference to be common in nature, and to be implicated in producing or maintaining floral character differences among sympatric plant species.

Many authors have documented floral character differences within or among plant species. Without detailed considerations of mechanisms of interaction, a number of them have argued that these character differences were produced or are maintained by some form of competition for pollination. In this chapter I will review the evidence that this is so. At the same time, I wish to point out relative strengths and weaknesses of the three different approaches available for the study of competition and community structure: inspection of patterns within single communities, comparison between communities, and experimental manipulations.

OBSERVATIONS OF CHARACTER DIFFERENCES WITHIN SINGLE COMMUNITIES

Differences in Flower Morphology and Color

One of the most striking features of natural plant communities is their high diversity in the morphology and color of flowers (Weevers, 1952; Guldberg and Atsatt, 1975; Heinrich, 1975a; Kevan, 1978). Floral diversity is correlated in general with morphological and behavioral diversity among flower-visiting animals. It seems possible that parallel adaptive radiation of flower and pollinator types represents a basic and pervasive evolutionary outcome of competition for pollination among plants, and concomitant competition for floral rewards among pollinators (cf. Heithaus, 1974). Indeed, it is difficult to imagine what biotic force *other* than competition, with its generally expected effect of producing differentiation and specialization among competitors, could have fostered overall diversity of pollination systems.

The view that specialization in pollination systems has increased through evolutionary time is widely held (e.g., van der Pijl, 1961; Stebbins, 1970; Faegri and van der Pijl, 1971; Heinrich, 1975b). Indeed, Grant (1949, 1952), Grant and Grant (1964, 1965), Sprague (1962), Macior (1970, 1971), and Schemske (1976) all describe intricate plant–pollinator relationships that they explain as the result of selection for ever-increasing specialization of plants on particular subsets of pollinators (see also Raven, 1972). One reason for specialization postulated by Grant (1949, 1952) is that it allows plants to avoid unfavorable hybridization events. As I have pointed out, this is the same as saying that specialization reduces competition for pollination through interspecific pollen transfer. Specialization is not a universal process, however, since many successful species retain broad "promiscuous" pollinator affinities. It also should be noted that the classification of species according to "pollination syndromes" (e.g., Grant 1949; van der Pijl, 1961; Baker and Hurd, 1968; Faegri and van der Pijl, 1971) will in many cases exaggerate specialization (cf. Macior, 1971). Among species I have studied, for example, are a typical "bumblebee flower" *(Delphinium nelsonii)* that is pollinated heavily by hummingbirds (Waser, 1978a) and a typical "hummingbird flower" *(Fouquieria splendens)* that is pollinated heavily by solitary bees (Waser, 1979).

Aside from allowing specialization for pollinator type, which may reduce both competition through pollinator preference and competition through interspecific pollen transfer, floral diversity might have evolved in part because species with unusual flower morphologies or colors are most able to induce pollinator constancy (see Waddington, this book), a behavior that will minimize competition through interspecific pollen transfer. Heinrich (1975a, 1976) proposes a similar hypothesis to explain the diverse shapes and colors of bee flowers in Maine bogs, and provides some evidence that constancy is based on flower morphology. There also have been several reports suggesting that color differences alone suffice to cause constancy (e.g., Grant, 1949; Bateman, 1956; McNaughton and Harper, 1960; Levin, 1968; Levin and Schaal, 1970; Faulkner, 1976; Jones, 1978), although almost as many suggest that this is not the case (e.g., Darwin, 1876, p. 416; Manning, 1957; Lewis, 1961; Macior, 1971; Heinrich, 1975b). Finally, Waddington (1979) discussed a behavior that can be called "foraging height constancy." He assumed that pollinators will tend to remain at a constant height above the ground as they fly between plants. If so, selection could lead to or maintain differences in inflorescence height among competing species. Waddington gave evidence for such differences within several small guilds of species sharing pollinator types and growing together in a single subalpine meadow. Unfortunately, while foraging height constancy has been reported for honeybees and butterflies (Levin and Kerster, 1973; Faulkner, 1976), he provided no evidence that the bumblebee pollinators in his system show this behavior.

It seems unlikely to me that the advantage of inducing constancy can provide a general explanation for floral diversity, since many pollinators, including hummingbirds (Grant and Grant, 1968; Waser, 1978a), butterflies and moths (Kislev et al., 1972; Levin and Berube, 1972), and even solitary bees and bumblebees (Clements and Long, 1923; Brittain and Newton, 1933; Grant, 1950; Free, 1966, 1970; Macior, 1970; Heinrich, 1976) often exhibit incomplete or no constancy. This should not necessarily be surprising. Constancy may not represent an adaptation for maximally efficient foraging, as is sometimes thought (e.g., Grant, 1949; Levin, 1971, 1978a; Oster and Heinrich, 1976), but rather a lack of adaptation (*sensu* Williams, 1966), because constancy often will cause a pollinator to skip over rewarding flowers and thus to exceed minimum flight costs. Such behavior is not predicted by optimal foraging theory (e.g., MacArthur and Pianka, 1966; MacArthur, 1972), unless one assumes some constraint such as inability of the pollinator to remember simultaneously how to handle flowers of different morphologies (see also Weaver, 1956; Heinrich, 1975a; Strickler, 1979). Indeed, many pollinator types appear to behave as optimal foragers (i.e., to forage in a way that maximizes net rate of energy intake; see Pyke, 1978a,b,c,d, 1979; Hartling and Plowright, 1979; Price and Waser 1979; Pyke and Waser, 1981; Waser and Price, this book), and it seems reasonable to expect those without memory constraints to minimize flight costs by being inconstant. The explanation for constancy that requires memory constraints rather than purely adaptive responses is in essence that hypothesized by Darwin (1876, p. 419). This explanation has been accepted widely (e.g., Proctor and Yeo, 1972; Heinrich, 1979), but to my knowledge remains without experimental verification.

Aside from allowing exclusive use of one or more pollinator species, or causing some form of constancy in individuals of shared species, there is one additional way in which morphological differences between flowers might reduce interspecific pollen transfer and thus might evolve to reduce at least one form of competition for pollination. If flowers of sympatric species sharing pollinators differ in placement of their sexual structures, each species may be able to use a unique part of the pollinator's body for transfer of gametes. Such a situation was noticed by Sprague (1962) in her study of pollination in the genus *Pedicularis*. Two species in the Sierra Nevada of California, *P. groenlandica* and *P. attolens,* share bumblebee pollinators but place pollen on the venter and head of a bee, respectively. Similarly, some degree of separation between species in pollen placement was reported for different communities of hummingbird-pollinated plants by Stiles (1975), Carpenter (1978), and Waser (1978a). Levin and Berube (1972) found that interspecific pollen transfer is low in natural mixed populations of *Phlox pilosa* and *P. glaberrima* and in laboratory feeding trials with hand-held butterflies, presumably in part because the two species differ in corolla length and exsertion of sexual parts. A final example comes from Brown and Kodric-Brown's (1979) study of hummingbird-pollinated plants in Arizona. Three common species that might compete for pollination in their study areas are *Penstemon barbatus, Castilleja integra,* and *Ipomopsis aggregata.* The anthers in *I. aggregata* flowers normally are radially arranged. However, Brown and Kodric-Brown propose that anthers and stigmas in their populations have become ventrally placed to avoid competition with *P. barbatus* and *C. integra,* whose sexual parts touch the head of a hummingbird. They present evidence (their Table 5) that most *I. aggregata* pollen indeed is carried on chins of hummingbirds. This pattern may not be as strong as their table suggests, however. If the birds examined were captured after foraging in a mixture of all three species, as appears to have been the case, it seems a necessary expectation that most *I. aggregata* pollen will be found on the chin. Any pollen deposited on the head will stand a chance of being removed by sexual parts of all three species, and thus should disappear more quickly than pollen deposited on the chin, which will be removed only by the sexual parts of *I. aggregata*. Brown and Kodric-Brown also present information on pollen loads taken from stigmas of *P. barbatus, C. integra,* and *I. aggregata* (their Table 6). Even though the first two species use the head of a hummingbird for pollen transport whereas *I. aggregata* is supposed to use the chin, stigmas of all species receive substantial loads of foreign pollen, and these loads do not differ consistently between species. From this observational evidence alone, then, it is not clear that placement of sexual parts in *I. aggregata* flowers has evolved in areas of sympatry with *P. barbatus* and *C. integra* to reduce competition through interspecific pollen transfer with these species.

From this discussion I conclude that the great diversity in flower color and morphology within communities generally reflects diversity among flower visitors and in the ways plants use visitors for pollen transfer. Unfortunately, this does not mean that studies of single communities have demonstrated, or are capable of demonstrating, that competition for pollination in any form promotes or maintains floral diversity. I will return later to this weakness of purely observational studies of single communities.

Differences in Flowering Time

Members of plant communities do not all flower at the same time. Charles Robertson was the first to propose that differences in flowering time might be produced or maintained by competition for pollination. In an early study (Robertson, 1895), he grouped members of a diverse Illinois woodland community according to major pollinator type, and argued in several cases that flowering times were staggered within such pollination "guilds." He also showed (Robertson, 1895, 1924) that introduced species had longer flowering times than natives. He ascribed this difference to a longer opportunity for evolutionary divergence among natives as a result of competition.

Many descriptive studies of flowering time

similar to Robertson's have appeared recently. For example, there have been surveys of unrelated species sharing pollinators in arctic (Hocking, 1968), temperate (Mosquin, 1971; Heinrich, 1975a; Reader, 1975; Carpenter, 1976; Waser, 1976, 1978a; Kodric-Brown and Brown, 1978; Parrish and Bazzaz, 1978; Schemske et al., 1978; Thomson, 1978), and tropical (Frankie et al., 1974; Heithaus, 1974; Stiles, 1975, 1977; Feinsinger, 1978) communities. There also have been reports of sequential flowering of sympatric congeners that share pollinators, and whose hybrids in several cases are known to be sterile or maladapted, including those of Whitaker (1944) for *Lactuca,* Clausen (1951) for *Madia,* Lewis (1961) for *Clarkia,* Grant and Grant (1964) for *Salvia,* Macior (1970) for *Pedicularis,* Heinrich (1975a) for several congeneric pairs in Maine bogs, and Lack (1976) for *Centaurea.* Finally, there have been reports of sequential flowering of wind-pollinated congeners or conspecifics that could suffer from interspecific pollen transfer, including those of Stebbins (1950) for three species of *Pinus* and of McNeilly and Antonovics (1968) for *Anthoxanthum* and *Agrostis.*

Mere inspection suggests fairly convincingly that flowering actually is sequential for some, but not all, of these studies. Sequential patterns are most evident where only a few species are involved (e.g., Lewis, 1961; Grant and Grant, 1964; Hocking, 1968; Reader, 1975; Carpenter, 1976; Lack 1976; Waser, 1978a). They are much less clear in complex communities (e.g., Mosquin, 1971; Heithaus, 1974; Heinrich, 1975a; Stiles, 1975, 1977; Feinsinger, 1978; Parrish and Bazzaz, 1978). In these cases it does not seem visually obvious that flowering times conform to any organized pattern, as we might expect if competition for pollination has played a role in their evolution.

In an attempt to test rigorously for organization of flowering times, Poole and Rathcke (1979) compared Stiles's (1977) data from a community of ten tropical hummingbird-pollinated plants with a statistically generated null hypothesis. They asked whether peak flowering dates for the species studied by Stiles were regularly spaced through the normal growing season, and concluded instead that peaks were significantly clumped during two yearly dry periods. From this, they argued that there was no clear reason to implicate competition for pollination as a force that had produced flowering time differences. A somewhat similar analysis was performed by Thomson (1978) for a large number of herbaceous species in subalpine meadows in the Colorado Rocky Mountains. He first grouped species into guilds according to major pollinator type, and then used a computer to distribute actual flowering time curves from each guild at random over the normal flowering season. Instead of examining spacing of flowering peaks, Thomson compared means and variances in overlaps of flowering curves from random and actual guilds. Although he examined six guilds at each of two sites, in no case was actual mean or variance in overlap significantly smaller than given by the appropriate random null model. In contrast to Poole and Rathcke (1979) and Thomson (1978), Pleasants (1980) was able to discern some uniform spacing of flowering times in complex communities. He examined Rocky Mountain meadow communities similar to those studied by Thomson (1978) and used similar computer methods for generating null expectations, but found that overlaps in flowering times within about half the pollination guilds studied were significantly smaller than expected at random.

Thus, although species in single communities always can be arranged from the earliest flowering to the latest flowering, mere inspection of such patterns cannot tell us whether they are different from random expectations. Comparison of actual patterns to those developed under suitable null hypotheses solves this problem in a statistical sense, but unfortunately does not tell us whether competition for pollination is or has been important. If competition acts only within subsets of pollination guilds, or if selective forces other than competition (e.g., parasitism, mutualism) have influenced flowering times within a guild, the overall pattern of flowering times may appear random, especially in large guilds (Thomson, 1978). Conversely, of course, finding that flowering times are nonrandomly spaced is not suf-

ficient in itself to demonstrate that competition for pollination rather than other forces has produced or maintains differences among species. This emphasizes again the difficulty of inferring causality from studies of single communities that I mentioned in discussing morphological differences between members of pollination guilds.

In drawing these conclusions I do not mean to suggest that the development and use of null expectations is a useless exercise. For example, I disagree with Stiles (1979), who discounted Poole and Rathcke's (1979) reanalysis of his data by implying that regularly spaced flowering times would not be expected by anybody familiar with his system. This caveat seems reasonable at first, but it implies that any kind of difference between species can be explained *a posteriori* as a result of competition for pollination. Obviously this leaves us with no ability to distinguish a random world from one organized by biotic interactions such as competition. In contrast, the rigorous approach of Poole, Rathcke, and others at least gives us great power to identify systems that merit further comparative or experimental examination of the kinds I will discuss below.

COMPARISONS BETWEEN COMMUNITIES

Geographic Variation in Flower Color and Morphology

Comparative studies are likely to yield more insight into the influence of competition for pollination on floral diversity than are studies of single communities in isolation (cf. Pittendrigh, 1958; Curio, 1973). By looking at the same species in different geographic locations, we can see whether traits such as flower color or morphology vary from place to place in ways that reflect the presence or absence of putative competitors for pollination. Such patterns, if consistent, will be much harder to ascribe to forces other than competition than are differences within single communities.

Comparative evidence for competitive effects on floral morphology appears restricted to several descriptions of character displacement (*sensu* Brown and Wilson, 1956) in zones of sympatry with potential competitors. Iltis (1958) reported that *Polanisia dodecandra* ssp. *dodecandra* has smaller flowers and shorter filaments where it is sympatric with *P. dodecandra* spp. *trachysperma* than in other areas. These differences might have evolved to reduce pollen transfer between subspecies, but Iltis gave no information on pollination with which to judge the likelihood of this explanation. Breedlove (1969) described a similar situation in two Mexican species of *Fucsia*. Where they are allopatric, both species appeared to be pollinated by hummingbirds and bumblebees, and both produce flowers of similar size and color, but abrupt differences in flower color and size occurred in areas of sympatry. These changes appeared to reduce pollinator sharing, with one species receiving mostly hummingbird visits and the other mostly bee visits. Some final examples come from Whalen's (1978) study of the genus *Solanum*. For several species pairs, differences in exsertion of sexual parts of flowers appeared abruptly in zones of sympatry. Whalen proposed that these differences limit effective pollinators of each species to bees of different size and thereby minimize interspecific pollen transfer. Unfortunately, he did not observe pollination in enough detail to confirm that morphological differences actually do correspond to changes in effective pollinators.

There are a few reports of character displacement involving flower color to add to those involving morphology. For some species, such as those studied by Iltis (1958) and Breedlove (1969), color changes accompany changes in flower size in zones of sympatry. In such cases it is possible that different colors are not themselves the basis for any pollinator discrimination that may exist, but instead act as cues that pollinators can associate with different floral morphologies. Apparent character displacement involving only color also occurs, however. Levin (1971) pointed out one case involving two *Clarkia* species studied by Lewis and Lewis (1955), and Levin and Kerster (1967) describe another case involving two *Phlox* species. Finally, a very detailed study of six species in the genus *Rudbeckia* by McCrea

(1981) describes several cases of displacement involving ultraviolet reflectance patterns of flowers. McCrea showed that these patterns have high heritability and that most of his species suffer reproductive losses when subjected to artificial interspecific pollen transfer. From these findings he argued that displacement would be expected to occur rapidly in zones of sympatry between most species. Unfortunately, he was unable to follow the small solitary bees that are major pollinators of his plants, and to show that their flower constancy was increased by divergence in ultraviolet patterns. To argue that color displacement in ultraviolet or human-visible wavelengths represents evolution to reduce competition for pollination, we must ultimately demonstrate that color changes themselves are able to induce pollinator constancy or shifts in pollinator affinities.

In summary, then, there appears to be little published evidence for displacement involving flower color or morphology that can be ascribed clearly to competition for pollination. Further searches for such patterns are needed, and they will be most valuable if they attempt to document effects of character displacement on identities and behavior of flower visitors.

Geographic Variation in Flowering Time

Flowering time obviously is a trait that varies greatly across the geographic range of species. For most species this variation cn be attributed at least in part to variation in latitude, elevation, or other factors that reflect climatic seasonality (e.g., Sørenson, 1941; Kucera, 1958; Jackson, 1966; Ray and Alexander, 1966; Hodgkinson and Quinn, 1978). In many cases, however, details of flowering time variation are not easily explained solely by climatic factors (e.g., Robertson, 1924; Olmsted, 1944; Bernström, 1952; Panje and Srinivasan, 1959; Sawamura, 1967). In some cases the patterns are suggestive of evolutionary divergence to reduce competition for pollination.

For instance, there are several reports of apparent character displacement involving flowering times. McNeilly and Antonovics (1968) examined two wind-pollinated grasses, *Agros-*

tis tenuis and *Anthoxanthum odoratum,* across a boundary between soils contaminated and those uncontaminated by heavy metals from mine tailings. In both species they found substantial differences in flowering times of individuals near the abrupt boundaries between soil types, but not further from boundaries. After growing plants under standard conditions they concluded that these differences were genetically determined. Although artificial crosses between individuals of each species from contaminated and uncontaminated soils uncovered no consistent incompatibility barriers, Jain and Bradshaw (1966) had shown previously that plants adapted to each soil type grew poorly on the other. Thus McNeilly and Antonovics (1968) reasoned that character displacement in flowering times could have evolved to reduce pollen transfer and resulting formation of hybrids of low viability in each soil type. Carpenter (1976) found similarly that two tree species on the island of Hawaii, *Metrosideros collina* and *Sophora chrysophylla,* flowered sequentially at elevations where they were sympatric. At other elevations where one or the other species grew alone, each had a substantially longer flowering time that encompassed much of the flowering period occupied by the other species in areas of sympatry. Both tree species are pollinated primarily by honeycreepers, and Carpenter hypothesized that character displacement in flowering times was an evolutionary response to competition for pollination. Finally, Stace and Fripp (1977) discussed an analogous situation for two races of *Epacris impressa* in Australia. Since different races have different colors and sizes of flowers, and since no information is given on flower visitors, however, it is not clear that the shifts in flowering time they reported from areas of sympatry are related to competition for pollination.

In retrospect, arguments that competition for pollination should lead to displacement in flowering time rest on the implicit assumption that this trait can be changed readily by selection. This seems very reasonable. Flowering time has been shown to change rapidly under various regimes of artificial and natural selection (e.g., Stanford et al., 1962; Snaydon,

1963a,b; Akemine and Kikuchi, in Allard and Hansche, 1964; McNeilly and Antonovics, 1968; Paterniani, 1969). Indeed, rates of change in some of these studies were so dramatic as to suggest that flowering time is an extremely flexible trait. If this is correct, we may expect character displacement to be widespread, even though there have been few reports of it to date. Similarly, we might expect a close examination of single species to reveal many differences in flowering time from place to place that reflect local competitive environments in a fine-tuned way.

I have some preliminary evidence for geographic variation in flowering time of *Ipomopsis aggregata* that is intriguing in light of this last prediction. This species is a common resident of dry montane meadows in the western United States, and is pollinated largely by hummingbirds (Grant and Grant, 1968; Waser, 1978a). Over much of its geographic range, populations of *I. aggregata* are restricted to isolated mountain "islands" separated by lower areas of Great Basin or Sonoran Desert vegetation.

I have observed flowering of *I. aggregata* systematically in two locations, the Pinaleño Mountains of southeastern Arizona and the Elk Mountains of west-central Colorado. These sites are separated in a north–south direction by about 700 km or 6 degrees of latitude. During the summer of 1974 I repeatedly censused flowering *I. aggregata* plants along a stretch of road that follows the crest of the Pinaleños and varies in elevation from about 1900 to 2900 meters. I have followed flowering of *I. aggregata* in Colorado for several years within permanent census plots at 2900 meters elevation (Waser, 1978a; Waser and Real, 1979). Figure 13-1 compares flowering times from the Pinaleños (1974) and Elks (1973, a fairly representative year). Despite the different census methods used, it is clear that *I. aggregata* flowers later in southern Arizona than it does in Colorado at comparable or even higher elevations. This difference is difficult to attribute to seasonal factors. It is most easily explained, in fact, by flowering times of other common species at each site that share hummingbird pollinators with *I. aggregata*. In the Elks these po-

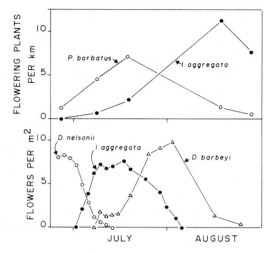

Figure 13-1. Flowering times of *Ipomopsis aggregata* and known or possible competitors for hummingbird pollination in two geographic locations. Top: Mean numbers of flowering plants of *I. aggregata* and *Penstemon barbatus* from surveys along a 30-km stretch of dirt road in the Pinaleño Mountains (32°40′ N. Latitude) in southeastern Arizona, from 1974. Bottom: Mean numbers of flowers of *Delphinium nelsonii*, *I. aggregata*, and *D. barbeyi* from counts in twenty 2 m × 2 m census plots at the Rocky Mountain Biological Laboratory, Elk Mountains (38°57′ N. Latitude) in Colorado, from 1973.

tential competitors for pollination are *Delphinium nelsonii* and *D. barbeyi*, which flower before and after *I. aggregata* in the same or nearby meadows, respectively (see Waser, 1978a). Experiments show that *D. nelsonii* indeed competes with *I. aggregata*, as I will discuss below. In the Pinaleños by far the most common and widespread species visited by hummingbirds are *I. aggregata* and *Penstemon barbatus*. Thus, these species are potential competitors for pollination, and *P. barbatus* flowers before *I. aggregata* (Fig. 13-1). A similar flowering sequence for *P. barbatus* and *I. aggregata* was found by Kodric-Brown and Brown (1978, their Fig. 1) in the White Mountains of Arizona, about 125 km north of the Pinaleños.

Stimulated by these patterns, I recorded from herbarium collections at the University of Arizona, University of Utah, and Museum of Northern Arizona the dates of all flowering specimens of *P. barbatus* and *I. aggregata* that had been collected between 1900 and 2900

meters elevation in six areas of Utah and Colorado. Figure 13-2 presents the results gleaned from this survey and from my direct observations in the Pinaleños (site 6 in Fig. 13-2) and Elks (site 2). The herbarium records suggest that *P. barbatus* flowers before *I. aggregata* in the White Mountains (site 5), which is in accord with direct observation. This gives some support to the use of herbarium records as crude substitutes for actual surveys in each area. In the Abajo Mountains of southeastern Utah (site 3) and the San Francisco Peaks of central Arizona (site 4), *P. barbatus* again appears to flower first. On the north and south rims of the Grand Canyon (sites 7 and 8), however, there seems to be a reversal in flowering times with *I. aggregata* flowering before *P. barbatus*. In these areas the flowering time of *I. aggregata* is similar to that in the Wasatch Mountains of Utah (site 1) and the Elk Mountains of Colorado (site 2) where *P. barbatus* is absent. The possible flip-flop in flowering times seems to me exactly the sort of pat-

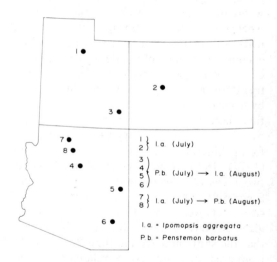

Figure 13-2. Flowering times of *I. aggregata* and *P. barbatus* at eight geographically distant sites between 1900 and 2900 meters elevation in Colorado, Utah, and Arizona, taken from direct observation or deduced from herbarium records. Sites are as follows: 1 = Wasatch Mountains above Salt Lake City, Utah; 2 = Elk Mountains, Colorado; 3 = Abajo Mountains above Monticello, Utah; 4 = San Francisco Peaks above Flagstaff, Arizona; 5 = White Mountains, Arizona; 6 = Pinaleño Mountains, Arizona; 7 = north rim of the Grand Canyon; and 8 = south rim of the Grand Canyon, Arizona.

tern to expect if different isolated montane areas are colonized by individuals of each species with slightly different original flowering times, and if subsequent divergence as a result of competition between species leads to alternate equilibrium states.

By this discussion I do not mean to imply that extensive perusal of herbarium specimens is in order. Surveys of this kind obviously fail to detect which plant species actually are pollinated by hummingbirds ("pollination syndromes" are not adequate to determine this, as I have mentioned), or which species actually grow together and thus potentially are subject to competition. The time I spent secluded in herbaria may at least suggest where to look for interesting geographic patterns, but the next step obviously is actually to go and count flowers. Unfortunately, a large allocation of funds and time would be needed to survey four or five isolated mountain islands in Arizona on a regular and frequent schedule, or to hire a field assistant to census each site. Seen in this light, detailed comparative studies seem costly and complicated as a means of demonstrating that competition for pollination occurs and influences diversity of flowering times, especially compared to experimental studies of single communities. A combination of the two techniques may be synergistic, however, as I will discuss later.

EXPERIMENTAL APPROACHES

Experiments were an integral part of many early investigations of pollination (e.g., Müller, 1873; Beal, 1880; Clements and Long, 1923). For some reason, however, this approach largely was abandoned by subsequent workers, and only recently has reappeared. This is surprising because field and laboratory experiments with plants and pollinators usually are straightforward. In addition, of course, experimentation is the logically correct method for studying causation, as Platt (1964) discusses clearly. Comparative studies sometimes are considered to be equivalent to experiments, since different populations of species under study have been "manipulated" by different natural environments. However, such "natural

experiments" cannot control for variables other than those whose causal importance are of interest, and thus are much less likely than true experiments to be conclusive (see Connell, 1975).

If our objective is to determine whether competition for pollination occurs between two sympatric species and suffices to maintain character differences, it is not hard to imagine a general experimental approach. Ideally, we should manipulate the degree of overlap or similarity between individuals of the two species in the character of interest, and measure any effect this has on pollinator behavior and reproductive success. If individuals of two species share a common character such as flower color or morphology or flowering time and suffer a net loss of fitness in each other's presence relative to individuals not sharing the character or not subjected to each other's presence, then competition represents a selective agent sufficient to maintain differences in the character in question (compare Curio, 1973; Connell, 1975; Thomson, 1981). If, in addition, we measure visitation rate and movement patterns of pollinators presented with mixtures of plants of the two species and observe where pollen is carried, we should be able to determine whether the interaction involves pollinator preference or interspecific pollen transfer. Actually to change the character of interest without changing other factors, of course, sometimes requires ingenuity. For example, one might attempt to investigate flower color displacement by transplanting individuals into flower pots and creating artificial populations in which potential competitors do and do not differ in color, and measuring subsequent seed set. In measuring reproductive success in this experiment, however, one would have to be careful to control for potential subtle effects of different genetic similarities between conspecific mates (Price and Waser, 1979; Waser and Price, this book). Adequately assessing reproductive success also is difficult. It seems reasonable to assume for now that competition will not affect quality of seeds produced, and thus to measure only seed set and not viability. On the other hand, competition may cause changes in fecundity through male sexual function alone, which will be very difficult to detect.

I have attempted a field experiment along the lines outlined above with two humming-bird-pollinated plants mentioned earlier that grow in the Elk Mountains of Colorado, *Ipomopsis aggregata* and *Delphinium nelsonii* (see Waser, 1978a for details). These species grow in much the same habitats and flower sequentially (e.g., Fig. 13-1, bottom). To determine whether different flowering times are maintained by competition for pollination, I compared seed set and pollinator behavior in small artificial arrays of potted plants placed outdoors under natural conditions. Arrays contained either each species alone or a mixture. By collecting potted plants from different elevations I was able to obtain simultaneous flowering of the two species in mixed arrays. In several replicate experiments, plants in mixed arrays had consistently and significantly lower seed sets than their matched controls in arrays containing each species alone. These fecundity reductions corresponded to interspecific flights and interspecific pollen transfer by humming-birds, judging from mixed pollen loads I observed on birds and stigmas. Furthermore, seed set reductions could be duplicated consistently, at least for *I. aggregata,* with artificial interspecific hand-pollinations in the laboratory. I concluded that competition for hummingbird pollination between *I. aggregata* and *D. nelsonii* suffices to maintain their sequential flowering in nature, and suggested that the interaction involves interspecific pollen transfer, at least in part. My study may be seen as incomplete, however, because observations of pollinator movements in artificial arrays were not extensive enough to determine the mechanism of interaction with great certainty.

Campbell and Motten (1981) recently carried out a study that avoids some of the inadequacies of my earlier work. They manipulated densities and degree of mixing of two spring wildflowers, *Claytonia virginica* and *Stellaria pubera,* and found effects on the latter species strongly suggestive of competition for pollination. By observing in detail the visitation behavior of bee flies, they were able to show that competitive effects involve loss of *S.*

pubera pollen during interspecific pollinator movements, rather than any reduction in pollinator visits to this species; indeed, visitation rate was inhanced by the presence of *C. virginica* flowers. Competition for pollination in this system may play a role in maintaining differences in peak flowering dates of the two plant species, but Campbell and Motten do not discuss this possibility.

An experimental approach also was taken by Levin (1978b) in an attempt to see if different inflorescence heights of two sympatric *Lythrum* species could be explained as an evolutionary response to induce foraging height constancy of pollinators. Levin placed plants in flower pots, constructed artificial mixed arrays with inflorescences at different relative heights, and observed honeybee foraging behavior. He found that interspecific flights indeed declined with increasing disparity in inflorescence heights. Unfortunately, he did not measure reproductive success of individuals under different degrees of interspecific overlap in this character, as would be required to demonstrate that competition for pollination is involved in maintaining observed differences.

Finally, Levin and Kerster (1967) used an experiment in a study of character displacement involving flower color. They reported that white-flowered plants are rare in most populations of the prairie herb *Phlox pilosa.* In some populations, however, *P. pilosa* grows mixed with *P. glaberrima* and flowers at the same time. In these cases, white-flowered *P. pilosa,* which contrast with the pink-flowered *P. glaberrima,* may predominate. Levin and Kerster transplanted normal pink-flowered *P. pilosa* into one such mixed population, and then compared loads of heterospecific pollen reaching stigmas of pink and white flowers of this species. They found that white flowers suffered much less contamination from *P. glaberrima* pollen than did pink flowers, which is expected from the tendency of the butterfly pollinators to discriminate between flowers on the basis of color. Pink-flowered transplants also appeared to be visited more frequently than white-flowered forms of *P. pilosa.* Thus, white-flowered forms received less conspecific as well as less heterospecific pollen. It is impossible to deduce whether they actually enjoyed a net fitness advantage relative to pink-flowered forms, since once again no attempt was made to measure reproductive success of individuals directly.

As far as I know, other experiments of this sort have not been attempted. They seem to me, however, the best suited of any single method of discerning whether competition for pollination serves as a general force maintaining character differences among sympatric plant species. I hope that others will be encouraged to undertake experiments in which they manipulate similarity between species in some character whose displacement is surmised to be related to competition. Alternative experimental methods for demonstrating that competition for pollination occurs or has occurred have been proposed, but I do not believe that they are logically sound. For example, Schemske et al. (1978) and Zimmerman (1980) imply that it is sufficient to look for correlations between seed set and pollinator numbers or pollen availability to infer competition. However, competition through interspecific pollen transfer could occur even if pollinators are abundant relative to flowers and even if seed set is not raised by addition of pollen, as discussed earlier. Conversely, finding that pollinator or pollen availability limits seed set does not suffice to demonstrate that competition through pollinator preference is occurring, since the removal of putative competitors of the same or another species could quite conceivably cause pollinators to abandon the plants in question for better pastures and thus lower seed set even further.

COMBINING APPROACHES

I have tried here to review evidence that competition for pollination occurs and that it is responsible for some of the great diversity we observe in floral traits. Most studies to date have used inspection of single communities to attack this problem. I have suggested that this technique in itself is not useful for inferring causation, although it may serve to detect interesting patterns and to direct us to systems that deserve further investigation. A smaller number of studies have used comparison between populations of single species in different geographic locations to explore competition. I have argued that comparative studies, even if they are considered to uncover results of so-called natural experiments, are much less powerful or productive than controlled experimental manipulations in inferring causation.

On the other hand, comparative studies and experimental manipulations may be combined to good advantage. A fundamental weakness of experiments is that they can tell us whether competition for pollination suffices to maintain character differences among sympatric species, but not whether competition was important in the original development of such differences. If we determine experimentally that competition maintains character differences among species in one place, however, comparison of the character from place to place can provide strong evidence that competition produced the differences as well. This is so if we can document that the character in question is flexible and varies in accord with the presumed competitive environment, as flowering time does in the example I discussed with *Ipomopsis aggregata*.

I conclude that competition for pollination, especially competition through interspecific pollen transfer, is likely on empirical and simple theoretical grounds to be common in nature. At this time, however, we have little strong evidence that competition actually is important in producing or maintaining floral diversity. To remedy this situation I think it will be necessary to concentrate on experimental approaches, but not to eschew entirely more traditional methods, especially those of comparison of populations of single species from place to place.

ACKNOWLEDGMENTS

I have discussed many of the ideas presented here with J. F. Addicott, J. H. Brown, C. Dunford, P. Feinsinger, S. Louda, J. M. Pleasants, H. R. Pulliam, M. V. Price, G. H. Pyke, B. J. Ratchcke, J. D. Thomson, and M. Zimmerman. M. V. Price also provided field assistance and editorial advice, and drew the figures. L. Arnow always made me welcome in the University of Utah herbarium. P. Lain promptly and patiently typed the manuscript.

The University of Arizona Foundation, Sigma Xi, the Chapman Memorial Fund, and the American Ornithologists' Union helped to finance my field studies in Colorado and Arizona. I thank all of these people and institutions.

LITERATURE CITED

Ågren, G. I. and T. Fagerström. 1980. Increased or decreased separation of flowering times? The joint effect of competition for space and pollination in plants. *Oikos* 35:161–164.

Allard, R. W. and P. E. Hansche. 1964. Some parameters of population variability and their implications in plant breeding. *Adv. Agron.* 16:281–325.

Baker, H. G. and P. D. Hurd. 1968. Intrafloral ecology. *Annu. Rev. Ent.* 13:385–414.

Bateman, A. J. 1956. Cryptic self-incompatibility in the wall flower: *Cheiranthus cheiri* L. *Heredity* 10:257–261.

Beal, W. J. 1880. Fertilization of flowers by hummingbirds. *Am. Nat.* 14:126–147.

Bernström, P. 1952. Cytogenetic intraspecific studies in *Lamium.* I. *Hereditas* 38:163–220.

Breedlove, D. E. 1969. The systematics of *Fuchsia* section *Encliandra* (Onagraceae). *Univ. Calif. Publ. Bot.* 53:1–69.

Brittain, W. H. and D. E. Newton. 1933. A study in the relative constancy of hive bees and wild bees in pollen gathering. *Can. J. Res.* 9:334–349.

Brown, J. H. and A. Kodric-Brown. 1979. Convergence, competition, and mimicry in a temperate community of hummingbird pollinated flowers. *Ecology* 60:1022–1035.

Brown, W. L. and E. O. Wilson. 1956. Character displacement. *Syst. Zool.* 5:49–64.

Campbell, D. R. and A. F. Motten. 1981. Competition for pollination between two spring wildflowers. *Bull. Ecol. Soc. Am.* 62:99.

Carpenter, F. L. 1976. Plant–pollinator interactions in Hawaii: pollination energetics of *Metrosideros collina* (Myrtaceae). *Ecology* 57:1125–1144.

———. 1978. A spectrum of nectar-eater communities. *Am. Zool.* 18:809–819.

Charnov, E. L. 1979. Simultaneous hermaphroditism and sexual selection. *Proc. Natl. Acad. Sci. U.S.A.* 76:2480–2484.

Clausen, J. 1951. *Stages in the Evolution of Plant Species.* Cornell University Press, Ithaca, N.Y.

Clements, F. E. and F. L. Long. 1923. *Experimental Pollination—An Outline of the Ecology of Flowers and Insects.* Carnegie Inst. Wash. Publ. No. 336.

Connell, J H. 1975. Some mechanisms producing structure in natural communities: a model and evidence from field experiments. *In* M. L. Cody and J. M. Diamond (eds.), *Ecology and Evolution of Communities.* Belknap Press of Harvard University Press, Cambridge, Mass.

Curio, E. 1973. Towards a methodology of teleonomy. *Experientia* 29:1045–1058.

Darwin, C. 1876. *The Effects of Cross and Self Fertilisation in the Vegetable Kingdom.* Murray, London.

———. 1890. *On the Origin of Species by Means of Natural Selection.* Appleton, New York.

Elton, C. 1927. *Animal Ecology.* Sidgwick and Jackson, London.

Emlen, J. M. 1973. *Ecology: An Evolutionary Approach.* Addison Wesley, Reading, Mass.

Faegri, K. and L. van der Pijl. 1971. *The Principles of Pollination Ecology,* 2nd rev. ed. Pergamon Press, Oxford.

Faulkner, G. J. 1976. Honeybee behaviour as affected by plant height and flower colour in brussels sprouts. *J. Apic. Res.* 15:15–18.

Feinsinger, P. 1978. Ecological interactions between plants and hummingbirds in a successional tropical community. *Ecol. Mon.* 48:269–287.

Frankie, G. W., H. G. Baker, and P. A. Opler. 1974. Comparative phenological studies of trees in tropical wet and dry forests in the lowlands of Costa Rica. *J. Ecol.* 62:881–913.

Free, J. B. 1966. The foraging behavior of bees and its effect on the isolation and speciation of plants. *In:* J. G. Hawkes (ed.), *Reproductive Biology and Taxonomy of Vascular Plants,* Pergamon Press, Oxford.

———. 1968. Dandelion as a competitor to fruit trees for bee visits. *J. Appl. Ecol.* 5:169–178.

———. 1970. The flower constancy of bumblebees. *J. Anim. Ecol.* 39:395–402.

Ganders, F. R. 1975. Fecundity in distylous and self-incompatible homostylous plants of *Mitchella repens* (Rubiaceae). *Evolution* 29:186–188.

Grant, K. A. and V. Grant. 1964. Mechanical isolation of *Salvia apiana* and *Salvia mellifera* (Labiatae). *Evolution* 18:196–212.

Grant, V. 1949. Pollination systems as isolating mechanisms. *Evolution* 3:82–97.

———. 1950. The flower constancy of bees. *Bot. Rev.* 16:379–398.

———. 1952. Isolation and hybridization between *Aquilegia formosa* and *A. pubescens. Aliso* 2:341–360.

Grant, V. and K. A. Grant. 1965. *Flower Pollination in the Phlox Family.* Columbia University Press, New York.

———. 1968. *Hummingbirds and Their Flowers.* Columbia University Press, New York.

Guldberg, L. D. and P. R. Atsatt. 1975. Frequency of reflection and absorption of ultraviolet light in flowering plants. *Am. Midl. Natur.* 93:35–43.

Hartling, L. K. and R. C. Plowright. 1979. Foraging by bumblebees on patches of artificial flowers: a laboratory study. *Can. J. Bot.* 57:1866–1870.

Heinrich, B. 1975a. Bee flowers: a hypothesis on flower variety and blooming times. *Evolution* 29:325–334.

———. 1975b. Energetics of pollination. *Annu. Rev. Ecol. Syst.* 6:139–170.

———. 1976. Bumblebee foraging and the economics of sociality. *Am. Sci.* 64:384–395.

———. 1979. *Bumblebee Economics.* Harvard University Press, Cambridge, Mass.

Heithaus, E. R. 1974. The role of plant–pollinator interactions in determining community structure. *Ann. Missouri Bot. Gard.* 61:675–691.

Hocking, B. 1968. Insect-flower associations in the high arctic, with special reference to nectar. *Oikos* 19:359–387.

Hodgkinson, K. C. and T. A. Quinn. 1978. Environmental and genetic control of reproduction in *Danthonia caespitosa* populations. *Aust. J. Bot.* 26:351–364.

Iltis, H. H. 1958. Studies in the Capparidaceae IV. *Polanisia* Raf. *Brittonia* 10:33–58.

Jackson, M. T. 1966. Effects of microclimate on spring flowering phenology. *Ecology* 47:407–415.

Jain, S. K. and A. D. Bradshaw. 1966. Evolutionary divergence among adjacent plant populations. *Heredity* 20:407–441.

Jones, C. E. 1978. Pollinator constancy as a pre-pollination isolating mechanism between sympatric species of *Cercidium*. *Evolution* 32:189–198.

Kevan, P. T. 1978. Floral coloration, its colorimetric analysis and significance in anthecology. *In* A. J. Richards (ed.), *The Pollination of Flowers by Insects*. Academic Press, London.

Kislev, M. E., Z. Kraviz, and J. Lorch. 1972. A study of hawkmoth pollination by a palynological analysis of the proboscis. *Israel J. Bot.* 21:57–75.

Kodric-Brown, A. and J. H. Brown. 1978. Influence of economics, interspecific competition, and sexual dimorphism on territoriality of migrant rufous hummingbirds. *Ecology* 59:285–296.

Kucera, C. L. 1958. Flowering variation in geographic selections of *Eupatorium rugosum* Houtt. *Bull. Torrey Bot. Club* 85:40–48.

Lack, D. 1947. *Darwin's Finches*. Cambridge University Press.

Lack, A. 1976. Competition for pollinators and evolution in *Centaurea*. *New Phytol.* 77:787–792.

Lawlor, L. R. and J. Maynard Smith. 1976. The coevolution and stability of competing species. *Am. Nat.* 110:79–99.

Levin, D. A. 1968. The effect of corolla color and outline on interspecific pollen flow in *Phlox*. *Evolution* 23:444–455.

———. 1971. The origin of reproductive isolating mechanisms in flowering plants. *Taxon* 20:91–113.

———. 1978a. Pollinator behaviour and the breeding structure of plant populations. *In* A. J. Richards (ed.), *The Pollination of Flowers by Insects*. Academic Press, London.

———. 1978b. The origin of isolating mechanisms in flowering plants. *Evol. Biol.* 11:185–317.

Levin, D. A. and W. W. Anderson. 1970. Competition for pollinators between simultaneously flowering species. *Am. Nat.* 104:455–467.

Levin, D. A. and D. E. Berube. 1972. *Phlox* and *Colias*: the efficiency of a pollination system. *Evolution* 26:242–250.

Levin, D. A. and H. W. Kerster. 1967. Natural selection for reproductive isolation in *Phlox*. *Evolution* 21:679–687.

———. 1973. Assortative pollination for stature in *Lythrum salicaria*. *Evolution* 27:144–152.

Levin, D. A. and B. A. Schaal. 1970. Corolla color as an inhibitor of interspecific hybridization in *Phlox*. *Am. Nat.* 104:273–283.

Lewis, H. 1961. Experimental sympatric populations of *Clarkia*. *Am. Nat.* 95:155–168.

Lewis, H. and M. E. Lewis. 1955. The genus *Clarkia*. *Univ. Calif. Publ. Bot.* 20:241–392.

MacArthur, R. H. 1972. *Geographical Ecology*. Harper and Row, New York.

MacArthur, R. H. and R. Levins. 1967. The limiting similarity, convergence and divergence of coexisting species. *Am. Nat.* 101:377–385.

MacArthur, R. H. and E. R. Pianka. 1966. On optimal use of a patchy environment. *Am. Nat.* 100:603–609.

Macior, L. W. 1970. The pollination ecology of *Pedicularis* in Colorado. *Am. J. Bot.* 57:716–728.

———. 1971. Co-evolution of plants and animals—systematic insights from plant–insect interactions. *Taxon* 20:17–28.

McCrea, K. D. 1981. Ultraviolet floral patterning, reproductive isolation and character displacement in the genus *Rudbeckia* (Compositae). Doctoral dissertation, Purdue University, West Lafayette, Ind.

McNaughton, I. H. and T. L. Harper. 1960. The comparative biology of closely related species living in the same area. I. External breeding barriers between *Papaver* species. *New Phytol.* 59:15–26.

McNeilly, T. and J. Antonovics. 1968. Evolution in closely adjacent plant populations. IV. Barriers to gene flow. *Heredity* 23:205–218.

Manning, A. 1957. Some evolutionary aspects of the flower constancy of bees. *Proc. Roy. Phys. Soc. Edinburgh* 25:67–71.

Mosquin, T. 1971. Competition for pollinators as a stimulus for the evolution of flowering time. *Oikos* 22:398–402.

Müller, H. 1873. On the fertilisation of flowers by insects and on the reciprocal adaptations of both. *Nature* 8:187–189.

Olmsted, C. E. 1944. Growth and development in range grasses. IV. Photoperiodic responses in twelve geographic strains of side-oats grama. *Bot. Gaz.* 106:46–74.

Oster, G. and B. Heinrich. 1976. Why do bumblebees major? A mathematical model. *Ecol. Mon.* 46:129–133.

Panje, R. R. and K. Srinivasan. 1959. Studies in *Saccharum spontaneum*. The flowering behavior of latitudinally displaced populations. *Bot. Gaz.* 120:193–202.

Parrish, J. A. D. and F. A. Bazzaz. 1978. Pollination niche separation in a winter annual community. *Oecologia* 35:133–140.

Paterniani, E. 1969. Selection for reproductive isolation between two populations of maize, *Zea mays* L. *Evolution* 23:534–547.

Pijl, L. van der. 1961. Ecological aspects of flower evolution. II. Zoophilous flower classes. *Evolution* 15:44–59.

Pittendrigh, C. S. 1958. Adaptation, natural selection, and behavior. *In* A. Roe and G. G. Simpson (eds.), *Behavior and Evolution*. Yale University Press, New Haven, Conn.

Platt, J. R. 1964. Strong inference. *Science* 146:347–353.

Pleasants, J. M. 1980. Competition for bumblebee pollinators in Rocky Mountain plant communities. *Ecology* 61:1446–1459.

Poole, R. W. and B. J. Rathcke. 1979. Regularity, randomness, and aggregation in flowering phenologies. *Science* 203:470–471.

Price, M. V. and N. M. Waser. 1979. Pollen dispersal and optimal outcrossing in *Delphinium nelsoni*. *Nature* 277:294–296.

Proctor, M. and P. Yeo. 1972. *The Pollination of Flowers*. Taplinger, New York.

Pyke, G. H. 1978a. Are animals efficient harvesters? *Anim. Behav.* 26:241–250.

———. 1978b. Optimal body size in bumblebees. *Oecologia* 34:255–266.

———. 1978c. Optimal foraging in bumblebees and co-evolution with their plants. *Oecologia* 36:281–293.

———. 1978d. Optimal foraging in hummingbirds: testing the marginal value theorem. *Am. Zool.* 18:739–752.

———. 1979. Optimal foraging in bumblebees: rules of movement between flowers within inflorescences. *Anim. Behav.* 27:1167–1181.

Pyke, G. H. and N. M. Waser. 1981. The production of dilute nectars by hummingbird and honeyeater flowers. *Biotropica*. 13:260–270.

Raven, P. H. 1972. Why are bird-visited flowers predominantly red? *Evolution* 26:674.

Ray, P. M. and W. E. Alexander. 1966. Photoperiodic adaptation to latitude in *Xanthium strumarium*. *Am. J. Bot.* 53:806–816.

Reader, R. J. 1975. Competitive relationships of some bog ericads for major insect pollinators. *Can. J. Bot.* 53:1300–1305.

Robertson, C. 1895. The philosophy of flower seasons, and the phaenological relations of the entomophilous flora and the anthophilous insect fauna. *Am. Nat.* 29:97–117.

———. 1924. Phenology of entomophilous flowers. *Ecology* 5:393–407.

Roughgarden, J. 1976. Resource partitioning among competing species—a coevolutionary approach. *Theor. Pop. Biol.* 9:388–424.

Sawamura, Y. 1967. An autecological study of the photoperiodic response of the geographic strains of *Polygonum thunbergii* Siebold et Zuccarini. *Jap. J. Bot.* 19:353–386.

Schemske, D. W. 1976. Pollinator specificity in *Lantana camara* and *L. trifolia* (Verbenaceae). *Biotropica* 8:260–264.

Schemske, D. W., M. F. Willson, M. M. Melampy, L. J. Miller, L. Verner, K. M. Schemske, and L. B. Best. 1978. Flowering ecology of some spring woodland herbs. *Ecology* 59:351–366.

Snaydon, R. W. 1963a. Morphological and physiological population differentiation of *Anthoxanthum odoratum* on the Park Grass Experiment, Rothamsted. *Heredity* 18:382.

———. 1963b. The diversity and complexity of ecotypic differentiation within plant species in response to soil factors. *Proc. 11th Int. Genet. Congr.*, The Hague, 1:143.

Sørensen, T. 1941. Temperature relations and phenology of the northeast Greenland flowering plants. *Medd. Grønland* 125:1–305.

Sprague, E. F. 1962. Pollination and evolution in *Pedicularis* (Scrophulariaceae). *Aliso* 5:181–209.

Stace, H. M. and Y. J. Fripp. 1977. Raciation in *Epacris impressa*. II. Habitat preferences and flowering times. *Aust. J. Bot.* 25:315–323.

Stanford, E. H., M. M. Laude, and P. de V. Booysen. 1962. Effects of advance in generation under different harvesting regimes on the genetic composition of Pilgrim Ladino clover. *Crop Sci.* 2:497–500.

Stebbins, G. L. 1950. *Variation and Evolution in Plants*. Columbia University Press, New York.

———. 1970. Adaptive radiation of reproductive characteristics in angiosperms, I. Pollination mechanisms. *Annu. Rev. Ecol. Syst.* 1:307–326.

Stiles, F. G. 1975. Ecology, flowering phenology, and hummingbird pollination of some Costa Rican *Heliconia* species. *Ecology* 56:285–301.

———. 1977. Coadapted competitors: the flowering seasons of hummingbird pollinated plants in a tropical forest. *Science* 198:1177–1178.

———. 1979. Reply to Poole and Rathcke. *Science* 203:471.

Straw, R. M. 1972. A Markov model for pollinator constancy and competition. *Am. Nat.* 106:597–620.

Strickler, K. 1979. Specialization and foraging efficiency of solitary bees. *Ecology* 60:998–1009.

Sukhada, K. and Jayachandra. 1980. Pollen allelopathy—a new phenomenon. *New Phytol.* 84:739–746.

Thomson, J. D. 1978. Competition and cooperation in plant–pollinator systems. Doctoral dissertation, University of Wisconsin, Madison.

———. 1981. Implications of different sorts of evidence for competition. *Am. Nat.*, 116:719–726.

Thomson, J. D., B. J. Andrews, and R. C. Plowright. 1981. The effect of a foreign pollen on ovule fertilization in *Diervilla lonicera*. (Caprifoliacaea). *New Phytol.* 90:777–783.

Waddington, K. D. 1979a. Divergence in inflorescence height: an evolutionary response to pollinator fidelity. *Oecologia* 40:43–50.

Waser, N. M. 1976. Food supply and nest timing of broad-tailed hummingbirds in the Rocky Mountains. *Condor* 78:133–135.

———. 1977. Competition for pollination and the evolution of flowering time. Doctoral dissertation, University of Arizona, Tucson.

———. 1978a. Competition for hummingbird pollination and sequential flowering in two Colorado wildflowers. *Ecology* 59:934–944.

————. 1978b. Interspecific pollen transfer and competition between co-occurring plant species. *Oecologia* 36:223–236.

————. 1979. Pollinator availability as a determinant of flowering time in ocotillo *(Fouqueria splendens)*. *Oecologia* 39:107–121.

Waser, N. M. and L. A. Real. 1979. Effective mutualism between sequentially flowering plant species. *Nature* 281:670–672.

Weaver, N. 1956. The foraging behavior of honeybees on hairy vetch: foraging methods and learning to forage. *Insectes Sociaux* 3:537–549.

Weevers, Th. 1952. Flower colours and their frequency. *Acta Bot. Neerl.* 1:81–92.

Whalen, M. D. 1978. Reproductive character displacement and floral diversity in *Solanum* section *Androceras*. *Syst. Bot.* 3:77–86.

Whitaker, T. W. 1944. The inheritance of certain characters in a cross of two American species of *Lactuca*. *Bull. Torrey Bot. Club* 71:347–355.

Williams, G. C. 1966. *Adaptation and Natural Selection*. Princeton University Press, Princeton, N.J.

Zimmerman, M. 1980. Reproduction in *Polemonium:* competition for pollinators. *Ecology* 61:497–501.

14
A REVIEW OF FLORAL FOOD DECEPTION MIMICRIES WITH COMMENTS ON FLORAL MUTUALISM

R. John Little

Rancho Santa Ana Botanic Garden
Claremont, California

ABSTRACT

Studies in floral mimicry are proliferating rather rapidly. This chapter begins with a consideration of basic principles, provides an overview of the more important published reports, and gives some insight into the historical development of this subject. Two basic types of floral mimicry are considered: food deception and floral mutualism. Several key studies of each type are discussed in some detail, followed by brief discussions of unresolved cases for which quantitative data are lacking. Characteristics common to food deception mimicries are summarized and grouped according to the categories of basic ecological conditions, pollination biology, and morphology. These serve as key points to analyze when investigating situations of deceptive mimicry.

KEY WORDS: Automimicry, Batesian, deceit mimicry, floral mimicry, floral mutualism, food deception mimicry, mimic, Müllerian, mutualistic mimicry, pollination, pseudocopulation, reproductive mimicry, signal-receiver.

CONTENTS

INTRODUCTION

The subject of floral mimicry is not new, but it has received comparatively little attention from biologists since it was recognized by Sprengel in 1793 (Nilsson, 1980; Heinrich, this book). Fortunately this situation is changing, and an increasing number of studies are being devoted to the subject (e.g., Powell and Jones, this book). In addition, there have been many anecdotal reports and casual observations cited in the literature that describe possible examples of floral mimicry, including several in this book.

Wiens (1978) was the first to attempt an overview of the topic of floral mimicry as part of a general review of plant mimicry. He proposed a class of plant mimicry known as reproductive mimicry in which he included all types of floral mimicries that are based on the utilization of false sensory cues to attract animals to effect pollen transfer. He recognized four categories: food source (food deception), territorial defense (pseudoantagonism), sexual response (pseudocopulation), and brood site selection.

The purpose of this chapter is to review and examine food deception mimicries and to comment on their relationship with floral mutualisms. Food deception mimicries are also referred to as deceit syndrome mimicries (e.g., Dafni and Ivri, 1981).

FLORAL MIMICRY

Floral mimicry is a subset of plant mimicry, which can be described in terms of the tripartite system established by Wickler (1968) and amplified by Vane-Wright (1976) for classifying mimetic phenomena. The tripartite system consists of two signal transmitters, namely, a model and a mimic, and a signal-receiver (or operator). In floral mimicry, flowers (or perhaps groups of flowers, e.g. umbels or capitula) of different taxa function as the model and mimic, and the signal-receiver is a pollinator.

Wiens (1978) and Vane-Wright (1980) have written similar definitions of mimicry (cited in Chapter 15, p. 311, this book). A significant feature of these definitions is the inclusion of the concept of fitness, for without it there is no test of the process, and putative examples must remain subjective (Wiens, 1978). These and other definitions were recently discussed by Endler (1981) in a very useful overview of mimicry definitions.

The expected biological result of a situation of floral mimicry is that the mimetic taxon achieves increased fitness over what it is capable of in the absence of the floral model and its pollinators. Fitness can be demonstrated by measuring an appropriate parameter related to fecundity.

Floral mimicry differs in an important and fundamental way from most other types of plant and animal mimicries. In a situation of floral mimicry, the "purpose" of the mimesis is to *attract* a signal-receiver (i.e., a pollinator). (Or, as a "result" of the mimesis, a signal-receiver is attracted.) In other types of plant and animal mimicries, mimesis serves to warn or repel or serves for camouflage (*sensu* Wiens, 1978; see also Endler, 1981). In other words, floral mimicries serve to elicit a "positive" response from the signal-receiver, whereas in many zoological mimicries a "negative" predator–prey relationship exists in which the mimic gains a survival advantage because of its similarity to a *nonattractive* model. Significant differences between plant and animal mimicries have been noted by a number of authors (Yeo, 1968; Wiens, 1978; Dodson, pers. comm., 1979 [unpub. ms]; C. E. Jones, pers. comm., 1979; Little, 1980; Powell and Jones, this book). Thus floral mimicries have characteristics that distinguish them from other types of plant and animal mimicries.

Floral mimicries are included by Wiens (1978) in a class of plant mimicry called reproductive mimicry. Specific types of floral mimicries are outlined below, all of them having as their basis the utilization of false sensory cues to attract animals to effect pollen (i.e., gamete) transfer:

Reproductive mimicry	Reproductive mimicry
Floral mimicry	Floral mimicry
Food source	Food deception
Territorial defense (Pseudoantagonism)	Antagonist deception
Sexual response (Pseudocopulation)	Sexual deception
Brood site selection	Odor deception
	Prey deception

The terms in the left column were proposed by Wiens (1978), and those on the right are from

Calaway H. Dodson (unpub. ms.; Dodson generously furnished myself and others with a copy of this ms. which was written prior to 1979). Dodson developed this classification system based on his studies of mimicry in orchids (van der Pijl and Dodson, 1966). Wiens integrated Dodson's work with examples from other plant families, modified the terminology slightly, and combined Dodson's categories of odor deception and prey deception into a single category of brood site selection. In this chapter, I concentrate on food deception (or food source) mimicries. Unfortunately, space does not permit a consideration of the other types.

Food deception can be said to occur when a signal-receiver (pollinator) is "mistakenly" attracted to one taxon (the mimic) in anticipation of obtaining a reward that is only, or more abundantly, available in another, similar-appearing taxon (the model). The mimic is pollinated as the signal-receiver attempts to gather food from it. Repeated mistakes, whatever their frequency, increase the fitness of the mimic in its environment. Similarity of model and mimic is usually gauged in terms of visual signals in food deception mimicries. However, olfactory signals may also be very important, either separately or in conjunction with visual signals.

A type of reproductive mimicry closely related to food deception mimicries is called floral mutualism. Mutualism is said to exist between two or more taxa when all provide a floral reward and are pollinated by a common pollinator. Mistakes are not made in the sense that they are in deceptive mimicries because the pollinators visit all of the flowers in the group indiscriminately. Also, in a mutualistic association there are no mimics, only models. For some this concept runs counter to what is normally expected to make up a mimetic relationship, namely, a tripartite association of model, mimic, and signal-receiver (Wiens, 1978). I agree with Wiens in principle that it is stretching the concept of mimicry to consider mutualistic associations as mimicry. However, as Endler (1981) has aptly argued, there are better ways to distinguish the phenomena of mimicry than we have had in the past, and recognition of this can alleviate much

of the confusion and argument about what constitutes mimicry and what does not.

Floral mutualism has also been called advertising mimicry (Proctor and Yeo, 1972) in recognition of the similarity of floral shape, presentation, phenology, proximity, and abundance of the taxa involved. The result of a mutualistic relationship is the convergence of floral characters and blooming phenology of *all* of the taxa. I hasten to add that convergence of characters is a common denominator regardless of whether the mimetic relationship is based on deceit or mutualism.

Floral mutualisms are as poorly studied as deception mimicries. One of the purposes of this chapter is to discuss examples of these two types from those that appear in literature. At the present time it appears that these are the two fundamental forms of floral mimicry that are based on food deception. It is not always clear, however, which type an author is referring to when citing an example. Future studies may show that these two forms are overly simplistic. Indeed, I suggest that deceit and mutualistic mimicries are parts of a continuum, the details of which remain to be worked out. Before illustrating the closeness of their relationship, let us first consider their main differences, which are summarized below:

Deceit mimicries	Mutualistic mimicries
Mimic lacks food (or "reward")	Food present in all taxa
Clearly defined model and mimic	No mimic; all taxa are models
Pollinators specialized and constant	Pollinators unspecialized or not constant
Model flowers are more abundant than mimics	Equal number of flowers of model taxa
Usually involves only two taxa	Often involves more than two taxa

I want to stress that all of these items can be prefaced with qualifiers, (e.g., most, usually, frequently, etc.), for I know of exceptions to each one. Nevertheless, the sum of these items serves to clarify the basis for choosing to classify an example as deceit or mutualistic.

Conceivably, floral mutualisms may be a

stage from which some deceit systems evolve. To illustrate the relationship, consider a mutualism involving only two taxa. If one of the taxa lost its ability to produce a food reward, the system would resemble deceit mimicry.

Before discussing examples of deceit and mutualistic mimicries, I will say a few words about terminology. Up to this point I have purposely avoided using the terms "Batesian" and "Müllerian" in connection with floral mimicry. Most biologists are familiar with these terms in connection with zoological mimicries, and I suppose for lack of anything better, some authors have used these terms to describe floral mimicries. Thus, a floral mimesis is called Batesian if some form of deception is involved where the putative mimic lacks a floral reward, and is called Müllerian if the flowers seem to be "advertising" as a group for pollinator attraction.

I have several objections to using Batesian and Müllerian terminology to describe floral mimicries, and suggest for the following reasons that these terms not be used in connection with them: First, as stated above, floral mimicries have characteristics that distinguish them from other types of plant and animal mimicries. For example, in floral mimicry the purpose of the mimesis is to attract a signal-receiver, whereas in animal mimicries mimesis serves to warn or repel or serves for camouflage. Second, Ford (1971) and Wiens (1978) have clearly pointed out that plant mimicry falls outside the scope of classical Batesian and Müllerian mimicry. Third, Wickler (1968) stated that "the distinctions between various forms of mimicry, such as Batesian, Peckhammian, Mertensian, show that one should *not* rely upon the original definitions that were based upon the concept of false warning coloration and on the presence of a predator as a signal-receiver" (my emphasis). Fourth, our present understanding of mimetic phenomena is more sophisticated than it was in the past, and rather than forcing the various overlapping phenomena into the general definition, it is better to retain specific terms to describe different types of mimicries (Endler, 1981). Finally, Endler (1981) has done an admirable job of categorizing various types of

mimetic relationships into six "classes." His third class, "Batesism," includes five types of mimicries including classical Batesian mimicry and reproductive mimicry. As stated earlier, reproductive mimicry encompasses all types of floral mimicry. Thus, in effect, Endler is saying that Batesian and reproductive mimicries are related, but their differences are significant enough to warrant that they be recognized as separate kinds of phenomena. Likewise, Müllerian mimicry is included in class 4, "Müllerism"; and although Endler did not discuss floral mutualisms, they nevertheless belong in this class, distinct from classical Müllerian mimicry. To summarize, it is no longer necessary, or appropriate, to lump all examples of floral mimicry into one of the two categories of classical mimicry.

FOOD DECEPTION MIMICRIES

Dodson (1962) was probably the first to treat food deception as a distinct category of floral mimicry. Various types of food deception mimicries have since been described or suggested, and can be grouped in three categories, as indicated below:

"Two taxa" deceit mimicries
    1. Noncongeneric model and mimic
    2. Congeneric model and mimic
Automimicry (conspecific mimicry)
    1. Dioecious/monoecious
    2. Others
Mimicry based on naiveté (instinct model)

Literature references dealing with deceit mimicries are of two kinds, research reports and anecdotal/casual observations. Some of the important studies are discussed below, whereas other potential cases not substantiated with experimental evidence are presented later under "Unresolved Food Deception Mimicries."

Two Taxa Deceit Mimicries

These are called two taxa because the model and mimic belong to two different species. Most of the examples that I would refer to this category consist of two taxa from different genera (i.e., noncongeneric pairs). There are even one or two cases where two or more models appear to be involved with one mimic. A second type of two taxa mimicry may be possible between species of the same genus, in which one species functions as the floral model and the other as the mimic. Although no solid cases have been made for this sort of relationship, it seems probable that some may exist. Carlquist (1979), in fact, may have uncovered possible examples of this type (see discussion below, under "Unresolved Food Deception Mimicries").

Relatively few studies investigate or test the deceit mimicry hypothesis through an observational or experimental approach. Some of the more important examples include Nierenberg (1972), Boyden (1980), Little (1980), Bierzychudek (1981), and Dafni and Ivri (1981).

Nierenberg (1972) studied food deception mimicry between yellow-flowered equitant Oncidiums (Orchidaceae) that occur in the Caribbean islands, and several species in two genera of Malphigiaceae (Stigmatophyllum and Malphigia). His examples include three species of Oncidium which occur on two different islands, O. lucayanum (Grand Bahamas), O. quadrilobum (Dominican Republic), and O. desertorum (Haiti). All of these species lack nectar and are believed to be mimics of the malphig species with which they occur. Pollination is by female bees of the genus Centris. Unfortunately, the identity of the species of malphigs involved is not revealed, and no quantitative data were presented. Evidence for mimicry was obtained by comparing ultraviolet (UV) photographs of the flowers of model and mimic.

Boyden (1980) in a brief paper described an example of food deception mimicry in Panama between Epidendrum ibaguense (Orchidaceae), the mimic, and Lantana camera (Verbenaceae) and Asclepias curassavica (Asclepiadaceae). Both of the latter species were reported to function as models. The mimic lacks nectar, whereas both models have it. An interesting feature of this situation was the discovery that the pollinaria of both models, and the pollen of the mimic, are deposited at dif-

ferent sites on the body of the pollinator, a monarch butterfly (Danaus plexippus). Macior (1971), Heinrich (1979), and Little (1980) described similar situations in which the pollen of the floral model and mimic was spatially separated on the bodies of their mutual pollinators. Such a situation may indicate an efficiency of the system, although Straw (1972) pointed out that even if the pollen of two species is intermingled, pollen function will not necessarily be totally depressed.

A situation similar to that described above was suggested by Dodson (unpub. ms.) in Costa Rica with the same models but a different species of Epidendrum, E. radicans. Interestingly, Bierzychudek (1981) recently reported on a study she conducted with these species in Costa Rica and concluded that mimicry alone cannot account for the similarities of their flowers.

While stressing that the sum of her data "provide no clear answer" as to whether or not mimicry is involved in this situation, she presents several alternative hypotheses to explain the similarities of their flowers. The value of this report is that it points to the need for a careful assessment of each potential situation of floral mimicry, as one cannot assume that floral resemblances are necessarily based on mimesis.

An extensive documentation of two taxa food deception mimicry between two desert annuals, Mohavea confertiflora (Scrophulariaceae) and Mentzelia involucrata (Loasaceae), was demonstrated by Little (1980; ms. in prep.). The model (Mentzelia) produces abundant pollen and nectar in open, actinomorphic flowers that are visited by several species of bees and beetles. In contrast, the mimic produces a minute amount of nectar, but this and its pollen are "hidden" within the closed floral tube of its zygomorphic corolla.

Evidence for floral mimicry is based upon morphological similarities of the flowers of model and mimic, and, more important, on observations of pollinator activity, which when coupled with information about the mimic's reproductive biology, demonstrated the increased fitness of the mimic. The primary pollinators of both model and mimic are two closely related species of *Xeralictus* and two closely related species of *Hesperapis* bees. These bees are essentially constant to the model and are the only known pollinators of the mimic. Analysis of every category of pollinator behavior (e.g., visits and pollinations) showed that the model consistently had a higher rate of activity than the mimic. In no case did the mimic outdo the model for pollinator service even in locations where the model was greatly outnumbered by the mimic in terms of available blossoms.

Pollination of the mimic by *Hesperapis* female bees appears entirely the result of chance mistakes. Females are believed to be more easily deceived than the males. The females forage with great intensity and for considerable periods of time (up to 11 minutes) on *Mentzelia* flowers, whereas only seconds are spent in a *Mohavea* blossom.

The pollination mechanism of *Mohavea confertiflora* is adapted to take advantage of occasional mistakes by its pollinators. For example: (1) The physical movements by which bees gain entry into a *Mentzelia* flower are identical with the movements required to enter the narrow floral tube of *Mohavea* flowers (e.g., the floral parts of both species are compact or closed, and the flower axis is vertical with the flowers erect); in order to enter these flowers, bees must push floral parts apart with their bodies and then descend headfirst into the floral tube. (2) *Mohavea* pollen deposition is nototribic and is spatially separated on the bee's body from *Mentzelia* pollen, which collects primarily on the head, abdomen, and ventrical surfaces. (3) *Mohavea* pollen is cohesive and sticks to the dorsal thorax of a bee pollinator where it accumulates, ready for another mistake; *Mentzelia* pollen is powdery and is combed into the corbiculae. (4) Bees carrying *Mohavea* pollen need enter the floral tube of another flower for only a split second in order to effect pollination.

Morphological evidence for the existence of floral mimicry consisted of comparing characters of model and mimic that were considered important for attracting visually orienting bees, such as plant height, flower size, degree of flower dissection, and analysis of floral spectral reflectance patterns. A high degree of similarity was found in the morphological characters examined, including a remarkable, but unexplained, similarity in pollen exine sculpturing (Little, 1980; ms. in prep.).

Dafni and Ivri (1981) reported an interesting three-year study of deceit mimicry between an orchid *(Orchis israelitica)* and *Bellevalia flexuosa* (Liliaceae). The criteria for mimicry that they investigated are similar to those used by Little (see above). For example, they studied the pollinators and their behavior, phenological dynamics, relative abundance, capsule production, and visual resemblance. Deceit mimicry was hypothesized because of the "striking similarity of coloration" between the orchid (the mimic) and *B. flexousa* (the model), and the fact that the mimic has no reward and shares the same pollinators as the model. The increased fitness of the mimic was demonstrated in terms of a much greater production of capsules in areas where it occurs sympatrically with the model.

The authors went on to discuss three "types" of floral mimicries that are based on the relative frequency of the model and the mimic and on their mutual relationships (all quotes from Dafni and Ivri, 1981): "Type A: By sharing the same search-image the two partners increase the chances for both to be pollinated. Such a situation could evolve and become adaptive under such selective pressure as a shortage of pollinators and/or high competition for them. Hence the deviating form is less adaptive by being less visited unless it becomes more similar to the model and/or abundant enough." Type A seems analagous to floral mutualism. "Type B: The abundant model rewards the pollinator while the uncommon mimic has less or no reward. The model is well visited and the less common species attracts the model's visitor by mimicry." Type B mimicries are all those types that I have included here under the heading of "food deception mimicries." I would add the following thought to their definition: In theory, the model "should" outnumber the mimic in a deceit system. However, in randomly distributed populations of annual species of model and mimic plants, the population mix in a given area may be such that the mimic outnumbers the model. This should be taken into consideration by the investigator. "Type C: The mimic has some reward, or a kind of deceit, and can attract at least a few pollinators when it appears alone, but increases its visitors considerably in the model's presence." Dafni and Ivri (1980) suggested that this be described as "facultative floral mimicry," which is the only mention of a "facultative" system that I am aware of. As an example, they cited the relationship between *Cephalanthera longifolia* which has pseudopollen and sweet fragrance and *Cistus salviifolius* which offers an abundance of edible pollen.

Automimicry (Conspecific Mimicry)

In this category are mimicries that occur within a single species. Two types are recognized, those that involve monoecious or dioecious plants and those that do not. The first type cannot be subdivided too finely because some monoecious plants are functionally dioecious. In addition, all sorts of different combinations of floral sexuality are possible. The concept of conspecific floral mimicries stretches the concept of mimicry somewhat because the model and mimic are not *different* organisms in the taxonomic sense. Indeed, with mimicries involving monoecious plants, in which the female flowers mimic the male, we are dealing with the same organism although there may be a temporal separation in blooming time. Such examples serve to illustrate how different plant mimicries can be from classical Batesian mimicry.

Monoecious/dioecious food deception mimicries have been noted by a number of workers (Gilbert, 1975; Baker, 1976; Bawa, 1977; Schmid, 1978; van der Pijl, 1978; Faegri and van der Pijl, 1979; Bawa, 1980; C. E. Jones, pers. comm. 1981), mostly in tropical plants. Baker (1976) suggested that this method of pollination may prove to be a widespread phenomenon and mentioned its occurrence in the Caricaceae, Buxaceae, Annonaceae, Fagaceae, and probably the Arecaceae (Palmae), Cucurbitaceae, and Eurphorbiaceae. To this list can be added Sapindaceae (Bawa, 1977). Several of these examples are discussed below.

Baker (1976) studied *Carica papaya* in Costa Rica, and because the floral structures of all the Caricaceae are "remarkably constant," he extended his results to include the entire family. He showed that in the dioecious *C. papaya,* cross-pollination still occurs even though the pistillate (female) flowers lack floral rewards (nectar and pollen), and are virtually odorless. In contrast, the staminate (male) flowers produce both nectar and pollen and have a "strong, sweet perfume" odor. In addition, there are many more staminate flowers produced than pistillate. These are the types of "characters" that one would expect to find in a two-species system. A study of the length of time spent by pollinators on staminate (model) and pistillate (mimic) flowers predictably showed that hawkmoths, the primary pollinating vector, spent considerably more time on the "model" flowers than on the "mimics." Even though stops at the pistillate flowers were very brief, pollinations occurred because of the placement of the pollen on the body of the hawkmoth relative to the position of the pistillate stigma.

In a study of another tropical tree, *Cupania guatemalensis* (Sapindaceae), Bawa (1977) showed that although this plant is morphologically monoecious, it is functionally dioecious because of a separation in flowering times of male and female flowers. The pistillate flowers contain anthers although they are nonfunctioning and do not dehisce. Thus, Bawa suggests that the anthers in pistillate flowers serve as "decoys." He reasoned that if the pistillate flowers were indeed mimics of the staminate, they should occur in low frequency and for only a short period of time. Both of these expectations were realized.

From these examples and others cited above, we may summarize some points that seem to apply to monoecious/dioecious automimicries: (1) female flowers lack nectar, pollen, or both, and mimic male flowers through similar floral shapes (petals, anthers/stigmas, etc.); (2) male flowers are more abundant than female flowers, and may last for a longer period of time; (3) pollinator visits and foraging times to the male flowers are of longer duration than the female; and (4) pollen of the model is positioned on the pollinator in such a manner that it can be rapidly transferred to the stigma of the mimic.

Two taxa mimicries and automimicries have a number of characteristics in common. These are outlined in Table 14-1, grouped according to three criteria: basic ecological conditions, pollination biology, and morphology. Ecologically, the model and mimic exist close enough to be within the range of mutual pollinating vectors, and are in bloom at the same time. The breeding system of the mimic can be anything except apomictic because such plants are not dependent on pollinators for reproductive success. Various categories of pollinator behavior, appropriate to the system being studied, show that the signal-receiver is constant to the model for a period of time, and that the model is preferred over the mimic by the signal-receiver in terms of the rate of pollinations and visits, length of foraging times, etc. From a study of the breeding system of the mimic, and of pollinator behavior in relation to it and the model, it can be demonstrated that the fitness of the mimic is increased because of its association with the model. Stuctural similarities in floral and plant morphology of the model and mimic may be expected but need not occur. A more important factor is that they are functionally similar viewed from the perspective of the signal-receivers. For example, similarities of floral spectral patterns would be expected if the signal-receivers normally depended on visual senses to locate sources of food, whereas similar floral odors would be expected if olfactory senses were of primary importance.

Mimicry Based on Naiveté

This category consists of food deception mimicries that occur because of the "mistakes" of naive pollinators. As used here, naive pollinators are newly emerged insects and perhaps juvenile migrant hummingbirds. However, just how long a pollinator remains naive and when it becomes "educated" must vary with the circumstances.

Only a few situations that have been described could be considered under this category, and those involve newly emerged bumblebee queens (Nilsson, 1980; Ackerman, 1981). Nilsson (1980), in a rather thorough study in Sweden of the pollination ecology of *Dactylorhiza sambucina* (Orchidaceae), concluded that "the species acts by deceit and obviously exploits a short stage of the *(Bombus)* queens' lives when they are newly emerged after hibernation and have not established foraging routes and are inexperienced on food-flowers." Ackerman (1981), in working with another orchid, *Calypso bulbosa* var. *occidentalis* in California, concluded that this plant "is a generalized food-flower mimic and must rely on exploratory visits of naive bees for pollination."

There are numerous similarities between the

Table 14-1. Concepts of floral mimicry based on food deception.

BASIC ECOLOGICAL CONDITIONS
  1. Model and mimic must be sympatric
  2. Blooming phenologies of model and mimic overlap
POLLINATION BIOLOGY
  A. Breeding system of floral mimic
    1. Mimic must require a pollinator to effect pollination (plant at least not total apomict)
    2. Floral rewards, if present at all, are in smaller quantities in mimic in comparison with model
  B. Pollinator behavior
    1. At least one pollinator must be constant to the model for a period of time
    2. Mimic is pollinated on the same foraging flight as the model
    3. Pollinators constant to the model must be observed to effectively pollinate the mimic
    4. Pollination rate and foraging times are greater for the model
    5. Pollen from the model could be spatially separated from pollen of the mimic on the body of the pollinator
MORPHOLOGY
  1. Floral morphologies of model and mimic functionally similar
  2. Similar habit of model and mimic
  3. Similar floral spectral patterns
  4. Similar floral odors (if odors present in model)

examples of Nilsson and Ackerman: the flowers are both allogamous, nectarless orchids; these orchids do not appear to mimic any other flowers in their community; *Bombus* queens are among the most important pollinators; flowering occurs simultaneously with the emergence of the queens; pollinations and fruit set decline shortly after the queens have emerged and have learned that the flowers are nonrewarding; pollinator interest in the flowers is "short"; solitary bees are present in the plant community, although they are not important pollinators; and variability occurs in the populations of the orchids in respect to certain floral characteristics, e.g. intensity of floral colors (*D. sambucina* and *C. bulbosa* var. *occidentalis*) and strength of floral odors (*C. bulbosa* var. *occidentalis*). Ackerman (1981) cited studies by Heinrich (1975a,b) showing that bumblebees visited *Calopogon tuberosus,* a nectarless orchid, an average of 5.4 times in quick succession. These orchids are also variable in color, and Heinrich suggested that this enhances the number of visits the pollinators require to learn to avoid the species.

There are, however, some conceptual problems with calling these particular situations mimicry, although this issue has not previously been addressed in the literature. The difficulty stems from the fact that a naive pollinator has no model to confuse with or discriminate from the mimic. Seemingly, one part of the tripartite system is missing—the floral model. In mimicry based on naiveté, the model (if it can be called that) is the *instinct* of the pollinator for a certain floral type, shape, color, odor, etc., or perhaps some combination of these. In fact, no current definition of mimicry provides for an instinct model. Currently acceptable models include living organisms, inanimate objects, and background material, although not all authors are in agreement even on these (see Endler, 1981).

Pseudocopulation is one of the classical examples of floral mimicry (Kullenberg, 1961; Stoutamire, 1975; Kullenberg and Bergstrom, 1976), although there is only one flower species involved—the "mimic." The signal-receivers in these situations are newly emerged male bees and wasps that are entirely naive in their first encounters with the mimetic flowers that resemble the females of these species. These males also act on instinct, as do the bumblebee queens mentioned above. Nevertheless, the former are eminently successful at effecting pollination, and few would deny that their response to the flowers does not increase the plants' reproductive success. In fact, once the female insects emerge, the males are completely captivated by them, and the flowers are no longer visited by the males. Thus pseudocopulatory flowers, limited primarily to the orchid genera *Ophrys* and *Cryptostylis,* capitalize on the instinctual behavior of the naive males by presenting them with visual and olfactory signals, and tactile stimuli that are perfectly sufficient to attract them. Although pseudocopulatory situations are based on sexual response and not food deceit, and are probably more finely tuned than the examples of naive mimicry cited above, the principles are exactly the same.

Would it be correct to say that pseudocopulation is not based on mimicry? I think not, for the following reason: A key characteristic of the male insects in pseudocopulatory mimicry is their nonrandom behavior in relation to the floral mimic. In other words, the males do not visit the mimetic flowers by chance, a factor that can be demonstrated. Herein lies the essence of criteria that help distinguish whether or not a potential case of naive mimicry is actually based on a mimetic (symbiotic) relationship: if the behavior of the pollinators is proved to be random with respect to their visits to the flower, the situation could not justifiably be considered mimetic; however, if the pollinators visit the flower in a nonrandom fashion, the situation may potentially be based on mimicry (C. E. Jones, pers. comm. 1982). Of course, it would also have to be shown that the pollinators actually had an affect on the reproductive success of the plant; i.e., that as a result of their visits pollination occurred, and seed was produced.

In the studies of Nilsson (1980) and Ackerman (1981), it was demonstrated that the floral mimics were almost completely dependent on naive bumblebees for pollination. Thus, in both of these examples, and in pseudocopulatory mimicry, naive pollinators appear to exert a strong evolutionary force on the floral characters and phenology of the "floral mimics" that they pollinate. Although one could argue that these flowers are simply opportunistic in respect to blooming when pollinators first appear, the evidence indicates clearly that *some* type of relationship or symbiosis is occurring based on the nonrandom behavior of their pollinators, and is affecting the reproductive success of these plants in a positive way. Except for the lack of a specific floral model, these situations are similar in every respect to other deceit mimicries.

Some authors have challenged the notion that mimicry can occur with naive pollinators (Williamson and Black, 1981). These authors stated that "naiveté itself is inconsistent with a mimicry hypothesis because the process of learning is fundamentally different from that of confusion of mimic for model." This was part of their criticism of Brown and Kodric-Brown's (1979) conclusion that "Batesian" mimicry occurred between *Lobelia cardinalis* and other hummingbird-pollinated species. One of the points that Williamson and Black attempted to make is that a naive organism cannot be a "sensitive signal-receiver" in the sense defined by Vane-Wright (1976). Wiens (1978), however, placed no such constraint on the mimetic relationship in his definition of mimicry. His only provision is that the signal-receiver be unable to consistently discriminate the mimic from the model. In other words, whether educated or naive, if the organism makes mistakes that increase the fitness of the mimic, it qualifies as a signal-receiver in a mimetic relationship. Of course, in Nilsson's (1980) and Ackerman's (1981) examples and those involving pseudocopulation, there are as we have said, no "models." However, if instinct is considered a "model," then there are no difficulties with considering naive pollinators as legitimate signal-receivers. Interestingly, Vane-Wright (1980) recently changed his previous definition and eliminated the "sensitive signal-receiver" criterion, a change apparently unkown to Williamson and Black.

Although it was not explicitly stated by Williamson and Black, I believe one of their concerns was that it is difficult to understand how indiscriminate pollinators can function as signal-receivers in a mimetic sense because many different types of flowers can be visited at random. I previously made a similar argument: "If there are no constant pollinators linking the mimic to the model, there can be no selection on the mimic to resemble the model" (Little, 1980). This was another way of stating that one cannot expect random pollinator behavior in a deceptive mimicry system. For reasons to be explained shortly, I agree with Williamson and Black that a deceptive mimicry interpretation of *L. cardinalis* is probably not warranted. However, for the reasons stated above, I disagree that naive pollinators cannot function in certain situations as signal-receivers.

In addition to Nilsson's and Ackerman's use of the term naive, Gentry (1974) described mature female euglossine bees as naive in respect to the nectarless bignoniaceous flowers that they visit in search of nectar. However, this use of the term is quite different from what is meant here, because the bees described by Gentry were experienced at gathering nectar. Although this situation was called mimicry, it is difficult from the brief description to tell what kind of mimicry this represents. Apparently the flowers are not mimics of any particular species, but instead bear a general resemblance to several other mass flowering species in their vicinity. The frequency of pollination is no doubt enhanced by the short bursts of flowering, plus the "conspicuous visual and olfactory stimuli" that characterize this species. This situation certainly merits further study, perhaps from the perspective of floral mutualism.

Some authors have inferred that the lack of nectar production per se is evidence for mimicry. However, as Williamson and Black (1981) have cautioned, other hypotheses should also be considered. There are situations in which the pollination of nectarless plants is not clearly related to mimetic symbiosis, e.g., some vibratile pollination systems (Buchmann, this book). It is important to recognize this in order to consider alternative hypotheses in situations where there is sufficient reason to doubt that mimicry is actually occurring. Perhaps the nectarless, nonmimetic situation (or "nectarless pollination syndrome") is simply another pollination strategy that should be recognized and dealt with. Some of these may be subtle variations on the mimicry theme. It has been suggested that nectarless or nectar-poor plants ("mimics") may follow nectar-bearing "models" in flowering time (Heinrich, 1975b; Schemske, et al., 1978).

Brown and Kodric-Brown (1979) described a complex situation involving hummingbirds and nine hummingbird-pollinated plants in Arizona. One part of their study focused on a population of nectarless plants *(Lobelia cardinalis)* that are pollinated primarily by juvenile migrant hummingbirds *(Selasphorus rufus.)* The authors suggested that *Lobelia* acted as a "Batesian" mimic of eight other nectar-producing species. Not only could this be interpreted as (1) an example of two taxa mimicry (with more than one model), it could also be interpreted as (2) mimicry based on naiveté of the juvenile hummingbirds, or (3) a potential case of automimicry (Williamson and Black, 1981). Thus, *L. cardinalis* may be functioning in one or more of these capacities at this particular site. However, the fact that it is pollinated is not surprising considering the foraging habits of hummingbirds, naive or otherwise. After all, hummers visit most anything that is a saturated color (e.g., long wavelengths of visible light)—including artificial feeders. *Lobelia cardinalis* would probably be pollinated even if it were not associated with other hummingbird plants because of the floral characters it shares with these types of plants. Thus, the fact that this particular population lacks nectar does not prove a food deception mimicry hypothesis. In addition to this nectarless population, Brown and Kodric-Brown (1979) cited another in Arizona. It would be interesting to know if nectarless populations of *L. cardinalis* were prevalent or exceptional. If prevalent, they could be interpreted in the context of a nectarless pollination strategy, or perhaps as some type of food deception mimicry. If not, this particular example could be interpreted as an unusual example of floral mutualism in which a nectarless species "functioned" in a deceptive manner.

Unresolved Food Deception Mimicries

A number of highly interesting, potential cases of food deception mimicry (and possibly floral mutualism) are discussed briefly below. All of these are based on observations alone, and are in need of quantitative and experimental verification.

Dodson (1962) reported a situation in Ecuador where nine species of orchids in five genera (*Elleanthus,* two spp.; *Odontoglossum,* three spp.; *Oncidium,* one sp.; and *Epidendrum,* three spp.) occur sympatrically with various species of ericaceous shrubs. The orchids appear to have converged to closely resemble these shrubs, which include the genus *Gaultheria.* The shrubs bloom continuously and have been observed to be pollinated by hummingbirds, which are believed by Dodson to be pollinators of the orchids. Unfortunately data are not included regarding whether any of these orchids produce nectar. This situation appears complex and begs for further study.

C. H. Dodson (unpub. ms.) described several examples of food deception mimicries involving orchids. For example, "in the genus *Calopogon,* false stamens in the form of fleshy cirrhae, with yellow, swollen tips are produced on the erect lip. At their base is an orange, false-nectar spot. The erect lip is flexible at its point of attachment to the column, and when bees land on the apex of the lip their weight causes the lip to depress rapidly. The bees fall on the column and thereby pollinate the flower." In Florida, the four species of *Calopogon* "are associated with plants which belong to other plant families and produce flowers in which a clear floral resemblance is evident. *Calopogon pulchellus* (Salisb.) R. Br. occurs frequently in the everglades prairies of south Florida associated with *Sabbatia grandiflora*" (Gentianaceae). "General color, size and overall shape are remarkably similar. *Calopogon barbatus* (Walt.) Chapm. occurs in pine-palmetto scrub in central Florida together with *Cuthbertia graminea*" (Commelinaceae). "In each case, floral similarity suggests that the pollinators of the plants which produce food are deceived by the orchid into visits which result in pollination." The basis of these mimicries is apparently pollen.

Dodson (unpub. ms.) also mentioned three other examples, which are apparently based on nectar: (1) *Oncidium* species are pollinated by female *Centris* bees, and show an overall similarity to members of the Malphigiaceae. This system was investigated by Nierenberg (1972) and is discussed above (see under "Two Taxa Deceit Mimicries"). (2) *Oncidium onustum* resembles the flowers of species of *Cassia* in the coastal deserts of Ecuador and Peru. The pollinator of *O. onustum* is *Xylocopa* cf. *transitoria,* which is attracted by the sweet floral scent and attempts to collect nectar from the false nectary at the base of the column. It is implied, but not specifically stated that this bee also pollinates *Cassia* species. (3) *Epidendrum radicans* is a mimic of *Lantana camera* and *Asclepias curassavica.* (See discussion above related to Boyden, 1980 and Bierzychudek, 1981.)

Schelpe (1966) reported a possible case of mimicry between an African orchid, *Orthopenthea fasciata,* and a species of *Adenandra,* a shrub of the Rutaceae family. The basis for the suggestion of mimicry is the general similarity of the flowers of these two species, the fact they are sympatric, and the observation that *Andenandra* flowers are more abundant than is *O. fasciata.* No information is given about the occurrence or lack of nectar in *O. fasciata.*

Yeo (1968; Proctor and Yeo, 1972) described an example of what may be food deception mimicry between *Euphrasia micrantha* (Scrophulariaceae) (the mimic) and *Calluna vulgaris* (Ericaceae) (the model), two species that occur sympatrically in England. The mimic has "scanty nectar and pollen supply," relative to the model, which in contrast produces large amounts of pollen and nectar. Although this case is an often cited example of actual or potential "Batesian" mimicry, quantitative data are completely lacking. Even though 14 years have elapsed since this was first reported, no one has bothered to investigate it further!

Carlquist (1979) reported a record number of potential examples of floral mimicry for a single genus—no fewer than a dozen. These are in the genus *Stylidium* (Stylidiaceae), all from Western Australia. His examples may be divided into two kinds: congeneric, in which mimicry between six pairs of *Stylidium* species is postulated; and noncongeneric, in which mimicry is suspected between a *Stylidium* species and a species of another genus. Unfortunately, no data are available that would allow one to determine the basis of the mimesis, i.e., floral mutualism or deceit. As a first rough cut, one may postulate that the congeneric examples are equivalent to floral mutualisms, whereas the noncongeners are examples of deceit. However, it is also possible that one or more pairs of the congeneric type may be based on food deception. A thorough study of this interesting and complex array of mimicries is certainly warranted.

Faegri and van der Pijl (1979) mentioned two potential examples of deceit automimicry involving herbaceous plants: (1) *Exacum* (Gentianaceae), in which the protandrous anthers remain fresh-looking but are empty by the time the stigmas ripen; and (2) the monoecious *Begonia* (Begoniaceae), in which the stigmas are remarkably similar to the anthers in shape and size, and the male flowers are much more abundant than the female.

Heinrich (1979, p. 189) described a situation involving a nectarless orchid, *Calopogon pulchellus,* which is pollinated by bumblebees that have not established their foraging specialties. His description suggests that mimicry based on naiveté is operating. In addition, *C. pulchellus* apparently mimics another orchid in the community that does provide nectar, *Pogonia ophioglossoides,* and this relationship seems analogous to two taxa deceit mimicries. Perhaps both types of mimicries are working at the same time in this situation. Further study is certainly needed.

FLORAL MUTUALISM

The subject of floral mutualism was introduced earlier (see under "Floral Mimicry") and discussed in comparison with deceit mimicries. To reiterate, a mutualistic association or symbiosis is said to occur when two or more taxa utilize the same pollinator(s) and each taxon benefits from the other's presence. In addition, all taxa produce some kind of floral reward. Thus, in essence, all taxa in a mutualistic association are models. Their pollinators tend to be unspecialized and/or are forms (e.g., hummingbirds) that are not normally constant to a given plant species. Situations of floral mutualism are referred to by some authors as "Müllerian mimicry," but for the reasons presented such usage is discouraged.

At present there are probably only three well-documented, published reports of floral mutualism and perhaps another half dozen potential cases consisting of descriptive or anecdotal information. The former include Brown and Kodric-Brown (1979), Schemske (1981), and Powell and Jones (this book); the latter include Pennell (1948), Grant and Grant (1968), Macior (1971, 1974), Proctor and Yeo (1972), and Watt et al. (1974).

The study of Brown and Kodric-Brown (1979) was cited above (see under "Food Deception Mimicries"). The thrust of their paper dealt, however, with the "Müllerian" association of eight or nine species of hummingbird-pollinated plants. They quantified observations made some 11 years earlier by Grant and Grant (1968) regarding the apparent convergence of floral shape and color in plants pollinated by migratory hummingbirds in western North America. In addition to verifying this striking convergence in floral shape and color, they concluded that the hummingbird pollinators visit the flowers randomly, but interspecific competition is reduced because pollen of the different species is usually deposited on different anatomical portions of the birds. Furthermore, nectar was secreted at similar rates in most of the species.

The relationship of these plants and their hummingbird pollinators has all of the requisite characteristics outlined earlier in this chapter for delineating a typical mutualistic association. For example, (1) food was present in eight of nine taxa in the population (*Lobelia cardinalis,* a ninth species, lacked nectar and was termed a "Batesian" mimic); (2) all of the species studied conform to a general hummingbird-pollinated, floral model; (3) the pollinators are not constant to individual species; (4) although all species were not present in the population in equal numbers, three were estimated as "very common" and, one as "common"; and (5) at least eight (and perhaps nine) species are involved. This association is one of the premiere examples of floral mutualism.

Schemske (1981) investigated the relationship between two bee-pollinated tropical herbs (*Costus allenii* and *C. laevis,* Zingiberaceae) that occupy similar habitats and flower synchronously. Analysis of the breeding systems of these species showed that both are self-compatible but not autogamous, and have strong barriers to hybridization. Nectar secretion patterns are identical. They share the same non-selective pollinator, which carries pollen from both species in exactly the same location. Their flowers are morphologically similar and are the same color.

Schemske (1981) demonstrated convergence of *C. allenii* and *C. laevis,* by emphasizing that these species neither partition pollinators through a temporal segregation of flowering times, nor through the use of constant pollinators. He suggests that both species function as "Müllerian mimics" (sic) with respect to food rewards. This study is exemplary in respect to the types of data that need to be obtained in order to demonstrate floral mutualism.

The study of Powell and Jones is presented in the following pages (Chapter 15). Only a brief comment will be made here to point out the highly interesting switch that occurs in the foraging habits of the pollinators in response to changing environmental parameters. It appears that this factor may need to be considered more often when analyzing patterns of pollinator behavior, at least when working with bumblebees.

Unresolved Mutualistic Mimicries

Pennell (1948), a noted student of the Scrophulariaceae, made a brief mention of the striking similarities in floral morphology between the genus *Collinsia* (Scrophulariaceae) and the flowers of papilionaceous legumes (Fabaceae). Yeo (1968) commented on this report and added that the flowers of *Collinsia parviflora* and *Lupinus bicolor* might be thought of as analogous to "Müllerian mimics [sic], both being attractive for their own sakes." These are annual species that exist over a wide range in similar habitats, from San Diego County, California to Oregon, in the United States. Even though there are no data at the present time to suggest a mutualistic relationship, this situation seems to merit further study.

Macior (1971, 1974) was one of the first American workers (along with C. H. Dodson) to promote the significance of floral mimicry as a factor governing the floral evolution of certain species. Macior (1971) described a putative example of mimicry between the pink-flowered *Erysimum amoenum* (Brassicaceae) and the pink-flowered *Primula angustifolia* (Primulaceae), which occur sympatrically, share the same pollinators, and bloom synchronously in the high alpine meadows of the Rocky Mountains. Interestingly, *E. nivale*, sometimes considered a yellow-flowered color variant of *E. amoenum*, is associated with the blue-flowered *Polemonium viscosum* (Polemoniaceae) at other alpine locations. Floral form is similar between these species, and they share the same pollinator. Although it was not specifically stated, apparently both species pairs mentioned above produce nectar. Therefore it is presumed that the type of mimicry occurring is based on mutualism. Clearly, a thorough study of these cases is needed.

As a result of these observations, Macior (1971) postulated an often cited precept of floral mimicry: "In situations where population size may be restricted by local topography, as it is in alpine meadows, mimicry may provide a means of sustaining pollinator interest in one area by enlarging the *functional* size of a plant population." This applies to both food deception and mutualistic mimicries.

Another apparent example of mutualistic mimicry was described by Macior (1971) between *Pedicularis groenlandica* (Scrophulariaceae) and *Dodecatheon pulchellum* (Primulaceae). Both species are nectarless and are pollinated by pollen-foraging *Bombus* bees. Because the pollen of these species is spatially separated on the bodies of the bees, Macior stated that this pollination association enjoys a high degree of efficiency.

Proctor and Yeo (1972, p. 375) mentioned three species of plants occurring sympatrically in the Alps that are possible mutualistic mimics, based on similar appearance and abundance: *Ranunculus alpestris, Dryas octopetala,* and *Chrysanthemum alpinum*. No pollinator information is provided. The authors

stated that this type of mimicry is more likely to be found among species with "fairly unspecialized and promiscuous pollination mechanisms."

Watt et al. (1974) suggested that a number of flowers commonly pollinated by *Colias* butterflies in Colorado "constitute at least a partial analogy to a Müllerian mimicry ring" (sic). This conclusion was based on the similar UV floral patterns of the flowers, which signal to UV-perceiving pollinators the presence of a dilute monosaccharide-rich nectar. This pattern is particularly attractive to butterflies and flies, primarily because of their preference for this kind of nectar. This situation appears to have a strong basis for floral mutualism because many of the appropriate criteria are met: e.g., food is present in all taxa; all taxa are apparently models; the pollinators are unspecialized and apparently not constant to any given species. Lacking are data on the plants' breeding systems and pollinator behavior, which are needed to test whether the plants experience increased fitness as a result of their association with the others. The authors themselves call for further study of this situation in terms of a floral mutualism context.

To summarize, situations of floral mutualism share many characteristics with deceit mimicries in terms of basic ecological conditions, types of breeding systems, and similarities in morphological features. Fundamental differences occur, however, with respect to types of pollinator behavior, presence of food rewards, and the relative number of plants (or blossoms) in the population. Finally, the various species in a putative mutualistic association must be shown to achieve increased fitness as a result of the association.

ACKNOWLEDGMENTS

I thank C. E. Jones for discussions, and criticisms of earlier drafts. I also thank Cynthia Little for typing and C. H. Dodson for the use of his unpublished materials.

LITERATURE CITED

Ackerman, J. D. 1981. Pollination biology of *Calypso bulbosa* var. *occidentalis* (Orchidaceae): a food-deception system. *Madrono* 28(3):101–110.

Baker, H. G. 1976. "Mistake" pollination as a reproductive system with special reference to the Caricaceae. *In* J. Burley and B. T. Styles (eds.), *Tropical Trees: Variation, Breeding and Conservation.* New York: Academic Press, pp. 161–169.

Bawa, K. S. 1977. The reproductive biology of *Cupania guatemalensis* Radlk. (Sapindaceae). *Evolution* 31:52–63.

———. 1980. Mimicry of male by female flowers and intrasexual competition for pollinators in *Jacaratia dolichaula* (D. Smith) Woodson (Caricaceae). *Evolution* 34(3):467–474.

Bierzychudek, P. 1981. *Asclepias, Lantana,* and *Epidendrum:* a floral mimicry complex? *Reprod. Bot.* 54–58.

Boyden, T. C. 1980. Floral mimicry by *Epidendrum ibaguense* (Orchidaceae) in Panama. *Evolution* 34(1):135–136.

Brown, J. H. and A. Kodric-Brown. 1979. Convergence, competition, and mimicry in a temperate community of hummingbird-pollinated flowers. *Ecology* 60(5):1022–1035.

Carlquist, S. 1979. *Stylidium* in Arnhem Land: new species, modes of speciation on the Sandstone Plateau, and comments on floral mimicry. *Aliso* 9(3):411–461.

Dafni, A. and Y. Ivri. 1980. Deceptive pollination syndromes in some orchids in Israel. *Acta Bot. Neerl.* 29:55. (Abstract)

———. 1981. Floral mimicry between *Orchis israelitica* Baumann and Dafni (Orchidaceae) and *Bellevalia flexuosa* Boiss. (Liliaceae). *Oecologia* 49:229–232.

Dodson, C. H. 1962. The importance of pollination in the evolution of the orchids of tropical America. *Am. Orchid Soc. Bull.* 31:525–534, 641–649, 731–735.

Endler, J. A. 1981. An overview of the relationships between mimicry and crypsis. *Biol. J. Linnean Soc.* 16:25–31.

Faegri, K. and L. van der Pijl. 1979. *The Principles of Pollination Ecology,* 3rd rev. ed. Pergamon Press, Oxford.

Ford, E. B. 1971. *Ecological Genetics,* 3rd ed. Chapman and Hall, London.

Gentry, A. H. 1974. Flowering phenology and diversity in tropical Bignoniaceae. *Biotropica* 6(1):64–68.

Gilbert, L. E. 1975. Ecological consequences of a coevolved mutualism between butterflies and plants. *In* L. E. Gilbert and P. H. Raven (eds.), *Coevolution of Animals and Plants.* University of Texas Press, Austin, pp. 210–240.

Grant, K. A. and V. Grant. 1968. *Hummingbirds and Their Flowers.* Columbia University Press, New York.

Heinrich, B. 1975a. The role of energetics in bumblebee-flower interrelationships. *In* L. E. Gilbert and P. H. Raven (eds.), *Coevolution of Animals and Plants.* University of Texas Press, Austin, pp. 141–158.

———. 1975b. Bee flowers: a hypothesis on flower variety and blooming times. *Evolution* 29:325–334.

———. 1979. *Bumblebee Economics.* Harvard University Press, Cambridge, Mass.

Kullenberg, B. 1961. *Studies in Ophrys Pollination.* Almqvist and Wiksells Boktryckeri Ab., Uppsala.

Kullenberg, B. and G. Bergstrom. 1976. The pollination of *Ophrys* orchids. *Bot. Notiser* 129:11–19.

Little, R. J. 1980. Floral mimicry between two desert annuals, *Mohavea confertiflora* (Scrophulariaceae) and *Mentzelia involucrata* (Loasaceae). Ph.D. dissertation, Claremont Graduate School, Claremont, Calif.

Macior, L. W. 1971. Co-evolution of plants and animals—systematic insights from plant–insect interactions. *Taxon* 20(1):17–28.

———. 1974. Behavioral aspects of coadaptations between flowers and insect pollinators. *Ann. Missouri Bot. Gard.* 61:760–769.

Nierenberg, L. 1972. The mechanism for the maintenance of species integrity in sympatrically occurring equitant *Oncidiums* in the Caribbean. *Am. Orchid Soc. Bull.* 41(10):873–882.

Nilsson, L. A. 1980. The pollination ecology of *Dactylorhiza sambucina* (Orchidaceae). *Bot. Notiser* 133:367–385.

Pennell, F. 1948. Taxonomic significance of an understanding of floral evolution. *Brittonia* 6:301–308.

Pijl, L. van der. 1978. Reproductive integration and sexual disharmony in floral functions. *In* A. J. Richards (ed.), *The Pollination of Flowers by Insects.* Academic Press, London, pp. 79–88.

Pijl, L. van der and C. H. Dodson. 1966. *Orchid Flowers/ Their Pollination and Evolution.* University of Miami Press, Coral Gables, Fla.

Proctor, J. and P. Yeo. 1972. *The Pollination of Flowers.* Taplinger Publishing Co., New York.

Schelpe, E. A. 1966. *An Introduction to the South African Orchids.* Macdonald, London.

Schemske, D. W. 1981. Floral convergence and pollinator sharing in two bee-pollinated tropical herbs. *Ecology* 62(4):946–954.

Schemske, D. W., M. F. Willson, M. N. Melampy, L. J. Miller, L. Verner, K. M. Schemske, and L. B. Best. 1978. Flowering ecology of some spring woodland herbs. *Ecology* 59(2):351–366.

Schmid, R. 1978. Reproductive anatomy of *Actinidia chinensis* (Actinidiaceae). *Bot. Jahrb. Syst.* 100:149–195.

Stoutamire, W. 1975. Pseudocopulation in Australian terrestrial orchids. *Am. Orchid Soc. Bull.* 44(3):226–233.

Straw, R. M. 1972. A Markov model for pollinator constancy and competition. *Am. Nat.* 106(951):597–620.

Vane-Wright, R. I. 1976. A unified classification of mimetic resemblances. *Biol. J. Linnean Soc.* 8:25–56.

———. 1980. On the definition of mimicry. *Biol. J. Linnean Soc.* 13:1–6.

Watt, W. B., P. Hoch, and S. Mills. 1974. Nectar resource by *Colias* butterflies. *Oecologia* 14:353–374.

Wickler, W. 1968. *Mimicry in Plants and Animals.* Trans. from German by R. D. Martin. McGraw-Hill, New York.

Wiens, D. 1978. Mimicry in plants. *In* M. K. Hecht, W. C. Steere, and B. Wallace (eds.), *Evolutionary Biology,* Vol. 11. Plenum Publishing, New York.

Williamson, G. B. and E. M. Black. 1981. Mimicry in hummingbird-pollinated plants? *Ecology* 62(2):494–496.

Yeo, P. F. 1968. The evolutionary significance of the speciation of *Euphrasia* in Europe. *Evolution* 22:736–747.

15
FLORAL MUTUALISM IN *LUPINUS BENTHAMII* (FABACEAE) AND *DELPHINIUM PARRYI* (RANUNCULACEAE)

Elizabeth A. Powell

Department of Botany
University of Hawaii at Manoa
Honolulu, Hawaii
and
C. Eugene Jones
Department of Biological Science
California State University
Fullerton, California

ABSTRACT

A sympatrically blooming population of *Lupinus benthamii* Heller (Fabaceae) and *Delphinium parryi* Gray (Ranunculaceae) was studied in the seasons of 1978, 1979, and 1980 in the Santa Lucia Mountains, Monterey County, California. These plants showed remarkable vegetative and floral similarities.

The major pollinators of the two plants were *Bombus caliginosus* (Frison) and *Bombus californicus* Smith workers. In 1979, *Bombus* spp. populations were low, and the plants were pollinated by solitary bees. Both plants were partially autogamous.

Although *Delphinium parryi* contained nectar, *Lupinus benthamii* was nectarless. Plots containing plants in mixed and pure species fields were observed for visitations by bumblebee workers. Bees visited both species of plants in mixed fields. Air temperature in 1978 and air temperature and humidity in 1980 were correlated to the ratio of bumblebee visitations each plant species received.

It was concluded that *Lupinus benthamii* and *Delphinium parryi* were involved in a mutualistic mimicry complex for the joint attraction and energetic support of pollinators.

KEY WORDS: Mimicry; mutualistic floral mimicry; *Lupinus; Delphinium;* Monterey County, California; pollination; *Bombus;* solitary bees.

CONTENTS

INTRODUCTION

Even with the abundance of literature on the subject, mimicry has evaded a universally acceptable definition. Controversy as to what constitutes mimicry and what resemblances are not mimetic continues as more cases of the phenomenon are described. Wickler (1968) described mimicry as a three-party interaction involving a model (signal-transmitter, S1), a mimic (signal-transmitter, S2), and a deceived animal (signal-receiver, E). The mimic is further defined as "that one of the two signal-transmitters to which the receiver directs a response which is not of advantage to the receiver itself." The "signal" that the mimic sends to the signal-receiver is a false one; the signal-receiver is deceived by the mimic. Mullerian mimicry is excluded by Wickler as a type of mimicry, "since no agent is deceived and no difference exists between the model and the mimic."

Vane-Wright (1976) classified all possible mimetic responses except crypsis based on the positive and negative interactions of the three parties. Inter- and intraspecific mimicry situations were classified. Mullerian mimicry was included in Vane-Wright's classification with the argument that there is a functional model and a functional mimic; "those individuals tested are the models, those shunned which might otherwise have been attacked are the mimics." He further stated that

A fundamental problem lies with the concept of "deception" which implies a value judgement on the relationship between mimic and operator [signal-receiver]. I consider an essence of mimicry lies not in deception as such, but in mistaken identification.

Crypsis is excluded from Vane-Wright's definition of mimicry, since it involves "a failure of awareness" and not mistaken identity.

Wiens (1978), in his review of mimicry in plants, sided with Wickler by rejecting Mullerian mimicry and including crypsis. Wiens defined mimicry in this manner:

Mimicry is the process whereby the sensory systems of one animal (operator) are unable to discriminate consistently a second organism or parts thereof (mimic) from either another organism or the physical environment (the models), thereby increasing the fitness of the mimic.

Vane-Wright (1980), in response to Wiens's definition, defended his original definition of mimicry and modified it to read:

Mimicry involves an organism (the mimic) which simulates signal properties of a second living organism (the model) which are perceived as signals of interest by a third living organism (the operator), such that the mimic gains in fitness as a result of the operator identifying it as an example of the model.

Vane-Wright maintained that it is the presence of signals of interest that separates mimicry from crypsis, a nonsignaling situation. Arguments for and against Mullerian mimicry and crypsis as types of mimicry appear to be purely semantic. Mimicry is a type of convergence in which the selecting factor, or operator, is a sensory acute animal. All forms of mimicry have this in common. We define mimicry as a type of convergent evolution in which one or more species have converged in morpology, chemistry, or behavior, on a part of itself or themselves (automimicry), another living entity, or an inanimate object as a result of selection by an animal.

Floral similarities in sympatrically blooming plants of unrelated taxa have prompted many authors to speculate on the existence of a mimicry complex for the attraction of pollinators (Pennell, 1948; Schelpe, 1966; van der Pijl and Dodson, 1966; Yeo, 1968; Proctor and Yeo, 1972; Macior, 1974; Heinrich, 1975; Schemske, 1978; Wiens, 1978; Boyden, 1980; Dafni and Ivri, 1981a,b; Schemske, 1981).

Macior (1971) cited an instance of possible floral mimicry between *Primula angustifolia* and *Erysimum amoenum*. He stated that "mimicry may provide a means of sustaining pollinator interest in one area by enlarging the functional size of the plant population." Heinrich (1979) indicated that mimicry could be a strategy used by unrewarding flowers to "double-cross" bees that have not established foraging specialties. The grass pink, *Calopogon pulchellus,* in his example, is a large, showy, and unrewarding orchid that blooms at the same time as the nectar-rich rose pogonia, *Pogonia ophioglossoides*. The two flowers resemble each other in visible color and ultraviolet reflectance patterns. The two species use the same pollinator.

Brown and Kodric-Brown (1979) described a community of hummingbird-pollinated flowers as eight "Mullerian" mimics and one "Batesian" mimic. *Lobelia cardinalis* (the "Batesian" mimic) does not secrete nectar in the population studied. However, it is visited by juvenile migrant hummingbirds that mistake the *Lobelia* for abundant and rewarding *Ipomopsis aggregata* and *Penstemon barbatus*. The use of "Mullerian" and "Batesian" mimicry to describe pollination strategies and plant-plant floral mimicry systems seems incorrect. Clearly, floral mimicry cannot be described in the same terms as predator–prey situations. Floral mimicry involves a mutualistic interaction that serves to attract pollinators and not to repulse predators. New terminology to describe floral mimicry interactions must be devised.

Floral mimicry systems in which one flower is not providing a reward but is receiving visits from either naive pollinators or pollinators specializing on a similar rewarding flower, could be called deceptive floral mimicry (Little, 1980). This type of mimicry is what was termed "Batesian" mimicry by Brown and Kodric-Brown (1979). The obvious difference from Batesian mimicry is that floral mimicry is based on the ability of the mimic to be attractive to the operator, rather than repulsive. Vane-Wright (1976) categorized this floral type of mimicry as "inviting mimicry."

Thomson (1981) described a situation of two similar plant species that flower together and use the same pollinator. He found that the presence of *Potentilla fruticosa* enhanced the visitation rate on *Potentilla gracilis*. Thomson called this situation "mutual facilitation." He also indicated that pollinators may "sum" the densities of the two flowers when making foraging site choices.

A mutualistic system of this kind could be called mutualistic floral mimicry. This is similar to what was called Mullerian mimicry by Brown and Kodric-Brown (1979). However, since the flowers are attracting operators, pollination vectors, that also gain positively from the interaction, it is difficult to provide evidence of deceit, to meet Wickler's (1968) definition of mimicry. It is difficult to also prove Vane-Wright's (1976) requirement for mimicry, "mistaken identification." It does not matter if the pollinator "thinks" the flowers are the same, when they are not to taxonomists. The most important feature of the flowers to the pollinator, the quality and quantity of rewards and the handling time (the time it takes a pollinator to reach rewards in a flower), may be the same or nearly the same. If the flowers consistently provide rewards that reinforce visitation to a generalized floral pattern, then the pollinator learns one pattern for all similarly rewarding flowers. The same is basically true of Mullerian mimicry, when a predator equally avoids all species in the complex after learning one pattern.

This chapter deals with a case of plant–plant floral mimicry in which the model and mimic roles are not clearly defined, and mistaken identify of a mimic for a model by an operator is not verified. However, floral and vegetative convergence in a sympatric population appears to be advantageous to the plants involved, resulting in mutual pollinator attraction.

METHODS

Study Plants

Lupinus benthamii and *Delphinium parryi* are native California wildflowers, whose ranges overlap in the Santa Lucia Mountains, Monterey County, California. *L. benthamii* is an annual that can be 3 to 6 dm tall. It is erect and villous with racemes 10 to 20 cm long. Flowers are light to deep blue with the spot on the banner petal turning from light violet to deep red after pollination. Leaves are divided into seven to ten leaflets. Flowering stalks are anthocyanous. *D. parryi* produces annual growth 3 to 9 dm tall from a perennial tuber. The leaves are mostly basal, three- to five-parted, and further divided. Flowers are born on racemes 6 to 18 cm long. Sepals are generally purplish-blue to deep blue, but can vary in color from white to violet. Upper petals are whitish to light violet, and the lower petals are similar in color to the sepals and are puberulent (Munz, 1968). The racemes and petioles are anthocyanous in some plants studied.

Study Site

This study was conducted in the Santa Lucia Mountains, Monterey County, during the spring of 1978, 1979, and 1980. Study sites bordered the Hunter-Ligget Military Reservation and the Los Padres National Forest in Nacimiento Canyon. The major portion of the study was conducted in an area 21 km from Jolon, California, on the Nacimiento-Fergusson Road. The elevation was approximately

500 meters. The study area was a disturbed mixed oak woodland bordering dense chamise chaparral. The area was dominated by introduced annual grasses and had been used to pasture cattle.

Floral Features

Flowers and inflorescences of *L. benthamii* and *D. parryi* were photographed in visible and ultraviolet (UV) light. Ultraviolet photographs were made using a Kodak Wrattan 18A ultraviolet transmitting filter. Total reflectance curves of floral parts were generated by a Beckman model DK-A Spectroreflectometer (Horovitz and Harding, 1972). Samples were analyzed by Dr. Stephen Buchmann at the Carl Hayden Bee Research Laboratory in Tucson, Arizona. Specimens were gathered fresh in the field and analyzed the following day. Voucher specimens of *L. benthamii* and *D. parryi* are in the Faye MacFadden Herbarium, California State University, Fullerton (MACF).

Fresh, untreated pollen of *L. benthamii* and *D. parryi* was coated with a mixture of palladium and gold before being viewed in a scanning electron microscope. The specimens were viewed and photographed by Mr. Mike Bell of Union Oil Research Center, Placentia, California. A voltage of 10 kilovolts and magnifications of 2,000 and 10,000 times were used.

Reproductive Biology

The flowering times of *L. benthamii* and *D. parryi* and the foraging behavior of their pollinators, *Bombus* queens and workers, were recorded during 1978, 1979, and 1980. Density of *L. benthamii* and *D. parryi* in a single field was determined by the number of plants encountered on five 30-meter line transects taken in the same site in May of the years 1978, 1979, and 1980.

Inflorescences of the study plants were bagged with insect-proof gauze while the flowers were in the unopened bud stage. Open flowers and fruits were removed from the inflorescences that were bagged. Bagged flowers were treated in the following ways: (1) Four inflorescences with a total of 71 flowers of *L. benthamii* and three inflorescences with 27 flowers of *D. parryi* were not manipulated in any way. These inflorescences were allowed to develop fruit. (2) Twenty *D. parryi* flowers on five inflorescences were emasculated in the bud. Nine of these flowers on two inflorescences were allowed to develop without further treatment. (3) Five flowers on one inflorescence were hand-pollinated with pollen from a dehiscing flower on the same plant. (4) Six flowers on two inflorescences were hand-pollinated with pollen from a plant 2 to 5 meters away. (5) Fourteen flowers on two inflorescences of *L. benthamii* were hand self-pollinated when the flower was open, approximately 2 days following bagging. Self-pollination was performed by depressing the keel of the flower in such a manner that the stigma came in contact with pollen shed into the beak of the keel by the dehiscing anthers. (6) Three flowers on one inflorescence were emasculated in the bud stage and hand-pollinated with pollen from another plant.

The pollen of *L. benthamii* was obtained by depressing the keel once and collecting the pollen presented. The pollen of *D. parryi* was collected by gently scraping the dehiscing anthers and placing the presented pollen on glycine paper. The pollen of ten flowers was weighed at a time.

Inflorescences of both species were bagged with insect-proof gauze overnight. *D. parryi* nectar was collected with 1-μl microcapillary tubes by inserting the capillary tube in both spurs of the flowers. Dried nectar samples on filter paper were analyzed by Dr. Irene Baker at the University of California at Berkeley. No flowers of *L. benthamii* yielded nectar.

Delphinium and lupine flowers in the study area were tagged and observed from bud to fruit over a 2-week period in May 1980.

Pollinator Behavior

A plot 5 meters by 2 meters was delimited in an area where *D. parryi* and *L. benthamii* grew in a mixed stand. This plot was studied in approximately the same area over the years 1978, 1979, and 1980. Bees visiting inflorescences in the plot were observed one at a time. Species of plant visited, order of visitation, time spent by a pollinator at a flower, and the number of flowers visited were recorded. Time of day, air temperature, general weather conditions, cloud cover, and wind were also recorded. Data on temperature, humidity, and precipitation for May 1978 were obtained from the U.S. Forest Service. These data were collected at a Nacimiento Canyon weather station about 3.2 km from the study site. However, owing to vandalism, no data were available from this station in 1979 and 1980. Temperature and humidity data for 1980 were obtained from the Hunter-Liggett Military Reservation air field weather station approximately 24 km from the study site. No weather data were available for 1979.

In the years 1979 and 1980, pure plots of 10 square meters of *L. benthamii* and *D. parryi* were delimited that were at least 10 meters from the nearest population of the other species. All three plots, the mixed and two pure plots, were in the same habitat, at the same altitude, and within approximately 150 meters of each other. On the morning of May 11, 1980, between 8:45 and 11:45 A.M., a mixed plot, a pure plot of *L. benthamii*, and a pure plot of *D. parryi* were watched simultaneously by three observers. On May 12, 13, and 14, one observer alternated between the three plots every half hour during the morning and afternoon hours that bees were the most active.

A visitation ratio was calculated for bumblebees *Bombus* spp. in the study plot by the following formula:

$$\frac{\dfrac{VL}{NL} - \dfrac{VD}{ND}}{\dfrac{VL + ND}{NL + ND}}$$

where VL = number of inflorescences of *L. benthamii* visited; VD = number of inflorescences of *D. parryi* visited; NL = number of inflorescences of *L. benthamii* in the plot; and ND = number of inflorescences of *D. parryi* in the plot.

A negative ratio indicated that *D. parryi*

was preferentially visited, whereas a positive ratio indicated that *L. benthamii* was preferentially visited. The larger that positive or negative number, the stronger the preference. A ratio of 0 indicates that equal numbers of *L. benthamii* and *D. parryi* inflorescences were visited on a single foraging flight in the study plot.

No attempt was made to distinguish between *Bombus caliginosus* and *Bombus californicus* workers when these bees were observed together in 1980. Different hives of *Bombus* spp. may have different foraging specialties (Free, 1970). This was not considered in this study. No attempt was made to mark individual bees, and since bumblebees tend to forage in the same areas, on the same species of plant, over several days (Manning, 1956; Free, 1970; Heinrich, 1976a), the same individuals may have been recorded in the study plots several times. This would not change the proportion of the behavioral spectrum exhibited by single individuals if the bumblebees did

not significantly change their individual foraging patterns, but it would bias the total sample.

Herbivore Behavior

Herbivorous beetles were found damaging flowers of *L. benthamii* and *D. parryi* in the study plots. One hundred forty flowers of *D. parryi* from a pure population and 102 flowers from a population growing mixed with *L. benthamii* were chosen at random and examined for damage. Flowers with holes in floral parts or missing floral parts were considered damaged. Damage was considered heavy if 50% or more of the essential reproductive parts, stamens, pistil, or spur were damaged or missing. Damage was considered slight if 10% of the petals and sepals were damaged or missing, but there was no damage to stamens, pistil, or spur. Damage was considered moderate if damage was intermediate between slight and heavy.

RESULTS

Vegetative and Flora Features

The general growth habit and floral arrangement of *L. benthamii* and *D. parryi* are similar. Both plants have basal leaves and flowers borne on upright racemes (Fig. 15-1). Field specimens of *L. benthamii* were generally taller than *D. parryi*. However, the heights of the two species overlap between 32 and 65 cm (Table 15-1). The leaves of the two plants can appear remarkably similar in size, shape, and presentation (Figs. 15-1 and 15-2). In many of the plants studied, the stems, petioles, and leaves of both species were anthocyanous. Both species have trichomes on the stems, petioles, and leaves. In *L. benthamii* the trichomes are long and stiff, and in *D. parryi* they are soft and appressed. The vegetative portions, pedicels, and unopened buds of both species are UV-absorptive.

The visible and UV floral patterns of both species are similar. The flowers are zygomorphic and have a UV pattern. The flower of *D. parryi* is twice as large as that of *L. ben-*

Figure 15-1. Greenhouse-grown specimens of *Delphinium parryi* and *Lupinus benthamii* (right) photographed in visible light. Scale is equal to 10 cm.

Table 15-1. A comparison of six anatomical features and one pollinator feature of *Lupinus benthamii* and *Delphinium parryi*.

Feature		*Lupinus benthamii*			*Delphinium parryi*			
	n	Range	$\bar{x}$	n	Range	$\bar{x}$	t
Height of inflorescence (cm)	63[1]	32–92	64.1	34[1]	29–65	46.4	6. 23***
Number of flowers visited by *Bombus* per inflorescence	83[1]	1–12	3.6	27[1]	1–5	1.6	3. 92***
Time spent at a flower by *Bombus* (seconds)	153[2]	0.5–3.5	1.04	58[2]	1.5–7.0	1.82	16.24***
Number of open flowers per inflorescence	69[1]	1–13	5.5	26[1]	1–7	2.9	4. 44***
Number of inflorescences per plant	16[4]	2–10	4.3	22[4]	1–3	1.1	3. 55***
Days of anthesis		4–7			9–16		
Weight of pollen available to *Bombus* per flower (µg)	30[3]	54–92	76	30[3]	47–176	108	0.83 NS

***Significant at .001 level. NS = not significant.
1 = number of inflorescenses measured; 2 = number of bees measured; 3 = number of flowers measured; 4 = number of plants measured.

thamii, but since *L. benthamii* has almost twice the number of flowers per inflorescence as *D. parryi*, the floral displays are similar (Fig. 15-3).

Reflectance curves of floral parts of both species are shown in Fig. 15-4. Both species have flowers that are blue or violet in the human visible spectrum (HVS). *D. parryi* se-

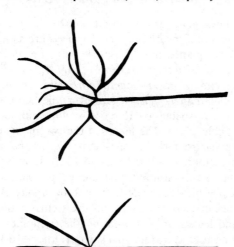

Figure 15-2. Outline of typical leaves of *Lupinus benthamii* (bottom) and *Delphinium parryi* (top). Leaves are ½ size.

pals and petals had a reflectance peak at 450 nm, whereas the peak reflectance in the blue range for floral parts of *L. benthamii* was at 400 nm. *D. parryi* had a smaller second peak at 600 nm and a larger one at 750 nm. The reflectance of *L. benthamii* flowers also increased in the red range that is invisible to bumblebees. The banner spot of *L. benthamii* and the upper petals of *D. parryi* are UV-absorptive and had a reflectance peak at approximately 440 nm. The upper petals of *D. parryi* appeared light violet in the HVS. The corresponding insect color would be yellow (Fig. 15-5 and Table 15-2) (Kevan, 1978). The banner spot of *L. benthamii* was light pink, but turned dark red after pollination. The ultraviolet absorptive character of the banner spot was unchanged by pollination. Only unpollinated *L. benthamii* flowers were tested with spectrophotometry.

Scanning electron microscopy of pollen grains of *L. benthamii* and *D. parryi* shows similarity in size and general morphology. The grains of both species were tricolpate and oblong. The grains of *L. benthamii* were 31 by 17 µ in size, rectangular in shape, and had a reticulated exine. The exine of *L. benthamii* grains were similar to the sculpturing in other genera of Fabaceae (Adams and Smith, 1977). The grains of *D. parryi* were 28 by 21 µ in size and were oval to rectangular-shaped. The

Figure 15-3. Individual flowers of *Delphinium parryi* (left) and *Lupinus benthamii* (right) photographed in visible light (a). Inflorescences of *Lupinus benthamii* (left) and *Delphinium parryi* (right) photographed in visible light (b). The same flowers and inflorescences as in (a) and (b) photographed with an ultraviolet transmitting filter (c) and (d). Scale is in centimeters.

exine on these grains was slightly granulose (Fig. 15-6).

Reproductive Biology

The density of *D. parryi* varied little over the three years of this study (Fig. 15-7). This would be expected, since *D. parryi* is a perennial whose density would probably be altered only by extremely favorable or unfavorable conditions. The density of *L. benthamii,* an annual, was variable (Fig. 15-7). The study area was burned by a brush fire in September of 1979, and this fire, the favorable rainfall of 1980, or a several-year cycle of seed production and germination may have been responsible for the 1980 increase in *L. benthamii* density.

The mean number of inflorescences per plant in a mixed field was 4.3 for *L. benthamii*

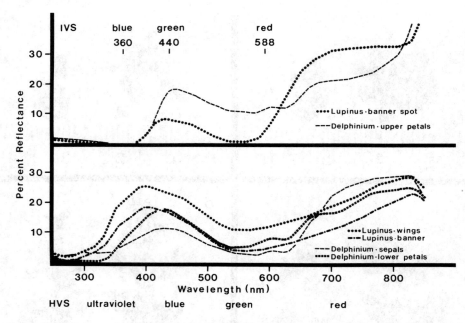

Figure 15-4. Reflectance curves in human visible spectrum (HVS) and insect visible spectrum (IVS) of floral parts of *Lupinus benthamii* and *Delphinium parryi*.

and 1.1 for *D. parryi*. The mean number of flowers per inflorescence was 5.5 for *L. benthamii*, compared to 2.9 for *D. parryi*. The density of plants in the field is four times greater for *L. benthamii* than *D. parryi*. A multiplication of these figures reveals that there were 29.65 lupine flowers per each delphinium flower in the mixed field studied (Table 15-1).

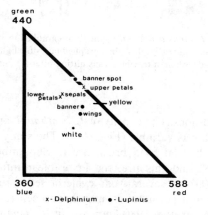

Figure 15-5. Reflectance proportions of *Delphinium parryi* and *Lupinus benthamii* floral parts plotted on a trichromatic scale based on three wavelengths to which bees are sensitive (Kevan, 1978).

Lupinus benthamii was observed blooming from mid-April to mid-June over the three years studied (Table 15-3), whereas the blooming season of *D. parryi* was restricted mostly to the month of May. Some individuals began their bloom in the last week of April, and some were still in bloom in the first week of June. By mid-June, any remaining delphinium flowers were dried on the racemes.

From mid-April to early May *Bombus* spp. queens were observed foraging on a variety of flowers, as well as *L. benthamii* and *D. parryi*. By mid-May, *Bombus* spp. queens were no longer observed in the field. *Bombus* spp. workers were observed foraging from late April until early June on *L. benthamii* and *D. parryi*. This coincided with the peak blooming period of the two flowers. By mid-June, however, *Bombus* spp. workers were foraging primarily on *Lotus scoparius* whose bloom began at the end of May as the bloom of *L. benthamii* and *D. parryi* declined (Table 15-3).

Delphinium parryi is protandrous. Individual flowers were observed to remain open from 9 to 16 days. After anthesis, all the flower's 30 anthers dehisced over a period of 7 to 13 days at a rate of approximately 3 anthers per day.

Table 15-2. Reflectance proportions of *Lupinus benthamii* and *Delphinium parryi* floral parts at three wavelengths corresponding to the visual sensitivity of bees and the insect color that these represent, following the system of Kevan (1978).

Species	Floral area	Reflectance proportions			Insect color
		360 nm	440 nm	588 nm	
Lupinus	Banner spot	0	67	33	Yellow
Delphinium	Upper petals	0	61	39	Yellow
Lupinus	Wings	37	42	21	Pale yellow
Lupinus	Banner	39	47	14	Pale yellow
Delphinium	Sepals	25	56	19	Pale green
Delphinium	Lower petals	21	55	24	Pale green

When the last set of anthers were dehiscing, the stigma pushed through them and opened. After 2 to 3 days in the female stage, the petals fell as the ovary began swelling. Significantly more nectar was produced by flowers in the female stage than those in the male stage ($t = 3.92$, Sig at 0.05, df 12, 20). This pattern of increasing nectar production in older flowers in the female stage may promote outcrossing (Pyke, 1978; Heinrich, 1979). The nectar of *D. parryi* is hexose-rich, whereas the *L. benthamii* plants in this study did not produce nectar. Pollen in both flowers was produced in nearly equal amounts. There was no significant difference between the two plant species in the weight of the pollen available to pollinators per visit (Table 15-1). The anthers of *L. benthamii* dehisce in the bud. The flowers remained open 4 to 7 days (Table 15-1). Within a day of pollination, the white banner spot turned deep red, and within 2 days of pollination, the flower withered as fruit development began. On bagged, unpollinated specimens, the banner remained white and the flowers were open for as long as 14 days.

Bagging experiments revealed that both species were partially autogamous. When flowers of *L. benthamii* were bagged in the bud with no manipulation, 32% set fruit. If flowers were hand self-pollinated, 57% set fruit (Table 15-4). No fruit development resulted from hand cross-pollination of *L. benthamii*, and 33% of the hand cross-pollinated *D. parryi* flowers set fruit. *D. parryi* produced fruit for all flowers hand self-pollinated and in 44% of the flowers bagged in bud and not manipulated. There was no fruit set in flowers that were emasculated in the bud. It is unknown what percent fruit set results on open pollinated plants of either species.

Pollinator Behavior

Three species of *Bombus* were observed foraging on *L. benthamii* and *D. parryi*. *Bombus caliginosus* (Frison) was the most common pollinator of the two flowers over the three years. *Bombus californicus* Smith was observed primarily in 1980, foraging on *D. parryi*. *Bombus edwardsii* was observed very infrequently in all years. In 1979, *Bombus* spp. were very scarce. Flowers were visited most frequently by several species of solitary bees (Table 15-5). In the years that *Bombus* spp. were abundant, 1978 and 1980, these solitary bees were not observed in the study plot with the exception of one *Emphoropsis* sp. that was observed in 1980. The more aggressive *Bombus* spp. will exclude these other bees from their foraging area (R. Snelling, pers. comm.). The solitary bees visited significantly fewer inflorescences while in the study plot ($\bar{x} = 2.7$) than *Bombus* spp. [$(\bar{x} = 6.7)(t = 10.88$, Sig at .01, $df = 76, 756)$]. Although there was an active feral hive of *Apis mellifera* approximately 10 meters from the study plot all three years, *Apis mellifera* was rarely observed on either flower species and appeared to be an infrequent and ineffectual visitor.

Trichodes ornata Say was captured on *D. parryi* and *L. benthamii* and frequently was seen dusted with pollen. These beetles may in-

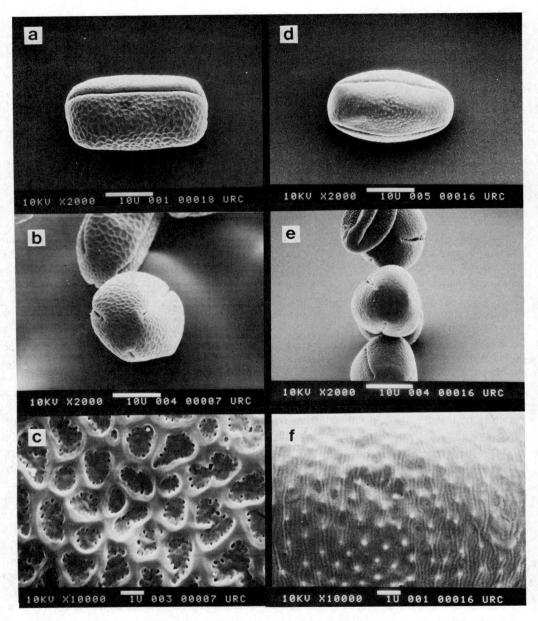

Figure 15-6. Scanning electron micrographs of pollen grains of *Lupinus benthamii* (a, b, and c); and *Delphinium parryi* (d, e, and f). Kilovolts (kV), magnification (×), and scale in microns (μ) are indicated for each micrograph. (a) and (d) are whole grain views. (b) and (e) are polar views. (c) and (f) are magnifications of the exine.

cidentally transfer pollen between inflorescences while laying eggs in the flowers (R. Snelling, pers. comm.). They were not considered active pollinators of the study plants.

Bombus caliginosus approaches and visits both *L. benthamii* and *D. parryi* in a similar manner (Fig. 15-8). The bee lands on the keel of *L. benthamii* or the lower petals of *D. parryi* facing the UV-absorptive nectar guide. This guide probably is the orientation cue for the bee for both flowers (Jones and Buchmann, 1974). *L. benthamii* pollen is deposited on the sternum of *Bombus* spp., and *D. parryi* pollen is deposited on the gular region of the bee. Sig-

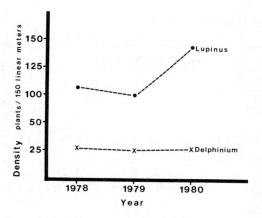

Figure 15-7. Density of sympatric *Lupinus benthamii* and *Delphinium parryi* in the same field over three years.

nificantly less time is spent by *Bombus* spp. on *L. benthamii* flowers than *D. parryi* flowers (Table 15-1). The increased time spent at *D. parryi* flowers is probably time spent in nectar collection and is not handling time.

One-way analysis of variance and Scheffe's multiple range test showed that the 1979 and

1980 seasons were significantly different from the 1978 season in the mean visitation ratio and temperature (at which data were collected). In 1978 a positive visitation ratio indicated that bees visited more lupines than delphiniums in the mixed study plot. Both 1979 and 1980 had negative visitation ratios, indicating that more delphinium flowers were visited than lupine flowers (Table 15-6).

A chi-square goodness-of-fit test of visitation in the mixed plot revealed that bees did not visit the two species randomly in the study plots. Bees visited more *L. benthamii* than expected in 1978 ($\chi^2 = 701.81$, Sig at .001, $df = 1$, $n = 438$), and more *D. parryi* than expected in 1979 ($\chi^2 = 7.05$, Sig at .01, $df = 1$, $n = 40$) and 1980 ($\chi^2 = 118.17$, Sig at .001, $df = 1$, $n = 105$).

The daily average temperature is the mean temperature over the 24-hour day measured every hour. The mean daily average temperature for days during which data were collected was significantly lower in 1980 than in 1978.

Table 15-3. Flowering phenology of *Lupinus benthamii* and *Delphinium parryi* and the foraging activity of *Bombus* queens and workers.

Event	Mid-April	Late April	Early May	Mid-May	Late May	Early June	Mid-June
Lupinus flowering	X	X	X	X	X	X	X
Delphinium flowering		X	X	X	X	X	
Bombus queens observed foraging	X	X	X				
Bombus workers observed foraging on *Lupinus* and *Delphinium*		X	X	X	X	X	
Bombus workers observed foraging on *Lotus scoparius*							X

Table 15-4. Percent fruit development of *Lupinus benthamii* and *Delphinium parryi* after maniuplations of bagged plants.

	Lupinus		*Dephinium*	
Treatment	Number of flowers	Percent fruit development	Number of flowers	Percent fruit development
No manipulation (control)	71	32	27	44
Hand self-pollinated	14	57	5	100
Emasculated only	—	—	9	0
Hand cross-pollinated	3	0	6	33

Table 15-5. A list of bee visitors to *Delphinium parryi* and *Lupinus benthamii*.

Species	Captured on	
	L. benthamii	D. parryi
Apidae		
Bombus caliginosus (Frison)	X	X
Bombus californicus Smith		X
Bombus edwardsii Cresson	X	
Apis mellifera Linnaeus	X	X
Anthophoridae		
Xylocopa tabaniformis orpifex Smith	X	
Anthophora crotchii Cresson		X
Emphoropsis rugosissima Cockerell	X	
Emphoropsis interspersa Cockerell	X	X
Synhalonia lunata Timberlake	X	
Megachilidae		
Hoplitis albifrons maura (Cresson)	X	X
Hoplitis sambuci Titus	X	
Osmia atrocyanea Cockerell	X	
Osmia calla Cockerell	X	
Megachile onobrychidis Cockerell	X	
Halicitae		
Lasioglossum sisymbrii Cockerell		X

The coolest of the three seasons was 1980, whereas 1979 was the warmest. Humidity was significantly higher in 1980 than in 1978 (Table 15-6). The mean monthly temperature is an average of the daily average temperatures for every day of the month of May. The mean monthly temperature of May 1978 was not significantly different from that of May 1980 (Table 15-6).

When all years, plots, and bees were considered, the mean visitation ratio for bees foraging at temperatures greater than 15.5°C ($\bar{x}$ = 0.40) was significantly higher than the ratio for bees foraging at temperatures equal to or less than 15.5°C (t = 5.75, Sig at 0.01, df = 832).

A chi-square goodness-of-fit test indicated that random behavior (behavior that appeared to be dependent on the number of inflorescences presented in the study plots) occurred at temperatures less than or equal to 12.7°C in 1978 (χ^2 = 2.16, NS, df = 1, n = 36), and at temperatures greater than 21.1°C in 1980 (χ^2 = 0.34, NS, df = 1, n = 16). Overall random behavior was observed when bees visited both species of flowers in the study plot in 1978 (χ^2 = 3.04, NS, df = 1, n = 88). These bees visited significantly more inflorescences ($\bar{x}$ = 9.7) than those bees that did not visit both species while foraging in the study plot ($\bar{x}$ = 6.38)(t = 4.39, Sig at .01, df = 87, n = 349). The sampling of a nectar source, *D. parryi*, in the plot may encourage bees to forage longer in the area (Hartling and Plowright, 1979).

It appears superficially, when the 1978 data are pooled, that the bees that visit both plant species in the study plot were totally random in their floral choices. However, a close examination of the visitation data of individual bees revealed that 52% (n = 46) of the bumblebees that were inconstant and visited both species of flowers had foraging specialties; they visited one species of flower more than twice as frequently as they visited the other species of flower. The remaining 48% (n = 42) of the bees that visited both species in the plot visited *L. benthamii* and *D. parryi* in approximately equal frequency. In terms of all the bees ob-

Figure 15-8. Photographs of *Bombus caliginosus* worker at a flower of *Lupinus benthamii* (a) and *Delphinium parryi* (b).

served in the 1978 season, 75.3% (n = 330) specialized on *L. benthamii,* and 15.1% (n = 66) specialized on *D. parryi.* The remaining 9.6% (n = 42) had no foraging specialty. The relative density of the whole field of flowers was 80% *L. benthamii* and 20% *D. parryi* in 1978. A chi-square goodness-of-fit test revealed that the frequency of visitation of specialist bees to either flower is related to the relative density in the entire mixed field (χ^2 = 2.54, NS, df = 1, n = 396). The bees that did not have an apparent foraging specialty foraged at significantly lower temperatures ($\bar{x}$ = 17°C) than those that did have a speciality ($\bar{x}$ = 19°C)(t = 2.33, Sig at .05, df = 42, n = 46). Those bumblebees that did not show a specialty, however, showed a preference for *D. parryi* (χ^2 = 25.1, Sig at 0.01, df = 1).

A stepwise multiple regression was used to compare the independent variables of temperature, humidity, wind speed, and hour with the dependent variable, visitation ratio for the 1978 and 1980 data. In 1978, temperature was significantly correlated with visitation ratio (r = 0.1613, F = 11.645, P = <.001). Temperature that year explained only 2.6% of the variation. Wind speed, humidity, and hour were not signficantly correlated with visitation ratio (r = 0.1048, F = 1.606, P = 0.187). Temperature, humidity, hour, and wind speed together explained 3.7% of the variation. For the 1980 data, temperature had a significant correlation with the visitation ratio (r =

Table 15-6. A comparison of climatic and behavioral features of the three study seasons.

Mean ($\bar{x}$)	1978	1979	1980	F/T
Visitation ratio	0.69	−0.81	−1.15	F = 54.9
n = number of bees	n = 438	n = 40	n = 105	df = 2
Temperature (°C)	19	27.3	17.4	F = 210.4*
n = number of bees	n = 438	n = 40	n = 105	df = 2
Humidity (relative %)	46.0	—	49.2	t = 3.8*
n = number of bees	n = 438	—	n = 31	df = 541
Monthly temperature (°C)	15.1	—	15	t = 1.28 NS
n = number of days	n = 31	—	n = 31	df = 60
Daily average temperature (°C)	15	—	12	t = 5.39*
n = number of days	n = 7	—	n = 4	df = 9

* = significant at .001 level. NS = not significant. — = data not available.

0.324, $F = 12.049$, $P < .001$). Temperature in 1980 explained 10.47% of the variation. Humidity was also significantly correlated with the visitation ratio ($r = 0.2023$, $F = 4.395$, $P = 0.039$). Hour was not significantly correlated with the visitaton ratio. Temperature, humidity, and hour jointly explained 18.1% of the variation in the visitation ratio. Wind speed data were not available for 1980.

Herbivore Behavior

Herbivorous beetles of the species *Lytta stygica* (LeC) were observed in the study plots eating the floral parts of both plant species. These beetles were probably responsible for the damage to *D. parryi* flowers (79% of which had some damage) examined in the pure stand. This is compared with the 25% damaged delphinium flowers examined in the mixed stand. The damage to delphinium flowers was more severe in the pure stand than in the mixed stand. There were no delphinium flowers examined in the mixed stand that sustained heavy herbivore damage. However, 42% of the delphinium flowers with damage in the pure stand were heavily damaged (Table 15-7).

Several individuals of the beetle *Trichochronellus stricticollis* Casey were found in some *D. parryi* flowers. It is probable that they are also herbivorous (Daly et al., 1978).

The mammalian herbivores of the study plants are unknown. However, in 1978, a pocket gopher *(Thomomys bottae)* was observed pulling a branch of *L. benthamii* into a hole. Although cattle walked through the site in 1980, no evidence was found of grazing on the study plants. Both *D. parryi* and *L. benthamii* have numerous generic relatives that are alkaloid-bearing (Kingsbury, 1964; Benn, 1966; Willaman and Li, 1970). There are sev-eral reported cases of delphinium and lupine poisoning of livestock (March and Clawson, 1916a,b; Shupe et al., 1970).

DISCUSSION

Both *L. benthamii* and *D. parryi* have flowers adapted to bumblebee pollination (Faegri and van der Pijl, 1979). The similarity in form, presentation of flowers, height, and dissection of leaves, also may be adapted to form recognition by bees (Manning, 1956). The difference in heights of the two plants may encourage pollinator fidelity (Levin and Kerster, 1973; Waddington, 1979). However, since the heights of the inflorescences overlap over a wide range (Table 15-1), height differences probably do not discourage inconstancy by bees foraging in mixed fields.

The flowers have similar reflectance curves, and have floral colors that appear very similar to the human eye. Bees, however, may be able to distinguish the subtle difference in the color of the two flowers (P. Kevan, pers. comm.). The banner spot of *L. benthamii* and the upper petals of *D. parryi* are the floral areas most similar in insect color. These parts may also be the most important for short-range orientation (Jones and Buchmann, 1974). The similarity in insect color of the orientation cue may override the differences in insect color of other floral parts. The similarity of size and shape of pollen grains of the two species (Fig. 15-6) may be an adaptation for pollen pickup by the same-size pollinators (Lee, 1978).

The two plant species share the same pollinator, *Bombus* spp. workers. These bees forage systematically in a field of flowers. Solitary bees pollinated these plants in 1979 when *Bombus* spp. populations were low. Solitary bees forage only for their own needs and to

Table 15-7. Herbivore damage in mixed and pure stands of *Delphinium parryi* flowers.

Stand	Number flowers examined	Number flowers damaged	Percent flowers damaged	Total damaged flowers					
				Slight		Moderate		Heavy	
				#	%	#	%	#	%
Pure	140	111	79	24	22	51	46	47	42
Mixed	102	26	25	22	84	4	16	0	0

provision their own brood. In contrast, *Bombus* spp. workers must forage for an entire hive, including individuals that do not forage. Bumblebees visited more flowers while foraging than solitary bees. The solitary bees captured on *L. benthamii* and *D. parryi* were all generalists (R. Snelling, pers. comm.).

Recent work on competition for pollinators would predict that flowers that are similar in color, form, height, floral presentation, and blooming time would be in competition for pollinators. Pleasants (1980) predicted that in guilds of flowers using the same pollinator, blooming times should diverge to reduce competition. However, the blooming peaks of the two flowers *L. benthamii* and *D. parryi* completely overlap and appear to be timed with the emergence of foraging *Bombus* workers. It also has been predicted that the least able competitor would need to diverge in some character in order to avoid extinction (Levin and Anderson, 1970; Mosquin, 1971; Pleasants, 1980). It appears, however, that *L. benthamii* and *D. parryi* both receive visitation from bees that do not appear constantly to reject one species. Straw (1972) predicted that similar flowers of two species that share a pollinator that is about equally constant to each, should survive and not be competitively excluded. It appears that convergence rather than divergence is operating in the floral characteristics of the two flowers.

Autogamy in *L. benthamii* and *D. parryi* may allow a more competitive situation to exist between the two flowers than could be tolerated in totally outcrossing species (Levin and Anderson, 1970). Further, autogamy in *L. benthamii* may be enhanced by the simple "working" of the flower by a heavy pollinator, such as *Bombus* spp., which need not even carry conspecific pollen (Wainwright, 1978). Although the pollen of *L. benthamii* and that of *D. parryi* were laced on different areas of the bee's body, these areas are close enough together that pollen mixing is not improbable. If there is interspecific pollen mixing on the bee's body, and pollen transfer to the proper stigmatic surface is imperfect, then it would be expected that the least autogamous species would either diverge to avoid this form of competition or be competitively excluded from the population (Waser, 1978). One way for plants to avoid this kind of mixing is to occur in separate clumps rather than mixed aggregations of flowers. This study did not examine the amount of interspecific pollen on stigmas of *L. benthamii* and *D. parryi* in mixed vs. pure stands. However, such a study would provide knowledge on this subject. It is also unknown what part grooming by bumblebees plays in reducing interspecific pollen transfer.

The effects of competition between the two species could be less for *D. parryi* because it is a tuberous perennial that may survive for decades (Epling and Lewis, 1952). *L. benthamii*, on the other hand, is an annual, and may therefore, be more immediately affected by serious changes in pollinator availability or effectiveness.

Environmental factors, as well as the needs of the individual bee or the hive, appear to play roles in flower choices by bumblebees. When the air temperature is cold and the honeypot is empty, a nectar source may be preferred. However, when the weather is warmer and the honeypot is full, foraging for pollen may be the most energetically efficient strategy for the hive.

We found that in both 1978 and 1980, when *Bombus* spp. were the major pollinators, visitation to *D. parryi* was higher at lower temperatures. The correlation between temperature and visitation ratio was significant at the 0.001 level in both 1978 and 1980, although temperature explains a higher percentage of the variation in the visitation ratio in 1980 than in 1978. All of the observations in 1980 were made at temperatures of less than 22°C, whereas only 65% of the observations in 1978 were made at temperatures less than that. Colder temperatures may be energetically more stressful for foraging bumblebees. As temperature decreases, air temperature may become more important to bumblebees in determining floral choices (Heinrich, 1972, 1973). *D. parryi* was the preferred species in 1979 and 1980, but *L. benthamii* was preferred by bumblebees in 1978. It would be expected that *D. parryi* would be preferred over *L. benthamii* at lower temperatures, since *D. parryi* provides both nectar and pollen, and *L. ben-thamii* provides only pollen. The solitary bees that were the major pollinators in 1979, preferred *D. parryi*. However, *Bombus* spp. did not uniformly prefer *D. parryi*.

It may be more energetically efficient for bumblebees to forage for nectar from *D. parryi* and pollen from *L. benthamii*. Although both *L. benthamii* and *D. parryi* provide almost the same amount of pollen per visit, there are approximately 30 times more *L. benthamii* flowers available than *D. parryi* flowers. Lupines grow with many flowers clustered on the racemes. This may make pollen gathering from *L. benthamii* more energetically efficient than from the few flowered *D. parryi* racemes. The difference in energy available to larval bees for each species of pollen is unknown. However, Colin and Jones (1980) found that the caloric energy values of several species of dicot pollen were similar.

In the bumblebee hive, some bees may specialize on pollen gathering and some on collecting nectar (Heinrich, 1976b). In 1978, most bumblebees did show definite specialties. Seventy-five percent of the bumblebees specialized on *L. benthamii*, and 15% specialized on *D. parryi*. Although it is unknown whether these specialist bees were from separate hives, it is possible that when the weather is warm and there is no energetic stress on the hive, most bees might be involved in pollen gathering, and only a few bees might be required to gather sufficient nectar supplies for the hive. An occasional visit to a nectar flower may allow a pollen specialist to track changing resources (Oster and Heinrich, 1976) or may provide an opportunity for the bee to gain sufficient energy in the field to continue uninterrupted foraging. The bees that visited both plant species in the mixed plot, in 1978, averaged 6.5 visits to *L. benthamii* inflorescences and 3.2 visits to *D. parryi* inflorescences. Of those bees, the bees that were identified as pollen specialists, that is, bees that visited *L. benthamii* more than twice as often as *D. parryi*, visited a mean of 11.9 *L. benthamii* inflorescences and 1.6 *D. parryi* inflorescences. This indicates that some bees foraged primarily on *L. benthamii* and visited *D. parryi* only occasionally.

It is possible that a bee that visits both *L. benthamii* and *D. parryi* does not need to distinguish the nectar from the pollen source if both needs are met in proper proportions as the bee forages in a mixed field. In 1978, visitations to the two flowers appeared to be related to the density of the two species in a mixed field. Chi square goodness-of-fit comparisons of the visitations of bumblebees to the density of the two flowers in the mixed plot showed a preference for *L. benthamii*. However, when the visitations of the two species were compared to the density of each floral species in the entire field of flowers, the bees appeared to be foraging randomly. This indicates that bees may visit the two flowers on a more or less "as they come to them" basis when pollen gathering is energetically feasible and the needs for nectar are low.

Bumblebees will take nectar from honeypots before starting pollen foraging (Heinrich, 1973). This indicates that access to a nectar source, either stored in the hive or available in the field, is important for bees before they commence pollen foraging. Free (1955) found that the proportion of pollen gatherers to nectar foragers increased during the day. This further indicates that nectar is necessary either at the lower temperatures of morning or early in the day before pollen foraging is begun.

Humidity may also play a role in pollen collection by bumblebees. Brian (1954) indicated that bumblebees collected nectar faster on damp days than on days that were dry. Humidity was significantly correlated with visitation ratio in 1980 but not in 1978, and may become important to bumblebees when temperature is not a critical factor.

Evidence is provided here to suggest that *L. benthamii* and *D. parryi* are involved in a mutualistic mimicry system for the joint attraction of pollinators. It appears that the *L. benthamii* and *D. parryi* mutualistic floral mimicry system is complex and dynamic. Delimiting which plant species is the model and which is the mimic in a mutualistic mimicry complex may be fruitless. It has been predicted that the model of a floral mimicry complex may: (1) be the more common species; (2) have a pollinator that is constant to this species (Little, 1980); (3) appear (bloom) first (Bobisud, 1978); (4) contain nectar (Brown and Ko-dric-Brown, 1979). Conversely, the mimic should: (1) be rare (Brown and Kodric-Brown, 1979); (2) be pollinated by a pollinator that is constant to the model (Little, 1980); (3) follow the model in blooming time; (4) be nectarless (Brown and Kodric-Brown, 1979). Neither *D. parryi* nor *L. benthamii* perfectly "fills the bill" for either role as model or mimic. *L. benthamii* is the more common of the two species, blooms before *D. parryi,* but does not contain nectar. *D. parryi* is preferred and has reasonably constant pollinators under some environmental conditions, whereas *L. benthamii* is preferred under other environmental conditions. On a warm, clear day, *D. parryi* may be preferred by solitary bees, and *L. benthamii* may be the flower of choice for *Bombus* spp. A visit to *L. benthamii* for a nectar-gathering bumblebee may be disadvantageous. However, a visit to *D. parryi* for a pollen-collecting bee may be a "bonus" energetically. The interaction between *L. benthamii* and *D. parryi* and their pollinators is complicated by the fact that at times bees appear to be able to distinguish between the two flowers and at other times do not show discrimination.

A mutualistic mimicry complex may come into being purely fortuitously, and be maintained by pollinators and herbivores. *L. benthamii* occurs in disturbed areas and grassy hills along Nacimiento-Fergusson Road in Nacimiento Canyon. *D. parryi* is patchily common thoughout Nacimiento Canyon along stream beds and in shady areas. The delphinium may also be established more readily in disturbed areas (Epling and Lewis, 1952). It seems possible that *L. benthamii* dispersed to the area after the area was disturbed by man and cattle. It may have been able to gain in numbers rapidly in the areas where it was associated with *D. parryi*. Since the two flowers are morphologically similar, they might both gain more visitations occurring together than either one alone (Bobisud and Neuhaus, 1975). However, since *L. benthamii* may be a good pollen source for bumblebees, its association with a nectar source may be even more advantageous for the plants involved. This nectar–pollen association could conceivably support a larger population of *Bombus* spp. than could *D. parryi* or *L. benthamii* alone. The two plants together may have been able to expand their numbers owing to their ability to mutually support the needs of beginning *Bombus* hives.

More work needs to be done in the area of mutualistic floral mimicry. Much about the interactions between *L. benthamii* and *D. parryi* and their pollinators needs to be explained. An examination of the pollinators, floral features, and phenology of these species in allopatry may provide knowledge on what kind of convergence or divergence has occurred as a result of their sympatry and pollinator-sharing association. The subtleties of possible competition for pollinators between similar sympatric species as it is related to the hive energetics of social bees need to be studied further. Also, a study of the difference in seed set in the mixed and pure plots compared to the seed set resulting from autogamy in these populations may provide an insight into the spatial arrangement of similar flowers that best enhances the mutualistic aspects of the association.

The ability of similar sympatric species to mutualistically deter herbivores also warrants investigation. It seems very probable that *D. parryi* and *L. benthamii* are involved in a genuine Mullerian mimicry complex for the repulsion of large herbivores, particularly cattle, and may even mutualistically dilute the deleterious effects of herbivorous beetles on floral structures in mixed stands.

ACKNOWLEDGMENTS

We especially thank Phil Adams, Joel Weintraub, Herbert and Irene Baker, Roy Snelling, Mike Bell, Steve Buchmann, Trudy Ericson, Leo Song, Jack Mixner, Tim Derby, and Ed Hall for their assistance during various phases of this study.

This study was partially funded by grants from the Departmental Associations Council and the Department of Biological Science, California State University, Fullerton. It was submitted in partial fulfillment of the requirements for the degree of Master of Arts in Biological Science for E. A. Powell at California State University, Fullerton.

LITERATURE CITED

Adams, R. J. and M. V. Smith. 1977. Scanning electron and light microscope studies of pollen of some Leguminosae. *J. Apic. Res.* 16(1):99–106.

Benn, M. H. 1966. Delphinium alkaloids. *Can. J. Chem.* 44:1–8.

Bobisud, L. E. 1978. Optimal time of appearance of mimics. *Am. Nat.* 112:962–965.

Bobisud, L. E. and R. J. Neuhaus. 1975. Pollinator constancy and the survival of rare species. *Oecologia* 21:263–272.

Boyden, T. C. 1980. Floral mimicry by *Epidendrum ibaguense* (Orchidaceae) in Panama. *Evolution* 34:135–136.

Brian, A. D. 1954. The foraging of bumblebees. Part I. Foraging behavior. *Bee World* 35:61–67.

Brown, J. and A. Kodric-Brown. 1979. Convergence, competition, and mimicry in a community of hummingbird-pollinated flowers. *Ecology* 60:1022–1035.

Colin, L. J. and C. E. Jones. 1980. Pollen energetics and pollination modes. *Am. J. Bot.* 67:210–215.

Dafni, A. and Y. Ivri. 1981a. Floral mimicry between *Orchis israelitica* Baumann and Safri (Orchidaceae) and *Bellevalia flexuosa* Boiss. (Liliaceae). *Oecologia* 49:229–232.

———. 1981b. The flower biology of *Cephalanthera longifolia* (Orchidaceae): pollen imitation and facultative floral mimicry. *Pl. Syst. Evol.* 137:229–240.

Daley, H. V., J. T. Doyen, and P. R. Ehrlich. 1978. *Introduction to Insect Biology and Diversity.* McGraw-Hill Book Co., New York, 564 pp.

Epling C. and H. Lewis. 1952. Increase of the adaptive range of the genus *Delphinium. Evolution* 6:253–267.

Faegri, K. and L. van der Pijl. 1979. *The Principles of Pollination Ecology,* 3rd ed. Pergamon Press, New York, 244 pp.

Free, J. B. 1955. The collection of food by bumblebees. *Insectes Sociaux* 2:303–311.

———. 1970. The flower constancy of bumblebees. *J. Anim. Ecol.* 39:395–402.

Hartling, L. K. and R. C. Plowright. 1979. Foraging by bumblebees on patches of artificial flowers: a laboratory study. *Can. J. Zool.* 57:1866–1870.

Heinrich, B. 1972. Energetics of temperature regulation and foraging in a bumblebee, *Bombus terricola* Kirby. *J. Comp. Physiol.* 77:49–64.

———. 1973. The role of energetics in bumblebee-flower interrelationships. *In* L. E. Gilbert and P. H. Raven (eds.), *Coevolution of Animals and Plants.* University of Texas Press, Austin, 246 pp.

———. 1975. Bee flowers: a hypothesis on flower variety and blooming times. *Evolution* 29:325–334.

———. 1976a. Resource partitioning among some eusocial insects: bumblebees. *Ecology* 57:874–889.

———. 1976b. The foraging specializations of individual bumblebees. *Ecol. Mon.* 46:105–128.

———. 1979. *Bumblebee Economics.* Harvard University Press, Cambridge, Mass., 245 pp.

Horovitz, A. and J. Harding. 1972. Genetics of *Lupinus* V. Intraspecific variability for reproductive traits in *Lupinus nanus. Bot. Gaz.* 133:155–165.

Jones, C. E. and S. L. Buchmann. 1974. Ultraviolet floral patterns as functional orientation cues in Hymenopterous pollination systems. *Anim. Behav.* 22:481–485.

Kevan, P. R. 1978. Floral coloration, its colorimetric analysis and significance in anthecology. *In* A. J. Richards (ed.), *The Pollination of Flowers by Insects.* Linnean Soc. Symp. Ser. No. 6. Academic Press, New York.

Kingsbury, J. M. 1964. *Poisonous Plants of the United States and Canada.* Prentice-Hall, Englewood Cliffs, N.J., 626 pp.

Lee, S. 1978. A factor analysis study of the functional significance of angiosperm pollen. *Syst. Bot.* 3:1–19.

Levin, D. A. and W. W. Anderson. 1970. Competition for pollinators between simultaneously flowering species. *Am. Nat.* 104:455–467.

Levin, D. A. and H. W. Kerster. 1973. Assortative pollination for stature in *Lythrum salicaria. Evolution* 27:144–152.

Little, R. J. 1980. Floral mimicry between two desert annuals, *Mohavea confertiflora* (Scrophulariaceae) and *Mentzelia involucrata* (Loasaceae). Ph.D. dissertation, Claremont Grad. School, California.

Macior, L. W. 1971. Co-evolution of plants and animals—systematic insights from plant–insect interactions. *Taxon* 20(1):17–28.

———. 1974. Behavioral aspects of coadaptations between flowers and their insect pollinators. *Ann. Missouri Bot. Gard.* 61:760–769.

Manning, A. 1956. Some aspects of the foraging behavior of bumblebees. *Behavior* 9:164–201.

Marsh, C. D. and A. B. Clawson. 1916a. Larkspur poisoning of livestock. United States Department of Agriculture Bull. No. 365.

———. 1916b. Lupines as poisonous plants. United States Department of Agriculture Bull. No. 405.

Mosquin, T. 1971. Competition for pollinators as a stimulus for the evolution of flowering time. *Oikos* 22:398–402.

Munz, P. A. 1968. *A California Flora.* University of California Press, Berkeley, 1681 pp.

Oster, G. and B. Heinrich. 1976. Why do bumblebees major? A mathematical model. *Ecol. Mon.* 46:129–133.

Pennell, F. W. 1948. The taxonomic significance of an understanding of floral evolution. *Brittonia* 6:301–308.

Pijl, L. van der and C. H. Dodson. 1966. *Orchid Flowers: Their Pollination and Evolution.* University of Miami Press, Coral Gables, Fla., 214 pp.

Pleasants, J. M. 1980. Competition for bumblebee pollinators in Rocky Mountain plant communities. *Ecology* 61:1446–1459.

Proctor, M. and P. Yeo. 1972. *The Pollination of Flowers.* Taplinger Publishing Co., New York, 418 pp.

Pyke, G. H. 1978. Optimal foraging in bumblebees and coevolution with their plants. *Oecologia* 36:281–293.

Schelpe, E. A. 1966. *An Introduction to the South African Orchids.* MacDonald Press, London.

Schemske, D. W. 1978. Flowering ecology of some woodland herbs. *Ecology* 59:351–366.

———. 1981. Floral convergence and pollinator sharing in two bee-pollinated tropical herbs. *Ecology* 62:946–954.

Shupe, J. L., L. D. Balls, and L. F. James. 1970. Changes in blood serum transaminase associated with lupine and larkspur poisoning in cattle. *Cornell Vet.* 58:129–135.

Straw, R. M. 1972. A Markov model for pollinator constancy and competition. *Am. Nat.* 106:597–620.

Thomson, J. D. 1981. Spatial and temporal components of resource assessment by flower-feeding insects. *J. Anim. Ecol.* 50:49–59.

Vane-Wright, R. I. 1976. A unified classification of mimetic resemblances. *Biol. J. Linnean Soc.* 8:25–56.

———. 1980. On the definition of mimicry. *Biol. J. Linnean Soc.* 13:1–6.

Waddington, K. D. 1979. Divergence in inflorescence height: an evolutionary response to pollinator fidelity. *Oecologia* 40:43–50.

Wainwright, M. 1978. The floral biology and pollination ecology of two desert lupines. *Bull. Torrey Bot. Club* 105:24–38.

Waser, N. M. 1978. Interspecific pollen transfer and competition between co-occurring plant species. *Oecologia* 36:223–236.

Wickler, W. 1968. *Mimicry in Plants and Animals.* World University Library. McGraw-Hill Book Co., New York, 255 pp.

Wiens, D. 1978. Mimicry in plants. *In* Max K. Hecht (ed.), *Evolutionary Biology,* Vol. 11. Plenum Press, New York.

Willaman, J. J. and Hui-lin Li. 1970. Alkaloid-bearing plants and their contained alkaloids. *J. Nat. Prod., Lloydia Suppl.,* Vol. 33, No. 3A. 286 pp.

Yeo, P. F. 1968. The evolutionary significance of the speciation of *Euphrasia* in Europe. *Evolution* 22:736–747.

16
POLLEN COMPETITION IN A NATURAL POPULATION

David L. Mulcahy,[1] Peter S. Curtis,[2] and Allison A. Snow[1]

[1]*Botany Department*
University of Massachusetts
Amherst, Massachusetts
and
[2]*Department of Ecology and Evolution*
S.U.N.Y./Stony Brook
Stony Brook, New York

ABSTRACT

This study indicates that, in a natural population of *Geranium maculatum,* pollen tube competition occurs between pollen grains that reach the stigma simultaneously, as well as between members of separate pollinations. Significant interactions exist between the rate of pollen arrival on the stigma and the rate and variance in pollen tube growth down the style. The study suggests that it is possible to quantify pollen tube competition under field conditions.

KEY WORDS: Fertilization, gametophytic selection, *Geranium maculatum,* pollen tube competition, pollen tube growth rate, stigma loading, seed set, style length.

CONTENTS

INTRODUCTION

One of the common assumptions made by pollination biologists is that the number of pollen grains arriving on stigmatic surfaces generally exceeds the number of ovules available for fertilization. This implies that there may be intense competition among microgametophytes, and fertilizations are thus accomplished most often by gametes from the most rapidly growing pollen tubes. This has led several investigators to ask what, if anything, would be the effect of pollen tube competition upon the sporophytic portion of the life cycle. This question can be approached at two different levels: one dealing with pollen competition as an interaction between sporophytes (as pollen sources) and another that involves selection among gametophytes (i.e., pollen grains) from a single pollen source. The first topic involves mixtures of pollen from two (or more) pollen sources, and, in this context, Stephens (1956) demonstrated that in an open-pollinated population, plants that produce above-average quantities of pollen do indeed leave above-average numbers of progeny. Somewhat less obvious (although no less important) was the finding that pollen tubes from heterotic F_1 plants of *Zea mays* grow more rapidly than do the pollen tubes from the inbred parents of these heterotic hybrids. This implies that the heterotic qualities of the sporophytic pollen source are expressed also in pollen tube growth rates (Murakami et al., 1972). Other studies have indicated that, generally speaking, those components of pollen mixtures that grow rapidly give rise to plants that themselves grow rapidly (see Mulcahy, 1971, 1974a; Ottaviano et al., 1980). A notable exception to this statement, however, sometimes appears when the pollen mixture includes a self pollination; that is, when one component of the pollen mixture is obtained from the pistillate parent. In these particular cases, observed in *Zea mays* (Jones, 1928) and *Lycopersicum esculentum* (Hornby and Shin-Chai, 1975), self pollen tubes out compete non-self pollen tubes, and the resultant seedlings exhibit classical symptoms of inbreeding depression. An explanation for these exceptional cases comes from the fact that they both involve highly inbred lines. When similar studies employ F_1 plants as staminate and pistillate parents, self pollen exhibits no advantage in pollen tube competition (Pfahler, 1967). This fact led Pfahler (1967) and Mulcahy (1974b) to suggest, and Johnson and Mulcahy (1978) to demonstrate, the following explanation. During the process of inbreeding, there is an intense selection for pollen tubes that penetrate the self style very rapidly. These pollen tubes thus represent specialists in one stylar environment, and when the pollen tube growth rate of this pollen type is compared with that of an unselected pollen type, the specialist predictably wins.

From the above studies, it is quite clear that if pollen grains from two (or more) pollen sources reach a stigma at the same time, pollen fom one source may greatly outpace the others in pollen tube growth rate and thus fertilize a disproportionate fraction of the eggs. These differential pollen tube growth rates have important implications in population biology, as has been pointed out by Willson (1979).

There is also another level of pollen competition, that between pollen grains from a single pollen source. A series of studies, initiated by Ter-Avanesian (1949, 1969, 1978a,b) and continued by others (Matthews, in Lewis, 1954; Mulcahy and Mulcahy, 1975; Mulcahy et al., 1975, 1978; Ottaviano et al., 1980), indicates that there are strong and positive correlations between the growth rates of pollen tubes and the growth rates of the resultant sporophytes. This conclusion has been reached in studies involving *Triticum aestivum, Gossypium hirsutum, Petunia hybrida,* and *Dianthus chinensis.* An ongoing study by Mary McKenna (S.U.N.Y., Stony Brook) has recently provided conclusive evidence that, in *Dianthus chinensis,* progeny from rapidly growing pollen tubes are competitively superior to other progeny.

The apparent benefits of pollen tube selection cannot help but call attention to the fact that the angiosperm style and mode of pollination should greatly accentuate pollen tube selection. This has led to the suggestion that intensified pollen tube selection may have been a significant factor in the rise of the angiosperms (Mulcahy, 1979). Insect pollination causes relatively large numbers of pollen grains to be deposited on the stigma simultaneously, and both the quantity and the timing of pollination will influence the intensity of pollen tube competition. Furthermore, the stylar tissue that separates the stigma from the ovules serves as an ideal structure for facilitating the separation of fast and slow pollen tubes. Thus, the angiosperms may be particularly well suited to benefit from a correlation between pollen and sporophyte qualities.

At this point, however, it is important to recall that although large quantities of pollen are produced by plants, the assumption that pollen tube competition is intense in natural populations rarely has been tested. Indeed, Bierzychudek (1981) recently reported that for several species (e.g., *Arisaema triphyllum* and *Erythronium albidum*) seed production in natural populations may be limited by inadequate pollinator activity, although apparently this is not so for *Geranium maculatum* or *Delphinium nelsonii* (Waser, 1978; Willson et al., 1979).

In the present chapter we investigate the possibility of pollen tube competition in a natural population and the factors that influence this phenomenon. Certainly, the number of pollen grains reaching the stigma is of major interest, but, by itself, gives surprisingly little insight into the possible intensity of pollen tube competition. The ovary of *Geranium maculatum* contains 10 ovules, which result in 5 seeds; because of this it may seem that more than 10 viable pollen grains in the stigmatic surface would indicate that at least some pollen tube competition had taken place. This assumption, however, could be wrong. If the first pollination deposited 10 pollen grains on the stigma, and their 10 pollen tubes reached all available ovules before other pollen grains were deposited, then any subsequent pollinations would

be superfluous. In such a case the ultimate presence of 100 pollen grains on a pistil that contains only 10 ovules would give a false indication of intense pollen tube competition, when, in fact, there had been none.

In the present chapter, therefore, we have attempted to estimate the rates of both pollination and fertilization in order to determine whether pollen tubes compete for ovules.

METHODS AND MATERIALS

Geranium maculatum was chosen as a species appropriate for the present study because it produces large flowers (up to 3 cm in diameter), bears a single pistil, and forms reasonably large populations. Furthermore, its pollen grains are large enough to be observed on the stigmatic surface with a good-quality 10× hand lens. Individual flowers open for 4–6 days and are protandrous. For all but the last day in the lifetime of a flower, the 5 stigmatic lobes of the pistil are pressed together, effectively preventing pollination. On the final day, the lobes of the stigma separate, thus initiating the

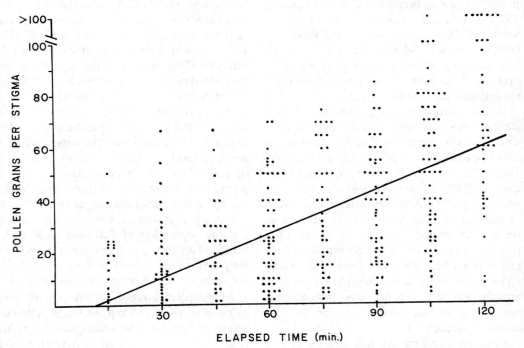

Figure 16-1. The numbers of pollen grains observed on stigmatic surfaces once pollinators initiate their visits. $Y = -5.05 + 0.55\,X$; $N = 396$.

period of stigma receptivity (Willson et al., 1979).

It was at the day of the stigma's receptivity that our field observations of flowers began. Individual plants were enclosed by cheesecloth-covered frames in order to exclude pollinators until we were ready to begin observations. Once the frames were removed, receptive stigmas were examined to ensure that no pollen was present. Then, at 15-minute intervals, the numbers of pollen grains on stigmatic surfaces were determined. These results are presented in Fig. 16-1.

In order to determine how many pollen grains must reach the stigma for full seed set to occur, a small number of hand pollinations were made using potted plants. Specific numbers of pollen grains from other individuals were placed on receptive stigmas. These flowers were observed until fruits had enlarged enough to indicate any successful fertilization. The results of this experiment are shown in Fig. 16-2.

A third part of the study was designed to determine how quickly pollen grains germinate and penetrate stylar tissues. To accomplish this, flowering branches of *Geranium maculatum* were cut and put in water, and receptive flowers were pollinated by hand with a small but uncounted number of pollen grains. The pistils sampled from these flowers were then fixed in 70% ethanol at 30-minute intervals.

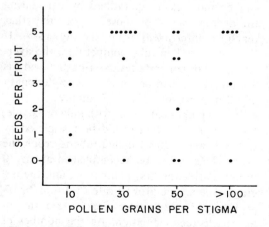

Figure 16-2. Effect of pollen load on seed set ($N = 26$). *Geranium maculatum* has ten ovules and sets a maximum of five seeds per fruit.

Three to six pistils were examined from each time period. At a later time the pistils were softened and cleared by placing them for 20 minutes in an 8 N solution of sodium hydroxide at 60°C. After being rinsed in several changes of distilled water and a final wash of 0.1 M Tris-glycine buffer, pH 8.4, the pistils were transferred to decolorized aniline blue, in which they were stored until observed under fluorescent microscopy. By this method it is possible to observe the pollen tubes of *Geranium maculatum* throughout the length of the style and also in the ovary.

The proportion of those pollen tubes observed in the stigma that had reached the ovary in each 30-minute time interval are presented in Fig. 16-3. Ideally, we should have compared the number of pollen grains on the stigma with the number of pollen tubes in the ovary, but with the method by which we cleared the style, it was quite difficult to determine how many pollen grains were originally on the stigma. Nevertheless, the data should serve to visualize some significant aspects of pollen tube competition.

RESULTS AND DISCUSSION

The data in Fig. 16-1 indicate that relatively large numbers of pollen grains are deposited on receptive stigmas in a short period of time. This regression predicts that 11.4 pollen grains will arrive within the first 30 minutes of observation, and another 16.4 arrive at each subsequent 30-minute period. These data can be combined with those in Fig. 16-2, which show that no additional seed set is achieved when more than 30 pollen grains are placed on the stigma of a flower. According to the regression line in Fig. 16-1, 30 pollen grains should have been deposited in approximately 63.7 minutes.

In themselves these data tell us very little about gametophytic competition. However, by combining the values of the first two figures with the data presented in Fig. 16-3, the possibility of gametophytic selection becomes clear. The data presented in Fig. 16-3 show that some pollen tubes traverse the distance between the upper style and the ovary in 30 minutes; others require more than 120 min-

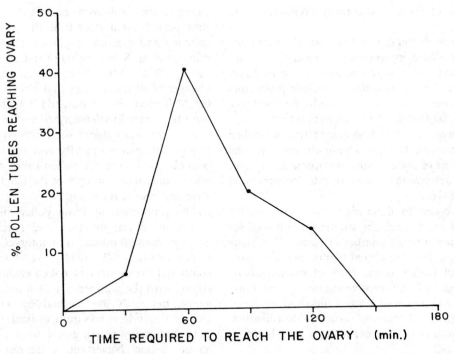

Figure 16-3. Frequency distribution of the times required for pollen tubes to grow from the base of the stigma to the ovary.

utes, with the greatest number requiring 60 minutes ($\bar{x}$ = 75.3 minutes). As will be further explained below, this variation in pollen tube growth rate allows the possibility of pollen tube selection. The extent to which this possibility is realized may be estimated by considering the time of arrival of pollen on the stigma.

Pollen obviously does not arrive continuously, but rather in a series of discontinuous pulses, despite the smooth regression shown in Fig. 16-1. These pulses have not yet been quantified, however, and for want of a better estimate, we have assumed that the pollinators arrived at 30-minute intervals, although the interval is more likely shorter. (We have also assumed, probably incorrectly, that pollen tube growth rates are unaffected by the previous passage of pollen tubes through the stylar tissues.) Given these assumptions, the rate of arrival of pollen tubes to the ovary may be represented as in Fig. 16-4. The utility of this figure is that it indicates that variance in pollen tube growth rate allows the fastest pollen tubes

from the second and still later pollinations to surpass the slowest pollen tubes of an earlier pollination. Assuming that pollinators are arriving at 30-minute intervals, it is possible to calculate the extent to which pollen competition selects for pollen tubes that grow rapidly (Fig. 16-5). We have found that this selection can be most easily quantified by calculating the average speed of those pollen tubes that reach an unfertilized ovule (see Fig. 16-5). If there is no pollen tube competition, the average speed of pollen tubes entering the ovary is 3.2 mm/hr, based on an average style length of 4.0 mm, traversed in 75.3 minutes.

In a hypothetical flower with only one ovule, however, fertilization would be completed by the first (and fastest) pollen tube to reach the ovary. This would be accomplished about 30 minutes after the first pollination, and the average pollen tube growth rate (8.0 mm/hr, $N = 1$) would be 2.51 times greater than that of an unselected population. As the number of pollen tubes required for full seed set increases, the average growth rate of the pollen tube pop-

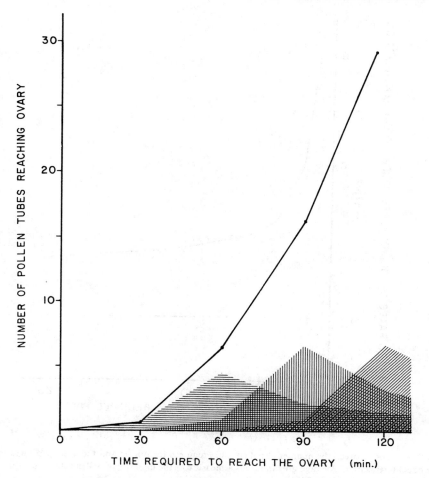

Figure 16-4. The arrival of pollen tubes in the ovary of *Geranium maculatum*. The solid line represents the total number of pollen tubes reaching the ovary, and the shaded areas represent proportional contributions of the first, second, and third pollinations.

ulation that effects fertilization decreases (Fig. 16-5). With multiple pollinations, the most rapid pollen tubes from later pollinations may pass the slower tubes of earlier pollinations (see Fig. 16-4), thus delaying the approach to the average speed of an unselected population of pollen tubes.

Pollen tube competition thus occurs not only between pollen grains that reach the stigma simultaneously, but also between members of separate pollinations. The intensity of pollen tube selection depends on the length of the style, the number of ovules available for fertilization, rate of pollination, and other factors.

If, in *Geranium maculatum*, the first 20 pollen grains to reach the ovary are sufficient to

produce full seed set, as we believe is the case, then, according to Fig. 16-5, these will exhibit an average speed of 4.3 mm/hr, 34% greater than that of an unselected population of pollen tubes. If only 10 pollen tubes are required, as may be the case, their average speed will be 41% greater than that of an unselected population.

The significance of pollen tube competition stems from the hypothesis that gametophytic and sporophytic qualities show a significant and positive correlation. Gametes from pollen tubes that grow rapidly give rise to sporophytes that grow more rapidly than do other plants (Ottaviano et al., 1980). Although this correlation may be both interesting and useful,

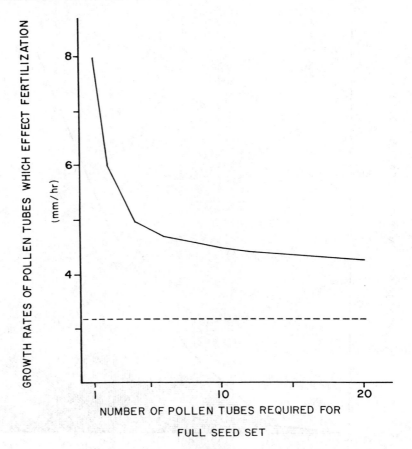

Figure 16-5. Relation between the number of pollen tubes required for full set and the average growth rate of pollen tubes that effect fertilization. Average growth rate is a measure of the intensity of pollen tube competition. At growth rates lower than 3.2 mm/hr (dashed line), there is no competition.

its ultimate significance in natural populations remains untested. The present study, albeit a preliminary one, does provide some measure of gametophytic selection in one natural population. Even when several pollinations are necessary for full seed set, competition can occur if the range in pollen tube growth rates is sufficient to allow pollen tubes from later pollinations to surpass some pollen tubes from earlier pollinations. Further quantifications of these phenomena are needed, but the data so far obtained suggest that gametophytic selection may be an adaptively significant phenomenon in some natural populations.

Future studies will clearly require extensive field observations of times between pollinator visits and the numbers of pollen grains left per visit. Once such data are available, however, it should be possible to learn something new about the selective significance of floral characteristics such as style length and ovule and seed numbers. A more distant, but not unattainable, goal could be a conceptual model of the functional basis for the angiosperm flower.

ACKNOWLEDGMENTS

This work was supported, in part, by NSP grant # DEB 7903685, to DLM, and by a grant from the O. C. Marsh fund, National Academy of Sciences, to PSC. The authors are grateful for the above financial support and also for the generous and enthusiastic assistance of Dr. Claire M. Johnson and Ms. Joyce Miller.

LITERATURE CITED

Bierzychudek, P. 1981. Pollinator limitation of plant reproductive effort. *Am. Nat.* 117:838–840.
Hornby, C. A. and L. Shin-Chai. 1975. Some effects of

multiparental pollination in tomato plants. *Can. J. Pl. Sci.* 55:127–132.

Johnson, C. M. and D. L. Mulcahy. 1978. Male gametophyte in maize. II. Pollen vigor in inbred plants. *Theor. Appl. Genet.* 51:211–215.

Jones, D. E. 1928. *Selective Fertilization.* University of Chicago Press, Chicago.

Lewis, D. 1954. Annual report of the department of genetics. *Annu. Rep. John Innes Hort. Inst.* 45:12–17.

Mulcahy, D. L. 1971. A correlation between gametophytic and sporophytic characteristics in *Zea mays* L. *Science* 171:1155–1156.

———. 1974a. Correlation between speed of pollen tube growth and seedling height in *Zea mays* L. *Nature* 249:491–493.

———. 1974b. Adaptive significance of gamete competition. *In* H. F. Linskens (ed.), *Fertilization in Higher Plants.* Elsevier Press, Amsterdam.

———. 1979. The rise of angiosperms: a genecological factor. *Science* 206:20–23.

Mulcahy, D. L. and G. B. Mulcahy. 1975. The influence of gametophytic competition on sporophytic quality of *Dianthus chinensis. Theor. Appl. Genet.* 46:277–280.

Mulcahy, D. L., G. B. Mulcahy, and E. Ottaviano. 1975. Sporophytic expression of gametophytic competition in *Petunia hybrida. In* D. L. Mulcahy (ed.), *Gamete Competition in Plants and Animals.* Elsevier Press, Amsterdam.

———. 1978. Further evidence that gametophytic selection modifies the genetic quality of the sporophyte. *Soc. Bot. Fr. Actualites botaniques* 1–2:57–60.

Murakami, I. I., M. Yamada, and K. Takayanagi. 1972.

Selective fertilization in maize, *Zea mays.* I. Advantage of pollen from F$_1$ plants. *Jap. J. Breeding* 22:203–208.

Ottaviano, E., M. Sari-Gorla, and D. L. Mulcahy. 1980. Pollen tube growth rates in *Zea mays*: implications for genetic improvement of crops. *Science* 210:437–438.

Pfahler, P. L. 1967. Fertilization ability of maize pollen grains. II. Pollen genotype, female sporophyte, and pollen storage. *Genetics* 57:513–521.

Stephens, S. G. 1956. The composition of an open pollinated segregating cotton population. *Am. Nat.* 90:25–39.

Ter-Avanesian, D. V. 1949. The influence of the number of pollen grains used in pollination. *Bull. Appl. Genet. Pl. Breeding, Leningrad* 28:119–133.

———. 1969. The significance of the quantity of pollen in the hybridization of wheat. *Genetika* 5(10):103–108.

———. 1978a. The effect of varying the number of pollen grains used in fertilization. *Theor. Appl. Genet.* 52:77–79

———. 1978b. Significance of pollen amount for fertilization. *Bull. Torrey Bot. Club* 105:2–8.

Waser, N. M. 1978. Competition for hummingbird pollination and sequential flowering in two Colorado wildflowers. *Ecology* 59:934–944.

Willson, M. 1979. Sexual selection in plants. *Am. Nat.* 113:777–790.

Willson, M. F., L. J. Miller, and B. J. Rathcke. 1979. Floral display in *Phlox* and *Geranium:* adaptive aspects. *Evolution* 33:52–63.

Section V
GENE FLOW AND POLLINATION

17
OPTIMAL AND ACTUAL OUTCROSSING IN PLANTS, AND THE NATURE OF PLANT–POLLINATOR INTERACTION*

Nickolas M. Waser
Mary V. Price

Department of Biology
University of California
Riverside, California
and
Rocky Mountain Biological Laboratory
Crested Butte
Colorado

ABSTRACT

One determinant of the success of mating between two plants, in terms of fecundity and off-spring viability, should be the genetic similarity of the mates. For certain plant species, the optimal genetic simiarity will correspond to a short outcrossing distance, which we call the *optimal outcrossing distance*. Experimental hand-pollinations of two montane perennials, *Delphinium nelsonii* and *Ipomopsis aggregata*, indicate that optimal outcrossing distances for these species are indeed short (between 1 and 100 meters), at least in terms of fecundity of a cross. We compare this finding with information available for other plants, and conclude that gene flow and population structure will be related to the optimal outcrossing distance for any species. In addition, several estimates suggest that actual outcrossing distances provided by humming-bird and insect pollinators of *D. nelsonii* and *I. aggregata* are even shorter than optimal distances. We discuss several possible explanations for this discrepancy, and stress that it is not unexpected, given that plant–pollinator interactions basically involve conflict of interest rather than cooperation.

KEY WORDS: Bumblebees, Colorado, *Delphinium nelsonii*, Fritz Müller's Law, genetic neigh-borhood, hand-pollination, hawkmoths, hummingbirds, inbreeding depression, *Ipomopsis aggregata*, microdifferentiation, *Mimulus guttatus*, optimal mating, outbreeding depression, plant–pollinator conflict, pollen dispersal, seed dispersal, seed set, seedling survivorship.

CONTENTS

*This chapter is dedicated to Fritz Müller (1821–1897).

INTRODUCTION

Not all possible matings are of equal value to individuals. Those that confer highest fitness in terms of the number and quality of resulting offspring can be called *optimal matings*. In considering what matings might be optimal, animal ecologists have concentrated on two factors: parental investment of resources that enhance fitness of a cross, and the quality of genes contributed by each parent to offspring (*e.g.*, Verner and Willson, 1966; Orians, 1969; Downhower and Armitage, 1971; Halliday, 1978; Thornhill, 1980).

It is hard to see how parental resource investment can be a factor in defining the optimal mating for plants, at least from the point of view of female sexual function, since pollen contributes no resource to developing seeds. On the other hand, Janzen (1977), Charnov (1979), and Willson (1979) have proposed that genetic quality of mates can influence the fitness of a cross. For example, Janzen (1977) predicts that female plants (or "female" tissue in monoecious or hermaphroditic plants) should discriminate in favor of pollen carrying genes conferring high fitness on progeny, such as genes for especially attractive flowers. For such discrimination to evolve, genes beneficial to the sporophyte must be associated with those expressed in the gametophyte during pollen germination and tube growth. In fact, Mulcahy (1971, 1974, this book) has shown for several species that such an association exists. It is not yet clear, however, how commonly discrimination of mates on the basis of their absolute genetic quality actually occurs in plants.

It seems to us that a third factor is more likely than resource investment or genetic quality to influence the fitness of a cross and thereby define the optimal mating for plants. This factor is the *genetic similarity* between mates. If mates are too similar, inbreeding depression is likely to occur in the form of reduced fecundity and offspring viability (Darwin, 1876; Grant, 1975). Conversely, if mates are too dissimilar, outbreeding depression may occur, at least at the level of crosses between closely related species or varieties (Müller, 1883, pp. 145–146, Moll et al., 1965; Grant, 1971) or between isolated populations of the same species (Kruckeberg, 1957; Martin, 1963; Hughes and Vickery, 1974; Hogenboom, 1975; Vickery, 1978). The fitness depressions following excessive inbreeding or outbreeding clearly define some *intermediate degree of outbreeding* as optimal. This conclusion is not novel; we propose that it be called "Fritz Müller's Law" in honor of the German naturalist who deduced it a century ago from his pollination experiments with the genus *Abutilon* (see Müller, 1883, p. 145).

We recently have suggested (Price and Waser, 1979) that Fritz Müller's Law can apply within populations. For reasons discussed later, we predict that outbreeding depression will begin to be expressed in matings between members of the *same* population for species experiencing localized gene flow, rather than being expressed only in matings between isolated populations or varieties. For such species, then, an intermediate optimal degree of outbreeding should correspond to a small physical distance between mates.

In this chapter we review and augment earlier evidence from hand-pollination experiments (Price and Waser, 1979) that short *optimal outcrossing distances* indeed exist for two herbaceous perennial wildflowers. We provide evidence for both species that pollen from donor plants growing a relatively short distance from recipients maximizes seed set of a cross, and we discuss some preliminary measures of the effects of outcrossing distances on seedling viability. We then compare our results to those being obtained by workers using plant species with apparently different population structures.

In addition, we compare our estimates of optimal outcrossing distances to estimates of actual pollen dispersal provided by animal pollinators. We conclude that *actual outcrossing distances* usually are shorter than optimal distances for the two species we have studied. This discrepancy leads us to consider how the fundamental nature of plant–pollinator interaction might involve conflict, rather than the attainment of simultaneously optimal conditions for both mutualistic partners.

FIELD EXPERIMENTS AND OBSERVATIONS

Our studies involve two plant species common to subalpine meadows near the Rocky Mountain Biological Laboratory (RMBL; 2900 meters elevation) in west-central Colorado. *Delphinium nelsonii* (Ranunculaceae) flowers after snowmelt in early summer (Waser, 1978; Waser and Real, 1979), reaching densities of about 10 plants/m² in some years. Near the RMBL it is pollinated primarily by broadtailed *(Selasphorus platycercus)* and rufous *(S. rufus)* hummingbirds and by queen bumblebees (*Bombus appositus, B. flavifrons, B. californicus,* and *B. nevadensis*) (Waser, 1978; Waser and Price, 1981). *Ipomopsis aggregata* (Polemoniaceae) flowers after *D. nelsonii* in the same meadows (Waser and Real, 1979), sometimes reaching densities of about 5 plants/m². Near the RMBL it is pollinated primarily by the two hummingbird species that visit *D. nelsonii*. In occasional years (e.g., 1973, 1978) hawkmoths *(Hyles lineata)* are common around the RMBL, visit *I. aggregata* intensively, and also carry its pollen to some extent (Waser, 1978; unpub. data). Neither plant species reproduces vegetatively. Flowers of both species are protandrous and set no seed when pollinators are excluded artificially (Waser, 1978).

To investigate the effect of outcrossing distance on plant fecundity, we hand-pollinated 1190 flowers on 377 *D. nelsonii* plants over four years (1976–1979) and 621 flowers on 60 *I. aggregata* plants over two years (1979–1980), with pollen brought from donor plants growing various distances from the recipients. We carried out hand-pollinations as follows (see also Price and Waser, 1979). We bagged all plants while in bud to exclude natural pollinators and assigned each at random to a single outcrossing treatment. As successive flowers on a plant became receptive, we pollinated each once with pollen from a donor plant growing the appropriate distance away. We used round wooden toothpicks to gather and transfer pollen. We standardized techniques to control for amounts of pollen collected, mechanical handling received, and the total time elapsed between collection and deposition. For each replicate experiment we chose a subset of the following outcrossing treatments: 0 (self-pollination), 1, 3, 10, 30, 100, and 1000 meters. We harvested fruits from hand-pollinated plants when seeds were mature or nearly so and could easily be counted. In all replicates, we left several flowers unpollinated to serve as controls for possible failure of bags to exclude pollinators. If control flowers on any plant set more seed than the average for pollinated flowers from the replicate as a whole, the plant was excluded from analysis.

For replicates with *D. nelsonii*, seeds were collected just at maturity so that they could be planted. Our procedure was as follows. In 1976 we lumped all pigmented, apparently mature seeds from each outcrossing treatment (i.e., selfed, 1, 10, 100, and 1000 meters) and planted them in separate subsections of a 10 m² plot 100 meters from the meadow containing female parents. In 1977, 1978, and 1979, we planted mature seeds from each individual separately within an 8 m² plot at the center of the female parental meadow, sowing them at exact spatial coordinates with the aid of a 1 m² wooden frame strung off into 10 cm by 10 cm quadrants. We then recorded coordinates and sizes of all seedlings appearing in subsequent years. There is no seed dormancy in this species (Price and Waser, unpub. data). We removed all flowering stalks within 2 meters of each plot, so that their seeds would not fall within plots and be confused with those we planted.

We used three methods to estimate actual pollen dispersal for *D. nelsonii* and *I. aggregata*. (1) We measured mean plant-to-plant and flower-to-flower flight distances of pollinators by following them as they foraged in natural meadows and recording distances flown between successively visited plants as well as number of flowers visited per plant. To be able to compare flight distances of different pollinator species, we matched observations of foraging bouts by meadow, date, and time of day. (2) We measured amounts of fluorescent dye powders deposited on successively visited stigmas by bumblebees and hummingbirds after they had visited single male flowers whose anthers had been dusted with dye. To do this we introduced pollinators into a small outdoor flight cage and exposed them to freshly picked flowering stalks of *D. nelsonii* or *I. aggregata* placed in vases. If transfer properties of dye particles approximate those of pollen grains, "dye carryover" is a reasonable estimate of "pollen carryover," and can be combined with observed flight distances to estimate pollen dispersal. Assumptions inherent in these calculations are discussed below. (3) We measured dispersal of dye to and from single plants in natural populations. To estimate dispersal *from* plants, we dusted their anthers with dye, waited 6–12 hours, and then harvested marked plants as well as known proportions of their neighbors growing within 0–1, 1–2, 2–4, 4–6, and 6–10 meters of them. We examined receptive stigmas from harvested plants under a dissecting microscope, and counted numbers of dye particles per stigma. We multiplied the mean number of particles per stigma at each distance interval (including particles transferred within subject plants) by

the number of potential recipients at that distance to estimate relative numbers of particles deposited in each interval. From this we calculated the mean distance reached by a dye particle, counting transfer within subject plants as zero and otherwise using the midpoint of each distance interval. To estimate dispersal *to* plants, we conducted the reciprocal experiment. We dusted anthers on all male flowers of a known proportion of plants growing within 0–1, 1–2, 2–4, and 4–6 meters of central subject plant, using a distinct dye color for each distance interval. After 6–12 hours we harvested subject plants and counted dye particles of each color on receptive stigmas. Because we had available only four distinguishable dye colors, we conducted separate experiments in the same meadow and at the same time to estimate the mean number of dye particles per stigma transferred within plants. We dusted dye on anthers of all male flowers on a plant and counted particles deposited after 6–12 hours on stigmas of its female flowers. We combined results from several such paired sets of experiments. We corrected for the number of potential pollen donors growing at each distance interval by dividing the mean number of particles of any color by the proportion of the population dusted at that distance. From this we could estimate the distance from which dye particles deposited on stigmas had come.

RESULTS

Optimal Outcrossing Distances for *Delphinium nelsonii* and *Ipomopsis aggregata*

In three replicate experiments in 1976 and 1977, an outcrossing distance between 1 and 100 meters appeared to maximize fecundity for *D. nelsonii* (Fig. 17-1). In these replicates as a whole, seed sets from 10-meter crosses were significantly greater than those from selfed ($X^2 = 24.20$, *d.f.* = 6, $P < 0.005$) or 1000-meter crosses ($X^2 = 12.59$, *d.f.* = 6, $P < 0.05$), combining probabilties from 1-tailed *t*-tests with each replicate (Sokal and Rohlf, 1969). Two replicates conducted in 1978 and

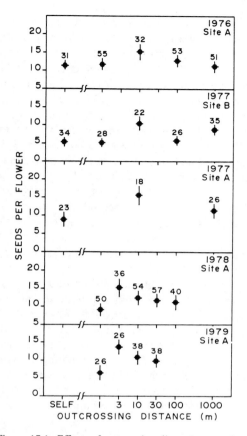

Figure 17-1. Effects of outcrossing distance on seed set from five replicate hand-pollination experiments with *D. nelsonii*. Values are means ± one standard error with sample sizes (number of flowers) shown. Sites A and B are two meadows, separated by approximately 100 meters, in which pollen recipients were located.

1979 narrow the optimal outcrossing distance to between 1 and 10 meters (Fig. 17-1). In these replicates, seed sets from 3-meter crosses were significantly greater than those from 1-meter crosses ($X^2 = 17.4$, $d.f. = 4$, $P < 0.005$), and greater than those from 30-meter crosses ($X^2 = 9.2$, $d.f. = 4$, $P = 0.058$), when probabilities were combined from separate *t*-tests as before. The pattern of peak fecundity with crosses in the interval 1–100 meters is repeated consistently in all five replicates conducted over 4 years and using two different populations of recipient plants. This suggests strongly that there is an intermediate optimal outcrossing distance for *D. nelsonii*. The probability of obtaining an intermediate optimum

at any point in the first replicate, followed by optima in the same 1–100-meter interval in all subsequent replicates, is only $\frac{3}{5} \times \frac{1}{3} \times \frac{1}{5} \times \frac{3}{5} \times \frac{3}{4} = 0.018$ by chance alone.

For *I. aggregata*, an optimal outcrossing distance, in terms of fecundity, again appears to lie between 1 and 100 meters (Fig. 17-2). Seed sets from 10-meter crosses were significantly greater than those from selfed crossed ($X^2 = 30.4$, $d.f. = 4$, $P < 0.001$), and greater than those from 100-meter crosses ($X^2 = 7.3$, $d.f. = 4$, $P \approx 0.15$), when probabilities were combined from separate *t*-tests. A two-way ANOVA with 10-meter and 100-meter seed sets from both 1979 and 1980 yields $F = 3.3$, $d.f. = 1, 183$, $P \approx 0.07$ for the effect of outcrossing treatment on seed set, again indicating that 10-meter crosses are more fecund than 100-meter crosses.

Results of our first few years of *D. nelsonii* survivorship data are summarized in Fig. 17-3. There is a tendency for the proportion of seeds germinating to increase as a function of outcrossing distance (Spearman's rank correlation $r_s = 0.72$, $n = 7$, $P \approx 0.08$), but for

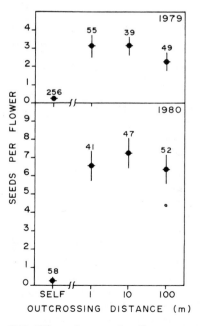

Figure 17-2. Effects of outcrossing distance on seed set from two replicate hand-pollination experiments with *I. aggregata*. Conventions are as in Fig. 17-1.

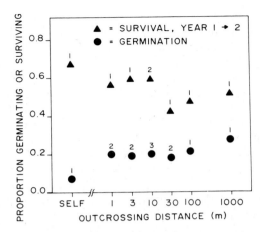

Figure 17-3. Summaries of germination and survival from year 1 to year 2 for hand-pollinated *D. nelsonii* seed progenies planted in parental meadows. Number of replicate experiments averaged to obtain each value is shown above each point.

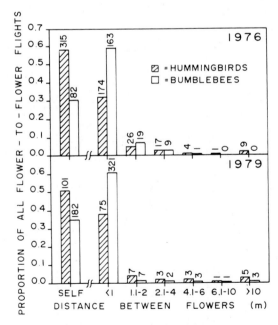

Figure 17-4. Distributions of flower-to-flower distances flown by hummingbirds and bumblebees visiting *D. nelsonii* in the same meadow (site B; see Fig. 17-1) in 1976 and 1979. Sample sizes (number of flights censused) appear above each histogram. Mean plant densities were 1.2 plants/m² in 1976 and 2.2 plants/m² in 1979.

seedling survivorship from first to second summers to decrease ($r_s = -0.73$, $n = 7$, $P \approx 0.07$). These contrasting patterns cause estimated overall survival of 10-meter seeds (12%) to exceed those of selfed (5%), 1-meter (11%), 3-meter (11%), 30-meter (8%), and 100-meter (10%) seeds, but not of 1000-meter seeds (16%). This illustrates that we may have to follow survivorship in both *D. nelsonii* and *I. aggregata* until plants reach reproductive maturity in order to discern whether there is a short optimal outcrossing in terms of survivorship. We estimate that reproductive maturity is reached only after many years for *D. nelsonii*. This species grows very slowly. Most of our second-year seedlings produced a single 2–3 lobed, 0.5-cm-radius leaf, and our third year seedlings a 3–4-lobed, 0.7-cm-radius leaf, whereas flowering adults produced two or more multilobed basal leaves of 1–2 cm radius. If the pattern of decreasing survivorship with increasing outcrossing distance persists past the second year of *D. nelsonii* seedling growth, then overall survivorship to reproductive maturity might indeed be highest for a short outcrossing distance; but our preliminary data are not conclusive.

Actual Outcrossing: Pollinator Flight Distances

Distributions of flight distances for hummingbirds and bumblebees visiting *D. nelsonii* differed significantly in 1976 and 1979 ($X^2 = 69.3$, $d.f. = 4$, $P < 0.005$; and $X^2 = 37.3$, $d.f. = 3$, $P < 0.005$, respectively), primarily because the two pollinator types had different tendencies to fly between flowers on the same plant and between neighboring plants less than 1 meter apart (Fig. 17-4). While hummingbirds flew less frequently between near neighbors than bumblebees, they also visited more flowers per plant [1976: birds = 2.5 ± 0.09 (296), bees = 1.4 ± 0.06 (502); 1979: birds = 2.2 ± 0.15 (97), bees = 1.5 ± 0.05 (344); values are mean flowers visited per plant ± one standard error with sample size in parentheses]. The net effect of these behavioral differences is for mean distances flown to be slightly longer for hummingbirds than for

Table 17-1. Various estimates of mean and median pollen transfer distances, compared to estimates of optimal outcrossing distances.

Species, year	Optimal outcrossing distance[1]	Pollen transfer distance		
		Hummingbird flights[2]	Bumblebee or hawkmoth flights[3]	Dye dispersal[4]
D. nelsonii				
1976	3 m	0.8 m (0 m)	0.6 m (0.3 m)	—
1979		0.9 m (0 m)	0.5 m (0.2 m)	—
1980		—	—	0.5 m (0.4 m)
I. aggregata				
1978	10 m	0.1 m (0 m)	0.1 m (0 m)	—
1980		0.3 m (0 m)	—	1.4 m (1.0 m)

1. These are distances at which we obtained peak seed sets in our most recent hand-pollination experiments.
2. Values are mean flower-to-flower flight distances, followed by median distances (in parentheses). A zero means that median flight was between flowers of a single plant.
3. Values are for bees visiting *D. nelsonii* or moths visiting *I. aggregata*.
4. Values are based on dispersal of dye both to and from plants in nature.

bumblebees, but for median distances flown to be slightly shorter for birds (Table 17-1). Another difference between these two pollinators is that bumblebees have a noticeable tendency to fly from lower, pistillate flowers of *D. nelsonii* to upper, staminate flowers, whereas hummingbirds do not (Epling and Lewis, 1952; Pyke, 1978a). Compared to hummingbird movements, therefore, relatively few bumblebee movements will cause selfing, and our analysis may slightly underestimate distances bees carry pollen by counting all their movements within plants as zeroes.

Distributions of flight distances for hummingbirds and hawkmoths visiting *I. aggregata* also differed significantly in 1978 ($X^2 = 22.0$ *d.f.* $= 4$, $P < 0.005$), again because of different tendencies to fly between flowers on the same plants or between neighboring plants (Fig. 17-5). Hummingbirds visited more flowers per plant than hawkmoths [birds: 7.4 $\pm$ 0.80 (213); moths: 4.7 $\pm$0.25 (239); conventions as above]. The net effect of this behavioral difference is for mean, but not median, distances flown by birds to be slightly less than those flown by hawkmoths (Table 17-1).

Finally, there was some year-to-year variation in distances flown by each pollinator. Distributions of flight distances on *D. nelsonii* differed significantly between 1976 and 1978 for bumblebees ($X^2 = 21.4$, *d.f.* $= 3$, $P < 0.005$), but not for hummingbirds ($X^2 = 6.5$, *d.f.* $=$

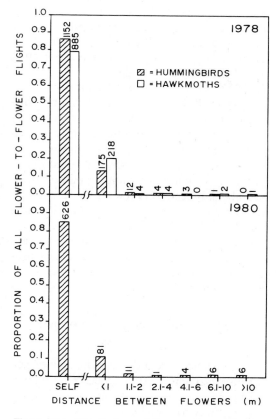

Figure 17-5. Distributions of flower-to-flower distances flown by hummingbirds and hawkmoths visiting *I. aggregata* in 1978 and by hummingbirds visiting *I. aggregata* in 1980, in two different meadows near the RMBL. Conventions are as in Fig. 17-4. Mean plant densities were 3.5 plants/m² in 1978 and 1.2 plants/m² in 1980.

Table 17-2. Various estimates of the fraction of pollen transferred into the optimal outcrossing interval.

| Species | Optimal outcrossing interval[1] | Pollen transferred optimal distance | | |
		Flights[2]	Flights plus carryover[3]	Dye dispersal[4]
D. nelsonii	1–10 m	8%/7%	38%	6%
I. aggregata	1–100 m	3%/1%	9%	38%

1. These are the intervals containing distances at which we obtained peak seed sets in our most recent experiments.
2. Values are based on flower-to-flower flight distances averaged across years. First value is for hummingbirds; second is for bumblebees *(D. nelsonii)* or hawkmoths *(I. aggregata)*.
3. Values are based on flight distances corrected by median dye carryover.
4. Values are based on dispersal of dye both to and from plants in nature.

$6, 0.5 < P < 0.1$). These differences were relatively slight, however, and did not clearly reflect the 25% change in mean plant density between years (1976: 1.2 plants/m²; 1979: 2.2 plants/m²). Distributions of hummingbird flights on *I. aggregata* contained relatively more long flights in 1980 than in 1978 ($X^2 = 23.9, d.f. = 6, P < 0.005$), probably reflecting a 65% decrease in plant density between years (1978: 3.5 plants/m² 1980: 1.2 plants/m²).

Despite all these differences in behavior, distributions of flower-to-flower distances flown by hummingbirds and insects visiting each plant species were remarkably similar. They were leptokurtic in all cases, with most flights occurring within single plants or between near neighbors. Indeed, very few flights carried pollen directly to optimal outcrossing distances (Table 17-2). Thus, if flight distances accurately reflect pollen transfer distances, it appears that optimal outcrossing distances usually exceed actual outcrossing distances for *D. nelsonii* and *I. aggregata*.

Actual Outcrossing: Dye Carryover Experiments

Flight distances are a direct indication of pollen dispersal only if all the pollen picked up at a flower is deposited at the first flower visited subsequently by a pollinator. They will underestimate dispersal if pollen is carried over past the first flower visited. Our estimates of pollen carryover, using dye powders as pollen analogues, are shown in Fig. 17-6. Some dye particles travel as far as the eighteenth successive flower after being picked up by bumblebees foraging on *D. nelsonii* inflorescences and the nineteenth successive flower after being picked up by hummingbirds foraging on *I. aggregata* inflorescences. For both plant species, however, the number of particles deposited on receptive stigmas declined quickly with the number of intervening flowers since dye pickup. If we express carryover as the expected flower reached by a dye particle following pickup, the average particle actually deposited on a stigma reached only the 3.3th *D. nelsonii* flower and the 5.5th *I. aggregata* flower. These values overestimate the typical number of flowers reached, however, because distributions are asymmetrical. The median dye particle reached only the 2.9th *D. nelsonii* flower and the 3.8th *I. aggregata* flower.

These estimates of dye carryover can be used to refine estimates of dye dispersal based on pollinator flight distances. If we assume that the median particle is deposited on the third and fourth flower for *D. nelsonii* and *I. aggregata*, respectively, we can calculate the probability that after traveling three or four unit flower-to-flower flights in a straight line away from the pollen door, the pollinator will have moved as far as the interval containing the optimal distance (1–10 meters for *D. nelsonii;* 1–100 meters for *I. aggregata*). To calculate this probability, we assume that pollinators can travel either 0, 0.5, 1.5, 3, 5, 8, or greater than 10 meters between flowers (these are midpoints of distance intervals given in Figs. 17-4 and 17-5), that the probability of a flight of any distance is independent of the dis-

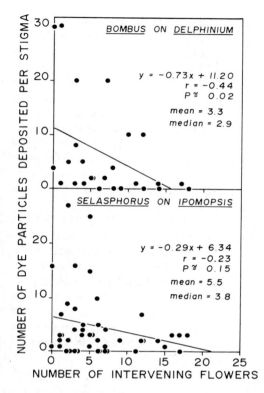

Figure 17-6. Carryover of dye by bumblebees visiting *D. nelsonii* flowers (top) and hummingbirds visiting *I. aggregata* flowers (bottom). Best straight-line fits are shown (in both cases they are as good as or better than exponential curve fits); they serve to show that the number of dye particles deposited on stigmas decreases as number of intervening flowers increases. Mean and median flowers reached by dye particles are shown in each case.

tances flown in previous flights, and that this probability is given by the proportion of all observed flights (averaged across years) that were between plants within the appropriate distance interval. We can then enumerate all possible three-way (or four-way) combinations of unit flights and their probabilities of occurrence, and estimate the probability the median pollen grain will be carried into the optimal outcrossing interval by summing probabilities of all appropriate unit flight combinations.

By these calculations we estimate that bumblebees visiting *D. nelsonii* and *I. aggregata* will carry a minority of pollen grains as far as the intervals containing optimal outcrossing distances (Table 17-2). Because our calculations make the assumption that pollinators

maintain the same direction of flight as they visit successive flowers, which is unlikely to be the case (G. H. Pyke, pers. comm.), this method of estimating pollen dispersal distances is likely to err on the high side. Once again, it appears that optimal outcrossing distances exceed actual distances.

Actual Outcrossing: Dye Dispersal in Natural Populations

By observing dispersal of dye powders in natural populations, we avoid having to make assumptions about pollinator movement patterns and dye carryover. Observed dispersal distances are the net result of natural patterns of visitation and dye transfer by all pollinators.

For both *D. nelsonii* (Fig. 17-7) and *I. ag-*

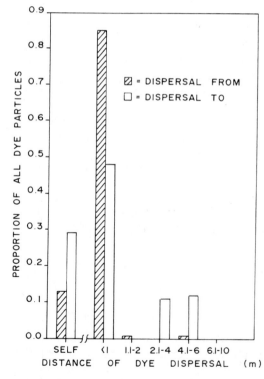

Figure 17-7. Distributions of dye dispersal distances of *D. nelsonii*. The "Dispersal From" distribution is an average from three 1980 replicates in different meadows in which a central individual was dyed and surrounding plants were examined for dye. The "Dispersal To" distribution is an average from ten 1980 replicates in different meadows in which surrounding plants were dyed and a central undyed individual was examined.

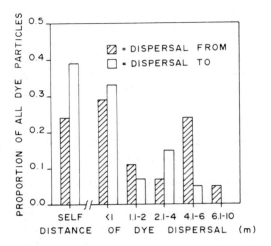

Figure 17-8. Distributions of dye dispersal distances for *I. aggregata*. Conventions are as in Fig. 17-7. "Dispersal From" is an average from three 1980 replicates, while "Dispersal To" is an average from eleven 1980 replicates.

gregata (Fig. 17-8) estimates of the numbers of dye particles that were carried various distances from single donor individuals are fairly similar to estimates of the numbers carried from various distances to recipient individuals. In most cases, dye dispersal was leptokurtic, with the number of deposited particles decreasing rapidly with distance. Mean and median dispersal distances were short for both species, below or in the lower end of intervals that contain optimal outcrossing distances (Table 17-1). Few dye particles were carried directly to or from distances within the optimal outcrossing intervals (Table 17-2).

DISCUSSION

Microdifferentiation of Populations and Optimal Outcrossing Distances

As far as we know, our finding of short optimal outcrossing distances is unprecedented. Outbreeding depression traditionally has been reported only from crosses between isolated populations of single species or from even wider crosses (e.g., Kruckeberg, 1957; Martin, 1963; Moll et al., 1965; Hughes and Vickery, 1974; Hogenboom, 1975). We suspect, however, that this is so partly because earlier workers did not look for outbreeding depression on a local

scale. We began with the expectation that outbreeding depression could occur over small distances and hence that optimal outcrossing distances could be short.

Our expectation was based first of all on the assumption that genetic differentiation can occur within single plant populations. In dense populations with restricted pollen and seed dispersal, heterogeneity in selective forces should indeed produce genetic differentiation on a small spatial scale. Even in the absence of selective gradients, genetic "isolation by distance" should occur in such populations by drift alone (Kimura and Weiss, 1964: Wright, 1969; Rohlf and Schnell, 1971; Turner et al., 1982). Considerable local genetic differentiation due to selection and/or drift has in fact been found in plant populations using electrophoresis (Schaal, 1974, 1975; Allard, 1975; Hamrick and Allard, 1975; Levin, 1977; Silander, 1979) and other techniques (Epling and Dobzhansky, 1942; McNeilly and Antonovics, 1968; Vickery, 1978; Silander and Antonovics, 1979). For example, Schaal (1974, 1975) documented differentiation on a scale of meters in *Liatris cylindracea*, a prairie herb with localized pollen and seed dispersal (Levin and Kerster, 1969b), and found genetic distance between plants (Nei, 1972) to be a linear function of physical distance between them.

Given that genetic differentiation can occur within populations and that genetic distance between plants is related to physical distance, our prediction of short optimal outcrossing distances follows from Fritz Müller's Law. If genetic similarity declines rapidly enough with distance, then optimal similarity could occur at a relatively short distance between mates. Notice that this prediction is entirely empirical; it is a logical extension of the observation that inbreeding and outbreeding depressions occur in nature, and it says nothing about ultimate causal explanations for inbreeding or outbreeding depressions in seed set. Unfortunately we have no space here to discuss these issues, and we will report elsewhere on our experimental investigation of possible causes of the apparent short optimal outcrossing distances for *D. nelsonii* and *I. aggregata*.

If the logic outlined above is correct, a simple expectation is that the optimal outcrossing distance for different species should depend on the spatial scale of differentiation within or between populations, which in turn will be sensitive to the scale of gene flow. We can use our study and others to explore this expectation.

First, both *D. nelsonii* and *I. aggregata* have highly localized gene flow and apparently short optimal outcrossing distances. As we have shown, various estimates suggest that most pollen dispersal is on the order of a few meters or less for both species. Furthermore, we have estimated seed dispersal by placing sheets of plastic around fruiting stalks, covering these with spray tanglefoot, and locating all seeds after seed drop. Primary dispersal from the central stalk was short and leptokurtic for *D.*

nelsoni ($\bar{x}$ = 11.1 cm, *s.d.* = 7.56 cm, *n* = 336 seeds from seven plants; see Fig. 17-9A and compare Epling and Lewis, 1952). We also have searched systematically for seedlings in sparse stands of *D. nelsonii*. Seedlings clearly were clumped around isolated mature plants, and the distribution of seedling distances from these putative parents was leptokurtic with mean and variance indistinguishable from that found for primary dispersal ($\bar{x}$ = 11.7 cm, *s.d.* = 7.64 cm, *n* = 46 seedlings). This correspondence suggests little secondary seed dispersal in *D. nelsonii*. Primary dispersal also was short, although not distinctly leptokurtic, for *I. aggregata* ($\bar{x}$ = 56.8 cm, *s.d.* = 25.33 cm, *n* = 124 seeds from two plants; see Fig. 17-9B). Thus we estimate populations of

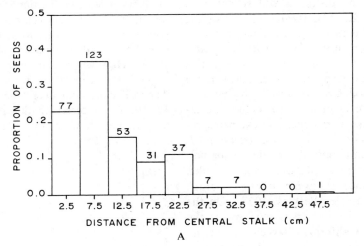

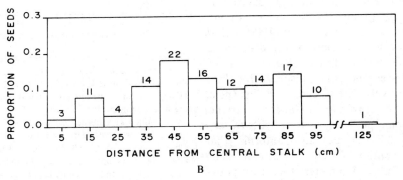

Figure 17-9. A. Seed dispersal in *D. nelsonii*, from seven plants measured in 1977 and 1978. Sample sizes are given above each histogram. Class limits are 0–5, 5.1–10, 10.1–15 cm, etc. Values are expressed as distance from central stalk of inflorescence, although dehiscing capsules were a mean distance of 2.6 cm from the central stalk and 11 cm from the ground. B. Seed dispersal in *I. aggregata*, from two plants measured in 1981. Conventions are as in (A), except that class boundaries differ. Dehiscing capsules were a mean of 2.0 cm from the central stalk and 32 cm from the ground.

both plant species to be effectively subdivided into small genetic "neighborhoods" (see Appendix to this chapter). As such they should meet conditions favoring local genetic differentiation through drift alone (Kimura and Weiss, 1964; Wright, 1969; Rohlf and Schnell, 1971; Levin and Kerster, 1974; Levin, 1979). It appears likely that there also could be considerable selective heterogeneity on a small spatial scale in the populations we have studied. Edaphic conditions and identity of neighboring plant species that could act as competitors change over distances of a few meters in meadows around the RMBL (Waser and Price, unpub.).

Recently, we have learned of other studies whose results bear on our expectation that optimal outcrossing distances will reflect gene flow distance. K. P. Lertzman (1981) of the University of British Columbia has obtained hand-pollination results for *Castilleja miniata* (Scrophulariaceae) similar to ours. For this small perennial herb, an outcrossing distance of 2 meters produces substantially higher seed set than distances of 1 or 10 meters. Lertzman's populations are dense and extensive, resembling our populations of *D. nelsonii* and *I. aggregata*. Hummingbirds are the major pollinators of his plants, and estimated pollen transfer distances between flowers average around 1 meter (see Perkins, 1977). On the other hand, S. Koptur (pers. comm.) of the University of California, Berkeley, studying *Inga brenesii* and *I. punctata* (Fabaceae) in Costa Rica, has found that seed sets with outcrossing distances of more than 1000 meters exceed those with an outcrossing distance of less that 500 meters. Indeed, Koptur has evidence that seed and pollen dispersal for these trees is relatively widespread. P. Fuller (pers. comm.) of the University of California, Irvine, has done some preliminary work with *Datura meteloides* (Solanaceae) in California that suggests the optimal outcrossing distance is on the order of a few kilometers for desert populations but not for coastal populations. No information is available yet on pollen or seed dispersal. It is interesting to speculate, however, that seeds in the desert might be widely dispersed by flash floods in the washes along

which plants grow. Finally, Coles and Fowler (1976; see their Figs. 1 and 2) present results that suggest an optimal outcrossing distance of several hundred meters for white spruce *(Picea glauca),* although they do not discuss this pattern themselves. This species is wind-pollinated, and it seems possible that pollen dispersal may be relatively widespread.

We also have obtained some pertinent information in a recent study of seed dispersal and microevolution of *Mimulus guttatus* (Scrophulariaceae) in Utah (Waser et al., 1982). This small perennial herb is restricted to mesic habitats and often grows in small populations isolated from one another by tens or hundreds of meters along watercourses draining the Wasatch Mountains near Salt Lake City (Vickery, 1978). Pollination is achieved mostly by small solitary bees (Harris, 1979), so that pollen dispersal is likely to be highly localized (see Levin and Kerster, 1974; Waddington, 1980). However, we have experimental evidence suggesting that water-mediated seed dispersal can occur over long distances (Fig. 17-10). In concert with this potentially long-distance gene flow, mean seed sets of hand-pol-

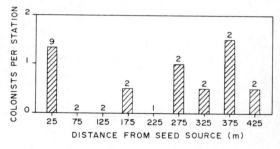

Figure 17-10. Summary of an experimental measurement of water-mediated seed dispersal of *M. guttatus,* conducted in Red Butte Canyon, Wasatch Mountains, Utah. Histograms show the mean number of seedlings colonizing a sampling station (consisting of three 23 cm by 30 cm flats filled with sterile soil) along the stream bank various distances downstream from two source populations. Class limits are 0–49, 50–99, 100–149 meters, etc. All flowering individuals were found in source populations, which were treated as replicates of one another in constructing the figure. Colonists in the interval 0–49 meters include those appearing in flats placed within one of the source populations. Numbers of sampling stations in each distance interval are shown above histograms. The median dispersal distance is 275 meters. Details are given in Waser et al. (1982).

linated crosses between populations connected by flowing streams (those sharing common canyons) are greater than those of crosses within populations or those of crosses between populations not connected by flowing water (those in separate canyons). Because variances in seed sets are high, this putative long optimal outcrossing distance for *M. guttatus* (Fig. 17-11) is not significant statistically (a one-way ANOVA gives $F = 0.7$; $d.f. = 2, 253$; $P \approx 0.5$ for the effect of "outcrossing level" on seed set). However, the pattern is suggestive and fits qualitatively with our expectation that optimal distance should depend on gene flow distance.

We hope that there will be more studies along these lines in the future, so that more extensive between-species comparisons can eventually be made of gene flow, population subdivision, and optimal outcrossing distance. If general patterns emerge from such comparisons, they may be a great aid to our under-standing of ultimate evolutionary causes of intermediate optimal distances.

Conflict in Pollination Systems: Do Plants Achieve Optimal Outcrossing?

We obtained reasonably consistent results using several methods to estimate actual pollen flow in natural populations of *D. nelsonii* and *I. aggregata*. There were only slight discrepancies between years in frequency distributions of flower-to-flower flights (Figs. 17-4 and 17-5), and in mean and median pollen dispersal distances estimated from these distributions (Table 17-1). There were larger discrepancies between years in distributions of dye dispersal (Figs. 17-7 and 17-8), as well as between methods in estimating the percentage of all pollen grains transported optimal outcrossing intervals (Table 17-2). All estimates of actual outcrossing distances, however, fell below those of optimal distances. Thus we conclude that there are discrepancies between optimal and actual outcrossing distances for *D. nelsonii* and *I. aggregata*. Three possible reasons for this come to mind (see also Price and Waser, 1979).

First, we acknowledge that discrepancies could be caused by inadequacies of method. We have attempted to determine optimal outcrossing distances by pollinating all flowers on a plant with pollen from a single distance. In nature, however, flowers and plants receive pollen from some array of distances. It is conceivable that more elaborate crosses with pollen brought from a mixture of distances would show that optimal matings involve shorter or longer distances than those we have identified. In addition, our estimates of actual outcrossing distances could be flawed. Although we have attempted to obtain estimates of pollen transfer in several different ways, we have chosen throughout to use dyes as pollen analogues. This technique is well established (e.g., Linhart, 1973; Stockhouse, 1976; Thomson, 1981) and has advantages of simplicity and speed relative to other methods of estimating pollen flow (see Schlising and Turpin, 1971; Reincke and Bloom, 1979; Schaal, 1980). However, the method is useful only insofar as mechanics of

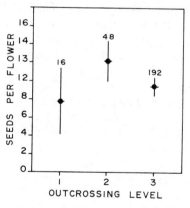

Figure 17-11. Effects of level of outcrossing on seed set from hand-pollinations within and between small isolated populations of *M. guttatus* in the Wasatch Mountains of Utah. Conventions are as in Fig. 17-1. Level 1 indicates crosses between plants from the same population (who often may be clone members); level 2 indicates crosses between plants from populations sharing the same side canyon of Big Cottonwood Canyon; and level 3 indicates crosses between plants from populations inhabiting different side canyons. Physical distances between mates were on the order of 1–10 meters in level 1, 100–1500 meters in level 2, and 300–4500 meters in level 3. Populations used in these experiments are described further in Fig. 7 and pp. 445–451 of Vickery (1978) and in Waser et al. (1982).

pickup, transfer, and deposition of dye follow those of pollen. Recently, J. D. Thomson (pers. comm.) has suggested that dye flow may overestimate pollen flow for some Canadian woodland herbs he has studied. In addition, Pyke (1981b) provided information on actual pollen carryover for *I. aggregata* at the RMBL, which we can compare to our Fig. 17-6. This comparison suggests again that dye flow overestimates pollen flow slightly. Pyke used a mummified hummingbird to probe a set number of flowers on two plants in a row. Since *I. aggregata* is self-incompatible (see Fig. 17-2), he attributed increment in seed set of probed flowers on the second plant relative to background pollination entirely to pollen picked up at the first plant. A linear regression from results of several such experiments showed that seed set of probed flowers declined depending on their position in the probing sequence, and fell to zero, relative to background pollination, by about the thirteenth flower probed. The best-fit linear regression for our dye carryover experiment with *I. aggregata,* however, predicts that dye deposition will fall to zero only at about the twenty-second flower probed by a hummingbird (Fig. 17-6). We conclude that there is as yet no strong evidence that dye dispersal is uncharacteristic of pollen dispersal. If there is a difference, however, it appears to be a direction that makes our estimate of actual pollen dispersal too liberal. That there are potential problems of method seems to us the least interesting potential cause of apparent discrepancies between optimal and actual outcrossing distances for *D. nelsonii* and *I. aggregata.*

A second possibility is that optimal matings for "pistillate" and "staminate" functions in an hermaphroditic flower might differ. This possibility is addressed by Janzen (1977), who discusses "rape" by pollen grains that confer low fitness to the zygote (see also Charnov, 1979). However, it is not clear to us how optimal *mean* outcrossing distance could differ for male and female sexual functions. Both sexual functions contribute genes equally to progeny. If progeny fitness depends in some way on outcrossing distance, selection should act equally on each sexual function to, on the

average, achieve matings over the optimal distance. On the other hand, pollen is energetically cheap to produce relative to seeds, and optimal *variance* in outcrossing distance could be greater for male than for female sexual function for the reasons discussed by Janzen (1977) and Charnov (1979). Our attempts to measure optimal outcrossing distance admittedly have been mainly in terms of fitness gained through female sexual function (seed set). At the same time, however, seed set variations in hand-pollinations represent realized differences in abilities of pollen from different outcrossing distances to contribute to mature seeds (with the female sexual function presumably exerting some degree of control over pollen germination and differential ovule abortion—see Brink, 1952; Dickinson and Lewis, 1973; Charnov, 1979). Thus, we can imagine that the female sexual function in our flowers might have a more precise outcrossing distance than is apparent from seed set results (i.e., that "rape," *sensu* Janzen, 1977, occurs), but not that male sexual function can achieve higher fertilization success at each distance than we observe. If this deduction is correct, what we observe suggests that conflict between sexual functions cannot explain discrepancies we see between actual and optimal outcrossing distances.

We are left with a third possible explanation for the discrepancies, which, in our view, is the most interesting of all. We would like to stress that the basic interaction between *D. nelsonii* and *I. aggregata* and their animal pollinators, and the basic interaction between all plants and pollinators, must involve *conflict* rather than *cooperation* (cf. Howe and Vande Kerckhove, 1979). The general point that cooperation between species cannot be a prediction of the theory of natural selection was made clearly by Darwin (1859, p. 100), and has been discussed recently for pollination systems by Pyke (1981a) and Pyke and Waser (1981). Along with recent interest in "coevolved" systems (see Janzen, 1980), however, some workers have implied that plant–pollinator interaction will lead to situations of ever increasing specialization and stability in which evolutionary "goals" of both mutualistic partners are achieved simultneously (e.g. Faegri and van der Pijl, 1971; Hainsworth and Wolf, 1976; Laverty, 1978; Howell, 1979; Brown and Kodric-Brown, 1979). It is hard for us to see how this can be so.

Consider the basic interaction between plants and pollinators. The optimal situation for the plant is to receive an unlimited quantity of high-quality gamete transfer at no metabolic cost to itself. The optimal situation for the pollinator is to receive an unlimited food reward at each flower. Clearly, these evolutionary "goals" are in direct conflict. The solution realized in nature is some compromise, with plants expending metabolic energy to produce floral advertisement and food reward and pollinators expending energy to extract their livelihood from flowers.

This bears on our discussion of actual and optimal outcrossing in *D. nelsonii* and *I. aggregata*. If bumblebees and hummingbirds are foraging optimally (cf. MacArthur and Pianka, 1966; Pyke et al., 1977), as seems to be the case (Whitham, 1977; Pyke, 1978a,b,c; Hartling and Plowright, 1979; Waddington, 1980), they should minimize flight costs by minimizing flight distances between successively visited flowers. When *D. nelsonii* and *I. aggregata* grow at such high densities that neighboring plants are separated by much less than the optimal outcrossing distance (as is the case around the RMBL), we cannot expect plant-to-plant flight distances (much less flower-to-flower distances) to average what is optimal for plants. Changes in plant density clearly can strongly influence actual outcrossing distance (see Levin and Kerster, 1969a,b; Schaal, 1978; Ellstrand et al.,1979; Waddington, 1979) but such changes are very unlikely to reflect adaptive responses of individuals. How then can individual selection acting on plant traits influence the chance that pollinators will fly distances as long as those optimal for plants? We can think of several possibilities.

First of all, pollinator behavior might be influences by changes in nectar (or pollen) reward available in flowers. This appears a common expectation, perhaps as a result of influential papers by Heinrich and Raven (1972) and Heinrich (1975a). In its simplist interpretation, however, the expectation seems unwarranted. For example, we can treat individual plants of a species as discrete resource patches and ask what predictions are made by optimal foraging theory about behavior of pollinators visiting patches of different quality (cf. Pyke, 1978c). Models of foraging behavior in a patchy environment (MacArthur and Pianka, 1966; Charnov, 1976) predict that pollinators should visit more flowers per plant on rare mutants with higher than average nectar production rate or concentration and/or should approach such plants more often than average plants (if they are able to remember the location of individuals; see Pyke and Waser, 1981). However, no prediction is made about pollinator flight distances. More complicated foraging models (see Pyke, 1981a) suggest that flight distances should decrease following visits to especially rewarding plants. However, all these models also predict that effects on flight distances of mutants producing different rewards disappear once mutants reach high frequencies in a population.

While alteration of floral rewards may not influence outcrossing distance in simple ways, a second possibility is that it might do so by leading to a change in the identity of pollinators attracted by a plant. This was stressed by Heinrich and Raven (1972) and Heinrich (1975a), who pointed out that species with rich floral rewards (and attendant changes in floral morphology) could specialize on pollinators with high energy requirements. One implication was that such "high energy" pollinators might provide greater outcrossing than "low energy" pollinators. It was noted that tropical bees and hummingbirds indeed may trapline over long distances while collecting nectar (Janzen, 1971; Linhart, 1973). Judging from our studies, however, such examples should not be taken to mean that pollinators differing in metabolic requirements necessarily differ in outcrossing services, as some people appear to have assumed (e.g., Brown et al., 1978; Stiles, 1978; Schmitt, 1980). Hummingbirds and bumblebees visiting *D. nelsonii* (which differ approximately fourfold in whole-animal metabolic rate during flight—see Hainsworth and

Wolf, 1972; Heinrich, 1975b), and humming-birds and hawkmoths visiting *I. aggregata* (which differ approximately fivefold in metabolic rate during flight—see Heinrich, 1971; Hainsworth and Wolf, 1972), provide strikingly similar outcrossing services to these plants. In retrospect this should not be surprising. Each pollinator, regardless of absolute energy requirement, may be expected to minimize flight costs by flying between plants that are near neighbors (unless foraging is not the only purpose of flight, as with the butterflies studied by Schmitt, 1980). This indeed is what all pollinators do in our system. We conclude that neither the plant species we have studied, nor others, are likely, in general, to realize substantial increases in outcrossing distances by switching pollinators.

A third possible way in which plants may be able to influence outcrossing involves placement or exsertion of sexual parts of flowers. If a population is polymorphic for placement of sexual parts, then carryover of pollen (and thus outcrossing distance) may be enhanced (Waser, unpub.; P. Feinsinger, pers. comm.). K. P. Lertzman of the University of British Columbia is now engaged in computer modeling and experimental tests of such effects (see Lertzman and Gass, this book). Heterostyly may be the most obvious example of such a polymorphism and may have evolved partly as a means of enhancing carryover. This suggestion seems to us related to Darwin's (1877) explanation for heterostyly as a mechanism for reducing pollen wastage (see also Ganders, 1979). Other than heterostyly, which is not especially widespread, polymorphisms in anther or stigma placement that might enhance pollen carryover have not been identified in the literature. This may mean that manipulation of pollen transfer by this means is uncommon. It may also mean, of course, that subtle polymorphisms exist but have escaped notice.

From these arguments it appears that variation in plant traits may have relatively little effect directly or indirectly on how far pollen is transferred. On the other hand, a female plant or female sexual function in a hermaphrodite may be able to influence *realized* outcrossing distance after pollination, by allowing only certain pollen grains to germinate and fertilize ovules or selectively aborting certain fertilized ovules (as mentioned above; see, e.g., Charnov, 1979). Thus, if excess pollen reaches stigmas, including a few grains from long distances, a plant may come closer to achieving optimal outcrossing from the female point of view than our measurements would suggest.

The optimal pollination for a plant must involve a complex sequence of pollinator behaviors before, during, and after the plant visitation. Despite this complexity, each detail of pollinator behavior will influence plant fitness through one of two fundamental components of pollination; the *quantity* of pollen transferred to, within, and from the plant, and the *quality* of the pollen transferred. In this chapter we have considered only one small part of the optimal pollination, the genetic quality of pollen transferred to a plant. We have presented experimental evidence that genetic similarity between mates is important to plant fitness, and we have suggested that patterns of gene flow and population subdivision could determine what outcrossing distance corresponds to the optimal genetic similarity. We have pointed out that pollinators transfer pollen over distances shorter than what appears to be optimal, and we argue that selection acting on individual plants is likely to have little effect on this intransigence of pollinators. We would argue, indeed, that there is little chance to find any component of pollination in nature optimized from the plant's point of view, much less to find all components optimized simultaneously. Thus, while it seems instructive to attempt a definition of the optimal pollination both theoretically and experimentally, it probably is unwise to take a Panglossian view of plant–pollinator interaction, despite the fact that mutualism and coevolution are involved.

ACKNOWLEDGMENTS

As this research progressed, we discussed it with many people, to our great benefit. We especially thank J. Antonovics, H. G. Baker, J. H. Brown, E. L. Charnov, N. C. Ellstrand, P. Feinsinger, M. Geber, K. P. Lertzman, D. G. Lloyd, H. R. Pulliam, D. A. Samson, L. L. Wolf, and R K. Vickery for their comments. P. Fuller, S. Koptur, and R. K. Vickery kindly let us discuss their unpublished hand-pollination results. Indispensable field help was provided by K. Donovan, D. Extine, and D. A. Samson, and welcome financial support was provided by the American Philosophical Society, Sigma Xi, the University of California at Riverside, and NSF grant DEB-8102774 to NMW. Finally, we thank P. Lain for patiently typing and retyping the manuscript.

LITERATURE CITED

Allard, R. W. 1975. The mating system and microevolution. *Genetics* 79:115–126.

Brink, R. A. 1952. Inbreeding and crossbreeding in seed development. *In* J. W. Gowen (ed.) *Heterosis.* Iowa State University Press, Ames.

Brown, J. H. and A. Kodric-Brown. 1979. Convergence, competition, and mimicry in a temperate community of hummingbird pollinated flowers. *Ecology* 60:1022–1035.

Brown, J. H., W. A. Calder, and A. Kodric-Brown. 1978. Correlates and consequences of body size in nectar-feeding birds. *Am. Zool.* 18:687–700.

Charnov, E. L. 1976. Optimal foraging, the marginal value theorem. *Theor. Pop. Biol.* 9:129–136.

———. 1979. Simultaneous hermaphroditism and sexual selection. *Proc. Natl. Acad. Sci. USA* 76:2480–2484.

Coles, J. F. and D. P. Fowler. 1976. Inbreeding in neighboring trees in two white spruce populations. *Silvae Genetica* 25:29–34.

Darwin, C. 1859. *On the Origin of Species by Means of Natural Selection.* Murray, London.

———. 1876. *On the Effects of Cross and Self Fertilisation in the Vegetable Kingdom.* Murray, London.

———. 1877. *The Different Forms of Flowers on the Plants of the Same Species.* Murray London.

Dickinson, H. G. and D. Lewis. 1973. Cytochemical and ultrastructural differences between intraspecific compatible and incompatible pollinations in *Raphanus.* *Proc. R. Soc. Lond. B* 183:21–38.

Downhower, J. F. and K. B. Armitage. 1971. The yellow-bellied marmot and the evolution of polygamy. *Am. Nat.* 105:355–370.

Ellstrand, N. C., A. M. Torres, and D. A. Levin. 1979. Density and the rate of apparent outcrossing in *Helianthus annuus* (Asteraceae). *Syst. Bot.* 3:403–407.

Epling, C. and Th. Dobzhansky. 1942. Genetics of natural populations. VI. Microgeographic races in *Linanthus parryae. Genetics* 27:317–332.

Epling, C. and H. Lewis. 1952. Increase of the adaptive range of the genus *Delphinium. Evolution* 6:253–267.

Faegri, K. and L. van der Pijl. 1971. *The Principles of Pollination Ecology.* Pergamon Press, Oxford.

Ganders. F. R. 1979. The biology of heterostyly. *N.Z. J. Bot.* 17:607–635.

Grant, V. 1971. *Plant Speciation.* Columbia University Press, New York.

———. 1975. *Genetics of Flowering Plants.* Columbia University Press, New York.

Hainsworth, F. F. and L. L. Wolf. 1972. Power for hovering flight in relation to body size in hummingbirds. *Am. Nat.* 106:589–596.

———. 1976. Nectar characteristics and food selection by hummingbirds. *Oecologia* 25:101–113.

Halliday, T. R. 1978. Sexual selection and mate choice. *In* J. R. Krebs and N. B. Davies (eds.), *Behavioural Ecology.* Sinauer, Sunderland, Mass.

Hamrick, J. L. and R. W. Allard. 1975. Correlations between quantitative characters and enzyme genotypes in *Avena barbata. Evolution* 29:438–442.

Harris, J. E. 1979. The pollination ecology and sexual reproductive biology of *Mimulus guttatus* along an elevational gradient. Master's thesis, University of Utah, Salt Lake City.

Hartling, L. K. and R. C. Plowright. 1979. Foraging by bumblebees on patches of artificial flowers: a laboratory study. *Can. J. Zool.* 57:1866–1870.

Heinrich, B. 1971. Temperature regulation in the sphinx moth, *Manduca sexta. J. Exp. Biol.* 54:141–152.

———. 1975a. Energetics of pollination. *Annu. Rev. Ecol. Syst.* 6:139–170.

———. 1975b. Thermoregulation in bumblebees. II. Energetics of warm-up and free flight. *J. Comp. Physiol.* 96:155–166.

Heinrich, B. and P. H. Raven. 1972. Energetics and pollination ecology. *Science* 176:597–602.

Hogenboom, N. G. 1975. Incompatibility and incongruity: two different mechanisms for the non-functioning of intimate partner relationships. *Proc. Roy. Soc. Lond. B* 188:361–375.

Howe. H. F. and G. A. Vande Kerckhove. 1979. Fecundity and seed dispersal of a tropical tree. *Ecology* 60:180–189.

Howell, D. J. 1979. Flock foraging in nectar-feeding bats: advantages to the bats and to the host plants. *Am. Nat.* 114:23–49.

Hughes, K. W. and R. K. Vickery, Jr. 1974. Patterns of heterosis and crossing barriers from increasing genetic distance between populations of the *Mimulus luteus* complex. *J. Genet.* 61:235–245.

Janzen, D. H. 1971. Euglossine bees as long-distance pollinators of tropical plants. *Science* 171:203–205.

———. 1977. A note on optimal mate selection by plants. *Am. Nat.* 111:365–371.

———. 1980. When is it coevolution? *Evolution* 34:611–612.

Kimura, M. and G. H. Weiss. 1964. The stepping-stone model of population structure and the decrease of genetic correlation with distance. *Genetics* 49:561–576.

Kruckeberg, A. 1957. Variation in infertility of hybrids between isolated populations of the serpentine species *Streptanthus glandulosus* Hook. *Evolution* 11:185–211.

Laverty, T. 1978. Flower visiting behavior of experienced and inexperienced bumblebees. Master's thesis, University of Alberta, Edmonton.

Lertzman, K. P. 1981. Pollen transfer: processes and consequences. Master's thesis, University of British Columbia, Vancouver.

Levin, D. A. 1977. The organization of genetic variability in *Phlox drummondii*. *Evolution* 31:477–494.

———. 1979. The nature of plant species. *Science* 204:381–384.

Levin, D. A. and H. W. Kerster. 1969a. The dependence of bee-mediated pollen and gene dispersal upon plant density. *Evolution* 23:560–571.

———. 1969b. Density dependent gene dispersal in *Liatris*. *Am. Nat.* 103:61–74.

———. 1974. Gene flow in seed plants. *Evol. Biol.* 7:139–220.

Linhart, Y. B. 1973. Ecological and behavioral determinants of pollen dispersal in hummingbird-pollinated *Heliconia*. *Am. Nat.* 107:511–523.

MacArthur, R. H. and E. R. Pianka. 1966. On optimal use of a patchy environment. *Am. Nat.* 100:603–609.

Martin, F. W. 1963. Distribution and interrelationships of incompatibility barriers in the *Lycopersicon hirsutum* Humb. and Bonpl. complex. *Evolution* 17:519–528.

McNeilly, T. and J. Antonovics. 1968. Evolution in closely adjacent plant populations. IV. Barriers to gene flow. *Heredity* 23:205–218.

Moll, R. H., J. H. Lonnquist, J. Velez Fortuno, and E. C. Johnson. 1965. The relationship of heterosis and genetic divergence in maize. *Genetics* 52:139–144.

Mulcahy, D. L. 1971. A correlation between gametophytic and sporophytic characteristics in *Zea mays* L. *Science* 171:1155–1156.

Mulcahy, D. L. 1974. Adaptive significance of gametic competition. *In* H. F. Linskens (ed.) *Fertilization in Higher Plants*. North-Holland Publishing Co., Amsterdam.

Müller, H. 1883. *The Fertilisation of Flowers*. Macmillan, London.

Nei, M. 1972. Genetic distance between populations. *Am. Nat.* 106:283–292.

Orians, G. H. 1969. On the evolution of mating systems in birds and mammals. *Am. Nat.* 103:589–603.

Perkins, M. D. C. 1977. Dynamics of hummingbird-mediated pollen flow in a subalpine meadow. Masters thesis, University of British Columbia, Vancouver.

Price, M. V. and N. M. Waser. 1979. Pollen dispersal and optimal outcrossing in *Delphinium nelsonii*. *Nature* 277:294–296.

Pyke, G. H. 1978a. Optimal foraging in bumblebees and coevolution with their plants. *Oecologia* 36:281–293.

———. 1978b. Optimal foraging: movement patterns of bumblebees between inflorescences. *Theor. Pop. Biol.* 13:72–98.

———. 1978c. Optimal foraging in hummingbirds: testing the marginal value theorem. *Am. Zool.* 18:739–752.

———. 1981a. Optimal foraging in nectar-feeding animals and coevolution with their plants. *In* A. C. Kamil and T. D. Sargent (eds.) *Foraging Behavior: Ecological, Ethological, and Psychological Approaches*, Garland, New York.

———. 1981b. Optimal nectar production in a hummingbird-pollinated plant. *Theor. Pop. Biol.* 20:326–343.

Pyke, G. H. and N. M. Waser. 1981. The production of dilute nectars by hummingbird and honeyeater flowers. *Biotropica* 13:260–270.

Pyke, G. H., H. R. Pulliam, and E. L. Charnov. 1977. Optimal foraging: a selective review of theory and tests. *Q. Rev. Biol.* 52:137–154.

Reincke, D. C. and W. L. Bloom. 1979. Pollen dispersal in natural populations: a method for tracking individual pollen grains. *Syst. Bot.* 4:223–229.

Rohlf, F. J. and G. D. Schnell. 1971. An investigation into the isolation-by-distance model. *Am. Nat.* 105:295–324.

Schaal, B. A. 1974. Isolation by distance in *Liatris cylindracea*. *Nature* 252:703.

———. 1975. Population structure and local differentiation in *Liatris cylindracea*. *Am. Nat.* 109:511–528.

———. 1978. Density dependent foraging on *Liatris pycnostachya*. *Evolution* 32:452–454.

———. 1980. Measurement of gene flow in *Lupinus texensis*. *Nature* 284:450–451.

Schlising, R. A. and R. A. Turpin. 1971. Hummingbird dispersal of *Delphinium cardinale* pollen treated with radioactive iodine. *Am. J. Bot.* 58:401–406.

Schmitt, J. 1980. Pollinator foraging behavior and gene dispersal in *Senecio* (Compositae). *Evolution* 34:934–943.

Silander, J. A. 1979. Microevolution and clone structure in *Spartina patens*. *Science* 203:658–660.

Silander, J. A. and J. Antonovics. 1979. The genetic basis of the ecological amplitude of *Spartina patens*. I. Morphometric and physiological traits. *Evolution* 33:1114–1127.

Sokal, R. R. and F. J. Rohlf. 1969. *Biometry*. Freeman, San Francisco.

Stiles, F. G. 1978. Ecological and evolutionary implications of bird pollination. *Am. Zool.* 18:715–727.

Stockhouse, R. E. 1976. A new method for studying pollen dispersal using micronized fluorescent dusts. *Am. Midl. Natur.* 96:241–245.

Thomson, J. D. 1981. Spatial and temporal components of resource assessment by flower-feeding insects. *J. Anim. Ecol.* 50:49–59.

Thornhill, R. 1980. Mate choice in *Hylobittacus apicalis* (Insecta: Mecoptera) and its relation to some models of female choice. *Evolution* 34:519–538.

Turner, M. E., J. C. Stephens, and W. W. Anderson. 1982. Homozygosity and patch structure in plant populations as a result of nearest-neighbor pollination. *Proc. Natl. Acad. Sci. U.S.A.* 79:203–207.

Verner, J. and M. F. Willson. 1966. The influence of habitats on mating systems of North American passerine birds. *Ecology* 47:143–147.

Vickery, R. K. 1978. Case studies in the evolution of species complexes in *Mimulus*. *Evol. Biol.* 11:405–507.

Waddington, K. D. 1979. Flight patterns of three species of sweat bees (Halictidae) foraging at *Convolvulus arvensis*. *J. Kansas Ent. Soc.* 52:751–759.

———. 1980. Flight patterns of foraging bees relative to density of artificial flowers and distribution of nectar. *Oecologia* 44:199–204.

Waser, N. M. 1978. Competition for hummingbird pollination and sequential flowering in two Colorado wild-flowers. *Ecology* 59:934–944.

Waser, N. M. and L. A. Real. 1979. Effective mutualism between sequentially flowering plant species. *Nature* 281:670–672.

Waser, N. M. and M. V. Price. 1981. Pollinator choice and stabilizing selection for flower color in *Delphinium nelsonii. Evolution* 35:376–390.

Waser, N. M., R. K. Vickery, and M. V. Price. 1982. Patterns of seed dispersal and population differentiation in *Mimulus guttatus. Evolution* 36:753–761.

Whitham, T. G. 1977. Coevolution of foraging in *Bombus* and nectar dispensing in *Chilopsis:* a last dreg theory. *Science* 197:593–596.

Willson, M. F. 1979. Sexual selection in plants. *Am. Nat.* 113:777–790.

Wright, S. 1969. *Evolution and the Genetics of Populations,* Vol. 2, *The Theory of Gene Frequencies.* University of Chicago Press, Chicago.

APPENDIX: ESTIMATES OF GENETIC NEIGHBORHOOD SIZE FOR *D. NELSONII* AND *I. AGGREGATA*

For plants with hermaphroditic flowers, which disperse genes through haploid pollen and diploid seeds, the genetic neighborhood size (Wright, 1969) is

$$N_e = 6.3d \, (\sigma_p^2 + \sigma_s^2)$$

where d is the genetically effective density, and σ_p^2 and σ_s^2 are variances in pollen and seed dispersal, respectively, around the parent plant (Levin and Kerster, 1974). Under the assumption that there is no net movement of pollen or seeds in any given direction, this becomes

$$N_e = 6.3d \left(\sum_i p_i^2/2N_p + \sum_i s_i^2/N_s \right)$$

where p_i and s_i are distances reached by the ith pollen grain and the ith seed, respectively, and N_p and N_s are total numbers of pollen grains and seeds, respectively.

For *D. nelsonii* we estimate from 1977 and 1978 seed dispersal experiments that $\sum_i s_i^2/N_s = 0.02$ m^2. We estimate from 1980 experiments with dye flow to and from plants that $\sum_i p_i^2/2N_p = 0.86$ m^2. Assuming that

genetic density equals actual density, and taking 5.0 plants/m^2 as a representative density of *D. nelsonii,* we arrive at

$$N_e = 27.7 \text{ plants}$$

This corresponds to a neighborhood area of 5.5 m^2 with radius of 1.3 m.

For *I. aggregata* we estimate from 1981 seed dispersal experiments that $\sum_i s_i^2/N_s = 0.38$m^2. We estimate from 1980 experiments with dye flow to and from plants that $\sum_i p_i^2/2N_p = 5.09$ m^2. Taking 2.0 plants/m^2 as a representative density, we arrive at

$$N_e = 68.9 \text{ plants}$$

This corresponds to a neighborhood area of 34.5 m^2 with radius of 3.3 m.

These estimates incorporate one novel feature, based on the assumption in the neighborhood size formula that all pollen grains and seeds considered actually transmit genes. From our hand-pollination studies we have some information on how distance from which a pollen grain comes influences its probability of fertilization. In calculating pollen dispersal variances, we used this information to weigh the probability of fertilizing; this weighing factor was most pronounced for self-pollination events. More accurate weighing factors might be derived if we knew how fertilization probability is influenced by a pollen grain occurring in a natural mixture of pollen from donors at various distances, but this information is not available, nor is information on seed or seedling success as a function of dispersal distance. Also, our estimates of neighborhood size may be biased on the high side, because actual densities usually exceed genetically effective densities (Levin and Kerster, 1974). For both *D. nelsonii* and *I. aggregata,* then, genetic neighborhood sizes appear to be small, which should facilitate local microdifferentiation (see, e.g., the theoretical results of Rohlf and Schnell, 1971, and Turner et al., 1982).

18
WHY BEES MOVE AMONG MASS-FLOWERING NEOTROPICAL TREES

Gordon W. Frankie and William A. Haber

Department of Entomological Sciences
University of California
Berkeley, California

ABSTRACT

In the Costa Rican dry forest, beginning nectar-monitoring studies revealed considerable inter-plant variation in the timing, quantity, and quality of nectar secretion in mass-flowering tree species pollinated by large bees. Recognition of slightly asynchronous flowering among different-sized conspecific individuals of most mass-flowering species added another variable to nectar secretion. Nectar flow patterns are presented for six tree species belonging to three different plant families, Bignoniaceae, Caesalpinaceae, and Fabaceae.

Limited studies suggested that large bees, especially the Anthophoridae, respond quickly to changes in nectar secretion. That is, as nectar production increased and decreased, bee numbers correspondingly increased and decreased.

It is hypothesized that intraspecific differences in flowering phenology, nectar flow, volume, and quality result in an ever-changing resources base that is continually monitored by large bees. Detection of changes, in turn, prompts the bees to make regular intertree movements, thereby effecting outcrossing.

KEY WORDS: Tropical trees, outcrossing, bee pollinators, neotropical forests, tropical ecosystems, intertree pollinator movement, interplant pollinator movement, traplining pollinators, mass-flowering plants, nectar monitoring and analysis, flowering phenology, Costa Rican dry forest, Anthophoridae, *Gaesischia exul, Centris* spp., *Xylocopa*, euglossine bees, Apidae, Costa Rican lowland wet forest, Fabaceae, mid-elevation cloud forest, bimodal nectar production, pollinator visitation rates, dioecy, androdioecy, hermaphroditism, climatic factors and nectar secretion.

CONTENTS

INTRODUCTION

In the recent past, biologists have questioned the ability of insect pollinators to move among widely dispersed tropical trees and thereby effect outcrossing. Recent findings from studies in tropical forests suggest that these pollinators are indeed able to move among trees (Gilbert, 1975; Frankie et al., 1976; Augspurger, 1980; Appanah, 1981, unpub. ms.). In fact, certain large bee pollinators are known to move long distances within neotropical forests (Janzen, 1971), and they are reported to be the primary pollinators of many plant groups (Frankie et al., this book). Further, studies on the breeding systems of many neotropical tree species (Bawa, 1974; Bawa and Opler, 1975; Bawa, in prep.) have demonstrated that a high percentage of them are obligate outcrossers.

Despite this new information about insect pollinators and plants in tropical ecosystems, numerous questions remain about the circumstances under which these pollinators make intertree movements (Frankie, 1976). One such question is, why do pollinators move at all among flowering trees in a forest? In the case of species (i.e., tree and other life forms) that produce relatively few flowers each day over an extended period (Frankie et al., 1974; Opler et al., 1980), it is understandable that for energetic reasons (Heinrich and Raven, 1972; Heinrich, 1975, 1976) a pollinator will move among plants to obtain a full complement of floral rewards; i.e., pollen or nectar. Janzen (1971, 1974) has characterized this pollinator behavior as traplining, and he reasons that pollinators apparently visit numerous plants of

this type along well-memorized routes in the forest. Other workers have provided additional information and insight on traplining behavior (Baker, 1973; Gilbert, 1975). However, in the case of mass-flowering tree species, which produce large quantities of flowers over a relatively short time, it is not immediately apparent why pollinators would move among the trees. A few workers (Frankie and Baker, 1974; Frankie, 1976; Gentry, 1978; Rausher and Fowler, 1979, unpub. data) have suggested that interactions (aggressive, territorial, etc.) among insect pollinators (and visitors) account for some of the movement. In general, these suggestions provide only a partial explanation of intertree pollinator movements that lead to effective outcrossing.

Since 1976 we have been monitoring nectar from trees and other life forms (treelets, shrubs, vines, and lianas) in the Costa Rican dry forest (Frankie et al., this book; Haber and Frankie, in prep.) and have learned that considerable variation in nectar flow patterns occurs within and between species in this forest. We now have sufficient data to hypothesize that these variations may be largely responsible for the movement of pollinators (especially bees) among the mass-flowering trees. In this chapter, we present evidence to support the hypothesis that is based primarily on flowering phenological information and on data concerned with the timing, quantity, and quality of nectar secretion for selected mass-flowering tree species that are pollinated by large bees. To a lesser extent, we present information on bee responses to temporal variation in nectar flow. Because of these limitations, we regard the data in this chapter as the starting point in testing our hypothesis.

METHODS

Study Sites

Most of our studies were conducted between the towns of Cañas and Liberia in the lowland dry forest of Guanacaste Province, Costa Rica. Selected flowering trees at two sites in particular, Hacienda La Pacifica (1200 ha) and Comelco Ranch (20,000 ha), were utilized, and at both locations the habitats ranged from slightly to greatly disturbed. These sites are described in Daubenmire (1972), Rockwood (1973, 1975), Frankie et al. (1974, this book), Heithaus et al. (1975), and Opler et al. (1980).

Tree Species

Intertree differences in nectar flow and quantity have been recorded for all large bee-pollinated tree species that have been intensively surveyed (three to six individuals for each of 11 species) (Frankie and Haber, unpub. data). For the purposes of this chapter, only those nectar patterns that exemplify the known range of differences are presented for the following species: *Tabebuia impetiginosa*, *T. ochracea*, *T. rosea* (Bignoniaceae); *Caesalpinia eriostachys* (Caesalpinaceae); and *Andira inermis*, *Machaerium biovulatum*, *Myrospermum frutescens* (Fabaceae).

Pollinators

The large bee pollinators (>1 cm in length) belonged primarily to the family Anthophoridae (Frankie et al., 1976, this book). The most commonly represented taxa were *Gaesischia exul* and about 12 species in the genus *Centris*. To a much lesser extent, 6 species of the genus *Xylocopa* (Anthophoridae) and several species of the euglossine group (Apidae) were considered effective pollinators of the study trees.

Nectar Monitoring

Nectar was systematically sampled from all accessible sides of study trees. With the exception of a very few, small-sized individuals, most trees were medium-sized and in full flower, and had several branches that could be reached with ladders (3.3 meters or less). Nectar was removed hourly* from the same open flowers, which were bagged and measured with precalibrated Drummond microcapillary tubes. Bagging was done with Pollen Tector bags around sunset on the day before monitoring; rarely, it was done prior to sunrise on the day of monitoring. With the exception of *Andira inermis* all species opened their one-day flowers prior to sunrise. From 20 to 83 flowers were monitored per tree, unless noted otherwise in the figure headings. Often, flower sample size decreased through time on a given tree, since flowers were periodically lost when they fell off in the bags or were blown out of the sampler's hand by wind. A mean amount of nectar per flower was calculated hourly for each tree. Variations around the means are represented by standard errors. (*Note:* Some standard errors were too low to be represented in the figures.)

Bee Monitoring

Most large bee species began visiting flowers around 0545 (= local time), when it was just becoming light enough to read. Bees were monitored on study trees in two different ways. Estimates were made of the number of bees observed on a certain portion of the crown twice an hour, or the number of bee visits to a small group of mapped flowers was recorded twice each hour. Bee numbers were recorded in only those areas where the forest was relatively undisturbed or had received only moderate disturbance.

---

*Continual removal of nectar by bees is a regular feature of the large bee-pollinated trees. Thus, we had no problems with nectar accumulation and evaporation.

RESULTS

Intertree Variation in Nectar Production due to Differences in Flowering Phenologies

Although members of most tree species populations have been generally described as flowering synchronously in the dry forest (Frankie, et al., 1974; Opler et al., 1980), careful examination of the phenology data shows that flowering of these individuals is actually staggered slightly (a few to several days) in any given area. When intensity of flowering (i.e., absolute number of flowers per tree) is considered along with staggered flowering, a more accurate representation of flowering among conspecific individuals in an area emerges, and this can be graphically displayed (Fig. 18-1). Thus, foraging bees are faced with a mosaic of conspecific flowering individuals: some just beginning to flower, some in full flower, and some producing a few flowers at the end of a flowering episode. As one source begins to give out, bees will shift their foraging activity to a nearby conspecific individual that is in good flower. Bee movement, which is thought to be a response to daily changes in flowering, was suggested for the large bees that visit the mass-flowering trees *Andira inermis* (Fabaceae) in the dry forest (Frankie et al., 1976) and *Dipteryx panamensis* (Fabaceae) in a lowland wet forest in Costa Rica (Perry and Starrett, 1980).

Intratree Variation in Quantity and Timing of Nectar Production

Nectar flow data from members of the following four species serve to exemplify the range in variation recorded within species in this forest: *Caesalpinia eriostachys* (Caesalpinaceae), *Myrospermum frutescens* (Fabaceae), *Tabebuia impetiginosa* (Bignoniaceae), and *Machaerium biovulatum* (Fabaceae).

In *Caesalpinia eriostachys,* the nectar secretion pattern was different in each of three individuals examined. In tree No. 3 (Fig. 18-2) the greatest amount of nectar was available at sunrise (0600 peak), with a small amount of

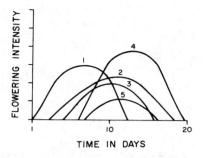

Figure 18-1. Hypothetical flowering intensities and phenologies of several conspecific trees at a particular location.

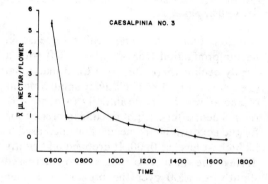

Figure 18-2. Nectar periodicity of *Caesalpinia eriostachys* No. 3 on 13 February 1981 at Hac. La Pacifica; *N* = 54–83 flowers.

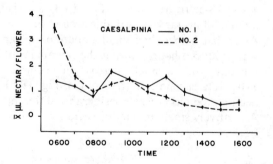

Figure 18-3. Nectar periodicities of *Caesalpinia eriostachys* Nos. 1 and 2 on 10 February 1981 at Hac. La Pacifica; *N* = 11–48 and 31–32 flowers for tree Nos. 1 and 2, respectively.

nectar being produced hourly throughout the rest of the day. There was also a slight increase in production around 0900 in this tree. In another individual of *Caesalpinia* (tree No. 2, Fig. 18-3) the amount of nectar available at 0600 was somewhat less than in tree No. 3. In

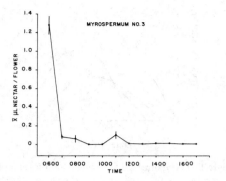

Figure 18-4. Nectar periodicity of *Myrospermum frutescens* No. 3 on 12 February 1981 at Hac. La Pacifica; N = 28–40 flowers.

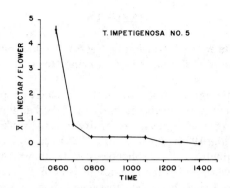

Figure 18-6. Nectar periodicity of *Tabebuia impetiginosa* No. 5 on 6 February 1981 at Hac. La Pacifica; N = 36–42 flowers.

addition, there was a noticeable increase in nectar production from 0900 to 1000 with a steady decline thereafter. In a third individual (tree No. 1, Fig. 18-3) relatively small amount of nectar was available initially (around sunrise). Then, after a slight decline in production for the next two hours, nectar flow increased to its highest level at 0900. It dropped off slightly during the next two hours and increased slightly at 1200 over the previous two-hour period.

Two different nectar flow patterns of *Myrospermum frutescens* are displayed in Figs. 18-4 and 18-5. In tree No. 3 (Fig. 18-4) a large amount of nectar was available around sunrise. After that time, only a small amount of nectar was produced hourly during the remainder of the day, with a small increase in nectar at 1100. In a second tree (tree No. 4, Fig. 18-5) nectar production was considerably different. A relatively small amount of nectar was avail-

able at sunrise, and production continued at a relatively modest level until 0900, after which time it dropped to almost zero. In other individuals of *Myrospermum*, described in detail below, it was not uncommon to observe one or rarely two large peaks of nectar produced after the initial peak.

Three individuals of *Tabebuia impetiginosa* serve to demonstrate the range in variation in nectar secretion in this species. The simplest pattern observed was that in tree No. 5 (Fig. 18-6) in which a relatively large amount of nectar was available at sunrise. Thereafter, production dropped off to very low levels and eventually to zero later in the day. In tree No. 2 (Fig. 18-7) nectar production did not drop off as rapidly as it did in tree No. 5. Furthermore, at 1200 and at 1400 two slight increases in production were recorded. In tree No. 1 (Fig. 18-8) two distinct, yet small, peaks of nectar were produced during the morning. The

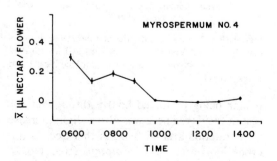

Figure 18-5. Nectar periodicity of *Myrospermum frutescens* No. 4 on 9 February 1981 at Hac. La Pacifica; N = 34–42 flowers.

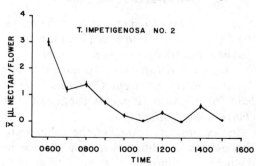

Figure 18-7. Nectar periodicity of *Tabebuia impetiginosa* No. 2 on 31 January 1981 at Bagaces; N = 30–61 flowers.

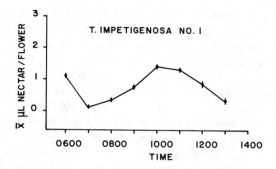

Figure 18-8. Nectar periodicity of *Tabebuia impetiginosa* No. 1 on 21 January 1980 at Chomes (Puntarenas Province); $N = 20$ flowers.

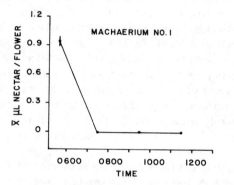

Figure 18-9. Nectar periodicity of *Machaerium biovulatum* No. 1 on 21 August 1979 at Hac. La Pacifica; $N = 8–11$ flowers.

first was available at sunrise, while the second was available between 1000 and 1100. Thereafter, nectar production dropped off to a low level in this tree.*

The nectar flow patterns of three individuals of *Machaerium biovulatum* are presented in Figs.18-9 and 18-10. One tree (No. 1) was in the lowland dry forest, while two individuals (Nos. 4 and 5) were monitored on the same day at a mid-elevation forest site near the city of Naranjo (about 1100 meters elevation). A relatively large amount of nectar was available in the flowers of tree No. 1 at sunrise. After that time, no nectar was produced during the day. The nectar flow patterns of tree Nos. 4 and 5 contrasted sharply with that of tree No.

*The nectar flow in this species is complicated somewhat, since a small proportion of the flowers open progressively during the day.

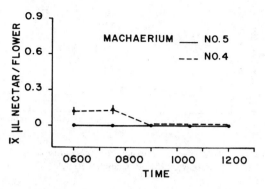

Figure 18-10. Nectar periodicities of *Machaerium biovulatum* Nos. 4 and 5 on 28 August 1980 in vicinity of Naranjo (Alajuela Province); $N = 31–39$ and 22 for trees 4 and 5, respectively.

1. Nectar of tree No. 4 was produced over a 1½-hour period. After that time, it dropped to zero. In tree No. 5, which was a small individual, none of the sampled flowers (about 7% of total flowers on tree) produced any nectar. On this tree, most bees merely probed the flowers, visiting perhaps three or four, and then left the tree almost immediately. In the process they picked up pollen on their abdomens and moved to other nearby trees (for example, tree No. 4) containing nectar. In Hawaii, Carpenter (1976) monitored nectar from the ohia tree, *Metrosideros collina* (Myrtaceae), and found that in 31 individuals sampled, two produced no nectar (Carpenter, pers. comm.).

It is not always clear what constitutes a genuine peak vs. an insignificant increase in nectar production. For example, in *Tabebuia rosea* (Fig. 18-11), tree No. 1 clearly produced only

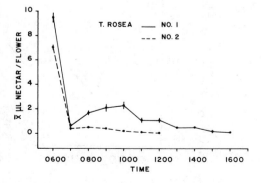

Figure 18-11. Nectar periodicities of *Tabebuia rosea* Nos. 1 and 2 on 6 March 1978 at Comelco Ranch; $N = 15$ flowers for each tree.

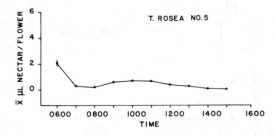

Figure 18-12. Nectar periodicity of *Tabebuia rosea* No. 5 on 23 February 1981 at Comelco Ranch; $N = 44$ flowers.

one peak of nectar during the day. However, tree No. 5 (Fig. 18-12) had one peak of nectar at 0600 and a prolonged increase in nectar production spanning the time period 0900 to 1100. Significant differences in the average amount of nectar produced per flower per day were also recorded among these trees. In tree No. 1, each flower averaged 19.6 $\pm$ S.E. 3.5 μl, whereas in trees 2 and 3 each flower averaged 8.8 $\pm$ 0.6 and 5.4 $\pm$ 0.5 μl, respectively. The differences between tree No. 1 and the other two trees were significant at the .05 level (one-way ANOVA followed by Duncan's Multiple range test).

We suggest that increases and decreases in bee numbers corresponding to measurable increases and decreases in nectar flow offer the best indication that these nectar patterns have ecological significance. This was documented

in the case of *Andira inermis* in 1972 (Fig. 18-13). It is well established in this species that two peaks of nectar are produced during the day (Frankie et al., 1976, this book). The first of these occurs at flower opening time between 0700 and 0800, whereas the second occurs somewhere between 1100 and 1300. When bees were carefully recorded from mapped flowers in the crown of one tree, it was observed that most visitation corresponded with the peaks in nectar production. A substantial decline in numbers of bees was also observed between the two peak periods (Fig. 18-13).

Andira inermis presents a unique example of variation in nectar production. Flowers on individual plants open synchronously between 0700 and 0800, depending on the individual and perhaps on the climatic conditions of a particular day. At that time, a relatively large amount of nectar is available to the bees. Afterward, nectar production drops off relatively sharply until about 1100. A second peak of nectar flow then develops. In some individuals this peak occurs as early as 1100 or 1130, whereas in others it may occur as late as 1200–1300. A theoretical composite of the various nectar production patterns in several individuals of *Andira* in a given forest situation would produce a mosaic setting of nectar availability through time (Fig. 18-14). In this species, flower opening time, time of the second peak of nectar flow, and nectar quantity from one

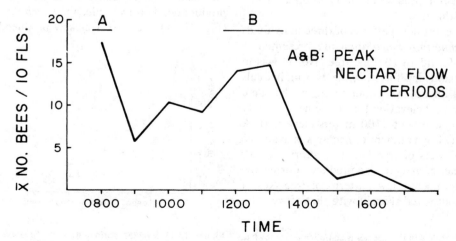

Figure 18-13. Hourly means of bees visiting 10 mapped flowers of one *Andira inermis* individual at Liberia, February 1974.

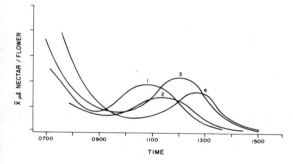

Figure 18-14. Hypothetical nectar periodicities of four individuals of *Andira inermis* that bloom on the same day at the same site.

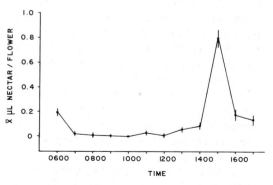

Figure 18-15. Nectar periodicity of *Myrospermum frutescens* No. 2 on 10 February 1981 at Hac. La Pacifica; $N = 42–53$ flowers. Mean number bees observed hourly in top half of crown.

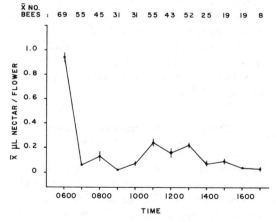

Figure 18-16. Nectar periodicity of *Myrospermum frutescens* No. 2 on 12 February 1981 at Hac. La Pacifica; $N = 41–46$ flowers. Mean number bees observed hourly in top half of crown.

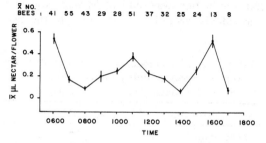

Figure 18-17. Nectar periodicity of *Myrospermum frutescens* No. 2 on 14 February 1981 at Hac. La Pacifica; $N = 42–49$ flowers. Mean number bees observed hourly in top half of crown.

day to the next, are variables that probably determine bee movement patterns among individuals. Bimodal nectar peaks have also been observed in some temperate plants (Frankie and Vinson, 1977; Brown et al., 1981; Carpenter, this book).

Changes in Nectar Production within Individuals Through Time

Although we have monitored nectar from only a limited number of individuals of selected species through time, two tentative conclusions have been reached on this effort to date. First, individuals of some species may be nearly invariate in their nectar flow patterns from one day to the next. They may, however, vary in their total nectar output per day (e.g., *Andira inermis* and *Myrospermum frutescens*). Second, individuals of some species can vary greatly from one day to the next, both in nectar secretion pattern and in total quantity produced during a day. For example, dramatic changes in nectar secretion were recorded in one individual of *Myrospermum* (No. 2, Figs. 18-15–18-17) over three sampling dates: 10, 12, and 14 February 1981. On 10 February very little nectar was produced until 1500, at which time a sharp rise in nectar production was observed (Fig. 18-15). On 12 February a large amount of nectar was available at sunrise, with a second, smaller peak that extended from 1100 to 1300 (Fig. 18-16). On the last sample date, the tree produced a modest amount of nectar early in the morning, and then a gradual increase in nectar flow was observed between 0900 and 1100, at which time a second large peak of nectar was recorded (Fig. 18-17). Nectar production dropped off substantially after that time and hit a low point at 1400, and then increased considerably to a third peak at 1600. Bees were monitored on the downwind side of the top half of the crown during these three days. Numbers of bees were counted twice hourly for five minutes just before and just after nectar was removed from the bagged flowers, and an average was computed. With the exception of a possible lag effect at 0600 on 10 and 14 February, and a relatively low bee visitation rate at the late afternoon peak on 14 February, bees generally increased and decreased with corresponding increases and decreases in nectar production (Figs. 18-15–18-17). In addition to daily differences in nectar secretion patterns, the average amount of nectar produced per flower varied each day. In this case, significant differences among increasing daily means were determined through a one-way ANOVA followed by Duncan's mean separation test (Table 18-1).

Possible Differences in Chemical Composition of Nectar

Some of our sampling suggests that chemical differences may be important in moving pollinators from one plant to the other. In 1978, two large individuals of *Tabebuia ochracea* were monitored for nectar production on the same day. The trees were separated by approximately 0.75 km. Results of the monitoring effort are presented in Fig. 18-18. Although nectar production was similar between the two individuals, visitation by the bees to the two

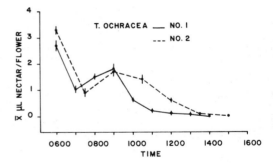

Figure 18-18. Nectar periodicities of *Tabebuia ochracea* Nos. 1 and 2 on 9 March 1978 in vicinity of Comelco Ranch; N = 15 and 17 flowers for trees 1 and 2, respectively.

trees was substantially different. With regard to tree No. 1, almost no bees were observed foraging at any time during the day. In tree No. 2, bee visitation was relatively high throughout the entire period when nectar flowed. Although no nectar samples were taken for chemical analysis, we suspect that there were differences in the chemical makeup of the trees (including sugar concentration) that would possibly explain differences in bee visitation. Furthermore, in 1981, we revisited tree No. 1 when it flowered, and again observed almost no bee visits to this tree; no nectar was sampled from this tree in 1981.

In 1978 we monitored nectar of two nearby, similar-sized individuals of *Tabebuia rosea* (Fig. 18-11). Throughout the period of nectar flow in the two trees, bee visitation rates were always significantly higher in tree No. 1 as compared to tree No. 2, even when the bees first arrived at 0600 (Frankie et al., 1981). Samples of nectar were taken from the two trees and analyzed chemically at Berkeley during the same year. In tree No. 2, relatively high levels of phenolics were detected in the nectar. We speculate that differences in plant secondary compounds in nectar may be, in part, responsible for differential foraging by the bees. Relevant to our observations on the *Tabebuia* nectars is the suggestion of Rhoades and Bergdahl (1981) that plant secondary compounds in nectar may function to sort out pollinator types.

Intraspecific variation in nectar sugar concentration has been recorded among the large

Table 18-1. Mean amounts of nectar produced in flowers of *Myrospermum frutescens* No. 2 on three sampling dates, Hacienda La Pacifica.

Sample date	$\bar{x}$ µl nectar/fl./day
10 Feb. 1981	1.27a ± S.D. 0.97
12 Feb. 1981	1.97b ± S.D. 1.14
14 Feb. 1981	2.69c ± S.D. 1.10

Means followed by different letters are significantly different at 0.05 level according to Duncan's mean separation test.

bee-pollinated tree species (Frankie et al., this book), and some of this variation may be responsible for certain bee visitation patterns that we have observed. In the case of *Andira inermis* sugar concentrations of trees growing in relatively undisturbed sites reached greater than 50%. In 1981, we examined the sugar concentrations of three *Andira* individuals that were growing in highly disturbed habitats along the Pan American highway near the town of Liberia. The average sugar concentration in these trees was 22, 24, and 32% respectively, which is considerably below the amount detected under more natural circumstances. Further, there was very little bee visitation to the trees, even though they occurred close to sites that were well represented by a diverse assortment of large bees. Almost all large bee-pollinated plant species in the dry forest, which offer nectar as the primary reward, have an average sugar concentration of 30% or more (Frankie et al., this book). Clearly, more work needs to be done on sampling for chemical differences that may exist among individuals of a population. Furthermore, observations and experimental manipulations need to be designed so as to assess how the bees respond to these chemical differences.

Differences in Nectar Production among Staminate and Pistillate Individuals of Dioecious Species

In a mid-elevation site (Monteverde, Puntarenas Province) we observed differences in nectar secretion among males and females of a dioecious species, *Mappia racemosa* (Rutaceae). In this species, nectar is available in male flowers about one hour earlier than in female flowers. Unsynchronized nectar flow between the sexes would facilitate movement of pollinators between male and female individuals by first providing a resource at the pollen station. When nectar gives out in male trees, pollinators will move to females and in the process transport pollen.

In a lowland mixed Dipterocarp forest, the Pasoh Forest Reserve in Negeri Sembilan, West Malaysia, S. Appanah (1981, unpub. ms.) found that males and hermaphrodites of the androdioecious tree *Xerospermum intermedium* (Sapindaceae) each produced characteristic nectar peaks during the day. Nectar peaks in males were unsynchronized with those of the hermaphrodites, and overall visitation by suspected pollinators corresponded to high and low periods of nectar production in both types of flowers. Further, Appanah presented data indicating that as visitors left males, they moved to hermaphrodites and in the process transferred pollen.

Other Relevant Bee Observations

In 1974, one of us (GWF) had the opportunity to observe how quickly large bees *Gaesischia exul* and several *Centris* species) respond to changes in nectar production on a given tree. During February of this year, two large trees of *Andira inermis* were monitored for nectar production on two separate days in the Costa Rican dry forest. Flowers at the top and bottom crown levels were monitored to see if nectar quantity varied between these two major regions. One tree was approximately 10 meters in height and was climbed to sample the nectar at the top level. The other tree was approximately 15 meters in height, and nectar from its high-level flowers was secured by standing on top of a platform that was supported by a hydraulic lift. Results of the study indicated no obvious difference in overall nectar quantity between top and bottom-level flowers. However, the study did indicate that during the second peak of nectar production (Frankie et al., 1976) nectar began flowing in the lower branches about one hour earlier than in the high-level branches. From the top of the lift truck, it was possible to characterize the response of the foraging bees to the changes in nectar flow in the following manner: As time progressed from the beginning of the second nectar peak to its end, bees formed a rather continuous ring (a relatively narrow band) around the outside of the crown. As nectar flowed in flowers from bottom to top, the bees, as a group, moved progressively up the tree in a ringlike fashion. Thus, the bees seemed to have no difficulty in tracking nectar availability as it changed through time.

DISCUSSION

Overall results of our studies on mass-flowering, large bee-pollinated tree species suggest that intraspecific differences in flowering phenology, nectar flow, and volume (and probably concentration of nectar) result in an ever-changing resource base that is continually monitored by large bees. Detection of changes, in turn, prompts the insects to make regular intertree movements, thereby effecting outcrossing. Despite the variation in nectar flow and volume, it is still possible to categorize species according to major secretion patterns, and this has been done for almost all of the bee-pollinated tree species in the dry forest study area (Frankie et al., this book; Haber and Frankie, in prep.).

Clearly, more data will be required to examine and test the hypothesis further. The major tasks that lie ahead are: (1) to more fully explore the range of variation within species, (2) to study the bases for the variation, and (3) to document relationships between visitors and the patterns of nectar variation.

The first task is a relatively simple one that will involve three to five workers monitoring nectar simultaneously from the same conspecific trees throughout their blooming periods. In addition, nectar must be sampled from these trees for later chemical analysis (see Baker and Baker, this book).

Variations in nectar production in flowers have been observed by numerous workers, especially in temperate regions. Factors known to influence nectar secretion and concentration include: age of plant, edaphic conditions, height and age of flower on plant, secretion and reabsorption activity of nectaries, insolation and exposure, air temperature, relative humidity, and removal rate of nectar by visitors (Baker, 1978; Pleasants and Zimmerman, 1979; Willmer, 1980; Plowright, 1981; Zimmerman, 1981; and numerous references in Percival, 1965; Corbet, 1978a,b; Feinsinger, 1978; Corbet et al., 1979). One of the few studies that deals with variation in nectar secretion in tropical plants is that of Feinsinger (1978). In his hummingbird plants he has recorded considerable interflower variation in

nectar secretion rates, and in Trinidad he experimentally demonstrated that hummingbirds spend less time visiting flowers that have variable vs. constant volumes of nectar. With regard to interflower differences of the hummingbird-pollinated plants at Monteverde (Puntarenas Province, Costa Rica), he speculated that the variable nectar secretion rates may arise from a complex of environmental and somatic factors (Feinsinger, 1978). A stimulating theoretical treatment of pollinator foraging behavior, changing floral resources, and breeding structure of plant populations is provided by Levin (1978).

In beginning to assess the possible origin(s) of nectar variation in the dry forest trees, those factors that relate directly to daily changes in climatic conditions may prove to be less important than other factors such as age of plant (Wadey, 1961), stage of flowering (Lloyd, 1980; Lloyd et al., 1980), edaphic conditions, etc. The middle of the dry season (mid-January to mid-March), when most of our nectar studies were conducted, is a time of predictable weather, i.e., warm-hot, sunny, and windy conditions. However, on occasional days there are small but noticeable changes in air temperature and wind speed (reduced or accelerated), and brief rain showers may occur very infrequently (Frankie et al., 1974; Frankie and Coville, 1979).

The occurrence of different nectar secretion patterns in nearby conspecific individuals on the same day (e.g., *Myrospermum* and *Tabebuia rosea*) also suggests that other than climatic factors may be influencing nectar production. Further, trees of some species (e.g., *Myrospermum*) may change while others remain constant in their nectar secretion patterns. In Berkeley, California, I. Baker (pers. comm.) simultaneously recorded a series of different nectar secretion patterns in several plants of *Silene alba* (Caryophyllaceae) that were growing under greenhouse-like conditions over a 30-day period.

Several relationships between bee visitors and nectar production patterns await investigation. Visitation studies should be conducted in relatively undisturbed forests that have large bee populations, since the bee populations were incredibly large before widespread disturbance of the dry forest occurred (Frankie et al., this book). This is an important feature of the large bee pollination system, since large numbers of bees remove nectar from flowers quickly, a phenomenon that in turn appears to lead to intertree movements. In the current study, only a limited effort was made to record bee visitation rates, owing to the fact that many study trees were in disturbed sites.

Our hypothesis assumes that large bees respond relatively quickly (probably within half an hour) to changes in nectar flow. Available data suggests that foraging bees do respond quickly to changes in nectar production. That bees move among conspecific individuals in a relatively predictable manner was documented in 1972 on bees visiting *Andira inermis* by Frankie et al. (1976). These workers observed an increasing tendency for bees to move to nearby conspecifics and at least one other mass-flowering tree, *Dalbergia retusa,* on successive days. Corresponding with this pattern was a reduction in the number of bees collected at the original release trees.

In a setting of continually changing nectar, foraging bees probably invest only a small portion of their time memorizing where and when nectar production will occur. Rather, they probably sample nectar regularly from several trees in order to appropriately monitor changes that may occur through time. This type of plant-provoked behavior largely contrasts with the traplining strategy described by Janzen (1971, 1974), which is dependent on the bees' memory to revisit the same plants over a relatively long time period. However, as Janzen points out, traplining bees probably devote a certain small percentage of time to exploring new floral resources, since most traplining plants do not flower year round.

Pollen viability remains a critical question in interpreting the importance of daily shifts by bees among conspecific individuals. In *Andira inermis,* pollen that had been removed from flowers 24 hours previously was found to have comparable viability to pollen that was taken on the current day (Frankie et al., 1976). In a cloud forest site in Costa Rica (Monteverde, Puntarenas Province) we have also observed that pollen retains its viability in some species for as long as five days (Haber and Frankie, in prep.). Extended pollen life would allow for movement of pollen between conspecific individuals over several days, provided the bees retain some pollen on their bodies. However, at this time we are uncertain about how efficiently and when these bees groom themselves. In the future, we will address this question by examining pollen contamination on the bodies of bees in their sleeping aggregations (Frankie, Coville, and Vinson, in prep.).

ACKNOWLEDGMENTS

The research was supported by several grants from the National Science Foundation. The Werner Hagnauer and Dave Stewart families kindly allowed us to use their properties, Hacienda La Pacifica and Comelco Ranch, respectively. We were assisted in the field by P. Bolstad, K. Timmerman, and W. Timmerman. S. Mandel assisted in analyzing the quantitative data, and figures were prepared by J. Fraser, J. Washburn, and C. Tibbetts. R. Thorp offered helpful suggestions on the possible ecological significance of staggered flowering phenologies in tropical forests. H. Baker, I. Baker, K. Bawa, L. Carpenter, S. Corbett, P. Feinsinger, B. Heinrich, S. Koptur, and J. Washburn kindly read an early draft of the manuscript.

LITERATURE CITED

Appanah, S. 1981. Pollination in Malaysian primary forests. *Malaysian Forester* 44:37–42.

Augspurger, C. K. 1980. Mass-flowering of a tropical shrub *(Hybanthus prunifolius):* influence on pollinator attraction and movements. *Evolution* 34:475–488.

Baker, H. G. 1973. Evolutionary relationships between flowering plants and animals in American and African Tropical Forests. *In* B. J. Meggers, E. S. Ayensu, and W. D. Duckworth (eds.), *Tropical Forest Ecosystems of Africa and South America: A Comparative Review.* Smithsonian Institution Press, Washington, D.C., pp. 145–159.

————. 1978. Chemical aspects of the pollination biology of woody plants in the tropics. *In* P. B. Tomlinson and M. H. Zimmermann (eds.), *Tropical Trees as Living Systems.* Cambridge University Press, Cambridge and New York, pp. 57–82.

Bawa, K. S. 1974. Breeding systems of tree species of a lowland tropical community. *Evolution* 28:85–92.

Bawa, K. S. and P. A. Opler. 1975. Dioecism in tropical trees. *Evolution* 29:167–179.

Brown, J. H., A. Kodric-Brown, T. G. Whitham, and H. W. Bond. 1981. Competition between hummingbirds and insects for the nectar of two species of shrubs. *Southwest Natur.* 26:133–145.

Carpenter, F. L. 1976. Plant–pollinator interactions in Hawaii: pollination energetics of *Metrosideros collina* (Myrtaceae). *Ecology* 57:1125–1144.

Corbert, S. A. 1978a. Bees and the nectar of *Echium vulgare*. *In* A. J. Richards (ed.), *The Pollination of Flowers by Insects*. Academic Press, London, pp. 21–29.

———. 1978b. Bee visits and the nectar of *Echium vulgare* L. and *Sinapis alba* L. *Ecol. Entomol.* 3:25–28.

Corbet, S. A., D. M. Unwin, and O. E. Prys-Jones. 1979. Humidity, nectar and insect visits to flowers, with special reference to *Crataegus, Tilia* and *Echium. Ecol. Entomol.* 4:9–22.

Daubenmire, R. 1972. Phenology and other characteristics of tropical semi-deciduous forest in north-western Costa Rica. *J. Ecol.* 60:147–170.

Feinsinger, P. 1978. Ecological interactions between plants and hummingbirds in a successional tropical community. *Ecol. Mon.* 48:269–287.

Frankie, G. W. 1976. Pollination of widely dispersed trees by animals in Central America, with an emphasis on bee pollination systems. *In* J. Burley and B. T. Styles (eds.). *Variation, Breeding and Conservation of Tropical Forest Trees*. Academic Press, London, pp. 151–159.

Frankie, G. W. and H. G. Baker. 1974. The importance of pollinator behavior in the reproductive biology of tropical trees. *An. Inst. Biol. Univ. Nac. Auton., Mexico* 45 *(ser. Bot.)*:1–10.

Frankie, G. W. and R. Coville. 1979. An experimental study on the foraging behavior of selected solitary bee species in the Costa Rican dry forest (Hymenoptera: Apoidea). *J. Kansas Ent. Soc.* 52:591–602.

Frankie, G. W. and S. B. Vinson. 1977. Scent marking of passion flowers in Texas by females of *Xylocopa virginica texana* (Hymenoptera: Anthophoridae). *J. Kansas Ent. Soc.* 50:613–625.

Frankie, G. W., H. G. Baker, and P. A. Opler. 1974. Comparative phenological studies of trees in tropical wet and dry forests in the lowlands of Costa Rica. *J. Ecol.* 62:881–919.

Frankie, G. W., P. A. Opler, and K. S. Bawa. 1976. Foraging behavior of solitary bees: implications for outcrossing of a neotropical forest tree species. *J. Ecol.* 64:1049–1057.

Frankie, G. W., W. A. Haber, H. G. Baker, and I. Baker. 1981. A possible chemical explanation for differential foraging by anthophorid bees on individuals of *Tabebuia rosea* in the Costa Rican dry forest. *Brenesia*. In press.

Gentry, A. H. 1978. Anti-pollinators for mass-flowering plants. *Biotropica* 10:68–69.

Gilbert, L. E. 1975. Ecological consequences of a coevolved mutualism between butterflies and plants. *In* L. E. Gilbert and P. H. Raven (eds.), *Coevolution of Animals and Plants*. University of Texas Press, Austin, pp. 210–240.

Heinrich, B. 1975. Energetics of pollination. *Annu. Rev. Ecol. Syst.* 6:139–170.

———. 1976. The foraging specializations of individual bumblebees. *Ecol. Mon.* 46:105–128.

Heinrich, B. and P. H. Raven. 1972. Energetics and pollination. *Science* 176:597–602.

Heithaus, E. R., T. H. Fleming, and P. A. Opler. 1975. Foraging patterns and resource utilization in seven species of bats in a seasonal tropical forest. *Ecology* 54:841–854.

Janzen, D. H. 1971. Euglossine bees as long-distance pollinators of tropical plants. *Science* 171:203–205.

———. 1974. The deflowering of Central America. *Nat. Hist.* 83:48–53.

Levin, D. A. 1978. Pollinator behavior and the breeding structure of plant populations. *In* A. J. Richards (ed.) *The Pollination of Flowers by Insects*. Linnean Soc. Symp. Ser. No. 6, Academic Press, New York, pp. 133–150.

Lloyd, D. G. 1980. Sexual strategies in plants. I. An hypothesis of serial adjustment of maternal investment during one reproductive session. *New Phytol.* 86:69–79.

Lloyd, D. G., C. J. Webb, and R. B. Primack. 1980. Sexual strategies in plants. II. Data on the temporal regulation of maternal investment. *New Phytol.* 86:81–92.

Opler, P. A., G. W. Frankie, and H. G. Baker. 1980. Comparative phenological studies of shrubs and treelets in wet and dry forests in the lowlands of Costa Rica. *J. Ecol.* 68:167–188.

Percival, M. S. 1965. *Floral Biology*. Pergamon Press, Oxford.

Perry, D. R. and A. Starrett. 1980. The pollination ecology and blooming strategy of a neotropical emergent tree, *Dipteryx panamensis. Biotropica* 12:307–313.

Pleasants, J. M. and M. Zimmerman. 1979. Patchiness in the dispersion of nectar resources: evidence for hot and cold spots. *Oecologia* 41:283–288.

Plowright, R. C. 1981. Nectar production in the boreal forest lily *Clintonia borealis. Can. J. Bot.* 59:156–160.

Rausher, M. D. and N. L. Fowler. 1979. Intersexual aggression and nectar defense in *Chauliognathus distinguendus* (Coleoptera: Cantharidae). *Biotropica* 11:96–100.

Rhoades, D. F. and J. C. Bergdahl. 1981. Adaptive significance of toxic nectar. *Am. Nat.* 117:798–803.

Rockwood, L. L. 1973. Distribution, density, and dispersion of two species of *Atta* (Hymenoptera: Formicidae) in Guanacaste Prov., Costa Rica. *J. Anim. Ecol.* 42:803–817.

———. 1975. The effects of seasonality of foraging in two species of leaf-cutting ants *(Atta)* in Guanacaste Province, Costa Rica. *Biotropica* 7:176–193.

Wadey, H. J. 1961. Nectar yield and age. *Bee Craft* 43:8, 86.

Willmer, P. G. 1980. The effects of insect visitors on nectar constituents in temperate plants. *Oecologia* 47:270–277.

Zimmerman, M. 1981. Patchiness in the dispersion of nectar resources: probable causes. *Oecologia* 49:154–157.

Section VI
TEMPERATE AND TROPICAL PHENOLOGICAL STUDIES

19
STRUCTURE OF PLANT AND POLLINATOR COMMUNITIES

John M. Pleasants

Department of Botany
Iowa State University
Ames, Iowa

ABSTRACT

The structure (pattern of niche relationships) of both plant and pollinator communities is discussed, with particular emphasis on bumblebee-dominated communities. The role of competition in producing structure is examined. In the first section I review the evidence that the differences among plants within a community are due to segregation to reduce competition for pollinators. Plants visited by bumblebees typically belong to one of three guilds. Each guild is associated with either a short-, a medium-, or a long-tongued bumblebee, and guild members have corolla lengths that generally match the tongue length of their principal visitor. The evidence for temporal segregation in flowering among plants in the same guild, as well as available methods for detecting regularity in flowering phenologies, is discussed. Another way a species can reduce loss of pollinator visits is by increasing its attractiveness relative to competitors. Increasing attractiveness, by increasing nectar production rates, may be an important option for plants and may diminish the necessity for divergence in flowering time. In the second section evidence for niche partitioning of floral resources by bumblebees is reviewed. Differences in tongue length among co-occurring bumblebees provide each species with relatively exclusive use of a set of plant species. The reasons why bumblebees of a particular tongue length visit only plants with matching corolla lengths are examined. The main factor that excludes bees from nonmatching flowers is the actions of more efficient bees, which lower the standing crop of nectar to unprofitable levels. Anomalies in the tongue-length:corolla-length relationship are discussed. Explanations for "too many" representatives of a particular tongue length class in certain communities are given. Evidence for other kinds of niche partitioning among bumblebees, including diurnal segregation, is reviewed.

KEY WORDS: Bumblebees, Colorado, competition, competitive release, community structure, corolla length, diurnal segregation, exploitation, flowering phenologies, foraging efficiency, guilds, interference, niche relationships, nectar production rates, null hyothesis, pollinator constancy, pollinator limitation, plant attractiveness, pollinator preference, Rocky Mountains, segregation, tongue length.

CONTENTS

PLANT COMMUNITY STRUCTURE IN RELATION TO POLLINATION

Introduction

Community structure can be defined as the niche relationships among species that use a particular set of resources. When defined in this way, every community will have structure no matter how haphazard. What is of interest, however, is whether there is a pattern to the structure, i.e.,some regularity to the relationships among species such that communities in different locales or with different species composition have similar structure. If such patterns of structure exist, it implies the presence of some structuring agent and the operation of an important and ubiquitous ecological principle. A theoretically likely candidate for a structure-producing agent is competition. Because the effects of competition place a limit on how similar species can be and still coexist, communities in which competition is important should display a pattern of resource partitioning. Species should be segregated from one another or overdispersed in niche space.

For plants, the activity of pollinating agents is a resource that is of critical importance for species that are neither autogamous, apomictic, nor exclusively vegetative reproducers. To what extent can the structure of plant communities be explained by competition for pollinators? I will discuss primarily plant communities whose pollinator fauna is dominated by bumblebees. This includes arctic, alpine, and bog plant communities. This choice reflects my own area of interest and that of many

others as well. Bumblebee-dominated communities have received more attention and are better understood than communities dominated by other pollinating agents. Bumblebees are also an important component of the pollinator fauna of other temperate zone ecosystems including prairies, woodlands, and some desert communities. Where bumblebees are of lesser importance, solitary bees typically are the major pollinators.

Competition can occur when two (or more) species share pollinators and time of flowering. The competitive effect can come about in two ways: (1) when pollinator visits to one species reduce the number of visits to the other (exploitation competition); (2) when pollinator moves between individuals of the two species reduce the effectiveness of pollinator visits (interference competition). Interference can occur when pollen from one species ends up on another species. Interference may affect male function by reducing the amount of pollen that reaches a conspecific. It can also affect female function by clogging stigmatic surfaces with foreign pollen that may limit the ability of conspecific pollen to fertilize ovules. These two types of competition are not mutually exclusive, and one, neither, or both may be operating. Interference and exploitation can be avoided in somewhat different ways (see Fig. 19-1). All the means of avoiding exploitation (by differences in pollinators or time of flow-

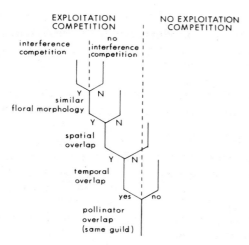

Figure 19-1. Model showing the relationship between exploitation and interference competition for pollinators and the possible means of avoiding competition. Although the model is dichotomous, implying either complete overlap or no overlap, in reality plants may overlap partially in one or more categories, giving rise to varying degrees of competition.

ering) also serve to reduce interference, but some means of avoiding interference (pollinator constancy, spatial segregation) do not reduce exploitation. Which type of competition is most important when competition does occur is a matter of some dispute (Silander and Primack, 1978, Waser, 1978a,b, and see also Chapter 13 in this book; Pleasants, 1980). Certainly the interference component would be reduced when the pollinators are generally constant, as is the case for bumblebees (Macior, 1974; Pleasants, 1980). It might be argued, of course, that the reason constancy exists is that the plants have evolved differences in such features as floral morphology, color, odor, or inflorescence height so as to promote pollinator constancy and reduce interference. In the discussion that follows, I will consider only exploitation competition, except where noted.

A number of investigations have merely suggested that competition for pollinators might be occurring based on circumstantial evidence, including studies of arctic communities (Hocking, 1968; Kevan, 1972), bog communities (Pojar, 1974; Reader, 1975; Heinrich, 1975, 1976b; Small, 1976), and meadow communities (Mosquin, 1971; Pojar,1974). Most of

these studies have shown at least a potential for competition in that plants cannot set seed without pollinator visits, and some species share pollinators. Some of the circumstantial evidence cited has been the presence of measurable standing crops of nectar, implying not many bee visits (Hocking, 1968), and displacement in flowering time of congeners (Heinrich, 1975) or confamilials (Reader, 1975).

Before addressing the question of competition, however, we first must determine whether pollinators are in fact a resource that is in limiting supply and thus a resource for which plants may compete. Unfortunately, data bearing on this question are very meager. For many species it can be shown by excluding pollinators that plants require pollinator service to set seed. Thus we are dealing with a potentially important resource. But is it a limiting resource? Typically seed and fruit set of plants open to ambient pollination fall below 100% (i.e., less than 100% of the ovules mature, or less than 100% of flowers set any fruit at all). However, many species may be energetically limited and thus unable to mature all the ovules that get fertilized (Sutherland, 1980; Udovic, 1981). Thus less than 100% seed set may not represent pollinator limitation. In some cases frequency of visitation to flowers on plants has been correlated with their seed set (Zimmerman, 1980), which is much better evidence for resource limitation. Pollinator limitation is also indicated in studies where higher seed set occurs with hand pollination than with natural pollination (Bierzychudek, 1981). More studies of this sort are needed.

Segregation by Pollinator Use

As previously mentioned, the two most important niche dimensions with regard to potential exploitation competition for pollinators are the type of pollinator used and the time of flowering. These niche axes are in most cases independent of one another so that the total overlap (competitive effect) between species is equal to the product of the overlaps on each axis. Thus if overlap on one dimension is zero, the total overlap is zero, and the species are not competitors. Considering the pollinator dimension first, we can inquire as to what extent plant species differ in the species or type of pollinator they use. (For convenience I employ the term "use" here although it is debatable as to who is using whom. It may be more appropriate to call pollinators the users, because they are mobile and can exhibit choice behavior. However, plants can certainly influence and manipulate that choice by their floral morphology and quantity of nectar reward as well as other attractants such as color and odor.) A number of studies of bumblebee-pollinated plants have shown differences among them in their principal bumblebee visitor(s) (Reader, 1975; Heinrich, 1976a: Pleasants, 1980). These differences appear to be due primarily to the fact that the plants differ in the corolla depth of their flowers. Because there is a correspondence between bumblebee tongue length and the corolla depth of the flowers they visit, plants with different corolla depths tend to have different visitors. The reason for this relationship will be discussed later. In meadows in the Rocky Mountais of Colorado the bumblebee fauna, in general, consists of three species, one with a short tongue (approximately 5.75 mm—worker tongue length), one with a medium-length tongue (7.7 mm), and one with a long tongue (10.5 mm). Each bumblebee is the primary visitor to a set of plant species (guild) with corolla depths that bracket its tongue length. Guilds are somewhat overlapping entities (Pleasants, 1980) such that some plant species may be visited to a large extent by two different bumblebees. In general, however, each plant species has a single major bumblebee visitor. Plants in different guilds will not be competitors. Thus segregation along the pollinator dimension represents the first level of niche separation in the community. However, members of the same guild are still potential competitors, and we can inquire to what extent competition among them is minimized by segregation along the second niche dimension, time of flowering.

Segregation by Flowering Time

First we need to determine whether two simutaneously flowering species that share pol-

linators are mutually depressing the number of visits, and more importantly the seed set, each receives. This is best studied by an experimental approach, although few studies of this type have been made to date. Those studies that have been done (Levin and Kerster, 1967; Levin, 1978; Waser, 1978a; Campbell and Motten, 1981) have concentrated on the interference aspect of competition. Their general approach has been to demonstrate a reduction in seed set when plants are grown in mixed-species plantings vs. monospecific plantings. While this can show a potential for interference, it must be combined with field observations to show that interference is important under natural conditions. For example, Campbell and Motten (1981) found a potential for interference by showing that for flowers allowed only one visit during their lifetime, seed set was lower if the only visit was preceded by a pollinator visit to a heterospecific rather than a conspecific. However, under field conditions they found flowers received enough visits for maximum seed set. When interference competition is thought to be responsible for sequential flowering in two species, seed set during the period of flowering overlap should be compared with seed set during nonoverlap (Waser, 1978a). However, one cannot say whether the decline in seed set observed during overlap is due to interference or due to a reduction in visitation, either because of exploitation competition or because pollinators are less attracted to species when their floral density is low (during the overlap period one species' floral abundance is declining, while the other's is just beginning to increase). Observations of pollinator visitation rates per flower before, during, and after the overlap period are needed to supplement seed set measurements. A second experimental approach, which has not been tried before (but a study is currently in progress—J. Pleasants and M. Zimmerman), involves performing experimental manipulations during the period when the two species overlap in flowering. If this is a period of competition, then if the density of one of the species is reduced (by bagging inflorescences), the other species should experience competitive release and have higher seed set than before or after the reduction. The reciprocal density reduction should also be done.

Whether plants that share pollinators exhibit temporal segregation is a question that has received quite a bit of attention. If process can be inferred from pattern, a regular spacing of blooming periods is inferential evidence that competition for pollinators exists. No alternative hypothesis would generate such a pattern. However, a weakness of many of the papers purporting to show a regular temporal spacing pattern is a lack of statistical verification. The conclusions are based on simple inspection of graphs showing the blooming periods of all species plotted over the entire flowering season. Several approaches have been taken to provide a statistical test for the regularity of flowering phenologies. In general these approaches compare the actual blooming dispersion pattern to expected patterns based on the null hypothesis of random arrangement of flowering times. One of them (Poole and Rathcke, 1979) examines the variance in the distance in time between the flowering peaks of temporally adjacent species. A simple expression gives the expected variance, based on random assignment of flowering peaks. The ratio of observed to expected variance is calculated. A ratio greater than 1.0 indicates aggregation of blooming, and a ratio less than 1.0 indicates a regular pattern. The significance of the ratio value can be determined with a X^2 test. This technique has the advantage of being easy to calculate, since it is strictly analytical (unlike the computer simulation techniques mentioned below). Consequently it has been employed in a number of studies (Poole and Rathcke, 1979; Anderson and Schelfhout, 1980; Pleasants, 1980; Rabbinowitz et al., 1981). The drawback of this technique is that it uses only information on the peak date of flowering for species and ignores the shapes and breadths of their flowering curves. As a result, it has less power to discern the presence of segregated blooming times when they do exist than have the simulation methods (Pleasants, 1980). In other words, this technique is conservative. It can be used to prove competition but not disprove it. When it shows a regular pattern, such a pattern does exist; but when it indicates a random pattern, there may actually be a regular one. A second analytical technique (Heithaus, 1974; Anderson and Schelfhout, 1980) determines if the same number of species is in flower at all points in time during the season (which is what a regular distribution would produce). The sequence of number of species in flower for each successive time interval is taken as the observed distribution and is compared with the expected flat distribution of an equal number of species in flower (based on the average number). Note that this method tests the competition hypothesis of regular blooming rather than the null hypothesis of random blooming. It also tends to be conservative in declaring a pattern to be regular.

Another approach to testing the null hypothesis uses computer simulation. In a technique that I developed (Pleasants, 1977, 1980), random communities are assembled by taking each of the actual species in the community and determining the time of initiation of flowering by choosing a random date between the beginning and end of the flowering season. The shape and breadth of each species flowering curve are retained. After all species have been randomly positioned, the average overlap in flowering for all species pairs is computed. A new average overlap is calculated after each simulation run. The average overlap for the actual community is tested to see if it could have come from this population of simulation values. Randomly constructed communities would be expected to have higher overlaps than communities in which competition has acted to minimize overlap. Another simulation technique (Thomson, 1978a) uses a different approach. Here the competition hypothesis is stated as follows: if competition has really molded blooming times, then a species' present temporal position is the one that results in the least overlap with and loss of pollinator visits to temporal neighbors. In the computer simulation each species slides along the temporal continuum, and the competitive load it experiences at each step of the way is determined. The potential load, if it bloomed at another time, is compared to the load it experiences in its present position. The potential load within the immediate vicinity of its present position is of greatest interest. This is so because if flowering time were to change, it would probably evolve as a gradual displacement from its present position. If a lower load were available some distance from its present position, flowering at that time would have to evolve in a leap-frog fashion to bypass intevening periods of heavier competition.

The conclusions reached in those studies of flowering phenologies where statistical procedures have been used are somewhat equivocal in their support for competition. Because of the ease of calculation the analytical techniques have been used most frequently, despite their lower sensitivity to nonrandom patterns. Tall grass prairie communities, where the major pollinator groups are Apidae (honeybees and bumblebees), Halictidae, and Anthophoridae (Parrish and Bazzaz, 1979), have received a lot of attention in this regard. Anderson and Schelfhout (1980) found a significant deviation from constant numbers of species in flower when the whole season was considered but not when smaller (40-day) intervals were used. Rabbinowitz et al. (1981) found no evidence of regularity in the blooming times of all entomophilous species using the Poole and Rathcke (1979) method. Parrish and Bazzaz (1979) found three seasonal aggregations of blooming times with gaps between them. While they did not use statistics, they reasoned that if competition were important, these gaps would have been filled by species seeking an escape from competition. Using computer simulations, Pleasants (1980) found that 10 of 11 guilds of bumblebee-visited plants in Rocky Mountain meadows had nonrandom flowering phenologies. Thomson (1978a) analyzed other Rocky Mountain plant communities using his technique and found that some plant species had a temporal position that minimized overlap, but others did not.

One problem with many of the studies that examine flowering phenologies is that the guild structure of the community is not identified. Usually all plants visited by bees are included in the phenological pattern. This is bound to include plant species that are not visited by the same species of bee and thus are not competitors. We would not expect such species to have

segregated blooming periods. If competition were important, each guild would have segregated blooming periods, but combining all guilds superimposes several regular patterns that could result in an overall random pattern. In addition, the resources (pollen or nectar) that the pollinators collect from each species often are not specified. Bumblebee pollinators, for example, often visit some species only for nectar, while others are visited only for pollen, and still others for both pollen and nectar. Since bees must collect both pollen and nectar from flowers to maintain their colonies, a visit to a plant for nectar does not reduce the supply of visits to pollen-providing plants, and vice versa. Thus competition between exclusively pollen- or nectar-providing species does not occur, and competition between either nectar or pollen producers and those providing both is reduced. Thus in examining flowering phenologies, it is improper to lump species that pollinators use for different resources.

Other Community Aspects of Competition

It is also useful to examine the validity of the competition hypothesis itself. What is often not realized is that sequentially flowering plants may be involved in mutualistic as well as competitive interactions (Waser and Real, 1979; Thomson, 1980, 1981). The mutualistic effect may be due to several factors. For example, when pollinator population size builds up over the season (as it does for social bees), the presence of an earlier blooming species benefits a later blooming species by providing a resource that contributes to the production of new worker bees, which then increases the pollinator availability for the second species. The second species may also benefit the first, albeit much more indirectly, by providing a resource that ultimately influences the number of new queens the colony produces and thus the number of bees available to the first species in the following year. Mutualism of this sort can also occur with pollinators that do not exhibit population increases over the flowering season. For hummingbirds, territory establishment and population density are determined by the floral abundance of early blooming species, and thus the availability of pollinators to late blooming species is dependent on these early species (Waser and Real, 1979). The effect of this kind of mutualism on segregation of flowering times is not clear. It would not appear to present any selection pressure for convergence in flowering time. A second type of mutualistic effect could promote convergence. When two species that share pollinators are patchily distributed or widely dispersed, there may be an advantage to flowering together (in both space and time) so as to provide a large floral display to attract pollinators (Thomson, 1978b, 1981; Brown and Kodric-Brown, 1979).

Even in a situation where competitive interactions override mutualistic ones, do we really expect competition to produce a regular dispersion of flowering times? In general, a perfectly regular dispersion of resource utilization curves along a resource continuum is expected, theoretically, when species have equal carrying capacities (K's) and when the competitive effect of species on one another is symmetrical ($\alpha_{AB} = \alpha_{BA}$) (Roughgarden, 1979). When K's and α's are different, the limiting similarity between successive pairs of species along the continuum will vary. However, the general pattern for the whole assemblage of species still may be somewhat regular, although not perfectly regular. One other point to consider is that there may be a difference between the temporal *overlap* between two species and their *competitive* effect on each other (α's). Competition theory states that competition should be minimized, not necessarily that overlap should be. If plants differ in their competitive ability, i.e., their ability to attract pollinators, then overlap may not be a good measure of competition. One way species may differ in their ability to draw visitors away from other species is through differences in floral abundance. When a rare species overlaps in blooming with an abundant species, the competitive effect will be asymmetrical. The rare species will experience a large loss of pollinator visits owing to the presence of the common species, but the common species suffers little from the presence of the rare one. If competition for pollinators is important, we might expect a species that experiences a large competitive load

to obtain more visits by (a) diverging in flowering time from competitors, or (b) increasing its attractiveness to pollinators (per flower) relative to competitors. This second option is rarely considered. It is equivalent to a species increasing its competitive ability (increasing α).

For animals, increasing competitive ability generally means developing interference mechanisms (such as aggression) to prevent the other species from getting resources. Selection for competitive ability has been considered theoretically (α-selection; Gill, 1974), although its community structure implications have not been explored. If this option is important to plants we would expect rare species, which potentially would have a competitively inferior position, to have evolved increased attractiveness per flower to enable it to persist in the competitive environment. Are rare species more attractive than common species? In several studies the investigators noted the presence of rare but highly attractive species (Beattie et al.,1973; Heinrich, 1976a). In Rocky Mountain meadows I found that a species' abundance was negatively correlated with its attractiveness (number of pollinators visits per flower) ($r = -.82$, $p < .01$, $n = 44$; Pleasants, 1980). The competitive load a species actually experienced (the number of visits it lost to competitors) was significantly reduced compared to what it would have been if it had been equal in attractiveness to its competitors (Pleasants, 1977). In addition, species that potentially stood to lose the most because of competitors were the most attractive ($r = .345$, $p < .05$, $n = 43$). The attractiveness of flowers to pollinators is primarily a function of the flower's nectar production rate (Pleasants, 1981). Thus by increasing its nectar production rate (npr) a rare or competitively disadvantaged species may be able to attract enough visitors to persist despite its overlap with other more common or competitively superior species. The fact that plants apparently can exercise the option of increasing competitive ability, which can serve as an alternative to diverging in flowering time, is an additional reason why a regular segregation of blooming times may not be observed. It also suggests that looking at flowering times alone to discern the importance of competition is inadequate.

In summary, the competition hypothesis can be stated as follows: To coexist with other species in a community a plant species must be able to acquire an adequate (which we leave undefined) supply of pollinator visits from the pool of available visits. A species can achieve this by differing from other species in the pollinators it uses and diverging in flowering time from species with which it shares pollinators. In addition, however, it can ensure visitation even when flowering with a competitor by increasing its attractiveness. If attractiveness can overcome a poor competitive situation, should we really expect a regular pattern of blooming among species within a guild? Couldn't species simply invade the guild at any point along the temporal continuum and become established by having a high attractiveness? This would result in a somewhat random flowering phenology sequence. This question is difficult to answer, but it should be pointed out that an attractiveness sufficient to invade or remain in a community must be bought at a high price. This is so because attractiveness is based on nectar production rate, and nectar production can be energetically costly (J. Pleasants and S. Chaplin, unpub. data). Thus it may be evolutionarily easier for a species to change its flowering time rather than increase its npr. I suspect, then, that most guilds will show a flowering pattern approaching regularity, but that a correlation between attractiveness and potential competitive position will also exist.

Interference Competition

I now will deal briefly with differences among species in a community that can be related to reducing the interference component of competition for pollinators. (See Waser, this book, for additional discussion of this topic.) Interference is potentially important for species that share both pollinators and time of flowering. Since interference can occur only when an individual pollinator moves interspecifically during a foraging bout, plant attributes that reduce this possibility will reduce interference. These attributes are discussed in the following paragraphs.

Spatial Isolation. During foraging bouts, many kinds of pollinators tend to move to neighbors of the inflorescences they have just left (Pyke, 1978; Zimmerman, 1979, 1981). As a result, when potentially interferring species are spatially isolated from one another, interspecific visits are unlikely. At one study site in the Colorado Rocky Mountains, I found that near the end of the flowering season five composite species (family Asteraceae) were in bloom simultaneously. The greater the temporal overlap between pairs of these species (and thus the greater the potential for interference), the larger was the degree of spatial isolation between them (Pleasants, 1980). Hurlbert (1970) found a similar relationship among several co-occurring goldenrod *(Solidago)* species.

Pollinator Constancy. Both bumblebees and honeybees are known to exhibit a fair degree of faithfulness to one species during a foraging bout (Free, 1963, 1970; Macior, 1974; Pleasants, 1980). Plants can promote pollinator constancy by differing in floral morphology. There are two reasons for this: (1) Morphological differences (including color and flower presentation differences) will allow a bee to discriminate between species while foraging. (2) The reason bees are constant in the first place is to increase foraging efficiency. Interspecific foraging will reduce efficiency by requiring repeated switching in the behavioral tactics (handling) involved in entering flowers. The greater the morphological difference between a plant species and a competitor, the less likely it is that a pollinator will include both during a foraging bout. Anderson and Schelfhout (1980) analyzed a group of composites blooming in tall grass prairie. They found a negative correlation between the morphological similarity of species and temporal overlap. Pollinators were not examined in this study, so we do not know if these species are in fact competitors, but composites do usually overlap in visitors (pers. obs.). Height of an inflorescence is also a morphological feature that can promote constancy, since some pollinators tend to remain in one horizontal zone while foraging (Levin and Kerster, 1973). Waddington (1979) found in Rocky Mountain meadows that species that share pollinators are more likely to be different in inflorescence height than those that do not share pollinators.

Differential Pollen Placement. Floral morphology differences may also contribute to reducing interference even when pollinators are not constant. The morphology of a flower will determine how a bee enters it, and what part of the bee's body contacts the anthers. For example, some flowers place pollen on a bee's back (nototribic), whereas others place it on a bee's underside (sternotribic). It is possible that two plant species could avoid interference by using different parts of a bee's body for pollen deposition and removal. Examples of this are scarce, but Macior (1971) found that pollinia from several co-occurring *Asclepias* species tended to be located in different places on a bumblebees body.

Diurnal Separation. Plant species with seasonal flowering overlap may reduce interference by having diurnal differences in time of flower opening or pollen and nectar presentation. In Rocky Mountain meadows two composites, *Erigeron speciosa* and *Viguera multiflora,* overlap broadly in pollinators and time of flowering. Bumblebee visitation to *viguera* is highest from the morning until midday, at which point it decreases markedly (J. Pleasants and B. Pleasants, unpub. data). Both species are visited only for nectar. Visitation to *Erigeron* does not begin until midday. Pollen shedding in *Viguera* begins in the morning, whereas it begins at midday for *Erigeron.* Presumably there is a diurnal nectar secretion pattern for each species corresponding to the times of highest visitation, but we have yet to examine this.

For all of the examples cited above, it is difficult to prove that the potential for interference competition was responsible for the evolution of the spatial, temporal, or morphological differences. All that can be said is that the patterns are consistent with this explanation. Careful studies will have to be done to verify this. For example, for two species that have little spatial overlap, apparently to avoid interference, the pollination fate of individual plants found in the proximity of individuals of the other species could be compared to the fate of well-isolated individuals. Naturally occurring pockets of spatial overlap between the two

species could be used for this experiment, or plants could be transplanted to create such areas (Waser, 1978a). Manipulating diurnal blooming patterns or floral morphology to test their effectiveness at reducing interference may not be possible.

Other Community Patterns

Another aspect of community structure of interest is the number of species that co-occur. In Rocky Mountain meadows the number of bumblebee-pollinated species is about 20 at 3000 meters in elevation and decreases to 7 at 4000 meters (Pleasants, 1977). This same trend of decreasing plant species diversity with increasing elevation is also found for bee-pollinated plants in California mountain communities (Moldenke, 1975). It is probably due to a shortened growing season at higher elevations. Within Rocky Mountain meadow communities the average number of plant species in a guild decreases, going from guilds visited by short-tongued bumblebees (8 plant species) to medium-tongued bees (6 plant species) to long-tongued bees (4 plant species) (Pleasants, 1977). This same trend of more plants with short corollas (visited by short-tongue bees) and fewer with long corollas is found in northern European bumblebee-pollinated plants (Ranta et al., 1981). Apparently more plants can be supported on the services provided by short-tongue bees. One reason for this is that short-tongue bees are more abundant, particularly later in the season, than long-tongue bees (Lundberg and Ranta, 1980; Pleasants, 1981). The more abundant short-tongued bees may be a resource that can be more finely subdivided, thus accounting for the higher number of species using them. Of course, it is not clear whether the higher floral abundance is due to higher bee abundance or vice versa, or, as is likely in this coevolved system, a bit of both. However, there are some indications that the intrinsic potential for population increase may be greater for shorter-tongued bees. The larvae of short-tongued bees develop more quickly into adults, and the number of eggs laid per brood cell is greater (Hobbs, 1966a,b, 1967, 1968).

Another possible consequence of the larger

resource base repesented by short-tongued bees is a greater degree of specialization among the plant species in this guild with regard to the type of resource(s) offered to the pollinator. In Rocky Mountain meadows this guild is the only one that has one or two species at a site that make no nectar and are visited only for pollen. Most of the other species in this guild are visited for nectar only. Going from the short- to the long-tongued bee guild, there is an increasing emphasis on species that are visited for both pollen and nectar (proportion of species: 0.16 short, 0.50 medium, 0.93 long; Pleasants, 1977).

COMMUNITY STRUCTURE OF POLLINATOR FAUNA

Introduction

Community-level studies of pollinators are relatively scarce. More commonly, the investigator is interested only in the pollinators associated with a particular plant species. The best-studied pollinator communities are those dominated by bumblebees although community studies of mixed-bee faunas (bumblebees and solitary bees) (Moldenke, 1975; Tepidino and Stanton, 1981) and hummingbird faunas (Brown and Kodric-Brown, 1979) also exist. Consequently, in the discussion that follows I will attempt only to summarize what is known about bumblebee communities. Bumblebees tend to predominate in cool environments because their ability to thermoregulate allows them to forage at low ambient temperatures when other bees cannot. Such environments occur at high altitudes, at high latitudes, and in certain coastal areas.

Bumblebees have an annual life cycle. Queens, already fertilized, emerge from their overwintering hibernaculae in the spring. Nests are established, typically in an abandoned rodent burrow, and the first brood is begun. After a sufficient number of worker bees have been produced, the queen ceases foraging and remains in the nest. Workers forage throughout the summer, and new workers are produced. In the late summer males and the new queens are produced by the colony, and

mating takes place. The new queens form the base for the next summer's population.

As with the discussion of plant community structure, we can inquire to what extent the niche relationships and differences among bumblebee species within a community have been the result of selection to avoid competition. First it must be shown that the resources that plants provide to pollinators (pollen and nectar) are in limiting supply. It can be argued that anytime a bee visits a flower that was previously visited by another bee it will get less nectar or pollen than it could have received. The presence of the first bee has a negative effect on the second; i.e., competition exists. The amount of pollen and nectar resources brought back to the nest determines the number of reproductives (queens and males) that the nest can produce. Thus the efficiency of workers in gathering resources will have a rather immediate effect on fitness. In any plant-pollinator system, even one with a low density of pollinators relative to flowers, competition can exist. However, the degree of competition, and the strength of selection pressures to evolve strategies to maximize foraging efficiency, will increase as the bee-to-flower ratio increases. The fact that competition is not an all-or-none phenomenon, but can exist in varying degrees, is often forgotten. If pollinator densities are low, such that there is competition among plants for pollinators, that does not necessarily imply a lack of competition among pollinators for flowers. More probably, there is competition on both sides but to varying degrees in different communities and at different times of year within a community.

There are several examples of competitive release of one bumblebee species following the removal of another that indicate that bumblebee species are resource-limited and in competition with one another. Inouye (1978) examined two plant species, *Aconitum columbianum* [(corolla depth) = 8.44 mm] and *Delphinium barbeyi* (C.D. = 13.96 mm). *Aconitum* is visited primarily by the medium-tongue-length bee *Bombus flavifrons,* but is also visited to a lesser extent by the long-tongued *B. appositus. Delphinium* is visited primarily by *B. appositus* and to a lesser extent by *B. flavifrons*. The primary visitor on each of these plant species has a faster foraging rate (flowers per second) than the secondary visitor (Inouye, 1978). In a monospecific patch of flowers, when the primary visitor was removed the abundance of the secondary visitor increased significantly. These results indicate that on each plant, the presence of the primary visitor lowered the standing crop of nectar so as to make foraging by the secondary visitor relatively unprofitable. In other words, under ordinary conditions the secondary visitor was competitively excluded, but it experienced competitive release when the primary visitor was removed. Pleasants (1977, 1981) found competitive release in a natural species removal experiment. In a year when the feral honeybee *Apis mellifera*, a short-tongued bee, was virtually eliminated from two study sites owing to a severe winter, the abundance of short- and medium-tongued bumblebees increased markedly over their abundance the previous year. The number of additional bumblebees seen in this year was amost exactly equal to the number of honeybees missing from the previous year. Density compensation of this sort is a common indicator of competitive release (Yeaton, 1974). The total number of bees (bumblebees and honeybees) seen during the summer was the same for the two years, indicating that the floral resources in a meadow can support only so many bees. This means that floral resources were limiting.

Segregation by Tongue Length

If interspecific competition does occur in bumblebee communities, what are its ramifications in terms of community structure? What are the differences among bumblebee species that reduce competition and allow them to coexist? Figure 19-2 gives a summary of the possible coexistence mechanisms, which will be discussed below. In meadows in the Rocky Mountains of Colorado the bumblebees fall into three tongue-length classes: short [Queens (Q) 8–8.5 mm, workers (w) 5.75mm], medium (Q 10.25 mm, w 7.8 mm) and long (Q 12–13 mm, w 9.5–10.5 mm) (Inouye, 1977; Pleasants, 1980). At any particular site there is usually

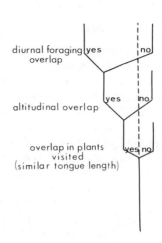

Figure 19-2. Model showing means of avoiding competition among bumblebee pollinators. Partial overlap may also occur.

one representative of each class. As was mentioned in the discussion of plant guilds, there is a morphological relationship between the corolla depth of flowers and the tongue length of the bees that visit them (see also Ranta and Lundberg, 1980). Thus bumblebees with different tongue lengths will tend to specialize on different plant species. This segregation will reduce competition and also suggests that competition is important. It is somewhat difficult, however, to rigorously test the hypothesis that the tongue-length differences among bees arose from selection pressures exerted by competition. How would one explicitly state the null hypothesis of noncompetition so that the observed pattern of tongue-length differences could be compared to it?

Why do bumblebees prefer flowers with corolla depths that are approximately equal to their tongue length? It is fairly easy to see why bees would avoid flowers with a corolla length longer than their tongue length. They would be unable to reach the nectar unless it was allowed to accumulate in the corolla tube. What is less obvious is why bees avoid flowers with corolla lengths shorter than their tongue length. One argument that has been made is that when a bee has a tongue that is too long for a flower, it is less efficient in foraging on the flower than a bee with the appropriate tongue length (Inouye, 1980; Ranta and Lundberg, 1980). This efficiency argument is summarized in Fig. 19-3.

What is the evidence for the efficiency relationships? First, are short-tongued bees the most efficient on short-corolla flowers? It is difficult to find long-tongue (l-t) bees foraging on short-corolla (s-c) flowers so as to compare their efficiency with bees of medium and short tongue length (m-t and s-t). However, I was able to overcome this difficulty by making observations at the end of the flowering season, when composites (which have short corollas) dominate the flora. I observed worker bees of all tongue-length classes foraging on *Aster bigelovii* [C.D. 6.05 mm (short)] and found the expected pattern C (Fig. 19-3). Inouye (1980) found only partial support for pattern C. Medium-tongued (m-t) bees foraged more slowly on s-c flowers than s-t bees, but l-t bees foraged as rapidly as s-t bees. Thus the efficiency argument is not sufficient to explain the avoidance of s-c flowers by l-t bees. Second, are medium-tongued bees the most efficient on medium-corolla flowers, as depicted hypothetically in pattern B, Fig. 19-3? Ranta and Lundberg (1980) analyzed the work of several authors and found U-shaped curves similar to pattern B. However, all the plant species they discuss are legumes that depend on a tripping mechanism for entry to the flower. Shorter handling time on such flowers by m-t (and thus medium body weight) bees may be more a function of the bees' being of the proper weight to depress the keel of the flower than of the bees' having the appropriate tongue length. For non-legumes, G. Pyke (pers. comm.) finds pattern B for bumblebees foraging on *Agastache urticifolia* [C.D. = 8.7 mm (medium)]. Further support, at least for the right half of pattern B, comes from the study of Inouye (1978) cited earlier, where the m-t bee, *B. flavifrons,* forages more rapidly than the l-t *B. appositus* on *Aconitum* [C.D. = 8.44 (medium)]. Third, are long-tongued bees the most efficient on long-corolla (l-c) flowers (pattern A, Fig. 19-3)? The superior efficiency of l-t bees on long-corolla flowers is well documented (Inouye, 1980).

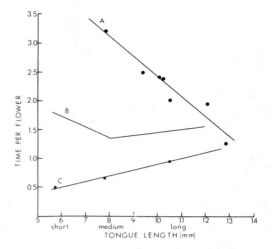

Figure 19-3. The foraging speed of bumblebees with different tongue length on flowers with different corolla depths. A. Bees foraging on a long-corolla flower, *Delphinium barbeyi.* Corolla depth (C.D.) = 13.96 mm. Data from Inouye (1980). B. Bees foraging on a medium-corolla flower (hypothetical, no data available). C. Bees foraging on a short-corolla flower, *Aster bigelovii.* C.D. = 6.05 mm.

In general, then, the efficiency argument appears to have some validity. However, efficiency is only a part of the explanation for the corolla-depth preferences of bees. Note in Fig. 19-3 that bees of any given tongue length can always visit s-c flowers more rapidly than medium- or long-corolla flowers. The relative positions of curves A, B, and C reflect the general pattern of increasing handling time for flowers of increasing corolla depth (Inouye, 1980). Thus if preference were based on minimizing handling time alone, then all bees, no matter what their tongue length, should prefer s-c flowers. Clearly more is involved. There is evidence that bumblebees prefer those plant species that provide the highest net energy intake rate (Pyke, 1980; Pleasants, 1981); i.e., that they are foraging optimally. The net energy intake rate per flower is equal to $E_g - E_l/t$ where E_g is the energy gained from the nectar in the flower, E_l is the energy lost while obtaining the nectar, and t is the handling time per flower. The efficiency argument only considers t, but clearly the net profit, $E_g - E_l$ must be considered as well. Long-corolla flowers typically have higher standing crops of nectar than s-c flowers (Heinrich, 1976c; Pleasants, 1981). This is due partly to the higher nectar production rate of l-c flowers (Heinrich, 1976c; Pleasants, 1981) and partly to the lower foraging activity of bees on them. Thus E_g increases with increasing corolla depth. E_l increases with increasing bee body size (Pyke, 1980), and body size and tongue length are well correlated.

Keeping these facts in mind, and given that bees prefer those species that maximize their net energy intake rate, we can make the following explanation for the observed pattern of bees' preference for flowers whose corolla length matches their tongue length. *Short-tongued bees for s-c flowers:* s-t bees are mechanically excluded from l-c flowers; on medium-corolla flowers they may not be able to reach the nectar when other bees are present, and, in addition, they forage more slowly on m-c flowers than on s-c flowers. This leaves the s-c flowers as their best choice. *Medium-tongued bees on medium-corolla flowers:* m-t bees are generally excluded from l-c flowers because l-t bees reduce the standing crop of nectar beyond their reach. Although m-t bees are capable of visiting s-c flowers and can visit them rapidly, s-t bees, because of their lower foraging costs, can lower standing crops of nectar in these flowers to a level that is still profitable for s-t bees but unprofitable for m-t bees. Thus m-c flowers will provide the highest energy intake rate for m-t bees. *Long-tongued bees on l-c flowers:* l-t bees can forage rapidly on medium- and short-corolla flowers, but the net profit from these flowers will be quite low because of the larger foraging costs for larger-sized bees. In addition, the activities of short- and medium-tongued bees on these flowers can reduce the net profit of these flowers to unprofitable levels for l-t bees. Thus a l-t bee is faced with a choice between short- and medium-corolla flowers, which it can visit more rapidly but which have a lower net profit because they must be shared with other bees, and l-c flowers which require greater handling time but provide a greater profit because they are not shared with other bees. Given this choice, the higher net energy intake rate can be found foraging on l-c flowers.

In summary, the general morphological concurrence between tongue length and corolla depth would appear to be based on the preference of bees with a particular tongue length for flowers with corollas of a similar length. The nonpreference of bees for flowers with corollas longer than their tongue is due primarily to an inability to reach the nectar, particularly where longer-tongued bees are present. The nonpreference of bees for flowers with corollas shorter than their tongues is due primarily to the actions of shorter-tongued bees, which, because of their lower foraging costs and higher foraging efficiency, can reduce the nectar rewards to unprofitable levels for longer-tongued bees. When a bee's tongue is very much longer than the corolla depth of the flower it is visiting, it might impede foraging, although data on this are lacking. Figure 19-4 is a pictorial representation of the points just made. Thus it is the presence of other bees, rather than some intrinsic foraging efficiency limitation that excludes bees from flowers. This conclusion is supported by the results of the species-removal experiment discussed earlier (Inouye, 1978). In that case medium-tongued bees visited l-c flowers only if l-t bees were removed, and l-t bees visited m-c flowers only if m-t bees were removed.

There are some anomalies in the preferences of bees that suggest that other factors, besides a morphological match-up between tongue length and corolla depth, may be involved. I will give three exmples. First, at certain elevations in the Rocky Mountains two long-tongued bumblebees, *B. appositus* and *B. kirbyellus,* are found together. *Castilleja sulphurea,* a long-corolla species (C.D. = 13.8 mm), often is present in the area of overlap. However, *B. kirbyellus* is the only bumblebee that visits this plant (Pyke, 1982). At lower elvations *B. californicus,* a rare bumblebee, is the only visitor to *Castilleja* (which is also rare), despite the presence of *B. appositus. Castilleja* is visited for both pollen and nectar. Second, three short-tongued bumblebees, *B. bifarius, B. frigidus,* and *B. sylvicola,* overlap at certain elevations. Each bumblebee appears to specialize on a different simultaneously flowering composite, although all the plants have short

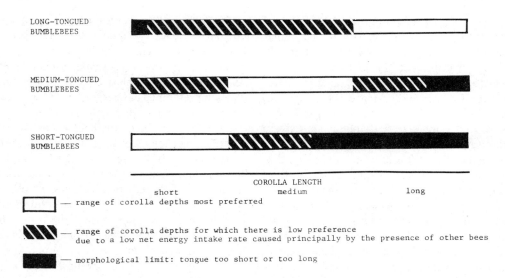

Figure 19-4. Model showing the processes that produce a correspondence between a bumblebee's tongue length and the range of corolla lengths of the flowers it visits.

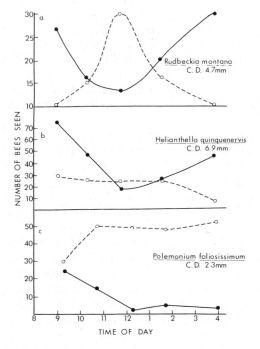

Figure 19-5. Durnal pattern of number of bees seen on shared species in bloom during July 26 and 30, 1978 at a 3070-meter-elevation meadow in the Rocky Mountains. *B. bifarius* [worker tongue length 5.75 mm (short)] = solid line; *B. flavifrons* [worker t.1. 7.81 mm (medium)] = dashed line. All plants are visited for nectar only except *Polemonium,* which is also visited for pollen. *Rudbeckia* and *Helianthella* are composites.

corollas (Pyke, 1982). These plants are visited for nectar only. Third, in Figs. 19-5 and 19-6 it can be seen that *B. flavifrons,* a medium-tongued bee, visits several short-corolla composites such as *Rudbeckia* (C.D. = 4.7 mm) and *Helianthella* (C.D. = 6.9 mm) but not *Helenium* (C.D. = 5.5 mm), although *Helenium* has a longer corolla than *Rudbeckia.* All these plants are visited for nectar.

Why do bees not visit flowers that they are capable of visiting? Or to put the question another way, why are some bumblebees willing to forage on species that other bumblebees, even of similar tongue length, avoid? For the third example cited, the answer may have to do with the fact that the nectar production rate of *Helenium* is much lower than *Rudbeckia* or *Helianthella* (Pleasants, 1981). *B. bifarius* with its smaller body and lower foraging costs may find *Helenium* profitable, whereas the larger-bodied *B. flavifrons* may not. For the other two examples it could be that different bumblebees have slightly different nutritional needs, and that bees tend to forage more heavily on those species whose pollen or nectar (amino acid composition) provides for those needs. It would be interesting to examine a number of bumblebee species throughout their ranges to see if, despite dfferences in the avail-

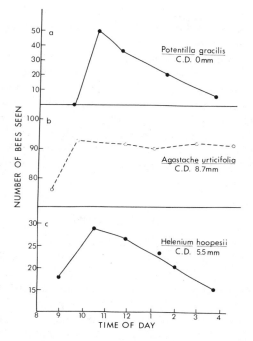

Figure 19-6. Diurnal pattern of bee visitation on un-shared species (see legend to Fig. 19-5). *Potentilla* is visited for pollen only, *Agastache* is visited for pollen and nectar, and *Helenium*, a composite, is visited for nectar only.

able plant species composition, a particular bee species always prefers plants whose nectar has a particular amino acid composition.

The picture of bumblebee communities as composed of just three species, a short-, a medium-, and a long-tongued type, is something of an oversimplification. In most bumblebee communities the species can be easily assigned to a tongue-length class (i.e., there are usually no bees with intermediate-length tongues—a tongue length differs from the next one in the size series by about a factor of 1.3; Inouye, 1976). However, there may be more than one representative of a particular tongue-length class. In Rocky Mountain meadows at elevations of 2700–3300 meters there are two s-t bumblebees, *B. bifarius* and *B. occidentalis*. One of them, *B. occidentalis*, has a unique role in the community as a nectar robber. It has heavier mandibles than other bees, that allow it to pierce the corolla of deep-corolla species and remove the nectar illegitimately (Inouye,

1977). This bee's having access to nectar that *B. bifarius* cannot use (although *B. bifarius* workers sometimes use the holes in the flowers made by *B. occidentalis*) may provide a means of coexistence between the species. The structure of European bumblebee communities is more complex. In some cases a fourth tongue-length class, superlong, is present. This class may be the European counterpart of the New World hummingbirds (Inouye, 1977). In a number of European bee communities there are several representatives of some classes, typically the short- and medium-length classes, such that 6 to 11 species may be found in a given locale (Ranta and Vepsalainen, 1981). However, when study localities are broken down into different habitats and elevations, the number of species within a given area is reduced to 5 or 6 (Lundberg and Ranta, 1980; Ranta et al., 1981). This still leaves several representatives of certain tongue-length classes. One explanation for the coexistence of bumblebees that use the same food plants has to do with the patchy nature of flowering species in space and time (Ranta and Vepsalainen, 1981). Flowering species in European communities appear as components of a mosaic of interdigitating habitats. The patchiness of habitats may be due in part to man-induced habitat disturbances. Consequently, within a given radius of some point there may not be a set of plant species that provides continuous floral resources through a season. This is in contrast to many meadows in the Rocky Mountains that are discrete entities, separated from other meadows by large expanses of trees. In such meadows, each community provides a rather predictable and continuous floral resource base. The success of a bumblebee colony in a European community may depend on whether it happened to get established within a reasonable proximity of continuous sources of nectar and pollen. Although worker bees may travel a considerable distance to obtain food (Alford, 1975), such travel entails a sizable energy cost. Colonies that occasionally find no suitable flowering species at economically profitable distances from the nest may face short-term energy crises. Because bumblebees are annual species, these crises may influence the number of reproductives that a colony can produce at the end of the season. Because of the spatial and temporal heterogeneity, the success of a colony may be unpredictable. This type of lottery system for success (Sale, 1976) may not allow any one species to competitively exclude another, permitting several similar species to coexist. Such a method of coexistence has also been suggested for the bee community of the North American high-altitude short-grass prairie (Tepidino and Stanton, 1981). It may also be applicable to many habitats in the Rocky Mountains and Sierras that are not of the discrete-meadow variety. (In fact, the discrete meadow may be the rarer habitat type.) More intense studies of species dynamics in these kinds of environments are needed.

One point should be mentioned: the designations short-, medium-, and long-tongue are relative terms and don't correspond to specific tongue lengths. For example, for short-tongue bees in communities in the Rocky Mountains, south Finland, and north Sweden the tongue lengths of queens are 8–8.5 mm, 8.5–9.5 mm, and 11–12 mm, respectively (Pleasants, 1981; Ranta et al., 1981). There is a general trend of increasing tongue length with increasing latitude. This may be so not because of selection for increasing tongue length per se but because of selection for the thermoregulatory benefits of large body size to withstand the rigors of high-latitude climates. Body size and tongue length are highly correlated. There does not appear to be a corresponding increase in corolla depths with high latitudes (Ranta et al., 1981), but as mentioned above, long-tongued bees are capable of visiting short-corolla flowers.

Altitudinal Segregation

A second means of resource partitioning in bumblebees, besides differences in tongue length, is altitudinal segregation. There is typically a replacement series of bumblebees with the same tongue length over an altitudinal gradient (Macior, 1974; Inouye, 1976; Pyke, 1982). For example, in the Rocky Mountains *B. bifarius* is the short-tongued representative

below about 3300 meters whereas from 3300 meters to 4000 meters *B. sylvicola* is the short-tongued bumblebee. Sometimes a third s-t bumblebee, *B. frigidus,* is sandwiched between these two. A similar replacement, at about the same elevation, is found between the two long-tongued bees *B. appositus* and *B. kirbyellus.* The altitudinal ranges of these replacement species do not abut but overlap to some extent. This can explain why, at certain elevations, the bee communities have "too many" representatives of certain tongue-length classes.

Diurnal Segregation

Although in many bumblebee communities species are separated by specializations based on tongue length, bees in adjacent tongue-length classes overlap in their use of some plant species (Pleasants, 1981). One way that competition can be reduced in such cases is if the two bee species forage on the shared plants at different times of the day. Several instances of diurnal foraging separation were found in bumblebee communities in Maine bogs (Heinrich, 1976a). I found evidence for this in Rocky Mountain meadows. Figure 19-5 shows the change in the abundance of the short-tongued *B. bifarius* (worker tongue length 5.75 mm) and the medium-tongued *B. flavifrons* (worker tongue length 7.81 mm) on three plant species during the course of the day . These species are visited for nectar only, except for *Polemonium,* which is also visited for pollen. As the abundance of *B. bifarius* on these plants increases or decreases throughout the day, the abundance of *B. flavifrons* changes in the opposite direction. What is the basis for this complementarity? Is one species excluding the other, or could the differences in abundance simply reflect the bees' preference for foraging at different ambient temperatures? To answer these questions I monitored the bee activity on all the other species blooming at the same time, particularly those used exclusively by one bumblebee (Fig. 19-6). In midmorning *B. bifarius* activity shifts to *Potentilla gracillis,* from which it collects pollen (flowers of this species open at about 1000 hours; it produces no nectar). *Bombus bifarius* activity also increases on *Helenium hoopesii,* from which it collects nectar. The implication is that because *B. bifarius* switches to *Helenium* and *Potentilla,* its activity on *Rudbeckia, Helianthella,* and *Polemonium* is reduced, and these species now become profitable for *B. flavifrons.* When *B. bifarius* returns to two of the species in Fig. 19-5 later in the day, *B. flavifrons* is once again excluded. The pattern in Fig. 19-5 probably is not due to a time of day (or temperature) foraging preference by either species. Overall, *B. bifarius* activity is still high at midday, and *B. flavifrons* maintains a relatively constant activity level throughout the day on *Agastache urticifolia,* where it is the only visitor.

In general, the fact that two species that utilize the same food items are separated in their time of feeding on those items does not necessarily mean they are no longer competing. The two species are still using and depleting the same resource, and thus each species will have a negative effect on the other. The strength of the competitive effect will depend on the renewability of the resources. With rapidly renewing resources, such as nectar, the species that gets first crack at them cannot completely eliminate them. If that first species cannot (perhaps for physiological reasons related to ambient temperature) or does not (because it is engaged in other activities) use those resources, later in the day there will still be some available for a second species. Neither can the second species competitively exclude the first by reduction of resource levels. Thus temporal separation can allow for the coexistence of similar species, but the reduction in competition that occurs will depend on resource renewability. In many ways temporal separation is analogous to interspecific territoriality. In the one case ecologically similar organisms stake out spatial refugia wherein they have exclusive use of resources; in the other it is a temporal refugia.

Segregation by Aggressive Interactions

A fourth means of resource partitioning among ecologically similar bees involves interference competition. Morse (1977) found that when bumblebees of a small-bodied species foraged alone on goldenrod, they utilized both proximal and distal portions of the inflorescence. When individuals of a large-bodied species were also present, the smaller bees restricted their foraging to primarily distal portions (which the larger bee was too heavy to visit). However, no direct behavioral interactions between the species were observed. Interference competition has not been observed in other studies of bumblebees. It would not be expected to occur except where bees of quite different size are foraging on the same species. Because of tongue-length foraging specialization this is not likely to happen, except perhaps at the end of the flowering season when primarily short-corolla composites, such as goldenrod, are in bloom.

In conclusion, the role of competition in determining both plant and pollinator community structure is fairly well documented, at least for bumblebee-dominated communities. However, we need more experimental studies directly demonstrating a fitness reduction for plants when they simultaneously use the same pollinator. We also need investigations of the mechanism of competition for pollinators (exploitation or interference), and an assessment of the extent of pollinator-limited plant reproduction in communities. Studies of plants and pollinators in communities where pollinator types other than bumblebees predominate are clearly needed. The role of other factors, besides competition, in regulating plant and pollinator species diversity needs to be examined as well.

ACKNOWLEDGMENTS

The ideas presented in this paper were clarified and sometimes even changed by discussions with a number of people, including D. Inouye, G. H. Pyke, J. D. Thomson, N. M. Waser, and M. J. Zimmerman. Barbara Pleasants assisted in all the field work and drew the figures. Field work was assisted by grants from Sigma Xi and the Steven A. Vavra Endowment to UCLA, and by the National Forest Service, which protected our study sites by issuing a special use permit.

LITERATURE CITED

Alford, D. V. 1975. *Bumblebees.* Davis-Poynter Ltd., London.

Anderson, R. C. and S. Schelfhout. 1980. Phenological patterns among tallgrass prairie plants and their implications for pollinator competition. *Am. Midl. Natur.* 104:253–263.

Beattie, A. J., D. E. Breedlove, and P. R. Ehrlich. 1973. The ecology of the pollinators and predators of *Frasera speciosa*. *Ecology* 54:81–91.

Bierzychudek, P. 1981. Pollinator limitation of plant reproductive effort. *Am. Nat.* 117:838–840.

Brown, J. H. and A. Kodric-Brown. 1979. Convergence, competition, and mimicry in a temperate community of hummingbird-pollinated flowers. *Ecology* 60:1022–1035.

Campbell, D. R. and A. F. Motten. 1981. Competition for pollination between two spring wildflowers. *Bull. Ecol. Soc. Am.* 62(2):99.

Free, J. B. 1963. The flower constancy of honeybees. *J. Anim. Ecol.* 32:119–131.

———. 1970. The flower constancy of bumblebees. *J. Anim. Ecol.* 39:395–402.

Gill, D. E. 1974. Intrinsic rate of increase, saturation density, and competitive ability. II. The evolution of competitive ability. *Am. Nat.* 108:103–116.

Heinrich, B. 1975. Bee flowers: a hypothesis on flower variety and blooming times. *Evolution* 29:325–334.

———. 1976a. Resource partitioning among eusocial insects: bumblebees. *Ecology* 57:874–889.

———. 1976b. Flowering phenologies: bog, woodland, and disturbed habitats. *Ecology* 57:890–899.

———. 1976c. The foraging specializations of individual bumblebees. *Ecol. Mon.* 46:105–128.

Heithaus, E. R. 1974. The role of plant–animal interactions in determining community structure. *Ann. Missouri Bot. Gard.* 61:675–691.

Hobbs, G. A. 1966a. Ecology of species of *Bombus* Latr. (Hymenoptera: Apidae) in southern Alberta. IV. Subgenus *Fervidobombus* Skorikov. *Can. Entomol.* 98:33–39.

———. 1966b. Ecology of species of *Bombus* Latr. (Hymenoptera: Apidae) in southern Alberta. V. Subgenus *Subterraneobombus* Vogt. *Can. Entomol.* 98:288–294.

———. 1967. Ecology of species of *Bombus* (Hymenoptera: Apidae) in southern Alberta. VI. Subgenus *Pyrobombus*. *Can. Entomol.* 99:1271–1292.

———. 1968. Ecology of species of *Bombus* (Hymenoptera: Apidae) in southern Alberta. VII. Subgenus *Bombus*. *Can. Entomol.* 100:156–164.

Hocking, B. 1968. Insect–flower associations in the high Arctic with special reference to nectar. *Oikos* 19:359–387.

Hurlbert, S. H. 1970. Flower number, flowering time, and reproductive isolation among ten species of *Solidago* (Compositae). *Bull. Torrey Bot. Club* 97:189–193.

Inouye, D. 1976. Resource partitioning and community structure: a study of bumblebees in the Colorado Rocky Mountains. Doctoral dissertation, University of North Carolina, Chapel Hill.

———. 1977. Species structure of bumblebee communities in North America and Europe. *In* W. I. Mattson (ed.), *The role of Arthropods in Forest Ecosystems.* Springer-Verlag, New York, pp. 35–40.

———. 1978. Resource partitioning in bumblebees: experimental studies of foraging behavior. *Ecology* 59:672–678.

———. 1980. The effect of proboscis and corolla tube lengths on pattern and rates of flower visitation by bumblebees. *Oecologia* 45:197–201.

Kevan, P. 1972. Insect pollination of high arctic flowers. *J. Ecol.* 60:831–848.

Levin, D. A. 1978. The origin of isolating mechanisms in flowering plants. *Evol. Biol.* 11:185–317.

Levin, D. A. and H. W. Kerster. 1967. Natural selection for reproductive isolation in *Phlox*. *Evolution* 21:679–687.

———. 1973. Assortative pollination for stature in *Lythrum salicaria*. *Evolution* 27:144–152.

Lundberg, H. and E. Ranta. 1980. Habitat and food utilization in a subarctic bumblebee community. *Oikos* 35:303–310.

Macior, L. W. 1971. Coevolution of plants and animals—systematic insights from plant–insect interactions. *Taxon* 20:17–28.

———. 1974. Pollination ecology of the Front Range of the Colorado Rocky Mountains. *Melanderia* 15:1–61.

Moldenke, A. R. 1975. Niche specialization and species diversity along a California transect. *Oecologia* 21:219–242.

Morse, D. 1977. Resource partitioning in bumblebees: the role of behavioral factors. *Science* 197:678–679.

Mosquin, T. 1971. Competition for pollinators as a stimulus for the evolution of flowering time. *Oikos* 22:398–402.

Parrish, J. A. D. and F. A. Bazzaz. 1979. Difference in pollination niche relationships in early and late successional plant communities. *Ecology* 60:597–610.

Pleasants, J. 1977. Competition in plant pollinator systems: an analysis of meadow communities in the Colorado Rocky Mountains. Doctoral dissertation. University of California, Los Angeles.

———. 1980. Competition for bumblebee pollinators in Rocky Mountain plant communities. *Ecology* 61:1446–1459.

———. 1981. Bumblebee response to variation in nectar availability. *Ecology*. 62:1648–1661.

Pojar, J. 1974. Reproductive dynamics of four plant communities of southwestern British Columbia. *Can. J. Bot.* 52:1819–1834.

Poole, R. W. and B. J. Rathcke. 1979. Regularity, randomness, and aggregation in flowering phenologies. *Science* 203:470–471.

Pyke, G. H. 1978. Optimal foraging: movement patterns of bumblebees between inflorescences. *Theor. Pop. Biol.* 13:72–98.

———. 1980. Optimal foraging in bumblebees: calculation and net rate of energy intake and optimal patch choice. *Theor. Pop. Biol.* 17:232–246.

————. 1982. Local geographic distributions of bumble-bees near Crested Butte, Colorado: competition and community structure. *Ecology.* 63:555–573.

Rabbinowitz, D., J. K. Rapp, V. L. Sork, B. J. Rathcke, G. A. Reese, and J. C. Weaver. 1981. Phenological properties of wind- and insect-pollinated prairie plants. *Ecology* 62:49–56.

Ranta, E. and H. Lundberg. 1980. Resource partitioning in bumblebees: the significance of differences in proboscis length. *Oikos* 35:298–302.

Ranta, E. and K. Vepsalainen. 1981. Why are there so many species? Spatio-temporal heterogeneity and northern bumblebee communities.*Oikos* 36:28–34.

Ranta, E., H. Lundberg, and I. Teras. 1981. Patterns of resource utilization in the Fennoscandian bumblebee communities. *Oikos* 36:1–11.

Reader, R. J. 1975. Competitive relationships of bog ericads for major insect pollinators. *Can. J. Bot.* 53:1300–1305.

Roughgarden, J. 1979. *Theory of Population Genetics and Evolutionary Ecology: An Introduction.* Macmillan, New York.

Sale, P. F. 1976. Reef fish lottery. *Nat. Hist.* 85(2):61–64.

Silander, J. A. and R. B. Primack. 1978. Pollination intensity and seed set in the evening primrose *(Oenothera fruticosa). Am. Midl. Natur.* 100:213–216.

Small, E. 1976. Insect pollinators of the Mer Bleue peat bog of Ottawa. *Can. Field Natur.* 90:22–28.

Sutherland, S. 1980. Energy limited fruit set in a paniculate agave: a test of the Bateman principle. *Bull. Ecol. Soc. Am.* 61(2):105.

Tepidino, V. J. and N. L. Stanton. 1981. Diversity and competition in bee–plant communities on short-grass prairie. *Oikos* 36:35–44.

Thomson, J. D. 1978a. Competition and cooperation in plant–pollinator systems. Doctoral dissertation. University of Wisconsin, Madison.

————. 1978b. Effects of stand competition on insect visitation in two-species mixtures of *Hieracium. Am. Midl. Natur.* 100:431–440.

————. 1980. Implications of different sorts of evidence for competition. *Am. Nat.* 16:719–726.

————. 1981. Spatial and temporal components of resource assessment by flower-feeding insects. *J. Anim. Ecol.* 50:49–59.

Udovic, D. 1981. Determinants of fruit set in *Yucca whipplei* reproductive expenditure vs. pollinator availability. *Oecologia* 48:389–399.

Waddington, K. D. 1979. Divergence in inflorescence height: an evolutionary response to pollinator fidelity. *Oecologia* 40:43–50.

Waser, N. M. 1978a. Competition for hummingbird pollination and sequential flowering in two Colorado wildflowers. *Ecology* 59:934–944.

————. 1978b. Interspecific pollen transfer and competition between co-occurring plant species. *Oecologia* 36:223–236.

Waser, N. M. and L. A. Real. 1979. Effective mutualism between sequentially flowering plant species. *Nature* 281:670–672.

Yeaton, R. I. 1974. An ecological analysis of chaparral and pine forest bird communities on Santa Cruz Island and mainland California. *Ecology* 55:959–973.

Zimmerman, M. 1979. Optimal foraging: a case for random movement. *Oecologia* 43:261–267.

————. 1980. Reproduction in *Polemonium:* competition for pollinators. *Ecology* 61:497–501.

————. 1981. Optimal foraging, plant diversity and the marginal value theorem. *Oecologia* 49:148–153.

20
PATTERNS OF FLOWERING IN TROPICAL PLANTS

Kamaljit S. Bawa

Department of Biology
University of Massachusetts
Boston, Massachusetts

ABSTRACT

Flowering pattern is determined by the timing, duration, and frequency of flowering. Plants display a wide variety of flowering patterns. In addition to intraspecific and interspecific differences, male and female plants of dioecious species also usually differ in their flowering patterns. Selective forces influencing the timing, duration, and frequency of flowering are varied and diverse. The timing of flowering is influenced by competition for pollinators among species, selection against interspecific gene flow, pollinator availability and nature of floral rewards, and selection for the optimization of life history traits other than flowering. The duration of flowering is evaluated in terms of control over relative investment in flowers and fruits, level of geitonogamy and outcrossing, predictability of pollination, competition for pollinators, selection against interspecific gene flow, flower herbivores, opportunities for colonization, and canopy position. Variation in the frequency of flowering is considered in the context of seed predation, unpredictable environment, and temporal pattern of resource availability. Differences between sexes are discussed in terms of mate selection and energetics of reproduction, while those among sexual systems are considered in reference to optimization of male and female reproductive success in dissimilar ways. Flowering patterns are apparently molded by a number of conflicting selective pressures. A clearer understanding of these pressures requires a knowledge of the way in which the flowering phenology interacts with the phenology of individual inflorescences, overall phenology of the plants, and the phenology of the pollinators. It also requires quantitative information on variation in timing, duration, and frequency of flowering at the population level over a number of flowering seasons.

KEY WORDS: Competition for pollinators, dioecious species, flowering pattern, flowering phenology, monoecious species, selection against interspecific gene flow, sexual selection, tropical forest trees.

CONTENTS

INTRODUCTION

The timing, duration, and the frequency of flowering largely describe the flowering pattern of a population and its constituents. Plants display a wide variety of flowering patterns. Individuals of a population may flower for periods as brief as one day or as long as one year; several times a year, once a year, or once every few years; once or many times in their lifetime (Frankie et al., 1974a; Gentry, 1974a,b; Janzen, 1976; Opler et al., 1980a). In addition there may be considerable variation between male and female plants of dioecious species (Bawa, 1980; Bullock and Bawa, 1981). The variation in flowering patterns has implications for many aspects of community ecology, e.g., organization and structure of communities (Janzen, 1967; Frankie et al., 1974b; Frankie, 1975; Stiles, 1978) and population biology, e.g., gene flow in plants as determined by the foraging behavior of pollinators in resource patches varying in time and space (Levin, 1979a), optimal foraging theory (Pyke et al., 1977), and the evolution of reproductive strategies (Schaffer and Gadgil, 1975; Giesel, 1976). However, the ecological basis of such variation has not received much attention (but see Janzen, 1967, 1974, 1978; Stiles, 1978). Here, I review the role of sexual selection, selection for outcrossing and reproductive isolation, pollinator availability and floral rewards, competition for pollinators, flower herbivores, and opportunities for colonization in influencing the timing, duration, and frequency of flowering.

The phenological studies conducted in tropical lowland ecosystems, particularly in Costa Rica (Fournier and Salás, 1966; Janzen, 1967, 1978; Frankie et al., 1974a; Heithaus, 1974; Stiles, 1975, 1978; Opler et al., 1980a), have contributed much to our understanding of the flowering patterns. This chapter is based on the aforementioned studies and the unpublished observations that I have made in Costa Rica during the last several years in a dry deciduous forest in the Guanacaste Province (for site description, see Frankie et al., 1974a) and in a wet evergreen forest in Heredia Province (for site description, see Stiles, 1978). Only trees and shrubs are considered although the conclusions also apply to lianas. Herbs are not considered because no information is available except for the species of *Heliconia* (Stiles, 1975), which are equivalent to large shrubs in terms of their size and the amount of floral resource offered to the pollinators. Two other points should be noted: (1) almost all species under consideration are animal-pollinated; (2) individual flowers in the vast majority of tropical species are receptive and attractive to pollinators for only one day.

Patterns in flowering phenology can be discerned at the level of communities, species-populations, and individuals. Since community-wide patterns have already been explored by many authors (e.g., Koelmeyer, 1959; Fournier and Salás, 1966; Janzen, 1967, 1974; Croat, 1969; Daubenmire, 1972; Medway, 1972; Frankie et al., 1974a; Heithaus, 1974; Putz, 1979; Hilty, 1980; Opler et al., 1980a; and others), only patterns prevalent at the species and infraspecific levels are considered below.

TIMING OF FLOWERING

The role of physical factors such as photoperiod in the induction of flowering (see Njoku, 1958 for tropical trees; Schwabe, 1971) and that of temperature and rainfall in stimulating flowering in those species that undergo dormancy following induction (Schwabe, 1971; Opler et al., 1976) are well known. But, as Janzen (1967) pointed out, proximate causes such as mechanisms that allow a particular species to flower at a particular time must be distinguished from ultimate causes that specify the advantages or disadvantages of flowering at a particular time. The discussion below is focused on the role of ultimate rather than proximate factors in the evolution of flowering times.

Variation among Species

In aseasonal tropical lowland wet forests, different species bloom at different times (Koelmeyer, 1959; Medway, 1972; Frankie et al., 1974a; Putz, 1979; Hilty, 1980; Opler et al., 1980a). Even in seasonal communities in which there is some synchronization in sexual reproduction (e.g., Janzen, 1967), there is no period without a significant number of species in bloom (Frankie et al., 1974a). Within the constraints imposed by physiological capacity, when should the individuals of a species bloom? Among the critical factors that determine the timing of flowering are: interspecific competition for pollinators and interspecific gene flow, pollinator availability and nature of floral rewards, and selection for the optimization of other life history traits.

Interspecific Competition for Pollinators and Interspecific Gene Flow. In tropical communities, as elsewhere, many species are serviced by the same group of pollen vectors. It is generally assumed that these species bloom at different times to avoid competition for pollinators (Heithaus, 1974; Frankie, 1975; Stiles, 1975, 1977). A critical test of the competition hypothesis, however, would eventually require the demonstration that temporal segregation of blooming period among competing species is not random (Pool and Rathcke, 1978; Cole, 1981) and that overlap in blooming results in reduction of seed set (Waser, 1978).

It is difficult to distinguish the effects of selection against interspecific gene flow on the temporal staggering of blooming periods from those of competition for pollinators. Thus the extent to which sequential blooming is the result of selection against interspecific gene flow is not known. In *Heliconia* (Musaceae), one of the few genera for which quantitative information on flowering phenology and pollination biology of several sympatric species is available, habitat segregation appears to be at least as important as temporal segregation in blooming times (Stiles, 1975).

The extent to which selection against gene flow and for avoidance of competition for pollinators may result in temporal segregation of blooming period may vary considerably. In some instances, if the floral syndromes (color, shape, odor, etc. of flowers) of several sympatric species increase each other's floral display, there actually may be selection for the convergence of blooming times of species serviced by the same group of pollinators. Brown and Kodric-Brown (1979) give examples of such convergence among hummingbird-pollinated plants of a temperate-zone community, and argue that disadvantages of interspecific pollen flow or competition may outweigh the advantages that accrue from increased display. However, the effects of competition and potential for interspecific gene flow could also be minimized by spatial segregation and/or partitioning of the pollinator resource by the time of the day. Gilbert (1975), for example, has shown that several species of the Passifloraceae that share the same pollen vector in a tropical wet forest, present pollen and nectar at different times of the day.

Pollinator Availability and the Nature of Floral Rewards. It is obvious that in zoophilous species, timing of flowering would coincide with the availability of pollinators (Waser, 1979). While in the temperate regions the prevailing temperature may largely influence the optimal conditions for pollinator activity (Schemske et al., 1978), in the tropical lowlands several biotic and abiotic factors may determine the availability of pollinators at a particular time (Janzen, 1967). For example, nectarivorous bats in the neotropics usually consume fruits and insects, in addition to nectar (Heithaus et al., 1975). Thus the temporal patterns of fruit and insect abundance could strongly influence the population densities of bats with concomitant effects on the flowering times of bat-pollinated plants (see also Janzen, 1967). In a dry deciduous forest, Frankie (1975) found that a disproportionately large number of moth-pollinated plants flower in the wet season. He suggested that this increase may be associated with an abundance of food for moth larvae in the form of increased foliage during the onset of rains. For the same forest, Janzen (1967) explained the simultaneous blooming of many species in the dry season in terms of pollinator availability. He listed a number of factors related to moisture and food resources that result in greater influence of pollinators on the dry- rather than the wet-season flowering. He also reasoned that the absence of leaves in the dry season may increase the floral display of bee- and bird-pollinated plants. Frankie et al. (this book) have indeed found that the vast majority of trees and lianas flowering in the dry season are bee-pollinated and mass-blooming (i.e., they produce a large number of flowers over a relatively short period). In contrast, the bee-pollinated plants in the wet season bear relatively few flowers per day, but for a long period of time. Mass blooming, apparently, does underlie the importance of visual cues. In short, although the availability of pollinators might explain the initiation of flowering at a particular time in a group of plants serviced by the same pollen vector, it does not predict the way in which flowering times of different species in the group might be distributed in time.

The role of floral rewards has also been implicated in the evolution of flowering times. Gentry (1974a) raised the possibility that certain nectarless bignoniaceous vines have been selected to bloom soon after or simultaneously with nectar-producing bignoniaceous trees and vines. The nectarless vine flowers are apparently successful visual mimics of the nectariferous tree flowers. By contrast, there may be selection in species with rich floral rewards for "escape-in-time" from species that offer little or no reward. Heinrich (1975b) has suggested this as a possible factor in the earlier blooming of nectar-rich species in a temperate-region herbaceous community.

Selection on Other Life History Traits. In explaining the blooming of a large number of species in the dry season in a deciduous forest, Janzen (1976) argued that there is selection for flowering at a time when there is no vegetative activity, because flowering and fruiting at the same time as vegetative growth could result in the loss of position in the canopy. Selection by seed dispersal agents acting on the timing of fruiting (Snow, 1966; Janzen, 1967; McKey, 1975) may influence the periodicity of flowering, but no examples involving such an effect have been documented fully. The closest case is that of mistletoes that have their pollen and seed dispersed by a flowerpecker (P. Davidar, pers. comm.). According to Davidar, the mistletoe flowers "mimic" the nutritionally rich fruits, and the phenology of current years' flowering is more or less contemporaneous with the fruiting of past years' flower crop.

Variation within Populations

Asynchronous blooming among individuals has been reported for several species-populations (Bawa, 1977; Primack, 1980; Bullock and Bawa, 1981; see also Opler et al., 1980a). The level of asynchrony has been quantified by Primack (1980) for three shrub species in New Zealand. Defining the limits for complete asynchrony and full synchrony within a population as 0 and 1 respectively, he obtained values from 0.34 to 0.74. The values varied from one year to another for the same population.

In outcrossing populations, an individual should be selected to bloom in synchrony with its conspecifics. Augspurger (1978) has indeed shown stabilizing selection for synchronous flowering in *Hybanthus prunifolius* (Violaceae) enforced by density-dependent pollinators and seed predators. In five other shrub species, she also noted that the level of intrapopulation synchrony increased with a decrease in the blooming time of the individuals. Presumably, when plants bloom over an extended period of time, opportunities for mating are not as severely limited in time as for plants with short blooming periods. Therefore when the flowering period is long, selection against a low level of asynchrony may be weak.

A high level of asynchronous blooming might also result in the reduction of the effective population size (Bawa, 1977; Primack, 1980), but the extent to which the reduction may have an appreciable effect on the level of inbreeding is uncertain in view of the suggestion that most matings are among close neighbors in plants and that neighborhood sizes are normally small (Levin, 1979b). Asynchronous blooming could in fact increase the diversity of matings by changing the close neighbors of a plant in time, if there is a poor year-to-year correlation in the blooming times of neighboring individuals.

Selection for asynchronous blooming may occur when there is intense intraspecific competition for pollinators. In a Costa Rican wet forest, two canopy species, *Terminalia lucida* (Combretaceae) and *Dipteryx panamense* (Fabaceae), have massive crowns, are relatively abundant, and have a fairly high level of asynchrony in the onset of flowering among individual plants (Bawa, unpub. obs.). Presumably populations of such species would require a very large number of pollen vectors if the flowering were fully synchronous among individuals. Thus, unless pollinators had their life cycles tied to the blooming of these species, or flowering was preceded by some other very common species that utilized the same pollen vectors, competition for pollinators could result in selection for asynchrony. Along the same lines, the number of pollinators would also be low for those species that temporally share pollinators with other species but are the first to bloom in the "season" when the population densities of pollinators might be low. Asynchronous blooming in figs (Janzen, 1979) constitutes a special case and is discussed in a later section.

In some large tropical trees, it is not unusual to find the parts of a crown out of phase, with some parts in flower, others in fruit, and others yet to flower (Medway, 1972; Hallé et al., 1978). It is not known whether intra-individual asynchrony is caused by poor genotypic canalization or has some adaptive significance. Certainly, a greater level of genetic diversity in the progeny is possible if different parts of the crown breed with different individuals. Asynchrony can also reduce the level of geitonogamy. In trees with very large crowns, asynchrony between branches may allow the individual to have an extended duration of flowering while pollinators perceive each crown part as blooming massively (see below).

Implicit in the discussion of the evolution of flowering times is the suggestion that there is genetic variation in populations with respect to flowering time on which selection can operate. Both the existence of genetic variability in flowering times and the action of selection on such variation have been demonstrated for several plants including fruit trees (Stanford et al., 1962; McNeilly and Antonovics, 1968; Janick and Moore, 1975).

DURATION OF FLOWERING

The duration of flowering refers to the length of the blooming period. In general, the longer the blooming period, the fewer flowers that are produced per day (Augspurger, 1978).

Variation among Species

There is a considerable variation among species with respect to the number of flowers produced per unit of time. At one extreme are species in which individuals produce a few flowers per day but bloom for a long period, lasting from several weeks to several months, and at the other extreme are species in which individuals produce a large number of flowers per day but bloom for a few days (Fig. 20-1). The first type of flowering is referred to here as extended blooming, and the second type as mass blooming (Gentry, 1974a; Opler et al., 1980a). An extreme example of the first type is *Muntingia calabura* (Elaeocarpaceae), in which plants flower more or less continuously throughout the year (Frankie et al., 1974a), and that of the second type is *Casearea praecox* (Flacourtiaceae), in which all the plants of a population flower synchronously for only one day (Opler et al., 1976). The two types represent only the end points of a spectrum that encompasses a broad range of variation—Gentry (1974a) has described five types of flowering patterns for the Bignoniaceae alone. Only two

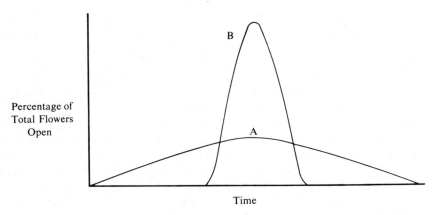

Figure 20-1. Two extreme patterns in the rate of flower production. Pattern A represents the production of a few flowers per day for many days, whereas B represents the production of many flowers per day for a few days.

extremes are considered here because that makes it easier to identify possible selective pressures. It should be emphasized that a large number of species can be distinctly classified as extended or mass bloomers only in relation to some other species. Further, extended or mass blooming is a property of the individuals and, by implication, of species containing such individuals. Asynchronous blooming of mass-flowering plants can result in extended blooming at the population level; however, such species-populations for the purpose of discussion below are referred to as mass bloomers.

Blooming over a long period of time (type A, Fig. 20-1) should confer several advantages: (1) It should allow better control of the relative investment in flowers and fruits than mass blooming because the rate of flower production can be adjusted to match the resources available for fruit production. (2) In outcrossing populations, extended blooming should increase the individual's chances of fertilizing a large number of mates and at the same time enable it to receive pollen donors from a large number of genotypes; the increased diversity of matings is simply a function of the greater amount of time available for matings. (3) The production of a few flowers per unit of time should reduce the level of geitonogamy and force pollinators to seek rewards from other conspecifics. (4) Extended blooming should entail less risk of reproductive failure resulting from bad weather or lack of pollinators than

mass blooming. Selection for extended blooming, however, could be influenced by the factors discussed in the following paragraphs.

Selfing. Selection for selfing could override benefits that accrue either from fertilizing a large number of embryos on different plants or from sampling pollen from a large number of genotypes. Of the five species in a dry deciduous forest that displayed high levels of self-compatibility (Bawa, 1974), individuals in three, *Ardisia revoluta* (Myrsinaceae), *Curatella americana* (Dilleniaceae), and *Sloanea terniflora* (Elaeocarpaceae), bloomed for 3 to 5 days. In a wet evergreen forest, there are indications of self-compatibility in *Hymenolobium* sp. and *Pterocarpus* sp. of the Fabaceae (Bawa and Perry, unpub. data). In both species, plants bloom for less than 4 days. *Hybanthus prunifolius* (Violaceae) also flowers for a few days and is self-compatible (Augspurger, 1978).

Competition for Pollinators and Selection against Interspecific Gene Flow. As the number of species competing for the same pollinator pool increases, selection against interspecific gene flow and reduced seed set during the overlap in blooming with other species should lead to a decrease in the blooming period. Discrete episodes of mass blooming among different species should result in greater pollinator fidelity and less interspecific gene flow than allowed by extended blooming (see Antonovics and Levin, 1980). Frankie (1975) found that many tree species in the Fabaceae flower during the dry season (January to April) and are pollinated by the same group of medium- to large-sized bees (Anthophoridae and Xylocopinae); however, each species flowers massively for a short period, and the flowering periods of different species are staggered, possibly minimizing competition and interspecific gene flow. Although all the species flowered massively, the duration of flowering at the individual and population levels varied from one species to another. Obviously, the extent to which competition for pollinators and selection against interspecific pollen flow may reduce the duration of blooming would depend upon the number of simultaneously competing species and the amount of potential gene flow among them. In another study, Frankie et al. (this book) noted that while most of the species pollinated by large bees in the dry season have short blooming periods, the few species that flower in the wet season bloom over an extended period of time. This is so presumably because potential for interspecific pollen flow is high in the dry season when a large number of species flower.

The effects of competition for pollinators, however, are not independent of the foraging behavior of pollinators and the relative abundance and dispersion of plants. The production of a few flowers per day over a long period of time could also alleviate the effects of interspecific competition if the pollinators forage in a trapline; that is, if they follow a feeding route of a fairly set sequence of many individual plants (Janzen, 1971a). On the other hand, mass flowering is associated with pollinators such as social or eusocial bees and flies that are attracted in large numbers to a resource patch to which they remain constant until the resource is depleted. Outcrossing in such massively blooming plants may be enhanced through intense competition among pollinators at the resource patch (Frankie and Baker, 1974; Frankie et al., 1976) or "anti-pollinators" (Gentry, 1978). Pollinators, of course, are capable of altering their foraging behavior in response to changes in plant resources so that the bees that generally forage in a trapline occasionally switch to mass-blooming plants

(A. H. Gentry, pers. comm.). The correlation between extended blooming and trapline foraging has been pointed out for several insect-pollinated plants by Frankie (1975) and Janzen (1971a). In addition, Stiles (1975) has shown that plants in three sympatric species of *Heliconia* pollinated by territorial hummingbirds have relatively short blooming periods and a high rate of flower production, whereas among species pollinated by nonterritorial, "traplining" hummingbirds, both the long continuous blooming and short discontinuous blooming patterns are encountered.

Plants in species that are rare in space or time (see below, under "Frequency of Flowering") should flower massively because the production of a small number of flowers per unit of time may not attract enough pollinators. On the other hand, plants in species that are patchily distributed could maintain extended blooming even in the face of intense interspecific competition for pollinators because a resource patch composed of a small number of flowers on a large number of closely spaced plants is equivalent to a large number of flowers on one or a few individuals. In a tropical lowland wet evergreen forest, two species of *Heliconia* (Musaceae) that are scattered in the forest bloom very synchronously (G. Stiles, pers. comm.). Although many "uncommon" species in the same wet evergreen forest have extended blooming, at least some of them are patchily distributed (J. Denslow, pers. comm.). Species with supra-annual blooming cycles, hence rare in time, tend to be mass-blooming (Frankie et al., 1974a; P. Ashton, pers. comm.). In general, data that correlate rarity and/or patchiness with flowering pattern are scarce and difficult to gather. In many cases, for example, it is difficult to determine whether a species is rare or patchily distributed on a general or a local scale.

Flower "Herbivores." Massive blooming should allow escape in time from bud and flower "herbivores" in plants not endowed with other means of defense, in the same way that mast fruiting (supra-annual production of large seed crops) is considered an escape from seed predators (Janzen, 1974). Flower "herbivores," however, may here not depress the reproductive potential of a plant to the same extent as seed predators. In many species, including hermaphroditic taxa, many flowers act by virtue of their structure, timing or position only as pollen donors; consumption of these flowers after the pollen has been dispersed should have little effect on fitness.

Opportunities for Colonization. Colonization by early successional and transient species is dependent upon the disturbance of the habitat. Recolonization following disturbance may occur from seeds dispersed to the site before or after disturbance (Ewel, 1980). Thus seed availability at the time of disturbance may be an important factor in colonization (Hartshorn, 1978). Since the disturbances may occur more or less continuously, as, for example, with gaps resulting from tree falls in the forest, long periods of seed production in early successional or transient species should be advantageous. To the extent that prolonged periods of seed production are associated with extended blooming, there may be indirect selection for long blooming periods in early successional species. Indeed, Opler et al. (1980a) found that early successional species generally have longer blooming periods than species in mature communities. In addition, a very large proportion of species in which the plants bloom more or less throughout the year (Opler et al., 1980b) occurs in early successional communities.

Additional or alternative explanations can also be advanced for the patterns observed by Opler et al. (1980 a,b). Since mass blooming may entail a greater risk of reproductive failure than extended blooming, the mass-blooming strategy may be particularly disadvantageous in ephemeral species that occupy a given habitat for only a limited period. Also, at least in some early successional communities there may be little or no coevolution among the biota owing to the ever-changing composition of the species. Thus if interspecific competition for pollinators is an important selective force in reducing the blooming period, then plants in early successional communities may not show such a trend.

Canopy Position. Long-distance visual cues to locate resources may be more important for pollinators in the canopy than in the understory. Mass blooming seems to be particularly common in canopy species (Bawa, unpub. obs.; P. Ashton, pers. comm.) and lianas (M. Grayum, pers. comm.). In addition, bromeliads and epiphytes tend to show a high degree of synchrony in flowering (G. Stiles, pers. comm). In the Bignoniaceae, the mass-flowering vines are predominantly canopy species, whereas the vines that bloom over long periods have flowers below the canopy (Gentry, 1974b). These observations may not appear to be in line with those of Opler et al. (1980a), who noted a greater length of blooming period for trees than for treelets and shrubs. Note, however, that the analysis by Opler et al. does not distinguish between the canopy and the understory trees.

In summary, the rate of flower production is influenced by a number of factors, such as the amount of selfing, competition for pollinators, selection against interspecific gene flow, foraging behavior of pollinators, flower herbivores, selection for colonization, and habit. In addition, one might expect selection on the timing and duration of vegetative growth and fruit production to influence the duration of flowering. However, to my knowledge, no attempts have been made to examine the constraints imposed by vegetative growth and fruit phenology on the length of blooming periods.

Although the curves displayed in Fig. 20-1 are symmetrical, flowering curves in natural populations are rarely, if ever, symmetrical. Recently Thomson (1980) has approached the question of skewness in flowering curves by considering the temporal distributions of flowers as resource utilization functions among species competing for pollinators. He has reasoned that species that are among the first to bloom, and are thus unfamiliar to pollinators, should have more positively skewed curves than the species that bloom later in the season.

Variation within Populations

Plants within a population have been shown to differ in the duration of flowering in a few species (Gentry, 1974b; Augspurger, 1978; Primack, 1980; Bullock and Bawa, 1981). This variation in part may be genetic, and in part may be correlated with size and the site-specific resources for reproduction. At least in mass-blooming species, stabilizing selection may reduce much of the genetic variance in the duration of flowering.

Differences between male and female plants of dioecious species in regard to the length of blooming period are discussed below (under "Intersexual Variation").

FREQUENCY OF FLOWERING

Variation among Species

Plants can be characterized as either polycarpic (many reproductive episodes during the life span) or monocarpic (single reproductive episode at the end of the life span). Within the polycarpic group are species with supra-annual blooming cycles (Frankie et al., 1974a; Janzen, 1974, 1978), several flowering episodes within one year (Gentry, 1974a,b; Opler et al., 1980a; Bullock et al., 1982), or a single flowering episode each year.

Among tropical plants, monocarpy has been reported in several taxa: bamboos (Janzen, 1976), palms (Moore and Uhl, 1973), and *Tachigalia versicolor* in the Fabaceae (Foster, 1977). Different explanations have been proposed for the evolution of monocarpy. In *Agave,* Schaffer and Schaffer (in Schaffer and Gadgil, 1975) found that bees prefer flowers on taller stalks. They suggested that a disproportionate gain in fitness with increasing reproductive effort could select for monocarpy. Janzen (1976) proposed the satiation of seed predators as the primary factor in the evolution of monocarpy in bamboos. Foster (1977) suggested that the death of *Tachigalia* trees following reproduction increases the probability of survival of their progeny in the space vacated by the adult plants.

In many polycarpic species flowering is supra-annual, the interval between successive flowering episodes varying from two years in most species (Janzen, 1970; Frankie et al., 1974a) to 21 years in some Dipterocarps (Jan-

zen, 1974). The interval may be fixed, or may vary between and within individuals (Janzen, 1978). The evolution of mast fruiting is presumed to result from selection exercised by seed predators (Salisbury, 1942; Janzen, 1971b).

A factor that has received no attention in the evolution of supra-annual blooming is the rate of fruit maturation. It is a common observation that in most plants flowering does not occur until the previous fruit crop has matured. In *Posoqueria grandiflora* (Rubiaceae), a small tree in the Costa Rican lowland wet forests, it takes 32 months for fruits to mature, and the flowering occurs every 36 months (i.e., soon after the dispersal of the fruit crop from the previous flowering cycle). The reason for the unusually long time for the maturation of fruits is not known.

In contrast to supra-annual blooming, many species flower more than once every year (Frankie et al., 1974; Gentry, 1974a; Opler et al., 1980a). This multiple blooming has been termed episodic flowering by Bullock et al. (1982). In episodic flowering, plants regularly display at least two discrete flowering (and fruiting) episodes per year (Fig. 20-2). The term is not extended to include those species in which a small minority of plants may flower more than once owing to poor canalization, or the species in which multiple flowering may result from asynchrony within the crown. In an episodic flowering species, the duration of each episode may be short [e.g., two to three days in bignoniaceous vines (Gentry, 1974a,b), in several Melastomaceae (Bawa, unpub. obs.), and in *Cassipourea elliptica* (Rhizophoraceae) and *Cestrum* sp. (Solanaceae) (S. H. Bullock and J. H. Beach, pers. comm.)] or long [e.g., several weeks in *Pentaclethra macroloba* (Mimosoideae) (G. H. Hartshorn, pers. comm.)]. The number of episodes may range from two in *Symphonia globulifera* (Guttiferae) to five in *Guarea rhopalocarpa* (Meliaceae) (Bullock and Bawa, unpub. obs.). The length of the interval between episodes also varies both among and within species. In *Guarea rhopalocarpa,* for example, it may range from 4 weeks to 20 weeks (Bullock et al., 1982).

Episodic flowering may be viewed as a bet-

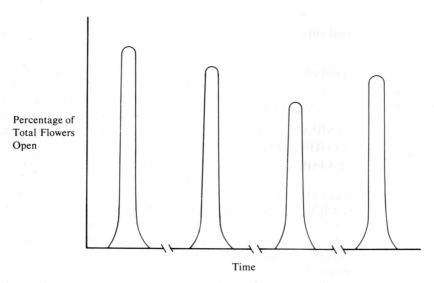

Figure 20-2. Multiple flowering: several flowering episodes within a year. The length of the interval between flowering episodes and the height of peaks may vary (see also Gentry, 1974a).

hedging strategy in the face of uncertainty in the environment (Gentry, 1974b). This uncertainty may stem from several factors. Pollination success may be unpredictable owing either to competition from other species or to inclement weather. The argument also can be extended to include seed dispersers. Indirect evidence to support the notion of a bet-hedging strategy comes from the cloud forests where the proportion of species that display episodic flowering is quite high (W. Haber, pers. comm). In these forests, the timing of optimal weather for pollinator, especially insect, activity is usually unpredictable. Gentry (1974a) has also argued that episodic flowering may evolve when plants offer little or no floral reward, but bloom repeatedly over short periods after species with rich rewards.

Episodic flowering may also be viewed as a strategy that allows the production of a few, relatively small "clutches" rather than a single, relatively large "clutch." This strategy might be adopted when resources for a certain phase of reproductive activity, though more or less continuously available over a long time, are somewhat limited at a given time (see Burley, 1980, for an analogous example).

Episodic flowering could easily evolve in species that have more or less continuous vegetative growth. This might explain the finding of Opler et al. (1980a) that episodic flowering is more common in the riparian (48%) than in the dry forest (17%). It is not obvious why species with episodic flowering have greater representation among shrubs and treelets than among trees, as observed by Frankie and his associates (Frankie et al., 1974a; Opler et al., 1980a).

Variation within Populations

Several types of variation in the frequency of flowering within populations have been noted. The most noteworthy case concerns species-populations in which some individuals skip reproduction (Bullock and Bawa, 1981; Bullock et al., 1982), but a detailed analysis of such a situation, unlike those for some vertebrates (Bull and Shine, 1979), is lacking.

In species that flower at intervals greater than one year, it is not uncommon to find plants in bloom in off years (Frankie et al., 1974a; Janzen, 1978). In species that display episodic flowering, individuals may differ with respect to the number of flowering episodes per year (Bullock et al., 1982), as well as the interval between episodes. The causes and the implications of such variation remain to be explored.

INTERSEXUAL VARIATION IN TIMING, DURATION, AND FREQUENCY OF FLOWERING

As emphasized by Darwin (1871), Bateman (1948), Trivers (1972), Williams (1975), and others, the reproductive goals of males and females usually are not identical. Males, because of the production of numerous energetically inexpensive gametes, tend to optimize the quantity of matings; females, because of the production of relatively few energetically expensive gametes, and because of the investment in parental care of the offspring, tend to optimize the quality of matings. Thus, there is usually intense competition among males for mates. In plants, such competition may result in the onset of flowering in males earlier than in females; a male that is already "entraining" pollinators slightly before the blooming of females may have an advantage in despositing its pollen on stigmas, over the males that bloom concurrently with females. This would be especially true for plants serviced by trapline pollinators because the earlier a plant gets incorporated into the feeding route of a pollinator, the more certain may be its access to females as they come into bloom (Gilbert, 1975; Bawa, 1980). It should be noted that male and female plants usually differ in flower number and floral rewards: male plants generally have many more flowers than the female plants (Opler and Bawa, 1978); male plants offer both pollen and nectar as food rewards, whereas female flowers offer only nectar (Bawa and Opler, 1975), and in some cases, female flowers lack food rewards and mimic male flowers (Baker, 1976; Bawa, 1980). Thus in dioecious species, pollinators presumably are generally cued to male plants, and female plants are visited later in the day (or night) after the resources in male flowers have been exhausted. Differences between male and female plants with respect to initiation of flowering have been documented for several species (Mulcahy, 1968; Godley, 1976; Webb, 1976; Lloyd and Webb, 1977; Bullock and Bawa,

1981). In all cases, males have been found to start flowering earlier than females (Fig. 20-3).

The male and female plants may be expected to differ also in the duration of flowering on the basis of sexual selection and the relative energy expenditure in the production of gametes. If, indeed, it is the quantity of matings that is being emphasized in the males, an extended period of flowering may be more advantageous for the males than for the females. Furthermore, the length of the blooming period in females may be curtailed if their reproductive success is limited by the resources available for fruit production rather than by their ability to get pollinated. This is based on the assumption that an increase in length of the blooming period occurs only at the expense of fruit production. Although there is no information concerning such a trade-off in wild populations, removal of young fruits to lengthen the blooming period in ornamental plants is a common horticultural practice. Average length of the blooming period in the two sexes has been documented for only one dioecious species: in *Jacaratia dolichaula* (Caricaceae), the blooming period of males is significantly longer than that of females (Bullock and Bawa, 1981).

Apart from the timing and duration of flowering, male and female plants might differ with respect to the frequency of flowering. It has been suggested that female plants may suffer a higher risk of mortality than the males because of a greater investment in reproductive structures (Lloyd and Webb, 1977). Higher reproductive costs in terms of biomass accumulated in reproductive structures of females have been shown for *Rumex acetosa* (Putwain and Harper, 1972), whereas mortality differences between males and females are indicated by data of Webb and Lloyd (1980) for Umbelliferae. If the greater investment in reproductive structures by the females were to reduce the probability of reproduction in future years, then the differences in reproductive costs could also be reflected in differences in the frequency of reproductive episodes between males and females. Information on intersexual differences in frequency of flowering

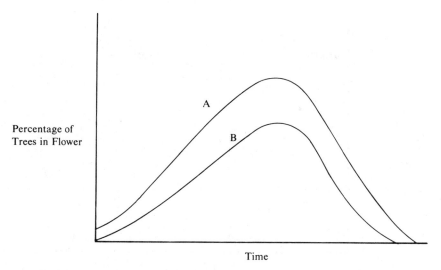

Figure 20-3. Differences in the flowering of male (A) and female (B) plants. A significantly greater proportion of male plants may start flowering earlier than female plants. (Adapted from Bawa, 1980.)

is available for one tropical tree species, *Jacaratia dolichaula,* and three temperate-region herbs, *Aciphylla aurea, Ac. subflabellata,* and *Aralia nudicaulis.* In *Jacaratia dolichaula,* the percentage of plants that flowered in one year but did not do so in the following year was significantly greater for females than for males (Bullock and Bawa, 1981). In *A. nudicaulis* (Bawa et al., 1982), *Ac. aurea,* and *Ac. subflabellata* (Lloyd and Webb, 1977), the male plants were found to flower more frequently than the female plants.

The intersexual differences mentioned above should also be displayed by monoecious plants that undergo distinct male and female phases during flowering. The male phase should precede the female phase, the female phase should be of shorter duration than the male phase, and there should be greater year-to-year variation in the frequency of the female than the male phase. These expectations are consistent with the observations made in a monoecious species, *Cupania guatemalensis* (Bawa, 1977).

VARIATION AMONG SEXUAL SYSTEMS

Monoecious plants in which male and female flowers mature at different times, e.g., Sapin-

daceae (Bawa, 1977) and Palmae (Schmid, 1970), have two distinctive features in their flowering phenology. First, intervening between male and female phases is a period during which no flowers are in anthesis (Fig. 20-4) although plants are essentially in the reproductive state in the sense that they bear old flowers and/or flower buds. Second, plants within a population display marked asynchrony in the onset of flowering.

In bisexual, outcrossing plants, the male pathway to reproductive success involves dispersal of pollen to other plants, whereas the female pathway involves receipt of pollen from other plants. Temporal separation of male and female phases not only reduces the possibility of selfing, but also permits the evolution of different adaptations associated with male and female reproductive success, thereby allowing the optimization of reproductive success via the two pathways in a dissimilar manner. The intervention of a nonflowering period eliminates altogether the possibility of overlap between the male and female phases. The nonflowering period also may be physiologically necessary for an orderly transformation of the male into the female phase.

Synchronous blooming in monoecious plants with distinct male and female phases would result in plants being in either the male or the

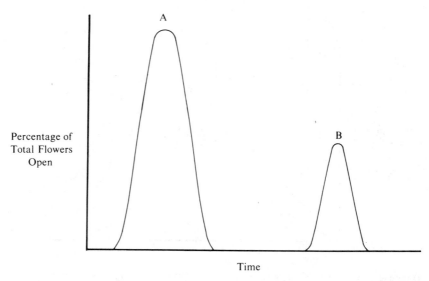

Figure 20-4. Temporal separation of male (A) and female (B) phases in monoecious species. The length of the interval between the phases may vary from one to several weeks in different species.

female phase at the same time (Fig. 20-5A). Asynchronous blooming, on the other hand, would lead to overlap of male and female phases among different plants (Fig. 20-5B). However, if the sequence of blooming were fixed, plants at the extremes would have lower fitness than plants in the middle of the flowering season because they would produce pollen or ovules when other plants were in the male or the female phase, respectively. This apparent paradox could be resolved in a wide variety of ways. For example, in *Cupania guatemalensis* the duration of male and female phases varies widely among individuals and probably between years; several plants in the population start with the female rather than the male phase; many plants undergo only the male phase in a particular year and have a long-enough blooming period to overlap with the female phase of other plants (Bawa, 1977). In palms, each plant may produce several inflorescences, one at a time, with each inflorescence undergoing a distinct male and female phase (Schmid, 1970), but the time interval between consecutive inflorescences may vary widely so that no particular plant consistently has a lower fitness due to asynchrony.

A high level of asynchronous blooming in figs constitutes a special case. Plants are in the female phase when wasps enter the syconia, and the male phase is not initiated until a new generation of wasps, which matures inside the figs, is ready to disperse pollen to other figs in the female phase (Wiebes, 1979). Synchronous blooming in figs would entail all plants being in the male phase soon after a seed crop has matured. Even if a new generation of figs could be developed to coincide with the maturation of the male phase, a considerable amount of selfing would occur by intratree movement of wasps from old to new figs. Asynchronous blooming in figs is further permitted by the fact that wasps are species-specific and seek out the proper species regardless of their rarity created by intrapopulation asynchrony (Janzen, 1979).

CONCLUDING REMARKS

Pollination of large woody plants requires the resolution of several conflicts. These plants must produce enough rewards to attract pollen vectors but not enough to satiate them so as to preclude interplant movement and therefore cross-pollination (Heinrich and Raven, 1972; Heinrich, 1975a). Cross-pollination in turn requires a certain degree of synchrony in the

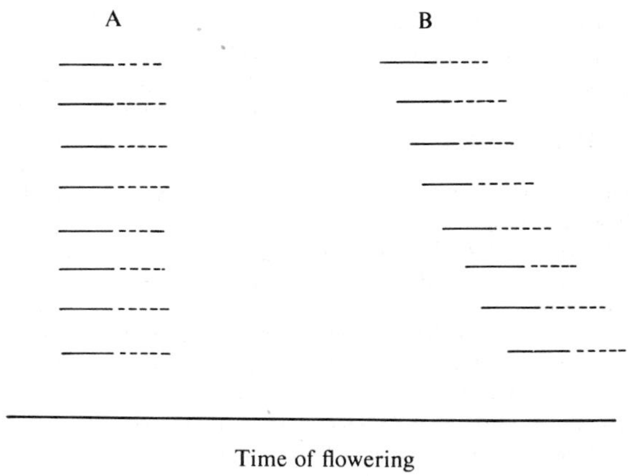

A   B

Time of flowering

Figure 20-5. Consequences of synchronous (A) and asynchronous (B) blooming on the overlap between male and female phases of different individuals in monoecious species. (Solid line—male phase; dashed line—female phase.)

onset and duration of flowering among conspecifics, but if the size of the pollinator pool is limited, the simultaneous blooming of all individuals could also result in intraspecific competition. Superimposed upon this conflict, in a species-rich community, is the competition for pollinator services among similar species. And, in addition to the conflict between plants and pollinators and among plants for pollinator services, there is the discrepancy between reproductive goals of male and female plants (Janzen, 1977; Willson, 1979; Bawa, 1980). The intersexual conflict is not confined to dioecious plants. Many monoecious plants, as pointed out above, undergo distinct male and female phases, and the flowering patterns of the two phases correspond to those displayed by the two sexes of dioecious plants. Even in hermaphroditic plants, the male and female reproductive functions may be optimized in different ways (Janzen, 1977; Charnov, 1979; Bawa and Beach, 1981; Lloyd, 1982). The pollen donation and pollen receipt may not be exactly symmetrical in time, and in some species there may be subtle differences between flowering patterns during the optimal "male phase" and the optimal "female phase" of the same plant.

The flowering patterns, to a very large extent, reflect different ways to resolve the conflicts involved in pollination, and represent compromises among different selective forces, making it difficult to assign a particular selective force to a specific flowering pattern. Furthermore, our understanding of different selective pressures is limited by a number of factors. First, although phenological data are available for a number of species on a community-wide basis, information on the phenology of pollinators and their relative abundance over space and time is almost nonexistent. Second, the way the phenology of flowers and inflorescences is integrated with the flowering phenology of the individual plant is not understood. This is important because the dynamics of pollinator movement depend not only upon the size of the resource patch (individual plant), but also upon spatial and temporal heterogeneity within and among patches. Third, the flowering phenology of individual plants has not been studied in the context of overall phenology; so constraints imposed by selection on the other life history traits are not adequately understood. Similarly, there is no information on the resources other than flowers that are crucial for the completion of pollinators' life cycles. Fourth, except for a few studies (Augspurger, 1978, 1980; Primack, 1980; Bullock and Bawa, 1981; Bullock et al., 1982), quantitative information on variation in timing, duration, and frequency of flowering at the individual and population levels for most species is lacking. Finally, there are no data on year-to-year variation in flowering patterns of a large number of marked individuals within the same species.

ACKNOWLEDGMENTS

The phenological observations reported here for the first time constituted a part of a wider study on the breeding systems of lowland tropical trees supported by grants DEB 7521018 and DEB 7725558 from the National Science Foundation. Jim Beach, Steve Bullock, Michael Grayum, and David Lloyd contributed to the development of several ideas and commented on the manuscript. Peter Ashton, Carol Augspurger, Julie Denslow, Gordon Frankie, Al Gentry, Bill Haber, Dan Janzen, Paul Opler, Richard Primack, Barry Tomlinson, Peter Stevens, Gary Stiles, and Colin Webb offered suggestions for improvements in the manuscript.

LITERATURE CITED

Antonovics, J. and D. A. Levin. 1980. The ecological and genetic consequences of density-dependent regulation in plants. *Annu. Rev. Ecol. Syst.* 11:411–452.

Augspurger, C. 1978. Reproductive consequences of flowering synchrony in *Hybanthus prunifolius* (Violaceae) and other shrub species in Panama. Ph.D. thesis, University of Michigan, Ann Arbor, 204 pp.

———. 1980 Mass flowering of a tropical shrub *(Hybanthus prunifolius). Evolution* 34:475–488.

Baker, H. G. 1976. "Mistake pollination" as a reproductive system with special reference to the Caricaceae. *In* J. Burley and B. T. Styles (eds.), *Tropical Trees: Variation Breeding and Conservation.* Academic Press, London, pp. 161–169.

Bateman, A. J. 1948. Intrasexual selection in *Drosophila. Heredity* 2:349–369.

Bawa, K. S. 1974. Breeding systems of tree species of a lowland tropical community. *Evolution* 28:85–92.

———. 1977. Reproductive biology of *Cupania guatemalensis* Radlk. (Sapindaceae). *Evolution* 31:52–63.

———. 1980. Mimicry of male by female flowers and intrasexual competition for pollinators in *Jacaratia dolichaula* (D. Smith) Woodson (Caricaceae). *Evolution* 34:467–474.

Bawa, K. S. and J. H. Beach 1981. Evolution of sexual systems in flowering plants. *Ann. Missouri Bot. Gard.* 68:254–274.

Bawa, K. S. and P. A. Opler. 1975. Dioecism in tropical forest trees. *Evolution* 29:167–179.

Bawa K. S., C. R. Keegan, and R. H. Voss. 1982. Sexual dimorphism in *Aralia nudicaulis* L. (Araliaceae). *Evolution* 36:371–378.

Brown, J. H. and A. Kodric-Brown. 1979 Convergence, competition, and mimicry in a temperate community of humming-bird pollinated flowers. *Ecology* 60:1022–1035.

Bull, J. J. and R. Shine. 1979. Iteroparous animals that skip opportunities for reproduction. *Am. Nat.* 114:296–303.

Bullock, S. H. and K. S. Bawa. 1981. Sexual dimorphism and the annual flowering pattern in *Jacaratia dolichaula* (D. Smith) Woodson (*Caricaceae*) in a Costa Rican rain forest. *Ecology* 62:1494–1504.

Bullock, S. H., J. H. Beach, and K. S. Bawa, 1982. Episodic flowering and sexual dimorphism in *Guarea rhopalcarpa* in a Costa Rican rain forest. *Ecology.* In press.

Burley, N. 1980. Clutch overlap and clutch size: alternative and complementary tactics. *Am. Nat.* 115:223–246.

Charnov, E. L. 1979. Simultaneous hermaphroditism and sexual selection. *Proc. Natl. Acad. Sci. U.S.A.* 76:2480–2484.

Cole, B. J. 1981. Overlap, regularity and flowering phenologies. *Am. Nat.* 117:993–997.

Croat, T. B. 1969. Seasonal flowering behavior in central Panama. *Ann. Missouri Bot. Gard.* 56:295–307.

Darwin, C. 1871. *The Descent of Man and Selection in Relation to Sex.* John Murray, London.

Daubenmire, R. 1972. Phenology and other characteristics of tropical semi-deciduous forest in northwestern Costa Rica. *J. Ecol.* 60:147–170.

Ewel, J. 1980. Tropical succession: manifold routes to maturity. *In* J. Ewel (ed.), *Tropical Succession. Biotropica,* Suppl. to Vol. 12, pp. 2–7.

Foster, R. B. 1977. *Tachigalia versicolor* is a suicidal neotropical tree. *Nature* 268:624–626.

Fournier, L. A. and S. Salás. 1966. Algunas observaciones sobre la dinamica de la floracion en el bosque tropical humedo de villa colon. *Revista Biol. Trop.* 14:75–85.

Frankie, G. W. 1975. Tropical forest phenology and pollinator–plant coevolution. *In* L. E. Gilbert and P. H. Raven (eds.), *Coevolution of Animals and Plants.* University of Texas Press, Austin, pp. 192–209.

Frankie, G. W. and H. G. Baker. 1974. The importance of pollinator behavior in the reproductive biology of tropical trees. *An. Inst. Biol. Univ. Nac. Auton. Mexico* 45, Ser. Bot. (1):1–10.

Frankie, G. W., H. G. Baker, and P. A. Opler. 1974a. Comparative phenological studies of trees in tropical wet and dry forests in the lowlands of Costa Rica. *J. Ecol.* 62:881–919.

———. 1974b. Tropical plant phenology: applications for studies in community ecology. *In* H. Lieth (ed.), *Phenology and Seasonality Modeling.* Springer-Verlag, Berlin, pp. 287–296.

Frankie, G. W., P. A. Opler, and K. S. Bawa. 1976. Foraging behavior of solitary bees: implications for outcrossing of a neotropical forest tree species. *J. Ecol.* 64:1049–1057.

Gentry, A. H. 1974a. Flowering phenology and diversity in tropical Bignoniaceae. *Biotropica* 6:64–68.

———. 1974b. Coevolutionary patterns in central American Bignoniaceae. *Ann. Missouri Bot. Gard.* 61:725–759.

———. 1978. Anti-pollinators for mass-flowering plants. *Biotropica* 10:68–69.

Giesel, J. T. 1976. Reproductive strategies as adaptations

to life in temporally heterogeneous environments. *Annu. Rev. Evol. Syst.* 7:57–79.

Gilbert, L. E. 1975. Ecological consequences of a co-evolved mutualism between butterflies and plants. *In* L. E. Gilbert and P. H. Raven (eds.), *Coevolution of Animals and Plants.* University of Texas Press, Austin, pp. 210–240.

Godley, E. J. 1976. Sex ratio in *Clematis gentianoides* DC. *N.Z. J. Bot.* 14:299–306.

Hallé, F., R. A. A. Oldeman, and P. B. Tomlinson. 1978. *Tropical Trees and Forests—An Architectural Analysis.* Springer-Verlag, New York.

Hartshorn, G. H. 1978. Tree falls and tropical forest dynamics. *In* P. B. Tomlinson and M. H. Zimmerman (eds.), *Tropical Trees as Living Systems,* Cambridge University Press, pp. 617–638.

Heinrich, B. 1975a. Energetics of pollination. *Annu. Rev. Ecol. Syst.* 6:139–170.

———. 1975b. Bee flowers: a hypothesis on flower variety and blooming times. *Evolution* 29:325–334.

Heinrich, B. and P. H. Raven. 1972. Energetics and population ecology. *Science* 176:597–602.

Heithaus, E. R. 1974. The role of plant–pollinator interactions in determining community structure. *Ann. Missouri Bot. Gard.* 61:675–691.

Heithaus, E. R., T. H. Fleming, and P. A. Opler. 1975. Foraging patterns and resource utilization in seven species of bats in a seasonal tropical forest. *Ecology* 56:841–854.

Hilty, S. L. 1980. Flowering and fruiting periodicity in a premontane rainforest in Pacific Columbia. *Biotropica* 12:292–306.

Janick, J. and J. N. Moore (eds.). 1975. *Advances in Fruit Breeding.* Purdue University Press, West Lafayette, Ind.

Janzen, D. H. 1967. Synchronization of sexual reproduction of trees within the dry season in Central America. *Evolution* 21:620–637.

———. 1970. Herbivores and the number of tree species in tropical forests. *Am. Nat.* 104:501–528.

———. 1971a. Euglossine bees as long-distance pollinators of tropical plants. *Science* 171:203–205.

———. 1971b. Seed predation by animals. *Annu. Rev. Ecol. Syst.* 2:465–492.

———. 1974. Tropical black-water rivers, animals and mast fruiting by the Dipterocarpaceae. *Biotropica* 6:69–103.

———. 1976. Why bamboos wait so long to flower. *Annu. Rev. Ecol. Syst.* 7:347–391.

———. 1977. A note on optimal mate selection by plants. *Am. Nat.* 111:365–371.

———. 1978. Seeding patterns of tropical trees. *In* P. B. Tomlinson and M. H. Zimmerman (eds.), *Tropical Trees as Living Systems.* Cambridge University Press, pp. 83–128.

———. 1979. How to be a fig. *Annu. Rev. Ecol. Syst.* 10:13–51.

Koelmeyer, K. O. 1959. The periodicity of leaf change and flowering in the principal forest communities of Ceylon. *Ceylon Forester* 4:157–189.

Levin, D. A. 1979a. Pollinator foraging behavior: genetic implications for plants. *In* O. T. Solbrig, S. Jain, G. B. Johnson and P. H. Raven (eds.), *Topics in Plant Population Biology.* Columbia University Press, New York, pp. 131–153.

———. 1979b. The nature of plant species. *Science* 204:381–384.

Lloyd, D. G. 1982. The selection for combined versus separate sexes in seed plants. *Am. Nat.* In press.

Lloyd, D. G. and C. J. Webb. 1977. Secondary sex characters in seed plants. *Bot. Rev.* 43:177–216.

McKey, D. 1975. The ecology of coevolved seed dispersal systems. *In* L. E. Gilbert and P. H. Raven (eds.), *Coevolution of Animals and Plants.* University of Texas Press, Austin, pp. 159–191.

McNeilly, T. and J. Antonovics. 1968. Evolution in closely adjacent plant populations. IV. Barriers to gene flow. *Heredity* 23:205–218.

Medway, Lord. 1972. Phenology of a tropical rain forest on Malaya. *Biol. J. Linnean Soc.* 4:117–146.

Moore, H. E., Jr. and N. W. Uhl. 1973. The monocotyledons: their evolution and comparative biology: VI. Palms and the origin and evolution of monocotyledons. *Q. Rev. Biol.* 48:414–436.

Mulcahy, D. L. 1968. The significance of delayed pistillate anthesis in *Silene alba. Bull. Torrey Bot. Club* 96:135–139.

Njoku, E. 1958. The photoperiodic response of some Nigerian plants. *J. West Afr. Sci. Assoc.* 4:99–111.

Opler, P. A. and K. S. Bawa. 1978. Sex ratios in tropical forest trees. *Evolution* 32:812–821.

Opler, P. A., G. W. Frankie, and H. G. Baker. 1976. Rainfall as a factor in the synchronization, release and timing of anthesis by tropical trees and shrubs. *J. Biogeog.* 3:231–236.

———. 1980a. Comparative phenological studies of treelet and shrub species in tropical wet and dry forests in the lowlands of Costa Rica. *J. Ecol.* 68:167–188.

Opler, P. A., H. G. Baker, and G. W. Frankie. 1980b. Plant reproductive characteristics during secondary succession in neotropical lowland forest ecosystems. *In* J. Ewel (ed.), *Tropical Succession. Biotropica,* Suppl. to Vol. 12, pp. 40–46.

Pool, R. W. and B. J. Rathcke. 1978. Regularity, randomness and aggregation in flowering phenologies. *Science* 203:470–471.

Primack, R. B. 1980. Phenological variation within natural populations I. Flowering in New Zealand montane shrubs. *J. Ecol.* 68:849–862.

Putwain, P. D. and J. L. Harper. 1972. Studies in the dynamics of plant populations. V. Mechanisms governing the sex ratio in *Rumex acetosa* and *Rumex acetosella. J. Ecol.* 60:113–129.

Putz, F. E. 1979. Aseasonality in Malaysian tree phenology. *Malaysian Forester* 42:1–28.

Pyke, G. H., H. R. Pulliam, and E. L. Charnov. 1977. Optimal foraging theory: a selective review of theory and tests. *Q. Rev. Biol.* 52:137–154.

Salisbury, E. J. 1942. *The Reproductive Capacity of Plants.* Bell, London.

Schaffer, W. M. and M. D. Gadgil. 1975. Selection for optimal life histories in plants. *In* M. L. Cody and J. M. Diamond (eds.), *Ecology and Evolution of Communities.* Harvard University Press, Cambridge, Mass., pp. 142–157.

Schemske, D. W., M. F. Willson, M. N. Melampy, L. J. Miller, L. Verner, K. M. Schemske, and L. B. Best. 1978. Flowering ecology of some spring woodland herbs. *Ecology* 59:351–366.

Schmid, R. 1970. Notes on the reproductive biology of *Asterogyne martiana* (Palmae). I. Inflorescence and floral morphology. Phenology *Principes* 14:3–9.

Schwabe, W. W. 1971. Physiology of vegetative reproduction and flowering. *In* F. C. Steward (ed.), *Plant Physiology: A Treatise,* VIA. Academic Press, New York, pp. 233–411.

Snow, D. W. 1966. A possible selective factor in the evolution of fruiting seasons in tropical forest. *Oikos* 15:274–281.

Stanford, E. H., M. M. Laude, and P. de V. Booysen. 1962. Effects of advance in generation under different harvesting regimes on the genetic composition of pilgrim ladino clover. *Crop Sci.* 2:497–500.

Stiles, F. G. 1975. Ecology, flowering phenology and hummingbird pollination of some Costa Rican *Heliconia* species. *Ecology* 56:285–310.

———. 1977. Coadapted competitors: the flowering seasons of hummingbird-pollinated plants in a tropical forest. *Science* 198:1177–1178.

———. 1978. Temporal organization of flowering among the hummingbird food plants of a tropical wet forest. *Biotropica* 10:194–210.

Thomson, J. D. 1980. Skewed flowering distributions and pollinator attraction. *Ecology* 61:572–579.

Trivers, R. L. 1972. Parental investment and sexual selection. *In* B. Campbell (ed.), *Sexual Selection and the Descent of Man.* Aldine, Chicago, pp. 136–179.

Waser, N. M. 1978. Competition for hummingbird pollination and sequential flowering in two Colorado wildflowers. *Ecology* 59:945–955.

———. 1979. Pollinator availability as a determinant of flowering time in ocotillo *(Fouquieria splendens).* *Oecologia* 39:107–121.

Webb, C. J. 1976. Flowering periods in the gynodioecious species *Gingidia decipiens* (Umbelliferae). *N.Z. J. Bot.* 14:207–210.

Webb, C. J. and D. G. Lloyd. 1980. Sex ratios in New Zealand apoid Umbelliferae. *N.Z. J. Bot.* 18:121–126.

Wiebes, J. T. 1979. Co-evolution of figs and their insect pollinators. *Annu. Rev. Ecol. Syst.* 10:1–12.

Williams, G. C. 1975. *Sex and Evolution.* Princeton University Press, Princeton, N.J.

Willson, M. F. 1979. Sexual selection in plants. *Am. Nat.* 113:777–790.

21
CHARACTERISTICS AND ORGANIZATION OF THE LARGE BEE POLLINATION SYSTEM IN THE COSTA RICAN DRY FOREST

G. W. Frankie,* W. A. Haber,* P. A. Opler,** and K. S. Bawa***

*Department of Entomological Sciences, University of California, Berkeley, California;
**Office of Endangered Species, Department of the Interior, Washington, D.C.; and ***Biology Department, University of Massachusetts, Boston, Massachusetts

ABSTRACT

Flowers that showed adaptation for pollination by large bees were studied in the highly seasonal dry forest of Costa Rica from 1969 to 1981. Characteristics of large bees that made them appropriate pollinators of these plants were also studied.

Plants with this adaptation were found primarily in the tree (35 of ca. 160 species) and vine-liana (39 of ca. 130 species) life forms. About 60% of the tree species belonged to the Bignoniaceae, Caesalpinaceae, or Fabaceae, and >85% of the vine-liana (or climber) species belonged to the Bignoniaceae, Fabaceae, Malpighiaceae, or Passifloraceae. Plants pollinated by large bees were generally absent in the shrub, herb, and epiphyte life forms.

Large bee flowers were diurnal, relatively large, and generally colorful, and usually lasted for only one day. Most were hermaphroditic, and many were zygomorphic (bilaterally symmetrical). Among the nectar-producing species, nectar sugar concentration averaged at least 24%. A seasonal sequence of flowering was observed during the year in the tree and climber species. In general, large bee plants producing massive displays of flowers bloomed in the dry season, whereas those producing a few flowers over extended periods bloomed commonly in the wet season.

Characteristics of small bee flowers in this forest contrasted sharply with those of large bee flowers. Small bee flowers were well represented in tree, shrub, some climber, and herb species. Flowers were generally small, white to cream-colored, and radially symmetrical. Most species produced considerably less nectar than the large bee flowers. The majority were hermaphroditic, but many were dioecious. Finally, a large number of the species flowered from March through May.

Two bee groups are believed to be the most important pollinators of large bee flowers. The first group consisted of *Centris* spp. and *Gaesischia exul* (Anthophoridae), which were almost exclusively confined in flight and nesting activities to the dry season, the period when most mass flowering species were in bloom. The second bee group, the euglossines (Apidae), were generally more abundant and diverse during the wet season, the period when most extended flowering species were in bloom. Both groups had wide host ranges.

Large bees were not attracted equally to all host species. In the more attractive nectar producers, some trees were known to draw up to 15,000 bees. Those bee levels quickly removed nectar from host trees such that none remained by the end of the day.

An examination of flowering times and floral reward presentation patterns suggested that the plants were not "actively" competing for the services of pollinators. However, the possibility that competition may have been one of the organizing (selective) forces in sorting out flowering periods was not ruled out.

Organization of the interactions between large bees and their plants was viewed at the community level. Previously published work on intertree movement patterns, self-incompatibility, and preference to forage in the canopy was considered with the current work to suggest that overall the large bee plant species and their bees are spatially and temporally structured, and this in turn is related to certain biological and behavioral traits of the two groups of organisms.

KEYWORDS: Costa Rica, large bees, Anthophoridae, Xylocopa, Euglossini, *Centris, Bombus,* Apidae, pollination ecology, Bignoniaceae, Fabaceae, Caesalpinaceae, Malpighiaceae, Passifloraceae, hermaphroditism, dioecy, nectar, pollen, traplining, competition, community ecology, plant reproductive biology, bee nesting biology, self-incompatibility, spatial and temporal structuring, anthophily, pollinator behavior, flowering phenology, anthesis, nectar secretion, production, visitation rate, nectar sugar concentration, floral sexuality, zygomorphy, outcrossing, mass flowering, vines, lianas, climbers, intraspecific variation, buzz pollination, bimodal nectar production, floral mimicry, assortive foraging, nectar monitoring, congeneric species, interplant pollinator movement.

CONTENTS

INTRODUCTION

In this chapter we characterize flowers in the Costa Rican dry forest that show adaptation for pollination by large-sized bees. We also describe characteristics of the large bees (vs. small bees) that make them appropriate pollinators of these plants. The large bee flowers aspects of the bees' biology and behavior are correlated with flowering and floral reward patterns. Overall, our approach is to examine empirically all plant species and their bees that interact to form this pollination type within the context of a large section of dry forest habitat.

This study represents one portion of a larger study aimed at characterizing all of the pollination types in the Costa Rican dry forest. The initial phase of the large study was conducted from 1969 to 1974. During this period numerous observations and collections were gathered primarily on arboreal anthophilous visitors. Selected aspects of the floral biology and associated pollinator behavior of numerous plant species also received attention (Bawa, 1974, 1977; Frankie and Baker, 1974; Bawa and Opler, 1975, 1977; Opler et al., 1975; Frankie et al., 1976; Opler, 1981). Further, general characteristics of most pollination types and speculation on their community organization in the dry forest were offered (Frankie, 1975). Using this work as a basis, an intensive effort to develop detailed information on the characteristics of large bee flowers and their bees was conducted between 1976 and 1981 by two of us (GWF and WAH).

STUDY SITES

Most of the research was conducted at Comelco ranch (ca. 20,200 ha) and Hacienda La Pacifica (ca. 1,200 ha) in Guanacaste Province. These sites, which are separated by about 20 km, are adjacent to the Pan American highway between the towns of Cañas and Liberia. In 1978, a large section of Comelco ranch was sold, but despite this change we were able to gain access to all sections of the property. The center of the original property was located about 14 km southwest of Bagaces (Fig. 21-1). Hacienda La Pacifica is located about 4 km north of Cañas. In addition to the two major sites, the general area between Comelco and La Pacifica was used to conduct pollination studies on selected plant species.

Mixtures of disturbed and mature lowland dry forest were found at all study sites. Climate and vegetation of the Comelco site are described in Frankie et al. (1974a), Vinson

and Frankie (1977), Heithaus (1979 a,b,c), and Opler et al. (1980). Climate and vegetation of Hacienda La Pacifica are described in Daubenmire (1972), Rockwood (1973, 1975), Heithaus et al. (1975), Turner (1975), and Glander (1978). A general description of the lowland dry forest in Costa Rica can be found in Holdridge et al. (1971) and Sawyer and Lindsey (1971).

There are two dry and two wet seasons in lowland Guanacaste. The first dry season, which is long and severe, begins in November and extends into May. The second dry season, or *veranillo,* is short and usually occurs during part of July and August. Most precipitation falls during the wet seasons (late May–early July and late August–early November); however, occasionally heavy showers fall during each dry season (Frankie et al., 1974a; Opler et al., 1976).

METHODS AND TERMINOLOGY

Investigation of the pollination ecology of large bee-adapted plants* focused on the following features of flowers and their anthophilous visitors:

1. Flowering phenology
2. Flower opening time (anthesis) and opening pattern
3. Flower color, odor, shape, and size
4. Flower life span
5. Principal reward(s): pollen and/or nectar
6. Pollen presentation pattern
7. Nectar presentation pattern
8. Sexual system
9. Visitors: type, relative numbers, activity periods, general behavior

Annual flowering period has been recorded for most plants (Frankie et al., 1974a; Opler et al., 1975, 1976, 1980; Opler, Frankie, and Baker, in prep.). Anthesis was initially recorded by spending early morning hours ob-

*Some scientific plant names used in our earlier papers (esp. Frankie et al., 1974a) have been updated. Currently recognized names and their synonyms are found in Janzen and Liesner (1980).

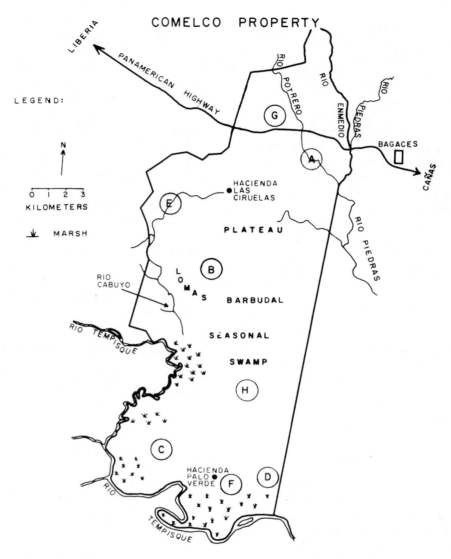

Figure 21-1. Original Comelco property in Guanacaste Province. Circled letters = specific study regions.

serving marked flowers in the field. Later, after representative times had been recorded, we merely noted whether marked flowers had opened before dawn. We also observed whether they opened synchronously or gradually through time. Features 3 through 5 are self-explanatory.

Several aspects of pollen and nectar rewards were examined. The presentation pattern of pollen was investigated by recording when and how anthers dehisced, and when and at what

rate pollen was removed by visitors. Nectar production was systematically monitored with precalibrated micropipettes (Drummond microcaps) from 12 to 40 flowers from accessible sides of medium-sized trees in full flower. Ladders (3.3 meters or less) were commonly used to reach flowers. One to three plants per species were monitored, depending on variation in nectar flow among flowers. Flowering branches were bagged the night before or just before dawn with Pollen Tector bags, and, in most cases, nectar was taken from the same flowers through time. Occasionally, when flowers were extremely delicate, new flowers were sampled through time, and thus a cumulative quantity of nectar was recorded. Nectar monitoring led to the determination of the daily flow pattern and reward structure, and it allowed for calculation of daily nectar production by a given flower. Sugar concentration was determined in the field with a pocket refractometer (Bellingham and Stanley) and by laboratory analysis by Irene Baker (cf. Baker, 1979) at the University of California, Berkeley.

Work on sexual systems was limited primarily to observations of floral sexuality (e.g., hermaphroditism, monoecy, dioecy, etc.).

Extensive collections of anthophilous visitors, especially bees, were gathered from 1970 to 1974 by one of us (PAO), and additional but smaller collections were made from 1975 to 1981 by two of us (GWF and WAH). These were deposited in the California Insect Survey collection at the University of California, Berkeley. The behavior of visitors between flowers on the same and different plants was also noted.

The term "large bee" is used throughout this paper for bees of medium–large body length ($\geq$ 1.2 cm) belonging to certain genera of Anthophorinae and Xylocopinae (Anthophoridae), Bombinae (Apidae), and Colletidae (see below). In contrast to these bees are the smaller bees (usually < 1.2 cm in length) that belong to the Halictidae, Megachilidae, and some species of the Meliponini (Apidae) and Anthophorinae. One anthophorid bee, *Gaesischia exul* (1.0–1.1 cm in length), could be considered as a small or a large bee. However, because of its abundance and apparent ability to cross-pollinate many large bee-adapted flowers (Frankie et al., 1976), it is placed in the large bee group.

The terms "large bee-adapted flower" or "large bee flower" (LBF) refer to those plants having flowers that are morphologically and/or chemically (in the case of pollen and nectar) adapted for pollination by medium–large bees. Implicit in this definition is the assumption that these plants are also primarily outcrossed by these bees.

RESULTS

Overview of "Large Bee Flowers"

Plant species that showed adaptation for pollination by large bees had the following floral characteristics. Flowers were diurnal, relatively large, and generally colorful, and usually lasted for only one day. Most were hermaphrodites, and many were zygomorphic or bilaterally symmetrical (Figs. 21-2–21-6). Among the nectar-producing species, nectar sugar concentration was on the average at least 24%. The flowers often attracted a great variety of bee taxa, especially large bees, during periods when pollen and nectar were presented. Large bees made regular contact with anthers and stigmas of the flowers, and in some cases only these bees were able to remove the well-protected pollen and nectar because of their large size and appropriate behavior. Further, large bees are known to move between trees (Frankie et al., 1976), and this behavior is important, since many LBF species are self-incompatible (Bawa, 1974). In contrast, other visitor types, including small bees, were much less effective in picking up pollen and transporting it to other flowers. Overall, other visitors usually occurred in smaller numbers or were smaller in size and therefore not as effective in opening flowers (e.g., some tight flag flowers of Fabaceae) or making contact with the anthers and stigmas.

Within the study area, 89 plant species (Tables 21-1–21-3) were found to be adapted for pollination primarily by large bees. Most were trees or climbers (vines/lianas). Plant species adapted to these bees were rarely found in understory vegetation (shrubs and herbs) or among the relatively few epiphytes in the study area.

Trees

Thirty-five of approximately 160 tree species (20%) found in the Guanacaste study area

Figure 21-2. Flowers of *Cassia grandis*; ~25 mm corolla diameter (widest pt.).

Figure 21-3. Extended flowering of *Cochlospermum vitifolium*; ~70 mm corolla diameter (widest pt.).

Figure 21-4. Flowers of *Gliricidia sepium*; ∼20 mm corolla diameter (widest pt.).

Figure 21-5. Mass-flowering of *Tabebuia impetiginosa*; ~60 mm corolla tube length.

Figure 21-6. Mass-flowering of *Arrabidaea patellifera*; ~45 mm corolla tube length.

showed adaptation for large bee pollination. About 60% of these belonged to the Bignoniaceae, Caesalpinaceae, or Fabaceae (Table 21-1). Their flowering characteristics are presented below.

Tree Phenology. An overall appraisal of vegetational responses in the dry forest (Frankie et al., 1974a) suggested that operationally there are only two seasons: the long dry season from mid-November to mid-May and the long wet period from mid-May to mid-November. Flowering phenologies of the common and occasional LBF tree species (31) are presented in Fig. 21-7 (and Table 21-1); four rare species are considered in a later section. Among the common or occasional species, 21 were characterized as mass flowering (= Type 3 or 4 of Gentry, 1974a,b). That is, individual plants produced large numbers of flowers over usually relatively short periods (3 days to 6 weeks). All of these species except *Eugenia salamensis* bloomed during the long dry season. Most flowering occurred from January through April, and during each of these months at least 8 species were in bloom. Ten species flowered over extended periods, each producing limited numbers of flowers daily (= Type 2 of Gentry, 1974a,b, and traplining type of Janzen, 1971). Most of these species bloomed during the wet season (Fig. 21-7). The extreme example of this behavior was observed in *Stemmadenia obovata,* which produced its few flowers daily over a 9-month period (Opler et al., 1980). *Byrsonima crassifolia* was intermediate in its flowering behavior. Although it produced moderate to large numbers of flowers daily, it did so for an extended period (Fig. 21-7 and Table 21-1).

Flowering of the species in Fig. 21-7 and for species of other life forms in this forest has been characterized as well synchronized (Frankie et al., 1974a,b; Opler et al., 1980). For the purpose of past work this was an accurate description of blooming activity. Further, despite some deviations from the main blooming patterns (Frankie et al., 1974a; Opler et al., 1976), the flowering sequence of the LBF trees in the general study area remained remarkably consistent year after year (Frankie, Haber, Opler, and Bawa, unpub.

notes, 1974–1980). However, in this chapter we recognize that for most mass blooming species, flowering among conspecifics in a given area is usually staggered by a few days to a week. Thus, for a particular species, one may expect many individuals to be in full flower while others will just be starting or ending their flowering episodes. The importance of slightly unsynchronized flowering in outcrossing is discussed in detail by Frankie and Haber (this book).

Climbers

Thirty-nine of approximately 130 vine and liana species (30%) showed adaptation for large bee pollination. The vast majority of LBF climbers (>85%) belonged to the Bignoniaceae, Fabaceae, Malpighiaceae, and Passifloraceae (Table 21-2). Most were woody perennials (lianas), although most Fabaceae were herbaceous annuals (vines).

Phenology of Climber Species. Yearly flowering activity of these species can be divided into two major periods: the long dry season and the long wet season. Flowering phenologies of common and occasional LBF climbers (35) are presented in Fig. 21-8 (and Table 21-2); four rare species are considered in a later section. In contrast to trees, 23% of the climbers (vs. 71% of the trees) flowered in the dry season (Fig. 21-7), whereas 11% (vs. 16%) flowered in the wet season, and 66% (vs. 13%) had their blooming periods extended over portions of both seasons. The number of climber species in flower at a given time varied from 8 (August–October) to 15 (May), and was generally higher during the dry season and the beginning of the wet season.

Because of their numerical dominance the Bignoniaceae play a major role in determining the seasonal flowering patterns. At least 2 bignoniaceous species were in flower each month, and 5 or more flowered in January and during the period from April to July. Most LBF Fabaceae and Malpighiaceae flowered during the dry season, whereas all Passifloraceae flowered during the wet season.

A bare majority, 19 species (54%), of LBF climbers had extended flowering; 16 (46%)

Table 21-1.   General floral characteristics of common and occasional LBF tree species. Order of seasonal flowering, beginning in December (Fig. 21-7), indicated in parenthesis.

Taxa	Sexuality, breeding system[a]	Prin. reward[b]	Color[c]	Fragrance[d]	Flower life (days)	FOT[e]	Flower opening pattern[f]
Apocynaceae							
(19) *Stemmadenia obovata* (Hook. & Arn.) K. Schum.	H	N	Y	Str	1	05-0830	*Syn,R/Asyn
(25) *Thevetia ovata* (Cav.) A.DC.	H	N	Y	Mod	1	before 0550	Syn,R
Bignoniaceae							
(24) *Godmania aesculifolia* (HBK.) Standl.	H.	N	Y & Pur	none	1?	Pk[g]1030-1230	Asyn
(2) *Tabebuia impetiginosa* (Mart. ex DC.) Standl.	H	N	Pur	Sli	1-2	Pk 0430-0800	Asyn
(18) *T. ochracea* (Cham.) Stand. spp. *neochrysantha* (A. Gentry) A. Gentry	H	N	Y	Mod	2	05-0600	*Syn,R/Asyn
(12) *T. rosea* (Bertol.) DC.	H	N	L–W	none	1	before 0530	Syn,R
Bixaceae							
(31) *Bixa orellana* L.	H	P	L	Sli	1	~0500	Syn,R
Caesalpinaceae							
(6) *Caesalpinia eriostachys* Benth.	H	N	Y	Str	1	before 0415	Syn,R
(7) *Cassia emarginata* L.	H	P	Y	Sli	>1?	before 0530	Syn,R
(16) *C. grandis* L.	H	P	R–L	Sli	1	before 0515	Syn,R
(30) *C. skinneri* Benth.	H	P	Y		>1?		
(11) *Haematoxylon brasiletto* Karst.	H	N	Y	none	1	before 0530	
(1) *Parkinsonia aculeata* L.	H	N	Y	Str	2	cont in AM	Asyn
Cochlospermaceae							
(3) *Cochlospermum vitifolium* (Willd.) Spreng.	H	P	Y	none	1	05-0600	Syn,R
Dilleniaceae							
(4) *Curatella americana*	H	P	W	Str	1	0430-?0730	Syn,R
Fabaceae							
(10) *Andira inermis* (Wright) HBK	H	N	Pur	Mod	1	0730-0830	Syn,R
(20) *Dalbergia retusa* Hemsl.	H	N	W	Str	1	0530 or later start 0800	Syn,R
(5) *Gliricidia sepium* (Jacq.) Walp.	H	N	L	none	1	Pk 13-1500	

Species		P/N	Color	Odor	No.	FOT	Syn/Asyn
(21) *Lonchocarpus costaricensis* Pittier	H	P/N*	R–Pur	none	1	1030–1130	Syn,R
(23) *L. eriocarinalis* Micheli	H	P/N*	Pur	none	1	12–1300	Syn,R
(17) *L. hondurensis* Benth.	H	N	L	none	1	most by 0600 few: 06–1500	Asyn
(14) *Myrospermum frutescens* Jacq.	H	N	L/W	Sli	1	0415	Syn,R
(22) *Piscidia carthagenensis* Jacq.	H	N	pale pink	none	2	start 0800 Pk 11–1300	Asyn
(8) *Platymiscium pleiostachyum* D. Sm.	H	N	Y–Or	Str	1	before 0630	
(15) *Pterocarpus rohrii* Vahl.	H	N	Y–Or	none	1	07–0800	Syn,R
Malpighiaceae							
(13) *Byrsonima crassifolia* (L.) HBK.	H	P/N	Y	none	1	0530–1300	Asyn
(29) *Malpighia emarginata* DC.	H	*P/?N	Pink	none	1	0545	Syn,R
Myrtaceae							
(27) *Eugenia salamensis* D.Sm. var. *hiraeifolia* (Standl.)McV.	H	P	cream W	Mod	1	0450	Syn,R
Rubiaceae							
(28) *Genipa americana* L. ♂	Dioecious	N	cream W	Sli	1	0615	Syn,R
(28) *G. americana* L. ♀	Dioecious	N	cream W	none	1?	07–0800	Syn,R
Styracaceae							
(9) *Styrax argentea* Presl. var. *argentea*	H	N	W	Str	1	0400	Syn,R
Tiliaceae							
(26) *Apeiba tibourbou* Aubl.	H	P	Y	Str	1	before 05–0800	Asyn

[a]H = hermaphrodite.
[b]N = nectar; P = pollen.
[c]Y = yellow; Pur = purple; W = white; L = lavender; Or = orange; R = red.
[d]Sli = slight; Mod = moderate; Str = strong.
[e]FOT = flower opening time.
[f]Syn = synchronous; R = rapid; Asyn = asynchronous.
[g]Pk = peak flowering time.
*Primary.

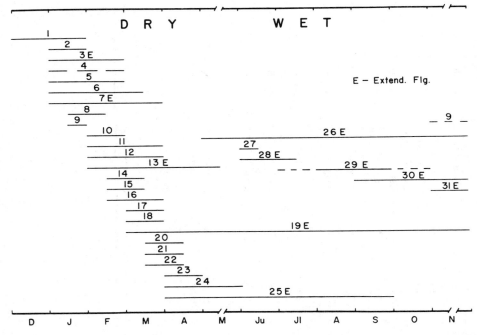

Figure 21-7. Phenology of common and occasional LBF tree species. Numbers refer to taxa given in Tables 21-1 and 21-5.

were generally described as mass flowering. The extended bloomers had at least 5 species in flower each month, with an apparent peak occurring in November (11 species). In direct contrast, mass bloomers, mostly Bignoniaceae, were predominant in the dry season, when from 4 to 7 species flowered in each month from January through May. No more than 3 mass bloomers were in flower from June through August, with none in flower from September through November.

Some LBF climbers had more than one brief, isolated flowering episode in the same season (= Type 5 flowering of Gentry, 1974a,b). *Hiraea reclinata* and *Xylophragma seemannianum* exhibited this trait and flowered in the dry season, whereas individual *Cydista heterophylla* flowered massively at intervals during the wet season. *Hiraea reclinata, X. seemannianum*, and *Macfadyena unguiscati* had their flowering stimulated by token rainfall in the latter half of the dry season (Opler et al., 1976), and *Cydista diversifolia* and *Securidaca sylvestris* may also be stimulated to flower by localized precipitation during the early dry season.

Other Floral Resources

In addition to the common and occasional tree and climber species, several other floral resources were used by the bees. These were separated into two groups: (1) rare tree and climber species and a few shrub, herbaceous, and epiphytic species that are adapted for large bees and (2) floral resources adapted for other pollinator types.

The rare LBF tree (4) and climber (4) species were totaled in the forest-wide figures provided earlier; however, they were not included in the phenology graphs (Figs. 21-7 and 21-8), since they represent only minor food resources within the study area. These species and the few shrub, herbaceous, and epiphytic species that show adaptation for large bee pollination are listed in Table 21-3.

Floral resources that are adapted for other pollinator types but were used to some extent by large bees, include many species. For example, the trees *Bombacopsis quinata* (Bombacaceae) and *Pithecellobium saman* (Mimosaceae), which are adapted to nocturnal pollinators, were commonly visited by carpen-

Table 21-2. General floral characteristics of common and occasional LBF climber species. Order of seasonal flowering, beginning in December (Fig. 21-8), indicated in parenthesis.

Taxa	Sexuality[a]	Prin. reward[b]	Flower color
Apocynaceae			
(31) *Fernaldia pandurata* (A.DC.) Woods	H	N	white
Bignoniaceae			
(23) *Adenocalymma apurense* (HBK.) Sandw.	H	N	yellow
(1) *Amphilophium paniculatum* (L.) HBK.	H	N	pink
(26) *Arrabidaea conjugata* (Vell.) Mart.	H	N	pink
(17) *A. corallina* (Jacq.) Sandw.	H	N	violet/purple
(4) *A. mollissima* (HBK.) Bur. & Schum.	H	N	lavender
(25) *A. patellifera* (Schlecht.) Sandw.	H	N	purple
(19) *Callichlamys latifolia* (L.Rich.) K. Schum.	H	N	yellow
(16) *Clytostoma binatum* (Thunb.) Sandw.	H	None[c]	pink
(13) *Cydista aequinoctialis* (L.) Miers	H	None[c]	pale pink
(9) *C. a.* var. *hirtella* (Benth.) A.Gentry	H	None[c]	pink
(8) *C. diversifolia* (HBK.) Miers	H	None	purple
(15) *C. heterophylla* Seib.	H	None[c]	purple
(18) *Macfadyena unguis-cati* (L.) A.Gentry	H	N	yellow
(5) *Mansoa hymenaea* (DC.) A.Gentry	H	N	pink
(22) *Pithecoctenium crucigerum* (L.) A.Gentry	H	N	white/yellow
(12) *Xylophragma seemannianum* (O.Ktze.) Sandw.	H	N	pink
Caesalpinaceae			
(33) *Bauhinia glabra* Jacq.	H	N	white
Dilleniaceae			
(3) *Davilla kunthii* St. Hil.	H	?	yellow
Fabaceae			
(2) *Centrosema pubescens* Benth.	H	N	pink
(34) *C. sagittatum* (H. & B.) Brand ex Riley	H	N	white/purple
(28) *Dioclea megacarpa* Rolfe	H	N	blue-purple
(32) *Mucuna pruriens* (L.) DC.	H	N	black-purple
(35) *Phaseolus atropurpureus* DC.	H	N	purple
Malpighiaceae			
(30) *Banisteriopsis muricata* (Cav.) Cuatr.	H	P/N	pink
(11) *Hiraea reclinata* Jacq.	H	P/N	yellow
(6) *Stigmaphyllon ellipticum* (HBK.) Adr. Juss.	H	P/N	yellow
(10) *S. lindenianum* Adr. Juss.	H	P/N	yellow
Passifloraceae			
(14) *Passiflora foetida* L.	H	N	white
(29) *P. platyloba* Killip	H	N	white
(24) *P. pulchella* HBK.	H	N	white/lavender
(20) P. sp. (PAO 967)	H	N	
(21) P. sp. #1	H	N	
(27) P. sp. #2	H	N	
Polygalaceae			
(7) *Securidaca sylvestris* Schlecht.	H	N	pink

[a]H = Hermaphrodite
[b]P = Pollen; N = Nectar
[c]Lacks nectar disc according to Gentry (1974a).

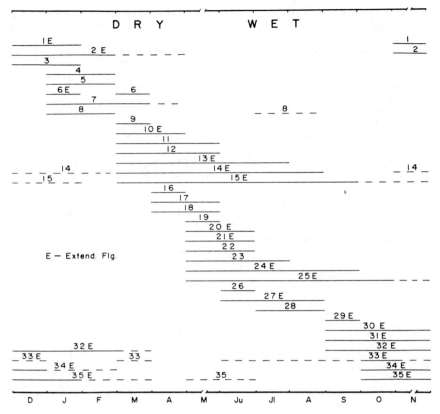

Figure 21-8. Phenology of common and occasional LBF climber species. Numbers refer to taxa given in Table 21-2. Flowering periods studied during 1972–1974 (Opler, unpub.).

ter bees around sunrise. A few of these plant species, e.g., the shrub *Stachytarpheta jamaicensis* (Verbenaceae), served as important food resources for several bee groups; however, the majority can be considered as minor resources. These are listed in Table 21-4.

Floral Biology: Trees

All but one species, *Genipa americana* (dioecious), had perfect flowers (Table 21-1).

Pollen was the principal or sole reward in 10 species (Table 21-1), and 6 of these were extended flowering species. Of the 4 remaining species, one, *Cassia grandis,* although not categorized as an extended bloomer, did produce pollen for about a 6-week period. Thus, as a group there was at least one pollen-producing tree species in flower during each month (Fig. 21-7 and Table 21-1). Nectar was the principal reward in 22 species (Table 21-1).

A wide variety of flower colors and odors was noted among the trees. However, the vast majority of colors were categorized as either yellow to orange or pink to purple; a few had white flowers (e.g., *Curatella americana*). All but two extended bloomers (*Bixa orellana* and *Malpighia emarginata*) had yellow flowers. Flower fragrances varied greatly from none detectable (as perceived by humans) to very strong. Detected odors were sweet and pleasant to humans.

Flowers of most trees were functional for only one day. In 4 species (*Parkinsonia aculeata, Tabebuia impetiginosa, T. ochracea,* and *Piscidia carthagenensis*) flowers lasted for up to 2 days. The stigma of *P. carthagenensis* was receptive on both days (Baker et al., 1982).

The flowers of most trees opened before sunrise (or at sunrise in a few cases), and the flower-opening pattern of these species was recorded as mostly synchronous and rapid. In 8

Table 21-3. Other LBF plant species.

Life form Taxa	Family	Prin. reward[a]	Season(s) of flowering Long dry	Long wet
Tree				
Stemmadenia donnell-smithii (Rose) Woodson	Apocynaceae	N	Portions of both	
Cassia hayesiana (B. & R.) Standl.	Caesalpinaceae	P		+
Schizolobium parahybum (Vell.) Blake	Caesalpinaceae	N	+	
Diphysa robinioides Benth.	Fabaceae	N	+	
Climber				
Mesechites trifida (Jacq.) M. Arg.	Apocynaceae	N	+	
Centrosema plumieri (Turp.) Benth.	Fabaceae	N	+	
Dalbergia glabra (Millsp.) Standl.	Fabaceae	N	+	
Adenopodia polystachya (L.) Dixon	Mimosaceae	N?	Portions of both	
Shrub				
**Cassia biflora* L.	Caesalpinaceae	P	+	
**Solanum flavescens* Dun.	Solanaceae	P	Both	
**S. hazenii* Britton	Solanaceae	P	Both	
**S. hirtum* Vahl.	Solanaceae	P	Both	
**S. ochraceo-ferrugineum* (Dunal.) Fern.	Solanaceae	P	Both	
Prockia crucis L.	Flacourtiaceae	P		+
Malpighia ? glabra L.	Malpighiaceae	P/N	Portions of both	
Ouratea lucens (HBK.) Engler	Ochnaceae	N	+	
Ouratea sp. (PAO 652)	Ochnaceae	N	+	
Herb				
Calathea allouia (Aubl.) Lind.	Marantaceae	N		+
Cyphomandra allophylla (Miers) Hemsl.	Solanaceae	P		+
Sesbania emerus (Aubl.) Urban	Fabaceae	N		+
Epiphyte				
Catasetum maculata HBK.	Orchidaceae	?		+
Encyclia cordigera (HBK.) Dressler	Orchidaceae	?	+	
Laelia rubescens Lindl.	Orchidaceae	?	+	

[a]P = pollen; N = nectar.
*Rare
**Important pollen resource for some large bee species in mostly disturbed habitats.

species (Nos. 5, 10, 15, 21–24, 28 in Table 21-1), flowers opened or began opening after sunrise, and in 3 (*Gliricidia sepium, Piscidia carthagenensis,* and *Godmania aesculifolia*) peak anthesis was at midday. Some species (Nos. 1, 2, 13, 17, 22, 24, 26) opened their flowers asynchronously, with more than half of these having a peak in anthesis during the day. Intraspecific variation in anthesis was observed in *Andira inermis, Curatella americana,* and *Dalbergia retusa.* A 1- to 3-hour difference in opening time occurred among some individuals of these species.

Anthers of most trees dehisced in the bud (Table 21-5). In *Byrsonima crassifolia, Cochlospermum vitifolium,* and *Tabebuia impetiginosa,* dehiscence occurred within an hour after anthesis (> Anth, Rapid). In 5 species,

(*Caesalpinia eriostachys, Dalbergia retusa, Haematoxylon brasiletto, Myrospermum frutescens,* and *Parkinsonia aculeata*) the anthers dehisced slowly after anthesis, over periods of more than an hour per flower (> Anth, Prog).

Among species that produced pollen as a reward, two patterns of removal by bees were observed. In *Apeiba tibourbou, Cochlospermum vitifolium, Cassia emarginata, Byrsonima crassifolia,* and *Lonchocarpus costaricensis,* pollen was removed gradually and this pattern in turn was related to anther structure (apical pores) in the cases of *A. tibourbou, C. vitifolium* and *C. emarginata.* In these 3 species, anthers were buzzed or vibrated for pollen by bees (Michener, 1962; Wille, 1963; Buchmann, 1974; Delgado and Sousa, 1977; Buchmann and Hurley, 1978). Gradual removal of

Table 21-4. Non–large bee floral resources that are used occasionally by large bees.

Plant taxa	Resource[a]	Large bee visitors
Alibertia edulis A. Rich.	N	*Euglossa*
Allophyllus occidentalis (Sw.) Radlk.	N	Anthophorinae
Bourreria quirosii Standl.	N	*Eulaema;* robbed by *Xylocopa*
Bursera tomentosa (Jacq.) Triana & Planch.	N	*Centris; Xylocopa*
Calliandra tapirorum Standl.	N	robbed by *Xylocopa*
Crataeva tapia L.	N	*Xylocopa*
Diospyros nicaraguensis Standl.	N	*Euglossa*
Erythroxylon havanense Jacq.	N	Anthophorinae
Inga vera Willd.	N	*Ptiloglossa, Xylocopa*
Ipomoea carnea Jacq.	N	robbed by *Xylocopa*
Laetia thamnia L.	N	*Centris; Xylocopa*
Lantana camara L.	N	*Centris*
Lonchocarpus minimiflorus D. Sm.	N/P	*Centris*
Luehea candida (DC) Mart.	N	*? Ptiloglossa; Xylocopa*
L. seemannii Triana & Planch.	N	*Ptiloglossa*
Maranta arundinacea L.	N	*? Euglossa*
Pisonia macranthocarpa D. Sm.	N	*Centris*
Pithecellobium mangense (Jacq.) Macbr.	N	*Xylocopa*
Psitacanthus calyculatus (DC.) G. Don.	N	*Xylocopa*
Spondias mombin L.	N	*Eulaema; Ptiloglossa; Xylocopa*
Stachytarpheta jamaicensis (L.) Vahl	N	*Eulaema; Xylocopa*
Ximenia americana L.	N	*Centris*

[a]P = pollen; N = nectar.

pollen in *B. crassifolia* and *L. costaricensis* was related mostly to relatively low visitation rates. The second pattern of removal was characterized as extremely rapid (ca. one hour) and did not involve unique anther structures. Rather, bees foraged intensively for the pollen of *Bixa orellana, Curatella americana, Cassia grandis,* and *Eugenia salamensis,* which was available at anthesis.

Three temporal patterns of nectar flow were observed. The most common pattern (11 spp.) was that of nectar production for a lengthy period (see "Cont" category in Table 21-5). Three variations on this pattern, depending on time of day of nectar flow, were also recognized. In the case of *B. orellana,* anthers were buzzed to remove the pollen.

The second pattern was that of bimodal nectar production in as many as 10 species (Table 21-5). Time elapsed between the first and second nectar peaks varied greatly within and between species. The greatest elapsed time was observed in *M. frutescens* (5 hr) with the shortest in *P. rohrii* (2 hr). Generally, the first nectar peak was approximately twice as great

as the second. A few individuals of some bimodal species produced three peaks of nectar, while others produced only one peak. Details on variations in nectar flow in these and other species are found in Frankie and Haber (this book). The third nectar flow pattern was uncommon. It involved a single prolonged period of nectar secretion by relatively few flowers on a tree *(Lonchocarpus costaricensis).*

Two general patterns of nectar reward structure were observed. The first and most common (Reward struct A, Table 21-5) was that of nectar production in all or nearly all of a tree's flowers. In most individuals of these species it was common to find considerable interflower variation in nectar production (see Frankie and Haber, this book). The second pattern (Reward struct B, Table 21-5) was that of nectar secretion in only a small proportion of the open flowers, and usually the quantity produced was small. This pattern was observed in *Godmania aesculifolia, Parkinsonia aculeata, Lonchocarpus costaricensis,* and most likely in *Lonchocarpus eriocarinalis.* Another species, *Piscidia carthagenensis,* had a

Table 21-5. Pollen and nectar characteristics of common and occasional LBF tree species. Order of seasonal flowering, beginning in December (Fig. 21-7), indicated in parentheses.

Taxa	Pollen			Nectar		
	Anther dehisce[a]	Pollen removal[b]	Flow pattern[c]	Reward struct[d]	$\bar{x}$ μl/fl./$d(N)$	$\bar{x}$ sugar conc % (range; N)
Apocynaceae						
(19) *Stemmadenia obovata*	Bud		Cont a	A		34(r29–40; 39)
(25) *Thevetia ovata*	Bud		Cont a	A	157?	32(r26–37; 21)
Bignoniaceae						
(24) *Godmania aesculifolia*	Bud		Cont a	B		
(2) *Tabebuia impetiginosa*	>Anth,Rapid		Bimod a	A	3.0(20)	48 1st pk; 30 2nd pk (N = sev.)
(18) *T. ochracea*	Bud		Bimod a	A	1st$d$:8.0(17)	30(r13–38; 10)
(12) *T. rosea*	Bud & >Anth		Bimod a	A	19.8(15)	39(r24–44; 23)
Bixaceae						
(31) *Bixa orellana*	Bud	Rapid	N/A			
Caesalpinaceae						
(6) *Caesalpinia eriostachys*	>Anth,Prog		Bimod a	A	14.9(15)	29(r23–36; 4)
(7) *Cassia emarginata*	?,Ap	Grad	N/A			
(16) *C. grandis*	Bud	Rapid	N/A			
(30) *C. skinneri*		Grad	N/A			
(11) *Haematoxylon brasiletto*	>Anth,Prog		Cont a	A		
(1) *Parkinsonia aculeata*	>Anth,Prog		Cont b	B	0.1(25)	31(N = 1)
Cochlospermaceae						
(3) *Cochlospermum vitifolium*	>Anth,Rapid,Ap	Grad	N/A			
Dilleniaceae						
(4) *Curatella americana*	? Bud	Rapid	N/A			
Fabaceae						
(10) *Andira inermis*	Bud		Bimod a	A	2.3(40)	50+(N = sev.)
(20) *Dalbergia retusa*	>Anth,Prog		Cont a	A	3.6(25)	41(r23–55; 9)
(5) *Gliricidia sepium*	Bud		*Cont c or Bimod b	A	3.3(20)	36(r25–51; 10)
(21) *Lonchocarpus costaricensis*	Bud	Grad	Midd	B		
(23) *L. eriocarinalis*	Bud			B		
(17) *L. hondurensis*	Bud		Cont a	A	7.5(22)	24(N = 3)
(14) *Myrospermum frutescens*	>Anth,Prog		Bimod a	A	3.0(46)	42(r37–53; 8)
(22) *Piscidia carthagenensis*	Bud		Bimod b	A/B		1d:55–59(N = 3)
						2d:14–50(N = 8)
(8) *Platymiscium pleiostachyum*			Cont b	A	0.3(25)	
(15) *Pterocarpus rohrii*	Bud		Bimod a	A	1.7(20)	

Table 21-5. Pollen and nectar characteristics of common and occasional LBF tree species. Order of seasonal flowering, beginning in December (Fig. 21-7), indicated in parentheses. (*Continued*)

| Taxa | Pollen | | | Nectar | |
	Anther dehisce[a]	Pollen removal[b]	Flow pattern[c]	Reward struct[d]	$\bar{x}\,\mu l$/fl./$d(N)$	$\bar{x}$ sugar conc % (range; N)
Malpighiaceae						
(13) *Byrsonima crassifolia*	>Anth,Rapid	Grad	Cont b	A		
(29) *Malpighia emarginata*	Bud	Grad		A		
Myrtaceae						
(27) *Eugenia salamensis*	Anth	Rapid	N/A			
Rubiaceae						
(28) *Genipa americana* ♂	Bud		Cont a	A		26(r24–29; 10)
(28) *G. americana* ♀			Cont a	A		24(r23–25; N = 2)
Styracaceae						
(9) *Styrax argentea*	Bud		*Bimod a or Cont a	A	10.1(13)	28(r26–35; 16)
Tiliaceae						
(26) *Apeiba tibourbou*	Bud,Ap	Grad	N/A			

[a]>Anth = after anthesis; Ap = apical; Bud = in the bud stage; Prog = progressively dehiscing through time.
[b]Applied only to pollen producers and refers to *Gradual* or *Rapid* removal of pollen by bees.
[c]Cont = Continuous; a: dawn → afternoon and usually declining through time; b: all morning and usually declining through time; c: afternoon and usually declining through time.
Bimod = Bimodal; a: dawn and midday; b: midday and late afternoon.
Midd = Midday peak (1000 ≃ 1300).
[d]A = Almost all flowers produce some nectar; B = only a small proportion of flowers produce some nectar.
*Primary.

reward structure that was somewhat intermediate between the two major patterns. Most individuals of *Parkinsonia aculeata** flowered at the beginning of the dry season, whereas the other 4 species were among the last to flower in the dry season (Fig. 21-7).

Daily quantities of secreted nectar varied greatly among species (Table 21-5). The flowers of some produced little or no nectar, whereas others, having flowers that each produced 1.0–8.0 µl of nectar daily, were arbitrarily considered as moderate nectar producers. A third group (4 spp.), in which flowers averaged > 8.0 µl nectar, were considered high nectar producers.

All tree species had mean nectar sugar concentrations of 24% or higher (sucrose equivalents, weight per total weight; Bolten et al., 1979) (Table 21-5 and see Baker and Baker, this book). As indicated by the ranges, there often was considerable variation in sugar concentration, and this reflects a combination of intra- and intertree variation.

Floral Biology: Climbers

Relatively little intensive study was devoted to the climber species. However, general information on sexuality, principal rewards, and flower color was gathered (Table 21-2), and detailed information on selected species (7) was also obtained (Table 21-6). All LBF climbers had hermaphroditic flowers of one of three colors: white, yellow, or pink to purple. Most offered nectar as the main visitor reward, whereas a few offered neither pollen nor nectar.

The 7 LBF climbers that received careful study (Table 21-6) differed generally in several respects from the LBF trees. First, 2- to 3-day flower life may be common, with the second- and third-day flowers producing little or no nectar (e.g., *A. mollissima, M. hymenaea,* and *S. sylvestris*). Second, flowers opened later in the LBF climbers. Third, nectar flow patterns were considerably different; for ex-

*Characteristically, the blooming period of *P. aculeata* was longer in Area F as compared with Area H (Fig. 21-1).

ample, in *A. mollissima* and *X. seemannianum* the greatest secretion of nectar occurred well after anthesis. Further, the pattern of a single early-morning burst of nectar in *D. megacarpa* and *S. sylvestris* was not observed in the trees. Finally, nectar appeared to concentrate in some LBF climbers, since it was not removed as quickly as in the LBF trees, and this pattern was related to the generally low visitation rates on climbers vs. trees (also see below).

Floral Mimicry

At least 3 sets of species appeared to be flower mimics. In each case flowering by a common tree was immediately followed by flowering of a less abundant (sometimes rare) liana with nearly identical-appearing flowers. *Tabebuia impetiginosa* was often followed by *Cydista diversifolia* (both Bignoniaceae) during the early dry season; *Tabebuia ochracea* was immediately followed by *Macfadyena unguiscati,* both Bignoniaceae, as is described by Opler et al. (1976), and the flowering of *Dalbergia retusa* was followed by *Dalbergia glabra*. It is presumed that the uncommon lianas received visits from large bees whose foraging search images had been shaped by flowering of the trees that had immediately preceded their own. Other lianas may exhibit similar behavior; for example, *Xylophragma seemannianum* has rose-pink flowers and blooms in the same habitat and time of year as *Tabebuia rosea*.

The Large Bees

The principal groups of large bees observed in the study area are listed in Table 21-7. Relative abundance levels, which are based on extensive bee collections accumulated since 1969, have been recorded for each taxon. The records were sorted according to dry and/or wet season. A list of all species of each bee group is provided in Appendix I (this chapter).

The general abundance of large bees was significantly greater on LBF plants, especially trees, during the long dry season as compared to host plants during the remainder of the year. During this season, several *Centris* spp. and *Gaesischia exul* were the primary visitors

Table 21-6. Selected floral characteristics of some common and occasional LBF climber species. Also, see Table 21-2 and Fig. 21-8.

Taxa[a]	Fragrance[b]	Flower life (days)	FOT[c]	Anther dehisce[d]	Flow pattern[e]	Nectar		
						Reward struct[f]	$\bar{x}$ μl/fl./ d(N)	$\bar{x}$ sugar conc % (range; N)
Arrabidaea mollissima	Sli	2+	06–0700	Bud	1st day: Cont a; Pk @ 0830	A	3.6(15)	28(r12–40; 18)
A. patellifera	none	1	0700	Bud	Cont b	A	5.3(13)	17(r10–25; 10)
Cydista diversifolia	none	3	most by 0600	Bud	—produces no nectar—			
Dioclea megacarpa	Mod	1	? dawn	Bud	Early AM	A	10.6(8)	44(r40–45; 7)
Mansoa hymenaea	none	2	<0550	Bud	1st d: Cont b	A	12.2(19)	39(r33–50; 26)
Securidaca sylvestris	none	2	06–0700	?Bud	1st d: Early AM	A	0.6(15)	
Xylophragma seemannianum	none	2+	most by 0540	Bud	1st d: Cont c; PK @ 1130	A	5.2(18)	18(r8–31; 6)

[a]All species except *D. megacarpa* are mass bloomers.
[b]Sli = slight; Mod = moderate.
[c]FOT = flower opening time.
[d]Bud = in the bud stage.
[e]Cont = continuous; a: all morning with peak as indicated; b: all morning and declining through time; c: morning and afternoon with peak as indicated. Early AM = one peak of nectar available only in early morning.
[f]A = almost all flowers produce some nectar.

Table 21-7. Major groups of large bees and primary (●) and secondary (○) seasons(s) of flight. D = dry; W = wet. Bee taxonomy according to Stephen et al. (1969).

Bee Taxa	Seasons[a]			
	D I	W I	D II	W II
Anthophoridae				
Anthophorinae				
9 *Centris* spp. (group I)	●			
4 *Centris* spp. (group II)	●	○	○	○
Gaesischia exul	●			
6 *Mesoplia* spp.	●			
Mesocheira sp.	●			
4 *Epicharis* spp.[b]	●	○	●	
Xylocopinae				
6 *Xylocopa* spp.	●	●	●	●
Apidae				
Bombinae				
2 *Bombus* spp.[b]	○		○	●
sev. *Euglossa* spp.	●	●	●	●
Eulaema nigrita and *E. polychroma*	●	●	●	●
3 *Eulaema* spp.			●	●
3 *Eufriesia* spp.[b]			○	●
2 *Exarete* spp.[b]	●			
Colletidae				
2 *Ptiloglossa* spp.	●	●	●	●

[a] *Dry I:* mid-Nov.-mid-May. *Dry II:* July-Aug. *Wet I:* mid-May-June. *Wet II:* Sep.-mid-Nov.
[b] Relatively rare species; few collections.

to LBF species. Overall, they accounted for ~95% of the large bee visits to most LBFs. For example, during normal rainfall years highly attractive nectar resources such as *Andira inermis* and *Dalbergia retusa* (Table 21-5) have yielded 500–700 large bees (mostly *Centris* spp. and *G. exul*) during a half-hour netting period on one tree to an experienced collector (Frankie et al., 1976). Less frequently encountered large bee groups during the dry season included several *Mesocheira, Mesoplia,* and *Xylocopa* spp., *Eulaema polychroma,* a few *Euglossa* spp., and 2 *Ptiloglossa* spp.

During the general wet period (mid-May–mid-November), numbers of large bees were very low compared to dry-season levels. The primary groups of wet-season large bees were the *Xylocopa, Eulaema,* and *Euglossa* spp. (Table 21-7). A high visitation rate at this time would be one *Eulaema* visiting flowers of one *Thevetia ovata* tree every 5 minutes. Less frequently encountered large bees at this time were *Bombus* and *Ptiloglossa* spp.

Overall, the four most frequently encountered groups of large bees were: (1) *Centris* spp. and *Gaesischia exul* (Anthophorinae), (2) Euglossini (Bombinae), (3) *Xylocopa* spp. (Xylocopinae), and (4) *Ptiloglossa* spp. (Colletidae).

Characteristics of each group, particularly as they relate to pollination of the LBF species, are presented below.

1. Centris and Gaesischia exul. Centris spp. and *Gaesischia exul* may be divided into two groups based on seasonal differences in collections from lowland Guanacaste Province, (Table 21-7 and Appendix I). The first group consisted of *G. exul* and several *Centris* spp.; these bees were collected only during the long dry season. In the second group were *C. analis, C. flavifrons, C. segregata,* and *C. vittata,* which flew primarily during the long dry season. In addition, most of these species had a second flight period during the second dry season, with limited flight for 2 species (*C. analis* and *C. flavifrons*) during the wet seasons.

Centris and *G. exul* used numerous host

plants during their respective flight periods. With regard to the nectar producers, almost all were mass bloomers. The host range of *Centris* bees ranged from 6 to 21 plant species, depending on bee species; *G. exul* visited 15 species. Numbers of hosts visited by these and other large bees are presented in Appendix I. The relative lack of specificity of neotropical bees has also been observed by others (Michener, 1954; Heithaus, 1979b). Differences in foraging behavior among *Centris* spp. and *G. exul* were noted. Some foraging differences were due to simple differences in structure among the various flower types. For example, *Cochlospermum vitifolium* and *Cassia emarginata* were vibrated to release pollen, whereas others such as the fabaceous flowers were pried open to remove hidden nectar. Rapid, darting flight between flowers, which were often at opposite sides of a flowering crown, was characteristic of most *Centris* species.

Although little is known about the nesting biology of the Anthophorinae in the study area, some specific information is available. Several *Centris* spp. and *Gaesischia exul* have been observed on many occasions excavating shallow earthen nests during the long dry season, and some *Centris* spp. nest in small tree holes (Frankie, Coville, and Vinson, in prep.). In a few species that are currently receiving study (Coville, Vinson, and Frankie, in prep.), several broods have been recorded for the long dry season. A detailed account of the nest of *C. aethyctera* was reported by Vinson and Frankie (1977). Our accumulated observations suggested that almost all nesting by these bees in lowland Guanacaste takes place during the long dry season. *Mesocheira* and *Mesoplia* spp., which parasitize nesting *Centris*, were commonly observed visiting the same nectar plants frequented by *Centris,* and all nesting activity of these parasites was observed only during the long dry season.

2. Euglossines. Euglossines appear to be traplining foragers (Janzen, 1971, 1974). Although they are found in the dry and wet seasons, most individuals were observed in the long wet season (May–November). Many observations of these bees were made within a kilometer of rivers with extensive evergreen forest or within extensive tracts of relatively undisturbed forest. Within the last two decades, as deforestation and pesticide use have increased dramatically in Guanacaste Province, euglossine populations have declined substantially (Janzen, pers. comm.). Most male euglossines that can be taken at chemical baits in the Guanacaste lowlands may actually breed and nest in evergreen forested slopes of nearby volcanos (Janzen, 1981; Janzen et al., 1982). However, nothing is known about the nesting biology of euglossines in lowland Guanacaste or in adjacent life zones. It is probable that euglossines were once more common in our study areas.

Euglossines were the most important pollinators of LBFs that bloomed in the wet season, including *Apeiba tibourbou, Thevetia ovata, Stemmadenia obovata, Cassia skinneri, Solanum hirtum,* wet-season Bignoniaceae, *Mesechites trifida, Centrosema* spp., and *Calathea allouia.* Among the common euglossines, e.g., *Euglossa* spp., *Eulaema polychroma,* and *E. nigrita,* relatively little interspecific difference in using pollen and nectar resources was observed. *Euglossa* and *Eulaema polychroma* females collected pollen throughout the year, whereas *E. nigrita* females were engaged in pollen collecting only during the wet season.

Euglossines used different flowers (especially for nectar) that required a wide array of approaches and handling. Flowers ranged from those requiring entrance into a tubular corolla (erect or lateral), hovering in front of the flower, hanging upside down on staminal masses, and manipulation of fabaceous "trip flowers." The complex behavior of male euglossines when visiting various orchids is well known (van der Pijl and Dodson, 1966; Dodson, 1975). No euglossines were observed at flowers of Malpighiaceae.

3. Xylocopa. The 6 species of carpenter bees found in our study area exhibited a wide range of foraging behavior and seasonality. With a few exceptions (see below), most can be characterized as opportunistic foragers. Depending on location, all but *X. strandi* (rare) were considered common to occasional species (Appendix I).

Xylocopa fimbriata and *X. gualanensis* usually exhibited traplining behavior (see Janzen, 1971). Because of their large size it was possible in some situations to follow individual bees (females) over 100 to 200 meters as they moved among pollen sources such as *Cassia biflora* and *Cochlospermum vitifolium*. Pollen collecting by females of both species was concentrated in the late wet season and early dry season, and nectar resources were drawn from a wide array of plant families (Appendix I). Although males of these species may be somewhat restricted seasonally to the wet–dry season interface (November), females were common throughout the year except for the dry–wet season interface (May).

Xylocopa subvirescens is typically a dry-season bee that usually exhibits opportunistic foraging behavior. Females collected pollen from *Apeiba tibourbou, Bixa orellana, Cassia biflora,* and *Cochlospermum vitifolium*, most of which have apical pores on the anthers, requiring the bees to vibrate their bodies in the pollen-collecting process. Nectar collecting included a wide variety of tree and other plant species with most of our records from the same fabaceous trees often used by *Centris* species. Both sexes were rarely observed during the long wet season.

Xylocopa muscaria and *X. barbatella* exhibited specific to opportunistic foraging, particularly for nectar sources. Male *X. muscaria* appeared unusually specific in their choice of nectar plants. Female *X. muscaria* and both sexes of *X. barbatella* were general in their choice of nectar plants. Pollen collection was limited to the dry season, and only female *X. muscaria* were found in appreciable numbers during the wet season.

Despite their activity year-round, most *Xylocopa* species have been observed nesting primarily during the long dry season (Frankie, unpub.). In a revealing study, Sage (1968) opened numerous *Xylocopa* nests in Guanacaste from June to August 1965 (Wet I and Dry II; Table 21-7) and found that of 70 occupied nests (with adults) only 2 contained provisioned cells. He suggested that at least during these months *Xylocopa* pass through a nonbreeding period.

4. Ptiloglossa. Two species of *Ptiloglossa*

are known from lowland Guanacaste (Janzen, pers. comm.). These bees are not often observed since they usually forage just prior to, and to a lesser extent during sunrise. In fact, they are often missed by collectors who begin their observations after the bees have finished foraging (Heithaus, 1979a).

These bees visited a variety of tree, shrub, and climber species for their floral rewards throughout the year. Among the LBF plants, the trees *Bixa orellana, Eugenia salamensis, Styrax argentea,* and certain *Cassia* species (Table 21-1) constituted important pollen and nectar resources. Commonly exploited shrubs included *Cassia biflora* (Frankie and Coville, 1979) and *Solanum* spp., both visited for pollen. The climbers *Ipomoea trifida* and *Passiflora foetida* were visited for their nectar. Other *Ptiloglossa* host plants, which are not typically adapted for large bees, are listed in Table 21-4. Nothing is known by us about their nesting biology in lowland Guanacaste.

Host Plant Preferences, Visitation Patterns, and Interplant Movements of Large Bees

As a group, the large bees were not attracted equally to all of their host plants. In general, LBF tree species and a very few climbers (e.g., *Securidaca sylvestris*) constituted the most attractive floral resources. Among the trees, the following were regarded as highly attractive: pollen producers—*Cassia grandis, Curatella americana,* and *Eugenia salamensis;* nectar producers—*Andira inermis, Dalbergia retusa, Haematoxylon brasiletto, Platymiscium pleiostachyum, Pterocarpus rohrii, Styrax argentea, Tabebuia impetiginosa,* and *T. rosea.* Because large numbers of bees were attracted to these species, the bees were able quickly to remove virtually all rewards from individual plants during presentation periods. This was verified by periodically monitoring nectar from numerous open flowers (unbagged). In general, high visitation rates were correlated with peak periods of pollen and/or nectar flow. A detailed account of bee visitation rates on some LBF tree species is provided in Frankie and Haber (this book).

The numbers of bees foraging on nectar

from an individual plant were in some cases staggering. For example, some moderate-sized flowering *Andira inermis* would attract up to 15,000 bees (Frankie, Opler, and Bawa, unpub. notes, 1972) during peak periods of nectar secretion (also see Frankie and Haber, this book). In 1972, a total of 70 different bee species (small and large) were collected from this tree (Frankie et al., 1976).

In 1969, we observed considerably larger numbers of bees visiting these host species as compared to more recent years (1977–1980). Undoubtedly, forest disturbances, which were mentioned earlier, have led to such a decline that it is common to find individuals of the above "attractive" host plants with residual reward at the end of the day. Further, D. H. Janzen (pers. comm.) informed us that bee levels we observed in 1969 were noticeably lower than those he observed in the early 1960s.

Large bees were commonly observed moving among nearby conspecific individuals (10–75 meters) of some LBF species. These observations were usually made on plants that did not attract large numbers of bees because of the ease in making observations on relatively few bees. The following bee taxa commonly moved: *Centris* spp. among *Byrsonima crassifolia* and *Cochlospermum vitifolium; Eulaema* spp. among *Apeiba tibourbou, Arrabidaea patellifera,* and *Stemmadenia obovata;* and *Xylocopa* spp. among *Cassia biflora, Cochlospermum vitifolium, Gliricidia sepium,* and *Parkinsonia aculeata.*

Other Visitors to Large Bee Flowers

A variety of visitors other than large bees were observed frequenting flowers of many LBF species; most were small bees and wasps. Heithaus (1979a,c) provides a lengthy list of anthophilous visitors he noted and collected on many LBF tree species listed in Table 21-1. In contrast to the large bees, most of these insects probably play a minor role in pollinating LBFs, for the following reasons. First, well-protected floral rewards, especially nectar, of several species preclude legal entry by visitors other than large bees, which force their way into the flower (e.g., *Gliricidia sepium, Tabe-*

buia spp., and *Thevetia ovata*). Second, regular contact of long anthers and/or stigmas of some LBFs (e.g., *Cochlospermum vitifolium*) can be achieved only by large-bodied bees. Third, several small bees, especially *Trigona,* rob flowers for their rewards rather than make legal visits (e.g., *Andira inermis* and *Tabebuia* spp.). Finally, although some small bees may serve as pollinators of the few LBF species where they make legal visits and proper contact with anthers and stigmas (e.g., the pollen producers *Bixa orellana, Byrsonima crassifolia,* and *Curatella americana,* and the nectar producers *Andira inermis, Dalbergia retusa, Myrospermum frutescens, Platymiscium pleiostachyum,* and *Pterocarpus rohrii,* which are all Fabaceae), their overall contribution in cross-pollinating these species appears small when large numbers of large bees visit these plants. In fact, when large bees are abundant, their foraging activity appears to repel small anthophilous visitors. On the contrary, if large bees are locally scarce in an area (which was common, for example, in subsites A and H of Comelco Ranch, Fig. 21-1) and small bees are abundant, the latter may indeed serve as pollinators.

Small Bee Pollinated Plants

Despite the above comments, the small bees in lowland Guanacaste are believed to be effective pollinators of a particular set of plant species (ca. 75 species), which include tree, shrub, and some climber and herbaceous species. The flowering phenology and floral biology of these species differs dramatically from that of the LBFs in the following ways. Although at least a few small bee flower (SBF) species bloom during each month, the vast majority bloom massively from March through May. Most species flower for a few days to a month. The flowers, which last one day, are small (3–8 mm in diameter), white to cream-colored, and generally short-tubed and radially symmetrical (Figs. 21-9–21-11). Dioecy is well represented in the SBF syndrome (Bawa and Opler, 1975). In the nectar-producing species, relatively little nectar is produced, and most is secreted in the early morning. A relatively high proportion

Figure 21-9. Flowers of *Coccoloba caracasana* (Polygonaceae).

Figure 21-10. Flowers of *Spondias mombin* (Anacardiaceae); ~6 mm corolla diameter (widest pt.).

Figure 21-11. Flowers of *Casearia aculeata* (Flacourtiaceae); ~10 mm corolla diameter (widest pt.).

of SBF species secrete no nectar, and most also offer pollen as a reward. Finally, large bees generally do not forage from SBF species.

Partitioning of LBF Resources

An overall appraisal of the temporal structuring of LBF floral rewards suggests that "active" competition among plants for large bees is at a minimum in the general study area. In the following section we examine flowering phenologies and reward presentation patterns of (1) congeneric species, (2) pollen producers, and (3) nectar producers from mostly the tree synusiae, since this is where most of the LBF resources and large bees are found.

1. Congeneric Species. Three species each of *Cassia, Tabebuia,* and *Lonchocarpus* flowered mostly at different times of the year (Fig. 21-7 and Table 21-5). Although *Cassia emarginata* and *C. grandis,* overlapped in flowering during a six-week period, their floral behaviors were distinctly different. *Cassia emarginata* flowered over an extended period, and its pollen was taken gradually through the day, whereas *C. grandis* was a mass bloomer that had most of its pollen removed rapidly early in the morning. The shrub, *Cassia biflora,* could be considered as a potential competitor of both *Cassia* species. However, because many large bee species prefer to forage from trees vs. shrubs (Frankie and Coville, 1979), and because the shrub only occurs in disturbed areas, *C. biflora* probably is not a serious competitor of tree *Cassia* species or other pollen-producing trees in relatively undisturbed forest.

2. Pollen Producers. Flowering times of these species (Fig. 21-7 and Table 21-5) were generally spread out during the year. Several overlapped to some extent in flowering; however, presentation of pollen by these species was largely sorted out based on whether they flowered massively or over an extended period, and whether the pollen was offered in a way that facilitated gradual or rapid removal. For example, *Cochlospermum vitifolium, Curatella americana,* and *Cassia emarginata* began flowering during the same general dry period of the year. However, *C. americana* differed from the others in that it flowered for a brief

period, and its pollen was presented in a manner that allowed for rapid removal (~1 hr) in early morning. The other 2 species produced a few flowers daily for extended periods, and removal of most or all the pollen required visitation throughout the morning. Potential competition among pollen-producing tree species in the wet season (Fig. 21-7) is reduced in ways similar to that of dry-season pollen producers.

3. Nectar Producers. As with the congeners and pollen producers, the same kind of temporal sorting was observed with the nectar producers. An examination of the behavior of the 5 nectar-producing tree species that began flowering during January (Fig. 21-7) provides an example. Flowering times were slightly different for most of these species. *Tabebuia impetiginosa** bloomed for only a few days in early January and presented its nectar during the morning. Flowers of *Gliricidia sepium* opened mostly at midday, and flowering extended over several weeks. In addition, *T. impetiginosa* attracted mostly *Centris* spp. and *Gaesischia exul,* whereas *G. sepium* was mostly attractive to *Xylocopa* spp. *Ceasalpinia eriostachys* also bloomed at this time, but nectar production began at dawn and continued at a usually declining rate into the afternoon (see also Frankie and Haber, this book). *Platymiscium pleiostachyum* and *Styrax argentea* began flowering a little later in the month, and each produced its nectar throughout much of the morning. However, *P. pleiostachyum* did not flower every year in the study area; individuals and/or small groups of *T. impetiginosa* also did not flower every year (Frankie et al., 1974a).

Parkinsonia aculeata (see above, under "Floral Biology: Trees") was one of the first trees to flower in the dry season, and *Godmania aesculifolia, Lonchocarpus costaricensis, L. eriocarinalis,* and *Piscidia carthagenensis,* which all flowered toward the end of the dry season, had a distinct nectar reward structure (Fig. 21-7 and Table 21-5). Whereas the other nectar producers had some nectar in

---

*Some individuals bloomed twice during January (Frankie et al., 1974a).

most flowers (Reward struct A), these 5 species had relatively few flowers with nectar (Reward struct B). Further, one congeneric species, *Lonchocarpus hondurensis,* flowered in the middle of the dry season, and all of its flowers secreted abundant nectar (Table 21-5). With the exception of *P. carthagenensis,* these species are regarded as relatively unattractive to most large bee species. It is tempting to speculate that if these low-reward species flowered during the middle of the dry season, they would be at a general disadvantage in relation to other nectar-producing trees.

DISCUSSION

Relatively few community pollination studies have been conducted in tropical life zones. The most extensive and complete studies known to us have been carried out in Costa Rica. Heithaus (1974) explored the interactions between butterflies and hummingbirds and their host plants in the Guanacaste dry forest. He and two other colleagues also studied nectarivorous bats and their host plants in the same forest (Heithaus et al., 1975). In a pollination-related study, Heithaus (1979a,b,c) examined species richness, temporal specialization, niche overlap, and flower-feeding specialization in bees and wasps in the Guanacaste dry forest. Stiles (1975, 1977, 1978, 1979) carefully worked out several interactions between hummingbirds and selected food plants in a lowland wet forest at La Selva, Heredia Province. Finally, Feinsinger (1976, 1978) provided an excellent account of hummingbirds and their host plants primarily in second-growth vegetation at a mid-elevation cloud forest at Monteverde, Puntarenas Province. Additional papers by Feinsinger on the hummingbirds and their plants in this habitat are expected as work on this system continues.

Numerically, the LBF species and their bees are important in the Guanacaste dry forest. A relatively high proportion of the tree species (20%) and climbers (30%) showed adaptation for pollination by these bees. Other important pollination systems in this forest include the small bee, diurnal generalist, hawkmoth, small or settling moth, and bat systems (Haber and Frankie, in prep.).

It is a composite of characteristics that actually defines an LBF species and separates it from the other pollination types. These characteristics may be viewed from the standpoint of the plant and the bees they interact with at the specific, population, and community levels.

Specific Level

In addition to numerous morphological and biological features previously mentioned, two other features of this pollination type are important. First, Bawa (1974) reported that in a large sample of tree species in this forest, most were obliged to outcrossing by virtue of their breeding system. Fourteen species in Table 21-2 were among those studied by Bawa, and 12 (Nos. 2, 3, 6, 10, 12, 15, 18, 20–24) were found to be self-incompatible. The remaining 2, *Byrsonima crassifolia* and *Curatella americana,* were self-compatible. Second, we have evidence that large bees regularly move among conspecific flowering individuals. The movement has been observed in selected cases previously mentioned, and it has been documented through mark-recapture studies. In 1976, Frankie et al. reported on the patterns of movement of several *Centris* species and *G. exul* among flowering trees of *Andira inermis* and *Pterocarpus rohrii.* Overall, low rates of intertree movements were recorded; however, these were adequate to account for average fruit set on the study trees. In a classic mark-recapture study, Janzen (1971) demonstrated that euglossine bees have the potential to visit widely spaced plants in tropical forests. One other study by Augspurger (1980) in Panama also demonstrated that bees readily moved among flowering plants. In this latter case, most bees were small, and the plants were shrubs.

Population Level

Large bee flowers and their large bees appear to interact in subtle ways that result in the cross-pollination of plants. Pollination of large trees involves the resolution of several conflicts

(Heinrich, 1975; Carpenter, 1976). One conflict relates to the production of floral resources that are large enough to attract pollinators but small enough to keep them moving between trees. In species that bloom over an extended period of time, a plant produces only a few flowers at a time, and their traplining pollinators move from one plant to another to bring about cross-pollination. In contrast, the mass flowering species concentrate a large amount of floral resource that also results in the concentration of pollinators in space and time (Frankie et al., 1976). However, slight variations in flowering phenology, daily nectar flow, and perhaps sugar concentration among conspecific mass bloomers appear to be extremely important in determining interplant movements of large bees. These variations and their apparent effects on large bees are examined in detail in Frankie and Haber (this book). The biomodal nectar flow pattern found in several LBF trees may also function as a subtle mechanism for promoting outcrossing in the following manner. Depletion of the food resource after the first nectar peak results in many bees leaving a tree. Presumably some bees search nearby conspecific individuals (and nonconspecifics) for food resources. When the second peak of nectar flows, high numbers of bees return. However, if some have searched widely during the low-nectar period, they may not return to the tree from which they originally foraged. Rather, they may take the one closest to them at the time of the second peak. Relevant to this scenario is the fact that large bees are known to move progressively away from original forage trees over a several-day period (Frankie et al., 1976). In summary, we hypothesize that the overall effect of this nectar pattern may be to promote increased mixing and interplant movement of bees within this forest, which in turn promotes outcrossing.

Community Level

At the community level there appear to be two dominant groups of large bees that serve as effective pollinators of LBFs. Separation of the groups is based on numerical dominance in their seasons (long dry and long wet) and on what appears to be a correlation with a partic-

ular flowering phenological pattern during these seasons. The first group consists of *Centris* species and *Gaesischia exul,* which are almost exclusively confined in flight and nesting activities to the dry season. Correlated with the seasonality of the bees is a pronounced tendency of mass-blooming, LBF species to flower in the dry season. In the case of the trees only one species, *Eugenia salamensis* (a pollen source), blooms in the wet season, and this flowering time may be unique, since its primary pollinators appear to be *Ptiloglossa.* Because of their lower numbers, the *Xylocopa, Mesoplia,* and *Mesocheira,* which nest almost exclusively during the dry season, are of lesser importance in the pollination of LBFs during this season.

The second group is composed of the euglossine bees, which were generally more abundant and diverse during the wet season (see also Janzen et al., 1982). Correlated with the predominance of euglossines in this season is the tendency for the extended-blooming, LBF species to flower in the wet season. Almost all flowering of extended-blooming nectar trees takes place during this season. Corresponding to this second pattern is the tendency for primarily euglossine bees to forage widely in the forest, visiting a few flowers on each plant. This behavior, which was first described by Janzen (1971, 1974) as traplining behavior, is believed to be responsible for the cross-pollination of plants that produce relatively few flowers each day. Possibly, the foraging behavior of the bees and the extended flowering pattern have coevolved with each other. However, there might be selection against mass flowering in the wet season because of the generally low density of bee populations that prevail at this time. It is also conceivable that, since there are relatively few LBF species in bloom in the wet season, these species have entered into a mutualistic relationship with each other by providing a continuous source of food (Baker, 1973; Gentry, 1974a) to the long-lived bees through extended and overlapping blooming periods of the involved plant species. In a relevant study in central Panama, Schemske (1981) studied 2 herbaceous *Costus* species (Zingiberaceae) that occupied the same understory habitat. He found that the 2 species converged in several floral characteristics, including flowering phenology, color, morphology, and nectar secretion pattern, and they shared the same pollinating euglossine bee. However, strong barriers to interspecific hybridization also existed. He concluded that these species benefited by the convergence through the overall increase in flower density and nectar quantity and through a probable increase in the regularity and rate of pollinator visitation.

The LBF species occurred predominantly in the canopy level of the forest, and were noticeably absent in the understory vegetation. Similar differentiation of LBFs has been observed in a lowland wet forest (La Selva) in Costa Rica (Bawa, unpub. data). During the dry seasons of 1974–1975 a study was carried out to investigate bee foraging behavior that was suspected of being related to this spatial pattern of flowering (Frankie and Coville, 1979). Bee visits were recorded at flowering individuals of the shrub *Cassia biflora* that were placed at top and bottom levels of wooden towers in second-growth forest. Overall, when given a choice of equal resources at each level, *Centris* bees preferred to forage at top tower levels. In fact, most *Centris* species were quite rigid in their tendency to forage at this level. *Xylocopa* and euglossines appeared to prefer top tower levels; however, they also displayed opportunistic foraging behavior when resources were plentiful in the understory. Finally, numerous observations of large bees foraging preferentially in the uppermost portions of many tree species in the dry forest have been made (Frankie et al., 1974b; Frankie, 1975). Assortative foraging or pollination has also been reported for bees visiting several temperate plant species (see references in Frankie and Coville, 1979).

It is difficult to determine whether LBF species occur in the canopy because bees forage there, or whether bees are found foraging in the canopy because bee-pollinated plants occur there. Whatever the case, the significant point that emerges from this analysis is that position of a species in the canopy, and consequently the structural organization of the communities, is undoubtedly subject to constraints imposed by pollinator–plant interactions. With

the exception of Hubbell's work (Hubbell, 1979), we are unaware of any other study that considers the structure of the whole community in relation to features of pollination biology.

The great number of tree and climber species that bloomed in the dry season had their flowering periods largely staggered through time. The staggering may be random (Heithaus, 1974; Poole and Rathcke, 1979), or it may reflect sorting of blooming periods due to competition (Robertson, 1895; Mosquin, 1971; Gentry, 1974b, 1976; Heithaus, 1974; Frankie, 1975; Heithaus et al., 1975; Stiles, 1975). There is, however, no direct evidence for competition of the type recently provided by Waser (1978) for 2 sequentially blooming herbaceous perennial species in the Colorado Rockies. The simultaneous blooming of several species at a given time during the dry season does not argue against an avoidance-of-competition hypothesis. As previously described, there is little overlap in reward presentation among pollen and nectar producers that flower during the same general period. It is also worth noting that flowering periodicities in Figs. 21-7 and 21-8 do not reflect peak flowering production in different species; so it is possible for more than one species to flower at the same time but avoid competition by having peak flower production staggered through time. Further, flowering phenologies and floral reward patterns indicate that those species having little reward to offer (nectar reward structure B) all bloomed at times that would not force them to compete with species having nectar reward structure A (most flowers have some nectar). Since LBF species do not attract bees with equal intensity and since bees are known to readily switch host plants as flowering conditions change (Frankie et al., 1976; Appendix I), the sorting of flowering times and/or floral reward presentations allows for numerous species to bloom during the dry season and still be serviced effectively by the same pollinating bee species. Finally, not all species and individuals of some species bloom every year (Frankie et al., 1974a).

At the opposite extreme of avoiding competition are those species that flower close together in time. We have observed that some climber species, which possess little or no nectar, come into flower soon after some of the mass-blooming tree species that produce substantial quantities of nectar. As first suggested by Gentry (1974a), the flowers of these climber species apparently mimic "nectar-producing mass flowering species, including other Bignoniaceae, which are dependent for pollination on exploratory visits by nectar-seeking bees investigating apparent new nectar sources." Supposedly, by mimicking nectar producers, the mimics occasionally attract the attention of naive pollinators and in the process are cross-pollinated. As pointed out by Gentry, many of these climber species bloom massively, but for a very short time of just three to four days, perhaps avoiding discrimination learning by the pollinators.

Up to this point we have discussed the characteristics that define or sort out an LBF species in the study area. It seems logical to ask about a species that does not quite fit the mold we have cast for this syndrome. Such a species is the tree *Machaerium biovulatum* (Fabaceae), which has all the morphological and biological characteristics of an LBF species, except it does not flower at the "right" time of the year.

Machaerium biovulatum is a rare to occasional mass-blooming nectar tree that flowers in the middle of the wet season and sets abundant fruit. No mass-blooming LBF nectar tree flowers at this time (Figure 21-7). At the beginning of our study we classified *M. biovulatum* as a unique LBF species. However, closer examination during the past two years suggests that its pollinators probably are not large bees (Frankie, Haber, H. Baker, and I. Baker, unpub.). With regard to visitors, which are relatively scarce on the tree at any time, we have made the following observations: (1) *Centris* and euglossine bees are not known to visit the flowers; (2) very few *Xylocopa* visit, and when they do, most make illegal visits; (3) *Trigona* mostly rob the flowers; and (4) only occasional megachilid bees make legal flower visits.

We have also studied the tree at mid-elevation (ca. 1,000 meters) sites around the central plateau of Costa Rica where it occurs more commonly (Holdridge and Poveda, 1975). At these sites, several large bee taxa, including

Centris spp., *Epicharis* sp., *Xylocopa* spp., and several megachilid species were commonly observed making legal flower visits. Based on comparative visitor observations and other tree distribution data (Frankie, Haber, H. Baker, and I. Baker, unpub.), we hypothesize that *M. biovulatum* has invaded the lowlands during recent time, and its few anthophilous visitors in the new habitat have yet to fully adjust to it as a host plant. Further, because most pollination in the lowlands seems to be due to megachilid bees, we have classified the tree there as a small bee flower species. However, in the mid-elevations, we recognize it as an LBF species. Theoretical considerations of invading tropical species and their potential pollination problems are discussed in Janzen (1967) and Frankie (1975).

CONCLUDING REMARKS

That pollinator–plant interactions may pose constraints to the organization of a community is an attractive and believable notion, but one that is extremely difficult to demonstrate experimentally for tropical forest ecosystems. Rather than attempt an experimental approach, we chose to gather large quantities of information on virtually all of the interacting players, large bees and their flowers, in order to define this pollination type within the context of the dry forest habitat. Our accumulated data strongly suggest that LBF species and their bees are spatially and temporally structured, a pattern that in turn is related to certain biological and behavioral traits of these two groups of organisms. However, we have no evidence to suggest that either group is more likely than the other to be responsible for driving these interactions. Other pollination types or systems are known to be uniquely structured in this forest owing to specific biological and behavioral features of the interacting pollinators and plants (Haber and Frankie, in prep.). In later papers we will complete the community picture by examining in detail the remaining pollination types.

ACKNOWLEDGMENTS

We thank the National Science Foundation for their continual support of this project from 1968 to 1980. P. D.

Hurd, Jr. (U.S. National Museum), L. Kimsey (University of California, Davis), R. R. Snelling (Los Angeles County Museum), and A. Wille (Universidad de Costa Rica) provided determinations on the bees. W. Burger (Chicago Field Museum), A. Gentry and R. Liesner (Missouri Botanical Garden), D. Austin (Florida Atlantic University), and L. Fournier (Universidad de Costa Rica) provided determinations on plant materials. During the development of this research, we benefited greatly from discussion with H. Baker, I. Baker, A. Gentry, S. Koptur, and R. Thorp. D. Janzen kindly supplied us with several unpublished observations. B. Heinrich, R. Heithaus, D. Inouye, D. Janzen, F. Stiles, and two anonymous individuals read an early draft of the chapter. We also thank the Organization for Tropical Studies for administrative facilities provided, and the Dave Stewart family (Comelco) and the Werner Hagnauer family (Hacienda La Pacifica) for their interest, patience, logistical support, and use of their land as study sites.

LITERATURE CITED

Augspurger, C. K. 1980. Mass-flowering of a tropical shrub *(Hybanthus prunifolius)*: influence on pollinator attraction and movement. *Evolution* 34:475–488.

Baker, H. G. 1973. Evolutionary relationships between flowering plants and animals in American and African tropical forests. *In* B. J. Meggars, E. S. Ayensu, and W. D. Duckworth (eds.), *Tropical Forest Ecosystems of Africa and South America: A Comparative Review.* Smithsonian Institution Press, Washington, D.C.

Baker, H. G., K. S. Bawa, G. W. Frankie, and P. A. Opler. 1982. Reproductive biology of plants in tropical forests. Ch. 12 *in* F. B. Golley and H. Lieth (eds.), *Tropical Rain Forest Ecosystems (Ecosystems of the World,* Vol. 14A). Elsevier Scientific Publishing Co., New York, pp. 183–215.

Baker, I. 1979. Methods for the determination of volumes and sugar concentrations from nectar spots on paper. *Phytochem. Bull.* 12:40–42.

Bawa, K. S. 1974. Breeding systems of tree species of a lowland tropical community. *Evolution* 28:85–92.

———. 1977. The reproductive biology of *Cupania guatemalensis* Radlk. (Sapindaceae). *Evolution* 31:52–63.

Bawa, K. S. and P. A. Opler. 1975. Dioecism in tropical trees. *Evolution* 29:167–179.

———. 1977. Spatial relationships between staminate and pistillate plants of dioecious tropical forest trees. *Evolution* 31:64–68.

Bolten, A. B., P. Feinsinger, H. G. Baker, and I. Baker. 1979. On the calculation of sugar concentration in flower nectar. *Oecologia* 41:301–304.

Buchmann, S. L. 1974. Buzz pollination of *Cassia quiedondilla* (Leguminosae) by bees of the genera *Centris* and *Melipona. Bull. So. Calif. Acad. Sci.* 73:171–173.

Buchmann, S. L. and J. P. Hurley. 1978. A biophysical model for buzz pollination in angiosperms. *J. Theor. Biol.* 72:639–657.

Carpenter, F. L. 1976. Plant–pollinator interactions in Hawaii: pollination energetics of *Metrosideros collina* (Myrtaceae). *Ecology* 57:1125–1144.

Daubenmire, R. 1972. Phenology and other characteristics of tropical semi-deciduous forest in north-western Costa Rica. *J. Ecol.* 60:147–170.

Delgado, A. O. and M. Sousa. 1977. Biologia floral del genero *Cassia* en la region de los Tuxtlas, Veracruz. *Bol. Soc. Bot. Mexico* 37:5–52.

Dodson, C. H. 1975. Coevolution of orchids and bees. *In* L. E. Gilbert and P. H. Raven (eds.), *Coevolution of Animals and Plants.* University of Texas Press, Austin, pp. 91–99.

Feinsinger, P. 1976. Organization of a tropical guild of nectarivorous birds. *Ecol. Mon.* 46:257–291.

———. 1978. Ecological interactions between plants and hummingbirds in a successional tropical community. *Ecol. Mon.* 48:269–287.

Frankie, G. W. 1975. Tropical forest phenology and pollinator plant coevolution. *In* L. E. Gilbert and P. H. Raven (eds.), *Coevolution of Animals and Plants.* University of Texas Press, Austin, pp. 192–209.

Frankie, G. W. and H. G. Baker. 1974. The importance of pollinator behavior in the reproductive biology of tropical trees. *An. Inst. Biol. Univ. Nac. Auton. Mexico* 45, *Ser. Bot.*1–10.

Frankie, G. W. and R. Coville. 1979. An experimental study on the foraging behavior of selected solitary bee species in the Costa Rican dry forest. *J. Kansas Ent. Soc.* 52:591–602.

Frankie, G. W., H. G. Baker, and P. A. Opler. 1974a. Comparative phenological studies of trees in tropical wet and dry forests in the lowlands of Costa Rica. *J. Ecol.* 62:881–919.

———. 1974b. Tropical plant phenology; applications for studies in community ecology. *In* H. Lieth (ed.), *Phenology and Seasonality Modeling.* Springer-Verlag, Berlin, pp. 287–296.

Frankie, G. W., P. A. Opler, and K. S. Bawa. 1976. Foraging behavior of solitary bees: implications for outcrossing of a neotropical forest tree species. *J. Ecol.* 64:1049–1057.

Gentry, A. H. 1974a. Coevolutionary patterns in Central American Bignoniaceae. *Ann. Missouri Bot. Gard.* 61:728–759.

———. 1974b. Flowering phenology and diversity in tropical Bignoniaceae. *Biotropica* 6:64–68.

———. 1976. Bignoniaceae in southern Central America: distribution and ecological specificity. *Biotropica* 8:117–131.

Glander, K. E. 1978. The feeding of howling monkeys and plant secondary compounds: a study of strategies. *In* G Montgomery (ed.), *Arboreal Folivores.* Symp. of the National Zoological Park. Smithsonian Institution, Washington, D.C.

Heinrich, B. 1975. Bee flowers: a hypothesis on flower variety and blooming times. *Evolution* 29:325–334.

Heithaus, E. R. 1974. The role of plant–pollinator interactions in determining community structure. *Ann. Missouri Bot. Gard.* 61:675–691.

———. 1979a. Community structure of neotropical

flower visiting bees and wasps: diversity and phenology. *Ecology* 60:190–202.

———. 1979b. Flower-feeding specialization in wild bee and wasp communities in seasonal neotropical habitats. *Oecologia* 42:179–194.

———. 1979c. Flower visitation records and resource overlap of bees and wasps in northwest Costa Rica. *Brenesia* 16:9–52.

Heithaus, E. R., T. H. Fleming, and P. A. Opler. 1975. Foraging patterns and resource utilization in seven species of bats in a seasonal tropical forest. *Ecology* 54:841–854.

Holdridge, L. R. and L. J. Poveda A. 1975. *Arboles de Costa Rica,* Vol. 1. Centro Cientifico Tropical, San Jose, Costa Rica.

Holdridge, L. R., W. C. Grenke, W. H. Hathaway, T. Liang, and J. A. Tosi, Jr. 1971. *Forest Environments in Tropical Life Zones. A Pilot Study.* Pergamon Press, Oxford.

Hubbell, S. P. 1979. Tree dispersion, abundance, and diversity in a tropical dry forest. *Science* 203:1299–1309.

Janzen, D. H. 1967. Synchronization of sexual reproduction of trees within the dry season in Central America. *Evolution* 21:620–637.

———. 1971. Euglossine bees as long-distance pollinators of tropical plants. *Science* 171:203–205.

———. 1974. The deflowering of Central America. *Nat. Hist.* 83:48–53.

———. 1981. Bee arrival at two Costa Rican female *Catasetum* orchid inflorescences, and a hypothesis on euglossine population structure. *Oikos.* 36:177–183.

Janzen, D. H. and R. Liesner. 1980. Annotated checklist of plants of lowland Guanacaste Province, Costa Rica, exclusive of grasses and non-vascular cryptogams. *Brenesia* 18:15–90.

Janzen, D. H., P. J. DeVries, M. L. Higgins, and L. S. Kimsey. 1982. Seasonal and site variation in Costa Rican euglossine bees at chemical baits in lowland deciduous and evergreen forests. *Ecology* 63:66–74.

Michener, C. D. 1954. Bees of Panama. *Bull. Am. Mus. Nat. Hist.* 104:1–176.

———. 1962. An interesting method of pollen collecting by bees from flowers with tubular anthers. *Rev. Biol. Trop.* 10:167–175.

Mosquin T. 1971. Competition for pollinators as a stimulus for the evolution of flowering time. *Oikos* 22:398–402.

Opler, P. A. 1981. Nectar production in a tropical ecosystem. *In* T. S. Elias and B. L. Bentley (eds.), *Biology of Nectaries.* Columbia University Press, New York.

Opler, P. A., H. G. Baker, and G. W. Frankie. 1975. Reproductive biology of some Costa Rican *Cordia* species (Boraginaceae). *Biotropica* 7:234–247.

Opler, P. A., G. W. Frankie, and H. G. Baker. 1976. Rainfall as a factor in the release, timing, and synchronization of anthesis by tropical trees and shrubs. *J. Biogeog.* 3:231–236.

———. 1980. Comparative phenological studies of shrubs and treelets in wet and dry forests in the lowlands of Costa Rica. *J. Ecol.* 68:167–188.

Pijl, L. van der and C. H. Dodson. 1966. *Orchid Flowers, their Pollination and Evolution.* University of Miami Press, Coral Gables, Fla.

Poole, R. W. and B. J. Rathcke. 1979. Regularity, randomness, and aggregation in flowering phenologies. *Science* 203:470–471.

Robertson, C. 1895. The philosophy of flower seasons, and the phaenological relations of the entomophilous flora and the anthophilous insect fauna. *Am. Nat.* 29:97–117.

Rockwood, L. L. 1973. Distribution, density, and dispersion of two species of *Atta* (Hymenoptera: Formicidae) in Guanacaste Prov., Costa Rica. *J. Anim. Ecol.* 42:803–817.

———. 1975. The effects of seasonality of foraging in two species of leaf-cutting ants *(Atta)* in Guanacaste Province, Costa Rica. *Biotropica* 7:176–193.

Sage, R. D. 1968. Observations on feeding, nesting, and territorial behavior of carpenter bees genus *Xylocopa* in Costa Rica. *Ann. Ent. Soc. Am.* 61:884–889.

Sawyer, J. O. and A. A. Lindsey. 1971. *Vegetation of the Life Zones of Costa Rica.* Monograph No. 2 of the Indiana Academy of Science, Brookville.

Schemske, D. W. 1981. Floral convergence and pollinator sharing in two bee-pollinated tropical herbs. *Ecology* 62:946–954.

Stephen, W. P., G. E. Bohart, and P. F. Torchio. 1969. *The biology and external morphology of bees.* Agr. Exp. Sta., Oregon State University, Corvallis.

Stiles, F. G. 1975. Ecology, flowering phenology and hummingbird pollination of some Costa Rican *Heliconia* species. *Ecology* 56:285–310.

———. 1977. Coadapted competitors: the flowering seasons of hummingbird-pollinated plants in a tropical forest. *Science* 198:1177–1178.

———. 1978. Temporal organization of flowering among the hummingbird foodplants of a tropical wet forest. *Biotropica* 10:194–210.

———. 1979. "Response to Poole and Rathcke." *Science* 203:471.

Turner, D. C. 1975. *The Vampire Bat—A Field Study in Behavior and Ecology.* Johns Hopkins University Press, Baltimore.

Vinson, S. B. and G. W. Frankie. 1977. Nests of *Centris aethyctera* (Hymenoptera; Apoidea: Anthophoridae) in the dry forest of Costa Rica. *J. Kansas Ent. Soc.* 50:301–311.

Waser, N. M. 1978. Competition for hummingbird pollination and sequential flowering in two Colorado wildflowers. *Ecology* 59:934–944.

Wille, A. 1963. Behavioral adaptations of bees for pollen collecting from *Cassia* flowers. *Rev. Biol. Trop.* 11:205–210.

Appendix I. Large bee species. Number of known host plants ( ) in Guanacaste.

Bee Taxa

Anthophoridae
  Anthophorinae
    Group I *Centris*
      C. adani (7), *C. aethyctera* (16), *C. bicornuta* (8), *C. fuscata* (14), *C. heithausi* (9), *C. inermis* (10), *C. lutea* (3), *C. nitida* (15), *C. trigonoides* (8)

    Group II *Centris*
      C. analis (6), *C. flavifrons* (9), *C. segregata* (21), *C. vittata* (6)

    Gaesischia exul (15)
    6 *Mesoplia* spp. (11)
    Mesocheira sp. (1)
    Epicharis albofasciata (2), *E. lunata* (2), *E. maculata* (6), *E. rustica* (2)

  Xylocopinae
    Xylocopa barbatella (11), *X. fimbriata* (15), *X. gualanensis* (20), *X. muscaria* (13), *X. strandi* (1), *X. subvirescens* (13)

Apidae
  Bombinae
    Bombus mexicanus (3), *B. pullatus* (1)
    Euglossa spp. (17)
    Eulaema bombiformis (?), *E. cingulata* (2), *E. meriana* (?), *E. nigrita* (10), *E. polychroma* (16)
    Eufriesia mexicana (1) *E. mussitans* (3), *E. surinamensis* (2)
    Exarete spp. (?)

Colletidae
  2 *Ptiloglossa* spp. (15+)

Section VII
SYSTEMS ECOLOGY IN POLLINATION BIOLOGY

22
COMPONENT ANALYSIS OF COMMUNITY-LEVEL INTERACTIONS IN POLLINATION SYSTEMS

James D. Thomson

Ecology and Evolution Department
State University of New York
Stony Brook, New York

ABSTRACT

In studies of communities, one almost always asks broad questions and is usually compelled to pursue these questions rather superficially. I suggest that dissecting pollination systems into components whose interactions can be modeled is the best insurance against making errors of interpretation in studies at this level. I give examples of some characteristics that make pollination systems resistant to coarse analysis, and try to demonstrate that component analysis does more than simply to provide insight into the mechanisms of community-level interactions; when pursued as a systematic research program, it reduces the probability of misinterpretation.

KEY WORDS: Community, competition, component analysis, pollen carryover, bees, foraging, constancy, inflorescence architecture, nectar, mimicry, time lag, scale, presentation.

CONTENTS

FOREWORD

This chapter was originally to be coauthored by Dr. R. C. Plowright of the Department of Zoology, University of Toronto. Much of the original work described was done by him and his students, and the editorial viewpoint herein is intended to represent his views as well as mine. Unfortunately, at the last minute he was unable to participate in the preparation of this chapter, and I have hastily compiled it from rough outlines that we had prepared jointly. Therefore, while Dr. Plowright deserves much of the credit for its contents, he cannot be held responsible for any misrepresentations or faults of exposition, which remain my responsibility.

INTRODUCTION

At least since the work of Robertson (1895), pollination ecologists have been interested in community-level interactions, i.e., ecological and evolutionary connections among species that extend beyond the direct relationship of a plant and its pollinator. Especially within the last decade, pollination community ecology has gained momentum as pollination ecologists have worked their way up from specific to general questions, and as community ecologists have adopted pollination systems as highly quantifiable arenas for study. The recognition that few plants and few pollinators are obligate specialists with regard to pollination compels the conclusion that "pollination webs," like food webs, will often be complex and cross-connected. The success of a particular species is likely to depend in part on its position in this matrix. In this chapter I wish to outline a framework for investigating such phenomena, and to flesh out the framework with a heterogeneous collection of small, but exemplary, case studies. I am concerned only with animal pollination, and my examples will reveal a bias toward bumblebees.

One can identify two extreme strategies for community studies, the "top-down" and "bottom-up" approaches. To illustrate the contrast, consider the case of competition between two plant species for pollinator services, reviewed by Waser (this book). At the broadest, "top-down" level, one may either compute simple measures, such as blooming-time overlaps, which may have something to do with competition, or conduct a large-scale experiment involving removals or additions of the potential competitors, as Waser (1978a) did. Although the experiment is likely to yield an unambiguously interpretable result (and the measurement of overlap is likely not to), neither ap-

proach gives information on the mechanism of the interaction. Unraveling mechanisms depends on a more detailed, "bottom-up" analysis of the *components* of the interaction in the style of C. S. Holling. By specifying the behavior of small system subunits, one can proceed to model system behavior, either in a logical, verbal way or explicitly mathematically, in a systems model such as those of Levin and Anderson (1970), Waser (1978b), or Lertzman and Gass (this book).

Lertzman and Gass discuss the general philosophical rationale for constructing this sort of model, and I will therefore not pursue this point. What I wish to stress is that pollination systems contain a number of quirks that seriously impede "top-down" analysis, and that dissection of components may often be necessary, not only to see the details of how the system works, but to avoid erroneous conclusions about the general nature of the interaction under study.

MAJOR COMPONENTS OF PLANT–POLLINATOR SYSTEMS

I first will separate the variables of interest into three major groups. The first of these, and the largest, is the collection of all plant attributes that affect pollination success. I lump all these characteristics under the general heading *plant presentation* (Fig. 22-1). Some of these are apparent to pollinators, whereas others are cryptic. A partial listing of the former would include: density; spatial pattern; timing of bloom; flower morphology, color, and scent; amount, type, and timing of floral rewards; plant height; inflorescence architecture (including flowers per inflorescence and inflorescences per plant, as well as the type of inflorescence); and variation in gender. Aspects that are not apparent to the pollinators but still determine reproductive success include such things as the clonal structure, if any; the genetic neighborhood size; the type of incompatibility system, if present, and the number of incompatibility alleles or mating types; the pollen production and ovule number; the functional response of seed set to varying amounts of pollen; and the seed-maturing resources

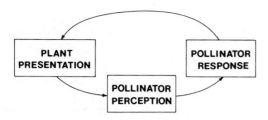

Figure 22-1. The major component categories of plant–pollinator interactions. Components within each category are detailed in the text.

available to the plant and the potential for competition among ovules within a fruit or among fruits within the plant.

A second major category is *pollinator response,* specifically meaning the ways in which pollinators respond to the apparent aspects of plant presentation. A number of these are concerns of optimal foraging theory, including: choice of species to visit and choice of place; foraging site fidelity; movement between plants (direction and distance); speed of working; flowers visited per inflorescence and per plant; degree of flower species constancy and foraging height constancy; extent of sampling of unfamiliar flowers, etc. Other aspects include: recruitment; traplining; territoriality; effectiveness at removing pollen from anthers and applying it to stigmas; and the extent of pollen carryover (treated by Lertzman and Gass elsewhere in this book).

Pollinator response will not, however, be a strict reflection of plant presentation because pollinators will seldom if ever have complete knowledge of all of the relevant characteristics of their potential food plants. Instead, response will depend on the way in which pollinators perceive the plants' presentations, and this perception may encompass a variety of distortions, including mistaken identities, time lags, nonlinearities in response to inflorescence size, and the like. Pollinators' assessments of flower density must involve some spatial scale, which might differ among pollinator species (Thomson, 1981). Structurally complex flowers require visitors to learn proper techniques for reward extraction; at least in bumblebees, some foragers never learn how to handle these flowers and therefore perceive them as nonrewarding, whereas other bees that have perfected

their techniques perceive the same flowers as highly rewarding (Heinrich, 1976; Laverty, 1980).

Ultimately, the pollinators' perceptions of the plants' presentations determine their foraging behavior and therefore the reproductive success of the plants. Differential reproductive success of plants with different presentations should in turn exert selective pressures on those aspects of plant presentation that are under genetic control. Pollinator response may also affect plant presentation directly in addition to long-term evolutionary feedbacks. An obvious example is the depletion of nectar by flower visitors, which immediately alters the presentation of the plant. The relationships among these major components are shown in Fig. 22-1.

HYPOTHETICAL AND REAL EXAMPLES OF COMPONENT INTERACTIONS

Clearly, assembling a formal systems model including all the components cited above is a forbiddingly complicated task, especially when one must consider not only a single plant species and single pollinator but a whole community. However, as the chapters in this book repeatedly demonstrate, many aspects of pollination systems can be quantified relatively easily in the field, and I feel that some simple communities can be largely "solved" by long-term studies. As matters stand, no such studies exist. Below, I will discuss a selection of single-component studies from a variety of systems, with the goal of showing how important some largely unstudied, "mechanical" details may turn out to be in lending character to large-scale interactions. The reader may regard the chapter as an annotated checklist of complications. To tie these elements together, I will present them in relation to a particular question: "What is the nature of the interaction between two plant species populations that overlap substantially in time of bloom and in pollinator use?"

The phrasing of the question suggests that interspecific competition for pollination service may be expected between the two species (see

Waser, this book). But the easy equation of ecological overlap and competition has been under attack for some time (e.g., Colwell and Futuyma, 1971; Connell, 1975, 1980; and many others), and certainly cannot be considered to be universally true. With specific reference to pollination systems, several authors now have suggested that similarities in bloom time and use of pollinators may in some cases indicate mutualism rather than competition for pollination, especially when flowers appear similar. In these cases, the similarity of the flowers has been explained as convergence driven by a sort of Müllerian mimicry in which a number of species may all receive better pollination by resembling each other (see Grant, 1966; Macior, 1971; Watt et al., 1974; Brown and Kodric-Brown, 1979; Parrish and Bazzaz, 1979; Thomson, 1980, 1981; Schemske, 1981; Little, this book; Powell and Jones, this book). Thus there have been two outcomes suggested for plant–plant interactions, and they are opposite in effect: competition and mutualism. I am aware of no critical tests of a mimicry hypothesis, but we can identify certain components that are necessary for such a relationship to work. If these component interactions are not present in nature, the mimicry hypothesis is weakened. Proceeding through a series of steps, I will demonstrate the style of the argument.

A hypothesis of Müllerian mimicry first of all depends on a density-dependent or frequency-dependent advantage, such that a dense population receives disproportionately more or better pollination when it comprises a larger fraction of the local flora. Do pollinators behave this way? Such relationships have been described from real systems. Levin and Anderson (1970) review the evidence for "Arnell's dominating flower phenomenon," one expression of frequency-dependent advantage. However, this phenomenon properly describes an increase in flower *constancy* with increasing frequency. More to the specific point is a demonstration that visitation rate on certain insect-pollinated species not only is higher where their density is higher, but increases further where similar flowers of different species coexist (Thomson, 1980, 1981, 1982a). Thus, at least some pollinators, including bumblebees,

do react positively to flower density, and "sum" the densities of similar flower species in their perception of density. This behavior satisfies one of the primary conditions for mimicry. Although necessary, this condition is not sufficient, owing to "minor details" of pollinator foraging and spatial dispersion of the plants. It is not certain that increased visitation causes increased fitness, especially in these circumstances, where different species are involved. If pollinators respond to mixed-species arrays of similar flowers by moving inconstantly from one plant species to another, the heterospecific transfer of pollen may actually reduce seed set by various mechanisms described by Waser (1978a,b, this book). Besides the wastage of pollen and the loss of stigmatic surface emphasized in Waser's (1978a) models, "allelopathic" pollen has been found in two genera of Compositae (*Parthenium*, Sukada and Jayachandra, 1980; *Hieracium*, Thomson et al., 1982b). These pollens inhibit successful fruiting in certain other species when applied in mixtures with conspecific pollen grains.

The degree of constancy displayed by the pollinators (an aspect of pollinator response) thus is clearly important to plant–plant interactions (Levin and Anderson, 1970; Straw, 1972; Bobisud and Neuhaus, 1975). It must be emphasized that many wild pollinators may be willing to forage inconstantly, and that the concept of pollinator constancy has perhaps been unduly influenced by the well-studied but unusually high constancy of honeybee workers (Free, 1970; Grant, 1950; Faegri and van der Pijl, 1979, p. 50). Individual bumblebees, for example, commonly visit two or more plant species on a single foraging trip. Such inconstancy may have a sampling function in tracking changes in resources (Heinrich, 1976, 1979; Oster and Heinrich, 1976); in other cases, because individual bees consistently return to the same individual plants in a mixed-species trapline (e.g., *Aralia hispida* and *Rubus* sp., Thomson, Maddison, and Plowright, unpub. ms.), the behavior seems less interpretable as sampling and probably represents simple harvesting. Indeed, in the hypothetical situation under consideration—similar flowers, intermingled to some extent—where in most cases the plants yield roughly equal rewards, optimal foraging models that include travel costs usually would predict an inconstant strategy.

When flower-constancy does occur in flexible animals like bumblebees or hummingbirds, it may partially or wholly reflect simple spatial aggregation of flower species ("passive constancy," Thomson, 1982a), which will be especially evident in monocultural crop plantings. If two plant species intermingle in a mosaic of monospecific patches, heterospecific visits will be comparatively rare even if visitors display no active constancy (Levin and Anderson, 1970; Waser, 1978a), especially if the visitors maintain restricted foraging areas that are small relative to the patch size of the plants involved (cf. Solbrig, 1975). However, as patch size increases, any "mimetic" advantages presumably will be dissipated because the pollinators no longer will sum the contributions of the two plant species in deciding where to forage (Thomson, 1981a).

The argument so far should suffice to show that the overall effect of pollinator sharing by plants is not easily predicted without paying careful attention to the fine structure of the interaction. A number of components are involved, and different components have opposite effects; i.e., some tend to produce mutualism and some competition. Figure 22-2 (from Thomson, 1982a) is an attempt to summarize the interplay among spatial pattern, visitation rate, and reproductive success, with the condition of decreasing constancy with increasing intermingling. No existing study has been thorough enough to determine whether any real plant–plant interaction actually fits the curves of Fig. 22-2, but experimental studies have shown that different natural systems occur in different regions of the "success space" depicted (Locations 1 and 2, Thomson, 1982a; location 3, Waser, 1978b).

Another phenomenon affecting plant relationships is pollen carryover, treated extensively by Lertzman and Gass in this book. All empirical treatments of carryover to date have considered one species only, but its importance in heterospecific visit sequences is readily apparent. When a pollinator visits a flower of species A, then B, then A, how much pollen

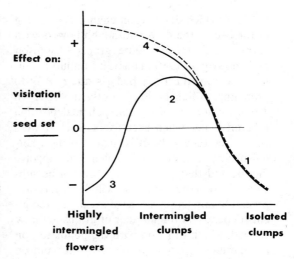

Effect on:

visitation
-----
seed set

Highly intermingled flowers   Intermingled clumps   Isolated clumps

Figure 22-2. A summary of the ways in which spatial intermingling may affect the interaction between two plant species that share pollinators. When plants occur in separate, discrete patches, the patches may compete for visitors. As the species intermingle more, visitation rate may rise as the pollinators respond to the greater density. However, seed set may drop off at extremes of intermingling owing to the increasing fraction of heterospecific visits.

will the second A receive from the first? Although data are lacking, assumptions of published competition models range from Straw's (1972) proposal that carryover is essentially complete, to Levin and Anderson's (1970) view that it is completely lacking. The Levin–Anderson model reflected the then-prevailing view that carryover within a species is generally low, with most pollen being placed on the next flower visited (Levin and Kerster, 1967, 1969). Recent experimental work suggests that carryover is often more extensive than this: In *Erythronium americanum, E. grandiflorum, Clintonia borealis,* and *Diervilla lonicera* (all flowers with open corollas, plentiful exposed pollen, and imprecise channeling of visitors over the reproductive parts), the pollen put on a nectar-feeding bumblebee by one flower declines roughly as an exponential decay with a "grain half-life" of between one and two flower visits. However, the tail of the fall-off curve usually is longer than predicted by exponential decay; i.e., a small fraction of grains travel considerable distances. There is great flower-to-flower variability in deposition, much of which depends on chance positioning of the bee on the flower, although flowers with extra nectar receive more grains (Thomson and Plowright, 1980; Thomson, unpub.). Pollen loss from the bee's body is due mostly to grooming, which is done mostly in flight. In *Erythronium grandiflorum,* pollen deposition falls off faster when bees fly between flowers than when they walk. It is clear that most heterospecific transfers of pollen will involve flights, and that pollen loss will often be substantial. Whether this loss will have any appreciable impact on the seed set of the recipients depends on the number of grains actually carried, which will vary with the pollinator; on the number of ovules to be fertilized; and on the functional response of seed set to pollination intensity. Despite being one of the more important components of any quantitative model of pollination success, this functional response apparently has been described for very few plant species (*Oenothera fruticosa,* Silander and Primack, 1978; *Passiflora vitifolia* and *Geranium maculatum,* A. Snow, pers. comm.). Extensive hand-pollinations in *Clintonia borealis* and *Medeola virginiana* have yielded only a very weak positive relationship between the number of outcross grains applied and the number or proportion of ovules developed (Plowright and Thomson, unpub. data). The weakness of the pollination/seed set relationship for these species is due largely to extensive fruit abortion, which may represent incompatible or semicompatible pollinations (cf. Bertin, 1981), or possibly deficiencies of technique. In view of the general lack of data, the extent to which the seed set response is coupled to pollination should become an active research topic; without such data, the significance of pollination rate and of pollen carryover is very difficult to treat.

The role of grooming in pollen carryover is another poorly understood relationship with important, sometimes elaborate ramifications. Hartling and Plowright (1979a,b; Plowright and Hartling, unpub. ms.) found that a bumblebee could pollinate as many as 50 florets after its arrival on a head of red clover *(Trifolium pratense),* a much larger number than

supposed previously. *T. pratense* is self-incompatible, and it had been thought that only a few florets could be fertilized on a new plant before self pollen overwhelmed the outcross pollen on the bee's body (Free and Butler, 1959; Michener, 1974). The outcross pollen that is deposited on clover florets apparently comes from the proboscidial fossa (Furgala et al., 1960; Spencer-Booth, 1965), which must not become overlaid by incompatible self pollen until the bee grooms, after leaving the flower head. Bees seem unable to groom the fossa clean, and they sweep pollen into it during other grooming movements. While not firmly demonstrated, this mechanism is the most likely explanation for the extensive within-head carryover. Bees also regulate the number of florets visited on a head according to the amount of nectar in the florets (Hartling, 1979; Hartling and Plowright, 1979a). If there is layering of pollen from successive heads in the proboscidial fossa, with the depth of each head's stratum proportional to the number of florets visited on the head, and if the amount of pollen removed from the superficial strata during a visit to a head also depends on the number of florets visited, interesting consequences follow. Since nectar-rich plants are the only ones whose stigmas will cut down to the deepest layers and the only ones whose pollen will subsequently be swept into those layers, there can be assortative mating for nectar levels without any discrimination by the pollinators. Nectar-rich plants could also have a larger neighborhood size in that some of their deeply deposited grains may travel considerable distances before being brought to light (see Lertzman and Gass, this book, on the effects of pollen layering).

The connections between nectar secretion and pollinator foraging have been made explicit in a systems model for pollination in red clover assembled by Plowright and Hartling (unpub. ms.) using parameter values from field experiments and the assumption of exponential decay of pollen. A flow chart for this model is given in Fig. 22-3. Although this form of the model does not incorporate the pollen-layering effects discussed above, it does allow one to ex-

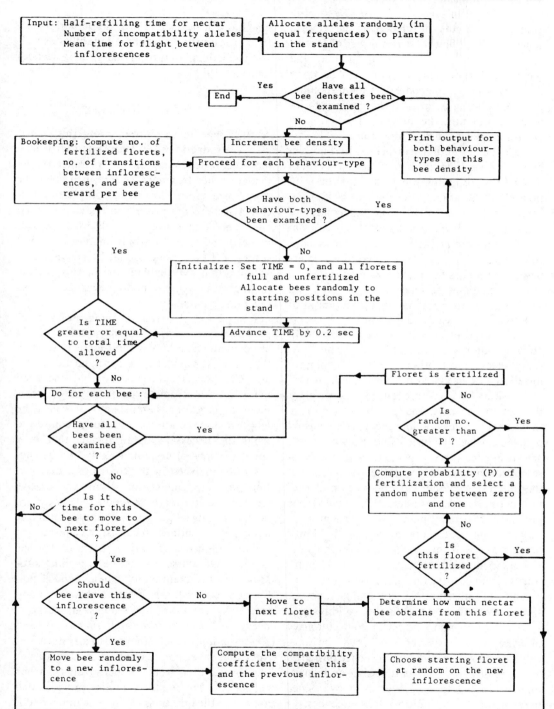

Figure 22-3. A flow chart of the *Bombus–Trifolium* model discussed in the text. Additional details may be found in Plowright and Hartling (unpub. ms.).

amine the effects of changing the reward schedule of the flowers, the number of incompatibility alleles in the clover population, and the rules of movement of the bees. Analyzing bee movement in the context of plant–pollinator interaction was, in fact, the original rationale for building the model. Plowright and Hartling used it to show that the flexible foraging behavior actually displayed by bumblebees also resulted in higher seed sets than behavior representing previous casual descriptions of bee behavior (Michener, 1974, p. 327). The model also points out the inadequacy of estimating seed set in a stand of clover from simply determining the frequency of interplant transitions. Behavior that maximizes interplant moves does not maximize seed set.

Manipulations of models of this sort can shed light on other pollination interactions. For example, it is possible to determine the nectar secretion rate that yields optimum seed set. In obligate outcrossers like red clover, this optimum clearly will often involve a "compromise" nectar production, rich enough to prevent visitors from visiting few flowers per plant but not so rich as to make them stay so long that their outcross pollen becomes exhausted or overlaid (see Heinrich and Raven, 1972). The residence time of pollinators on an inflorescence may also vary with the architecture of the inflorescence and with the pattern of nectar secretion. For instance, Pyke (1978) has proposed that many plants with vertical inflorescences present more nectar in the lower flowers, and that this presentation prompts the visitors to move upward and to depart when rewards diminish. Again, analysis of the components of presentation, perception, and response economically reveals the workings of a system subunit and suggests a coevolutionary interpretation. However, one must be very cautious in interpreting inflorescence design as a means of optimizing outcrossing by balancing visitation and carryover as above, since many plants have multiple inflorescences. In such cases, a spike or head that is optimally suited for receiving a pollinator's first visit to the plant will be progressively less suited for successive visits made within the same plant. Theories of inflorescence evolution (e.g., Willson and Rathcke, 1974; Willson and Price, 1977) must reflect this unpredictability as well as the modular nature of most inflorescences (Harper, 1977).

SUMMING UP

As I warned in the introduction, this is primarily a didactic essay. I concede that it also may seem, depending on the tolerance of the reader, either fragmentary and incomplete or simply premature. My main point is to advocate a systems approach, yet I have presented only one real systems model (the clover model), and even that only includes a small subset of the components that I have identified as important. My explanation of this embarrassing situation, as might be expected, has several subunits.

First, I feel that the advantages of a component-analysis approach *in revealing detail* are evident enough in subunit models (such as the red clover model cited here and Lertzman and Gass's carryover models) that one need not wait for a complete community-level treatment to proclaim them. The wait would be long because the intensity of study required to yield the necessary detail means that only small portions of systems usually can be successfully analyzed in the course of a dissertation project. Unless long-term cooperative studies become more common than they are now, our knowledge of community-level interactions will continue to consist of a number of isolated glimpses of various parts of various systems. Of course, this is the prevailing situation in all of community ecology, and if pollination ecologists are forced to draw too many inferences from too few studies—if bumblebees, and *Delphinium* and *Ipomopsis* and *Erythronium,* reach the talisman status of intertidal invertebrates and West Indian anoles—we are no worse or better off than our colleagues. This state of affairs provides one answer to the question, "Why is so much detail needed?" Because we will probably continue to be heavily dependent on a small number of studies, and continue to "solve" other systems by analogy and extension, it is critical to realize that small differences in structure may make otherwise similar systems behave differently. Breaking interactions into components sensitizes one to the importance of these differences.

A second, more practical need for detail is in most applied studies where the goal usually is to manipulate the pollination system so as to maximize a particular quantity such as fruit or seed set. It is obvious that an understanding of the relevant components will often be helpful in arriving at an effective program with a minimum of trial and error (see Tepedino, 1981).

One still might object that certain straightforward questions can be settled by single experiments without investigating intricacies. However, there are easily imagined circumstances in which ignorance of detail may result in undetected flaws in experimental design. Consider again the "simple" question of whether or not two coexisting plant species compete for pollination. Elsewhere in this book, Waser argues cogently for experiments to settle this question. However, the simplest conceivable experiment—removing or bagging the flowers of one of the potential competitors (see Thomson, 1980, 1982a)—ignores the "detail" that the pollinator community does not reshuffle itself randomly every day. Important pollinators, including honeybees (Ribbands, 1949; Singh, 1950), bumblebees (Manning, 1956; Heinrich, 1976), and birds (at least territory holders, Linhart, 1973), tend to return to the same foraging areas each day. The response of such animals to the sudden disappearance of a formerly favored flower type may be to switch to other species in the area, which would be interpreted as a confirmation of competition for visitors between the plants. However, it is possible that the animals would not even be foraging in the area if the removed species had not been present initially to attract them (see Thomson, 1978). If so, the experiment reveals nothing about competition between the species under natural circumstances.

Examples like those given above convince me that finding the truth about pollination systems usually will require dissection and some amount of model building. My emphasis on detail must not be taken to imply that I find small-scale mechanical processes more interesting or more important than large-scale ecological and evolutionary processes. However, seemingly trivial details of plant presentation and pollinator response evidently can determine whether a given ecological interaction is competitive or mutualistic, and select for or against certain evolutionary directions. They cannot be safely ignored.

ACKNOWLEDGMENT

The contribution of R. C. Plowright to this chapter already has been mentioned. I am also particularly grateful to the following members of the Plowright lab for their discussion of these ideas: Lester Hartling, Lawrence Harder, and Terence Laverty. Many conversations with scientists too numerous to list helped shape this chapter; talks with Nick Waser and Mary Price were especially helpful. Some of the work cited was supported by the U. S. National Academy of Sciences and some by the National Sciences and Engineering Research Council of Canada.

LITERATURE CITED

Bertin, R. 1981. Paternity and fruit production in trumpet creeper *(Campsis radicans). Bull. Ecol. Soc. Am.* 62:172. (Abstract)

Bobisud, L. B. and R. J. Neuhaus. 1975. Pollinator constancy and the survival of rare species. *Oecologia* 21:263–272.

Brown, J. H. and A. Kodric-Brown. 1979. Convergence, competition, and mimicry in a temperate community of hummingbird-pollinated flowers. *Ecology* 60:1022–1035.

Colwell, R. K. and D. J. Futuyma. 1971. On the measurement of niche breadth and overlap. *Ecology* 52:567–576.

Connell, J. H. 1975. Some mechanisms producing structure in natural communities: a model and evidence from field experiments. *In* M. L. Cody and J. M. Diamond (eds.), *Ecology and Evolution of Communities.* Harvard University Press, Cambridge, Mass., pp. 460–490.

———. 1980. Diversity and the coevolution of competitors, or the ghost of competition past. *Oikos* 35:131–138.

Faegri, K. and L. van der Pijl. 1979. *The Principles of Pollination Ecology.* Pergamon Press, Oxford.

Free, J. B. 1970. *Insect Pollination of Crops.* Academic Press, London.

Free, J. B. and C. G. Butler. 1959. *Bumblebees.* Collins, London.

Furgala, B., K. W. Tucker, and F. G. Holdaway. 1960. Pollens in the proboscis fossae of honeybees foraging certain legumes. *Bee World* 41:210–213.

Grant, K. A. 1966. A hypothesis concerning the prevalence of red coloration in California hummingbird flowers. *Am. Nat.* 100:85–98.

Grant, V. 1950. The flower constancy of bees. *Bot. Rev.* 16:379–398.

Harper, J. L. 1977. *Population Biology of Plants.* Academic Press, New York.

Hartling, L. K. 1979. An investigation of the relationship between bumblebee foraging behavior and the pollination of red clover: a component analysis approach. M. Sc. thesis, University of Toronto, 92 pp.

Hartling, L. K. and R. C. Plowright. 1979a. An investigation of inter- and intra-inflorescence visitation rates by bumblebees on red clover with special reference to seed set. *Proc. IVth Int. Symp. on Pollination.* Maryland Agric. Exp. Sta. Spec. Misc. Publ. No. 1:457–460.

———. 1979b. Foraging by bumblebees on patches of artificial flowers: a laboratory study. *Can. J. Zool.* 57:1866–1870.

Heinrich, B. 1976. The foraging specializations of individual bumblebees. *Ecol. Mon.* 46:105–128.

———. 1979. "Majoring" and "minoring" by foraging bumblebees, *Bombus vagans:* an experimental analysis. *Ecology* 60:245–255.

Heinrich, B. and P. H. Raven. 1972. Energetics and pollination ecology. *Science* 176:597–602.

Laverty, T. M. 1980. The flower-visiting behavior of bumblebees: floral complexity and learning. *Can. J. Zool.* 58:1324–1335.

Levin, D. A. and W. W. Anderson. 1970. Competition for pollinators between simultaneously flowering species. *Am. Nat.* 104:455–467.

Levin, D. A. and H. W. Kerster. 1967. Natural selection for reproductive isolation in *Phlox. Evolution* 21:679–687.

———. 1969. Density-dependent gene dispersal in *Liatris. Am. Nat.* 103:61–74.

Linhart, Y. B. 1973. Ecological and behavioral determinants of pollen dispersal in hummingbird-pollinated *Heliconia. Am. Nat.* 107:511–523.

Macior, L. W. 1971. Coevolution of plants and animals—systematic insights from plant–insect interactions. *Taxon* 20:17–28.

Manning, A. 1956. Some aspects of the foraging behaviour of bumblebees. *Behaviour* 9:164–201.

Michener, C. D. 1974. *The Social Behavior of the Bees.* Belknap Press of Harvard University Press, Cambridge, Mass.

Oster, G. and B. Heinrich. 1976. Why do bumblebees major? A mathematical model. *Ecol. Mon.* 46:129–133.

Parrish, J. A. D. and F. A. Bazzaz. 1979. Difference in pollination niche relationships in early and late successional plant communities. *Ecology* 60:597–610.

Pyke, G. H. 1978. Optimal foraging in bumblebees and coevolution with their plants. *Oecologia* 36:281–294.

Ribbands, C. R. 1949. The foraging method of individual honeybees. *J. Anim. Ecol.* 18:47–66.

Robertson, C. 1895. The philosophy of flower seasons, and the phaenological relations of the entomophilous flora and the anthophilous insect fauna. *Am. Nat.* 29:97–117.

Schemske, D. W. 1981. Floral convergence and pollination sharing in two bee-pollinated tropical herbs. *Ecology* 62:946–954.

Silander, J. A. and R. B. Primack. 1978. Pollination intensity and seed set in the evening primrose *(Oenothera fruticosa). Am. Midl. Natur.* 100:213–216.

Singh, S. 1950. Behavior studies of honeybees in gathering nectar and pollen. *Cornell Agr. Exp. Sta.* 288:1–59.

Solbrig, O. T. 1975. On the relative advantages of cross- and self-fertilization. *Ann. Missouri Bot. Gard.* 63:262–276.

Spencer-Booth, Y. 1965. The collection of pollen by bumblebees, and its transportation in the corbiculae and the proboscidial fossa. *J. Apic. Res.* 4:185–190.

Straw, R. M. 1972. A Markov model for pollinator constancy and competition. *Am. Nat.* 106:597–620.

Sukada, K. and Jayachandra. 1980. Pollen allelopathy—a new phenomenon. *New Phytol.* 84:739–746.

Tepedino, V. J. 1981. The pollination efficiency of the squash bee *(Peponapis pruinosa)* and the honeybee *(Apis mellifera)* on summer squash *(Cucurbita pepo). J. Kansas Ent. Soc.* 54:359–377.

Thomson, J. D. 1978. Effects of stand composition on insect visitation in two-species mixtures of *Hieracium. Am. Midl. Natur.* 100:431–440.

———. 1980. Implications of different sorts of evidence for competition. *Am. Nat.* 116:719–726.

———. 1981. Spatial and temporal components of resource assessment by flower-feeding insects. *J. Anim. Ecol.* 50:49–59.

———. 1982a. Visitation rate patterns in animal-pollinated plants. *Oikos.* 39:241–250.

Thomson, J. D., B. J. Andrews, and R. C. Plowright. 1982b. The effect of a foreign pollen on ovule fertilization in *Diervilla lonicera* Mill. (Caprifoliaceae). *New Phytol.* 90:777–783.

Thomson, J. D. and R. C. Plowright. 1980. Pollen carryover, nectar rewards, and pollinator behavior with special reference to *Diervilla lonicera. Oecologia* 46:68–74.

Waser, N. M. 1978a. Interspecific pollen transfer and competition between co-occurring plant species. *Oecologia* 36:223–236.

———. 1978b. Competition for hummingbird pollination and sequential flowering in two Colorado wildflowers. *Ecology* 59:934–944.

Watt, W. B., P. C. Hoch and S. G. Mills. 1974. Nectar resource use by *Colias* butterflies. *Oecologia* 14:353–374.

Willson, M. F. and P. W. Price. 1977. The evolution of inflorescence size in *Asclepias* (Asclepiadaceae). *Evolution* 31:495–511.

Willson, M. F. and B. J. Rathcke. 1974. Adaptive design of the floral display in *Asclepias syriaca* L. *Am. Midl. Natur.* 92:47–57.

23
FLORAL-VISITATION-SEQUENCES BY BEES: MODELS AND EXPERIMENTS

Keith D. Waddington

Department of Biology
University of Miami
Coral Gables, Florida

ABSTRACT

The term "flower-constancy," which has been used to denote the floral choice behavior of bees, is discussed. Reasons are given for abandoning the term and replacing it with "floral-visitation-sequences" (F-V-S). New approaches for studying and better understanding the bees' F-V-S are suggested. Two predictive models that may form a framework on which future work can be based are presented, along with some tests of these models.

KEY WORDS: Bees, foraging behavior, foraging theory, flower-constancy, optimization, artificial flowers, flower choice, behavioral ecology.

CONTENTS

INTRODUCTION

The literature is replete with investigations on the behavior of numerous species of bees foraging for pollen and nectar from the flowers of angiosperms (von Frisch, 1967, Proctor and Yeo, 1972, and Waddington and Heinrich, 1981 offer overviews of this work). Of this literature, one aspect of bee foraging behavior will be pursued in this chapter—that of the bees' patterns of floral choice.

The following will be discussed: (1) The literature on the floral choices of bees, generally referred to as flower-constancy, will be reviewed from a historical perspective. (2) I will show cause for revising the terminology associated with studies on bee floral choices and make a revision. (3) Novel approaches to the investigation of floral choice patterns through modeling, field observations, and controlled laboratory experiments will be exemplified by description of two current independent lines of research.

FIELD STUDIES AND FLOWER-CONSTANCY

The floral choices of individual bees have been investigated in the field by examining the constituents of the bees' pollen loads or by directly observing the bees' paths of movement between plants. In order to analyze the pollen load, pollen pellets are collected from bees in the field or at the hive entrance (in the case of the honeybee), smeared onto a glass slide, and examined microscopically. It is usually possible to identify pollen to genus and sometimes to species (Free, 1963; Macior, 1970). In the majority of studies the pollen loads are classified as pure (i.e., they contain the pollen of a single plant species) or mixed (the pollen of two or more plant species is present) (Betts, 1920; Clements and Long, 1923; Brittain and Newton, 1933; Macior, 1967, 1968, 1970, 1974, 1975a,b). The results of some of these studies, encompassing work on various bee taxa, are compiled and presented in Table 23-1. Although there is considerable variation in the proportion of pure pollen loads, both between and within bee taxa, a rather high proportion of the pollen loads are pure. Furthermore, mixed loads are generally comprised primarily of the pollen of one species and only small amounts of pollen from one or more other species (Grant, 1950). However, caution is advised in interpreting these results. In order for the proportions of the pollen load constituents accurately to reflect the proportion of the bees' visits to the respective plant species, the amount of pollen per plant must be the same for all species, and the pollen must be equally available to the bees (i.e., it must have the same properties for being picked up and lost, and the bees must collect pollen from all plants visited). It is likely that these assumptions are not usually met, and that the data are biased in some unknown ways.

There are also limitations on what can be

Table 23-1. Results of analyses of bee pollen loads.

Genus	Relative Frequency*		Authority
	Pure loads	Mixed loads	
Andrena	.44	.56	Brittain and Newton (1933)
	.64	.36	Clements and Long (1923)
Halictus	.84	.16	Brittain and Newton (1933)
	.75	.25	Clements and Long (1923)
Megachile	.75	.25	Brittain and Newton (1933)
	.55	.45	Clements and Long (1923)
Anthophora	.20	.80	Clements and Long (1923)
Bombus	.59	.41	Brittain and Newton (1933)
	.49	.51	Clements and Long (1923)
	.57	.43	Betts (1920)
	.49	.51	Macior (1970)
	.50	.50	Macior (1975b)
	.69	.31	Macior (1975a)
B. lucorum	.66	.34	Free (1970a)
B. agrorum	.37	.63	Free (1970a)
Apis	.62	.38	Brittain and Newton (1933)
	.88	.12	Clements and Long (1923)
	.93	.07	Betts (1920)
	.94	.06	Free (1970a)

*Many of the compilations from the authorities' work are taken from Grant (1950).

Table 23-2. Results from direct field observations of foraging bees.

	Number of individuals observed	Number of flowers visited	Proportion that are intraspecific visits*	Authority
Honeybees	8	258	1.0	Christy (1883)
	6	45	0.83	Bennett (1883)
Bumblebees	51	1660	0.51	Christy (1883)
	33	17	0.70	Bennett (1883)

*A visit to a plant of the same species as the previous visit.

learned by using this method. It is impossible, for example, to determine the bees' sequence of visits to the flowers, and the method can shed no light on the foraging patterns of nectar-feeding bees (Free, 1970a). Despite the weaknesses inherent in this method, the data in Table 23-1 certainly seem to indicate that bees often visit a single plant species during a foraging bout (i.e., one round trip from hive to floral patch). This result is generally confirmed by direct observation of the bees' visits in the field (Bennett, 1883; Christy, 1883; Bateman, 1951; Heinrich, 1976a). As an example, Grant (1950) has summarized the data of Christy (1883) and Bennett (1883) for bumblebees and honeybees; these data are presented here in Table 23-2.

This apparent propensity of bees to visit a single plant species during a foraging bout was noted more than 2000 years ago by Aristotle (Darwin, 1883), and the behavior has been referred to by floral biologists and melittologists for at least a century as *flower-constancy*. The literature on flower-constancy has been reviewed several times (Free, 1963, 1970a,b; Grant, 1950). The term has also been used to describe the behavior of bees on successive foraging trips and on successive days of foraging (Free, 1963, 1970a). Flower-constancy is defined in this chapter in terms of the behavior of bees during single foraging bouts.

A REVISION OF TERMINOLOGY

There are problems associated with referring to the floral visitation behavior of bees and the study of this behavior as flower-constancy. Below, reasons are given for abandoning the term as a general description of the bees' behavior.

The term flower-constancy does not always accurately describe the behavior of bees in the field. Constancy is defined in *The American Heritage Dictionary of the English Language* (Morris, 1969) as "1. Steadfastness in purpose or loyalty; faithfulness. 2. An unchanging quality or state." Employing these definitions, flower-constancy should be reserved for describing the singular situation of a bee visiting only a single plant species throughout a foraging bout; the term should not be used to describe generally the bees' choice-patterns. The data in Tables 23-1 and 23-2 suggest that bees exhibit flower-fidelity, or flower-constancy. However, this behavior is by no means exhibited exclusively by individual bees. Fidelity of bees to flowers of a single species is simply one outcome of a whole array of possible behavioral patterns.

The above argument may appear trivial, or perhaps one of semantics; however, I feel that choice of the term flower-constancy was truly unfortunate in that it has tended to make many biologists place undue emphasis upon the bee's flower species fidelity and may have stifled possibilities for novel productive lines of research. This conceptual framework, of defining the result (flower-constancy) *a priori,* has not been conducive to asking interesting questions such as: Why aren't bees constant? What parameters are important in predicting the bees' choice patterns? And so on. Perhaps this is why, aside from a few studies, the same kinds of observational and empirical work have been carried out during the last decade as were used a century of more ago. Most investigators have been content to verify that bees are more or less flower-constant.

Heinrich (1976a) took the first step in leaving behind the shibboleth of flower-constancy.

He has spent numerous hours observing floral visits of individually marked bumblebees in order to describe their *foraging specializations*. He found that bumblebees have individually distinct multiple specializations; they *major* on some plant species (visit them most frequently) and *minor* on others (their secondarily preferred species). This system of nomenclature has the advantage of permitting a more complete and accurate representation of the bees' behavior and gives the flexibiity required to describe the variable patterns of visitation that have been observed. A result of this work was the construction of some theoretical considerations, which will be described below (Oster and Heinrich, 1976). Perhaps these considerations were not stimulated or brought to light solely by Heinrich's system of description. However, the system does emphasize both common and rare visits to floral species and thereby provides a basis for such theoretical models. A problem with Heinrich's system of description is that it provides a qualitative and not quantitative description of the bees' behavior.

I suggest that the term *floral-visitation-sequence* (F-V-S) is more appropriate and should be used as a general qualitative heading for describing a bee's flower-visiting behavior and as a heading to designate such studies. The term is "neutral" in that it does not connote a specific behavioral pattern. The important part of this term is "sequence," for it is the actual sequence or order of visits to flowers that most completely describes a bee's behavior. Investigators have determined more or less precisely the proportion of visits to the flowers of one or more plant species, and this is all that has been necessary for loose discussions of flower-constancy. However, proportions provide an incomplete description of the bee's behavior, so that potentially valuable information is lost.

The F-V-S can be described qualitatively by an ordered list of flower visits. This is cumbersome and, except for some obvious patterns, difficult to interpret. Statistical methods such as the runs test (Sokal and Rohlf, 1969) can be of use for determining whether visits to flowers of different species are a random sequence, or whether a visit is infuenced by the outcome of previous visits. This method permits a qualitative statement about a sequence (e.g., a random sequence) with some probability of assurance. This information, coupled with the proportion of visits to flowers of alternative species, gives a rather complete picture of the bee's F-V-S.

The best description of the F-V-S is a wholly quantitative one, since a quantitative description provides the maximum information on a bee's behavior. Perhaps the best means of quantitatively describing the bees' F-V-S is by determining the probabilities associated with the possible transitions between the plant species on successive interplant flights. Thus the transition probabilities can be cast into an $n \times n$ (n being the number of available plant species) probability matrix:

$$P = \begin{matrix} p_{11} & p_{12} & \cdots & p_{1n} \\ p_{21} & p_{22} & \cdots & p_{2n} \\ \cdot & \cdot & & \cdot \\ \cdot & \cdot & & \cdot \\ \cdot & \cdot & & \cdot \\ p_{n1} & p_{n2} & \cdots & p_{nn} \end{matrix}$$

As examples, p_{11} is the probability of a bee flying from a plant of species 1 to another plant of species 1, and p_{21} is the probability of making a species-2 visit followed by a visit to species 1.

The probability matrix is the basis of the theory of Markov chains, and as Straw (1972) has indicated, the movements of pollinators between flowers "fall naturally into the category of phenomena amendable to description as Markov processes." The advantage of this method is that the F-V-S can be subjected to all the rigors of the theory (Howard, 1971). When the Markov model is applied, it is generally assumed that the matrix of transition probabilities is known, although questions of hypothesis testing and maximum-likelihood estimation have been investigated and are summarized by Billingsley (1961). In short, this general route of investigating the visitation sequences of bees may generate new ideas and understanding. A first step was taken along these lines by Bateman (1951). Markov models of pollinator activity also have been developed by Straw (1972) and Bobisud and Neuhaus (1975) for studying the effects of pollinator foraging patterns on the maintenance of intermingled simultaneously blooming species of plants.

APPROACHES FOR INVESTIGATING THE F-V-S

To date, most investigations of the F-V-S of bees have been descriptive and largely unaccompanied by any meaningful predictive or unifying theory. Until a few years ago, with few exceptions (see Bateman, 1951), the usual scientific method of formulating and testing hypotheses was generally not used. This trend was not unique to the study of foraging in bees; the same appears to apply generally to studies of feeding behavior (Schoener, 1971).

During the past 15 years, since the papers by Emlen (1966) and MacArthur and Pianka (1966), there has been increased interest in constructing theoretical models for the purpose of predicting the foraging behavior of animals, and, in particular, their food-choice patterns. This accumulated body of theoretical literature is generally known as optimal foraging theory and has been reviewed by Schoener (1971) and Pyke et al. (1977). The assumption that forms the basis for these models is that the fitness of animals is positively correlated with feeding efficiency; thus, within certain constraints (behavioral, physiological, etc.), natural selection will result in foraging behavior that maximizes foraging efficiency. Once the relationship between foraging efficiency (generally defined in terms of intake of some measure of food value per unit time, or net energy intake), or fitness, and all possible foraging behaviors has been established, it is possible to determine the "best" behavioral pattern. These studies have enjoyed some success in predicting the foraging behavior of various animals and provide a useful framework for future theoretical and empirical investigations.

Hints that the F-V-S of bees may be influenced by energetics have been made on numerous occasions. Darwin (1883) speculated

that the reason bees generally make repeated visits to a single plant species is that they are "thus enabled to work quicker; they have just learnt how to stand in the best position on the flower, and how far and in what direction to insert their proboscides." Heinrich and Raven (1972) were first to suggest, and extensively support with data from the literature, an energetic basis for studying the foraging behavior of bees and other pollinators. Several papers on various pollinator taxa soon followed (e.g., bees, Heinrich, 1972, 1975a,b; general, Covich, 1974; bats, Heithaus et al., 1974, 1975; birds, Gill and Wolf, 1975; Hainsworth and Wolf, 1976).

Below I will present two examples of research currently in progress that illustrate starting points from which I believe work on the F-V-S of bees can proceed. The methodology employed in these studies is new for investigating bee behavior. First, models are constructed based on the major assumption that bees are foraging so as to maximize foraging efficiency. Oster and Heinrich (1976) developed a model for the purpose of constructing a logical framework for understanding current field data on foraging bumblebees. Theirs is a schematic model that yields some interesting qualitative predictions based on the predictability of the resource-environment. In a second study, Waddington and Holden (1979) present a model that yields a quantitative description of the F-V-S and a new method for testing the predictions.

The Models

A. Oster and Heinrich (1976) developed a game-theoretical model for determining the best of two possible foraging strategies for a bumblebee faced with a field of numerous simultaneously blooming plant species. A bee can either visit only a single flower species or it can visit, in some frequencies, two or more of the available species. It is assumed that the floral species can be ranked as to their net profit to a bee (i.e., the difference between energy intake and costs of foraging). The authors consider first the problem of predicting the best F-V-S in a predictable, unchanging floral environment and then consider the case for a changing, unpredictable environment, i.e., where the best flower species is changing.

Three parameters are included in their model. The *state of nature* is the floral species that will indeed net the highest energy-profit to a foraging bee. A bee is assumed to have no information regarding the state of nature prior to reaching a floral patch. The bee must employ a "necessary imperfect sampling" process to gain information about the floral patch. The result is the bee's *perception* of the state of nature which is some fraction of the sampling time during which the bee will perceive one or more of the floral species as being best. Finally, the *possible foraging strategies* are stated. In order to illustrate the general situation, the simplest case of just two species of available flowers is considered, and they use a graphical proof (see Figs. 1 and 2 of their paper).

The results of the model are intriguing. In a predictable resource-environment, regardless of a bee's perception of which floral species will yield the greatest profits of energy, foraging on a single species of flowers is *always* better than random foraging among two or more floral species. The species on which the bee should major is determined by the bee's perception of the true state of nature. Under conditions of a changing resource-environment it is no longer advantageous to forage at one species of plant. Rather, a higher average reward can be attained by employing a "mixed strategy"; that is, foraging with some frequency at more than one floral species. For the two-species case, as the environment becomes less predictable, the bee should become less and less specialized on a single species.

Real (1980) has predicted the same outcome in another model. His is an evolutionary model that includes the variance of fitness (rather than mean fitness) that accompanies environmental uncertainty for making the maximization principle. When uncertainty in the floral environment is high, the model predicts evolutionary strategies consisting of diverse behavioral patterns ("mixed strategy"). Only when the fitness variance is zero, in a predictable environment, should a bee forage on a single species of plant.

B. Waddington and Holden (1979) considered the F-V-S of honeybees foraging for nectar at arrays of two randomly and independently patterned floral "types." The types may be two species of flowers or two morpological variants of the same floral species. Each floral type has associated with it a density and an expected caloric reward that can be obtained from each flower.

It is assumed that once a bee arrives at the floral patch, it forages in such a way as to maximize the intake of calories per unit of foraging time, given the behavioral constraints mentioned below. So, while at the patch, a bee incurs both costs (time) and benefits (calories from nectar) while foraging. The costs, u, to move from one plant to the next include the time to make the flight and, once arrived at the next plant, the time to gain access to the nectar (handling time). The gains on the move, C, are the calories acquired at the next chosen flower. $Q = (C/u)$ is the benefit–cost ratio for a single move.

The following behavior is assumed operative in the honeybee. A bee can estimate the caloric reward and handling time for flowers of each type and time to fly to a chosen flower. Assuming the bee can plan ahead one move, the best it can do is to choose *the* one flower, among all others sighted, that will result in the highest benefit–cost ratio, Q. If repeated over each move, the behavior should result in maximum intake of calories per time at the patch.

The behavioral pattern is illustrated in Fig. 23-1. A bee flies to flower b and there, after imbibing nectar, must "decide" which flower to fly to next. For simplicity, the bee's behavior is viewed at first in terms of one floral type, type 1. The bee's speed of flight is assumed unity, and the handling times for the types are equal. The bee, at b, scans the patch and sights the closest type 1 flower at d, which is X distance and time units away. Since the expected reward of type 2 flowers is twice that of type 1 flowers, the bee should fly to the closest type 2 flower if one occurs within distance $2X$. If a type 2 flower does not occur within $2X$ units, the bee should fly to d. The appropriate choice in Fig. 23-1 is e, the closest type 2 flower. Given the handling times, densities, and ex-

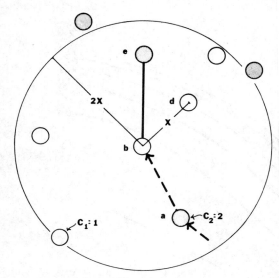

Figure 23-1. Open circles represent type 1 flowers, each equaling 1 calorie; shaded circles represent type 2 flowers, each worth 2 calories. Heavy solid and dashed lines indicate the path of movement of a bee between flowers. See text for complete explanation.

C_i = expected caloric rewards of type i flowers

D_i = flower density

h_i = handling times

v = speed of the bee in flight

Examples of the kinds of predictions that the model makes in relation to the relative densities and expected caloric rewards of the two floral types are illustrated in Fig. 23-2. As the values of D_1/D_2 and/or C increase, the curves asymptotically approach $p^* = 1$.

Also, the model predicts a random sequence of visits to the two floral types for the species under consideration, where the types are randomly and independently patterned. Thus, the F-V-S is rather precisely defined.

Experiments and Results

A. It is not possible to test precisely the predictions of Oster and Heinrich's (1976) model. Their model is schematic and as such does not yield quantitative answers. The model does, however, delineate the two extreme cases of the pure and mixed foraging strategies. It should be possible to relate these qualitative predictions to observations in the field.

Heinrich (1976a) has recorded the F-V-S of numerous individual bumblebees, in the forests, bogs, and hayfields of Franklin County, Maine. Individual bees were recognized by gluing a small, numbered, plastic disc to the dorsal thorax, or by the use of fast-drying paint. Nectar samples were taken from most species to determine the flowers' daily sugar production. Microcap capillary tubes were used to extract nectar, and sugar concentrations were measured with a hand refractometer (Heinrich, 1972).

There is clear evidence that a floral environment is constantly in a state of flux. Plants of the same species and of different species pass in and out of bloom (Mosquin, 1971; Heinrich, 1975c, 1976b,c), and nectar volumes change throughout the day (Waddington, unpub. data; Frankie, this book). These observations meet the criteria of an unpredictable resource-environment as defined in Oster and Heinrich's (1976) model. Thus they predict that

pected caloric rewards of the floral types, the problem of predicting the F-V-S comes down to determining whether a type 2 flower lies within some radius that is determined by the closest type 1 flower, and this is different on each move. Through manipulation of probability theory, Waddington and Holden (1979) derived the expected frequency of visits to the flower types. Thus, p^* is the optimal frequency of visits to type 1 flowers, and $q^* = 1 - p^*$ is the optimal frequency of visits to type 2 flowers (when the types are randomly and independently patterned).

$$p^* = \frac{\tilde{C}^2 D_1}{\tilde{C}^2 D_1 + B^2 D_2}, \quad C_2 > 0$$

$$p^* = 1, \qquad\qquad C_2 = 0$$

where:

$$B = \frac{(1 + v h_1)}{(1 + v h_2)}$$

$$\tilde{C} = \frac{C_1}{C_2}$$

i = flower types, 1 or 2

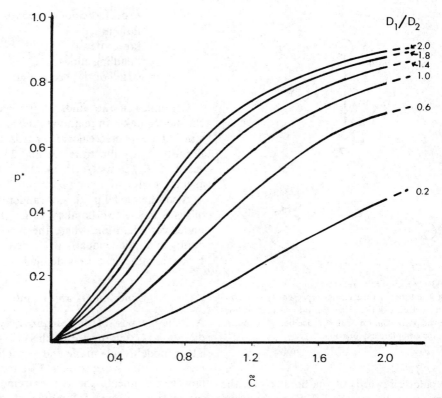

Figure 23-2. The prediction of a bee's optimal foraging behavior, p^*, is illustrated in relation to C (relative caloric rewards) for several values of D_1/D_2 (relative flower densities). It is assumed that $h_1 = h_2$ (handling times).

bumblebees should exhibit a mixed foraging strategy. This is generally borne out by Heinrich's field results. Most bees do not forage on a single plant species, nor do they visit flowers at random and in proportion to their relative abundance. Marked bees primarily visited flowers of one species (their major) and secondarily, in low frequency, visited one or more other species (their minors). It is interesting that individual bees did not have the same F-V-S. This too is consistent with predictions of their model; each bee, constrained by its limited sampling of the flowers, obtained a different perception of which species was best in terms of net caloric rewards. It is also likely that since the reward spectrum is always shifting, the actual best floral species is also always changing.

Differences in the foraging behavior of queens and workers were also consistent with the predictions. Workers have a shorter life span than queens and would therefore be expected to experience fewer changes in the resource-environment during their lives. The model predicts that workers should specialize more than queens on single species, and Heinrich's (1976a) data support this prediction.

Heinrich et al. (1977) studied the foraging behavior of bumblebees in the laboratory in an enclosed arena of artificial flowers. Blue and white patterns that each had a hole in the center were placed on a template of green Plexiglas that, in turn, covered a table containing a series of wells. Into these wells precise volumes of 50% sugar syrup could be dispensed. Several interesting experiments were performed.

In one set of experiments the bees gained a reward only from flowers of one color. Whether blue or white, the rewarding flowers were visited in high frequency (bees' major), whereas the the nonrewarding flowers were visited infrequently (their minor). Despite the predictability of the floral resource in this experiment, the bees generally maintained the minor. These results are contrary to predictions of the model.

In another experiment they examined the bees' behavior in response to a changing resource environment. Individual bees were rewarded at either blue or white flowers for ten days. The bees established the majoring strategy to the appropriately colored flowers. Subsequently, nonrewarding flowers of the previously rewarding color were added to the arena, thus "diluting" their reward value. The results varied with the color of flowers to which the bees were initially trained. When trained to blue, the bees did not change their major to the rewarding white flowers as the empty blue flowers were added, even though the more rewarding behavior was to major on the white flowers. After training to white flowers for 10 days, as empty white flowers were added, the bees visited the blue flowers at a rather high frequency, and in some instances switched majors. From these data, Heinrich et al. (1977) suggest that the variable strategies of bees observed in the field (Heinrich, 1976a, Fig. 5) "could in part be due to different foraging histories."

B. The model of Waddington and Holden (1979) makes quantitative predictions based on several assumptions of the bees' behavior and the values of several environmental variables (including the floral density and distribution, and the caloric rewards available). In order to determine whether a bee is employing p^* (the predicted F-V-S at a particular floral array), it is essential to know values of these variables for the patch so that the null hypothesis (F-V-S equals p^*) can be formulated. Because these measurements are difficult to make in the field, a method employing artificial flowers was developed (Waddington, 1979). The method permits control and manipulation of the variables used in the model. Because the details of the methods have already been reported, they will be stated briefly here (Waddington, 1979).

The patch of artificial flowers is made of a square (1.5 m²) piece of Plexiglas laid over a sheet of white paper (of the same dimensions) on which are arrayed, in grid-fashion, 2209 spatially fixed points at which flowers can be positioned. Wells were drilled in the top surface of the Plexiglas and centered over each of the possible flower positions. The wells served as "nectaries" into which small quantities of sugar syrup were dispensed with a syringe dispenser assembly (Model PB600-1, Hamilton Co.). A computer algorithm was used to select floral positions, and these positions were output on white computer paper. The output was pieced together, and glued, to form the 1.5 m² array. Rubber stamps were used to apply the floral shapes to the paper. Bees foraged singly for sugar syrup from the wells on the Plexiglas surface.

For purposes of testing the model, seven floral patches were prepared. Each patch had 200 type 1 flowers (blue, gentian-like shape) and 200 type 2 flowers (yellow, composite-like shape). The floral positions were generated randomly. The ratios of mean caloric rewards ($\bar{C}$; see model) were varied among the patches by varying the proportion of the flowers of each type that received 2 μl of 25% sucrose solution. The relative caloric rewards used yielded a wide range of predictions of p^*. Five to eight bees were permitted to forage at each patch up to ten times (each time is a foraging bout). The sequence of visits to flowers was recorded.

Since the flowers were randomly distributed on the seven patches, the bees' visits should be sequences of independent trials in which the probabilities of visiting flowers of the two types are p_j^* and q_j^* (jth patch). Thus, the F-V-S is defined. Two of the predictions of this model were tested. As expected, the bees' visits to the two floral types did represent a sequence of independent trials. The second prediction, that the probability of a bee visiting a type 1 (type 2) flower should remain constant over all foraging bouts and should equal the prediction [p_j^* (and q_j^*), for the jth patch], was rejected statistically (binomial test). However, by relaxing some assumptions, the model was improved. It appears that each bee that tested a patch employed the same F-V-S over each foraging bout, but different bees displayed slightly different probabilities (of visiting the two floral types) which varied around the predicted values (p_j^* and q_j^*). When the model was modified to permit the probability estimates of individuals to randomly deviate from the optimal prediction, then the results were consistent with the predictions (Fig. 23-3). This model remains to be field-tested, so its true value is currently unknown. Waddington and Holden (1979) have enumerated some problems expected in attempting to predict the F-V-S of honeybees in the field; however, the model does provide a basis for making field predictions.

Like any model, this one will profit from further testing of the predictions and the assumptions on which it is formulated. The work by Waddington et al. (1981) has already demonstrated this. Waddington and Holden (1979) assumed that it is the relative mean caloric val-

ues of flowers that are important in the bees' decision-making processes. By not having considered it, they assume the variance in nectar rewards to be unimportant in this regard. Waddington et al. (1981) found that the bumblebee, *Bombus edwardsii,* a close relative of the honeybee, preferably fed from artificial flowers yielding the same reward on each visit (low variance) rather than from flowers yielding variable rewards (higher variance), even though the long-term expectation (the mean) of reward was the same at each type of flower. Reward variance should be considered as a constraint in future models of bee foraging behavior.

THE FUTURE

The path, or sequence of floral visits, that a bee makes through a particular floral array is a re-

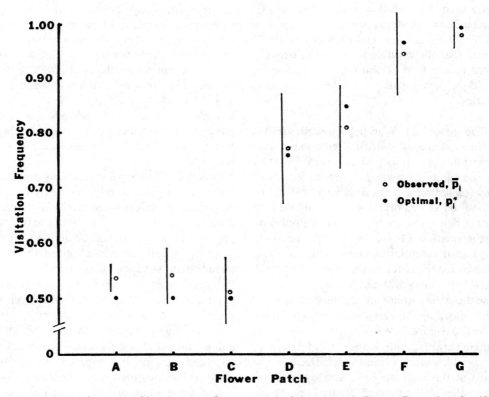

Figure 23-3. Visitation frequency of bees to type 1 flowers at each of the j (j = A, B, . . . , G) patches of artificial flowers. Open circles indicate the mean visitation frequency, $\bar{p}_1$, of bees tested on each patch, and the vertical lines indicate 95% confidence intervals. Dark circles show p_j^*, the optimal visitation frequency predicted by the model. (Modified with permission from the University of Chicago Press; from Waddington and Holden, 1979.)

sult of interaction between (1) the set of "rules" the bee employs while foraging and (2) those important parameters of the environment (floral patch) that when varied change the bee's F-V-S. Examples of environmental input might include the number of plant species available, the expected caloric rewards of the species, the distributions and densities of the plants, and their floral morphologies. If the goal is to predict the F-V-S of bees at any floral array, these two confounding sets of input must be unraveled and the set of foraging rules deciphered. Determination of the rules must be the major thrust of future studies of bee choice patterns. It is likely that this work will not be accomplished quickly, since it may mean determining a vast array of behavioral, physiological, and morphological constraints under which the bees function while foraging. Examples of such constraints include: the kinds of environmental cues (e.g., floral odors, nectar concentrations, plant spacing, etc.) that are utilized and the fashion in which they are utilized, the kinds of decisions that are based on these cues, the effects of past experience, rates of learning, the completeness of learning, the powers of discriminating between inputs, and the effects of inherited behavioral responses. To add to the complexity of the task is the likelihood that bees of different species employ different sets of rules, and that maybe different individuals of the same species employ different rules, and it is quite likely that the set of rules for any organism is not static with respect to time and physiological state. Clearly there is room for work on the F-V-S by biologists from most disciplines, and it appears that years will pass before we will fully understand and confidently predict the F-V-S of various bee species.

The question remains: What is the most efficient and reliable means for obtaining information on the bees' rules for foraging? There are numerous routes of research to take and many quantitative methods to utilize. For the behavioral studies, I think programs that combine controlled, manipulative laboratory experiments (following von Frisch, 1967; Nunez, 1970; Menzel and Erber, 1972; Heinrich et al., 1977; Waddington, 1979; and Waddington et

al., 1981) with field investigations will prove most rewarding. By formulating hypotheses that can be tested using the above combination, the set can be constructed rule by rule. The rules can then form the building blocks (assumptions) of models that can be used for predicting the F-V-S. C. S. Holling (e.g., 1959, 1965) has used this scheme (component analysis) successfully for predicting predator–prey interactions.

Students of bee foraging have their work cut out. However, the rewards are likely to be worth the efforts. Information on bee choice patterns will be of use to people performing applied research, e.g., crop pollination (Free, 1970b), and biologists interested in basic research. The mechanisms of bee choice patterns may provide additional insight into matters such as plant community composition and plant speciation.

ACKNOWLEDGMENTS

Thanks to Tracy Allen, Sydney Cameron, Bernd Heinrich, Julian Lee, and Charles Michener for their comments on a draft of the manuscript. My research on bee foraging behavior has been supported by NSF, NIH, The American Philosophical Society, The Theodore Roosevelt Fund, and The University of Kansas. Their support is gratefully acknowledged.

LITERATURE CITED

Bateman, A. J. 1951. The taxonomic discrimination of bees. *Heredity* 5:271–278.

Bennett, A. W. 1883. On the constancy of insects in their visits to flowers. *J. Linnean Soc. (Zool.)* 17:175–185.

Betts, A. D. 1920. The constancy of the pollen-collecting bee. *Bee World* 2:10–11.

Billingsley, P. 1961. Statistical methods in Markov chains. *Am. Math. Stat.* 32:12–40.

Bobisud, L. B. and R. J. Neuhaus. 1975. Pollinator constancy and survival of rare species. *Oecologia* 21:263–272.

Brittain, W. H. and D. E. Newton. 1933. A study of the relative constancy of hive bees and wild bees in pollen gathering. *Can. J. Res.* 9:334–349.

Christy, R. M. 1883. On the methodic habits of insects when visiting flowers. *J. Linnean Soc. London (Zool.)* 17:186–194.

Clements, F. E. and F. L. Long. 1923. *Experimental Pollination. An Outline of the Ecology of Flowers and Insects.* Carnegie Inst. Wash., Publ. No. 336.

Covich, A. 1974. Ecological economics of foraging among coevolving animals and plants. *Ann. Missouri Bot. Gard.* 61(3):794–805.

Darwin, C. R. 1883. *The Effects of Cross and Self Fertilization in the Vegetable Kingdom.* D. Appleton and Company, New York.

Emlen, J.M. 1966. The role of time and energy in food preferences. *Am. Nat.* 100:611–617.

Free, J. B. 1963. The flower constancy of honeybees. *J. Anim. Ecol.* 32:119–131.

———. 1970a. The flower constancy of bumblebees. *J. Anim. Ecol.* 39:395–402.

———. 1970b. *Insect Pollination of Crops.* Academic Press, London and New York.

Frisch, K. von. 1967. *The Dance Language and Orientation of Bees.* Belknap Press of Harvard University Press, Cambridge, Mass.

Gill, F. B. and L. L. Wolf. 1975. Foraging strategies and energetics of East African sunbirds at mistletoe flowers. *Am. Nat.* 109:491–510.

Grant, V. 1950. The flower constancy of bees. *Bot. Rev.* 16:379–398.

Hainsworth, F. R. and L. L. Wolf. 1976. Nectar characteristics and food selection by hummingbirds. *Oecologia* 25:101–113.

Heinrich, B. 1972. Energetics of temperature regulation and foraging in a bumblebee, *Bombus terricola* Kirby. *J. Comp. Physiol.* 77:49–64.

———. 1975a. The role of energetics in bumblebee–flower interrelationships. *In* L. E. Gilbert and P. H. Raven (eds.), *Coevolution of Animals and Plants.* University of Texas Press, Austin.

———. 1975b. Energetics of pollination. *Annu. Rev. Ecol. Syst.* 6:139–170.

———. 1975c. Bee flowers: a hypothesis on flower variety and blooming times. *Evolution* 29:325–334.

———. 1976a. The foraging specializations of individual bumblebees. *Ecol. Mon.* 46:105–128.

———. 1976b. Resource partitioning among some eusocial insects: bumblebees. *Ecology* 57:874–889.

———. 1976c. Flowering phenologies: bog, woodland, and disturbed habitats. *Ecology* 57:890–899.

Heinrich, B. and P. H. Raven. 1972. Energetics and pollination. *Science* 176:597–602.

Heinrich, B., P. R. Mudge, and P. G. Deringis. 1977. A laboratory analysis of flower constancy in foraging bumblebees: *Bombus ternarius* and *B. terricola*. *Behav. Ecol. Sociobiol.* 2:247–265.

Heithaus, E. R., P. A. Opler, and H. G. Baker. 1974. Bat activity and pollination of *Bauhinia pauletia:* plant-pollinator coevolution. *Ecology* 55:412–419.

Heithaus, E. R., T. H. Fleming, and P. A. Opler. 1975. Foraging patterns and resource utilization in seven species of bats in a seasonal tropical forest. *Ecology* 56:841–854.

Holling, C. S. 1959. Some characteristics of simple types of predation and parasitism. *Can. Entomol.* 92:385–398.

———. 1965. The functional response of predators to prey density and its role in mimicry and population regulation. *Mem. Ent. Soc. Can.* 45:1–60.

Howard, R. A. 1971. *Dynamic Probabilistic Systems.* Vol. 1, *Markov Models.* Wiley, New York, 576 pp.

MacArthur, R. H. and E. R. Pianka. 1966. On optimal use of a patchy environment. *Am. Nat.* 100:603–609.

Macior, L. W. 1967. Pollen-foraging behavior of *Bombus* in relation of pollination of nototribic flowers. *Am. J. Bot.* 54(3):359–364.

———. 1968. Pollination adaptation in *Pedicularis canadensis. Am. J. Bot.* 55(9):716–728.

———. 1970. The pollination ecology of *Pedicularis* in Colorado. *Am. J. Bot.* 57(6):716–728.

———. 1974. Pollination ecology of the front range of the Colorado Rocky Mountains. *Melanderia* 15:1–59.

———. 1975a. The pollination ecology of *Delphinium tricorne* (Ranunculaceae). *Am. J. Bot.* 62(10):1009–1016.

———. 1975b. The pollination ecology of *Pedicularis* (Scrophulariaceae) in the Yukon Territory. *Am. J. Bot.* 62(10):1065–1072.

Menzel, R. and J. Erber. 1972. The influence of the quantity of reward on the learning performance in honeybees. *Behaviour* 16:27–42.

Morris, W. (ed.) 1969. *The American Heritage Dictionary of the English Language.* American Heritage Publishing Co., Inc. and Houghton Mifflin Company, Boston.

Mosquin, T. 1971. Competition for pollinators as a stimulus for the evolution of flowering time. *Oikos* 22:398–402.

Nunez, J. A. 1970. The relationship between sugar flow and foraging and recruiting behaviour of honeybees (*Apis mellifera* L.). *Anim. Behav.* 18:527–538.

Oster, G. and B. Heinrich. 1976. Why do bumblebees major? A mathematical model. *Ecol. Mon.* 46:129–133.

Proctor, M. and P. Yeo. 1972. *The Pollination of Flowers.* Taplinger Publishing Co., New York.

Pyke, G. H., H. R. Pulliam, and E. L. Charnov. 1977. Optimal foraging: a selective review of theory and tests. *Q. Rev. Biol.* 52:137–154.

Real, L. A. 1980. Fitness, uncertainty, and the role of diversification in evolution and behavior. *Am. Nat.* 115:623–638.

Schoener, T. W. 1971. Theory of feeding strategies. *Annu. Rev. Ecol. Syst.* 2:369–404.

Sokal, R. R. and F. J. Rohlf. 1969. *Biometry.* W. H. Freeman and Co., San Francisco.

Straw, R. M. 1972. A Markov model for pollinator constancy and competition. *Am. Nat.* 106:597–620.

Waddington, K. D. 1979. Quantification of the movement patterns of bees: a novel method. *Am. Midl. Natur.* 101:278–285.

Waddington, K. D. and B. Heinrich. 1981. Patterns of movement and floral choice by foraging bees. *In: Foraging Behavior: Ecological, Ethological and Psychological Approaches.* Garland Press, New York.

Waddington, K. D. and L. R. Holden. 1979. Optimal for-

aging: on flower selection by bees. *Am. Nat.* 114:179–196.

Waddington, K. D., T. Allen, and B. Heinrich. 1981. Floral preferences of bumblebees *(Bombus edwardsii)* in relation to intermittent versus continuous rewards. *Anim. Behav.* 29:779–784.

24
ALTERNATIVE MODELS OF POLLEN TRANSFER

Kenneth P. Lertzman and Clifton Lee Gass

Department of Zoology
and Institute of Animal Resource Ecology
University of British Columbia
Vancouver, British Columbia

ABSTRACT

Pollen is often assumed to be dispersed by pollinators only a few flowers beyond its source. This assumption is not supported by recent experimental results, nor is it plausible in terms of mechanisms that generate pollen flow.

Process-oriented models that include the distribution of individual variability are more likely to be useful in understanding pollination systems than deterministic pattern-oriented models that are primarily sensitive to mean values.

As components of pollen transfer are modeled more explicitly, pollen transfer is less predictable from flower visitation sequences alone, and pollen is carried farther and more variably past its source.

Comparison of these models suggests that increasing structural complexity can provide increasing heuristic power, but we caution against making the conclusion that this also implies increasing realism.

Critical experiments in natural systems are needed to assess the contribution of rare events (i.e., pollen transferred in the tails of the dispersal distribution) to plant reproductive success.

KEY WORDS: Pollen carryover, pollen pool, process models, exponential decay, layering, stigmal capacity, floral variability, rare events.

CONTENTS

... even when we have correct premises, it may be very difficult to discover what they imply. All correct reasoning is a grand system of tautologies, but only God can make direct use of that fact. The rest of us must painstakingly and fallibly tease out the consequences of our assumptions.

Simon (1981)

INTRODUCTION

Transfer of pollen among flowers is a fundamental component of gene flow in many plant species. In the literature, pollen dispersal is often described by three components: pollinator flight distances, pollinator directionality, and pollen carryover. Pollinator flight distances have received much attention as estimators of pollen dispersal (Levin and Kerster, 1969a,b; Levin et al., 1971; Beattie, 1979; Beattie and Culver, 1979; Schmitt, 1980; Waser and Price, this book), and various aspects of pollinator behavior, including interflight directionality of pollinators, have been studied by those interested in foraging (e.g., Pyke, 1978; Heinrich, 1979; Zimmerman, 1979; Waddington, 1980; Gass and Montgomerie, 1981). However, little is known about carryover, and assumptions about carryover are weak links in the chain of reasoning about gene flow.

For this discussion we will define pollen carryover as the number of subsequent flowers to which the pollen of one flower is carried. Carryover could also be defined in terms of numbers of plants. If pollen from one flower is carried over and deposited at more flowers than just the next one the pollinator visits, then pollinator flight distances will underestimate pollen transfer distances. Many times researchers have made relatively unsupported assumptions about the extent of pollen carryover (Levin et al., 1971; Frankie et al., 1976; Feinsinger, 1978; Richards and Ibrahim, 1979; Augspurger, 1980; Schmitt, 1980).

Many current ideas about pollen dispersal and the factors affecting it stem from the work of Levin and his colleagues (Levin and Kerster, 1967, 1968, 1969 a,b, 1974; Kerster and Levin, 1968; Levin and Berube, 1972; Levin et al., 1971). In general, they observed that pollinator flight distances are leptokurtic and closely related to plant spacing distances, and inferred or assumed low pollen carryover. Some of these results appear to be general; for instance, a leptokurtic distribution of pollinator flight distances is common in hummingbirds, bumblebees, hawkmoths, and butterflies (e.g., Perkins, 1977; Beattie, 1979; Price and Waser, 1979; Schmitt, 1980; Gass and Montgomerie, 1981; Waser and Price, this book), although there may be significant differences in the degree of leptokurtosis between different pollinators (Schmitt, 1980).

Recent work on hummingbirds (Perkins, 1977; Lertzman, 1981; Waser and Price, this book) and bumblebees (Thomson and Plowright, 1980; Waser and Price, this book) suggests that pollen carryover may be longer and more variable than was indicated previously. This may be due to variation in both floral morphology (Lertzman, 1981) and pollinator behavior (Thomson and Plowright, 1980). If nuances of floral morphology or pollinator behavior do strongly affect whether pollen is picked up by or dropped off from pollinators at any given flower, then little generalization between systems may be possible without taking these factors into account.

This chapter will describe alternative sets of assumptions about the processes involved in pollen transfer and will discuss their consequences in terms of pollen carryover. We have three objectives: to organize and state explicitly the common assumptions about pollen carryover, to assess the validity of these assumptions, and to suggest some alternatives. Our backgrounds are in experimental field biology, and this discussion is based on a series of simulation modeling exercises we have conducted to clarify our own ideas about pollen transfer. Before discussing specific models in detail, it is important to discuss our approach to modeling in general. The following section does this, and also introduces the structural framework of the models.

ON BUILDING MODELS OF POLLINATION

Physical scientists and molecular biologists have repeatedly demonstrated that formal deduction is one of the most powerful tools of experimental science (Platt, 1964). In this chapter we illustrate an approach to pollination ecology in which models of biological processes are used to explore the hypotheses that generate the models; i.e., we use models as deductive tools. This will illustrate several points about systems approaches in pollination ecology. First, modeling is a useful tool for field biologists to organize existing knowledge and focus attention on critical areas of ignorance. In this sense this chapter is a review of the theory of pollination. Second, we want to show how these tools can be used both to suggest experiments and to generate predictions of their outcomes. Third, we want to reinforce the idea that our use of models to explore ideas cannot be separated from our use of experiments to evaluate those ideas.

In our attempts to understand complexity, we often construct models of reality. Physical models, like model airplanes to be tested in wind tunnels, and abstract models, like the simulations described in this chapter, are useful for exploring perceived relationships. Although models vary in the degree of realism, precision, and generality they embody (Walters, 1971), building them forces careful description of both what we know and what we don't know. This description is one of the most valuable benefits of modeling.

Models vary in their intended purposes. Rubber-band-powered airplanes mimic important aspects of the overall behavior of real airplanes without attempting to portray how they achieve it and may only vaguely resemble real airplanes in appearance. Other models portray details of the mechanical function of airplanes without attempting to achieve flight, and they may or may not look like the real thing. Thus we recognize both *pattern-oriented* models and *process-oriented* models. Some models may attempt description of both overall behavior and its underlying processes, but few can achieve both.

Pattern-oriented models provide useful descriptions of broad relationships expected from theory or observed in nature, and to the extent that the patterns they generate are accurate, may be useful aids to understanding higher-level phenomena. For example, models that generate patterns of pollen distribution can be used to examine the implications of these patterns for the neighborhood structure of plant populations (Levin and Kerster, 1969a; Levin et al., 1971; Schmitt, 1980). In our view, however, models can provide only limited insight into how patterns are generated unless they embody realistic biological processes. That is, the usefulness of models increases when they generate realistic biological patterns at one level through the simulation of realistic biological processes at a more detailed level (Clark and Holling, 1979). Such hierarchically structured models are doubly vulnerable to logical or experimental invalidation, and therefore doubly powerful as heuristic tools. Not only can their output fail to show reasonable patterns, but their explicit assumptions about the underlying mechanisms can be shown to be wrong.

There are three essential components to process models of pollination: pickup of pollen by a pollinator from the anthers of flowers, transport of pollen in a pool of pollen on the pollinator, and deposition of pollen on the stigmas (or anthers!) of other flowers. Alternative models may be distinguished by the number of subcomponents into which these basic three are disassociated, by the assumptions that are made about those subcomponents, and by the ways in which interactions between these components are organized.

The central feature of each model is a visit by a pollinator to a single flower: the "unit interaction." Population-level consequences of assumptions about the unit interaction are generated by iterating through it many times, collecting information from each iteration about pollen picked up or deposited, or about the structure of the pollen pool. By allowing variation in the factors that affect the components of this interaction, for instance, variation in floral morphology, we are able to study how the distribution of traits in populations of flowers is reflected in population-level patterns of pollen dispersal (Lertzman, 1981).

ALTERNATIVE MODELS

Models 1 through 3 were programmed in BASIC on Apple microcomputers. Model 4 was written in "C" and run on a Digital PDP-11/45. Copies of all programs are available on request.

The models are arranged in order of increasing biological realism of both process and pattern and increasing structural complexity (Table 24-1). Model 1 is pattern-oriented, with little underlying realism. It is simple, is historically significant, and represents commonly held assumptions. We treat it as an initial null hypothesis for comparison with other models and experimental data. Model 2 is a simple modification of model 1 that produces more reasonable *patterns* of pollen dispersal than model 1, but still lacks a realistic *mechanism*. Models 3 and 4 make more explicit statements than the other two models about pollen pool structure and how it affects pollen deposition. Model 4 incorporates stochastic variation in floral morphology that determines from where in the pollen pool pollen will be deposited or picked up.

These models assume that the pollinator deposits pollen on a flower's stigma prior to picking up pollen from its anthers. Though this refers to the temporal sequence of the mechanics of pollen transfer, it implies an assumption

about plant breeding systems. Thus the models will apply most directly to systems where flowers rarely, if ever, receive their own pollen.

Spatial structure is absent from these models. Flowers are unique, and are visited only once. They have no predefined association as plants. Superposition of spatial factors and pollinator movement on top of these models is the next step in the modeling of pollen transfer, but it is not discussed here.

We used 50 flower sequences (i.e., 50 iterations through the unit interaction per run of the model) to generate output for comparison of the four models. Data presented are from up to 60 runs of a model in order to generate a distribution of possible outcomes from a given set of assumptions. The restriction to a maximum of 50 flowers per run was set by space limitations on the Apple microcomputers. Model 4 could have been discussed with more flowers per run but was also limited to 50 flowers so as to be more directly comparable to the other three models. The behavior of model 4 with longer flower sequences (85 flowers) is discussed elsewhere (Lertzman, 1981).

The output variables are mean and maximum carryover. Mean carryover is calculated by summing the sequence numbers of the flowers that received pollen and dividing that total by the number of flowers that received pollen. For instance, a sequence in which pollen was deposited on the 5th, 16th, and 24th flowers past the source would have a mean carryover of: $(5 + 16 + 24)/3 = 15$ flowers. The max-

Table 24-1. General characters of the models.

Character	Model 1	Model 2	Model 3	Model 4
stigmal capacity	unlimited	unlimited	limited	limited
anther capacity	limited	limited	limited	limited
pollen picked up from every flower	yes	no	yes	no
pollen from each source deposited at every flower	yes	no	no	no
number of pollen grains picked up, when not 0	constant	constant	constant	variable
number of pollen grains deposited, when not 0	constant proportion of the pollen remaining from each source	constant	constant number randomly chosen from the pollen pool	variable: function of number and distribution of pollen remaining from each source

imum carryover is the sequence number of the last flower in a sequence to receive pollen. Unless otherwise specified, "carryover" refers to both of these statistics. Grand means and mean maxima are the means of mean and maximum carryover for a number of sequences.

Model 1: Constant Proportion of Pollen Grains Deposited: Exponential Decay

To study gene flow or to make inferences about the genetic structure of plant populations often requies a knowledge of pollen dispersal distances (e.g., Levin and Kerster, 1968, 1969a,b; Richards and Ibrahim, 1979; Schmitt, 1980; Waser and Price, this book). Because this variable is difficult to measure, observed pollinator movement distances are often used to estimate pollen transfer distances (e.g., Beattie and Culver, 1979). This correspondence is perfect if there is no pollen carryover. Levin and Kerster (1969a) used exponential decay to model carryover to produce a correction factor for pollinator flight distances so as to better estimate pollen transfer distances. They found that, given the assumption of exponential decay, their estimates of pollen transfer were insensitive to their assumptions about carryover (i.e., the parameters of the model). Since this approach characterizes much of the literature, we will start with exponential decay as our null hypothesis for pollen transfer.

Exponential decay is produced if a pollen pool is decremented by a constant proportion of its current size at each flower visited. When this proportion is large, the decay curve is steep. Thus, in this model, a constant number of pollen grains is picked up at each flower visited by the pollinator, and a constant proportion of the pollen remaining from each previously visited source is deposited at each subsequent flower. Pollen from different sources is assumed not to interact in the pollen pool on the pollinator, and is equally available for deposition. There is no limit to the number of grains that can be deposited on stigmas (one kind of interaction among pollen grains that is

disallowed is layering, which will be discussed below).

The amount of pollen deposited declines rapidly with the number of flowers beyond the pollen source (Fig. 24-1). When the proportion of the remaining pollen deposited from each source is high (0.4), no pollen is deposited after 11 flowers. When the proportion is very low (0.05), small amounts of pollen are deposited up to 46 flowers beyond the pollen source.

There are three key qualitative attributes of this model that we can compare with data: (1) all flowers receive pollen and the curve is thus smooth; (2) a constant proportion of the remaining pollen is deposited at each flower, so that we should observe deposition as a necessarily decreasing function of the flower visitation sequence; and (3) if the proportion of pollen that is deposited is large, the curve drops steeply. Based on experimental results from bumblebees (Thomson and Plowright, 1980; Waser and Price, this book) and hummingbirds (Perkins, 1977; Lertzman, 1981), each of these points can be rejected. In each of these experiments pollen deposition was variable; some flowers received no pollen, and later flowers in a sequence often received more pollen than earlier flowers. However, some of these results must be accepted cautiously and qualitatively. Lertzman (1981) and Waser and Price (this book) used fluorescent powders as pollen mimics, and they assumed that the dyes mimic pollen at least qualitatively; however, the correspondence between pollen carryover and dye carryover is not well established (Thomson and Plowright, 1980; Lertzman, 1981; Waser and Price, this book).

While this model may seem something of a "straw man," it is implicit in much current thinking about pollen carryover. It is corroborated by Levin and Berube's (1972) data on *Phlox* and *Colias* butterflies, but their results are probably not general. For instance, Levin et al. (1971) cautioned that systems in which pollen is transported on a proboscis will show lower carryover than those in which pollen is carried on the body of the pollinator because probosci are more efficient at depositing pollen.

Because the paper by Levin and Berube

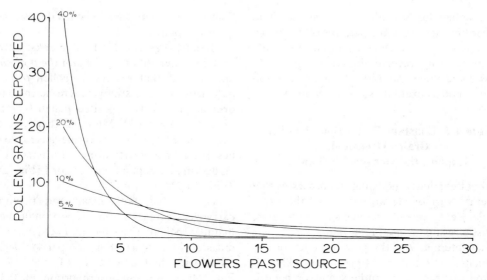

Figure 24-1. Pollen deposition curves from model 1: constant proportion deposited, simple exponential decay. Each curve indicates the pattern of pollen deposition from source flowers expected under model 1 if 40%, 20%, 10%, or 5% of the pollen from each source remaining on the pollinator were deposited on each flower visited. The ordinate scale is arbitrary, indicating that 100 grains were picked up from the source flower.

(1972) has had a large impact on assumptions about carryover, and because pollen flow is probably sensitive to these assumptions (though Levin and Kerster, 1969a, showed that pollen flow was insensitive to parameter changes given the exponential decay model, this does not necessarily hold for the assumptions about the underlying model itself), it is valuable to discuss their results in detail. One factor that may have led to the view of very efficient pollen deposition from probosci is that Levin and Berube (1972) experimentally prevented the butterflies in their experiments from retracting their probosci between flowers. Normal coiling and uncoiling could displace pollen from its sites of initial deposition, which would lead to a more patchy distribution of pollen. A spatially patchy distribution of pollen on the pollinator probably results in longer carryover than a uniform and contiguous distribution (Thomson and Plowright, 1980; Lertzman, 1981). Additionally, Levin and Berube (1972) used short (5-flower) sequences in their experiments; therefore they could not have observed long carryover even had it occurred. Observed carryover is very sensitive to

the length of the sequence of flowers used to measure it. For instance, when we increased the number of flowers in a sequence from 30 to 100 (in model 4), the proportion of pollen transferred more than 19 flowers past the source increased from about 10% to about 30%.

The effect of sequence length on observed carryover seems an obvious point, but there are *no* experimental measures of carryover with realistically large numbers of flowers. In the wild, some hummingbirds need to feed from an average of 3700 flowers per day (Gass et al., 1976), and we have no idea whether short to moderate (e.g., 30 flowers) experimental sequences (e.g., Thomson and Plowright, 1980; Lertzman, 1981, Waser and Price, this book) give anywhere near a realistic picture of pollen transfer when such large numbers of flowers are involved.

What can be inferred about carryover from the pollen loads on stigmas observed in nature? Based on the highly variable pollen loads found on the stigmas of adjacent flowers and plants, Levin and Kerster (1967, 1968) inferred low carryover. Alternatively, these variable pollen loads could have resulted from longer but more variable patterns of pollen deposition, in which case adjacent flowers would not necessarily receive similar pollen loads. How could this variability originate? The following models examine some possibilities.

Model 2: Constant Proportion of Pollen Grains Deposited: Exponential Decay with Blanks

Without explicitly modeling the processes that generate variability, we can make the exponential decay model produce somewhat more realistic results by assuming that some flowers do not interact with the pool of pollen on the pollinator (Lertzman, 1981): no pollen is picked up from their anthers and/or none is dropped off onto their stigmas. In nature, this could be produced by variation in floral morphology, in the reproductive state of the anthers and stigmas, and in pollinator behavior (e.g., the orientation of relevant parts of the body while contacting the flowers). These factors could act in either or both of two ways. They could influence the probability that flowers will contact the pollinator, or they could influence the probability of successful pollen transfer if that contact is made. We call these

flowers that create gaps in pollen deposition sequences "blanks."

Model 2 has two kinds of blanks: anthers that don't deposit any pollen *on the pollinator,* and stigmas that receive no pollen *from the pollinator.* Missed stigmas increase the carryover of previously deposited pollen (compare Figs. 24-1 and 24-2). Missed anthers decrease the observed mean carryover because zero values from "nonparticipating" anthers are included in the population statistics (Fig. 24-3; Table 24-2).

As expected, both mean and maximum carryover decrease as the proportion of the remaining pollen deposited at each flower increases (Fig. 24-4). For comparison of this model with the following two (Table 24-2 and Fig. 24-3), we use a proportion of 0.3 as representative.

This model assumes that there is no interaction between pollen from different sources, and that stigmas have unlimited capacity for pollen. Although the deposition of pollen is still a decreasing function of the number of flowers visited previously, the number of grains that stigmas receive is variable because not all previously visited flowers contribute pollen to the current flower.

This model of exponential decay with blanks

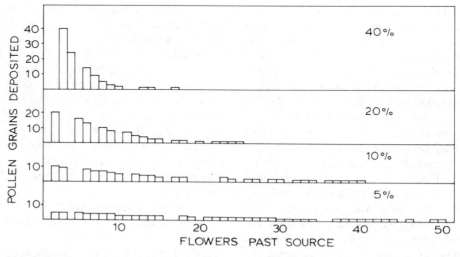

Figure 24-2. Pollen deposition curves from model 2: constant proportion deposited, exponential decay with blanks. Symbols and scales are as in Fig. 24-1, except that the abscissa extends to 50 flowers past the source flower. Based on field measurements of reproductive state in *Castilleja miniata* (Lertzman, 1981), 26% of the stigmas and 27.5% of the anthers, randomly chosen, did not interact with the pollen pool on the pollinator, i.e., were blanks.

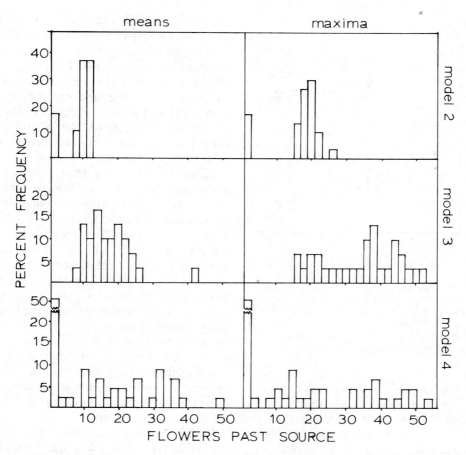

Figure 24-3. Frequency distributions of mean and maximum pollen deposition from models 2, 3, and 4. Sample sizes (number of runs of the model) are given in Table 24-2.

Table 24-2. Carryover statistics for models 2, 3, and 4.*

	Model 2		Model 3	Model 4	
	Including zero deposition	Not including zero deposition	Including zero deposition	Including zero deposition	Not including zero deposition
grand mean carryover	7	9	16	5	14
median mean carryover	9	9	15	0	14
variance	14.2	1.2	43.3	85.5	98.2
mean extreme carryover	14	17	32	7	17
median extreme carryover	17	18	35	0	16
variance	49	4.5	116.4	145.4	135.4
n	30	25	30	60	40

*There is only one category for model 3 because under the conditions that these data were generated, model 3 had no zero deposition values.

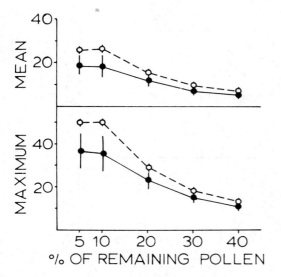

Figure 24-4. Mean and maximum pollen deposition distances from model 2. Closed circles and solid lines indicate means of 30 runs. Open circles and broken lines indicate the 25 of those runs in which some pollen was dropped off (i.e., excluding zero deposition values). Vertical lines indicate 95% confidence intervals.

is primarily of value as a more reasonable null hypothesis than simple exponential decay; it mimics the variability of the real world without specifying causal mechanisms. Thus it provides a basis for comparison with the following models, which make explicit statements about the processes that generate variability in pollen deposition.

Model 3: Constant Number of Pollen Grains Deposited: Limited Stigmal Capacity

In his models of competition between plants for pollinators, Waser (1978) assumed that a constant number of pollen grains was picked up from each flower visited, and that a constant number of grains, randomly chosen from the pool, was deposited at each flower. Each flower had a limited number of grains on its anthers, and a limited amount of space available on its stigma. Model 3, based on Waser's assumptions, assumes that pollen grains from early flowers in visitation sequences are just as likely to be chosen for deposition as pollen from more recently visited flowers. Thus there

is again no interaction between pollen from different source flowers (in contrast, model 4 explicitly models interactions among pollen grains in the pollen pool). Following Waser (1978), in model 3 the pollinator deposits 3 randomly chosen pollen grains on the stigma of each flower, and picks up 10 pollen grains from the anthers of each flower. Grains are drawn independently and without replacement from the pool of pollen on the pollinator.

There are two differences between this model and Waser's (1978). Whereas he simulated a pollinator moving through a spatial array of flowers, and flowers could be revisited, in this model flowers have no spatial identity and are visited only once. Similarly, while a total of 3 grains were deposited per flower per visit, these were counted toward a total of 5 per flower as a maximum stigmal capacity. In model 3, since revisits do not occur, the 3 grains deposited on one visit are equivalent to the stigmal capacity. The difference between 3 and 5 grains would be important if we were also directly modeling gene flow, but because we are primarily interested in the consequences of the pollen deposition rules, the distinction is not important.

Model 3 produces long and irregular carryover (Fig. 24-3, Fig. 24-5, and Table 24-2) that is similar to some experimental results (Thomson and Plowright, 1980; Lertzman, 1981). This variability most likely is due to the limited stigmal capacity. Stigmal capacity is less than the number of pollen grains available for deposition, and even though each stigma is saturated, only a small proportion of the pollen grains available can be deposited on any given stigma. Thus for a given pollen source, many blanks exist. When the model is driven with increasing stigmal capacity, mean carryover decreases and maximum carryover shows a peak at between 3 and 5 grains per stigma (Fig. 24-6). This peak of maximum carryover occurs in the range of pollen pickup-to-deposition ratios commonly observed in real systems (Waser, 1978).

The shorter carryover at the two extremes of stigmal capacity has different causes. The only situation in which there were zero dropoff values (i.e., when all pollen from a particular

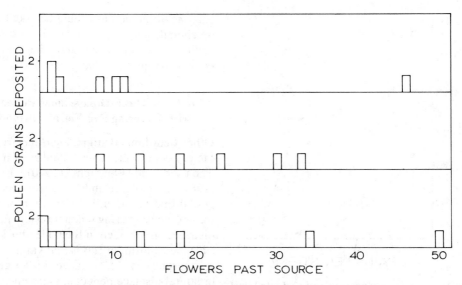

Figure 24-5. Pollen deposition curves from model 3: constant number of grains deposited. One, two, or three pollen grains could be deposited per stigma.

source remained on the pollinator and was never deposited) occurred when only one grain was deposited per stigma. Thus at the lowest stigmal capacity, carryover was lowered because pollen was remaining on the pollinator. The probability of a pollen grain being deposited is sensitive to the current size of the pollen pool on the pollinator. After many flowers have contributed pollen to the pool, the probability that pollen from any given flower will be cho-

sen is quite small: $p = 1/(10N - N)$, where N is the number of flowers fed on previously. The reason that carryover decreases at *high* ratios of deposition to pickup is that as the number of pollen grains deposited per stigma approaches 10, pollen is being deposited from the pollinator almost as fast as it is being picked up. When 10 grains were deposited by anthers and 10 picked up by stigmas, carryover was to one flower past the source.

Carryover (grand means and mean maxima) is both longer and more variable in model 3 than in model 2 (Table 24-2, Fig. 24-3). Carryover in model 2 could be made longer by having a higher proportion of "blank" stigmas, but we suspect that the variance would not increase. It is important to note that this is variance not only in pollen deposition in individual sequences of flowers (i.e., Fig. 24-5), but in the population parameters associated with the statistical distribution of that variability (i.e., Table 24-2, Fig. 24-3). Variability in model 3 output is highest at low stigma-deposition-to-pickup-from-anther ratios (Fig. 24-7), which is consistent with the decreased probability that a given pollen grain will be deposited under these conditions.

The pollen pool itself is assumed to be a randomly organized bin of pollen grains: selection

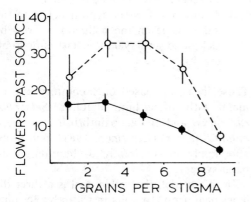

Figure 24-6. Mean (closed circles and solid lines) and maximum (open circles and broken lines) pollen deposition from 30 runs of model 3 at each of several levels of stigmal capacity. Vertical lines indicate 95% confidence intervals.

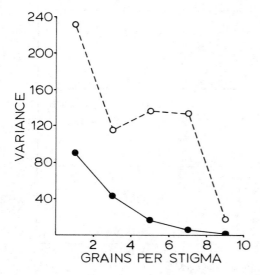

Figure 24-7. Variances of mean (closed circles and solid lines) and maximum (open circles and broken lines) pollen deposition distances from the same 30 runs of model 3 as in Fig. 24-6.

of pollen for transfer to stigmas is random, without regard for either the characteristics of the current flower or the sequence in which pollen was deposited in the pool. This is difficult to imagine in terms of underlying biological processes. Although grooming by bumblebees or preening by hummingbirds could conceivably randomize pollen in the pool with respect to visitation sequence (or obliterate it altogether), we would still expect some nonrandom pollen pool structure. In other words, we would expect precedence for deposition from the pollinator based on the time since last grooming or preening.

An alternative model might retain limited stigmal capacity but incorporate precedence for deposition based on visitation sequence. For instance, the probability that pollen will be deposited could be made dependent on the number of grains remaining from a flower and the number of flowers visited since they were picked up. This probability could be calculated using arbitrary scaling factors to produce reasonable patterns, but we chose instead to model a hypothetical causal mechanism. The next step, then, is to explicitly model pollen pool structure and its input/output processes. When comparing model 4 with the present

one, the question that should be asked is: How much understanding is gained by adding these additional components and their attendant complexity, and at what cost?

Model 4: Three-Dimensional Pollen Pool with Layering and Floral Variability

Other than limited stigmal surface area, what factor could result in the "blanks" of model 2? Thomson and Plowright (1980), Lertzman (1981), and Waser and Price (this book) suggested that variation in orientation of pollinator and flower during contact results in patchy distribution of pollen in the pollen pool. A possible mechanism for this is that variation in floral morphology, for instance, in flower length or lateral displacement of stigmas and anthers, would result in pollen from different flowers being deposited in different locations in the pollen pool. Flowers of different lengths would contact pollinators in different places, and hence pollen transfer between them would be unlikely.

This model rests on the following assumptions:

1. Pollinator–flower geometry during contact determines where pollen will be deposited on pollinators.
2. Pollinator–flower geometry is dependent on variability in floral morphology (one of several possible mechanisms).
3. If different flowers deposit pollen in the same place on the pollinator, it will be deposited in layers, or last-in-first-out stacks.

These ideas are based on laboratory experiments with rufous hummingbirds *(Selasphorus rufus)* and Indian paintbrush *(Castilleja miniata)* flowers (Lertzman, 1981), and probably apply most directly to hummingbird–plant systems.

In this model, the pollen pool is a three-dimensional array 14 units in width by 36 units in length (the *x,y* plane). The dimensions are based on measurements of pollen loads on freshly netted wild *Selasphorus rufus* that held territories in *Castilleja miniata* meadows,

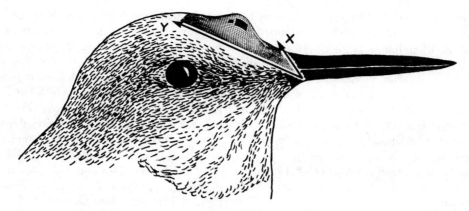

Figure 24-8. Drawing of an immature male rufous hummingbird (traced from a photograph) showing the hypothetical bivariate normal distribution of contact between the pollinator and floral reproductive organs. The two dark rectangles indicate the relative sizes of *Castilleja miniata* stigmas (small) and anthers (large). Bivariate normal distribution from *Biometry: The Principles and Practice of Statistics in Biological Research,* Second Edition, by Robert R. Sokal and F. James Rohlf. W. H. Freeman and Company. Copyright © 1981.

and on laboratory birds that had fed from known numbers of flowers. The third dimension, depth, is allocated dynamically as space is needed (Fig. 24-8). Whether model anthers and stigmas contact this pool or not, and where they contact it, if they do, depends on their length and lateral displacement from the midline of the flower. Both anthers and stigmas are constant in size, but anthers have an area six times larger than stigmas (4 × 6 vs. 2 × 2 pollen pool map units, based on measurements of *Castilleja miniata*).

We assume that variation in floral morphology (lateral displacement and corolla tube length, respectively) is normally distributed (Fig. 24-8). Real *Castilleja miniata* lengths deviate significantly from normality (they are both skewed and leptokurtic; Lertzman, 1981), but the lateral displacement of stigmas and anthers does not deviate significantly from normality.

In unit interactions, anthers and stigmas are chosen randomly from normal distributions with a given mean and variance. They are thus most likely to contact the center of the pool rather than the edges. Flowers in the tails of the distributions are likely to contact the pool only partially, if at all, and will have little surface area available for pollen transfer. If there is no pollen at a location in the pool when a stigma contacts it, then no pollen is deposited

on that stigma. Stigmas receive pollen from the top layer of the pool only, and once the top layer at a location is removed, the remaining layers are pushed up one step closer to the surface. They are pushed back down if pollen is added on top of them.

This model produces highly variable carryover (Fig. 24-3, Table 24-2). Many flowers receive no pollen, and those that do, receive variable amounts. In these 50-flower runs, there are many cases in which no pollen from a flower is deposited on other flowers. The frequency distributions of mean and maximum carryover show a high proportion of such zero deposition values, and have long tails representing less frequent long carryover (Fig. 24-3).

Zero deposition could result from either no pollen being picked up, pollen being buried by other pollen and never being uncovered, or pollen being deposited at locations where other flowers are unlikely to contact it. Because this model assumes that the pollen pool is rectangular, and flower variability follows a bivariate normal distribution, the corners of the pollen pool have a low probability of contact, even when compared to the edges of the pool in the middle of its length. If pollen is picked up here (a low-probability event), then it will be unlikely to be deposited. If a uniform distribution of flower variability is assumed, and zero deposition values are still present, then zero deposition is not a result of the distribution of flower variation, but more likely a result of the covering up of pollen. This appears to be the case; when we run model 4 assuming uniformly distributed variation in floral morphology, many zero values still occur. If the runs had been longer, say 120 flowers, pollen from these sources eventually might have been deposited, but we have not observed this.

Two kinds of modifications to model 4 would make it more likely that buried pollen would resurface:

1. The "coal scoop" hypothesis suggests that stigmas are dragged *through* the pollen pool, rather than touching on its surface as we have assumed so far. This has two new consequences: pollen from more than the surface layer is available for deposition on stigmas, and unit interactions create three-dimensional grooves through the pollen pool rather than simply lifting the top layer of pollen from an area of their own size. Both of these consequences would make it much more likely that stigmas would contact pollen in a patchy or sparsely filled pool, and that buried pollen would resurface. This would decrease the incidence of zero deposition and increase mean carryover.
2. In nature, differential stickiness of pollen to other pollen, to stigmas, and to anthers may keep pollen pools from building up as much as in this model. Additionally, periodic sloughing of patches of pollen, induced by rain, preeing, wind resistance during flight, or weak binding of pollen in the pool, may "reset" the pool and expose deeply buried pollen.

Overall mean carryover for model 4 is less than for model 3 (Table 24-2), and the incidence of zero deposition is much greater (Fig. 24-3). However, if mean carryover is calculated excluding the zero deposition values (i.e., if we only consider carryover of pollen that *was* deposited on stigmas), then carryover in model 4 has increased substantially (Table 24-2). Both mean and maximum carryover in model

4 have greater variances than in model 3. This difference is maintained even when zero deposition values are excluded from the model 4 calculations (Table 24-2).

Thus models 3 and 4 both show regular long carryover (Fig. 24-3), but model 4 has a higher variance associated with carryover, and a much higher frequency of zero deposition. When we decreased the ratio of stigmal pickup to anther deposition in model 3 from 3:10 toward 1:10 (Figs. 24-6 and 24-7), the behavior of model 3 approached that of model 4. However, it remained intermediate and did not reach the level of model 4. For instance, model 3 (Fig. 24-9) produced 13% zero deposition when 1 grain was deposited per stigma, compared to 65% in model 4 [variable number (between 1 and 4) of grains deposited per stigma]. Model 4 has potentially a much smaller stigma: anther capacity ratio, and the remaining differences between model 3 and model 4 could be due to this. However, we suspect that they are due to a combination of the large number of grains in the pollen pool relative to stigmal capacity *and* the likelihood that many of those grains will be permanently covered by other pollen. As long as the rate of covering is greater than the rate of uncovering, the possibility of layering will enhance the effect of limited stigmal capacity relative to anther deposition and produce this effect of many zero deposition values.

Though there is some experimental evidence suggesting layering of pollen on hummingbirds (Lertzman, 1981), it is far from conclusive. Because of the potential that layering profoundly affects carryover by increasing the likelihood of zero deposition (which may just be *very long* carryover), it is important to assess the occurrence of layering in real systems.

GENERAL DISCUSSION

Implications for Mechanisms

Much pollination research has attempted to study broad population-level patterns of pollen and gene flow, but it has been limited by incomplete knowledge of pollen transfer pro-

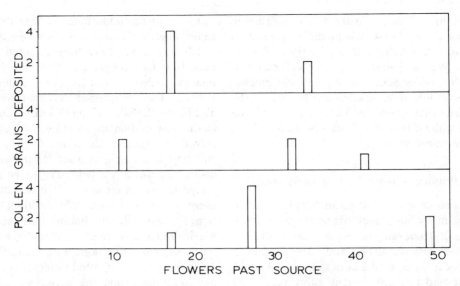

Figure 24-9. Pollen deposition curves from model 3: three-dimensional pollen pool. Zero, 1, 2, 3, or 4 pollen grains could be deposited per stigma.

cesses at the individual level. Historically, as in this discussion, the focus has moved from description of patterns toward the consideration of processes, with some accompanying increase in our understanding of how the patterns are generated. Each alternative model added components to the determination of pollen transfer: "blanks," limited stigmal surface area, variability in flower morphology (producing a spatially structured pollen pool), and the third dimension of the pollen pool—depth. These additional components and the ways that they interact profoundly affect the conceptual context in which we conduct our research, but we need critical experiments to assess their importance and generality.

The most important and general conclusion to be drawn from these models is that simple and often only implicit assumptions about pollen pool structure and the rules for pickup and deposition of pollen can have major consequences in terms of pollen carryover. One example of this is the complex pattern of pollen deposition that resulted from assuming random selection of pollen grains and limited stigmal capacity in model 3. This kind of result illustrates and reinforces the importance of making assumptions about carryover as explicit as possible.

Models 3 and 4, which make explicit statements about the processes that generate variability in pollen deposition, predict longer carryover than has often been assumed. If carryover is indeed lower (or higher) than we have suggested here, either the processes we modeled are invalid, or there are other processes acting that limit carryover. Some possibilities are: (1) that pollen grains stick together and come off in large clumps; (2) that the flowers and the pollinators contact each other in more stereotyped ways than we assumed; or (3) that stigmal capacity is high relative to the amount of pollen picked up by the pollinator. Any of these could lower flower-to-flower carryover, even if the processes described by these models were in operation. Note that the first possibility reflects a *structural* characteristic of the natural system and would require a structural change in the model, whereas the others are merely quantitative (i.e., they would be modeled simply by changing parameter values in the model).

Increasing the complexity of model structure resulted in an increasingly complex pattern of model output. While this has great heuristic value in terms of understanding the consequences of our assumptions, increasing complexity does not necessarily imply increasing realism. This is illustrated by comparison of models 3 and 4, which differ greatly in structure but behave in qualitatively similar ways. We need more knowledge of real systems to assess both the hypothetical mechanisms on which our models are based and the patterns they generate. These models will be worthwhile if their results can be used to focus future research.

Implications for Plant Populations

Based on an average of about 5 flowers visited per plant by hummingbirds on single bouts of foraging (*Selasphorus rufus* and *Castilleja miniata;* Perkins, 1977), mean and maximum pollen carryover values of both models 3 and 4 would lead to plant-to-plant carryover in the range suggested by Levin and Kerster (1969a) and Levin et al. (1971)—about 3 to 10 plants, depending on the number of flowers fed from per plant. However, these models suggest that while much of the pollen will be deposited near its source plant, a substantial proportion may be carried many flowers past its source. The tail of the distribution of carryover distances is long; both models 3 and 4 show infrequent but regular deposition 40 to 50 flowers past the source. To the extent that the processes described in these models represent real sytems, we need to be asking about the importance of such rare events in nature.

If, after pollination, real pollen grains from different sources compete for a limited number of ovules, or if reproductive success of ovules (i.e., seed number, seed quality, or seedling quality) varies with dispersal distance in the way suggested by Price and Waser (1979) and Waser and Price (this book), then a small proportion of the pollen transfers could contribute disproportionately to reproductive success. This possibility can be assessed only with reference to the pollen transfer mechanisms and optimal outcrossing distances of particular systems; so it would be premature to conclude that long-distance pollen transfers do generally have disproportionate fitness consequences. However, we can conclude that mean values of pollen carryover are not sufficient for assessing the importance of rare events. In fact, it is logically impossible in general to understand anything about the importance of rare events by examining measures of central tendency.

These analyses have been oriented toward male function; the primary focus has been on how many flowers past its source pollen is deposited. The complementary analysis would also be worthwhile, asking this question: What is the distribution of identities of pollen received by stigmas? Certainly, for any understanding of how selection acts on factors influencing carryover we will have to be able to integrate consequences of pollen-dispersal events in terms of both male and female function. For example, the finding that seed set is maximized at a particular pollen-dispersal distance (female function; Price and Waser, 1979) must be interpreted in terms of the factors that determine the actual distances to which pollen is carried (male function; this study). These two kinds of factors clearly are not independent. For instance, flooding a pollen pool with one's own pollen should increase success as a male, but decrease success as a female (Schemske, 1980; Lertzman, 1981). There is a growing literature on the tradeoffs and constraints involved in male vs. female expression and the tactics appropriate for each (e.g., Willson and Rathcke, 1974; Janzen, 1975; Charnov et al., 1976; Charnov and Bull, 1977; Charnov, 1979; Willson, 1979; Hancock and Bringhurst, 1980; Lertzman, 1981). It will be most important in the future to tighten the link between our interest in fitness consequences of different patterns of pollen dispersal such as these and our knowledge of the processes that act in producing these patterns.

ACKNOWLEDGMENTS

This chapter is dedicated to Dixon Jones, who originally got us thinking about pollen being layered on a hummingbird's forehead. We would also like to acknowledge the influence that Buzz Holling has had on how we think about systems and their components. Nick Waser has given his ideas and support freely from the inception of this project and deserves more credit than we can give him here.

Steve Oberski, Dave Zittin, Dave Marmorek, and Bill Webb were tremendously helpful with the programming, and Scott Carley, Larry Dill, Glenn Sutherland, Mary Price, Nick Waser, Monica Geber, and Gene Jones reviewed the manuscript. Mary McGechaen typed the tables. Thanks to all.

This project was submitted in partial fulfillment of the requirements for the M.Sc. degree of K. P. Lertzman and was supported financially by the National Sciences and Engineering Research Council of Canada (by a Postgraduate Research Fellowship to KPL and an operating grant 67-9876 to CLG).

LITERATURE CITED

Augspurger, C. K. 1980. Mass-flowering of a tropical shrub *(Hybanthus prunifolius):* influence on pollinator attraction and movement. *Evolution* 34:457–488.

Beattie. A. 1979. Plant–animal interactions affecting gene flow in *Viola. In* A. J. Richards (ed.), *The Pollination of Flowers by Insects.* Linnean Soc. Symp. Ser. No. 6. Academic Press, New York, pp. 151–164.

Beattie, A. J. and C. D. Culver. 1979. Neighborhood size in *Viola. Evolution* 33:1226–1229.

Charnov, E. L. 1979. Simultaneous hermaphroditism and sexual selection. *Proc. Natl. Acad. Sci. U.S.A.* 76:2480–2484.

Charnov. E. L. and J. Bull. 1977. When is sex environmentally determined? *Nature* 226:828–830.

Charnov, E. L., J. M. Smith, and J. J. Bull. 1976. Why be an hermaphrodite? *Nature* 263:125–126.

Clark, W. C. and C. S. Holling. 1979. Process models, equilibrium structures, and population dynamics: on the formation and testing of realistic theory in ecology. *Fortschr. Zool.* 25:29–52.

Feinsinger, P. 1978. Ecological interactions between plants and hummingbirds in a successional tropical community. *Ecol. Mon.* 48:269–287.

Frankie, G. W., P. A. Opler, and K. S. Bawa. 1976. Foraging behaviour of solitary bees: implications for outcrossing of a neotropical forest tree species. *J. Ecol.* 64:1049–1057.

Gass, C. L. and R. D. Montgomerie. 1981. Hummingbird foraging behaviour: decision-making and energy regulation. *In* A. C. Kamil and T. D. Sargent (eds.), *Foraging Behaviour: Ecological, Ethological and Psychological Approaches.* Garland STPM Press, New York.

Gass, C. L., G. Angehr, and J. Centa. 1976. Regulation of food supply by feeding territoriality in the rufous hummingbird. *Can. J. Zool.* 54:2046–2054.

Hancock, J. F. and R. S. Bringhurst. 1980. Sexual dimorphism in the strawberry *Fragaria chiloensis. Evolution* 34:762–768.

Heinrich, B. 1979. Resource heterogeneity and patterns of movement in foraging bumblebees. *Oecologia* 140:235–245.

Janzen, D. H. 1975. A note on optimal mate selection by plants. *Am. Nat.* 111:365–371.

Kerster, H. W. and D. A. Levin. 1968. Neighborhood size in *Lithospermum caroliniense. Genetics* 60:577–587.

Lertzman, K. P. 1981. Pollen transfer: processes and consequences. Unpublished M.Sc. thesis, University of British Columbia, Vancouver.

Levin, D. A., and D. C. Berube. 1972. *Phlox* and *Colias:* the efficiency of a pollination system. *Evolution* 26:242–250.

Levin, D. A. and H. W. Kerster. 1967. An analysis of interspecific pollen exchange in *Phlox. Am. Nat.* 101:387–399.

———. 1968. Local gene dispersal in *Phlox. Evolution* 22:130–139.

———. 1969a. The dependence of bee-mediated pollen and gene dispersal upon plant density. *Evolution* 23:560–571.

———. 1969b. Density dependent gene dispersal in *Liatris. Am. Nat.* 103:61–74.

———. 1974. Gene flow in seed plants. *Evol. Biol.* 7:139–220.

Levin, D. A., H. W. Kerster, and M. Niedzlek. 1971. Pollinator flight directionality and its effect on pollen flow. *Evolution* 25:113–118.

Perkins, M. D. C. 1977. Dynamics of hummingbird mediated pollen flow in a subalpine meadow. M.Sc. thesis, University of British Columbia, Vancouver.

Platt, J. 1964. Strong inference. *Science* 146:347–353.

Price, M. V. and N. M. Waser. 1979. Pollen dispersal and optimal outcrossing in *Delphinium nelsonii. Nature* 277:294–297.

Pyke, G. H. 1978. Optimal foraging in bumblebees and coevolution with their plants. *Oecologia* 36:281–293.

Richards, A. J. and H. Ibrahim. 1979. Estimation of neighbourhood size in two populations of *Primula veris. In* A. J. Richards (ed.), *The Pollination of Flowers by Insects.* Linnean Soc. Symp. Ser. No. 6. Academic Press, New York, pp. 165–174.

Schemske, D. W. 1980. Floral ecology and hummingbird pollination of *Combretum farinosum* in Costa Rica. *Biotropica* 12:169–181.

Schmitt, J. 1980. Pollinator foraging behaviour and gene dispersal in *Senecio* (Compositae). *Evolution* 34:934–943.

Simon, H. A. 1981. *The Sciences of the Artificial.* The M.I.T. Press, Cambridge, Mass.

Sokal, R. R. and F. J. Rohlf. 1969. *Biometry.* W. H. Freeman, San Francisco, 749 pp.

Thomson, J. D. and R. C. Plowright. 1980. Pollen carryover, nectar rewards, and pollinator behavior with special reference to *Diervilla lonicera. Oecologia* 46:68–74.

Waddington, K. D. 1980. Flight patterns of foraging bumblebees relative to density of artificial flowers and distribution of nectar. *Oecologia* 44:199–204.

Walters, C. J. 1971. The systems approach: mathematical models in ecology. *In* E. P. Odum (ed.), *Fundamentals of Ecology,* 3rd ed. W. B. Saunders, New York.

Waser, N. M. 1978. Interspecific pollen transfer and competition between co-occurring plant species *Oecologia* 36:223–236.

Willson, M. F. 1979. Sexual selection in plants. *Am. Nat.* 113:777–790.

Willson, M. F. and B. J. Rathcke. 1974. Adaptive design of the floral display in *Asclepias syriaca* L. *Am. Midl. Natur.* 92:47–57.

Zimmerman, M. 1979. Optimal foraging: a case for random movement. *Oecologia* 43:261–267.

Section: VIII
APPLIED POLLINATION ECOLOGY

25
POLLINATION OF ENTOMOPHILOUS HYBRID SEED PARENTS

Eric H. Erickson, Jr.

North Central States Bee Research Laboratory
USDA-ARS
University of Wisconsin
Madison, Wisconsin

ABSTRACT

The purpose of this chapter is to expand awareness, particularly among plant breeders, of the problems associated with the pollination of entomophilous hybrid seed parents and their F_1 progeny. The cause(s) of inadequate hybrid pollination are presented along with a discussion of the relative merit of the various techniques that have been suggested to ensure seed set. Past experiences with the development of hybrids and the results of current pollination research in a number of commercially produced hybrid crops are presented.

KEY WORDS: Bees, honeybees, pollination, hybrids, alfalfa, carrots, cotton, crucifers, onions, sunflower.

CONTENTS

INTRODUCTION

Perception of the advantages to be gained from improvement of crop plants is probably as old as civilization. A slow but steady improvement in the productivity and quality of crops has occurred since prehistoric times (Emsweller, 1961). Early efforts undoubtedly centered on selection for desirable traits, whereas, later on, improved cultivars were developed using various plant breeding methods. Most recently, modern plant breeding technology has provided a vast potential for improving crop quality and yield through commercial production of hybrid seed.

Hybridization of entomophilous crops has its roots in the need to overcome autogamy, since 86% of all phanerogamic seed plants are bisexual and 72% are hermaphroditic (Frankel and Galun, 1977). Outcrossing occurs with greater frequency in allogamous species although, even here, it cannot be assured. Coadaptative evolution of plants and their pollinators has established specific patterns of behavior within both systems that have con-

tributed significantly to the survival of each. These are based primarily upon the presence of floral visual cues, olfactory cues, and nectar cues that stimulate or reward pollen gathering. Pollination is an incidental consequence. Profitability incentives (e.g., nectar/pollen quantity and quality) benefit the pollinator, while successive intraspecific visits, assured by floral isolating mechanisms, ensure perpetuation (fitness) of the plant species (Grant, 1949; Percival, 1965; Levin, 1978).

Floral isolating mechanisms that limit interspecific and interpopulation foraging by insect pollinators are well known, particularly in autogamous species. Isolation may be "imposed by structural contrivances of the flowers which prevent the pollen of one species from being conveyed to the stigmas of the other (mechanical isolation), or imposed by the constancy of the pollinators to one kind of flower (ethological isolation)" (Grant, 1949). In either case interspecific hybridization is inhibited because of heritable differences among species. As hybrid crop systems are developed among entomophilous plant species, the need for solution of attendant pollination problems inevitably follows because of genetically induced floral differences between seed parents, e.g., intraspecific isolating mechanisms. Successful production of hybrids involves an understanding of the bionomics of both plant and pollinator; too often the latter is overlooked (Rubis, 1970a).

Commercial hybrids, most of which are produced via hand or wind pollination, genetic male sterility (heritable absence of viable pollen), or incompatibility mechanisms, or vegetatively, already exist in a number of flower and vegetable garden species. However, large-scale commercial production of horticultural and agronomic hybrids is limited; as late as 1961 only hybrid corn and onions were produced commercially (Emsweller, 1961). Today cultivated anemophilous hybrids in corn and sorghum reach 85% and 90% (respectively) of the total acreage planted to each, but comparatively few acres are devoted to entomophilous hybrids such as those in onions and carrots, 27% and 7% respectively (Frankel and Galun, 1977). Complete conversion to hybrids in a number of field crops would have taken place by now if seed could have been reliably produced (Franklin, 1970; Grant, 1971; Frankel and Galun, 1977).

Hybrid seed production is currently one of the most critical problem areas in entomophilous crop pollination. Although in some crops commercial hybrids have been available for over 30 years, there have been few successes and numerous failures. Financial losses have been commonplace among hybrid seed producers and growers. Those plant breeders, who have experienced the problem of inadequate pollination first hand, fully appreciate its impact. Pollination specialists are unable to keep pace with the development of commercial hybrids. There is a great need to bring together existing knowledge of the biologies of flowers and insects and to present an interpretation that focuses attention on problems related to pollination that occur in the development of hybrids.

In addition, recognition of the differing pollination needs of hybrid seed parents and their F_1 progeny is essential. Among the former there is always the need to effect pollen transfer between parent lines. Conversely, some F_1 progeny may be used as forage or food, e.g., alfalfa, carrots, and onions, wherein seed production is unimportant. Yet, in other hybrids like cotton and sunflower, marketable seed or related tissue is the cash crop. In these, plant breeders must ensure normal floral development in the F_1, since abnormalities may lessen the chances for insect pollination and resultant seed set.

McGregor (1976) noted that pollination terminology, often misunderstood or misapplied, may also contribute to the problems of hybrid development. Pollination is simply the physical transfer of pollen from anther to stigma. Fertilization includes all subsequent events from germination of the pollen and growth of the pollen tube to union of the gametes, and may take place wherever cross-compatibility exists. In the strictest sense, self-pollination involves transfer of pollen within a single flower (autogamy), but genetically speaking, geitonogamy, which refers to pollen transfer within the same plant but not the same flower, leads to the same thing, namely, selfing. By contrast, xenogamy, which refers to pollen transfer from one plant to another within the same genotype, race, or species, leads to outcrossing. Hybridization results from both intra- and interspecific cross-pollination (allogamy) (Percival, 1965; McGregor, 1976; Frankel and Galun, 1977). A plant that is self-fertile may or may not self-pollinate. Self-pollination and self-fertility are not synonymous.

Entomophily vs. Melittophily

Few problems are encountered in the pollination of anemophilous (wind-pollinated) hybrids. Of greater importance here are the difficulties encountered in the development of plant breeding systems and in maintenance of heterogenous populations (Frankel and Galun, 1977).

Insect pollination is often less reliable than anemophily owing to the instinctive foraging behavior of insects. Pollination syndromes among entomophilous plants are diverse and involve a variety of insect species. Nevertheless, nearly all insect-pollinated plants are pollinated by bees (Free, 1970; McGregor, 1976), the result of coadaptive melittophily (bee pollination). Insects such as flies are often used to advantage in achieving pollination within cages used in research and in breeding nurseries. However, their usefulness is limited to these circumstances. In commercial seed production, entomophily is really melittophily.

Commercial seed production relies upon the presence and dispersal of large numbers of pollinators. Honeybees (*Apis mellifera* L.) are almost always the bees of choice (Bohart and Todd, 1961; Haragsim, 1974): except under unusual circumstances (alfalfa is a notable exception) 90 to 95% of the pollinators for seed fields are honeybees. Although less efficient pollinators than other bees, honeybees predominate because they are easily propagated and can be moved from location to location as needed to ensure adequate pollinator populations. This is particularly important in areas where monoculture, complete cultivation, or other cropping practices have reduced or eliminated wild bee populations or nest habitats.

Presently, only three other bee species are used extensively in commercial pollination, and two of these only in the production of alfalfa seed. Management methods developed for the alfalfa leafcutter bee (*Megachile rotundata* Fabr.) permit propagation of large populations in nests that can be moved from field to field to accommodate the needs of seed producers. The alkali bee (*Nomia melanderi* Ckll.), a ground nesting species, can be encouraged in permanent nest sites or propagated in artificial nest sites in the vicinity of seed production fields. However, such permanent sites are advantageous only when, from year to year, the fields in seed production are within about a mile of the nest. The third species, *Osmia cornifrons* (Radoszkowski), is used extensively for apple pollination in Japan. Continuing research may demonstrate that still other species of native bees can be successfully managed for commercial seed production (Free, 1970; McGregor, 1976). Even so, most of these will be of limited usefulness

in hybrid seed fields, since they are principally mono- or oligolectic. There is little evidence demonstrating that the level of flower constancy practiced by other species of bees (having somewhat narrower ranges) is significantly different from that of honeybees, but opinions differ (Akerberg, 1974; Tasei, 1974; Parker, 1981b). Furthermore, since native bees are principally pollen collectors, we must presume that the likelihood of their readily crossing over to male sterile lines would be greatly reduced although perhaps not in every instance (Tasei, 1974; Parker, 1981a). Thus, for all practical purposes the honeybee is the only universal polylectic pollinator. Hence further consideration of pollinator foraging behavior in this chapter will be limited to that known to be characteristic of honeybees.

Mechanisms for Hybridization

The term hybrid, as used in commercial seed production, is restricted to first generation, F_1, progeny obtained by crossing two or more genotypes maintained by inbreeding or cloning (Weiss and Little, 1961). The objective is to combine desirable type characteristics to improve plant vigor, uniformity, and productivity. To be sure, other considerations are stimulating interest in hybridization, not the least of which are the economics of proprietary germplasm control because it provides protection of the plant breeders' investment.

According to the Federal Seed Act Regulations (amended in 1968), when "95 percent or more of the seed is hybrid then the term 'hybrid' may be shown on the label." If the percentage of hybrid seed is less than 95, but greater than 75, then the statement "contains 75–95 percent hybrid seed" must be shown on the label. The term "hybrid may not be used if the pure seed contains less than 75% hybrid seed" (Childers and Barnes, 1972).

Extensive use of hybrids, particularly among field crops, is dependent upon the large-scale production of seed via controlled pollination. In hybrid corn, this is achieved via hand detasseling, a procedure whereby field workers are employed to pull the tassels manually from immature plants, or via heritable male sterility. Hybridization in entomophilous species is facilitated by using the incompatibility systems common in allogamous species, hand emasculation, genetic male sterility, or cytoplasmic male sterility (Frankel and Galun, 1977; Forsberg and Smith, 1980). Other methodologies aimed at regulating sex epression in plants include manipulation of environmental factors as well as the use of chemical agents (plant growth regulators). However, these techniques generally have not been successful.

The following brief description of hybrid plant breeding systems is provided to enhance the readers' perception of the hybrid pollination problem. Frankel and Galun (1977) and Fehr and Hadley (1980) should be consulted for additional detail.

Control of Sex Expression. Hybrid cucumber production, and that of other Cucurbitaceae, differs somewhat from most other hybrid systems. Basically, spatial separation of sex organs and subsequent outcrossing is achieved in cucumbers via genetic regulation of sexual expression. Here, homozygous female (gynoecious) lines that produce many more female than male flowers are used as seed parents, and normal (monoecious) lines that usually produce more male than female flowers are used as pollinators (Frankel and Galun, 1977). Hence, crossing is assured, but the integrity of seed parents and of F_1 hybrid seed is not. This system is somewhat unique because cucurbits are attractive to bees, and pollination is rarely a problem (Akerberg, 1974).

Unfortunately, climatic and edaphic factors induce variance in flowering among gynoecious cucumber lines (Tiedjens, 1928; Whitaker, 1931; Nitsch et al., 1952; Shifris and Galun, 1956; Galun, 1959; Matsuo et al., 1969; N. M. Kauffeld, unpub. data) and hence affect hybridity. Shifris and Galun (1956) showed that the array of flowers along a vine could be influenced by environmental conditions, and Kaziev and Seidova (1965) found that female cucumber flowers had higher volumes of nectar than did male flowers. Hence, even with control of sex expression, the production of hybrid cucumbers has encountered difficulties.

Incompatibility. This phenomenon, best described by Frankel and Galun (1977) as the "inability of plants having functional gametes to set seed when either self-pollinated or crossed with some of their genetic relatives" is most common in crops that naturally cross-pollinate (allogamous species). Incompatibility systems are diverse and may involve morphological or physiological mechanisms. Self-incompatibility maintains a significant level of heterozygosity in plant populations. However, few commercial hybrids utilize incompatibility systems, and those that do have difficulty in achieving adequate pollination (Frankel and Galun, 1977; Faulkner, 1978). These problems are nearly identical to those enumerated for male sterile systems (Rubis, 1970b).

Male Sterility. Heritable male sterility in plants eliminates the need for hand emasculation. It may be under nuclear control (genetic male sterility, GMS) or cytoplasmic control (cytoplasmic male sterility, CMS). In GMS the alleles for sterility are recessive; hence, male sterility is influenced by both parents and is not complete. Since it is difficult to identify and eliminate male fertile plants from the male sterile populations, GMS is not suitable for large-scale production of hybrid seed. However, like hand emasculation, it is a useful plant breeding tool widely utilized in breeding nurseries or in small-scale seed production.

Cytoplasmic male sterility, on the other hand, is the most reliable method for large-scale production of commercial hybrids. Unidentified factors in the "maternal" cytoplasm induce complete sterility that may be restored either by recessive alleles contributing to the cytoplasm or by nuclear dominance (genecytoplasmic sterility). Once a CMS system has been established, pollination becomes the major obstacle (McGregor, 1976). Isogenic male sterile and male fertile lines are presumed to result from extensive backcrossing and may well exist for selected characteristics. However, experience has shown that other factors such as floral characteristics, not followed in the selection process, differ widely in so-called isogenic lines. In either case, pollination problems stem from the differing physiologies of seed parent genotypes and the production of pollinator foraging cues characteristic of each.

Frequently these differences are great enough to preclude random foraging and hence crossing between seed parents. This subject is discussed in detail in the next section.

Pollination Problems

Most problems encountered in the production of F_1 seed from hybrid seed parents can be classified as those problems resulting from (1) the influence of climatic or edaphic factors, (2) incompatibility or the lack of general combining ability, or (3) inadequate pollination. Care is often needed in identifying which factor or combination of factors is responsible for inadequate seed set. The first two circumstances will not be considered further, as they go beyond the intent of this chapter and have been examined extensively in other works (Frankel and Galun, 1977; Fehr and Hadley, 1980). The biological factors that contribute to the lack of adequate pollination are discussed below.

Less well-known than the advantages of heterosis are the innumerable problems encountered in assuring adequate pollination between entomophilous hybrid seed parents and occasionally in their F_1 progeny. Natural hybridization occurs infrequently among bisexual plant species. So it should be surprising that hybrids are expected in breeding nurseries, and especially in extensive seed production fields. The fact that entomophilous hybrids are obtained only with difficulty is hardly noteworthy.

Several authors have alluded to difficulties encountered in the pollination of hybrid seed parents (Franklin, 1970; Rubis, 1970b; Grant, 1971; Frankel and Galun, 1977; Faulkner, 1978; Mesquida and Renard, 1978; Forsberg and Smith, 1980; Stuber, 1980). Some have suggested that pollination problems can be overcome by adopting complex planting schemes, by saturating an area with pollinators, or by manipulating pollinator foraging activity. Unfortunately, these suggestion are misleading and far too simplistic, as they do not account for the instinctive and learned behaviors of pollinators (Grant, 1949; Franklin, 1970; Free, 1970; Gary, 1975, Levin, 1978;

Waters, 1978; Erickson and Peterson, 1979). These strategies may contribute varying levels of success in breeding nurseries or in the confinement of cages and greenhouses. Cross-pollination is most likely to occur when patch size is small and the resource is unpredictable (Levin, 1978), as in a crossing block. Fortuitous pairing of similar genotypes may also be contributory. However, there is a need to differentiate between the solutions found satisfactory for experimental plots and those that might work for the large-scale problems encountered in commercial seed production fields. Experience has shown that the two breeding situations have little in common.

Popular concepts of honeybee activities are misleading: bees are not exceptionally industrious (busy), nor are they merely incidental pollen vectors (Stanley and Linskens, 1974). When foraging they are opportunistic and may or may not pursue optimal foraging strategies. They have evolved with specific behavior modes that contribute to reproductive isolation in plants and to their own welfare. Most important is the intraspecific floral constancy that individual honeybees exercise while foraging. Gregor Mendel was an early observer of this behavior, known since the time of Aristotle (Grant, 1949; Orel, 1974). Apiculturalists recognize that discrimination between and fidelity to particular foraging cues normally limits interspecific floral contacts to less than 3%, on rare occasions interspecific contacts may exceed 10% (see also Free, 1970; Gary, 1975; McGregor, 1976). Thus, ethological isolation owing to pollinator foraging behavior is the principal deterrent to hybridization. Levin (1978) points out that "by definition, the frequency of hybridization will be an inverse function of flower specialization and constancy, since both constrain the wanderings of pollinators." A preponderance of floral characteristics contribute to foraging efficiency and maximize floral isolation (species integrity). Suggestions that tailor-made bees be bred for pollination of certain hybrids ignore the complexities of pollinator foraging behavior, the result of coadaptive evolution.

Experience has now shown that plant breeding procedures leading to the development of hybrid seed parents frequently alter flower nectar and aroma chemistry, the synchrony of floral events, and visual cues (including those in the ultraviolet portion of the bees' visual spectrum; Frisch, 1950, 1967) (Franklin, 1970; Baker and Baker, 1976; Frankel and Galun, 1977; Erickson and Peterson, 1978; Faulkner, 1978; Fick, 1978; Kevan, 1978, 1979; Erickson et al., 1979a; F. D. Parker, unpub. data). These alterations preclude nicking (Erickson and Peterson, 1978) and random foraging by bees (Rubis, 1970a; Loper et al., 1974; Waller et al., 1974) and hence significantly reduce or eliminate hybrid seed set (Franklin, 1970). Floral differences between lines presumed to be isogenic are of such magnitude that honeybees respond to them as they do interspecific differences (Percival, 1965; Erickson and Peterson, 1978, 1979; Erickson et al., 1979a). Bees are sensitive to both quantitative and qualitative differences in floral cues and rewards. They learn to associate specific levels of reward with certain cues (see von Frisch, 1967; Free, 1970; Johnson and Wenner, 1970). Avoidance learning in honeybees is also well documented (see McGregor, 1976).

Hybrid seed parents often appear as sympatric populations having acquired differing floral characteristics through selection and inbreeding rather than via geographical isolation (see Grant, 1949). Floral isolating mechanisms operate between genotypes otherwise described as isogenic. Both mechanical isolation (male sterility, morphological abnormalities, and asynchrony of floral events) and ethological isolation (floral cues and rewards) are evident between hybrid seed parents. Moreover, these types of isolation appear to reinforce one another (Grant, 1949). Nectary function and floral attractiveness to pollinators can be restored through selection and back-crossing (see section on "Hybrid Crucifers").

Frequently, but not always, male sterile flowers have less nectar than those on male fertile lines (Frankel and Galun, 1977; Mesquida and Renard, 1978; Erickson and Peterson, 1979), thus suggesting a pleiotrophic effect resulting from selection for male sterility. Foraging honeybees often exhibit a preference for male fertile lines (Akerberg, 1974; Tasei, 1974; McGregor, 1976; Erickson et al., 1979a). Since floral aromas are usually associated with the nectar production, aromaticity may be changed along with the nectar. Bees may exhibit a preference for, and will forage on male sterile lines when nectar is present. However, in most instances careful observation will show that few bees readily cross over between male sterile and male fertile lines (Erickson and Peterson, 1979). Altered nectar and aroma production in male steriles may occur because either the glands or the vascular bundles that feed them are altered, reduced, nonfunctional, or nonexistent (Mesquida and Renard, 1978; Erickson and Peterson, 1979; see also section on "Hybrid Crucifers"). A logical hypothesis is that genotypic differences in plant biochemistry are responsible for differences in the glandular products.

Nectaries may be vasculated by phloem branching from staminal or gynoecial traces (Waddle and Lersten, 1973; Erickson and Garment, 1979), or they may be independent organs (Fahn, 1974). In some plants it is thought that the nectaries and sexual organs share a common merstematic origin. This concept is supported by the concurrent diminution of nectary function that usually occurs with the loss of viable pollen on hybrid seed parents (see above). It now appears likely that this situation varies among plant species, thus adding to the need for a better understanding of the effects of genetic manipulation on the origin and function of floral organs.

Attempts to manipulate instinctive bee foraging behavior by means of colony placement or chemical attractants and repellents have been unsuccessful. Foraging instinct and learning go hand in hand; so, for example, when attractants are applied to a crop, the bee merely learns to associate the altered or new aroma with the same inadequate reward. Any resultant change in foraging activity is usually temporary. Similarly, row planting schemes are ineffective where cross-pollination is essential because bees tend to identify and work up and down rows. If rows are interplanted, experience has shown that the bees may forage selectively within the row (see sections on alfalfa and carrots). Certainly, the relative effectiveness of such attempts depends in part on the degree of similarity in foraging cues and rewards that exist between parental lines. These concepts have long been employed successfully in orchard crops such as apples and almonds, wherein varieties are self-incompatible, and high yields are attained through a program of interplanting with varieties of similar floral phenotype (Free, 1970; Blazek, 1974; Martin, 1975; McGregor, 1976; Thorp, 1978). So-called synthetic alfalfa hybrids are produced by a similar approach (see section on alfalfa).

Autogamous species, along with some allogamous species, are tolerant of inbreeding (Allard, 1960). However, a nearly imperceptible inbreeding depression has been noted among carrot floral characteristics (Erickson, unpub. obs.). Here, a genetic drift toward corolla phyllody occurs, along with a gradual loss of nectar and aroma quantity and quality. Certainly, problems such as this should be watched for in the development of new hybrid systems.

In retrospect, it must be pointed out that, until recently, there has been little, if any, understanding of the floral abnormalities induced by selection for male sterility or their consequences. In the past many crops have been intentionally or unintentionally bred for autogamy and self-fertility. The development of hybrids may require that the selective trends of many years (or centuries) be reversed. We now know that pollinator foraging cues must be included along with other characteristics in any selection index if a functional hybrid system is to be assured (Akerberg, 1974; Kubisova-Kropacova, 1974; Loper et al., 1974; Waller et al., 1974; McGregor, 1976; Erickson and Peterson, 1979; Stuber, 1980). Climatic and edaphic factors also contribute to floral response and must be considered (Rubis, 1970a). Unfortunately, here there are few meaningful data.

There are instances in which F_1 hybrids do not compete well with open-pollinated cultivars or surrounding plant species for the attention of pollinators. Heritability may significantly alter or eliminate pollinator profitability (flower nectar and pollen content) among F_1 progeny. When plants are unable to compete effectively for the attention of bees, yields are reduced. Plant breeders must watch for such problems developing among F_1 progeny.

Remedial Measures

It is obvious from the preceding discussion that plant breeding methods aimed at the production of hybrid seed often precipitate abnormalities in floral biology and thus defeat the normal pollination syndrome (Franklin, 1970; Frankel and Galun, 1977). As Rubis (1970a) so aptly pointed out, entomophilous hybrids cannot be produced successfully when one parent has no pollen and functionless nectaries. This concept can be expanded to include the change in the composition of nectar and aroma now known to occur among certain hybrid seed parent genotypes. Suggestions that bee breeders develop strains of honeybees that are tailor-made for such circumstances ignore the complexities of insect behavior.

Several avenues can be pursued in efforts to resolve the problems of pollination of hybrid seed parents as well as their F_1 progeny.

1. The floral biologies of normal autogamous and allogamous species should be understood prior to embarking on a breeding program.
2. Hybrid seed parents should be selected in the presence of honeybees if the bees are to be used in production fields. When bee pollination of F_1 progeny is required, field testing for pollinator activity should precede varietal release.
3. New breeding programs should include evaluation of and selection for floral characteristics.
4. Pairing of seed parents should be based on measured compatibility of pollinator foraging cues (i.e., similarity of nectar, aroma, and color).
5. Whenever possible, the effect of climatic and edaphic factors upon principal seed parents as well as F_1 progeny should be evaluated prior to release of a new variety.
6. Ideally, close cooperation between plant

breeders and apiculturalists should be established early and maintained throughout the developmental period of hybrids.

HYBRID CASE HISTORIES

The following sections develop a perspective for the problems encountered in attempting to successfully cross-pollinate hybrid seed parents and hence produce hybrid seed in selected hybrid systems. Experiences as well as experimental results of several principal investigators working in diverse cropping systems are presented. Problems experienced in achieving pollination of hybrid seed parents were remarkably similar in each case. As new hybrid systems are developed, additional knowledge and technology will be required to assure their pollination.

HYBRID ALFALFA

William H. Davis
Ring Around Products Inc., Plainview, Texas

Alfalfa (*Medicago sativa* L.),* "the most important forage legume grown in North America" (Childers and Barnes, 1972) is a perennial pasture and hay crop with an evolutionary adaptation for cross-pollination (McGregor, 1976). Anthesis in alfalfa has been described by Barnes et al. (1972). An average of about ten flowers are borne on each of many racemes per plant. Flower color is highly variable and ranges from purple, blue, or yellow-green, to yellow and white. Alfalfa flowers are pentamerous and zygomorphic; a large dorsal standard petal faces opposing wing petals, and a small pair of ventral keel petals encloses the sexual column. The pistil is surrounded by a column of nine fused stamens; a tenth free stamen arises from the receptacle of the flower just above the ovary base. Nectar is produced at the base of the ovary and free stamen (Teuber et al., 1980). Alfalfa produces large quantities of nectar. Nectar quality (sugar content) is acceptable to bees. Both nectar quantity and

*Alfalfa, derived from the Persian *aspasti*, which means horse fodder, is the oldest of all crops grown for forage.

quality are highly variable among alfalfa genotypes (Barnes and Furgala, 1978).

Pollination takes place after the flower is "tripped." Pollinator foraging activities cause the keel petals to separate and the restrained sexual column to spring forward. This syndrome (tripping), while not uncommon botanically, is unique among cultivated crops. The pollination ecology of alfalfa probably has been studied more thoroughly than that of any other plant species. However, owing to unique bee–plant interactions, the results of alfalfa pollinator studies may not be applicable to other crops.

At the time of seed set among alfalfa cultivars, both self-pollination and cross-pollination occur (Barnes, 1980). Self-pollination is undesirable because it reduces both seed production and viability (Weihing et al., 1943; see also Barnes et al., 1972; Childers and Barnes, 1972). Self-incompatability is not "a reliable method of pollen control for hybrid production" (Childers and Barnes, 1972). Seed yields approaching 2000 pounds per acre can be obtained when best cultivars (cvs) and good crop management is practiced, and when adequate numbers of pollinators are present (McGregor, 1976). Best yields are achieved in western states where low relative humidity and moderate to high temperatures are expected during pollination and seed maturation.

According to Hunt et al. (1976), "the discovery of cytoplasmic male sterility in alfalfa has not completely solved the dilemma of hybrid alfalfa. Pollen dispersal in alfalfa is by insect pollinators." Significant differences have been shown to exist in the preferences of pollinators for different alfalfa cultivars and clones (Hanson et al, 1964; Kauffeld et al., 1967, 1969). Heritable differences in nectar production occur among alfalfa clones (Teuber et al., 1980). Similarly, differences can exist in the amount of crossing (32–96%) between alfalfa populations because of varietal differences, planting schemes, and environmental factors (Kehr, 1973). Based on pollinator foraging cues and rewards (flower color, aroma, nectar and pollen), honeybees and perhaps other pollinators recognize intraspecific differences and discriminate during foraging between alfalfa genotypes (Clement, 1965; Kauf-feld and Sorensen, 1971, 1980; Childers and Barnes, 1972).

Given the importance of alfalfa as a forage crop, it "is not surprising that alfalfa breeders have been seeking new ways to improve the yield of alfalfa varieties" (Childers and Barnes, 1972). Cytoplasmic male sterility (CMS), a tool for hybridizing alfalfa, was discovered in alfalfa in 1963 (Davis and Greenblatt, 1967). It was later confirmed by Pederson and Stucker, (1969), and Bradner and Childers (1968). A second CMS system was described in Europe in 1978 (Barnes, 1980).

Following discovery of CMS, a team was assembled by the L. Teweles Seed Co. of Milwaukee, Wisconsin, to develop and produce hybrid alfalfa. The seed production phase was undertaken in California, while the forage testing program was conducted in Wisconsin and other midwestern states (Fig. 25-1).

In 1963, we clonally propagated ten partially male sterile plants (T-2). These were sent to California where, in 1964, one acre was planted to this line in the San Joaquin Valley. Unfortunately, seed production was poor because this male sterile clone proved to be highly unattractive to foraging bees, primarily honeybees.

In 1965 we began selecting several hundred new sterile clones. Cuttings were transplanted in a large commercial seed field in groups of five to ten cuttings per clone. Seed was harvested separately from each clonal group. It soon became apparent that honeybees and leafcutter bees did not prefer all clones to the same degree. Seed yields varied from one gram to one hundred grams per clone. In the fall of 1965, sufficient cuttings were made from the five clones most preferred by bees to plant approximately one acre with these selected A lines and one-half acre with a known male fertile (B line) called D-2. In 1966 we could see that this field was not going to be successful. Bees were attracted to the new A lines but would not visit the B line; seed set was ex-

Figure 25-1. Hybrid alfalfa forage trials—Clinton, Wisconsin, 1973.

tremely poor. Fortunately, at this time, there was available a large number of S2 B lines for top crossing in a commercial seed field of alfalfa. Several of these B lines set high amounts of seed (150+ grams per plant), while others set lesser amounts, as low as one gram per plant.

In 1967, A and B line cuttings were planted to approximately five acres. Two rows of A lines were used for every row of B line. The A lines were progeny derived from T1 MS; the B line was an S2 plant from the variety Buffalo, called D-31. In 1967 sufficient A line seed from the T1 MS × D-31 was produced to plant more than 20 acres of commercial hybrid seed.

Also in 1967, in California, we ran a replicated row pattern of A and B lines using the highly preferred A and B clones. Patterns of 2 to 2, 2 to 4, 2 to 6, and 2 to 8 rows were planted in 40-inch rows. The data from this replicated study indicated that bees do follow patterns. On 2 rows of A with 2 rows of B, we noted no yield differences in A lines vs. B lines. The same results were obtained with 4 rows of A with 2 rows of B; however, in the 6 rows of A with 2 rows of B, the inner A rows showed a reduction in yield. On the 8 rows of A with 2 rows B, there was a major reduction on the inner rows of the A line. The selection of the A and B lines for seed set and insect visitation had proved successful, since A line seed yields from rows near the B lines were equal to the open-pollinated B lines.

A review of the amount of A line seed available and the planting pattern data indicated that a ratio of 2A lines to 1B line was best for the first commercial attempt to produce hybrid alfalfa from direct seeding. However, commercial combines do not harvest less than six 40-inch rows. Thus, we tried planting 12 rows of A line and 6 rows of B line on both sides of this 12-row A block. A total of 28 acres was planted in early spring of 1968 using 2 pounds of each type of seed per acre. Both lines germinated very well, and stands were uniform. At flowering, observations of honeybee foraging activity were made (no leafcutter bees were used on this field). Seed set definitely diminished toward the inner 4 rows of the A line

with the outer 4 rows setting seed equal to the B line. Approximately 300 pounds of seed per acre on the A line and 400 pounds per acre on the fertile B line were harvested. The first commercial three-way hybrid alfalfa had been produced using a cytoplasmic male sterile system. Hybrid alfalfa appeared to be a success.

"The first commercial alfalfa hybrids utilizing cytplasmic-male sterility were marketed in the Midwest by the L. Teweles Seed Co., Milwaukee, Wis., in 1968. On March 16, 1971, a U.S. Patent, no. 3,570,181 was issued to the L. Teweles Seed Co. . . ." (Childers and Barnes, 1972). Unfortunately, in the two years following 1967, seed production on the A line fell steadily, with 1970 production going under 200 pounds per acre. We could not determine whether or not the honeybees were visiting the A and B lines indiscriminately. In order to see if alfalfa leafcutter bees could do a better job, we moved our hybrid seed production into Idaho where these bees were abundant.

Because of our concern that the wide block of A line plants was too far from the source of the pollen, we established a study to blend the B line at rates of 5–10–15% into the A line. In this way rows were eliminated, and bees visiting flowers in the A line were thought more likely to pick up pollen from nearby B line pollen plants in the mixture.

Discriminative pollinator foraging between clones greatly affected our ability to produce hybrid alfalfa seed. Some effort was made to understand why certain A and B lines were highly attractive to bees. Only two factors were apparent. One was aroma. Highly attractive A and B lines were very highly aromatic when nectar flow was induced following irrigation. Loper et al. (1974) and Waller et al. (1974) also noted significant differences in the aromas of certain alfalfa clones and in the response of foraging bees to these aromas. The second factor was easy tripping. This proved to be an important factor when honeybees were used to pollinate male sterile clones. Moreover, Pederson and Barnes (1973) and Staszewski (1979) have reported significant differences in pollen dehiscence among alfalfa genotypes. No nectar quality studies were undertaken on alfalfa hybrid seed parents.

Since alfalfa hybrids can be vegetatively propagated, "no concern needs to be given to the fertility of hybrid plants" (Childers and Barnes, 1972).

The Current Status of Hybrid Alfalfa Seed Production Research

S. T. (Steve) Yen and Paul L. F. Sun
Dairyland Research International, Clinton, Wisconsin

Hybrid alfalfa seed production is dependent upon bee visitation. Bee preference for certain alfalfa genotypes has been well documented. Hence, each year alfalfa scientists at Dairyland Research International, a division of Dairyland Seed Company, Inc., screen thousands of plants and clones at their western research station in Sloughhouse, California. Individual plants with 100 to 300 grams of seed are not uncommon under spaced planting (40 inches between rows and 18 inches between plants) for each A line (cytoplasmic male sterile), B line (maintainer, fertile), and C line (pollinizer, fertile).

Seed setting ability of experimental hybrid lines is evaluated at each step of the breeding program, such as at $A \times B$, $(A \times B_1) \times B_2$, $(A \times B_1 \times B_2) \times C$, etc. Only the best seed yielders are saved for forage yield tests and PPI (pollen production index)* readings. Seed production among hybrid seed lines was found to be highly correlated with that of the pollinizer. For example, correlation coefficients of 0.95 and 0.86 were observed between two $A_i \times B_j \times B_k$ lines and ten cultivars including "Iroquois" and "Saranac AR" under row planting patterns at Sloughhouse station in 1980. Both lines had PPIs of about 10.

The PPI of our breeding lines is routinely checked. The stability of the PPI at different years and locations as well as its relation to seed production is also routinely investigated.

$$*PPI = \frac{0(ms) + 0.1(pms) + 0.6(pf) + 1(f)}{\text{Total no. of plants}} \times 100$$

where *ms, pms, pf,* and *f* are the number of male sterile, partially male sterile, partially male fertile, and male fertile plants respectively.

Pollen production indices in a majority of our breeding lines are stable, excepting some variance from one hybrid seed category to the other. However, no conclusive trend between PPI and seed production has been established.

Several ways of overcoming the limitations of producing hybrid alfalfa seed are under study. One method is the blending of seed of $(A \times B)$ and C lines prior to planting; over 1200 pounds of commercial hybrid seed per acre has been obtained by this technique. Here hybridity exceeds 75%. This is but one of the many techniques for producing modified hybrid alfalfa seed proposed by Sun in his 1977 United States Patent (see Staszewski, 1979). Another means of reducing the cost of hybrid seed production while maintaining the benefit of hybrid seed is the certification of seed from female rows as hybrid and that from male rows as a different cultivar when they meet the requirements and regulations of the Alfalfa Review Board. Here, there is no seed production problem at the experimental level. Studies for commercial seed production are under way. Preliminary results are promising.

Acknowledgments

The author (Davis) wishes to acknowledge Ralph Arthur, Tom Pilling, Wells Oppel, Bob Teweles, and Rudy Zungia, who cooperated in the development of the first commercial hybrid alfalfa.

HYBRID CARROTS

Eric H. Erickson, Jr. and Clinton E. Peterson
North Central States Bee Research Laboratory and Vegetable Crop Production Research, USDA-ARS, University of Wisconsin, Madison, Wisconsin

Open-pollinated carrot (*Daucus carota* L. var. *sativa* DC.) cultivars are highly attractive to a variety of insects; hence, pollination is easily achieved in commercial seed fields. Although they are self-fertile, autogamy is rare because carrots are protandrous; normally, pollen dehiscence on an umbel is complete before the first stigma becomes receptive. Flowering of the entire umbel (particularly the primary umbel) may take two weeks or more, pollen dehiscence requires 6–7 days, and the period of stigma receptivity may last a week or longer. Several orders of umbels are produced, and these open in waves at 9- to 13-day intervals. The first three orders of umbels produce 90% of the seed harvested (Hawthorn et al., 1956, 1960; Franklin, 1958, 1970; Bohart and Nye, 1960).

Pollen produced on epigynous carrot florets is exposed, abundant, and readily gathered by pollinators. Open-pollinated carrot cvs usually produce ample quantities of nectar from a pore-covered spongy disc above the ovary (Fig. 25-2). Because the petals are small and flared at anthesis, the nectar produced is exposed and, like carrot pollen, relfects both visible and uv light (Thorp et al., 1975; Erickson et al., 1979a). The nectariferous area appears to cover the entire disc surface, including the underside, where, in the vicinity of the anther bases, the pores (nectar stomata) are much larger and different in appearance from those above (Erickson et al., unpub. ms.). This leads to speculation that different functions may be associated with these two types of stomata. Carrot umbels are very fragrant although the aroma complex differs among cultivars (Erickson and Peterson, 1979).

Development of commercial hybrid carrot seed production followed the discovery in 1946 of a CMS plant at Charleston, South Carolina (Welch and Grimball, 1947). Discovery of

Figure 25-2. Typical male fertile carrot floret. A, × 31; B, nectary, × 50; C, longitudinal section, nectary (upper arrow), filament base (lower arrow), × 50.

other CMS steriles followed. Although numerous carrot hybrids have since been released, hybrid seed production has been beset by disasters. Presently, hybrid carrots constitute only 7% of the acreage devoted to carrot seed production (Frankel and Galun, 1977). This fact reflects the inability to produce hybrid seed reliably.

As in most commercial seed crops, honeybees are the principal pollinators (Bohart and Nye, 1960) and should average eight bees per square yard to assure adequate pollination (Hawthorn et al., 1956; McGregor, 1976). However, the ineffectiveness of honeybees in hybrid carrot seed fields has been notable and is due to the inherent floral diversity between seed parents. It is commonplace for carrot seed producers to attempt to locate their fields adjacent to uncultivated areas where other, "less discriminating," insect pollinators exist in abundance. While limited success has been reported from this approach to carrot pollination, such areas are difficult to find, and reliance upon them as a source of pollinators will limit development of the hybrid carrot seed industry.

In male sterile carrots the anthers are either distorted and brown (domestic cytoplasm) or absent (wild cytoplasm) with petaloid structures in their place (Fig. 25-3). Most commercial F_1 hybrids are from the petaloid type, since the brown anther steriles are less stable. This lack of stability may account for their greater uniformity in nectar and aroma production (E. H. Erickson, unpub. obs.). Flower color among seed parents varies from white in normal male fertiles to off-white and phylloidial green in petaloid male steriles. Plant height between seed parents is often variable as well (Erickson and Peterson, 1979; Erickson et al., 1979a).

Wide differences in the synchrony of flowering events influence pollinator foraging among hybrid carrot seed parents, especially among the male sterile lines (Fig. 25-4). Some of these extremes undoubtedly result from cytoplasmically induced morphological abnormalities characteristic of flowers from many but not all CMS lines. Nectar production is significantly reduced in many petaloid male

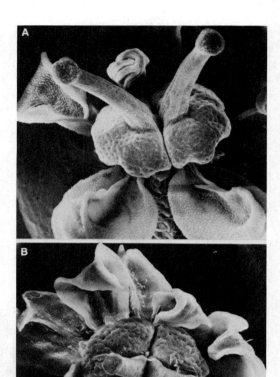

Figure 25-3. Comparison of a white-corolla, brown anther floret with a white-corolla, petaloid floret. A, brown anther male sterile, × 50; B, petaloid male sterile, × 50.

steriles (see also section on "Hybrid Crucifers"); sugar content of the nectar also appears to be somewhat reduced (fertiles, $\bar{x} = 30\%$; white corolla steriles, $\bar{x} = 11\%$; and green corolla steriles, $\bar{x} = 20\%$ dissolved solids). Other nectar components that might influence bee foraging have not been evaluated but are also presumed to differ (Erickson et al., 1979a,b). Occasionally male steriles are nectarless, a condition associated with and perhaps the result of petal carpeloidy, which is evident in some inbred lines (Fig. 25-5). Aroma production in male sterile carrot lines differs significantly both in quantity and quality from that in male fertiles (Erickson and Peterson, 1979).

Asynchrony of pollen dehiscence and stigma

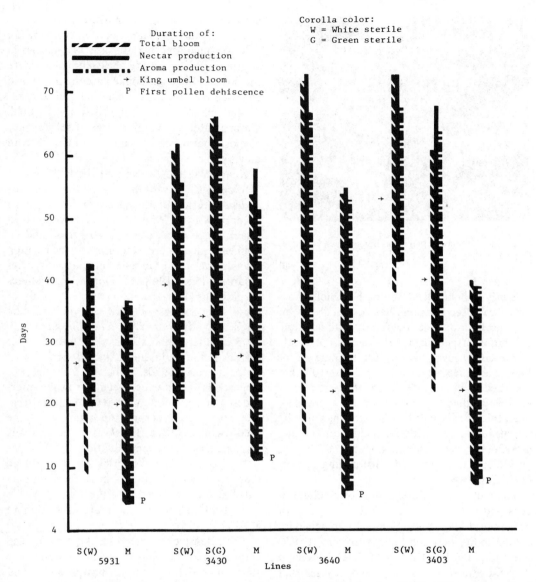

Figure 25-4. Anthesis of some male sterile (S) and fertile (M) hybrid seed parent lines (1974–1975 greenhouse data).

receptivity, beyond that of normal protandry, is evident among many seed parents. Some male sterile lines may bloom as many as 30 days later than male fertiles. Even within genotypes segregating for corolla color, green and white flowered plants may not bloom simultaneously. The level of variability for any characteristic is far greater in male steriles than in male fertiles. Further, it appears that the loss of pollen precipitates a loss of aroma and nectar production during the period when pollen usually dehisces. Nectar and aroma production may be delayed 5 to 10 days. This hiatus becomes extended in some lines (Fig. 25-4); in other lines these floral cues are nonexistent. The general tendency is for male sterile lines to secrete nectar and release volatiles much later than male fertile lines (Erickson et al., 1979a; Erickson and Peterson, 1979). As much as a sixfold preference for male fertile lines can exist when measured by the percent of foraging bees.

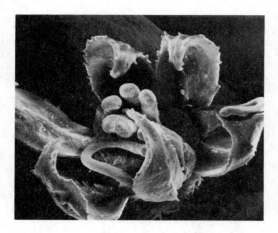

Figure 25-5. Abnormal green-corolla petaloid (carpeloid) carrot floret, × 50.

Study of honeybee foraging behavior on hybrid carrot seed parents has revealed that, as in other hybrids, bees readily discriminate and exhibit intraspecific fidelity (Erickson, et al., 1979a). Crossover from fertile to male sterile lines is least frequent—usually 3.3 to 10%, but recorded as high as 23.6% for the best combinations studied. At lower levels of crossing-over, these deviations from a pattern of fidelity might well be considered incidental. Higher crossing percentages (e.g., above 20%) may be an indication of similarity in foraging cues or rewards between seed parents.

The numerous differences in floral pollinator cues and rewards are responsible, in part, for poor hybrid seed yields. Some physiological incompatibility between lines is evident (Erickson et al., 1979a). Rubis's (1970a) statement that little seed can be expected in lines that have no pollen or nectar is particularly meaningful in carrots. Additional breeding and selection must be done if hybrid carrots are to be widely utilized in the future.

HYBRID COTTON

Joseph O. Moffett
Forage Crop Production Research, USDA-ARS, University of Oklahoma, Stillwater, Oklahoma

Hybrid cotton has been known since the 1890s when Mell (1894), and later Cook (1909), reported hybrid vigor from interspecific crosses between long staple cotton, *Gossypium barbadense* L., and short staple cotton, *G. hirsutum* L. Cook suggested planting the two species in alternate rows and using bees to cross-pollinate the flowers. By using unlike parents, the hybrid could be identified after planting and self-pollinated plants easily removed from the field. Unfortunately, the high degree of self-fertility of most commercial cottons made Cook's plan unworkable.

Kearney (1924) wrote that the F_1 hybrid of "Holden" *(G. hirsutum)* and "Pima" *(G. barbadense)* cotton produced larger bolls and an exceptionally fine-fibered cotton in Arizona. In Israel, crosses between *G. barbadense* and *G. hirsutum* gave more heterosis for seed cotton than intraspecific crosses of either species (Marani, 1967, 1968a,b). During early stages of plant growth, the interspecific hybrids of these two species outgrew the intraspecific crosses by a wide margin (Marani and Avieli, 1973). Other authors have reported large increases in lint yield due to hybrid vigor.

In the early 1970s, India, with the world's largest cotton acreage, began growing interspecific hybrid cotton on a large scale (ca. 2,000,000 acres) despite the high cost of the seed (Srinivasan, et al., 1972). This seed was produced by hand emasculation and pollination. Some of these hybrids have produced more than twice as much lint as the high parent and have outproduced the best available nonhybrids by significant margins (Singh et al., 1972).

The extensive line breeding of cotton by seed companies in the United States, as well as by state and federal scientists, has produced and is continuing to produce excellent commercial cotton cultivars. Hence, it may be difficult to obtain the large increases in yield that have occurred with hybrids in India. Davis (1978) notes that several authors have found the percentage of heterotic increase to be greater at low yield levels than at maximum yields. In six of seven studies summarized by Davis, interspecific hybrids between *G. hirsutum* and *G. barbadense* showed a yield increase of 20% or more over the best-adapted commercial varieties compared to similar results in only three of nine tests involving intraspecific crosses with *G. hirsutum*. The range of increase for the intraspecific crosses was from 10 to 138%, compared to 7 to 50% for the interspecific crosses. Overall, interspecific crosses between long and short staple cotton have given greater heterosis and yield advantage than intraspecific crosses within *G. hirsutum*.

In an extensive review of the vigor of hybrid cotton, Loden and Richmond (1951) concluded that male sterile lines offered the best possibilities for the development of hybrid cottons. They further concluded that the two most promising methods of producing these steriles were by either genetic or cytoplasmic male sterility.

Figure 25-6. A male sterile cotton flower (left) and a male fertile flower (right) with petals, calyx, and bract removed.

Genetic Male Sterility

Justus and Leinweber (1960) described a recessive gene (*ms* 1) in short staple cotton that causes partial or complete male sterility when it is homozygous. Richmond and Kohel (1961) found a recessive gene (*ms* 2) that gave complete male sterility. A third recessive gene (*ms* 3) that causes partial to complete male sterility in cotton was reported by Justus et al. (1963). A dominant gene (*ms* 4) that gives complete male sterility was described by Allison and Fisher (1964).

Weaver (1968) discovered two pairs of recessive genes (*ms* 5 and *ms* 6) that gave complete male sterility. These two genes were stable at Coimbatore, India under all seasonal conditions (Srinivasan et al., 1972). Crosses involving 27 varieties of *G. hirsutum* were made with the male sterile Gregg line carrying *ms* 5 and *ms* 6. Several of these 27 crosses combined favorably with Gregg. Both the yields and fiber quality of these Indian crosses were good. A second dominant gene for male sterility (*ms* 7) was reported by Weaver and Ashley (1971). However, Davis (1978) notes that all serious work on the genetic steriles has been dropped since the development of a good cytoplasmic male sterile.

Cytoplasmic Male Sterility

Meyer and Meyer (1961) first reported cytoplasmic male sterility in cotton from a breeding program started in 1948. After several false starts Meyer (1973) reported development of a stable cytoplasmic male sterile in cotton. The sterility was obtained by combining the chromosomes of the "Rowden" cultivar of a tetraploid, *G. hirsutum*, with the cytoplasm of a diploid, *G. harknessii* Brandegee.

At the present time almost all of the hybrid cotton breeding programs are using the male sterile cytoplasm developed from *G. harknessii* by Meyer (Fig. 25-6). Some difficulty has been encountered in obtaining a reliable fertility restorer line and in consistently achieving adequate pollination.

Chemical Sterility

After Eaton (1957) reported that FW-450 (sodium A,B-dichloro-isobutyrate) produces male sterility, an unsuccessful attempt to produce hybrid cotton seed commercially using FW-450 was tried. Repeated sprayings were needed to maintain sterility. Inconsistent results, phytotoxicity, and other problems developed. The program was eventually discontinued.

Insect Pollination Needed

Meyer (1969) wrote that one of the main obstacles to the practical production of hybrid cottonseed in the United States is an adequate and economical method of pollinating the male steriles.

Cotton flowers are large (2–4 inches long), pentamerous, self-fertile, and partially self-pollinating. One- to two-inch-long styles terminate in an elongate stigma and are surrounded by a full-length staminal column (Fig. 25-6). Both floral and extrafloral nectaries are present. However, the flowers of *G. hirsutum* and *G. barbadense* differ substantially in appearance and in their presentation of pollinator foraging cues. Those of *G. hirsutum* are cream-colored, secrete a low volume of nectar, and have cream-colored pollen (20% of existing cultivars have yellow pollen), while *G. barbadense* flowers are yellow with a maroon nectar guide, produce more nectar (of lesser sugar concentration than *G. hirsutum*), and have orange pollen (McGregor, 1976; B. W. Hanny, pers. comm.). I am unaware of reported differences in floral aroma between cultivars or hybrid seed parents.

Long staple cotton *(G. barbadense)* is characterized by high levels of gossypol, an insect toxicant and repellent. Breeding for insect resistance in *G. hirsutum* has increased the gossypol content of certain cultivars, some of which are identifiable by the yellow pollen inherited from *G. barbadense*. Increased levels of gossypol in the pollen of long staple and some short staple cultivars may be responsible for the lack of pollinator interest in the pollen. Chemical analysis of developing anthers of *G. hirsutum* and *G. barbadense* has shown not only that the two species differ quantitatively and qualitatively in terpenoid aldehydes (gossypol and related analogs), but that they also differ qualitatively in carotenoid and flavonoid constituents (B. W. Hanny, pers. comm.). It is therefore probable that mature pollens of these two species differ in their attractant and nutritive properties and thereby affect honeybee foraging activity. Moreover, genotypes lacking pollen may well have a lower gossypol (repellent) fraction in the flower.

Cotton pollen is heavy and sticky. It must be transferred from pollen-producing flowers to male sterile flowers by pollinating agents other than wind currents if male sterile flowers are to produce seed. The pollinating agent is normally an insect. A *Collops* beetle, *Collops vittatus* (Say), is common in cotton flowers in Arizona. These beetles have a smooth cuticle, so large quantities of pollen do not adhere to their bodies. Even so, when present in large numbers, *Collops* may contribute to the pollination of male sterile flowers. Several other nonhymenopterous insects are also found in cotton flowers (Moffett et al., 1976a). However, only bees and wasps normally have noticeable amounts of cotton pollen clinging to their bodies. These hymenopterans are probably the only effective pollinators of male sterile cotton. Occasionally bees and wasps visit cotton flowers in sufficient numbers to be valuable pollinators, but at other times they are scarce (Ware, 1927; McGregor et al., 1955; McGregor, 1959; Butler et al., 1960; Moffett et al., 1976b).

Moffett et al. (1976b) reported a wide variation between years in the visits of wild bees and wasps in the same cotton fields and in the same month. The number of bee visits required to set seed on a male sterile flower is not known, but a minimum of 50 pollen grains per stigma is required to fertilize all of the ovules (McGregor, 1976). G. D. Waller and associates (unpub. data) found that eight separate visits by honeybees to male sterile flowers resulted in a 69% boll set, compared to 38% when four visits were made and 13% for two visits. It is generally believed that in the field a 1% visitation (one bee per 100 flowers) is sufficient. However, the number of bee visits per flower necessary to set a hybrid seed crop varies according to:

1. The planting pattern of the male sterile and fertile plants.
2. The frequency with which the bees cross between the male sterile flowers and the flowers containing pollen.
3. The ratio of pollen flowers blooming compared to the male sterile flowers.
4. The speed with which the bees move from flower to flower.
5. The species of bees involved.
6. The length of time the pollen is viable.

Honeybees

The visits of honeybees to cotton flowers (Fig. 25-7A,B) vary greatly according to season, day, location, and year (Moffett et al., 1975a). Honeybees appear to exhibit some preference for the extra floral nectaries that function before the flower opens and continue functioning after flowering. However, this may be only the result of habituation. G. D. Waller and associates (unpub. data) noted that honeybees preferred and maintained a fidelity to male sterile flowers in their 1980 study.

Honeybees often do not collect cotton pollen even when suffering from a pollen deficiency and surrounded by thousands of acres of cotton. In Arizona, a bee emerging from a cotton flower *(G. barbadense)* coated with pollen alights elsewhere, and brushes off much of the adhering pollen (McGregor, 1976). G. D. Waller and associates (unpub. data) found only two cotton pollen masses in 20,000 examined in their 1980 studies on *G. hirsutum.* Unconfirmed reports indicate that this behavior is not the same for *G. hirsutum* in other areas. However, at times honeybees will collect large amounts of cotton pollen (Waller et al.,

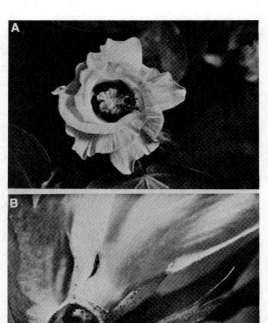

Figure 25-7A. A honeybee emerging from a flower of *G. barbadense.*
Figure 25-7B. A honeybee collecting nectar from a cotton flower. The bee's proboscis is protruding between two petals near their base.

1981). Foraging differences may be the results of environmental factors or genotype. Lack of preference for some cotton pollen may be the result of the presence of high levels of gossypol (see previous section). Simpson (1954) noted that among other things pollinators are sensitive to pollen type, and hence intervarietal crossing is affected "by the abundance of unlike pollen." Nevertheless, pollen usually adheres to the body hairs of honeybees, so that they are effective pollinators when they visit male sterile flowers.

In Arizona honeybees sometimes fail to visit cotton flowers even when large numbers of colonies are present (Fig. 25-8). However, adequate floral visitation has always occurred when an area is saturated with honeybees. When 50,000 colonies were moved into the Mettler area of California to pollinate 20,000 acres of alfalfa, McGregor (1976) found 20 or more honeybees per 100 cotton flowers within one-fourth mile of the alfalfa fields. Moffett et al. (1978a) found that when 900 colonies were moved in, adequate honeybee visits were made to cotton flowers on more than 1000 acres of cotton.

In 1977, 500 colonies were moved into a hybrid seed field in the Aquila area of Arizona. The bees set bolls well on 89 acres of male sterile flowers until applications of insecticides killed most of the field force of the colonies (Moffett et al, 1978b). After this, bee visits and boll set dropped more than 90%. Before the spray application, bee visits averaged 3.5%.

G. D. Waller and associates (unpub. data) in 1980 obtained yields on male sterile lines equal to those of male fertiles in the South Plains of Texas when five or more colonies per acre were moved into the area. These colonies had been badly weakened by insecticides before being moved. Many of the colonies had less than four frames of bees and brood. Bee visits to cotton flowers averaged about 1% in this Texas study.

Wild Bees and Wasps

Bumblebees (*Bombus* spp.) are very effective pollinators of cotton flowers in many areas in

Figure 25-8. Honeybee colony placement near a hybrid cottonseed field.

the eastern United States. By contrast, in the western half of the cotton belt one may spend an entire day in a large cotton field without seeing a single bumblebee (McGregor, 1976). In Georgia, *Melissodes* and honey bees were abundant in some cotton fields, while bumblebees and a large wasp, *Campsomeris plumipes* Drury (Swliidae), were present in smaller numbers (Allard, 1911a,b). Weaver (1978) reported large numbers of bumblebees visiting flowers in his Georgia experimental cotton plots. In Pakistan, *Anthophora* and *Apis* and a wasp *Myzinum* (= *Elis*), were the most important cotton flower visitors (Khan and Afzal, 1950). Balls (1929) wrote that *Nomia, Anthophora, Halictus, Xylocopoda,* and *Campsomeris* were important cross-pollinators of cotton. Moffett et al. (1976a) found more than 70 species of insects visiting cotton flowers in Arizona. Butler et al. (1960) reported that honeybees and three species of *Melissodes* were the only abundant, active, and efficient pollinators of cotton in Arizona. They found that bumblebees and other wild bees were not abundant.

Because of the large acreages of cotton, the numerous seed companies in the area, and the absence of large-scale applications of insecticides to cotton (hence the presence of extensive populations of native pollinators), much of the breeding work on developing hybrid cotton is occurring in the South Plains of Texas. Here, in 1974, Moffett et al. (1977) found that bee visits (mostly wild bees), averaged 1.59% to male sterile flowers. The most common wild bees were three species of *Melissodes,* two species of *Svastra,* one species of *Agapostemon,* and *Halictus ligatus* Say. Bumblebees were common off the caprock, but were scarce on the South Plains. Bee visits varied greatly between counties, but were adequate for good seed set on the male sterile flowers except in areas of low bee visitation. Even so, in 1980 several breeders who relied on wild bees to pollinate their male sterile cotton flowers had very low seed yields. Wild bee populations were abnormally low in most areas of the South Plains. These low wild bee populations were apparently caused by either the unusually hot dry weather or the increased use of pesticides, or a combination of both.

A more extensive study of the bees visiting cotton in this area was made in 1979 (Moffett et al., 1980). Thirty-five species of bees were

Table 25-1. The ten most numerous species of bees found in cotton flowers in the High Plains of Texas, 1979 (Moffett et al., 1980).

Species[a]	Number collected	Percent of total bees collected
1. *Agapostemon angelicus* Ckll.	65	24.6
2. *Apis mellifera* L.[b]	39	14.8
3. *Halictus ligatus* Say	16	6.1
4. *Melissodes thelypodii* Ckll.	15	5.7
5. *Bombus fraternus* (F. Sm.)	13	4.9
6. *Svastra atripes* (Cress.)	13	4.9
7. *Evylaeus* (Dialictus) sp. #3	13	4.9
8. *Triepeolus*[c] *helianthi* Root.	11	4.2
9. *Bombus americanorum* (F.)	8	3.0
10. *Nomada*[c] *texana* Cress	8	3.0

[a]Nine scoliid wasps, *Campsomeris pilipes* Sauss., were also collected in the cotton flowers, and eight had grains of cotton pollen on their bodies.
[b]Bees are from feral populations plus preexisting apiaries. No colonies were moved into the cotton.
[c]*Triepeolus* and *Nomada* are parasitic on other bees in their larval stage.

collected visiting cotton flowers. Twenty-five percent of the visits were made by *Agapostemon angelicus* Cockerell, while honeybees made 15% of the total visits to the inside of the cotton flower (Table 25-1). Frequently, honeybees visited the extrafloral nectaries to the exclusion of going inside the flowers. Bumblebees made 8% of the total visits in 2 counties off the caprock, but were rare in the other 11 counties studied.

Large numbers of the alfalfa leafcutter bee, *Megachile rotundata* F., have been brought into several cotton fields in bee boards. Yet only four of these bees were seen in cotton flowers. This species appears to have little promise as a pollinator of field cotton grown for hybrid seed.

Planted Row Ratio

Knowledge of the most efficient planting row ratios of pollen plants (B lines) to male sterile plants (A lines) is important for the economical production of hybrid cotton seed. In Arizona in 1974, a planting pattern of 2 fertile rows alternated with 6 male sterile rows gave good seed yields. The bee visits averaged 2%. The third male sterile row from the pollen plants produced 95% as much seed per flower

as the male sterile row adjacent to the pollen rows (Moffett et al., 1976d).

In 1977 a large field in west central Arizona was planted to the following pattern: 2 pollen rows, a skip row, 4 male sterile rows, then another skip row. Honeybees only crossed between the A and B lines on every eighth floral visit, or half as often as in solid row plantings. This was apparently because of the skip row between the A and B lines. Also, the inside two rows in the A line plantings produced much less in comparison to the outside rows than occurred in similar planting patterns of open pollinated cotton. The bee visits averaged 3.5% prior to application of insecticides (Moffett et al., 1978b).

Bees Prefer Some Genotypes

Genotypes of cotton vary widely in their floral attractiveness to honeybees. In a study of almost 100 genotypes in Arizona, Moffett et al. (1975b) found large and consistent differences in their floral attractiveness to honeybees. Flowers of genotypes containing cytoplasm from species other than *G. hirsutum* and *G. barbadense* were often more attractive to honeybees than flowers from lines developed from these two species.

Moffett et al. (1976c) reported a correlation between sugar concentration of their nectar and floral attractiveness of genotypes to honeybees. It was concluded that increasing the sugar concentration of selected cotton genotypes would increase their attractiveness to honeybees.

Nectarless cottons do not have extrafloral nectaries, but the nectaries at the base of the ovary in the flower still exist. By increasing the attractiveness of pollen in a nectarless cultivar, two things could be accomplished (B. W. Hanny, pers. comm.):

1. Nectar and pollen would be available to pollinators, but would not be in competition with extrafloral nectaries.
2. The elimination of extrafloral nectaries has been proved (Meridith, 1976) to be very effective in host plant resistance (HPR), particularly for *Lygus* and *Heliothis* control, but does not have a significant effect on beneficial insects. This HPR character could reduce the use of insecticide, thereby benefiting pollinators.

Problems Facing Hybrid Cotton Seed Production

Some problems remain in the development of hybrid cotton. One is the discovery of an R line that will restore 100% fertility to the F_1. Plant breeders have encountered inconsistent fertility restoration in cotton: fertility restoration in one F_1 was almost 100% for several years. Then, fertility dropped sharply during the hot summer of 1980. Some other breeders have had consistent restoration of over 90%, but have had difficulty reaching 100%. Still other cotton breeders believe they have solved or are solving the restoration problem (Wm. H. Davis, pers. comm.). Adequate restoration is much easier in interspecific crosses between short and long staple cotton than between intraspecific crosses with *G. hirsutum* (Davis, 1978). Evidently *G. barbadense* carries a gene for the enhancement of fertility.

A consistent and economical method of pollinating male sterile flowers needs to be developed. At times wild bees that visit cotton are abundant and provide excellent pollination. At other times they are scarce. Wild bee populations are greatly reduced by cultivation and insecticides. Often, almost none can be found in intensively sprayed areas.

Honeybees are erratic visitors to cotton flowers. They may often visit the cotton flowers in large numbers and pollinate the male sterile flowers adequately. Then, they suddenly start visiting other more attractive flowers that have started to bloom and cease visiting the cotton flowers. At other times honeybees visit only the extrafloral nectaries.

I obtained excellent visitation and good set of male sterile flowers with honeybees some years when only a few colonies were in the field vicinity. At other times honeybees have visited the flowers poorly, e.g. when 150 or more colonies were adjacent to a field (15 colonies/acre). This poor visitation resulted in very poor seed set in the male steriles. In this instance there were cotton fields and other plants competing for the attention of the bees.

We have always obtained honeybee visitation and pollination of male sterile flowers by saturating a wide area with 500 or more colonies of honeybees (one colony/two acres). However, cost factors make this impractical in commercial fields unless large acreages of hybrid seed are being produced in a small area. There is now no completely reliable and economical method of pollinating smaller fields being used to grow hybrid cottonseed.

Methods for controlling harmful insects may be the greatest obstacle to pollination of hybrid cotton.

HYBRID CRUCIFERS

Paul H. Williams and Hei Leung
Department of Plant Pathology, University of
    Wisconsin, Madison, Wisconsin

Cruciferous crops are represented primarily by species of *Brassica* and *Raphanus*. The brassicas consist of six interrelated species: the diploids—*B. nigra* (L.) Koch (b genome, $n = 8$) (black mustard), *B. oleracea* L. (c genome, $n = 9$) (cole crops), and *B. campestris* L. (a genome, $n = 10$) (turnip, oilseed rape, and oriental greens); and the amphidiploids—*B. carinata* A. Braun (bc genome, $n = 17$), *B. juncea* (L.) Czerniak (ab genome $n, = 18$), and *B. napus* L. (ac genome, $n = 19$). Radish is *R. sativus* L. (r genome, $n = 9$) (Tsunoda et al., 1980). Within these crucifers there exists a range of morphological diversity and utilization from edible seed and oils to leafy and swollen rooted forms used widely as vegetables and animal fodder. Morphotypes used for their edible leaves, buds or flowers, swollen stems, bulbous roots, and seed oils are found in *R. sativus, B. campestris, B. oleracea,* and *B. napus.*

Distinguishing taxonomic features of cruciferous crops are the characteristic floral structure of four sepals, four equal cruciform petals, six stamens (four long and two short), and an ovary with two parietal placentae. *Brassica* and radish flowers are produced in terminal branching racemes. Plants are annual or biennial, frequently being grown as winter annuals for seed production. Various forms and cultivars require from 2 to 16 weeks of temperatures from 4 to 12°C for floral initiation. Those requiring vernalization are normally grown in areas with cool mild winters and dry summers. Plants from late-summer- or fall-sown seed grow through the winter in the rosette stage before elongating into a many-branched flowering axis in early spring. The flowering period is 2 to 4 weeks, with each flower remaining open for 3 to 4 days. Flowers are 1 to 2 cm in size, and in brassicas are bright yellow or occasionally cream or white. Radish flowers are commonly pink but can range from white through yellow to pink or deep purple. Flowers of radish and brassica are highly attractive to pollen- and nectar-gathering insects, particularly the honeybee. Abundant production of nectar from four nectaries located at the base of the stamens (Fig. 25-9) and profuse production of heavy pollen promote insect pollination. Fertilization occurs within 24 hours of pollination.

The diploid brassicas and radish are strongly allogamous, with pollen germination and tube growth controlled by a sporophytically determined incompatibility system of multiple alleles at the *S* locus. Amphidiploid

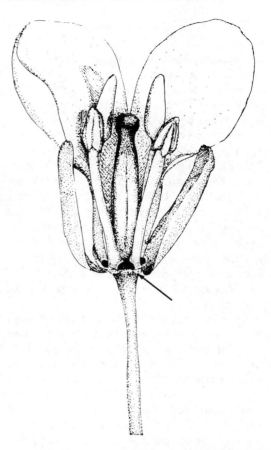

Figure 25-9. Cutaway diagram of a cabbage flower *(B. campestris)* showing location of the nectaries (dark areas, arrow).

species are autogamous and exhibit varying degrees of allogamy.

Hybrids via *S*-allele Incompatibility

The advantages of hybrid uniformity and vigor have long been recognized by breeders of crucifer vegetables and forages. *S*-allele incompatibility has been widely used in the production of superior hybrid cole crops (cabbage, broccoli, brussels sprouts, cauliflowers, chinese cabbage, oriental radish, and fodder kales) (Crisp, 1976). Hybrids currently occupy over 80% of the commercial cole crop production in the United States, parts of northern Europe, and Japan. By carefully selecting appropriate *S*-alleles, two-, three-, and four-way cross hybrids have been produced. In the production of hybrids, inbred parents are normally produced via bud pollination, a process in which self and sib incompatibility is overcome by placing conspecific pollen on the immature stigmas of hand-opened buds (Pearson, 1929). This labor-intensive process largely accounts for the high cost of hybrid crucifer seed.

Production of hybrid crucifers is dependent upon bee movement between parental lines. Though brassica and radish flowers are highly attractive to bees, it is apparent that a number of factors may contribute to seed set. Differential flowering times of the inbred parents can result in poor nick. In regions where crucifer seeds are produced, cool or wet weather frequently restricts bee movement. Bee preference for certain inbred parents is also a factor in reducing the production of hybrid seed. Though both parents may produce abundant pollen and nectar, other undocumented factors involved in bee foraging behavior may favor bee visitation to a particular inbred parent. In crucifers, as with many crops, the evaluation of bee visitation between potential inbred parents is rarely carried out at the time combining ability tests are being run. Thus, the ability of the inbreds to support seed production in the field is left until the first field-scale production of the crop. Evaluations of nicking and bee visitation between potential inbreds should be carried out with overall combining ability evaluations and should not be limited to hand pollinations or bee-assisted cage production.

Use of Bees in Inbred Production

Bees are now being used in the production of inbred crucifer seed. The self incompatibility of certain lines becomes nonfunctional after plants have been exposed to approximately 5% carbon dioxide for 0.5 hour to 2 hours. Inbreds are grown in polyethylene-covered cages. Each night, between 3:00 and 4:00 A.M., CO_2 is injected into the cage for the appropriate duration, followed by flushing with air. Each day bees enter the cage through a tube, which can be blocked off at night, during the CO_2 treatment period. Pollinations occurring 4 to 6 hours following the exposure to CO_2 may be compatible, depending on the genotype of the

line. Bee-aided inbred production has been shown to be considerably less costly than bud pollination.

Hybrids via Genic Male Sterility

Genic male steriles exist as single recessives and commonly appear after selfing of diploid brassicas and radishes. Genic male steriles are widely used in China, primarily in the production of hybrid chinese cabbage. The male sterile (MS) phenotype is maintained in the inbred seed parent by hand roguing fertile heterozygotes from the line. Inbred lines are maintained by recurrent bud crossing to a pollen source heterozygous for the *ms* gene. Highly uniform hybrids can be produced with careful surveillance and roguing. The production of genic MS hybrids also requires bud pollination to overcome *S*-allele incompatibility inherent in the production of the inbred.

In the production of hybrids using genic male sterility, parental attraction to pollinators may be further restricted by the absence of pollen in the seed parent. The Chinese have noted that in hybrid seed production fields, bees visit the male rows in the morning prior to 9:00 A.M. primarily seeking pollen. After about 9:00, bees move more randomly, seeking nectar in both male and MS–female rows, thus effecting cross-pollination. Normally three female rows are alternated with each male row. Care must be taken in the early stages of inbred development to ensure adequate nectary function and bee attractiveness. This can be done by evaluating at each generation of inbreeding, the male steriles for seed set in the field.

Cytoplasmic Male Sterility (CMS)

The high cost of production of incompatible inbred seed together with continual testing to ensure purity and stability of the *S*-alleles in each inbred parent has prompted breeders to seek CMS as an alternative to incompatibility in the production of crucifer hybrids. Though a number of CMS sources have been identified in the Cruciferae, none has proved to be without complications that have mitigated against

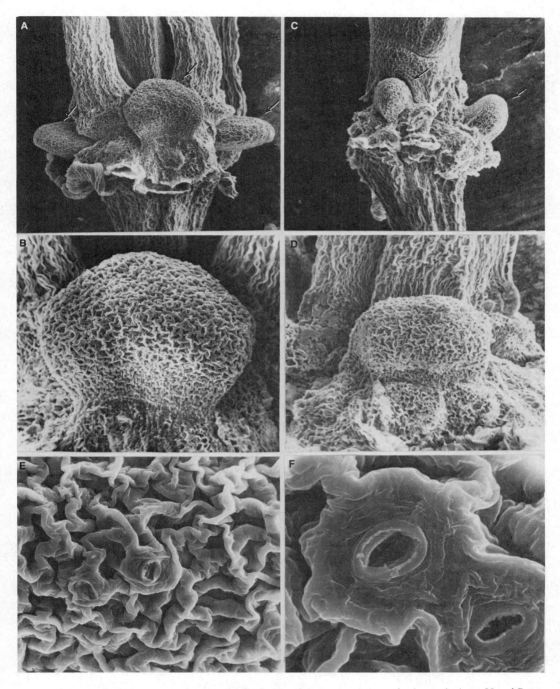

Figure 25-10. Scanning electron photographs of fully developed *B. campestris* nectaries (arrows), A, × 88 and B, × 220, and partially developed nectaries, C, × 88 and D, × 220, found in some CMS *B. campestris*. Nectar stomata, E, × 1100 and F, × 3740, are also shown.

widespread exploitation. Cytoplasmic male sterility *B. napus* and *B. oleracea* and radish have proved to be environmentally unstable and thus of limited value in hybrid production.

Cytoplasmic male sterility, derived by substituting the nucleus of one crucifer species in the cytoplasm of another, has given highly stable male steriles. However, various abnormalities associated with the wide cross have yet to be fully eliminated from the lines. The highly stable R_1 CMS developed by Bannerot by substituting the nuclei of *B. oleracea* and *B. napus* in the CMS cytoplasm of *Raphanus* has shown promise despite chlorosis of leaf tissues at low temperatures and reduced nectary function (Fig. 25-10) associated with the male sterility (Williams, 1980).

Early generation backcrosses of *B. oleracea* and *B. campestris* to the R_1 cytoplasm were examined in the greenhouse for secretion of nectar and scored on a 0–9 scale where 0 is the absence of nectar and 9 represents abundant production similar to that found in plants with normal cytoplasm (Fig. 25-11). During the summer growing season at Madison, Wisconsin, plants that flower in the field are observed for both nectar secretion and the presence of honeybee activity. By selecting for normal nectaries, nectar secretion, bee visitation, and seed set in the field, nectary function has been partially restored over six backcross generations.

A survey of the change in the average nectary function of backcross families in three *B. campestris* pedigrees suggested that the recovery of nectary function was favored by certain genotypic combinations. Over three generations of selection, pedigree C (recurrent parent P1419106) and pedigree B (recurrent parent PHW64040) showed faster gain toward normal nectaries than pedigree A (recurrent parent PHW64033) (Fig. 25-12).

Despite the demonstration that rapid progress can be made in restoring nectary function to CMS flowers, the question of relative attractiveness of CMS flowers compared to those of the male inbred remains a most important one for the effective exploitation of CMS in hybrid brassica production. Before relatively inexpensive hybrid crucifer seed can be produced, the problem of ensuring a high degree of bee attractiveness in both the seed parent and pollen parent, together with cross-compatibility between the two, will have to be overcome.

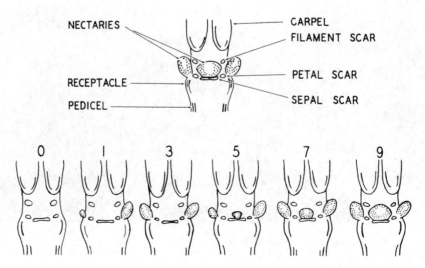

Figure 25-11. Top: Diagram of flower base of *B. campestris* indicating the position of nectaries. Bottom: Scale of nectary morphology for the assessment of nectary function in CMS *B. campestris*. 0 = no nectaries, 1 = one or two partially developed nectaries, usually small; 3 = two partially developed nectaries, close to normal size, 5 = three partially developed nectaries, one usually smaller; 7 = four partially developed nectaries, from small to nearly normal size; 9 = four fully developed nectaries.

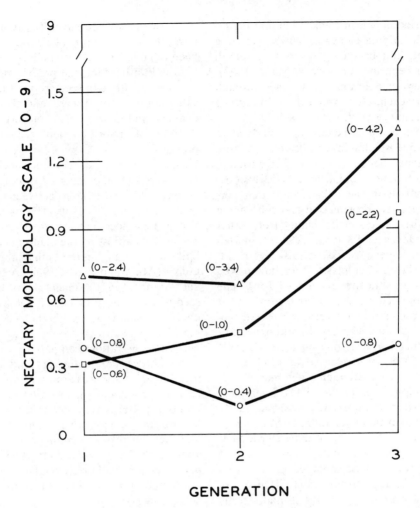

Figure 25-12. Developmental response of *B. campestris* nectaries to three generations of selection in three pedigrees, A (o—o), B (□—□), and C (△—△). Morphology scale is an assessment of nectary function in CMS *B. campestris* with 0 being the absence of nectaries and 9 being normal. Each point represents the average scale of ten plants examined in a family. Numbers in brackets indicate the range of scales found among plants in each family.

HYBRID ONIONS

Gordon D. Waller
USDA-ARS Carl Hayden Bee Research Center,
  Tucson, Arizona

The common onion (*Allium cepa* L.) is biennial and produces an umbel that varies from 6 to 20 cm in diameter with 50 to 2000 florets. These florets are both perfect and self-fertile, but because they are protandrous (Jones and Emsweller, 1933; Agati, 1952), cross-pollination predominates (van der Meer and van Ben-

nekom, 1968). Flowers open irregularly over the umbel for a period of 3 to 4 weeks; a field will bloom for one month or more, depending on the abundance of secondary seed stalks.

Each tiny floret has six stamens in two whorls and a simple pistil with three carpels, each containing two ovules. Pollen is produced at irregular intervals for two days following anthesis—first by the inner stamens and then the outer stamens (Jones and Emsweller, 1933). Pollen decreases in viability as the flower ages (Mann and Woodbury, 1969) and also may be less viable during the latter part of the period of bloom (Ockendon and Gates, 1976). The stigma elongates slowly, reaching full length and becoming receptive after all pollen has been shed (Jones and Mann, 1963). Receptivity then gradually declines through day six or seven, after which the stigma is no longer receptive (Moll, 1953, 1954).

In 1925 a plant in the cultivar (cv) "Italian Red" (13-53) was found with flowers that bore no functional pollen (Jones and Emsweller, 1936). Fortunately Italian Red 13-53 also developed bulbils in the inflorescence that permitted vegetative propagation and subsequent crosses with pollen-fertile plants from other cultivars. Jones and Clarke (1943) reported that a single-gene recessive *(ms ms)* maintained sterility in cytoplasm from Italian Red 13-53. This was the first example of using male sterility for seed parents in any crop, and it made the production of hybrid onion seed economically feasible. The first hybrid onion cultivar produced with cytoplasmic-genetic male sterility (CMS) was released in 1944, along with a method for producing commercial hybrid seed (Jones and Davis, 1944). Eventually the recessive maintainer was found in nearly all cultivars (Little et al., 1944, cited in Davis, 1957), and a wide variety of male sterile lines were developed by simply back-crossing to substitute the selected genotypes into sterile-inducing cytoplasm from Italian Red 13-53 (Jones and Clarke, 1947). When male sterile plants are found in commercial cultivars, they can be used to test for fertile plants having the *ms* gene; then little backcrossing is needed, since A and B lines are genetically similar (Jones and Mann, 1963). The low amount of selfing within male sterile onion lines (Clarke and Pollard, 1949) and the stability of male sterility (Barham and Munger, 1950) results in a uniform hybrid F_1 cross.

Gabelman (1974) reported that only 27% of onion seed produced in the United States during a two-year period was hybrid. However, the percentage of hybrid seed used for production of mature bulbs was much higher—over 90% in Wisconsin, for example. Conversely, onions used to produce dry sets are seldom hybrid. The complete conversion from open pollinated to hybrid cultivars did not occur, de-spite the early success with hybrid onions, because the production of hybrid seed has not been reliable. Inadequate pollination is probably the limiting factor in production of hybrid onion seed, although seed crop failures also result from drought, disease, and thermal stress (Tanner and Goltz, 1972). Nevertheless, production of hybrid onion seed has been profitable.

One of the first criteria for successful production of hybrid onion seed is to achieve synchrony of bloom (nicking) between male fertile and male sterile plants (Franklin, 1970). When this does not occur naturally, it is achieved by adjusting the planting dates of the bulbs or by changing the bulb storage temperatures (Atkin and Davis, 1954, cited in Jones and Mann, 1963). When plants are grown seed-to-seed, the problem of not having synchronous bloom can also be helped by adjusting planting dates (Franklin, 1970). Planting patterns (row ratios) compatible with the foraging patterns of pollinators and the movement of pollen by bees are also important (Erickson and Gabelman, 1956; Franklin, 1958). Planting patterns vary somewhat for different hybrids, but 4, 6, or 8 male sterile rows with 2 male fertile rows are common when there is abundant pollen production by the male fertile plants. Pollen availability might be reduced by low temperatures (van der Meer and van Bennekom, 1969).

Given synchronous bloom and adequate production of pollen having high viability, a pollinator that moves freely between male fertile plants and male sterile plants is essential (Kordakova, 1956; Sakharov, 1956; Bohart et al., 1970; Benedek and Gaal, 1972; Ewies and El-Sahhar, 1977). Cultivars of onion vary in their attractiveness to pollinators (Nye et al., 1973; Shasha'a et al., 1973; Carlson, 1974). Honeybees (*Apis mellifera* L.) and other bees tend to remain on either the male fertile flowers or the male sterile flowers (Lederhouse et al., 1972; Waters, pers. comm.). However, Gary et al. (1977a) showed "that few bees were discriminating between male sterile and male fertile lines during foraging." Numerous bees, flies, and wasps have been collected from onion flowers (Bohart et al., 1970; Caron et al.,

1975), and their pollination effectiveness has been established (Parker and Hatley, 1979), but the honeybee remains the only pollinator available in adequate numbers for use on a commercial basis. An alternative might be to locate hybrid seed fields near uncultivated areas having an abundance of native pollinators; however, this generally is not practical.

Nectar accumulates at the base of the stamens, with each floret producing 1 mg or more of nectar (Kordakova, 1956; Screbtzov and Screbtzova, 1977). A commercial onion seed field may yield up to 100 kg honey per ha (Kordakova, 1956; Julia et al., 1965; Bykov, 1969). Onion nectar generally contains relatively high concentrations of sugar (30 to 60%) (Lederhouse et al., 1968; Gary et al., 1972, 1977b; Waller, 1972a; Waters, 1972). Thus, the reward of sugar-rich nectar should attract and maintain adequate levels of bee activity to provide for pollination. However, Gary et al. (1972, 1977a) showed that the mean distance honeybees traveled to visit onion fields in California was less than they had traveled to visit carrots (*Daucus carota* L.) or safflower (*Carthamnus tinctoris* L.), and thus concluded that onions are decidedly less attractive to honeybees than these two crops. Because onion nectar is fully exposed, it is readily collected by bees, wasps, flies, and numerous other insects (Bohart et al., 1970), and the sugar concentration rapidly increases or decreases according to environmental conditions such as low humidity or rain. Onion nectar is fully visible and produces an intense fluorescence that may serve as a visual cue to pollinators (Thorp et al., 1975).

Because onion nectar is low in sucrose (hexose-dominant) (Waller, 1974) and high in potassium (Waller et al., 1972), it is marginal in acceptability to honeybees (Waller, 1972b). Hence bee visitation of onions is sporadic and unpredictable. In the absence of honeybees, nectar accumulates in the flowers and becomes highly viscous following the evaporation of water (Waters, 1972). Highly concentrated nectar is often avoided by bees.

Onion nectar can be detected in returning foragers using either the high potassium level (Waller et al., 1976) or its bright blue fluorescence (Waller and Martin, 1978). This fact has been used to show that most nectar foragers from colonies near onion seed fields in Arizona have foraged on onion flowers (Waller, unpub. data).

Onion pollen, the only other reward provided, is often brushed off and discarded by foraging honeybees (Benedek and Gaal, 1972; Williams and Free, 1974). Moreover, pollen-collecting honeybees generally make up less than one-tenth of the foragers observed (Gary et al., 1972, 1977a; Waller, 1972a; Williams and Free, 1974; Benedek, 1977). However, active collection of onion pollen does occur at times (Atkins et al., 1970). Pollen trapped from colonies near onion seed fields show that honeybees collected other types of pollen in preference to that of onion (Nye et al., 1971; Waller, unpub. data). Thus, it appears that onion pollen collection is usually incidental, and honeybees are principally nectar collectors when visiting onion flowers.

The preponderance of nectar collectors in a population of honeybees foraging on a hybrid onion seed field should enhance the movement of bees between the male fertile and male sterile flowers. What is needed for random, nondiscriminating bee visitation is for the male steriles and male fertiles to have comparable nectar rewards and similar visual and olfactory cues. Lederhouse et al. (1968) reported slightly higher nectar sugar concentrations in male fertile onion flowers than in male steriles; however, Gary et al. (1977b) reported no differences between fertiles and steriles for either nectar volume or sugar concentration. I know of no effort to evaluate foraging cues (color or odor) among hybrid onion seed parents.

At the onset of flowering it is a common practice for growers to place groups of 10 or 12 colonies of honeybees at intervals about the perimeter of each field at a rate of 3 to 5 colonies per acre. (Very few growers attempt to place colonies within their fields.) Populations of honeybee colonies (evidence of colony quality) are rarely determined, and visitation to the crop is often a matter of looking near the edge of the field to see whether the bees are working the crop. Often, the beekeeper views the pollination service provided as a diversion from the business of honey production. As a result, the colonies may be given a minimum of attention during the period of onion bloom. Bees rarely produce a crop of honey from onion, and when they do, the quality (flavor) is unpleasant (to most people). When competing crops are available near the onion field, the bees often visit them more intensely than they visit the onions (Waller et al., 1976). On rare occasions, honeybees have been observed to do almost no foraging when only onions were available, and colonies may actually decline in the absence of other forage. Humid, cloudy, and rainy weather may increase the attractiveness of onion nectar to bees by diluting highly concentrated and viscous nectar that occurs during hot, dry weather in arid climates.

Conversion to hybrid onions has been more successful than some other commercial hybrid crops requiring insects for pollination; for example, only 7% of carrots produced in the United States in the early seventies were hybrid (Gabelman, 1974). Nevertheless, the acreage planted for onion seed production in the United States has doubled since 1960, while the yield per acre has been cut nearly in half, and annual seed production has remained constant at about one million pounds. Some growers are reluctant to contract for production of hybrid onion seed because of the low yields (ca. 300 pounds/acre) and the likelihood of a crop failure (Campbell et al., 1968). As a result, some seed firms have purchased land to grow their own hybrid seed. Yields up to 1000 pounds/acre and more can be expected from open pollinated varieties; thus even with a lower price per pound the probability of a profit is greater because crop failures are less frequent.

The production of a new hybrid begins with pollen transfer by hand between individual plants or by caging flies (Jones and Emsweller, 1933, 1934) or bees (Moffett, 1965) with pairs of sterile and fertile heads. Seed increases of promising crosses are made in progressively larger cages. Eventually the area requirements force production to move to small open plots, in isolated areas to prevent contamination (Sparks and Binkley, 1946; Franklin, 1970). The above conditions provide little opportunity

to evaluate attractiveness to bees and seed production potential for a new onion cultivar. As a result, breeders tend to blame production people for seed failures in commercial-size fields of new cultivars, and production people accuse breeders of not discarding inbreds that are poor seeders.

During the 1970s, representatives of the major seed-producing firms held regular meetings with research scientists to discuss problems of onion seed production—especially the pollination of hybrids. New information about onion pollination problems was discussed, but little progress was made toward assuring adequate pollination. Some have even speculated that the trend toward hybrid onions has reversed. Major seed firms now seek better areas in which to produce hybrid onion seed, viz., Mexico, South Africa, and Australia. Hybrid onion production problems are complex, and solutions have evaded many scientists who have searched for them.

HYBRID SUNFLOWERS

Wayne A. Griffiths and Eric H. Erickson, Jr.
North Central States Bee Research Laboratory, USDA-ARS, University of Wisconsin, Madison, Wisconsin

The cultivated sunflower [*Helianthus annuus* v. *macrocarpus* (D.C.) (Ckll)] is an important oilseed and confectionary crop that is benefited by insect pollination (Furgala, 1970; McGregor, 1976; Dedio and Putt, 1980; Parker, 1981a). Sunflowers are highly cross-pollinated (Fick, 1978; Dedio and Putt, 1980); Hurd et al. (1980) report that all species of *Helianthus* in North America are obligate outcrossers [with two exceptions, *H. agrestis* (Pollard) and an uncultivated strain of *H. annuus*]. The sunflower capitulum (flower head) is an excellent example of an evolutionary adaptation for insect entomophilous pollination as the crowding of flowers (florets) "ensures conspicuousness" and the pollination of a maximum number by a single insect visit (Hurd, et al., 1980).

The inflorescence is composed of many individual florets in a large disc subtended by prominent ray flowers. The ray flowers are pis-

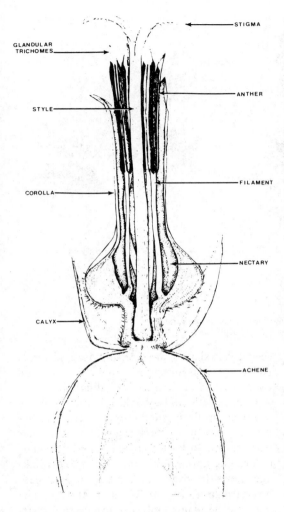

Figure 25-13. Sunflower disc floret.

Labels on figure: STIGMA, GLANDULAR TRICHOMES, ANTHER, STYLE, FILAMENT, COROLLA, NECTARY, CALYX, ACHENE

surfaces of the stigma are in close contact medially. The proximal end of the style is attached to the base of the corolla tube superior to the basal ovary. Nectar is secreted at the base of the floret. Sugar concentration of sunflower nectar is normally about 50%, but significant variation in nectar volume and sugar percentage may occur (Free, 1970; Furgala, 1974; McGregor, 1976; Fick, 1978).

The flowering sequence has been described by a number of researchers (Putt, 1940; Heiser, 1976; Jain and Dubey, 1979; Forsberg and Smith, 1980). It occurs in centripetal succession starting at the periphery of the head and proceeding at a rate of one to four rows of florets daily (Fig. 25-14). A typical head in bloom will have withered florets toward the periphery, then a ring of female-stage florets (forked stigmas), next a ring of male-stage florets (shedding pollen), and finally unopened florets toward the center (Free, 1964; McGregor, 1976). The entire flowering process for a head may last from 5 to 16 days (McGregor, 1976). The duration of flowering time is dependent upon pollination, environmental conditions, and the head diameter.

Anthesis of a single floret usually commences very early in the morning. The anther tube is exserted fully by the rapid elongation of the filaments. Subsequently the pollen is discharged into the anther tube. The style elongates and pushes the stigma through the anther tube. Pollen grains that have been discharged in the anther tube become attached to the hispidulous (covered with stiff hairs), nonreceptive outer surface of the stigma and are pushed out by the elongating pistil. A loss of turgidity in the filaments causes a recession of the anther into the corolla. Separation of the stigma lobes to expose the receptive surfaces for pollination continues throughout the following day (Putt, 1940; McGregor, 1976).

Pollination and fertilization may occur soon thereafter. Once pollination/fertilization has occurred, the stigma withers and recedes. If cross-pollination fails to occur, the stigma lobes remain receptive and continue to grow until they form a coil that allows the receptive surface to contact pollen adhering to the outer surface of its style (Hurt, 1948; Free, 1970).

tillate (Hurd et al., 1980), but the disc florets (Fig. 25-13) are hermaphroditic and protandrous (McGregor, 1976; Frankel and Galun, 1977). The number of florets per head (1000–4000) varies, depending upon the cultivar and the head size (McGregor, 1976). The florets are arranged in arcs radiating from the center of the head, and each represents a potential seed (achene).

The corolla tube is formed by five petals that are fused except at the tip. The five anthers are united and form a tube supported by five free and flattened filaments that are attached to the base of the corolla tube. Enclosed in the anther tube is the style, which terminates distally in a bilobed stigma. Prior to opening, the receptive

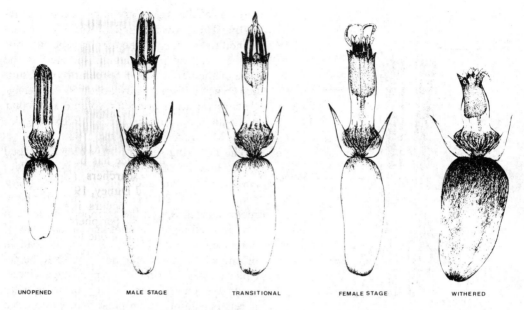

UNOPENED   MALE STAGE   TRANSITIONAL   FEMALE STAGE   WITHERED

Figure 25-14. Flowering sequence of sunflowers.

Self-pollination in sunflowers must be regarded as an exception and a rather inefficient evolutionary strategy that ensures survival of the species when cross-pollination has failed. The ability of a floret to set seed decreases over time because either stigma receptivity declines, or pollen on the style becomes inviable (Free, 1970). The achene of sunflower develops a hull regardless of whether a seed develops or not (Knowles, 1978). Empty achenes have been attributed to genetic makeup, environmental stress (Knowles, 1978), and insufficient pollination (Deshmukh and Nachane, 1977).

Hybridization

The development of hybrid sunflower production was stimulated by two events. The first was the introduction in about 1960 of high-oil cultivars from Russia. The second was the discovery of cytoplasmic male sterility (CMS) and the subsequent finding of fertility restoring genes (RHA) (Smith, 1978; Beard, 1981). The major objective of ongoing sunflower breeding programs since about 1970 has been to develop hybrid sunflowers with improved seed yield; higher oil content; uniformity in

seed color, shape, and size; disease and insect resistance; and autogamy. Generally, seed oil content is higher with cross-pollination.

Both genetic and cytoplasmic male sterility occur naturally in sunflower. Genetic sterility is controlled by a recessive gene (*ms*) that has been shown to be linked to a dominant gene controlling anthocyanin production. This linkage allows the grower to rogue out purple seedlings (male fertile *MsMs* and *Msms*). The male sterile plants lack the anthocyanin characteristic (*msms*).

Leclercq in 1970 discovered cytoplasmic male sterility in sunflower (Heiser, 1978). This discovery combined with the subsequent discovery of fertility restoring genes mainly from wild species of sunflower (Enns et al., 1970; Kinman, 1970; Vranceanu and Stoenescu, 1971) has been used extensively in hybrid seed production (see Forsberg and Smith, 1980 for complete explanation of CMS development and maintenance). Cytoplasmic male sterility and fertility restoration gene discoveries provided a mechanism for overcoming the problem of utilizing heterosis in sunflower (Putt, 1978). The first hybrids produced by the use of CMS and RHA were made available for commercial production in the United States in

1972. By 1976 it was estimated that they accounted for 80% of the sunflower production (Fick, 1978). Although a new source of CMS in sunflowers was recently discovered by Whelan (1980), all CMS sunflowers presently in use are related to the Leclercq line.

CMS plants usually have anthers that do not project out of the corolla; the anthers are fused only at their base not at the tip. In some CMS plants the anthers are normal but lack pollen. The female flower parts of CMS plants appear normal and are fertile when pollinated (Knowles, 1978). However, obvious floral abnormalities, pollinator preferences, and differences in ability to set seed have been observed among many CMS genotypes (E. H. Erickson, unpub.obs.; B. Furgala, unpub. obs.; F. D. Parker, unpub. obs.). These are the bases for most of the difficulties encountered in pollinating CMS lines.

In hybrid seed production CMS lines are interplanted with fertility restorer lines in ratios from 2:1 to 10:1 (Dedio and Putt, 1980). Time difference in flowering must be synchronized by manipulating planting dates, planting depths, and soil fertility (Dedio and Putt, 1980), although the use of multiple-head wild type RHA lines has recently mitigated this problem. Fick (1978) reported that no significant problems have been encountered using CMS and RHA system for hybrid seed production. However, recently hybrid seed producers and growers have experienced substantial losses that have precipitated litigation.

Sunflower Pollination

Sunflower florets are monoclinous; many are self-compatible, and some cultivars self-pollinate. Considerable variation in the ability to self-pollinate has been demonstrated among self-compatible cultivars. Seed yields can be significantly improved with insect pollination (D. E. Freund, B. Furgala, and M. A. Sugden, unpub. data).

Recently Hurd et al. (1980) reported 284 species of native bees that collected pollen from flowers of *Helianthus*. In addition, 72 species of the pollen-collecting bees are known to visit *Helianthus* for nectar only, as do about

56 species parasitic bees. Parker (1981a) reported 400 species of bees native to the United States visited *Helianthus* for nectar and pollen: many are oligoleges of sunflower.

Ratios of honeybees to wild bees on sunflower are variable. For example, Furgala (1974) observed in their Minnesota studies the ratio of 4.4 to 1.3. In North Dakota the ratio was 5.3 to 5.6. More recently D. E. Freund and B. Furgala (Unpub. data) found that 83% of all pollinators were honeybees. In Wyoming, 80% of the bees working sunflowers were honeybees (Krause and Wilson, 1981). Differences in such data undoubtedly reflect the relative densities of bee populations in those areas and seasonal patterns of foraging. As monoculture and pesticide use increase and nesting sites are destroyed, native bee populations decline dramatically.

Bees gathering nectar are the most efficient pollinators (Hurd et al., 1980). They insert their tongues and head between the petals and anther tube in order to reach the nectar at the base of the floret. This behavior dusts the bee with pollen.

The importance of honeybees, the principal pollinators of cultivated sunflowers, is well documented in the literature. Sunflowers are an important nectar source for beekeepers. Honeybees have also been observed by numerous researchers to groom sunflower pollen from their bodies while hovering over the plants (Guynn and Jaycox, 1973; McGregor, 1976; Robinson, 1978; Hurd et al., 1980; Parker, 1981b). This practice may contribute to increased pollination via aerial dusting of the florets (Free, 1970). Conversely, this behavior has also been viewed as contributing to reduced seed set, the result of reduced numbers of available pollen grains.

The recent trend in sunflower breeding programs has been to select self-pollinating, self-fertile lines (Fick and Rehder, 1977). This has been attributed to what has been perceived as a lack of adequate information on pollination. Furgala (1970) points out that "What precludes adequate pollination is not so much a lack of knowledge *per se* but the lack of relevant knowledge" and further concluded that "If we closely examine the results of . . . published studies we cannot arrive at a meaningful recommendation for adequate pollination of sunflowers." To be meaningfully interpreted, future studies of sunflower pollination should document the pollinator resource.

Pollination of Sunflower Hybrid Seed Parents

Insect pollination is essential in the production of sunflower seed from male fertile and CMS hybrid seed parents. However, as in other hybrid systems, considerable difficulty has been encountered in achieving adequate pollination. And, as with other hybrids, these problems may stem from heritable differences in pollinator foraging cues and rewards among parental genotypes.

Morphological differences in flower type, which may serve as visual foraging cues, have been reported to be genetically controlled in sunflower, as are ray and disc flower coloration (Fick, 1978). Genetic variation for nectar production has been reported by Furgala et al. (1976). Stigma coloration, tubular corolla length, and diameter in the disc floret (a morphological characteristic that may influence the accessibility of nectar to pollinators) have also been shown to be heritable (Fick, 1978).

In studies of 40 cultivars, CMS, HA, and RHA lines at the USDA Bee Investigation Lab., Madison, Wisconsin, it has been observed that ultraviolet radiating and absorbing color patterns among them range from distinct floral patterns at the base of the ray florets to no distinguishable floral patterns. In addition, definite stigma color differences and aberrations including extremely long or short styles and multiple stigma lobes (4–12) have been observed. Also noted was variation in resin production (which fluoresces in the UV and has a characteristic aroma), ranging from no discernible resin in some cultivars to copious amounts in others (E. H. Erickson and W. A. Griffiths, pers. obs.). Kreitner and Gershenzon (1980) have shown similar differences in the number of glandular trichomes on the anthertips among various wild and domestic sunflowers (Fig. 25-15A). Although pollinator foraging activity removes these trichomes (Fig. 25-

15B), their function is unknown. Many of these sunflower floral differences parallel differences reported among carrot hybrid seed parents and their F_1 progeny and may significantly affect the relative bee-attractiveness of each sunflower genotype as noted by F. D. Parker (unpub. data).

Recent studies (Parker, 1981b; B. Furgala and M. A. Sugden, unpub. data) have shown that honeybees forage discriminately on male sterile genotypes even though sterile and restorer lines, compared to fertile lines, secrete less nectar with less dissolved solids. Some native bees appear to prefer male fertile rows while still others show no preference. However, documentation of foraging preferences tends to obfuscate the principal issue. Row planting patterns, essential for cost-effective harvesting, predispose pollinators to preferentially select one or another seed parent and exercise considerable fidelity, foraging up and down the row. Hence, row crossing and coincidental cross pollination can be largely inhibited.

In his 1974 studies, Furgala (1974) demonstrated that only 6% of the honeybees foraging on the male sterile row were dusted with pollen. An additional 17% had a spot of pollen on the frons (face of the bee). Thus, fewer than 25% of these bees crossed over from male fertile to male sterile rows. These data are quite comparable to data presented for other hybrids (for example, see "Hybrid Carrots"). One hundred percent of the bees foraging on the fertile rows were dusted with pollen. Yields reflected this foraging behavior, as bee-attractive hybrids outyield those that are unattractive (George and Shein, 1980; Parker, 1981a). Bee-attractive lines should be thoroughly researched so that essential factors can be incorporated into new hybrid systems (Parker, 1981a).

Pollination of F_1 Hybrid Sunflowers

Great differences exist in the self-compatibility (autogamy and physiological self-compatibility) of F_1 hybrids (George and Shein, 1980). Care must be taken when interpreting self-compatibility data, as they are dependent upon the concomitant yield of open pollinated plots.

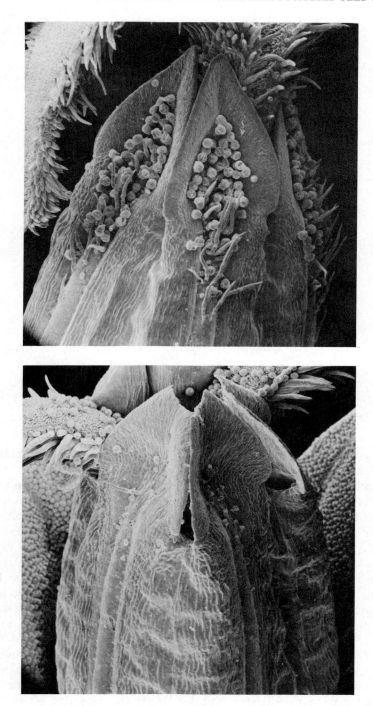

Figure 25-15. Glandular trichomes (spherical) and nonglandular hairs at the tip of sunflower anthers. A, cv USDA 903, no exposure to pollinators (×60); B, cv Interstate 903, after exposure to pollinators.

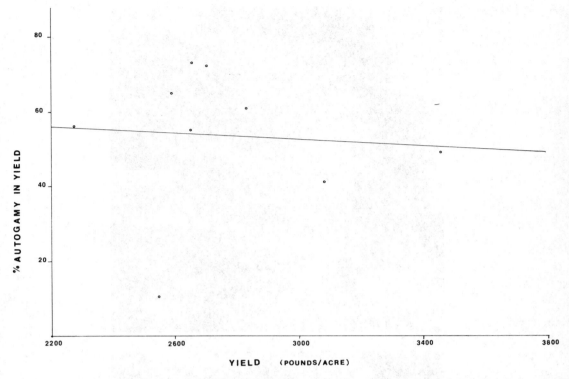

Figure 25-16. Autogamy in yield (as used by Robinson, 1980) vs. seed yields for 9 sunflower cvs, Waseca, MN 1979 (from unpub. data, D. E. Freund and B. Furgala).

These yields are in turn dependent upon the relative floral characteristics of the respective genotype, available pollinator resource, and climatic and edaphic factors. Obviously, in sunflowers as in other hybrids, there is a need to assure floral attractiveness if insect pollination is to be depended upon.

Differences exist in the attractiveness to bees of various hybrids (George and Shein, 1980). Furgala et al. (1976) determined that F_1 hybrids from crosses of cytoplasmically male sterile and fertility restorer lines produced higher levels of nectar than either parent. They further stated that pistillate florets have more nectar and nectar solids than do staminate or post-pistillate stages (previous reports have presented the opposite view; probably the result of sampling errors) (Furgala, 1974). Thus, they concluded that seed production crossing fields may be less attractive to bees than F_1 hybrid fields. If varieties of F_1 hybrids are minimally attractive to bees because

of sunflower genotype, bees may forage elsewhere.

The recent trend has been to select for self-fertile (autogamous) hybrids which, according to published reports, exhibit increased seed yields in the absence of pollinators (Robinson, 1980). However, analysis of unpublished data (D. E. Freund and B. Furgala; B. Furgala and M. A. Sugden) suggests that when "autogamy in yield"* is plotted against seed yield/acre, there is a tendency toward a decrease in yield (see Figs. 25-16 and 25-17). In Fig. 25-17 the decrease was significant ($r = 0.68$), probably the result of hot dry conditions during anthesis. (Note that here levels of autogamy in yield exceed 100%—an unrealistic circumstance. Hence the derivation of the formula is suspect.) Nevertheless, these data demonstrate that while selfing does occur, the plant still de-

---

* $\dfrac{\text{Weight of bagged heads}}{\text{Weight of unbagged heads}} \times 100$ (Robinson, 1980).

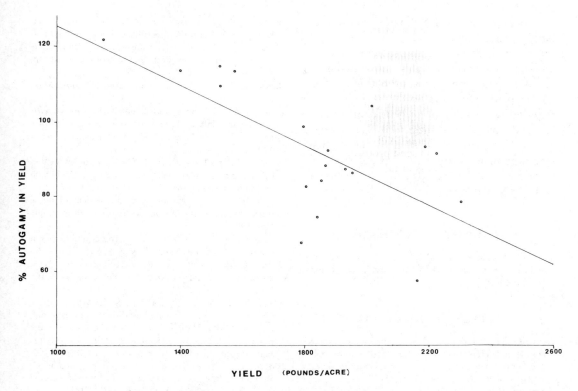

Figure 25-17. Autogamy in yield (as used by Robinson, 1980) vs. seed yields for 19 sunflower cvs, Waseca, MN 1980 (from unpub. data, B. Furgala and M. A. Sugden).

pends upon insect pollination for maximum yield.

Self-compatibility is usually 40 to 60% with some F_1 hybrids as low as 10% and others as high as 87% (Kessler, 1980). In a test of three hybrids with three self-compatibility ratings the following yield data were obtained by Furgala et al. (1979):

Self-compatibility	Caged without bees	Caged with bees	Uncaged
16%	1019 lb	2850	2801
41%	2192	2858	3168
87%	2510	2508	2960

Obviously autogamy did not contribute to increased yield. Similarly, Krause and Wilson (1981) reported 50% higher yields among three hybrids exposed to insect pollinators. Insect-pollinated cultivars outproduce autogamous varieties (Parker, 1981a).

Robinson (1980) coined the term "artifact autogamy" to describe the tendency of sunflower plants to produce larger and heavier achenes when achene set is limited as it may be with inadequate pollination (Forsberg and Smith, 1980). This phenomenon is known to exist in many horticultural and agronomic crops. Robinson (1980) considered 17 of 60 F_1 hybrids in 1978 and 26 of 73 hybrids in 1979 entirely autogamous when he compared yields with open pollinated controls. But no pollinator resource level was given. The pitfalls in this approach have already been described.

Pesticides and Pollinators

One final point should be presented here. Sunflowers are somewhat unique in that there may be a great need, especially in the southern regions, to control insect pests while the plants are in bloom and pollinators are present (Krause and Wilson, 1981). Either the pest insects or the absence of pollinators can substantially reduce yields. Highly autogamous hybrids could alleviate this problem, and it is apparent that this is a consideration in ongoing breeding programs. Obviously, this goal has not been achieved, and until it is, judgment is required when integrating pest control methods and pollinator management. The issue is quite different for hybrid seed production fields where pollinators are essential. Perhaps solutions eventually developed for the latter circumstance will be applicable to the former.

ACKNOWLEDGMENTS

The authors wish to acknowledge E. J. Garvens who provided the illustrations used and B. Furgala for his many helpful suggestions.

LITERATURE CITED

Agati, G. 1952. Indagini ed osservazioni sulla biologia fiorale della cipolla [Investigations and observations on the floral biology of the onion]. *Rivista della Ortoflorofrutticoltura Italiana* 36(3/4):67–77.

Akerberg, E. 1974. Insect pollination of hybrid varieties from cultivated plants. *Proc. III Int. Symp. Pollination,* Prague, pp. 83–89.

Allard, H. A. 1911a. Some experimental observations concerning the behavior of various bees in their visits to cotton blossoms. I. *Am. Nat.* 45:607–622.

———. 1911b. Some experimental observations concerning the behavior of various bees in their visits to cotton blossoms. II. *Am Nat.* 45:668–685.

Allard, R. W. 1960. *Principles of Plant Breeding.* John Wiley and Sons, Inc. New York, 485 pp.

Allison, D. C. and W. D. Fisher. 1964. A dominant gene for male-sterility in Upland cotton. *Crop Sci.* 4:548–549.

Atkins, E. L., Jr., L. D. Anderson, and E. A. Greywood. 1970. Research on the effect of pesticides on honeybees 1968–69. *Am. Bee J.* 110:387–389.

Baker, I. and H. G. Baker. 1976. Analyses of amino acids in flower nectars of hybrids and their parents, with phylogenetic implications. *New Phytol.* 76:87–98.

Balls, W. L. 1929. The natural crossing of cotton flowers in Egypt. *Egyptian Min. Agr. Tech. and Sci. Serv. Bull.* 89:1–27.

Barham, W. S. and A. M. Munger. 1950. The stability of male sterility in onions. *Proc. Am. Soc. Hort. Sci.* 56:401–409.

Barnes, D. K. 1980. Alfalfa. *In: Hybridization of Crop Plants.* Am. Soc. of Agron. and Crop Sci. Soc. of Am., Madison, Wis., pp. 177–187.

Barnes, D. K. and B. Furgala. 1978. Nectar characteristics associated with sources of alfalfa germplasm. *Crop Sci.* 18:1087–1089.

Barnes, D. K., E. T. Bingham, J. D. Axtell, and W. H. Davis. 1972. The flower, sterility mechanisms, and pollination control. *In* C. H. Hanson (ed.), *Alfalfa Science and Technology.* Am. Soc. of Agron., Madison, Wis., pp. 123–141.

Beard, B. H. 1981. The sunflower crop. *Sci. Am.,* May, pp. 150–161.

Benedek, P. 1977. Behaviour of honeybees (*Apis mellifera* L.) in relation to the pollination of onion (*Allium cepa* L.) inflorescences. *Z. Angew. Entomol.* 82:414–420.

Benedek, P. and E. Gaal. 1972. The effect of insect pollination on seed onion, with observations on the behavior of honeybees on the crop. *J. Apic. Res.* 11:175–180.

Blazek, J. 1974. Pollination in variety blocks of apple trees. *Proc. III Int. Symp. Pollination,* Prague, pp. 129–139.

Bohart, G. E. and W. P. Nye. 1960. Insect pollinators of carrots in Utah. *Utah Agr. Exp. Sta. Bull.* 419:1–16.

Bohart, G. E., and F. E. Todd. 1961. Pollination of seed crops by insects. *In: Seeds.* The Yearbook of Agric. USDA, pp. 240–246.

Bohart, G. E., W. P. Nye, and L. R. Hawthorn. 1970. Onion pollination as affected by different levels of pollinator activity. *Utah Agr. Exp. Sta. Bull.* 482, 57 pp.

Bradner, N. R. and W. R. Childers. 1968. Cytoplasmic male-sterility in alfalfa. *Can. J. Plant Sci.* 48:111–112.

Butler, G. D., F. E. Todd, S. E. McGregor, and F. G. Werner. 1960. *Melissodes* bees in Arizona cotton fields. *Ariz. Agr. Exp. Sta. Tech. Bull.* 139:11.

Bykov, B. 1969. [Onions as nectar plants.] *Pchelovdstovo* 3:24–26.

Campbell, W. F., S. D. Cotner, and B. M. Pollock. 1968. Preliminary analysis of the onion seed (*Allium cepa* L.) production problem, 1966 growing season. *HortScience* 3:40–41.

Carlson, E. C. 1974. Onion varieties, honeybee visitations, and seed yield. *Calif. Agr.* 28:16–17.

Caron, D. M., R. C. Lederhouse, and R. A. Morse. 1975. Insect pollinators of onion in New York State. *HortScience* 10(3):273–274.

Childers, W. R. and D. K. Barnes. 1972. Evolution of hybrid alfalfa. *Agr. Sci. Rev.* (3rd Quarter) 10(3):11–18.

Clarke, A. E. and L. H. Pollard. 1949. The amounts of self-pollination in male-sterile onion lines. *Proc. Am. Soc. Hort. Sci.* 53:299–301.

Clement, B. A. 1965. Flower color, a factor in attractiveness of alfalfa clones for honeybees. *Crop Sci.* 5:267–268.

Cook, O. F. 1909. Suppressed and intensified characters in cotton hybrids. *USDA Bur. Pl. Ind. Bull.* 147, 27 pp.

Crisp, P. 1976. Trends in breeding and cultivation of cruciferous crops. *In* J. G. Vaughan, A. J. MacLeod, and B. M. G. Jones (eds.), *The Biology and Chemistry of the Cruciferae.* Academic Press, London, 355 pp.

Davis, D. D. 1978. Hybrid cotton: specific problems and potentials. *Adv. Agron.* 30:129–157.

Davis, E. W. 1957. The distribution of the male-sterility gene in onion. *Proc. Am. Soc. Hort. Sci.* 70:316–318.

Davis, W. H. and I. M. Greenblatt. 1967. Cytoplasmic male-sterility in alfalfa. *Heredity* 58:301–305.

Dedio, W. and E. D. Putt. 1980. Sunflower. *In: Hybridization of Crop Plants*. Am. Soc. of Agron. and Crop Sci. Soc. of Am., Madison, Wis., pp. 631–644.

Deshmukh, S. D. and M. N. Nachane. 1977. Effect of bee visits and climate on sunflower pollination. *Indian J. Ent*. 39(3):281–283.

Eaton, F. M. 1957. Selective gametocide opens way to hybrid cotton. *Science* 126:1174–1175.

Emsweller, S. L. 1961. Fundamental procedures in breeding crops. *In: Seeds*. The Yearbook of Agric. USDA, pp. 127–133.

Enns, H., D. G. Dorrell, J. A. Hoes, and W. O. Chubb. 1970. Sunflower research, a progress report. *Proc. Fourth Int. Sunflower Conf*. Memphis, Tenn., pp. 162–167.

Erickson, E. H. and M. B. Garment. 1979. Soya-bean flowers: nectary ultrastructure, nectar guides, and orientation on the flower by foraging honeybees. *J. Apic. Res*. 18(1):3–11.

Erickson, E. H. and C. E. Peterson. 1978. Problems encountered in the pollination of cytoplasmically male sterile hybrid carrot seed parents. *Proc. IVth Int. Symp. Pollination*, College Park, Maryland, pp. 59–63.

———. 1979. Asynchrony of floral events and other differences in pollinator foraging stimuli between fertile and male sterile carrot inbreds. *J. Am. Soc. Hort. Sci*. 104(5):639–643.

Erickson, E. H., C. E. Peterson, and P. Werner. 1979a. Honeybee foraging and resultant seed set among male-fertile and cytoplasmically male-sterile carrot inbreds and hybrid seed parents. *J. Am. Soc. Hort. Sci*. 104(5):538–635.

Erickson, E. H., R. W. Thorp, D. L. Briggs, J. R. Estes, R. J. Daun, M. Marks, and C. H. Schroeder. 1979b. Characterization of floral nectar by high-performance liquid chromatography. *J. Apic. Res*. 18(2):148–152.

Erickson, H. T. and W. H. Gabelman. 1956. The effect of distance and direction on cross-pollination in onions. *Proc. Am. Soc. Hort. Sci*. 68:351–357.

Ewies, M. A. and K. F. El-Sahhar. 1977. Observations on the behavior of honeybees on onion and their effects on seed yield. *J. Apic. Res*. 16:194–196.

Fahn, A. 1974. *Plant Anatomy*. 2nd ed., Pergamon Press, London, 611 pp.

Faulkner, G. J. 1978. The effect of insect behaviour on hybrid seed production of brussels sprouts. A. J. Richards (ed.), *In The Pollination of Flowers by Insects*. Linnean Soc. Symp. Ser. No. 6, Academic Press, New York, pp. 201–202.

Fehr, W. R. and H. H. Hadley (eds.) 1980. *Hybridization of Crop Plants*. Am. Soc of Agron. and Crop Sci. Soc. of Am., Madison, Wis., 765 pp.

Fick, G. N. 1978. Breeding and genetics. *In: Sunflower Science and Technology*. Am. Soc. of Agron. and Crop Sci. Soc. of Am., Madison, Wis., pp. 279–338.

Fick, G. N., and D. A. Rehder. 1977. Selection criteria in development of high oil sunflower hybrids. *Proc. II Sunflower Forum*, Fargo, N. Dak., pp. 26–27.

Forsberg, R. A. and R. R. Smith. 1980. Sources, maintainance, and utilization of parental material. *In: Hybridization of Crop Plants*. Am. Soc. of Agron. and Crop Sci. Soc. of Am., Madison, Wis., pp. 65–81.

Frankel, R. and E. Galun. 1977. *Pollination Mechanisms, Reproduction and Plant Breeding*. Springer-Verlag, Berlin, Heidelberg, New York, 281 pp.

Franklin, D. F. 1958. Effects on hybrid onion seed production of using different ratios of male sterile and pollen rows. *Proc. Am. Soc. Hort. Sci*. 71:435–439.

———. 1970. Problems in the production of vegetable seed. *In: The Indispensable Pollinators*. Ninth Poll. Conf., Hot Springs, Arkansas, pp. 112–140.

Free, J. B. 1964. The behaviour of honeybees on sunflowers (*Helianthus annuus* L.) *J. Appl. Ecol*. 1(1):18–27.

———. 1970. *Insect Pollination of Crops*. Academic Press, New York, 544 pp.

Frisch, K. von. 1950. *Bees, their Vision, Chemical Sense, and Language*. Cornell University Press, Ithaca, N.Y., 119 pp.

———. 1967. *The Dance Language and Orientation of Bees*. Harvard University Press, Cambridge, Mass., 566 pp.

Furgala, B. 1970. Sunflower pollination—a neglected research problem area. *In: The Indispensable Pollinators*. Ninth Poll. Conf., Hot Springs, Arkansas, pp. 37–42.

———. 1974. Honeybee pollination of sunflowers: evaluation of limiting factors. Research Agreement No. 12-14-100-10880(33), 17 pp. (Cooperative project between University of Minnesota and the USDA, ARS, unpub.)

Furgala, B., E. C. Mussen, D. M. Noetzel, and R. G. Robinson. 1976. Observations on nectar secretion in sunflowers. *Proc. 1st Sunflower Forum*, Fargo, N. Dak., pp. 11–12.

Furgala, B., D. M. Noetzel, and R. G. Robinson. 1979. Observations on the pollination of hybrid sunflowers. *Proc. IVth Int. Symp. on Pollination*, Maryland Agric. Exp. Sta. Spec. Misc. Publ. 1:45–48.

Gabelman, W. H. 1974. F₁ hybrids in vegetable production. *Proc. 19th Int. Hort. Congr.*, Warsaw 3:419–428.

Galun, E. 1959. The role of auxins in sex expression of the cucumber. *Physiol. Plantarum* 12:48–61.

Gary, N. E. 1975. Activities and behavior of honeybees. Ch. 7 *in: The Hive and the Honeybee*. Dadant and Sons, Inc. Hamilton, Ill., pp. 185–264.

Gary, N. E., P. C. Witherell, and J. Marston. 1972. Foraging range and distribution of honeybees used for carrot and onion pollination. *Environ. Ent*. 1:71–78.

Gary, N. E., P. C. Witherell, K. Lorenzen, and J. M. Marston. 1977a. The interfield distribution of honeybees foraging on carrots, onions and safflower. *Environ. Ent*. 6:637–640.

———. 1977b. Area fidelity and intra-field distribution of honeybees during the pollination of onions. *Environ. Ent*. 6:303–310.

George, D., and S. Shein. 1980. Effect of pollination and compatibility on seed set. *Sunflower*, May/June, pp. 13–14.

Grant, V. 1949. Pollination systems as isolating mechanisms in angiosperms. *Evolution,* 3:82–97.

———. 1971. *Plant Speciation.* Columbia University Press, New York, 435 pp.

Guynn, G. and E. R. Jaycox. 1973. Observations on sunflower pollination in Illinois. *Am. Bee J.* May, pp. 168–169.

Hanson, C. H., H. O. Graumann, L. J. Elling, J. W. Dudley, H. L. Carnahan, W. R. Kehr, R. L. Davis, F. I. Fresheisser, and A. W. Hovin. 1964. Performance of 2-clone crosses of alfalfa and an unanticipated self-pollination problem. USDA Tech Bull. 1300.

Haragsim, O. 1974. Bees as pollinators of entomophilous plants. *Proc. III Int. Symp. Pollination,* Prague, pp. 31–37.

Hawthorn, L. R., G. E. Bohart, and E. H. Toole. 1956. Carrot seed yield and germination as affected by different levels of insect pollination. *Proc. Am. Soc. Hort. Sci.* 67:384–389.

Hawthorn, L. R., G. E. Bohart, E. H. Toole, W. P. Nye, and M. D. Levin. 1960. Carrot seed production as affected by insect pollination. *Utah Agr. Exp. Sta. Bull.* 422:1–18.

Heiser, C. B. 1976. *The Sunflower.* University of Oklahoma Press, Norman, Okla., 198 pp.

———. 1978. Taxonomy of *Helianthus* and origin of domesticated sunflower. *In: Sunflower Science and Techonology,* Am. Soc. of Agron. and Crop Sci. Soc. of Am., Madison, Wis., pp. 31–53.

Hunt, O. J., D. K. Barnes, R. N. Peaden, and B. J. Hartman. 1976. Inheritance of black seed character and its use in alfalfa improvement. Report of the Twenty-Fifth Alfalfa Improvement Conference, Cornell University, July 13–15. ARS-NC-52, 33 pp.

Hurd, P. D., W. E. La Berge, and E. G. Linsley, 1980. Principle sunflower bees of North America with emphasis on the southwestern United States *(Hymenoptera: Apoidea). Smithsonian Contrib. Zool.* No. 310, 158 pp.

Hurt, E. F. 1948. *Sunflower for Food, Fodder and Fertility.* Faber and Faber Ltd., London, 175 pp.

Jain, K. K. and C. S. Dubey. 1979. Anthesis studies in sunflower (*Helianthus annuus* L.). *Sunflower Newslett.* 3(3):15–18.

Johnson, D. L., and A. M. Wenner. 1970. Recruitment efficiency in honeybees: studies on the role of olfaction. *J. Apic. Res.* 9(1):13–18.

Jones, H. A. and A. E. Clarke. 1943. Inheritance of male sterility in the onion and the production of hybrid seed. *Proc. Am. Soc. Hort. Sci.* 43:189–194.

———. 1947. The story of hybrid onions. *In: Yearbook of Agriculture,* 1943–1947. U. S. Dept. Agr., Washington, D.C., 944 pp. (pp. 320–326).

Jones, H. A. and G. N. Davis. 1944. Inbreeding and heterosis and their relation to the development of new varieties of onions. USDA Tech. Bull. 874, pp. 1–28.

Jones, H. A., and S. L. Emsweller. 1933. Methods of breeding onions. *Hilgardia* 7(16):625–642.

———. 1934. The use of flies as onion pollinators. *Proc. Am. Soc. Hort. Sci.* 31:160–164.

———. 1936. A male sterile onion. *Proc. Am. Soc. Hort. Sci.* 34:582–585.

Jones, H. A., and L. K. Mann. 1963. *Onions and Their Allies.* Interscience, New York, 286 pp.

Julia, F., F. Pirvu, and G. Illyes. 1965. The melliferous value of esparsette *(Onobrychys viciifolia)* and onion (*Allium cepa* L.) under the pedoclimatic conditions around Cluj (in Rumanian). *Lucr. stiint. Inst. agron. Cluj.* 21:99–106.

Justus, N. and C. L. Leinweber. 1960. A heritable partially male-sterile character in cotton. *J. Hered.* 51:191–192.

Justus, N., J. R. Meyer, and J. B. Roux. 1963. A partially male-sterile character in Upland cotton. *Crop Sci.* 3:428–429.

Kauffeld, N. M. and E. L. Sorensen. 1971. Interrelations of honeybee preference of alfalfa clones and flower color, aroma, nectar volume, and sugar concentration. Kansas St. Agr. Exp. Sta. Res. Publ. 163, 14 pp.

———. 1980. Foraging behavior of honeybees on two-clone alfalfa combinations differing in bee attractiveness. *J. Econ. Ent.* 73:434–435.

Kauffeld, N. M., E. L. Sorensen, and R. H. Painter. 1967. Preferences of two races of honeybees for selected alfalfa clones under competitive and noncompetitive foraging. *J. Econ. Ent.* 60:1135–1139.

———. 1969. Stability of attractiveness of alfalfa clones to honeybees under varying locations, seasons, and years. *Crop Sci.* 9:225–228.

Kaziev, I. P. and S. S. Seidova. 1965. The nectar yield of flowers of some Cucurbitaceae under Azerbaidjan conditions. *XX Int. Beekeeping Jubilee Cong.* II:31.

Kearney, T. H. 1924. A hybrid between different species of cotton. *J. Hered.* 15:309–320.

Kehr, W. R. 1973. Cross-fertilization of alfalfa as affected by genetic markers, planting methods, locations, and pollinator species. *Crop Sci.* 13:296–298.

Kessler, Karl. 1980. Honeybees sweeten sunflower profits. *The Furrow.* 85(1): 14–15.

Kevan, P. G. 1978. Floral coloration: its colorimetric analysis and significance in anthecology. *In* A. J. Richards (ed.), *The Pollination of Flowers by Insects.* Linnean Soc. Symp. Ser. No. 6. Academic Press, New York, pp. 51–82.

———. 1979. Vegetation and floral colors revealed by ultraviolet light: interpretational difficulties for functional significance. *Am. J. Bot.* 66(6):749–751.

Khan, A. H. and M. Afzal. 1950. Vicinism in cotton. *Indian Cotton Growing Rev.* 4:227–239.

Kinman, M. L. 1970. New developments in the U.S.D.A. and state experiment station sunflower breeding programs. *Proc. Fourth Int. Sunflower Conf.* Memphis, Tenn., pp. 181–183.

Knowles, P. F. 1978. Morphology and anatomy. *In: Sunflower Science and Technology.* Am. Soc. of Agron. and Crop. Sci. Soc. of Am., Madison, Wis., pp. 55–87.

Kordakova, Z. M. 1956. [Honeybees and pollination of seed plants of the common onion] (in Russian). *In* I. V. Krishehunas and A. F. Gubin (eds.), *Pollination of*

Agricultural Plants. State Publishing House for Agricultural Literature, pp. 163–171. Moscow.

Krause, G. L. and W. T. Wilson. 1981. Insect pollination and honeybee visitation patterns on hybrid oilseed sunflowers in central Wyoming *(Hymenoptera: Apidae).* *J. Kansas Ent. Soc.* 54(1):75–82.

Kreitner, G. L. and J. Gershenzon. 1980. Secretory structures of disk florets as possible sources of chemical resistance to sunflower moth larvae. Paper #639 presented before the Entomological Society of America, Atlanta, Ga.

Kubisova-Kropacova, Sylvie. 1974. Adaptation of plants and bees for pollination purposes. *Proc. III Int. Symp. Pollination,* Prague, pp. 183–190.

Lederhouse, R. C., D. M. Caron, and R. A. Morse. 1968. Onion pollination in New York. *New York's Food and Life Sci.* 1:8–9.

———. 1972. Distribution and behavior of honeybees on onion. *Environ. Ent.* 1:127–129.

Levin, D. A. 1978. Pollinator behaviour and the breeding structure of plant populations. *In* A. J. Richards (ed.), *The Pollination of Flowers by Insects.* Linnean Soc. Symp. Ser. No. 6. Academic Press, New York. pp. 133–150.

Loden, H. D. and T. R. Richmond. 1951. Hybrid vigor in cotton—cytogenetic aspects and practical applications. *Econ. Bot.* 5:387–408.

Loper, G. M., G. D. Waller, and R. L. Berdel. 1974. Olfactory screening of alfalfa clones for uniform honeybee selection. *Crop Sci.* 14:120–122.

Mann, L. D. and G. W. Woodbury. 1969. The effect of flower age, time of day, and variety of pollen germination of onion, *Allium cepa* L. *J. Am. Soc. Hort. Sci.* 94(2):102–104.

Marani, A. 1967. Heterosis and combining ability in intraspecific and interspecific crosses of cotton. *Crop Sci.* 7:519–522.

———. 1968a. Heterosis and F_2 performance in intraspecific crosses among varieties of *Gossypium hirsutum* L. and of *G. barbadense* L. *Crop Sci.* 8:111–113.

———. 1968b. Heterosis and inheritance of quantitative characters in interspecific crosses of cotton. *Crop Sci.* 8:299–303.

Marani, A. and S. E. Avieli. 1973. Heterosis during the early phases of growth in intraspecific and interspecific crosses of cotton. *Crop Sci.* 13:15–18.

Martin, E. C. 1975. The use of bees for crop pollination. Ch. 20 *in: The Hive and the Honeybee.* Dandant and Sons, Inc., Hamilton, Ill., pp. 579–614.

Matsuo, E., S. Uemoto, and E. Fakushima. 1969. Studies on the photoperiodic sex differentiation in cucumbers, *Cucumis sativus* L.: aging effect on the photoperiodic dependency of sex differentiation. *J. Fac. Agr. Kyushu Univ.* 15:287–303.

McGregor, S. E. 1959. Cotton-flower visitation and pollen distribution by honeybees. *Science* 12:97–98.

———. 1976. *Insect Pollination of Cultivated Crop Plants.* Agr. Handbook No. 496, USDA, 411 pp.

McGregor, S. E., C. Rhyne, S. Worley, Jr., and F. E.

Todd. 1955. The role of honeybees in cotton pollination. *Agron. J.* 47:23–25.

Meer, Q. P. van der and J. L. van Bennekom. 1968. Research in pollen distribution in onion seed fields. *Euphytica* 17:216–219.

———. 1969. Effect of temperature on the occurrence of male sterility in onion (*Allium cepa* L.). *Euphytica* 18:389–394.

Mell, P. H. 1894. Experiments in crossing for the purpose of improving the cotton fiber. Ala. Agr. Exp. Sta. Bull. 56, 49 pp.

Meredith, W. R., Jr. 1976. Nectariless cottons. *Proc. Beltwide Cotton Production–Mechanization Conf.* Las Vegas, Nevada, pp. 34–37.

Mesquida, J. and M. Renard. 1978. Entomophilous pollination of male-sterile strains of winter rapeseed (*Brassica napus* L. Metzger var. *Oleifera*) and a preliminary study of alternating devices. *Proc. IV Int. Symp. Pollination,* College Park, Maryland, pp. 49–57.

Meyer, J. R. and V. G. Meyer. 1961. Cytoplasmic male sterility in cotton. *Genetics* 46:883. (Abstract)

Meyer, V. G. 1969. Some effects of genes, cytoplasm, and environment on male sterility of cotton *(Gossypium).* *Crop Sci.* 9:237–242.

———. 1973. A study of reciprocal hybrids between Upland cotton *(Gossypium hirsutum* L.) and experimental lines with cytoplasm from seven other species. *Crop Sci.* 13:439–444.

Moffett, J. O. 1965. Pollinating experimental onion varities. *Am. Bee J.* 105:378.

Moffett, J. O., L. S. Stith, C. C. Burkhardt, and C. W. Shipman. 1975a. Honeybee visits to cotton flowers. *Environ. Ent.* 4:203–206.

———. 1975b. The influence of cotton genotypes on the floral visits of honeybees. *Crop Sci.* 15:782–784.

———. 1976a. Insects visitors to cotton flowers. *J. Ariz. Acad. Sci.* 21:47–48.

———. 1976b. Fluctuation of wild bee and wasp visits to cotton flowers. *J. Ariz. Acad. Sci.* 11:64–68.

———. 1976c. Nectar secretion in cotton flowers and its relation to floral visits by honeybees. *Am. Bee J.* 116:32–34, 36.

Moffett, J. O., L. S. Stith, and C. W. Shipman. 1976d. Influence of distance from pollen parent on seed produced on male-sterile cotton. *Crop Sci.* 16:765–766.

———. 1977. Producing hybrid cotton on the high plains of Texas. *Proc. 1977 Beltwide Cotton Prod. Res. Conf.* Atlanta, Georgia, pp. 90–92.

———. 1978a. Effect of honeybee visits on boll set, seeds produced and yield of Pima cotton. *Proc. 1978 Beltwide Cotton Prod. Res. Conf.* Dallas, Texas, pp. 80–82.

———. 1978b. Producing hybrid cotton seed on a field scale using honeybees as pollinators. *Proc. 1978 Beltwide Cotton Prod. Res. Conf.,* Dallas, Texas, pp 77–80.

Moffett, J. O., H. B. Cobb, and D. R. Rummel. 1980. Bees of potential value as pollinators in the production of hybrid cottonseed on the High Plains of Texas. *Proc.*

Beltwide Cotton Prod. Res. Conf. St. Louis, Missouri, pp. 268–270.

Moll, R. H. 1953. A study of receptivity in the onion flower, *Allium cepa* Linn. and some factors affecting its duration. M.S. thesis, University of Idaho, Moscow, 22 pp.

———. 1954. A study of receptivity in the onion flower, *Allium cepa* L., and some factors affecting its duration. *Proc. Am. Soc. Hort. Sci.* 64:399–404.

Nitsch, J., E. B. Kurtz, J. L. Livermann, and F. U. Went. 1952. The development of sex expression in cucurbit flowers. *Am. J. Bot.* 39:32–43.

Nye, W. P., G. D. Waller, and N. D. Waters. 1971. Factors affecting pollination of onion in Idaho during 1969. *J. Am. Soc. Hort. Sci.* 96(3):330–332.

Nye, W. P., V. S. Shasha'a, W. F. Campbell, and A. R. Hamson. 1973. Insect pollination and seed set of onions (*Allium cepa* L.). Utah Agr. Exp. Sta. Res. Rep. 6, 15 pp.

Ockendon, D. J. and P. J. Gates. 1976. Variation in pollen viability in the onion (*Allium cepa* L.). *Euphytica* 25:753–759.

Orel, V. 1974. Mendel's contribution to the scientific basis of pollination research. *Proc. III Int. Symp. Pollination,* Prague, pp. 39–43.

Parker, F. D. 1981a. Sunflower pollination: abundance, diversity and seasonality of bees and their effect on seed yields. *J. Apic. Res.* 20:49–61.

———. 1981b. How efficient are bees in pollinating sunflowers? *J. Kansas Ent. Soc.* 54(1):61–67.

Parker, F. D. and C. L. Hatley. 1979. Onion pollination: viability of onion pollen and pollen diversity on insect body hairs. *Proc. IV Int. Symp. Pollination,* College Park, Md., 1978, pp. 201–206.

Pearson, O. H. 1929. Observations on the type of sterility in *Brassica oleracea* var. *capitata. Proc. Am. Soc. Hort Sci.* 26:34–38.

Pedersen, M. W. and D. K. Barnes. 1973. Alfalfa pollen production in relation to percentage of hybrid seed produced. *Crop Sci.* 13:652–656.

Pedersen, M. W. and R. E. Stucker. 1969. Evidence of cytoplasmic male sterility in alfalfa. *Crop Sci.* 9:767–770.

Percival, M. S. 1965. *Floral Biology.* Pergamon Press, London, 243 pp.

Putt, E. D. 1940. Observations on the morphological characters and flowering processes in the sunflower (*Helianthus annuus* L.) *Sci. Agr.* 21(4):167–179.

———. 1978. History and present world status. *In: Sunflower Science and Technology.* Am. Soc. of Agron. and Crop Science of Am. Madison, Wis., pp. 1–29.

Richmond, T. R. and R. J. Kohel. 1961. Analysis of a completely male-sterile character in American Upland cotton. *Crop Sci.* 1:397–401.

Robinson, R. G. 1978. Production and culture. *In: Sunflower Science and Technology.* Amer. Soc. of Agron. and Crop Sci. Soc. of Am., Madison, Wis., pp. 89–143.

———. 1980. Artifact autogamy in sunflower. *Crop. Sci.* 20:814–815.

Rubis, D. 1970a. Breeding insect-pollinated crops. *In: The Indispensable Pollinators.* Ninth Poll. Conf., Hot Springs, Arkansas, pp. 19–24.

———. 1970b. Bee-pollination in the production of hybrid safflower. *In: The Indispensable Pollinators.* Ninth Poll. Conf., Hot Springs, Arkansas, pp. 43–54.

Sakharov, M. K. 1956. Bees and onion seed production (in Russian). *In* I. V. Krishehunas, and A. F. Gubin (eds.), *Pollination of Agricultural Plants.* State Publishing House for Agricultural Literature, pp. 172–173. Moscow.

Screbtzov, M. F. and N. D. Screbtzova. 1977. Some peculiarities of onion (*Allium cepa* L.) flowering, nectar secretion and pollination. *Proc. Int. Symp. on Melliferous Flora,* Budapest, 1976. Apimondia Publ. House, Otaniemi-Helsinki, Finland, pp. 210–215.

Shasha'a, N. S., W. P. Nye, and W. F. Campbell. 1973. Path-coefficient analysis of correlation between honeybee activity and seed yield in *Allium cepa* L. *J. Am. Soc. Hort. Sci.* 98:341–347.

Shifris, O. and E. Galun. 1956. Sex expression in the cucumber. *Proc. Am. Soc. Hort. Sci.* 67:479–486.

Simpson, D. M. 1954. Natural cross-pollination in cotton. U. S. Dept. Agr. Tech. Bull. 1094, 17 pp.

Singh, A., M. S. Kairon, and B. R. Mor. 1972. Review of efforts made to develop hybrid cotton in Haryana. *Cotton Develop.* 2:33–36.

Smith, D. L. 1978. Planting seed production. *In: Sunflower Science and Technology.* Am. Soc. of Agron. and Crop Sci. of Am., Madison, Wis., pp. 371–386.

Sparks, W. C. and A. M. Binkley. 1946. Natural crossing in Sweet Spanish onions as related to distance and directions. Proc. Amer. Soc. Hort. Sci. 47:320–322.

Srinivasan, K., V. Santhanan, and S. Rajasekaran. 1972. Development of hybrid cotton utilizing male-sterile line. *Cotton Develop.* 2:27–39.

Stanley, R. G. and H. F. Linskens. 1974. *Pollen. Biology Biochemistry Management.* Springer-Verlag, New York, 307 pp.

Staszewski, Z. W. 1979. *Cytoplasmic Male Sterility and Heterosis in Alfalfa.* Warsaw, Poland, 95 pp.

Stuber, C. W. 1980. Mating designs, field nursery layouts, and breeding records. *In: Hybridization of Crop Plants.* Am. Soc. of Agron. and Crop Sci. Soc. of Am., Madison, Wis., pp. 83–104.

Tanner, C. B. and S. M. Goltz. 1972. Excessively high temperatures of seed onion umbels. *J. Am. Soc. Hort. Sci.* 97:5–9.

Tasei, J. N. 1974. Pollinator insects of the field bean (*Vicia faba equina*) which is producing hybrid seed. *Proc. III Int. Symp. Pollination,* Prague, pp. 115–119.

Teuber, L. R., M. C. Albertsen, D. K. Barnes, and G. H. Heichel. 1980. Structure of floral nectaries of alfalfa (*Medicago sativa* L.) in relation to nectar production *Am. J. Bot.* 67(4):433–439.

Thorp, R. W. 1978. Honeybee foraging behavior in California almond orchards. *Proc. IV Int. Symp. Pollination,* College Park, Maryland, pp. 385–392.

Thorp, R. W., D. L. Briggs, J. R. Estes and E. H. Erickson. 1975. Nectar fluorescence and ultraviolet irradiation. *Science* 189:476–478.

Tiedjens, V. A. 1928. Sex ratio and sex expression in the cultivated cucurbits. *Am. J. Bot.* 18:359–366.

Tsunoda, S., K. Hinata, and C. Gomez-Campo (eds.). 1980. *Brassica Crops and Wild Allies.* Japan Sci. Soc. Press, Tokyo, 354 pp.

Vranceanu, V. A. and F. M. Stoenescu. 1971. Pollen fertility restorer gene from cultivated sunflower (*Helianthus annuus* L.) *Euphytica* 20:536–541.

Waddle, R. M. and N. R. Lersten. 1973. Morphology of discoid floral nectaries in Leguminosae, especially tribe Phaseoleae (Papilionoideae). *Phytomorphology* 23:152–161.

Waller, G. D. 1972a. Evaluating responses of honeybees to sugar solutions using an artificial-flower feeder. *Ann. Ent. Soc. Am.* 65(4):857–862.

———. 1972b. Chemical differences between nectar of onion and competing plant species and probable effects on attractiveness to pollinators. Ph.D. dissertation, Utah State University, Logan, 132 pp.

———. 1974. Evaluating the foraging behavior of honeybees as pollinators of hybrid onions. *Proc. III Int. Symp. Pollination,* Prague, pp. 75–180.

Waller, G. D. and J. H. Martin. 1978. Fluorescence for identification of onion nectar in foraging honeybees. *Environ. Ent.* 7:766–768.

Waller, G. D., E. W. Carpenter, and O. A. Ziehl. 1972. Potassium in onion nectar and its probable effect on attractiveness of onion flowers to honeybees. *J. Am. Soc. Hort. Sci.* 97(4):535–539.

Waller, G. D., G. M. Loper, and R. L. Berdel. 1974. Olfactory discrimination by honeybees of terpenes identified from volatiles of alfalfa flowers. *J. Apic. Res.* 13(3):191–197.

Waller, G. D., N. D. Waters, E. H. Erickson, and J. H. Martin. 1976. The use of potassium to identify onion nectar-collecting honeybees. *Environ. Ent.* 5:780–782.

Waller, G. D., F. D. Wilson, and J. H. Martin. 1981. Influence of phenotype, season, and time of day on nectar production in cotton. *Crop Sci.* 21:507–511.

Ware, J. O. 1927. The inheritance of red plant color in cotton. Arkansas Agr. Exp. Sta. Bull. 220, 80 pp.

Waters, N. D. 1972. Honeybee activity in blooming onion fields in Idaho. *Am. Bee J.* 112:218–219.

———. 1978. Pollination problems in hybrid onion seed production using honeybees. *Proc. IV Int. Symp. Pollination,* College Park, Maryland, pp. 65–72.

Weaver, J. B., Jr. 1968. Analysis of a genetic double recessive completely male-sterile cotton. *Crop. Sci.* 8:597–600.

———. 1978. Observations on bee activity in several genotypes of cotton. *Proc. 1978 Beltwide Cotton Prod. Res. Conf.* Dallas, Texas, pp. 76–77.

Weaver, J. B. and T. Ashley. 1971. Analysis of a dominant gene for male-sterility in Upland cotton, *Gossypium hirsutum* L. *Crop Sci.* 11:596–598.

Weihing, R. M., D. N. Robertson, O. H. Coleman, and R. Gardner. 1943. Growing alfalfa in Colorado. Colorado Agr. Exp. Sta. Bull. 480, 3 pp.

Weiss, M. G. and Elbert L. Little, Jr. 1961. Variety is a key word. *In* Seeds. The Yearbook of Agric. USDA, pp. 359–364.

Welch, J. E. and E. L. Grimball, Jr. 1947. Male sterility in the carrot. *Science* 106:594.

Whelan, E. D. P. 1980. A new source of cytoplasmic male sterility in sunflower. *Euphytica* 29:33–46.

Whitaker, T. W. 1931. Sex ratio and sex expression in the cultivated cucurbits. *Am. J. Bot.* 18:359–366.

Williams, I. H. and J. B. Free. 1974. The pollination of onion (*Allium cepa* L.) to produce hybrid seed. *J. Appl. Ecol.* 11(2):409–418.

Williams, P. H. 1980. Chemistry and breeding of cruciferous vegetables *In* T. Swain and R. Kleiman (eds.), *The Resource Potential in Phytochemistry. Rec. Adv. Phytochem.,* Vol. 14. Plenum Press, New York, 215 pp.

26
POLLINATION FROM TWO PERSPECTIVES: THE AGRICULTURAL AND BIOLOGICAL SCIENCES

James R. Estes, Bonnie B. Amos,* and Janet R. Sullivan
Robert Bebb Herbarium
University of Oklahoma
Norman, Oklahoma

ABSTRACT

Pollination biology is an integral subdiscipline of each of the fields of ecology, evolution, and systematics, providing a unique medium for the testing of models and theories. In agriculture, pollination biology holds promise for increasing yields through the management of components of the pollination-reproductive system. The foraging component is currently the area of most intense research and greatest potential. Suggestions are made for improving terminology pertinent to breeding systems and for improving interdisciplinary communication.

KEY WORDS: Agronomy, apiculture, foraging, honeybees, horticulture, native bees, nectar, pollen, pollination, oligolecty, reproductive biology.

CONTENTS

*Present address: Department of Biology, Baylor University, Waco, Texas.

INTRODUCTION

Understanding diversity is the central theme among the disciplines of ecology, evolution, and systematics. A useful synthesis contributing to our understanding of the origin and distribution of the diverse elements of the earth's biota is based on energy and energy flow (Hardin, 1966, p. 241). Pollination biology provides an excellent opportunity to explore various aspects of energetics in terms of resource productivity (Baker and Baker, 1973), foraging strategy (Pyke, 1979), adaptive radiation (Grant and Grant, 1965), community structure (Parrish and Bazzaz, 1978), competition (Schaffer et al, 1979), reproductive strategy (Cruden, 1977), coadaptation (Macior, 1974), and energetics itself (Heinrich, 1975). The utility of pollination systems in probing for basic information and in testing models and hypotheses concerning energetics, or other fields of endeavor, arises from the simplicity of those systems: (1) interactions involve relatively few organisms; (2) interdependence among the organisms is often high, or even absolute; and (3) interrelationships between the producers and the consumers involve fundamental phenomena, such as plant reproduction and resource collection by animals. In addition, the observations are often straightforward.

On the other hand, the emphasis of agronomy, apiculture, and horticulture with respect to pollination is manipulative: (1) development of cultivars requires controlled pollination (Frankel and Galun, 1977); (2) maintenance of cultivars requires isolation from contaminating pollen sources (Frankel and Galun, 1977); (3) yield of many horticultural crops is dependent upon successful insect pollination, which may require management of hive bees (Martin and McGregor, 1973) and of competing weeds (Free, 1968); and (4) maximum hive development, and hence honey and brood production, requires access to nectar and pollen resources throughout the growing season (Thorp et al., 1974). Consequently agricultural scientists are concerned with precisely the same issues as population biologists, but

from an engineering perspective—namely, the increase of net yield of seed or nectar through the management of the reproductive-pollination system.

Ecology and systematics are synthetic fields of investigation, both of which require gathering and comparing data from a variety of disciplines. Pollination biology adds yet another synthetic tier of knowledge with most, or even all, its theories and paradigms rooted in ecological and evolutionary thought. Pollination biology, then, does not have a unique, discrete literature—the pollination process involves the study of animal behavior, ecology, and physiology and of plant ecology, genetics, physiology, and reproduction, all of which impinge on zoological and botanical classification. Thus, even studies dealing explicitly with pollination ecology (e.g., Barber and Estes, 1978: Macior, 1978; Cruden and Hermann-Parker, 1979) relate directly to ecology, systematics, and evolution. Pollination biology is, however, one of the basic sciences providing a theoretical framework for plant breeding and bee rearing (Frankel and Galun, 1977).

The study of pollination systems is therefore an active interface, uniting fields of knowledge. But within this burgeoning literature some data, perhaps inevitably, are not readily retrievable. For instance, publications on the ecology of insects most likely include data relative to floral records, pollen loads, and intrafloral behaviour (e.g., LaBerge, 1967; Eickwort, 1977; Buchmann and Jones, 1980; Johnson, 1981). Yet plant scientists unfamiliar with the entomological literature might fail to realize the significance of the work to plant biology because they cannot find plant taxa listed in the title or abstract. Even the basic reference for bees of the eastern United States (Mitchell, 1960, 1962) fails to cross-index plant taxa, though plant records are included in the descriptions of the bee species. Of course the reverse is equally true for many studies in plant systematics (e.g., Raven and Gregory, 1972; Towner, 1977), though lamentably plant systematists seem less likely to provide insect records, resource productivity, and intrafloral behavior of pollinators as well as floral predators than their zoological counterparts (for instance, see most floras and phytotaxonomic monographs).

It appears to us that an index—perhaps similar to *Index to Plant Chromosome Numbers*—in which plant–pollinator records and references would be cross-indexed with respect to animal and plant taxa would be a useful enterprise, materially benefiting pollination biologists.

Communication among pollination biologists in the agricultural and biological sciences is also restricted, especially from the agricultural fields to biology. This is in spite of the extensive observations and experiments relating to topics such as the biology of nectar (Raw, 1953; Waller, 1972; Waller et al., 1972; McLellan, 1975; Corbet, 1978; Waller and Martin, 1978) and foraging (Waller, 1970; Benedek and Goal, 1972; Gary et al., 1972; Erickson et al., 1973; Faulkner, 1976; Collison and Martin, 1979; Parker, 1981a). These two sets of references are representative of the research reported in agriculturally oriented journals relative to pollination biology. For obvious reasons agricultural pollination studies focus on a limited array of plant taxa—most notably from the families Cucurbitaceae, Ericaceae, Fabaceae, Malvaceae, Rosaceae, Rutaceae, and Solanaceae—and the single pollen vector *Apis mellifera*. Partially as a result of this extreme concentration and partially as a result of the enterprise of the United States Department of Agriculture and its scientists, the pollination of many crop systems is thoroughly studied in comparison to most nonagricultural systems. An example of the fine analysis of reproduction, pollination, and intrafloral ecology is the extensive work with almond and the foraging of honeybees (Tufts, 1919; Griggs et al., 1952; Griggs, 1953; Kester and Griggs, 1959; Griggs and Iwakiri, 1960, 1964; Kester, 1965; Thorp et al., 1967; Sheesley and Poduska, 1970; Thorp et al., 1973; Thorp et al., 1974; Griggs and Iwakiri, 1975; Thorp et al., 1975; Gary et al., 1976; Erickson et al., 1977; Gary et al., 1978; Thorp, 1978; Thorp and Mussen, 1978; Erickson et al., 1979).

The agricultural literature is also rich with innovative and useful techniques for the investigation of the biology of pollination (Gary, 1971; Waller, 1972; Gary and Lorenzen, 1976; Erickson et al., 1977; cf. Frankel and Galun, 1977).

Faced with this overwhelming diversity of themes and topics, we decided to concentrate on only two: foraging behavior and plant reproductive systems. This selection stems from the extensive literature in both areas, because both lend themselves to experimentation, the centrality of the issues, and our own interests. Further these two topics are intimately correlated and central to both population biology and agriculture. Thus population structure, dynamics, and evolution are directly affected by the rate of gene migration, and the distance genes move via pollen migration in a single generation determines the distance required to maintain agronomic lines.

FORAGING BEHAVIOR

Agricultural Sciences

The various components of the foraging system—attractive elements, resource productivity and presentation, nutritional levels of the foraged crop, sensory and recall capacity of the forager, intrafloral and interfloral behavior patterns, forager physiology and morphology—interact in a compelling fashion, strongly influencing reproduction of the host plant and the available energy for the anthophilous forager. It seems reasonable to assume that the exercise of control over appropriate aspects of the system could lead, in mellitophilous flowers, to increases in yield of seed, fruit, honey, and/or brood. The goals of the bee scientist and plant scientist may, however, be incompatible; for instance, increasing the nutrient and productive levels of nectar without a concomitant increase in the relative density of pollinating bees might diminish interfloral and interplant flights, resulting in reduced seed set, even though honey yields might increase. Similarly, the use of pollen-sterile lines is a valuable tool in the production of hybrid seed; however, the lack of pollen production affects foraging behavior (Erickson, this book). Not surprisingly, agricultural scientists have long understood the interactive nature of the components of the system and the potential value of an in-depth comprehension of the foraging system (Martin and McGregor, 1973; Frankel and Galun, 1977; Caron, 1979). A series of examples stress the potential for manipulation.

Genetic Control of Foraging. Alfalfa seed production is a major agricultural enterprise in the western United States of America. Honeybees routinely forage alfalfa; however, most (up to 99–100%) nectar foragers do not trip the keel and the inclusive reproductive structures (Reinhardt, 1952; Bohart, 1957, 1960). Further, the proportion of nectar foragers is relatively high, 50–85% nectar collectors in Arizona, and the percentage tends to increase with higher latitudes and cooler temperatures (Bohart, 1960). Presumably, the increase in percentage of pollen collectors in Arizona and California, at least partially, explains the higher seed set in the southern region. Therefore, selection of bee strains that either forage primarily for pollen or that do not bypass the anthers and stigma when collecting nectar seems to offer a method of improving seed productivity. To this end, Nye and Mackenson (1965, 1968; Mackenson and Nye, 1966, 1969) initiated a selective breeding program to develop honeybee strains with a propensity for pollen foraging rather than nectar collection. Within six generations of the initial selection a clear difference between the high alfalfa pollen collection line (high-APC) (86% pollen collectors in the colony) and the low line (8% pollen collectors) was detectable. Thus not only was the heritability of this significant component of the foraging system demonstrated, but a technique for the increase in yield in alfalfa seed production seemed feasible. The high-APC strain was also demonstrated to prefer pollen collection in areas where alfalfa had not been previously foraged by honeybees (Nye and Mackenson, 1970). Honey yield was, however, predictably lower in the high-APC strain. At least partially as a result of lack of interest among commercial bookeepers, the APC-line is no longer maintained, however.

Not unexpectedly, foraging efficiency for nectar may also be under genetic control and thus subject to artificial selection (Rothenbuhler et al., 1979). By the second generation significant differences were noted by these workers between a cage-reared lineage selected for rapid collection of syrup from feeders and a lineage of slow feeders, and between each and the initial population's rate of feeding. Furthermore, a positive correlation between weight gain and rate of feeding had been demonstrated earlier (Kulincevic and Rothenbuhler, 1973; Kulincevic et al., 1974). Nevertheless, field tests reported in 1979 were disappointing. Variation was high in the second generation, but the fast-line still outgained the slow-line feeders; however, by the fourth and fifth generations, either weight gains were not significantly different between the lines, or the slow-line actually gained more weight. A number of potential explanations might account for this unexpected turn; perhaps it was related to the selection of increased consumption of nectar and not actually foraging efficiency. Whatever the cause, the results serve to illustrate the overall complexity of the system and make extrapolation to field conditions based on perturbations of individual components difficult at best.

Even obvious manipulations of the pollination system and its individual subsets may fail to bear fruit. For example, commercial sweet almonds yield considerable supplies of nectar (Erickson et al., 1979) and pollen. Although the flowers are protandrous, pollen foraging bees are generally expected to be the most efficient pollinators (Thorp et al., 1974). Erickson et al., (1975, 1977) tested the efficacy of disposable pollination units (DPUs) for increasing productivity of almond orchards. Foragers from DPUs predominately collect nectar (only about 10% pollen collectors per DPU). Yet DPU colonies were more effective as pollinators than overwintered colonies even though about 55% of the foraging workers from the latter colonies collect pollen. Thus it appears that nectar foraging bees may be the most effective pollinators, at least in almonds.

Prunus dulcis (sweet almond) is a self-incompatible species; because commercial varieties are vegetatively propagated, orchards must include at least two and preferably several varieties for cross-pollination and hence fertilization to occur (Griggs, 1953; Kester, 1965). The varieties also mature at differing times and have differing, and commercially significant, features of the fruit; thus harvesting is more efficient if the varieties are grown in rows rather than randomly placed throughout the orchard. Under those circumstances foraging patterns of *Apis mellifera,* which is utilized as the prime pollinator, are of importance. Most notably, almond breeders and growers need to know intertree flight potentials and whether an increase in intertree flights can be induced—either genetically or through apiary management—in the hope that interrow flights, and subsequently fruit set, will increase significantly. As previously noted, increasing nectar and pollen productivity is likely to be counterproductive in terms of fruit production. Whether reduced nectar flow with slow replenishment would benefit growers is nevertheless unclear. Thorp and Mussen (1978) did find, however, that the first flowers produced for each variety—when rewards are at a low for each tree—set fruit most frequently, presumably because intervarietal flights were favored by the foragers.

Manipulation of the Attractive Components of the System. For crops that are less attractive to honeybees, increasing the visual or olfactory effects of the blossom or the quality or quantity of the forage (typically pollen or nectar) through plant breeding or mechanical means, might have a strong positive effect on seed and fruit set. The latter form of management via artificial addition of odors (Waller, 1970), pollen (Griggs et al., 1952; Griggs and Iwakiri, 1960), and sugar water (Free, 1965) has been attempted, but the results are somewhat mixed, and the engineering is still in the formative stages. The potential for plant breeding to increase attractiveness is largely unknown; unfortunately the data sets are inadequate to permit estimates of heritability and intraspecific variation for most plant species; however, alfalfa has been studied rather extensively, and it appears that selection for increased nectar yield is feasible (Teuber and Barnes, 1978).

Considerable diversity of opinion even exists concerning the relative attractiveness of the three primary sugars present in nectar (Wykes, 1952; Jamieson and Austin, 1956; Percival, 1961; Krapacova, 1965 as cited in Bachman and Waller, 1977). Bachman and Waller (1977), employing feeding-preference tests, demonstrated most convincingly, however, that honeybees exhibit preference for sugar water that is more heavily sucrose-laden, in contrast to sugar water that has more fructose or glucose. If this is an accurate reflection of the attractiveness of nectar, then selective breeding for a higher percentage of sucrose might be advantageous. Other constituents of nectar that show promise include amino acids (Baker and Baker, 1973), cations (Waller et al., 1972), and fluorescing compounds (Thorp et al., 1975).

The attractiveness and nutrient content of pollen is even less well known than that of nectar, but is undoubtedly much more complex (Martin and McGregor, 1973); nonetheless pollen is an essential component of the diet of bees (Stanley and Linskens, 1974). Just as with nectar, differences exist with respect to the preference of honeybees for differing pollens (Levin and Bohart, 1955; Hugel, 1962; Standifer et al., 1970). Campana and Moeller (1977), for instance, compared the relative nutritive and attractive properties of pollen of willow, sweet clover, boxelder, blackberry, and mixed pear and plum. The control pollen consisted of a mixture of various pollens. Twenty-four colonies were fed sugar water, and one of the test pollens presented in cake-form with the cakes composed of pollen:sugar:water (34.4:60:5.6 by percentage). The total pollen consumed, number of cells of brood, and number of cells of brood per gram of pollen consumed were all determined or calculated. More bees were reared by the colonies on sweet clover pollen than any other; however, the most nutritive pollen based on brood cells/gram of pollen was boxelder and mixed pear and plum. Interestingly, there was a positive correlation between amount of pollen consumed and the number of bees reared. Campana and Moeller (1977) thus concluded that (1) attractiveness of the tested pollen is a more significant feature than nutritive level; (2) all the tested pollens were of sufficient nutritional levels to rear brood. Although it is clear that the development of an attractive and highly nutritive pollen source would be advantageous to the apiculturists, its advantage to agronomists and horticulturists is less clear, and the issues of intraspecific variation on which selection might be based and whether all crop pollens are nutritious to honeybees have not been answered. In fact, Colin and Jones (1980) revealed that energy levels of pollen from anemophilous and entomophilous dicots are statistically indistinguishable. One implication from these studies is that nutritional levels, at least caloric levels, in most pollens are adequate for brood-rearing. Apparently either the minimal levels of energy, protein, lipids, and minerals necessary for pollen function are appropriate levels for bees, or the coincidence is the result of the long-coevolved history of bees and angiosperms.

Intraspecific and intervarietal variation in features of the pollen is known to occur in cotton; however, the distinction is that some varieties produce pollen that appears to be antagonistic to honeybees, whereas the pollen of other varieties is collected by foraging workers (Eric Erickson, pers. comm.; Barbara (Hanny) Ross, pers. comm.). The chemical nature (amino acids, flavonoids, tannins, and gossypol) of cream and yellow anthers and pollen of cotton has been documented by (Hanny) Ross and coworkers (Hanny, 1980; Hanny and Elmore, 1980), and the higher percentage of gossypol in yellow anthers presumably accounts for the growth suppression of larvae of the anther-feeding tobacco budworm (Hanny et al., 1979; Hanny and Elmore, 1980). Gossypol may therefore be the basis for the rejection of the pollen of some cotton varieties, but this requires experimental verification.

Biological Sciences

Foraging behavior has received considerable interest following publication of the watershed papers by Grant (1949) concerning foraging behavior and floral constancy, and Heinrich and Raven (1972) regarding the energetics of foraging. Various aspects of foraging have recently been reviewed—for example, foraging pattern (Levin and Kerster, 1969; Levin et al., 1971; Waddington, 1979a,b; Waddington and Heinrich, 1979), optimal foraging strategy (Pyke, 1978a,b,c, 1979; Pleasants and Zimmerman, 1979; Waddington and Heinrich, 1979; Waddington, 1980; Real, 1981), and breeding structure of the plant populations as influenced by foraging behavior (Grant, 1971; Levin, 1978). Rather than review these interesting, well discussed areas, we chose to consider various aspects of foraging for pollen, especially by solitary bees, and oligolectic behavior. The two will be jointly discussed.

A reasonable estimate of the worldwide number of species of bees is 20,000 (Stephen et al., 1969). In their treatment of the bees of northwestern North America, these authors estimate that perhaps 10% of the bees of the Northwest are social or semisocial, another 10% are parasitic, and the remainder are solitary. These figures are probably reasonable for all the temperate zone bee faunas although solitary bees may be even more abundant in warm, arid regions. Species of social bees are more numerous (on a percentage basis) in both the tropic and arctic zones. Linsley (1958, 1978) has reviewed the ecology of the solitary bees; Michener (1979) has treated their biogeography; and Thorp (1979) has provided an overview of the structural, physiological, and behavioral adaptions of bees in relation to pollination. From these and other papers it is apparent that the natural history of the solitary bees differs dramatically from that of members of the Apidae (the truly social bees). Other than preparing the nest, provisioning the nest with pollen, and depositing a fertilized egg, females of solitary species are dissociated from the offspring. Recruitment of workers for the exploitation of plentiful forage does not occur among nonsocial bees, and a worker caste is lacking.

Individual foragers of the eusocial genera tend to exhibit floral constancy—the visitation of only one floral resource per foray (Grant, 1950; Free, 1963, 1970; Heinrich, 1976)—with constancy presumably based on either nectar or pollen or both. A colony as a whole, however, collects forage from a diversity of crops. Honeybees and bumblebees are therefore anthophilous generalists. In addition, individual foragers may shift from one forage source to another as one crop wanes and another waxes, and the shifts may be diurnal or seasonal. Pollen and nectar, the caloric source for foragers, may also be stored in the colony.

Species of solitary bees, in contrast, may exhibit oligolecty, the collection of pollen from only one group of closely related species of a single family, genus, or section, even when other sources are abundant and available. Examples of oligolectic bee taxa, taken from the genus *Andrena*, are *Andrena (Onagendrena)*, in which all species are pollen collectors on the Onagraceae (Linsley et al., 1963a,b, 1964, 1973; MacSwain et al., 1973); *A. vercunda* and related species on *Pyrrhopappus* (Shinners, 1958; LaBerge, 1967; Barber and Estes, 1978); and *A. erigeniae* on the single species *Claytonia virginica* (Schemske, 1977; Schemske et al., 1978; Reese and Barrows, 1980). (The latter case may be considered monolecty.) Oligolecty is most significant in warm, temperate, arid regions of the world where solitary bees are most abundant (Michener, 1979).

Bees that are pollen specialists are typically closely coadapted to the host plant (Linsley, 1958, 1978), more so than is typical among members of the Apidae. Specialization includes seasonal and diurnal periodicity, structural modification for the collection and transport of pollen, and intrafloral behavior, all of which are often intimately related with the floral features of the plant (Thorp, 1979). (We hasten to add that many taxa of solitary bees are polylectic, foraging a variety of unrelated taxa for pollen.)

The foraging strategy employed by generalist bees, whether or not they are constant on a single foray, is based on a more protracted foraging season than for specialists. Thus maximum total levels of pollen may be collected during a season, increasing the total brood that can be reared. With extreme specialization for the pollen of only a single group, oliogolectic bees yield the greater available supplies of pollen for the advantage of greater efficiency involved in the collection of pollen from only a single source. This strategy may be more profitable if the oligoletic bee is the sole pollen collector for the plant species. This latter pattern is rather common and we present three cases, merely as examples:

Callirhoe and Diadasia. Callirhoe is a member of the Malvaceae, and, typical of genera of the family, it produces a prodigious pollen crop (over 25,000 grains/flower and up to about 150 flowers/plant for *C. scabriuscula*), with the individual grains among the largest within the flowering plants (about 100 μm in diameter). A wide array of social and solitary visitors collect nectar from the flowers; however, only bees of the genus *Diadasia* actively collect pollen (plus another suspected oligoleg, *Melissodes intorta*). All 23 Anglo-American species of *Diadasia*, with the exception of six, are oligolectic on members of the mallow family. *Diadasia* is restricted to western North and South America. The other foragers on *Callirhoe* actively discard the copious amounts of pollen grains that adhere to their bodies during nectar collection. Apparently this rejection of *Callirhoe* pollen is similar to that described for cotton, also a member of the Malvaceae. *Diadasia* have been rarely observed taking pollen from *Kallstroemia*, the pollen of which is rejected by many other bees. (References: Linsley et al., 1956; Linsley and MacSwain, 1957, 1958; Cazier and Linsley, 1974; Adlakha, 1969; Eickwort, et al., 1977; Amos, 1981.)

Pyrrhopappus and Hemihalictus. Pyrrhopappus is a ligulate composite with all species but one occurring in the southeastern and south-central United States. *Hemihalictus lustrans*, the only species of the genus, takes pollen only from species of *Pyrrhopappus*. They do not, however, collect nectar; rather they have been noted taking nectar from *Tamarix* (Tamaricaceae), *Ruellia* (Acanthaceae), and *Ludwigia* (Onagraceae). This lack of fidelity for nectar sources is not uncommon among pollen specialists (Michener, 1979). Characteristic of many oligoleges, the mode of pollen collection is specific, and the anthers are opened by the bees prior to dehiscence; this, in combination with the early morning foraging, excludes many potential competitors for the pollen—pollen tends to be completely removed by 0900 hours Central Daylight Time—except for females of the *Andrena verecunda*-complex. The seasonal and daily cycles of activity of *H. lustrans* and *P. carolinianus* are highly coincident. However, on *P. geiseri,* bees of the *Andrena verecunda*-species complex replace *Hemihalictus* prior to the closure of the flowering head. The geographic range of *Hemihalictus* and the genus are more tightly correlated than the examples given by Michener (1979), even to the extent that the disjunct *P. rothrockii,* a plant of the southern Rocky Mountains, also supports *H. lustrans.* (References: Michener, 1947; Shinners, 1951, 1953; Daly, 1961; Northington, 1974; Estes and Thorp, 1975; Barber and Estes, 1978.)

Ipomoea pandurata-complex and Cemolobus, Melitoma, and Ptilothrix. Ipomoea pandurata, I. leptophylla, and *I. longifolia* are closely related species, and, with the exception of an area of overlap in eastern Oklahoma and north-central Texas, the two former taxa are allopatric. The monotypic anthophorid genus *Cemolobus* forages for both nectar and pollen in the early morning from *I. pandurata.* *Melitoma taurea* is also a matinal visitor to *I. pandurata* for nectar and pollen, and visits other members of the genus *Ipomoea* and the closely related genus *Calystegia* as well. Unfortunately, population demographics are unavailable for these two visitors. Other bees may forage *I. pandurata* for nectar, but most of them are oligolectic visitors to unrelated plant genera with the exception of *Bombus* spp. *Bombus* appears to pick up casual pollen rather than carry the pollen in the scopae. *Ipomoea leptophylla* is foraged for both primary attractants only by *Melitoma grisella* (Anthophoridae); all other anthophilous visitors collect only nectar. The southwestern U. S. and Mexican *I. longifolia* has two oligoleges—*Melitoma segmentaria* and *Ptilothrix sumichrasti.* As the flowering season for *I. longifolia* concludes, these bees may switch to other morning glory species. (References: Robertson, 1891, 1925a,b, 1928, 1929; Linsley et al., 1956; Linsley, 1960; Austin, 1978.)

Within these examples, *Hemihalictus lustrans* on *Pyrrhopappus carolinianus, Melitoma grisella* on *Ipomoea longifolia,* and *Diadasia* on *Callirhoe,* interspecific competition for pollen is restricted, as they are the only collectors. In each circumstance the bees are also capable of extending their foraging season, via the collection of pollen from closely related species with overlapping flowering seasons. Whether each becomes inactive following daily closure of the host flowers, as *H. lustrans* does, is unknown, but that is typical behavior (Eickwort and Ginsberg, 1980).

The biology of pollen foraging has received less attention than has nectar foraging, and the pollen foraging patterns of solitary bees even less (Eickwort and Ginsberg, 1980). There is no reason to believe, however, that foraging patterns of pollen-collecting solitary bees would deviate significantly from models based on optimal foraging strategy (Levin, 1978; Pyke, 1978a,b,c, 1979; Waddington and Holden, 1979; Waddington, 1980). In fact, for solitary bees such as *Hemihalictus* (Estes and Thorp, 1975) that forage for pollen (1) in the cool early morning, (2) immediately upon departing the burrow without first taking nectar, and (3) for two to three hours before ultimately collecting nectar, the pressure to increase foraging efficiency may be even greater. Although Estes and Thorp did not determine flight distances of individual bees, circumstantial evidence from the transfer of marker dyes indicates that flight distances of *H. lustrans* are indeed minimal. They also noted that intercapitular flights tend to be to the next encountered capitulum along the flight path. Flight direction of *H. lustrans* is heavily skewed from randomness and oriented to the west, toward the heliotropic blossom faces.

The same workers with Briggs (unpub.) studied foraging patterns of *Lasioglossum titusii,* a closely related matinal species—*Hemihalictus* is a segregate of *Lasioglossum*—foraging another ligulate composite, *Agoseris heterophylla,* which produces both nectar and pollen. Average flight distance coincides with mean distances between nearest neighbors for the crop (Table 26-1). Therefore these bees must be selecting the shortest flight pattern.

Table 26-1. Comparison of flight distance of *Lasioglossum titusii* with nearest neighbor spacing of *Agoseris heterophylla*.

	N	Mean	t-value	s²	X² value
FNN	88	19.96		53.22	
			5.30[b]		80.3[b]
FD	107	9.65		7.88	
			0.51		0.0007
SNN	22	8.76		7.81	

[a]FD—flight distance; FNN—four nearest neighbors; SNN—single nearest neighbor. All measurements are in centimeters. Statistics employed were the nonparied t-test and the Bartlett test for homogeneity of variance. Flight distance differed from the four nearest neighbors at the 0.001 level of probability, but was not significantly different from the single nearest neighbor.
[b]Differ at the 0.001 level.

The most elegant study of foraging patterns of solitary bees is that of Waddington (1979a), involving *Agapostemon texanus*, *Augochlorella striata*, and an unnamed species of *Lasioglossum* on artificial populations of *Convolvulus arvensis* at three densities. The bees were collecting both nectar and pollen. The results for all three tested bee species were similar, with flights tending to be very short, but increasing significantly with increase in interplant spacing. Flight distances of *Lasioglossum* were, however, shorter than those of the other two bee species. In addition, the bees tend to forage on a direct path through the population.

The results of these three studies of foraging distance are similar and comparable to patterns reported for honeybees and bumblebees. One distinction concerning flight direction was found for *Hemihalictus*, in which practically all the flights were away from the rising sun, or more accurately toward the east-facing capitulae of *Pyrrhopappus*. Apparently this unidirectionality is uncommon, but the remaining studies in which flight angle was determined demonstrate a typical conformity to a base-line of flight.

Production and secretion of nectar requires an expenditure of photosynthate or plant energy. Consequently, the cessation of nectar flow would be an advantage if, and only if, reproductive effort were not hampered. For strictly anemophilous species the nectar is inconsequential in the pollination process, and most such flowers are nectarless. For entomophilous species, however, foraging insects must be induced to visit the resource-depleted pistillate phase. As expected, Stanley and Linskens (1974, p. 108) report that plants that supply only pollen are typically anemophilous; however, a number of bee-pollinated, dry flowers are known to us from the southern prairie [and adjacent woodlands] (*Cassia fasciculata, Commelina erecta, Helianthemum rosmarinifolium, Pyrrhopappus carolinianus, Sabatia campestris,* and *Solanum rostratum*). Grant and Grant (1979) report that *Echinocereus fasciculata* is probably dry as well, though other members of the genus produce nectar (Leuck, 1980). Of the five pollen flowers other than *Echinocereus*, only *P. carolinianus* is known to be associated with a specific, oligolectic pollinator. A congener of this species, *P. geiseri*, does yield minute (less than 0.1 ml/floret) amounts of nectar (Barber and Estes, 1978). Even though *Hemihalictus lustrans* collects pollen from *P. geiseri* as well as *P. carolinianus*, it does not collect nectar, but this results, at least partially, from the timing of nectar secretion, which occurs after the pollen has been largely collected and andrenid bees have replaced the halictids on the heads. The andrenids collect the nectar, the availability of which is coordinated with stigma opening; thus they are the pollinators of *P. geiseri*. In all likelihood the production of nectar is the primitive state in *Pyrrhopappus*, but the remaining three species of outcrossers need to be tested. *Helianthemum* (Dugan and Estes, unpub.) and *Sabatia* (Taylor, unpub.) are pollinated exclusively by solitary bees, whereas flowers of *Cassia fasciculata* and *Solanum rostratum* are pollinated primarily by bumblebees that vibrate the anthers to release pollen from apical pores (Bowers, 1975; Thorp and Estes, 1975). Both *C. fasciculata* and *S. rostratum* are also visited by solitary bees, some of which serve as pollinators of *Cassia*. *Commelina erecta* has three sterile anthers that apparently serve as visual attractants, one feeding anther, and two obscure pollinating anthers (McCollum et al., unpub.). Flowers or florets of all the taxa except *Sabatia* are open for just one day, and there is considerable overlap in the time of pollen presentation and stigma receptivity. Only *Sabatia* is strongly dichogamous, and bees are apparently deceptively induced to visit the pistillate flowers in search of pollen. We are unaware of estimates of the number of pollen-only flowers for any flora, and monographs and floristic treatments seldom, if ever, list the condition of nectar secretion or its absence. We suspect, however, that the two strategies employed by these five taxa in Oklahoma and Texas to elicit visits to the stigma are the most common—that is, lack of dichogamy-herkogamy or the employment of deception. However, for *Hepatica,* Motten (unpub.) reports that the species is protogynous, but this may be a special case as the visitors are male andrenids, and the plant is highly autogamous. The protandrous *Claytonia virginica* is not a dry flower, but only limited quantities of nectar are produced during the staminate phase, with more produced in the largely pollen-free pistillate phase (Eickwort et al., 1977).

Based on our literature search and on personal observations, solitary bees generally collect both pollen and nectar on each floral visit, and often collect both simultaneously, as for the ipomoea bees described above. [Honeybees tend to specialize with respect to either pollen or nectar collection; e.g., only 17% were reported to collect both on single forays (Parker, 1927, as cited in Stanley and Linskens, 1974), but pollen foragers can draw on the colony's provision of nectar.] The advantage for a solitary bee to forage only for pollen on forays, for example, *Osmia sparsiflora* (Rust and Clement, 1977), and to specialize on pollen flowers, for example, *Hemihalictus lustrans* (Estes and Thorp, 1975), is obscure, although, as Michener (1979) states, specialization is typically based on pollen and not on nectar. Perhaps the often used, but rarely tested, hypothesis that foraging specialization is advantageous is appropriate (Eickwort and Ginsberg, 1980). It also seems reasonable to us that, in some instances, bees first became oligolectic on a nectariferous plant species that then lost the ability to exude nectar, but the bees remained faithful to the plant.

The mode(s) of origin of oligolecty remain problematic even though the subject has been

the source of speculative writing since the early part of the century (Robertson, 1914). The most recent views are those of Linsley (1958, 1978), Cruden (1972), and Thorp (1979). Linsley (1978) suggests that the feeding patterns associated with oligolecty are learned each generation, conditioned by the pollen stored in the natal cell—the only feasible alternative is that the selection of the pollen-host plant is inherent. He notes that most solitary bees emerge prior to pollen dehiscence and for several days prior to nest building. This time, suggests Linsley, is spent in collecting nectar, and thus becoming familiar with floral location and timing. Clearly that is not the situation for pollen flowers or for bees that take pollen from one plant and nectar from others. Also, apparently not all bees emerge in time for preconditioning exercises (Rust et al., 1974; Rust and Clement, 1977; Batra, 1980); however, the presence of bees prior to opening of the specific flower host might well go unnoticed. The example cited by Linsley, *Andrena (Onagandrena) linsleyi,* as an example of learned response cannot unequivocally be attributed to learning. In some seasons the *A. linsleyi* must forage on an alternative host, until the preferred host emerges; the bees then switch. Genetic predisposition, though perhaps unlikely, cannot be excluded as a rational explanation without rigorous testing.

Linsley's model circumscribes selection of the host each year; it does not attempt to describe switches from one host to a second, or the onset of specialization from a nonoligolectic ancestor. Thorp (1969) considers stress in association with a paucity of preferred host-pollen to be a significant stage in the former process. Certainly if populations of bees are to survive stress conditions, they must either remain inactive or forage from alternate hosts. Both situations are known (Thorp, 1979). If switches occur, the bees may later resume oligolecty on the first host or become oligolectic on the new host. Experimental evidence is strongly called for but unfortunately unavailable.

The origin of the Apoidea is obscure and the early fossil record of this taxon even more incomplete than that of the flowering plants; presumably, however, the bees evolved in response to the rise of angiospermy, or flowering plants and bees coevolved (Michener, 1979). Based on the relative appearance of the two groups in the fossil record, the former is more likely. Thus prototype bees would have become dependent upon the pollen of the early flowering plants. The pollen of most entomophilous plants is oily and often spiny, whereas anemophilous plants produce smooth, dry pollen. It is inconceivable that the wind-pollinated ancestors of the insect-pollinated angiosperms or pro-angiosperms produced sticky pollen: the detriment to anemophily would have simply been too great. We presume therefore that the first pollen-collecting insects must have gathered buoyant, nonsticky pollen of wind-pollinated plants. With increasing foraging and entomophilous pollination, selection would have favored a pollen that more readily adhered to the body of the insect. Bees ultimately responded via the development of elaborate means of collecting the pollen, packing the pollen in pollen-baskets on the hind tibia or ventral abdomen, and transferring it to their burrows—assuming the ancestral bees were mining bees. Some members of the Colletidae, however, carry pollen in the crop rather than externally in a scopa; these bees, segregated in two subfamilies, Euryglossinae and Hylaeninae, are considered the most primitive extant members of the Apoidea (Michener, 1979). The Apidae, including the honeybee, have evolved even more elaborate corbiculae for pollen transfer as well as elongate mouth-parts for taking nectar from deep sources. The origin of oligolecty from polylectic taxa is however, largely unexplored, even theoretically. Some families and genera (examples are Panurgidae and *Diadasia*) exhibit a propensity for oligolecty and presumably arose from oligolectic ancestors, or oligolecty was a factor in their origin.

PLANT REPRODUCTIVE SYSTEMS

Rather than discuss breeding systems in general, we wish to make two brief appeals—the first is for plant systematists to consider more carefully potential variation in breeding systems and to explore pollination systems in greater detail, and the second is for greater care in the use of the descriptive terminology of reproductive biology.

Variation in Breeding Systems

With some notable exceptions exemplified by the excellent work of Robert Ornduff, Otto Solbrig, Herbert Baker, Robert Cruden, *inter alios,* and their students, plant systematists have too often failed to observe intrafloral ecology or to conduct sufficiently sophisticated experiments with breeding systems to comprehend the pollination-reproductive system, and hence have often missed valuable ecological and evolutionary evidence bearing on the phylogeny and classification of the plants under investigation. The examples expressed below are intended merely to represent a general tendency and not to be critical of the authors. In fact, the examples chosen are taken from outstanding taxonomic monographs.

All treatments of the genus *Cassia* prior to 1975, in floras and monographs we examined, fail to describe the variation of the corolla and androecium of *C. fasciculata* that Thorp and Estes (1975) consider crucial in the pollination syndrome of the species. The anthers are in two groups, one with nine anthers and the other with only one. The group of nine is erect and shielded by the erect upper petal, the cucullus. Large bees, the pollen vectors, must therefore approach the upper nine—feeding anthers—from slightly below. The pollinating anther and style are diverged and roughly parallel to the larger, lower petal and both are in the approach flight-path of the bees. It appears that as the bees vibrate pollen onto their venters from the feeding anthers, pollen from the pollinating anther is deposited along the side of the bee. Results from preliminary experimental floral manipulation substantiate this hypothesis. Flowers from which either the cucullus or the lower petal had been removed failed to set seed, whereas fruit-set in control plants was almost full (Stauffer, Estes, and Sullivan, unpub.). The situation is similar to that elegantly tested and described by Bowers (1975) for *Solanum rostratum*. Examination of the pollination system thus drew attention to a set of characters that are difficult to assess on herbarium specimens, even with foreknowledge of the petal and stamen arrangements.

The features taken with the pollination system may also be useful in phylogenetic inference. Thus the family of *Cassia*, the Fabaceae, exhibits a pronounced evolutionary trend from actinomorphic to papilionaceous flowers. The advantage, or adaptive strategy, of strongly bilateral flowers has been commented upon by Stebbins (1974, p. 63). The situation in *Cassia*, subfamily Caesalpinioideae, reveals that slight deviations from radial symmetry may be selectively advantageous and in somewhat the same fashion as for the subfamily Lotoideae; therefore, it is not unreasonable to assume that selection could induce a change from actinomorphy to incipient zygomorphy to strong zygomorphy.

In his excellent monographic treatment of *Pyrrhopappus*, Northington (1974) correctly noted the diversity of anthophilous visitors—beetles and bees—to heads of *Pyrrhopappus*, though he failed to observe the oligotrophic noctuid moth, *Schinia mitis* (Zwick and Estes, 1981). He concluded that all the visitors were pollinators; however, as previously discussed, *Pyrrhopappus carolinianus* and *P. geiseri* have specialized and different pollination systems, lending credence to Shinners' (1951, 1953) recognition of *P. geiseri* as a discrete taxon, rather than the treatments of Correll and Johnston (1970) and Northington (1974).

In an insightful study by Anderson (1979), what had been previously described as two hermaphroditic species in the genus *Solanum* were discovered to be different sexes of the same species. *Solanum appendiculatum* and *S. inscendens* were found to function as staminate and pistillate phases, respectively. Anderson documented the dioecious nature of this taxon with evidence from reciprocal crosses; pollen tube growth; and pollen quantity, morphology, and stainability. Recently, Anderson and Symon (pers. comm.) found that a similar situation exists in several androdioecious solanums of Australia. In these species, flowers that appear hermaphroditic are functionally female. It appears in both instances that pollen

is retained in pistillate flowers to attract and reward pollinators, since flowers in the genus typically lack nectar. Anderson and Symon believe that the development of dioecy in the cases described is a broad convergence for an apparently advantageous breeding system, since the Australian taxa are placed in a different subgenus from the Central American species.

Diversity in breeding and pollination systems has long been recognized as a major element in the evolution of the flowering plants (Fryxell, 1957; Percival, 1965; Faegri and van der Pijl, 1966). Fryxell, Percival, and Faegri and van der Pijl realized full well that the reproductive and pollination syndromes they described were but end-points along continuously varying spectra. The variation might take one of three modes: (1) a switch of reproductive or pollination schemes in response to an environmental factor; (2) genetically incomplete systems; or (3) interpopulational variation with respect to pollination and/or reproduction. Examples of each are discussed below.

Environmentally Induced Change. The phenomenon in which plants alternatively produce cleistogamous and chasmogamous flowers is well known and has been reviewed in reasonable detail by Faegri and van der Pijl (1966). In the case of *Lamium amplexicaule* cleistogamous flowers are produced early and at the close of the season, and chasmogamous ones during midseason. Floral visitation may also be influenced by environmental fluctuations; in the case already described, *Andrena (Onagandrena) linsleyi* is typically a visitor of *Oenothera deltiodes*, but in certain years *Oe. clavaeformis* is available as a pollen-host when the bees emerge. The bees forage this species until *Oe. deltoides* emerges later in the season (Linsley, 1978).

Genetically Incomplete Systems. Self-compatibility and self-incompatibility are the well-defined and reasonably well-understood extremes of a continuum; they are not absolutes. For instance, Perkins and co-workers (Perkins et al., 1975) found that seed set under exclosures for four species of *Verbena* deviated from the extremes of 0 and 100% and were not equivalent to percentages of seed set under conditions of open pollination. This situation is certainly far from unique.

Interpopulational Variation. Acer pensylvanicum occurs in the eastern deciduous forest of North America. Duncan Porter (pers. comm.) reports that apomixis is an important factor in the reproduction of this small tree in the southern states, but the species produces seed via outcrossing in New England (Sullivan, 1980). Also, recall that alfalfa is visited extensively by honeybees in most of its range, but not in parts of Idaho (Nye and Mac-Kenson 1970). Honeybees are capable pollinators in Arizona, but less efficient in Canada, except when abundant (Bohart, 1960). Thus the pollination syndrome of alfalfa varies. In addition, Whalen (1978) found evidence for character displacement and hence interpopulation variation in *Solanum grayi* and *S. lumholtzianum.* Flower size is similar for the two species over much of their respective ranges, but in the zone of overlap the flowers differ in size, with a small-flowered race of *S. grayi* displacing the more typical race. This size differential is apparently sufficient to affect intrafloral behavior and to mechanically isolate the two in the area of sympatry.

These papers reveal only a small portion of the diversity encompassed by many species of flowering plants. In addition, when comprehensive, detailed studies of pollination-reproductive systems are conducted, anomalous systems may be revealed: as examples, entomophilous intrafloral pollination (Estes and Brown, 1973); temporal dioecy (Cruden and Hermann-Parker, 1979; Amos, 1981); pollen sterile flowers and duodichogamy (Sullivan, 1980); transition from perfect to functionally staminate flowers in a single inflorescence (McCollum et al., unpub.); etc.

If a biosystematic study includes an investigation of breeding and pollination systems, the investigator should be aware of atypical modes of reproduction and of the potential for variation in the system. Experiments and observations should be planned accordingly: (1) Observations of the anthophilous fauna should be conducted at several sites and preferably across the range of the plant taxon. (2) Detailed observations of intrafloral behavior and pollen load, to include purity of the load and its position in relation to the stigma, must be made. (3) Floral phenology and visitor demography should also be included; most important are the period of anther dehiscence and the period of stigmatic receptivity. (4) Breeding experiments should include insect exclusion tests, artificial cross-pollinations, and self-pollinations—including crosses within and between flowers on single plants. Notice should be made of spatial distances between plants involved in cross-pollinations. The experiments should also involve numerous individuals, not merely numerous flowers on a few individuals, and numerous populations. Further interpopulation crosses should be scheduled. (5) The observer should be cognizant of such potential phenomena as nondehiscent or sterile anthers, deformity of the style or pistil, self-pollination via insect or wind action, or temporal dioecy—that is, the separation of sexual phases on a plant with perfect flowers via timing.

If resources permit, the scientist should also (1) note quality and quantity of nectar and pollen; (2) observe flight patterns of pollinators; and (3) monitor gene flow via genetic markers, chemical or physical markers, or genetic fine-analysis.

Variation in breeding systems has frequently been employed by agronomists, especially the use of male-sterility factors in the production of hybrid seed crops, for example, carrots (Erickson and Peterson, 1979).

Descriptive Terminology

The terminology, jargon if you wish, of pollination-reproduction has become confused through the use of significant terms from the fields of pollination and reproduction as homonyms. For instance, unmodified use of the term "autogamy" could imply self-pollination, self-fertilization, or a breeding system. Frequently in our literature survey we encountered the term used so loosely we could not ascertain the intended meaning. We believe authors dealing with pollination-reproduction should excercise extreme care in employing terms such as these. Another solution is to establish an alternative terminology. Table 26-2

Table 26-2. A proposed system of nomenclature for the fields of pollination biology and plant reproductive biology. The terms as proposed refer to gamic systems only.

		Relationship	
		Pollination	Fertilization
One plant	One flower	Autogamy	Automixis
	Two flowers	Geitonogamy[a]	
Two plants		Xenogamy[a]	Allomixis[b]

[a]*Geitonogamy* and *xenogamy* collectively are *allogamy*.
[b]*Panmixis* and *vicinism* are special cases of *allomixis*.

provides a matrix of relationships and a proposed usage with two new terms. The system employs the pollination terms already proposed by Faegri and van der Pijl (1966) and Percival (1965). In addition the suffix "mixis," which has been employed in terms denoting lack of fertilization (apomixis) or broadly based fertilization (panmixis), is also widely accepted. Automixis then would be self-fertilization, and allomixis, crosses between individuals. Panmixis and vicinism (Grant, 1971) would be special cases of allomixis. Utilization of this nomenclatural system would resolve ambiguity in the description of modes of pollination and reproduction. We fully appreciate that for many, or perhaps most, reproductive-pollination systems the differences are inconsequential. Thus automixis is typically preceded by autogamy and allomixis by allogamy. However, in species that are self-compatible, in which a single genotype produces massive numbers of attractive blooms (e.g., trees or rhizomatous perennials), most pollination events will be zoophilous and geitonogamous, but the reproduction will be automictic. The incidence of geitonogamy is largely uninvestigated, but the phenomenon does occur. *Aristolochia californica,* for instance, is a rhizomatous vine with each individual genotype producing tens of thousands of flowers. The flowers deviate somewhat from the typical trap-blossom, but they are strongly dichogamous, thus inhibiting autogamy. The pollinating flies effect automixis via geitonogamy as well as allomixis via xenogamy. The flight patterns of the dipteran visitors (predominately midges, gnats, sandflies, and fungus flies) are,

however, unknown (Estes, Thorp, and Briggs, unpub.). A similar situation is known for the bee-pollinated *Ludwigia peploides* ssp. *glabrescens,* populations of which are often initiated via a single rhizome washed downstream (Estes and Thorp, 1974). Estes and Brown (1973) even argue that mutations resulting in self-compatibility would be favored in species with this set of reproductive-pollination features.

SUMMARY

Pollination biology is to us a unified field of science but comprised of scientists with diverse interests, from theoretical to applied and from botanical to zoological. Any division is artificial and seems to be based only on place of publication—we wish to stress that the four-way schism based on type organism (primary producer vs. consumer) and field (agriculture vs. biology) is routinely crossed by many, or even most, pollination biologists. We believe a serialized *Index* would be of most benefit to individuals who are not practicing pollination biologists, but rather employ the results of pollination research—the systematists, ecologists, and phylogeneticists.

In terms of the cross-pollination of ideas from agricultural and biological disciplines, the general field of foraging behavior as related to reproduction (Levin, 1978) seems most fruitful. What is required from theoretical population biology is a general model or paradigm concerning optimal foraging strategy that permits agricultural scientists to test the effects of perturbations, singly and in tandem. Certainly theorists, such as Pyke and Levin; experimentalists, such as Waddington and Gary; and naturalists, such as Thorp and Macior, have made great strides in accomplishing this goal. Still lacking are important data-sets concerning levels of resources, attractive units, and heritability, among others.

We appreciate the role of *Apis mellifera* in the pollination of crops and realize that honeybees will continue to be the most significant management tool for seed production. However, probably reflecting our research bias, we also believe that for certain bee-pollinated, out-breeding crops with problems associated with pollination, the coevolved plant-native bee systems should be explored more thoroughly. The potential seems clear: oligolectic bees tend to be the most efficient pollinators—more efficient than wild polyleges which frequently have been studied, and more efficient than honeybees. For many crops the ease of managing honeybees overshadows any potential reduction in yield, but perhaps not for crops where the yield loss is economically significant. Parker (1981a,b), for instance, has clearly shown that native oligoleges are far more proficient at pollinating sunflowers by up to one order of magnitude. The oligoleges *Andrena helianthi* and *Melissodes agilis* and *Helianthus annuus* have evolved in North America in a mutualistic relationship; so those results are hardly surprising. The use of wild bees as pollinators is not without problems, however; the potential problems are local abundance, seasonality, maintenance of nesting sites, and pesticide kills. Nevertheless, for crops such as cotton, native bees from the center of origin of cotton or *Diadasia* could perhaps, through selection, become as specific as oligoleges on native mallows. The potential seems significant. The bees would also emerge in coincidence with cotton blossoms and become inactive as the season closes, reducing the chances of contact with pesticides. What is required is more evidence concerning the mode of origin of oligolecty—albeit agricultural scientists working to institute the system would likely contribute to knowledge of the origin—and knowledge of foraging, nesting, and reproductive behavior of the bees. For some crops the effort might not

be worthwhile; for others, however, the effort could well result in significant improvements in yield.

ACKNOWLEDGMENTS

We wish to express our appreciation to R. W. Thorp, University of California at Davis; E. H. Erickson, United States Department of Agriculture, Madison, Wisconsin; and G. E. Uno, University of Oklahoma, for their contributions to this manuscript. Thorp and Erickson provided helpful suggestions as we planned the scope of the chapter. Uno and Erickson critically read, criticized, and edited the manuscript; their thoughtful comments were instrumental in improving the discussion.

We also wish to recognize Mary Mason, who patiently and cheerfully typed the manuscript through several revisions.

LITERATURE CITED

Adlakha, R. L. 1969. A systematic revision of the bee genus *Diadasia* Patton in America north of Mexico (Hymenoptera: Anthophoridae). Unpublished doctoral dissertation, University of California, Davis.

Amos, B. B. 1981. Reproductive studies in the genus *Callirhoe* (Malvaceae). Unpublished doctoral dissertation, University of Oklahoma, Norman.

Anderson, G. J. 1979. Dieocious *Solanum* species of hermaphroditic origin is an example of a broad convergence. *Nature* 282:836–838.

Austin, D. F. 1978. Morning glory bees and the *Ipomoea pandurata* complex (Hymenoptera: Anthophoridae). *Proc. Ent. Soc. Washington* 80:397–402.

Bachman, W. W. and G. D. Waller. 1977. Honeybee responses to sugar solutions of different compositions. *J. Apic. Res.* 16:165–169.

Baker, H. G. and I. Baker. 1973. Amino acids in nectar and their evolutionary significance. *Nature* 241:543–545.

Barber, S. C. and J. R. Estes. 1978. Comparative pollination ecology of *Pyrrhopappus geiseri* and *Pyrrhopappus carolinianus*. *Am. J. Bot.* 65:562–566.

Batra, S. W. T. 1980. Ecology, behavior, pheromones, parasites, and management of the sympatric vernal bees, *Colletes inaequalis, C. thoracicus,* and *C. validus. J. Kansas Ent. Soc.* 53:509–538.

Benedek, P. and E. Goal. 1972. The effect of insect pollination on seed onion with observations on the behavior of honeybees on the crop. *J. of Apic. Res.* 11:175–180.

Bohart, G. E. 1957. Pollination of alfalfa and red clover. *Annu. Rev. Ent.* 2:355–380.

———. 1960. Insect pollination of forage legumes. *Bee World* 41:57–64, 85–97.

Bowers, K. A. W. 1975. The pollination ecology of *Solanum rostratum* (Solanaceae). *Am. J. Bot.* 62:633–638.

Buchmann, S. L. and C. E. Jones. 1980. Observations on the nesting biology of *Melissodes persimilis* CKll. (Hymenoptera: Anthophoridae). *Pan-Pacific Entomol.* 56:200–206.

Campana, B. J. and F. E. Moeller. 1977. Honeybees: Preference for and nutritive value of pollen from five plant sources. *J. Econ. Ent.* 70:39–41.

Caron, D. M. 1979. New research on bee pollination of crops: a report of the Fourth International Symposium on Pollination. *Bee World* 60:128–136.

Cazier, M. A. and E. G. Linsley. 1974. Foraging behavior of some bees and wasps at *Kallstroemia grandiflora* flowers in southern Arizona and New Mexico. *Am. Mus. Novit.* 2546:1–20.

Colin, L. J. and C. E. Jones. 1980. Pollen energetics and pollination modes. *Am. J. Bot.* 67:210–215.

Collison, C. H. and C. E. Martin. 1979. Behaviour of honeybees foraging on male and female flowers of *Cucumis sativus*. *J. Apic. Res.* 18:184–190.

Corbet, S. A. 1978. A bee's view of nectar. *Bee World* 59:25–32.

Correll, D. S. and M. C. Johnston. 1970. *Manual of the Vascular Plants of Texas*. Texas Research Foundation, Renner.

Cruden, R. W. 1972. Pollination biology of *Nemophila menziezii* (Hydrophyllaceae) with comments on the evolution of oligolectic bees. *Evolution* 26:373–389.

———. 1977. Pollen–ovule ratios: a conservative indicator of breeding systems in plants. *Evolution* 31:32–46.

Cruden, R. W. and S. M. Hermann-Parker. 1979. Butterfly pollination of *Caesalpinia pulcherrima*, with observations on a psychophilous syndrome. *J. Ecol.* 67:155–168.

Daly, H. V. 1961. Biological observations on *Hemihalictus lustrans*, with a description of the larva (Hymenoptera: Halicitidae). *J. Kansas Ent. Soc.* 34:134–141.

Eickwort, G. C. 1977. Aspects of the nesting biology and descriptions of immature stages of *Perdita octomaculata* and *P. halictoides* (Hymenoptera: Andrenidae). *J. Kansas Ent. Soc.* 50:577–599.

Eickwort, G. C. and H. S. Ginsberg. 1980. Foraging and mating behavior in Apoideae. *Annu. Rev. Ent.* 25:421–446.

Eickwort, G. C., K. R. Eickwort, and E. G. Linsely. 1977. Observations on nest aggregations of the bees *Diadasia olivacea* and *D. diminuta* (Hymenoptera: Anthophoridae). *J. Kansas Ent. Soc.* 50:1–16.

Erickson, E. H. and C. E. Peterson. 1979. A synchrony of floral events and other differences in pollination foraging stimuli between fertile and male-sterile carrot inbreds. *J. Am. Soc. Hort. Sci.* 104:639–643.

Erickson, E. H., L. O. Whitefoot, and W. A. Kissenger. 1973. Honeybees: a method of delimiting the complete profile of foraging from colonies. *Environ. Ent.* 2:531–535.

Erickson, E. H., R. W. Thorp, and D. L. Briggs. 1975. Comparisons of foraging patterns among honeybees in disposable pollination units and in overwintered colonies. *Environ. Ent.* 4:527–530.

———. 1977. The use of disposable pollination units in almond orchards. *J. Apic. Res.* 16:107–111.

Erickson, E. H., R. W. Thorp, D. L. Briggs, J. R. Estes, R. J. Daun, M. Marks, and C. H. Schroeder. 1979. Characterization of floral nectars by high-performance liquid chromatography. *J. Apic. Res.* 18:148–152.

Estes, J. R. and L. S. Brown. 1973. Entomophilous, intrafloral pollination in *Phyla incisa*. *Am. J. Bot.* 60:228–230.

Estes, J. R. and R. W. Thorp. 1974. Pollination in *Ludwigia peploides* ssp. *glabrescens* (Onagraceae). *Bull. Torrey Bot. Club* 101:272–276.

———. 1975. Pollination ecology of *Pyrrhopappus carolinianus* (Compositae). *Am. J. Bot.* 62:148–159.

Faegri, K. and L. van der Pijl. 1966. *The Principles of Pollination Ecology*. Pergamon Press, New York.

Faulkner, G. J. 1976. Honeybee behaviour as affected by plant height and flower colour in brussels sprouts. *J. Apic. Res.* 15:15–18.

Frankel, R. and E. Galun. 1977. *Pollination Mechanisms, Reproduction and Plant Breeding*. Springer-Verlag. New York. Monographs on Theoretical and Applied Genetics.

Free, J. B. 1963. Flower constancy of honeybees. *J. Anim. Ecol.* 32:119–131.

———. 1965. Attempts to increase pollination by spraying crops with sugar syrup. *J. Api. Res.* 4:61–64.

———. 1968. Dandelion as a competitor to fruit trees for bee visitors. *J. Appl. Ecol.* 5:161–178.

———. 1970. The flower constancy of bumblebees. *J. Anim. Ecol.* 39:395–405.

Fryxell, P. A. 1957. Mode of reproduction of higher plants. *Bot. Rev.* 23:135–233.

Gary, N. E. 1971. Magnetic retrieval of ferrous labels in a capture–recapture system for honeybees and other insects. *J. Econ. Ent.* 64:961–965.

Gary, N. E. and K. Lorenzen. 1976. A method for collecting the honey-sac contents from honeybees. *J. Apic. Res.* 15:73–79.

Gary, N. E., P. C. Witherell, and J. Marston. 1972. Foraging range and distribution of honeybees used for carrot and onion pollination. *Environ. Ent.* 1:71–78.

———. 1976. The inter- and intra-orchard distribution of honeybees during almond pollination. *J. Apic. Res.* 15:43–50.

———. 1978. Distribution and foraging activities of honeybees during almond pollination. *J. Apic. Res.* 17:188–194.

Grant, V. 1949. Pollination systems as isolating mechanisms in angiosperms. *Evolution* 3:82–97.

———. 1950. The flower constancy of bees. *Bot. Rev.* 16:379–398.

———. 1971. *Plant Speciation*. Columbia University Press, New York.

Grant, V. and K. A. Grant. 1965. *Flower Pollination in the Phlox Family*. Columbia University Press, New York.

———. 1979. Pollination of *Echinocereus fasciculatus* and *Ferocactus wislizenii*. *Pl. Syst. Evol.* 132:85–90.

Griggs, W. H. 1953. *Pollination requirements of fruits*

and nuts. Calif. Agr. Exp. Sta., Davis, Extension Service Circular No. 424.

Griggs, W. H. and B. T. Iwakiri. 1960. Orchard tests of beehive pollen dispensers for cross-pollination of almonds, sweet cherries, and apples. *Proc. Am. Soc. Hort. Sci.* 75:114–128.

———. 1964. Timing is critical for effective cross-pollination of almond flowers. *Calif. Agr.* 18:6–7.

———. 1975. Pollen tube growth in almond flowers. *Calif. Agr.* 29:4–7.

Griggs, W. H., G. H. Vansell, and B. T. Iwakiri. 1952. The use of beehive pollen dispensers in pollination of almonds and sweet cherries. *Proc. Am. Soc. Hort. Sci.* 60:146–150.

Hanny, B. W. 1980. Gossypol, flavonoid, and condensed tannin content of cream and yellow anthers of five cotton (*Gossypium hirsutum* L.) cultivars. *J. Agr. Food Chem.* 28:504–506.

Hanny, B. W. and C. D. Elmore. 1980. Amino acids in cream and yellow anthers of *Gossypium hirsutum. Phytochemistry* 19:137–138.

Hanny, B. W., J. C. Bailey, and W. R. Meredith. 1979. Yellow cotton pollen suppresses growth of larvae of tobacco budworm. *Environ. Ent.* 8:706–707.

Hardin, G. 1966. *Biology: Its Principles and Implications,* 2nd ed. W. H. Freeman, San Francisco.

Heinrich, B. 1975. Energetics of pollination. *Annu. Rev. Ecol. Syst.* 6:139–170.

———. 1976. The foraging specializations of individual bumblebees. *Ecol. Mon.* 46:105–128.

Heinrich, B. and P. H. Raven. 1972. Energetics and pollination ecology. *Science* 176:597–602.

Hugel, M. F. 1962. Etude de quelque constituants du pollen. *Ann. Abeille* 5:97–133.

Jamieson, C. A. and G. H. Austin. 1956. Preference of honeybees for sugar solutions. *Proc. Tenth Int. Cong. Ent.,* Montreal, 4:1059–1062.

Johnson, M. D. 1981. Observations on the biology of *Andrea (Melandrena) dunningi* Cockerell (Hymenoptera: Andrenidae). *J. Kansas Ent. Soc.* 54:32–40.

Kester, D. E. 1965. *A Review of Almond Varieties.* University of California Agricultural Extension Service, Davis.

Kester, D. E. and W. H. Griggs. 1959. Fruit setting in the almond: the effect of cross-pollinating various percentages of flowers. *Proc. Am. Soc. Hort. Sci.* 74:206–213.

Kulincevic, J. M. and W. C. Rothenbuhler. 1973. Laboratory and field measurements of hoarding behaviour in the honeybee. *J. Apic. Res.* 12:179–182.

Kulincevic, J. M., V. C. Thompson, and W. C. Rothenbuhler. 1974. Relationship between laboratory tests of hoarding behaviour and weight gained by honeybee colonies in the field. *Am. Bee J.* 114:93–94.

LaBerge, W. E. 1967. A revision of the bees of the genus *Andrena* of the Western Hemisphere. Part I. *Callandrena.* (Hymenoptera: Andrenidae). *Bull. Univ. Nebraska State Mus.* 7:1–318.

Leuck, E. E., II. 1980. Biosystematic studies in the *Echinocereus viridiflorus* complex. Unpublished doctoral dissertation, University of Oklahoma, Norman.

Levin, D. A. 1978. Pollinator behavior and the breeding structure of plant populations. *In* A. J. Richards (ed.), *Pollination of Flowers by Insects.* Linnean Soc. Symp. Ser. No. 6. Academic Press, London, pp. 133–150.

Levin, D. A. and H. W. Kerster. 1969. The dependence of bee-mediated pollen and gene dispersal upon plant density. *Evolution* 23:560–571.

Levin, D. A., H. W. Kerster, and M. Niedzlek. 1971. Pollinator flight directionality and its effect on pollen flow. *Evolution* 25:113–118.

Levin, M. D. and G. E. Bohart. 1955. Selection of pollens by honeybees. *Am. Bee J.* 95:392–393.

Linsley, E. G. 1958. The ecology of solitary bees. *Hilgardia* 27:543–599.

———. 1960. Observations on some matinal bees of *Cucurbita, Ipomoea* and *Datura* in desert areas of New Mexico and southeastern Arizona. *J. N.Y. Ent. Soc.* 58:13–20.

———. 1978. Temporal patterns of flower visitation by solitary bees, wtih particular reference to the southwestern United States. *J. Kansas Ent. Soc.* 51:531–546.

Linsley, E. G. and J. W. MacSwain. 1957. The nesting habits, flower relationships, and parasites of some North American species of *Diadasia* (Hymenoptera: Anthophoridae). *Wasmann J. Biol.* 15:199–235.

———. 1958. The significance of floral constancy among bees of the genus *Diadasia* (Hymenoptera: Anthophoridae). *Evolution* 12:219–223.

Linsley, E. G., J. W. MacSwain, and R. F. Smith. 1956. Biological observations on *Ptilothrix sumichrasti* (Cresson) and related groups of Emphorine bees. *Bull. So. Calif. Acad. Sci.* 55:83–101.

Linsley, E. G., J. W. MacSwain, and P. H. Raven. 1963a. Comparative behavior of bees and Onagraceae. I. Oenothera bees of the Colorado Desert. *Univ. Calif. Publ. Ent.* 33:1–24.

———. 1963b. Comparative behavior of bees and Onagraceae. II. Oenothera bees of the Great Basin. *Univ. Calif. Publ. Ent.* 33:25–58.

———. 1964. Comparative behavior of bees and Onagraceae. III. Oenothera bees of the Mojave Desert, California. *Univ. Calif. Publ. Ent.* 33:59–98.

Linsley, E. G., J. W. MacSwain, P. H. Raven and R. W. Thorp. 1973. Comparative behavior of bees and Onagraceae. V. *Camissonia* and *Oenothera* bees of Cismontane California and Baja California. *Univ. Calif. Publ. Ent.* 71:1–68.

Macior, L. W. 1974. Behavioral aspects of coadaptations between flowers and insect pollinators. *Ann. Missouri Bot. Gard.* 61:760–769.

———. 1978. The pollination ecology and endemic adaptation of *Pedicularis furbishiae* S. Wats. *Bull. Torrey Bot. Club* 105:268–277.

Mackensen, O. and W. P. Nye. 1966. Selecting and breeding honeybees for collecting alfalfa pollen. *J. Apic. Res.* 5:79–86.

———. 1969. Selective breeding of honeybees for alfalfa pollen collections: sixth generation and outcrosses. *J. Apic. Res.* 8:9–12.

MacSwain, J. W., P. H. Raven, and R. W. Thorp. 1973.

Comparative behavior of bees and Onagraceae. IV. *Clarkia* bees of the western United States. *Univ. Calif. Publ. Ent.* 70:1–80.

Martin, E. C. and S. E. McGregor. 1973. Changing trends in insect pollination of commercial crops. *Annu. Rev. Ent.* 18:207–226.

McLellan, G. R. 1975. Calcium, magnesium, potassium and sodium in honey and in nectar secretion. *J. Apic. Res.* 14:57–61.

Michener, C. D. 1947. Some observations on *Lasioglossum (Hemihalictus) lustrans* (Hymenoptera, Halictidae). *J. N.Y. Ent. Soc.* 45:49–50.

———. 1979. Biogeography of the bees. *Ann. Missouri Bot. Gard.* 66:277–347.

Mitchell, T. B. 1960. *Bees of the Eastern United States,* Vol. I. N.C. Agr. Exp. Sta., North Carolina State University, Raleigh.

———. 1962. *Bees of the Eastern United States,* Vol. II. N.C. Agr. Exp. Sta., North Carolina State University, Raleigh.

Northington, D. K. 1974. *Systematic Studies of the Genus Pyrrhopappus (Compositae: Cichorieae).* Special Publications of the Museum, No. 6, Texas Tech University. Texas Tech Press, Lubbock.

Nye, W. P. and O. Mackensen. 1965. Preliminary report on selection and breeding of honeybees for alfalfa pollen collection. *J. Apic. Res.* 4:43–48.

———. 1968. Selective breeding of honeybees for alfalfa pollination: fifth generation and backcross. *J. Apic. Res.* 7:21–27.

———. 1970. Selective breeding of honeybees for alfalfa pollen collection: with tests in high and low alfalfa pollen collection regions. *J. Apic. Res.* 9:61–64.

Parker, F. D. 1981a. How efficient are bees in pollinating sunflowers? *J. Kansas Ent. Soc.* 54:61–67.

———. 1981b. Sunflower pollination: abundance, diversity and seasonality of bees and their effect on seed yields. *J. Apic. Res.* 20:49–61.

Parrish, J. A. D. and F. A. Bazzaz. 1978. Pollination niche separation in a winter annual community. *Oecologia (Berl.)* 35:133–140.

Percival, M. S. 1961. Types of nectar in angiosperms. *New Phytol.* 60:235–281.

———. 1965. *Floral Biology.* Pergamon Press, New York.

Perkins, W. E., J. R. Estes, and R. W. Thorp. 1975. Pollination ecology of interspecific hybridization in *Verbena. Bull. Torrey Bot. Club* 102:194–198.

Pleasants, J. M. and M. Zimmerman. 1979. Patchiness in the dispersion of nectar resources: evidence for hot and cold spots. *Oecologia (Berl.)* 41:283–288.

Pyke, G. H. 1978a. Optimal foraging: movement patterns of bumblebees. *Theor. Popl Biol.* 13:72–98.

———. 1978b. Optimal foraging in bumblebees and coevolution with their plants. *Oecologia (Berl.)* 36:281–293.

———. 1978c. Are animals efficient harvesters? *Anim. Behav.* 26:241–250.

———. 1979. Optimal foraging in bumblebees: rule of movement between flowers within inflorescences. *Anim. Behav.* 27:1167–1181.

Raven, P. H. and D. P. Gregory. 1972. A revision of the genus *Gaura* (Onagraceae). *Mem. Torrey Bot. Club* 23:1–96.

Raw, G. R. 1953. The effect on nectar secretion of removing nectar from flowers. *Bee World* 34:23–25.

Real, L. A. 1981. Uncertainty and pollinator–plant interactions: the foraging behavior of bees and wasps on artificial flowers. *Ecology* 62:20–26.

Reese, C. S. L. and E. M. Barrows. 1980. Co-evolution of *Claytonia virginica* (Portulacaceae) and its main native pollinator, *Andrena erigeniae* (Andrenidae). *Proc. Ent. Soc. Washington* 82:685–694.

Reinhardt, T. F. 1952. Some responses of honeybees to alfalfa flowers. *Am. Nat.* 86:257–275.

Robertson, C. 1891. Flowers and insects, Asclepiadaceae to Scrophulariaceae. *Trans. St. Louis Acad. Sci.* 5:569–598.

———. 1914. Origin of oligotrophic bees. *Ent. News* 23:457–460.

———. 1925a. Habits of the *Hibiscus* bee, *Emphor bombiformis. Psyche* 32:278–282.

———. 1925b. Heterotropic bees. *Ecology* 6:412–436.

———. 1928. *Flowers and Insects.* Published by the author, Carlinville, Ill.

———. 1929. Phenology of oligolectic bees and favorite flowers. *Psyche* 36:112–118.

Rothenbuhler, W. C., J. M. Kulincevic, and V. C. Thompson. 1979. Successful selection of honeybees for fast and slow hoarding of sugar syrup in the laboratory. *J. Apic. Res.* 18:272–278.

Rust, R. W. and S. L. Clement. 1977. Entomophilous pollination of the self-compatible species *Collinsia sparsiflora* Fisher and Meyer. *J. Kansas Ent. Soc.* 50:37–48.

Rust. R. W., R. W. Thorp, and P. F. Torchio. 1974. The ecology of *Osmia nigrifrons* with a comparison to other *Acanthosmiodes. J. Nat. Hist.* 8:29–47.

Schaffer, W. M., D. B. Jensen, D. E. Hobbs, J. Gurevitch, J. R. Todd, and M. V. Schaffer. 1979. Competition, foraging energetics, and the cost of sociality in three species of bees. *Ecology* 60:976–987.

Schemske, D. W. 1977. Flowering phenology and seed set in *Claytonia virginica* (Portulacaceae). *Bull. Torrey Bot. Club* 104:254–263.

Schemske, D. W., M. F. Willson, M. N. Melampy, L. J. Miller, L. Verner, K. M. Schemske, and L. B. Best. 1978. Flowering ecology of some spring woodland herbs. *Ecology* 59:351–366.

Sheesley, B. and B. Poduska. 1970. Strong honeybee colonies prove value in almond pollination. *Calif. Agr.* 24:4–6.

Shinners, L. H. 1951. Notes on Texas Compositae. VII. *Field and Lab.* 19:81–82.

———. 1953. Notes on Texas Compositae. VIII. *Field and Lab.* 21:93–94.

———. 1958. *Spring Flora of the Dallas-Fort Worth Area Texas.* Published by the author, Dallas, Tex.

Standifer, L. N., R. H. MacDonald, and M. D. Levin. 1970. Influence of the quality of protein in pollens and of a pollen substitute on the development of the hypopharyngeal glands of honeybees. *Ann. Ent. Soc. Am.* 63:909–910.

Stanley, R. G. and H. F. Linskens. 1974. *Pollen: Biology, Biochemistry, Management.* Springer-Verlag, New York.

Stebbins, G. L. 1974. *Flowering Plants: Evolution above the Species Level.* Belknap Press of Harvard University Press, Cambridge, Mass.

Stephen, W. P., G. E. Bohart, and P. F. Torchio. 1969. *The Biology and External Morphology of Bees: With a Snyopsis of the Genera of Northwestern America.* Agr. Exp. Sta., Oregon State University, Corvallis.

Sullivan, J. R. 1980. Comparative reproductive biology of *Acer pensylvanicum* and *A. spicatum* (Aceraceae). Unpublished master's thesis, University of Connecticut, Storrs.

Teuber, L. R. and D. K. Barnes. 1978. Breeding alfalfa for increased nectar production. *Maryland Agric. Exp. Sta. Spec. Misc. Publ.* 1:109–116.

Thorp, R. W. 1969. Systematics and ecology of bees of the subgenus *Diandrena* (Hymenoptera: Andrenidae). *Univ. Calif. Pub. Ent.* 52:1–145.

———. 1978. Bee management for almond pollination. *In* W. Micke and D. E. Kester (eds.), *Almond Orchard Management.* University of California, Davis, Division of Agricultural Sciences Manual.

———. 1979. Structural, behavioral, and physiological adaptations of bees (Apoidea) for collecting pollen. *Ann. Missouri Bot. Gard.* 66:788–812.

Thorp, R. W. and J. R. Estes. 1975. Intrafloral behavior of bees on flowers of *Cassia fasciculata. J. Kansas Ent. Soc.* 48:175–184.

Thorp, R. W. and E. Mussen. 1978. Honeybees in almond pollination. University of California, Davis, Division of Agricultural Sciences, Leaflet No. 2465.

Thorp, R. W., W. Stanger, and T. Aldrich. 1967. Effects of artificial pollination on yield of Nonpareil almond trees. *Calif. Agr.* 21:14–15.

Thorp, R. W., E. H. Erickson, F. E. Moeller, M. D. Levin, W. Stanger, and D. L. Briggs. 1973. Flight activity and uniformity comparisons between honeybees in disposable pollination units (DPU's) and overwintered colonies. *Environ. Ent.* 2:525–529.

———. 1974. Disposable pollination units tested for almond pollination in California. *Am. Bee J.* 114:58–60.

Thorp, R. W., D. L. Briggs, J. R. Estes, and E. H. Erickson. 1975. Nectar fluorescence under ultraviolet irradiation. *Science* 189:476–478.

Towner, H. F. 1977. The biosystematics of *Calylophus* (Onagraceae). *Ann. Missouri Bot. Gard.* 64:48–120.

Tufts, W. P. 1919. Almond pollination. *Calif. Agr. Exp. Sta. Bull.* 306:337–366.

Waddington, K. D. 1979a. Flight patterns of three species of sweat bees (Halictidae) foraging at *Convolvulus arvensis. J. Kansas Ent. Soc.* 52:751–758.

———. 1979b. Quantification of the movement patterns of bees: A novel method. *Am. Midl. Natur.* 101:278–285.

———. 1980. Flight patterns of foraging bees relative to density of artificial flowers and distribution of nectar. *Oecologia (Berl.)* 44:199–204.

Waddington, K. D. and B. Heinrich. 1979. The foraging movements of bumblebees on vertical "inflorescences": an experimental analysis. *J. Comp. Physiol.* 134:113–117.

Waddington, K. D. and L. R. Holden. 1979. Optimal foraging: on flower selection by bees. *Am. Nat.* 114:179–196.

Waller, G. D. 1970. Attracting honeybees to alfalfa with citral, geraniol and anise. *J. Apic. Res.* 9:9–12.

———. 1972. Evaluating responses of honeybees to sugar solutions using an artificial-flower feeder. *Ann. Ent. Soc. Am.* 65:857–862.

Waller, G. D. and J. H. Martin. 1978. Fluorescence for identification of onion nectar in foraging honeybees. *Environ. Ent.* 7:766–768.

Waller, G. D., E. W. Carpenter, and O. A. Ziehl. 1972. Potassium in onion nectar and its probable effect on attractiveness of onion flowers to honeybees. *J. Am. Soc. Hort. Sci.* 97:535–539.

Whalen, M. D. 1978. Reproductive character displacement and floral diversity in *Solanum* section Androceras. *Syst. Bot.* 3:77–86.

Wykes, G. R. 1952. The preferences of honeybees for solutions of various sugars which occur in nectar. *J. Exp. Biol.* 29:511–518.

Zwick, F. B. and J. R. Estes. 1981. Adaptation of the life history of *Schinia mitis* (Lepidoptera: Noctuidae) to its host-plant, *Pyrrhopappus. J. Kansas Ent. Soc.* 54:416–432.

INDEX